Visualizing Nutrition
Everyday Choices

Fourth Edition

MARY B. GROSVENOR, M.S., RD

LORI A. SMOLIN, PH.D.
University of Connecticut

WILEY

VICE PRESIDENT & DIRECTOR	Petra Recter
SENIOR EDITOR	Alan Halfen
PROJECT MANAGER	Lauren Elfers
DEVELOPMENT EDITOR	Melissa Edwards Whelan
SENIOR PRODUCT DESIGNER	Linda Muriello
MARKET DEVELOPMENT MANAGER	Kristine Ruff
EDITORIAL ASSISTANT	MaryAlice Skidmore
MARKETING ASSISTANT	Maggie Joest
SENIOR CONTENT MANAGER	Svetlana Barskaya
SENIOR PHOTO EDITOR	MaryAnn Price
SENIOR PRODUCTION EDITOR	Trish McFadden
SENIOR DESIGNER	Wendy Lai
COVER PHOTO CREDITS	Vegetables: © Stolyevych Yulia/Shutterstock; Choices: © Kiselev Andrey Valerevich/Shutterstock; Family: © Spotmatik Ltd/Shutterstock; Laboratory: © Dusan Petkovic/Shutterstock; Fruit: © Andriy Blokhin/Shutterstock

This book was set in 9.5/12.5 Source Sans Pro by CodeMantra. Printed and bound by Quad Graphics.

Founded in 1807, John Wiley & Sons, Inc. has been a valued source of knowledge and understanding for more than 200 years, helping people around the world meet their needs and fulfill their aspirations. Our company is built on a foundation of principles that include responsibility to the communities we serve and where we live and work. In 2008, we launched a Corporate Citizenship Initiative, a global effort to address the environmental, social, economic, and ethical challenges we face in our business. Among the issues we are addressing are carbon impact, paper specifications and procurement, ethical conduct within our business and among our vendors, and community and charitable support. For more information, please visit our website: www.wiley.com/go/citizenship.

This book is printed on acid-free paper. ∞

Copyright © 2018 John Wiley & Sons, Inc. All rights reserved.

No part of this publication may be reproduced, stored in a retrieval system, or transmitted in any form or by any means, electronic, mechanical, photocopying, recording, scanning or otherwise, except as permitted under Sections 107 or 108 of the 1976 United States Copyright Act, without either the prior written permission of the Publisher, or authorization through payment of the appropriate percopy fee to the Copyright Clearance Center, Inc., 222 Rosewood Drive, Danvers, MA 01923 (website: www.copyright.com). Requests to the Publisher for permission should be addressed to the Permissions Department, John Wiley & Sons, Inc., 111 River Street, Hoboken, NJ 07030-5774, (201) 748-6011, fax (201) 748-6008, or online at: www.wiley.com/go/permissions.

Evaluation copies are provided to qualified academics and professionals for review purposes only, for use in their courses during the next academic year. These copies are licensed and may not be sold or transferred to a third party. Upon completion of the review period, please return the evaluation copy to Wiley. Return instructions and a free-of-charge return shipping label are available at: www.wiley.com/go/returnlabel. If you have chosen to adopt this textbook for use in your course, please accept this book as your complimentary desk copy. Outside of the United States, please contact your local representative.

ePUB ISBN 13 978-1119-39553-9

The inside back cover will contain printing identification and country of origin if omitted from this page. In addition, if the ISBN on the back cover differs from the ISBN on this page, the one on the back cover is correct.

Printed in the United States of America.

V10017710_022520

Preface

How Is Wiley Visualizing Different?

Wiley Visualizing is based on decades of research on the use of visuals in learning.[1] The visuals teach key concepts and are pedagogically designed to **explain, present, and organize** new information. The figures are tightly integrated with accompanying text; the visuals are conceived with the text in ways that clarify and reinforce major concepts, while allowing students to understand the details. This commitment to distinctive and consistent visual pedagogy sets Wiley Visualizing apart from other textbooks.

Wiley Visualizing texts offer an array of remarkable photographs, maps, media, and film from photo collections around the world. Visualizing images are not decorative, which can often be distracting to students, but purposeful and the primary driver of the content. These authentic materials immerse the student in real-life issues and experiences and support thinking, comprehension, and application.

Together these elements deliver a level of rigor in ways that maximize student learning and involvement. Wiley Visualizing has proven to increase student learning through its unique combination of text, photographs, and illustrations, with online video, animations, simulations, and assessments.

1. **Visual Pedagogy.** Using the Cognitive Theory of Multimedia Learning, which is backed up by hundreds of empirical research studies, Wiley's authors create visualizations for their texts that specifically support students' thinking and learning. For example, visuals help students identify important topics, organize new information, and integrate new material with prior knowledge.

2. **Authentic Situations and Problems.** *Visualizing Nutrition: Everyday Choices, 4e* offers an array of remarkable photographs, maps, and media. These materials immerse the student in real-life issues related to nutrition and thereby enhance interest, learning, and retention.[2]

3. **Designed with Interactive Multimedia.** *Visualizing Nutrition: Everyday Choices, 4e* is tightly integrated with WileyPLUS, our online learning environment that provides interactive multimedia activities in which learners can actively engage with the materials. The combination of textbook and *WileyPLUS* provides learners with multiple entry points to the content, giving them greater opportunity to explore concepts and assess their understanding as they progress through the course. *WileyPLUS* is a key component of the Wiley Visualizing learning and problem-solving experience. This sets Wiley Visualizing apart from other textbooks whose online component are mere drill-and-practice.

Visualizing Nutrition and *WileyPLUS* are designed to be a natural extension of how we learn

To understand why the Visualizing approach is effective, it is first helpful to understand how we learn.

1. Our brain processes information using two main channels: visual and verbal. Our *working memory* holds information that our minds process as we learn. This "mental workbench" helps us with decisions, problem-solving, and making sense of words and pictures by building verbal and visual models of the information.

2. When the verbal and visual models of corresponding information are integrated in working memory, we form more comprehensive, lasting, mental models.

3. When we link these integrated mental models to our prior knowledge, stored in our *long-term memory*, we build even stronger mental models. When an integrated (visual plus verbal) mental model is formed and stored in long-term memory, real learning begins.

The effort our brains put forth to make sense of instructional information is called *cognitive load*. There are two kinds of cognitive load: productive cognitive load, such as when we're engaged in learning or exert positive effort to create mental models; and unproductive cognitive load, which occurs when the brain is trying to make sense of needlessly complex content or when information is not presented well. The learning process can be impaired when the information to be processed exceeds the capacity of working memory. Well-designed visuals and text with effective pedagogical guidance can reduce the unproductive cognitive load in our working memory.

---

[1] Mayer, R.E. (Ed) (2005). *The Cambridge Handbook of Multimedia Learning*. Cambridge University Press.
[2] Donovan, M.S., & Bransford, J. (Eds.) (2005). *How Students Learn: Science in the Classroom*. The National Academy Press. Available at http://www.nap.edu/openbook.php?record_id=11102&page=1

New To This Edition

As the science of nutrition evolves, *Visualizing Nutrition: Everyday Choices* continues to evolve with it. This fourth edition includes the most up-to-date nutrition topics and recommendations along with many new photographs, new and improved illustrations, and enhanced critical thinking pedagogy.

- **Updated information:** The entire text has been updated and re-referenced to reflect the most recent nutrition science and guidelines. Topics at the forefront of today's nutrition landscape such as the impact of our food environment on diet, the role of the gut microbiome, newly recognized gluten sensitivities, and the protein needs of athletes have been added or expanded. New feature topics, reflecting the most current issues, have replaced older ones. For example, a new *Debate* in Chapter 4 addresses the impact of sugar on health and new *What a Scientist Sees* features in Chapters 5, 8, 13, and 14 discuss how eggs fit into a healthy diet, hidden sources of dietary sodium, the impact of cured meat on cancer risk, and growing concerns about food waste, respectively.

- **Organization of Micronutrients:** The presentation of the vitamins in Chapter 7 and the minerals in Chapter 8 has been reorganized into functional groupings. Although vitamins and minerals are still discussed in separate chapters, vitamins have been grouped based on their role in energy metabolism, blood health, antioxidant function, and gene expression and minerals have been grouped based on their roles as electrolytes, in bone health, as antioxidants, in blood health, and in energy metabolism. This organization makes it easier for students to identify, understand, and recall nutrient functions.

- **New Guidelines for Nutrition and Athletic Performance:** Recently published guidelines on nutrition and athletic performance have replaced older recommendations in Chapter 10. Students are guided through recommendations on the type, amount, and timing of intake of food, fluids, and supplements to support health and optimize exercise performance in casual and competitive sport.

- **New Food Label Legislation:** The Food Labeling Modernization Act of 2013 has been passed by Congress but has not yet been fully implemented. Therefore, both the current and the planned food labeling information are presented. Coverage of this topic in Chapter 2 compares the current Nutrition Facts label with the planned labels; they are compared and contrasted in discussions of specific nutrients in subsequent chapters.

- **Improved Thinking it Through Exercises:** These critical thinking case study exercises have been updated to be more effective in promoting critical thinking. The questions have been refined to make them more specific and to better emphasize current nutrition goals and the cases have been modified to help students more effectively navigate the obstacles in our unhealthy food environment.

- **Enhanced Visual Leaning:** To optimize the use of visuals that stimulate and enhance learning, many photos have been replaced and illustrations replaced or revised. Students see the most up-to-date information in new and updated graphs, charts, and maps. New photos in the *What Is Happening in This Picture* features in Chapters 1 and 12 focus on our food environment. In Chapter 3 new art helps students better understand the intestinal microbiota and how it affects health. In Chapter 5 a *Nutrition Insight* about fatty acid structures has been reorganized to more effectively show the relationship between different categories of fatty acids, in Chapter 6 a new illustrated table helps students visually distinguish different types of vegetarian diets, and in Chapter 10 a new figure illustrates relative energy deficiency in sport and its relationship to the female athlete triad.

- **Online Features:** Each chapter is supported by a variety of online features. *Food for Thought* videos by Laura Christoph accompany each chapter; these include chapter opening clips and summaries, as well as chapter-based topics, which help students to *Understand the Science* behind nutrition and how *Everyday Choices* impact their health. The *Video Bites* series, developed in collaboration with the authors, features student-focused sketches that explore the impact of day-to-day nutrition choices using real-life scenarios. Each video examines a topic germane to one or more chapters and blends up-to-date information with humor to pique student interest, address misconceptions, and spark discussion. Animations, Games, Interactivities, Audio clips and iProfile with accompanying Case Studies support each chapter and enhance the visual learning aspect of this title. End of chapter student support materials, such as the Critical and Creative Thinking Questions, Self-Tests, Appendices and links to Additional Resources have also been moved to an online format. Gradable Concept Check questions are a new online feature for this edition. Concept Questions apply to each learning objective and provide students with instant feedback on their understanding of the content.

- **Nutrition and Disease Presentation:** Nutrition-related disease topics, such as diabetes and heart disease, have always been integrated throughout the *Visualizing Nutrition* texts. For those who require slightly more in-depth coverage of disease topics, want to cover it as a single topic, or who want help integrating the role of nutrition in promoting health and preventing disease, a new separate chapter called *Focus on Nutrition and Disease* is available online. As with Metabolism (see online chapter *Metabolism: Energy for Life*), material presented in the text has been consolidated and topics such as nutrition and the immune response and the pathophysiology of various non-communicable diseases have been expanded.

How Does Wiley Visualizing Support Instructors?

Wiley Visualizing Site

Wiley Visualizing The Wiley Visualizing site hosts a wealth of information for instructors using Wiley Visualizing, including ways to maximize the visual approach in the classroom and a white paper titled "How Visuals Can Help Students Learn," by Matt Leavitt, instructional design consultant. Visit Wiley Visualizing at www.wiley.com/college/visualizing.

Wiley Custom Select

Wiley Custom Select Wiley Custom Select gives you the freedom to build your course materials exactly the way you want them, offering your students a cost-efficient alternative to traditional texts. In a simple three-step process create a solution containing the content you want, in the sequence you want, delivered how you want. Visit Wiley Custom Select at http://customselect.wiley.com.

PowerPoint Presentations

(available in *WileyPLUS* and on the book companion site)

A complete set of highly visual PowerPoint presentations—one per chapter—by Stephanie Colavita, Montclair State University, is available online and in WileyPLUS *Learning Space* to enhance classroom presentations. Tailored to the text's topical coverage and learning objectives, these presentations are designed to convey key text concepts, illustrated by embedded text art.

Test Bank

(available in *WileyPLUS* and on the book companion site)

The visuals from the textbook are also included in the Test Bank by Jennifer Zimmerman, Tallahassee Community College. The Test Bank has approximately 80 questions per chapter, many of which incorporate visuals from the book. The test items include multiple-choice and essay questions testing a variety of comprehension levels. The test bank is available online in MS Word files, as a Respondus Test Bank, and within *WileyPLUS*. The easy-to-use test-generation program fully supports graphics, print tests, student answer sheets, and answer keys. The software's advanced features allow you to produce an exam to your exact specifications.

Instructor's Manual

(available in *WileyPLUS* and on the book companion site)

The Instructor's Manual includes in class activities, an outline of WileyPLUS resources to utilize for each chapter, and discussion questions.

Nutrition Bytes Blog

(available in *WileyPLUS* and on the book companion site)

The Nutrition Bytes Blog provides an ongoing dialogue of trending topics and controversies in nutrition that spark discussion, highlight the relevance of nutrition in our lives, and encourage critical thinking. Nutrition Bytes is accessible on mobile devices and available from both the student and instructor companion sites, as well as within *WileyPLUS*. The blog is written by Katie Ferraro, University of California, San Francisco, and updated on a weekly basis, ensuring that discussions focus on the most current and relevant issues in nutrition. Blogs are searchable for topics of interest and students and instructors can join the discussion by posting their own comments. Users can subscribe to the newsfeed, which will automatically add it to their Favorites Center and be kept up to date.

Book Companion Site

All instructor resources (the Test Bank, Instructor's Manual, PowerPoint presentations, and all textbook illustrations and photos in jpeg format) are housed on the book companion site (www.wiley.com/college/grosvenor). Student resources include self-quizzes, glossary and flashcards, Nutrition Bytes Blog, and reference materials.

How Does *WileyPLUS* Support Instructors and Students?

WileyPLUS is designed for personalized, active learning. Several resources are available for instructors and students within *WileyPLUS*.

Hear This Illustration Audio Tutorials

Select figures in each chapter are accompanied by audio that narrates and discusses the important elements of that particular illustration. All audio files are accompanied by downloadable scripts of the narration.

Nutrition Interactivities

The fourth edition of *Visualizing Nutrition* includes nutrition activities for student practice. Each activity is embedded within the e-book so that students can practice as they are learning the content. Activities include food source identification drag-n-drop exercises for the macro- and micro-nutrient chapters and calculating and critical thinking activities that ask students to compute caloric intake and percent RDA. These types of exercises include informative feedback about the health consequences of specific nutrient toxicities or deficiencies.

Create-a-Plate and Revise-a-Recipe Activities

Each chapter includes an activity that asks students to create a balanced meal or snack to meet specific nutrient recommendations, incorporating MyPlate guidelines. Students can also practice altering meals by substituting foods with healthier choices.

Videos

Visualizing Nutrition contains multiple libraries of videos (such as the *Video Bites, Food for Thought,* and CBS/BBC video series), which focus on topics that pique student interest. Videos address current topics and discussions in nutrition, such as gluten-related disorders and our obesogenic environment. Many of these videos can be used as discussion tools, or can be assigned through *WileyPLUS* with gradable accompanying assessments. All videos have closed captioning for the hearing-impaired.

iProfile Mobile

The iProfile dietary analysis program now contains a database of over 50,000 foods and is available as a mobile-enabled website. Students can enter their food intakes and activities into their journal on the go via their smartphones and tablets.

iProfile is also available, with additional functionality, fully integrated with *WileyPLUS*. *WileyPLUS* includes a few types of assessments around iProfile, including computer graded iProfile Dietary Analysis Exercises in Chapters 4 through 9, written by Lori A. Smolin and Mary B. Grosvenor. These exercises ask students to analyze and modify a diet in relation to the specific nutrients discussed in the chapter. iProfile Case Study Assignments are also available in every chapter, which have students focus on a specific nutritional concept related to that chapter's content. Students can analyze the impact of different food and activity choices using iProfile reports.

Personalized Practice

ORION adaptive practice exercises meet students at just above their level in order to keep them challenged, but not frustrated. ORION adaptive practice assesses student understanding at the objective level. All students begin with a unique, short diagnostic quiz that establishes a baseline from which each student develops his own unique path. Adaptive practice includes extensive actionable reports that focus student study in key areas individual to each learner. Our rich question database contains nearly 6000 questions at every level of difficulty and all Bloom's levels.

Mobile Enabled Assets

All of the key resources students need to succeed in their nutrition course are now accessible on mobile devices. These include the How It Works, Estimating Portion Sizes, and MyPlate animations. Also mobile are the Nutrition Interactivities, Create-a-Plate and Revise-a-Recipe activities, and Interactive Process Diagrams.

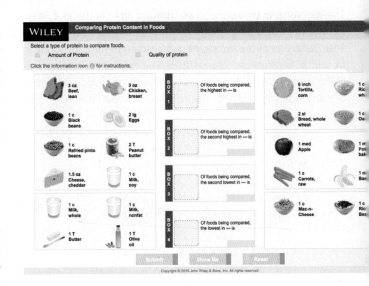

How Has Wiley Visualizing Been Shaped by Contributors?

Instructor Contributions

Throughout the process of developing this edition, we benefited from the comments and constructive criticism provided by the instructors and colleagues listed below. We offer our sincere appreciation to these individuals for their helpful reviews, invaluable contributions to the online resources, and continued involvement with the Wiley nutrition program:

Katherine Alaimo, Michigan State University
Celine Santiago Bass, Kaplan University
William Banz, Southern Illinois University
Abbe Breiter-Fineberg, Kaplan University
Amy Allen-Chabot, Anne Arundel Community College
Elizabeth Chu, San Diego Mesa College
Kara Egan, Indiana University-Purdue University Indianapolis
Elizabeth Eilender, Montclair State University
Nicolle Fernandes, Ball State University
Katie Ferraro, University of California, San Francisco
Melany Gilbert, Ozarks Technical Community College
Mindy Haar, New York Institute of Technology
Donna Huisenga, Illinois Central College
Karen Friedman-Kester, Harrisburg Area Community College
Rose Martin, Iowa State University
Deborah Murray, Ohio University
Melissa Silva, Bryant University
Nancy Berkoff, Los Angeles Trade Technical College
Jayne Byrne, College of St. Benedict
Linda Brown, Florida State University

Melanie Burns, Easting Illinois University
Laura Christoph, Holyoke Community College
James F. Collins, University of Florida
Sylvia Crixell, Texas State University
Barbara Goldman, Palm Beach State College
Timaree Hagenburger, Consumnes River College
Donna Handley, University of Rhode Island
Karen Israel, Anne Arundel Community College
Shanil Juma, Texas Women's University
Younghee Kim, Bowling Green State University
George Liepa, Eastern Michigan University
Owen Murphy, Central Oregon Community College
Judy Myhand, Louisiana State University
Cheryl Neudauer, Minneapolis Community and Technical College
Kristen Hilbert, SUNY Oneonta
Priya Venkatesan, Pasadena City College
Shahla Wunderlich, Montclair State University
Stephanie Colavita, Montclair State University
Jennifer Zimmerman, Tallahassee Community College
Jennifer Koslo, Maricopa Community Colleges

Special Thanks

- Special thanks to John Ambrose for providing a fresh eye to the material through the *Video Bites* series and updated chapter introductions for this new edition. We appreciate his "non-nutritionist" view of the science we know too well.

- We are extremely grateful to the many members of the editorial and production staff at John Wiley & Sons who guided us through the challenging steps of developing this text. Their tireless enthusiasm, professional assistance, and endless patience smoothed the path as we found our way. We thank in particular Senior Editor, Alan Halfen, for his support and for developing new ways to ensure this edition's success; and Trish McFadden, Senior Production Editor, who guided the production process. Our sincere thanks also go to Petra Recter, Vice President and Director, who oversaw the entire project; Kristine Ruff, Market Development Manager, who adeptly represents the Visualizing imprint; and Lauren Elfers for her continued support in the many roles she has played throughout this project. We appreciate the expertise of Mary Ann Price, Senior Photo Editor, in managing and researching our photo program. We are grateful to MaryAlice Skidmore, Editorial Assistant for helping to bring this project to fruition. And thanks to Development Editor, Melissa Edwards Whelan, for guiding us through the development of this text and its ancillary materials and to Jeanine Furino for keeping us on schedule through the perils of production.

About the Authors

MARY B. GROSVENOR holds a bachelor of arts in English and a master of science in Nutrition Science, affording her an ideal background for nutrition writing. She is a Registered Dietitian and has worked in clinical as well as research nutrition, in hospitals and communities large and small in the western United States. She has taught at the community college level and has published articles in peer-reviewed journals in nutritional assessment and nutrition and cancer. Her training and experience provide practical insights into the application and presentation of the science in this text.

LORI A. SMOLIN received a bachelor of science degree from Cornell University, where she studied human nutrition and food science. She received a doctorate from the University of Wisconsin at Madison, where her doctoral research focused on B vitamins, homocysteine accumulation, and genetic defects in homocysteine metabolism. She completed postdoctoral training both at the Harbor–UCLA Medical Center, where she studied human obesity, and at the University of California—San Diego, where she studied genetic defects in amino acid metabolism. She has published articles in these areas in peer-reviewed journals. Dr. Smolin is currently at the University of Connecticut, where she has taught both in the Department of Nutritional Science and in the Department of Molecular and Cell Biology. Courses she has taught include introductory nutrition, life cycle nutrition, food preparation, nutritional biochemistry, general biochemistry, and introductory biology.

Dedication

To my sons, David and John, and my husband, Peter. In the beginning, their contribution was support and patience with my long hours but over the years it has grown to include editing and writing as well. Thanks for helping me keep balance in my life.

(from Mary Grosvenor)

To my sons, Zachary and Max, who have grown up along with my textbooks, helping me to keep a healthy perspective on the important things in life. To my husband, David, who has continuously provided his love and support and is always there to assist with computer, technological, and life issues that arise.

(from Lori Smolin)

Letter to Students

Dear Students,

Everybody eats. So, we all have some basic understanding of nutrition: We need food to survive. We see that different foods affect our bodies in different ways and that not everybody responds to the same diet in the same way. And we see nutrition information all around us.

Never has the discussion of nutrition been more present in our own lives. It seems like we have access to an endless well of stirring documentaries about food, major news stories focus on nutrition, and advertisements about food and nutrition litter all forms of media. More and more young people are making conscious choices about their diets, whether it's to boycott GMOs or choose an atypical diet like veganism. Regardless of your personal choices one thing is clear: People are asking more questions than ever before. This is where we step in.

Our goal in writing *Visualizing Nutrition: Everyday Choices* is to provide you with a context to understand nutrition science as well as the sociocultural impact of food and nutrition. We are not here to provide simple answers to complex questions. We want to be part of the conversation. Our approach, which uses understandable explanations, debates, critical thinking exercises, visuals, videos, and online assessment tools, supports all learning styles. The clear, concise writing style—reinforced with colorful, engaging illustrations and photographs—makes the science user-friendly. And our critical thinking approach avoids vague hypotheticals and invites you to engage with the text by placing discussed topics into real-world situations.

Visualizing Nutrition: Everyday Choices helps you not only learn about how certain foods affect your health, but also gain an understanding of how your body processes the foods you choose to consume and how those choices present themselves in your day-to-day life. By helping you understand the "whys" and "hows," we hope we can satisfy and encourage your growing curiosity about nutrition in whatever direction it takes you, whether that means making more informed choices about your own diet and lifestyle, helping others do the same, or even pursuing a career within the field.

Brief Contents

PREFACE iii

1. Nutrition: Everyday Choices 1
2. Guidelines for a Healthy Diet 25
3. Digestion: From Meals to Molecules 51
4. Carbohydrates: Sugars, Starches, and Fibers 82
5. Lipids: Fats, Phospholipids, and Sterols 113
6. Proteins and Amino Acids 143
7. Vitamins 170
8. Water and Minerals 219
9. Energy Balance and Weight Management 265
10. Nutrition, Fitness, and Physical Activity 304
11. Nutrition During Pregnancy and Infancy 339
12. Nutrition from 1 to 100 371
13. How Safe Is Our Food Supply? 405
14. Feeding the World 440

APPENDIX A Dietary Reference Intakes A-1

ONLINE APPENDICES

APPENDIX A Dietary Reference Intakes
APPENDIX B Healthy Eating Patterns
APPENDIX C U.S. Nutrition Guidelines and Recommendations
APPENDIX D Energy Expenditure for Various Activities
APPENDIX E Standards for Body Size
APPENDIX F Normal Physiological Standards of Nutritional Relevance
APPENDIX G Choice (Exchange) Lists
APPENDIX H Food and Supplement Labels
APPENDIX I World Health Organization Nutrition Recommendations
APPENDIX J Calculations and Conversions
APPENDIX K Popular Dietary Supplements
APPENDIX L Answers to Thinking it Through

GLOSSARY GL-1

REFERENCES R-1

INDEX I-1

Contents

1 Nutrition: Everyday Choices 1

1.1 Food Choices and Nutrient Intake 2
1.2 Nutrients and Their Functions 6
1.3 Nutrition in Health and Disease 9
1.4 Choosing a Healthy Diet 13
1.5 Evaluating Nutrition Information 15

2 Guidelines for a Healthy Diet 25

2.1 Nutrition Recommendations 26
2.2 Dietary Reference Intakes (DRIs) 31
2.3 The Dietary Guidelines for Americans 33
2.4 MyPlate: Putting the Guidelines into Practice 38
2.5 Food and Supplement Labels 42

3 Digestion: From Meals to Molecules 51

3.1 The Organization of Life 52
3.2 The Digestive System 55
3.3 Digestion and Absorption of Nutrients 57
3.4 The Digestive System in Health and Disease 65
3.5 Delivering Nutrients and Eliminating Wastes 73
3.6 An Overview of Metabolism 77

4 Carbohydrates: Sugars, Starches, and Fibers 82

4.1 Carbohydrates in Our Food 83
4.2 Types of Carbohydrates 85
4.3 Carbohydrate Digestion and Absorption 88
4.4 Carbohydrate Functions 92
4.5 Carbohydrates in Health and Disease 96
4.6 Meeting Carbohydrate Needs 102

5 Lipids: Fats, Phospholipids, and Sterols 113

5.1 Fats in Our Food 114
5.2 Types of Lipids 115
5.3 Absorbing and Transporting Lipids 121
5.4 Lipid Functions 125
5.5 Lipids in Health and Disease 128
5.6 Meeting Lipid Needs 133

6 Proteins and Amino Acids 143

6.1 Proteins in Our Food 144
6.2 The Structure of Amino Acids and Proteins 145
6.3 Protein Digestion and Absorption 147
6.4 Protein Synthesis and Functions 149
6.5 Protein in Health and Disease 153
6.6 Meeting Protein Needs 157
6.7 Vegetarian Diets 163

7 Vitamins 170

7.1 A Vitamin Primer 171
7.2 Vitamins and Energy Metabolism 179
7.3 Vitamins and Healthy Blood 186
7.4 Antioxidant Vitamins 196
7.5 Vitamins in Gene Expression 200
7.6 Meeting Needs with Dietary Supplements 211

8 Water and Minerals 219

8.1 Water 220
8.2 An Overview of Minerals 226
8.3 Electrolytes: Sodium, Potassium, and Chloride 232
8.4 Minerals and Bone Health 240
8.5 Minerals and Healthy Blood 249
8.6 Antioxidant Minerals 253
8.7 Minerals and Energy Metabolism 257

9 Energy Balance and Weight Management 265

9.1 Body Weight and Health 266
9.2 Energy Balance 273
9.3 What Determines Body Size and Shape? 278
9.4 Managing Body Weight 283
9.5 Medications and Surgery for Weight Loss 290
9.6 Eating Disorders 293

10 Nutrition, Fitness, and Physical Activity 304

10.1 Food, Physical Activity, and Health 305
10.2 The Four Components of Fitness 307
10.3 Physical Activity Recommendations 309
10.4 Fueling Activity 313

- 10.5 Energy and Nutrient Needs for Physical Activity 319
- 10.6 Food and Drink to Optimize Performance 325
- 10.7 Ergogenic Aids 329

11 Nutrition During Pregnancy and Infancy 339

- 11.1 Changes in the Body During Pregnancy 340
- 11.2 Nutritional Needs During Pregnancy 345
- 11.3 Factors That Increase the Risks Associated with Pregnancy 351
- 11.4 Lactation 356
- 11.5 Nutrition for Infants 359

12 Nutrition from 1 to 100 371

- 12.1 The Nutritional Health of America's Youth 372
- 12.2 Nutrition for Children 376
- 12.3 Nutrition for Adolescents 383
- 12.4 Nutrition for the Adult Years 388
- 12.5 The Impact of Alcohol Throughout Life 397

13 How Safe Is Our Food Supply? 405

- 13.1 Keeping Food Safe 406
- 13.2 Pathogens in Food 410
- 13.3 Preventing Microbial Food-Borne Illness 417
- 13.4 Agricultural and Industrial Chemicals in Food 421
- 13.5 Technology for Keeping Food Safe 427
- 13.6 Biotechnology 432

14 Feeding the World 440

- 14.1 The Two Faces of Malnutrition 441
- 14.2 Causes of Hunger Around the World 443
- 14.3 Causes of Hunger in the United States 448
- 14.4 Eliminating World Hunger 450
- 14.5 Eliminating Food Insecurity in the United States 457

APPENDIX A Dietary Reference Intakes A-1

ONLINE APPENDICES

APPENDIX A Dietary Reference Intakes
APPENDIX B Healthy Eating Patterns
APPENDIX C U.S. Nutrition Guidelines and Recommendations
APPENDIX D Energy Expenditure for Various Activities
APPENDIX E Standards for Body Size
APPENDIX F Normal Physiological Standards of Nutritional Relevance
APPENDIX G Choice (Exchange) Lists
APPENDIX H Food and Supplement Labels
APPENDIX I World Health Organization Nutrition Recommendations
APPENDIX J Calculations and Conversions
APPENDIX K Popular Dietary Supplements
APPENDIX L Answers to Thinking it Through

GLOSSARY GL-1

REFERENCES R-1

INDEX I-1

CHAPTER 1

Nutrition: Everyday Choices

How do you choose what to eat? For most of the world's population, the answer is simple: You eat what you can grow, raise, catch, kill, or purchase. Subsistence is the principal motivator of food consumption: If you don't eat, you die. Historically, the game or crops people could kill or cultivate successfully became staples of their diet. As food production became more sophisticated, a greater array of food choices became available. As people explored and migrated across continents, new foods were discovered: Corn became part of the diet of European settlers in North America, and the potato was brought to the Old World from the New. Today, in our global society, you may literally choose from the world's dinner table.

Biological, social, economic, and cultural factors as well as personal tastes affect what you choose from this plethora of foods. And what you choose affects how healthy you are. Because the nutrients in the foods you eat form and maintain the structure of your body, you really are what you eat. The challenge is to find a satisfying balance between what you like and what optimizes your health. The choice is yours.

CHAPTER OUTLINE

Food Choices and Nutrient Intake 2
- Nutrients from Foods, Fortified Foods, and Supplements
- Food Provides More Than Nutrients
- What Determines Food Choices?

Nutrients and Their Functions 6
- The Six Classes of Nutrients
- What Nutrients Do

Nutrition in Health and Disease 9
- Undernutrition and Overnutrition
- Diet–Gene Interactions

Debate: Is There a Best Diet for You?

Choosing a Healthy Diet 13
- Eat a Variety of Foods
- Balance Your Choices

What Should I Eat? A Healthy Diet
- Practice Moderation

Thinking It Through: A Case Study on Choosing a Healthy Diet

Evaluating Nutrition Information 15
- The Science Behind Nutrition
- How Scientists Study Nutrition
- Judging for Yourself

What a Scientist Sees: Behind the Claims

1.1 Food Choices and Nutrient Intake

LEARNING OBJECTIVES

1. **Define** nutrient density.
2. **Compare** fortified foods with dietary supplements as sources of nutrients.
3. **Distinguish** essential nutrients from phytochemicals.
4. **Identify** factors in your food environment that influence your food choices.

What are you going to eat today? Will breakfast be a vegetable omelet or a bowl of sugar-coated cereal? How about lunch—a burger or a bean burrito? The foods we choose determine the **nutrients** we consume. To stay healthy, humans need more than 40 **essential nutrients**. Because the foods we eat vary from day to day, so do the amounts and types of nutrients and the number of **calories** we consume.

Nutrients from Foods, Fortified Foods, and Supplements

Any food you eat adds some nutrients to your diet, but to make your diet healthy, it is important to choose nutrient-dense foods. Foods with a high **nutrient density** contain more nutrients per calorie than do foods with a lower nutrient density because foods with a lower nutrient-density are higher in **empty calories** (Figure 1.1). Empty calories are calories from unhealthy fats and added sugars. If a large proportion of your diet consists of foods that are low in nutrient density and high in empty calories, such as soft drinks, chips, and candy, you could have a hard time meeting your nutrient needs without exceeding your calorie needs. By choosing nutrient-dense foods, you can meet all your nutrient needs and have calories left over for occasional treats that are lower in nutrients and higher in calories.

FIGURE 1.1 **Nutrient density** Nutrient density is important in choosing a healthy diet. Nutrient-dense foods provide more nutrients in fewer calories.

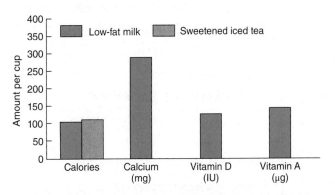

a. An 8-ounce glass of low-fat milk provides you with about the same number of calories as 8 ounces of bottled iced tea, but the milk also provides calcium, vitamin D, vitamin A, and other nutrients, including protein. The calories in the iced tea are empty calories from added sugar.

b. Typically, less processed foods provide more nutrients per calorie. For example, a roasted chicken breast is more nutrient dense than a highly processed form of chicken such as chicken nuggets; a baked potato is more nutrient dense than French fries; and apples are more nutrient dense than apple pie.

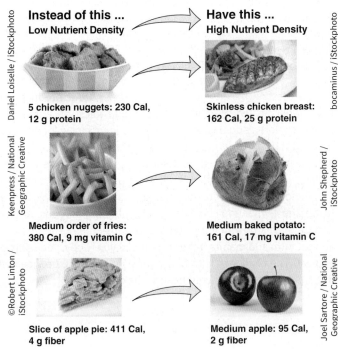

---

nutrient A substance in food that provides energy and structure to the body and regulates body processes.

essential nutrient A nutrient that must be consumed in the diet because it cannot be made by the body or cannot be made in sufficient quantities to maintain body functions.

calorie A unit of measure used to express the amount of energy provided by food.

nutrient density A measure of the nutrients provided by a food relative to its calorie content.

FIGURE 1.2 **Foods that are high in phytochemicals** Fruits, vegetables, and whole grains provide a variety of phytochemicals, such as those highlighted here. Supplements of individual phytochemicals are available, but there is little evidence that they provide the health benefits obtained from foods that are high in phytochemicals.[2]

Garlic, broccoli, and onions provide sulfur-containing phytochemicals that help protect us from some forms of cancer by inactivating carcinogens or stimulating the body's natural defenses.[3, 4]

Yellow–orange fruits and vegetables, such as peaches, apricots, carrots, and cantaloupe, as well as leafy greens, are rich in carotenoids, which are phytochemicals that may prevent oxygen from damaging our cells.[7]

Soybeans are a source of phytoestrogens, hormone-like compounds found in plants that may affect the risk of certain types of cancer and delay the progression of heart disease.[5, 6]

Purple grapes, berries, and onions provide red, purple, and pale yellow pigments called flavonoids, which prevent oxygen damage and may reduce the risk of cancer and heart disease.[8, 9]

Todd Gipstein / National Geographic Creative

In addition to nutrients that occur naturally in foods, we obtain nutrients from fortified foods. The **fortification** of foods was begun to help eliminate nutrient deficiencies in the population, with the federal government mandating that certain nutrients be added to certain foods. Foods such as milk with added vitamin D and grain products with added B vitamins and iron are examples of this mandated fortification that have been part of the U.S. food supply for decades.

Voluntary fortification of foods is also now common practice. Vitamins and minerals are routinely added to a variety of foods, including breakfast cereals and snack foods. The amounts and types of nutrients added to these voluntarily fortified foods are at the discretion of the manufacturer. These added nutrients contribute to the diet but are not necessarily designed to address deficiencies and may increase the likelihood of consuming an excess of some nutrients (see *Debate* in Chapter 7).

Dietary supplements are another source of nutrients, but they do not offer all the benefits of food (see Chapters 2 and 7). More than half of U.S. adults take some sort of daily dietary supplement.[1]

Food Provides More Than Nutrients

In addition to nutrients, food contains substances that, though not essential to life, can be beneficial for health. In plants, these health-promoting substances are called **phytochemicals** (**Figure 1.2**). Although fewer such substances have been identified in animal foods, animal foods also contain substances with health-promoting properties. These are called **zoochemicals**.

Some foods, because of the complex mixtures of nutrients and other chemicals they contain, provide health benefits that extend beyond basic nutrition. Such foods have been termed **functional foods**. The simplest functional foods are unmodified whole foods, such as broccoli and fish, that naturally contain substances that promote health and protect against disease, but some foods fortified with nutrients or enhanced with phytochemicals or other substances are also classified as functional foods (**Table 1.1**).[21] These modified foods, such as water with added vitamins, oatmeal with added soy protein, and orange juice with added calcium, have also been called **designer foods**. The term **nutraceutical** refers to any food or supplement that delivers a health benefit.

What Determines Food Choices?

Do you eat oranges to boost your vitamin C intake or ice cream to add a little calcium to your diet? Probably not. We need these nutrients to survive, but we generally choose foods for reasons other than the nutrients they contain.

The factors around us that influence what we choose to eat are referred to as our **food environment**. This includes the geographic, social, and economic factors that affect eating habits and patterns (**Figure 1.3**). It involves factors such as

---

fortification The addition of nutrients to foods.

phytochemical A substance found in plant foods that is not an essential nutrient but may have health-promoting properties.

functional food A food that has health-promoting properties beyond basic nutritional functions.

food environment The physical, economic, and social factors that affect eating habits and patterns.

CHAPTER 1 Nutrition: Everyday Choices

TABLE 1.1 Functional foods provide benefits beyond their nutrients

Food	Potential health benefit
Blueberries	May reduce the risk of heart disease and cancer.[9,10]
Breakfast cereal with added flaxseed	Helps reduce blood cholesterol levels and the overall risk of heart disease.[11]
Chocolate	May help reduce blood pressure and other risk factors for heart disease.[12]
Garlic	Helps reduce blood cholesterol levels and the overall risk of heart disease.[13]
Kale	May reduce the risk of age-related blindness (macular degeneration).[14]
Margarine with added plant sterols	Reduces blood cholesterol levels.[15]
Nuts	May reduce the risk of heart disease.[16]
Oatmeal	Helps reduce blood cholesterol.[17]
Orange juice with added calcium	Helps prevent osteoporosis.
Salmon	Reduces the risk of heart disease.[18]
Green tea	May reduce the risk of certain types of cancer.[19]
Whole-grain bread	Helps reduce the risk of cancer, heart disease, obesity, and diabetes.[20]

NUTRITION INSIGHT

FIGURE 1.3 Food environment affects food choices Our food environment influences what we eat and can make it challenging to choose a healthy diet.

What we choose to eat is influenced by what is available as well as what we are enticed to eat. Across much of America, fast-food restaurants dominate the landscape, and advertisements for these foods bombard us on TV and in other media. The availability, affordability, and familiarity of these foods result in increased consumption.

In the inner cities, supermarkets are scarce, and affordable perishable food is limited in small corner stores. This makes it difficult to choose a varied diet and to include healthy amounts of fresh fruits and vegetables.

David McNew / Staff / Getty Images

Richard B Levine / Newscom

(continues)

FIGURE 1.3 *(continued)*

Food preferences and eating habits learned as part of an individual's family, cultural, national, and social background are part of our food environment. In many parts of the world, insects, such as these cicadas and grasshoppers, are considered a treat, but in U.S. culture, insects are considered food contaminants, and most people would refuse to eat them.

Food can provide comfort and security, but what foods you choose when you are sick, cold, tired, depressed, or lonely depends on family, culture, and tradition. Some of us have a cup of hot tea, while others choose chicken soup, chocolate, or ice cream to help us feel better.

In the United States, food imports enhance the variety of foods available, but in many parts of the developing world, food choices are restricted to foods produced locally. In times of crisis, such as famine or a natural disaster, food choices may be limited to what humanitarian aid programs deliver.

Access to farmer's markets allows people even in the inner city to choose fresh locally grown produce.

access to grocery stores and restaurants as well as the products and pricing of food available in those stores and restaurants. It also involves family and cultural traditions; if we were never exposed to a food as a child, it would not be part of our food environment. What we choose to eat is also affected by our personal convictions, such as environmental consciousness or vegetarianism, as well as our personal preferences for taste, smell, appearance, and texture.

> **Concept Check**
>
> 1. **Which** has a higher nutrient density: a soda or a glass of milk?
> 2. **Why** are foods fortified?
> 3. **Why** is it better to meet your vitamin C needs by eating an orange than by taking a dietary supplement?
> 4. **What** factors determine what you will eat for lunch?

1.2 Nutrients and Their Functions

LEARNING OBJECTIVES

1. **List** the six classes of nutrients.
2. **Discuss** the three functions of nutrients in the body.

There are six classes of nutrients: carbohydrates, lipids, proteins, water, vitamins, and minerals. Carbohydrates, lipids, proteins, and water are considered **macronutrients** because they are needed in large amounts. Vitamins and minerals are referred to as **micronutrients** because they are needed in small amounts. Together, the macronutrients and micronutrients in our diet provide us with energy, contribute to the structure of our bodies, and regulate the biological processes that go on inside us. Each nutrient provides one or more of these functions, but all nutrients together are needed to provide for growth, maintain and repair the body, and support reproduction.

The Six Classes of Nutrients

Carbohydrates, lipids (commonly called fats), and proteins are all **organic compounds** that provide energy to the body. Although we tend to think of each of them as a single nutrient, there are actually many different types of molecules in each of these classes. **Carbohydrates** include starches, sugars, and **fiber** (**Figure 1.4a**). Several types of **lipids** play important roles in nutrition (**Figure 1.4b**). The most recognizable of these are **cholesterol**, **saturated fats**, and **unsaturated fats**. There are thousands of different **proteins** in our bodies and our diets. All proteins are made up of units called **amino acids** that are linked together in different combinations to form different proteins (**Figure 1.4c**).

Water, unlike the other classes of nutrients, is only a single substance. Water makes up about 60% of an adult's body weight. Because we can't store water, the water the body loses must constantly be replaced by water obtained from the diet. In the body, water acts as a lubricant, a transport fluid, and a regulator of body temperature.

Vitamins are organic molecules that are needed in small amounts to maintain health. There are 13 vitamins, which perform a variety of unique functions in the body, such as regulating energy metabolism, maintaining vision, protecting cell membranes, and helping blood to clot. **Minerals** are **elements** that are essential nutrients needed in small amounts to provide a variety of diverse functions in the body. For example, iron is an element needed for the transport of oxygen in the blood, calcium is an element important in keeping bones strong. We consume vitamins and minerals in almost all the foods we eat.

---

organic compound A substance that contains carbon bonded to hydrogen.

carbohydrates A class of nutrients that includes sugars, starches, and fibers. Chemically, they all contain carbon, along with hydrogen and oxygen, in the same proportions as in water (H_2O).

fiber A type of carbohydrate that cannot be broken down by human digestive enzymes.

lipids A class of nutrients, commonly called fats, that includes saturated and unsaturated fats and cholesterol; most do not dissolve in water.

cholesterol A type of lipid that is found in the diet and in the blood. High blood levels increase the risk of heart disease.

saturated fat A type of lipid that is most abundant in solid animal fats and is associated with an increased risk of heart disease.

unsaturated fat A type of lipid that is most abundant in plant oils and is associated with a reduced risk of heart disease.

protein A class of nutrients that includes molecules made up of one or more intertwining chains of amino acids.

FIGURE 1.4 **Carbohydrates, lipids, and proteins** Varying combinations of carbohydrates, lipids, and proteins provide the energy in the foods we eat.

a. Some high-carbohydrate foods, such as rice, pasta, and bread, contain mostly starch; some, such as berries, kidney beans, and broccoli, are high in fiber; and others, such as cookies, cakes, and carbonated beverages, are high in empty calories from added sugars.

b. High-fat plant foods such as vegetable oils, avocados, olives, and nuts are high in healthy unsaturated fats. High-fat animal foods such as cream, butter, meat, and whole milk are sources of cholesterol and are high in saturated fat, which increases the risk of heart disease.

c. The proteins we obtain from animal foods, such as meat, fish, and eggs, better match our amino acid needs than do most individual plant proteins, such as those in grains, nuts, and beans. However, when plant sources of protein are combined, they can provide all the amino acids we need.

Some are natural sources: Oranges contain vitamin C, milk provides calcium, and carrots give us vitamin A. Other foods are fortified with vitamins and minerals; a serving of fortified breakfast cereal often has 100% of the recommended intake of many vitamins and minerals. Dietary supplements are another source of vitamins and minerals for some people.

What Nutrients Do

Nutrients are involved in providing energy, forming body structures, and regulating physiological processes (**Figure 1.5**). Carbohydrates, lipids, and proteins are often referred to as **energy-yielding nutrients**; they provide energy that can be measured in calories. The calories people talk about and see listed on food labels are actually **kilocalories** (abbreviated kcalorie or kcal), units of 1000 calories. When spelled with a capital C, Calorie means kilocalorie. Carbohydrates provide 4 Calories/gram; they are the most immediate source of energy for the body. Lipids also help fuel our activities and are the major form of stored energy in the body. One gram of fat provides 9 Calories. Protein can supply 4 Calories/gram but is not the body's first choice for meeting energy needs because protein has other roles that take priority. Alcohol, though it is not a nutrient because it is not needed for life, provides about 7 Calories/gram. Water, vitamins, and minerals do not provide energy (calories).

With the exception of vitamins, all the classes of nutrients are involved in forming and maintaining the body's structure. Fat deposited under the skin contributes to our body shape, for instance, and proteins form the ligaments and tendons that hold our bones together and attach our muscles to our bones. Minerals harden bone. Proteins and water make up the structure of the muscles, which help define our body contours, and proteins and carbohydrates form the cartilage that cushions our joints. On a smaller scale, lipids, proteins, and water form the structure of individual cells. Lipids and proteins make up

8 CHAPTER 1 Nutrition: Everyday Choices

NUTRITION INSIGHT

FIGURE 1.5 Nutrient functions The nutrients we consume in our diet provide energy, form body structures, and regulate body processes.

Energy Whether riding a bike through the fall foliage, walking to the mailbox, or gardening, physical activity is fueled by the carbohydrates, fat, and protein in the food we eat.

Structure Proteins, lipids, carbohydrates, minerals, and water all contribute to the shape and structure of our bodies.

Regulation Water helps regulate body temperature. When body temperature increases, sweat is produced, cooling the body as it evaporates from the skin.

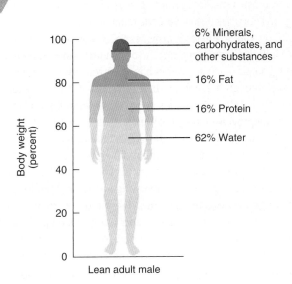

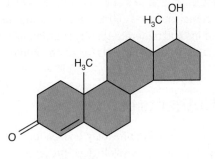

Regulation Lipids, such as the hormone testosterone, illustrated here, help regulate body processes. Testosterone is made from cholesterol. In men, it stimulates sperm production and the development of secondary sex characteristics, such as body and facial hair, a deep voice, and increased muscle mass.

the membranes that surround each cell, and water and dissolved substances fill the cells and the spaces around them.

All six classes of nutrients play important roles in regulating body processes. Keeping body temperature, blood pressure, blood sugar level, and hundreds of other parameters relatively constant involves thousands of chemical reactions and physiological processes. Proteins, vitamins, and minerals are regulatory nutrients that help control how quickly chemical reactions take place throughout the body. Lipids and proteins are needed to make regulatory molecules called **hormones** that stimulate or inhibit various body processes.

Figure 1.5 illustrates some of the ways various nutrients are involved in providing energy, forming body structures, and regulating physiological processes.

Concept Check

1. **Which** classes of nutrients provide energy?
2. **What** are the three overall functions that nutrients provide?

1.3 Nutrition in Health and Disease

LEARNING OBJECTIVES

1. **Describe** the causes of malnutrition.
2. **Explain** ways in which nutrient intake can affect health in both the short term and the long term.
3. **Discuss** how the genes you inherit affect the impact your diet has on your health.

What we eat has an enormous impact on how healthy we are now and how likely we are to develop chronic diseases such as heart disease, obesity, and diabetes. Consuming either too much or too little of one or more nutrients or energy will result in **malnutrition**. Malnutrition can affect your health not just today but 20, 30, or 40 years from now. The impact of your diet on your health is also affected by your genetic makeup.

Undernutrition and Overnutrition

Undernutrition occurs when intake doesn't meet the body's needs: The more severe the deficiency, the more dramatic the symptoms (**Figure 1.6**). Some nutrient deficiencies occur quickly. Dehydration, a deficiency of water, can cause symptoms

FIGURE 1.6 **Undernutrition** Nutrient deficiencies may be mild enough to cause no obvious symptoms or severe enough to cause debilitating illness and death.

a. Even though this child looks normal and healthy, she has low iron stores. If the iron content of her diet is not increased, she will eventually develop iron deficiency anemia. Mild nutrient deficiencies like hers may go unnoticed because the symptoms either are not immediately apparent or are nonspecific. Two common nonspecific symptoms of iron depletion are fatigue and decreased ability to fight infection.

b. The symptoms of starvation, the most obvious form of undernutrition, occur gradually over time when the energy provided by the diet is too low to meet the body's needs. Body tissues are broken down to provide the energy to support vital functions, resulting in loss of body fat and wasting of muscles.

malnutrition A condition resulting from an energy or nutrient intake either above or below that which is optimal.

in a matter of hours. Drinking water can relieve the headache, fatigue, and dizziness caused by dehydration almost as rapidly as these symptoms appeared. Other nutritional deficiencies may take much longer to become evident. Symptoms of scurvy, a disease caused by a deficiency of vitamin C, appear after months of deficient intake; **osteoporosis**, a condition in which the bones become weak and break easily, occurs after years of consuming a calcium-deficient diet.

We typically think of malnutrition as undernutrition, but **overnutrition**, an excess intake of nutrients or calories, is also a concern. An overdose of iron can cause liver failure, for example, and too much vitamin B_6 can cause nerve damage. These nutrient toxicities usually result from taking large doses of vitamin and mineral supplements because foods generally do not contain high enough concentrations of nutrients to be toxic. However, chronic overconsumption of calories and certain nutrients from foods can also cause health problems. The typical U.S. diet, which provides more calories than are needed, has resulted in an epidemic of obesity in which more than 70% of adults are overweight or obese (**Figure 1.7a**).[22] Diets that are high in sodium contribute to high blood pressure; an excess intake of saturated fat contributes to heart disease; and a dietary pattern that is high in red meat and saturated fat and low in fruits, vegetables, and fiber may increase the risk of certain cancers.[23] It has been estimated that a poor diet is associated with 26% of deaths in the United States, more than smoking and alcohol combined (**Figure 1.7b**).[24]

Diet–Gene Interactions

What you eat affects your health, but diet alone does not determine whether you will develop a particular disease. Each of us inherits a unique combination of **genes**. Some of these genes affect your risk of developing chronic diseases, such as heart disease, cancer, high blood pressure, and diabetes, but their impact is affected by what you eat. Your genetic makeup determines the impact a certain nutrient will have on you (see *Debate: Is There a Best Diet for You?*). For example, some people inherit a combination of genes that makes their blood pressure sensitive to sodium intake. When these individuals consume even an average amount of sodium, they may develop high blood pressure (discussed further in Chapter 8). Others inherit

FIGURE 1.7 **Overnutrition** Obesity is a form of overnutrition that increases the risk of a variety of other nutrition-related chronic diseases that contribute to death.

a. Obesity occurs when energy intake surpasses energy expenditure over a long period, causing the accumulation of an excessive amount of body fat. Adults are not the only ones who are getting fatter; 17% of U.S. children and adolescents, ages 2 to 19 years, are obese.[25]

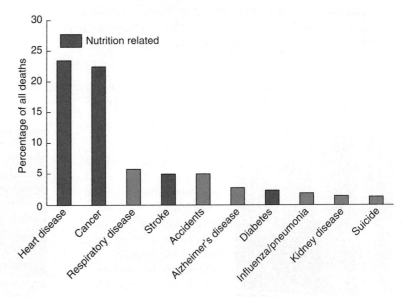

b. Four of the top 10 causes of death in the United States are nutrition related.[26] They are all thought to be exacerbated by obesity.

Interpret the Data

Based on this graph showing the leading causes of death in the United States, about what percentage of all deaths are due to nutrition-related diseases?

a. 5%   c. 50%
b. 10%   d. 90%

genes Units of a larger molecule called DNA that are responsible for inherited traits.

Debate

Is There a Best Diet for You?

The Issue

The genes you inherit affect your risk of developing a variety of chronic diseases. If analysis of your genes can reveal these risks, then a personalized diet prescription might be made to prevent, moderate, or cure the chronic diseases for which you are at risk. Is this individualized approach better than following the overall recommendations for a healthy diet?

Our health is determined by many factors, including environment, lifestyle, and genetics. We have control over lifestyle factors that affect our disease risk, such as what we eat and how much we exercise. We don't have control over our genetic background. The science of nutritional genomics suggests that we can reduce our risk of disease by tailoring our diets to our individual genetic makeup.[28] It predicts that someday we may go to the doctor's office, have our genes analyzed, and then have specific foods and dietary supplements prescribed to optimize our health and prevent diseases to which we are susceptible (see figure).

PeopleImages / iStockphoto

Current nutrition guidelines are designed to improve and maintain the health of almost all healthy people in the population. Yet, we know that different people respond differently to the same diet, so dietary advice that is good for the majority of people may not be optimum for everyone. Modern medicine is already practicing nutritional genomics at a very basic level; dietitians design special diets based on patients' existing medical conditions. People with elevated blood lipid levels are instructed to reduce their intake of saturated fat and increase their fiber intake, those with high blood pressure are shown how to reduce their sodium intake, and people with diabetes are taught how to manage blood sugar levels by modifying their carbohydrate intake.

Proponents of nutritional genomics suggest that reviewing an individual's genetic analysis will permit the development of more personalized dietary recommendations, which will prevent or improve outcomes for a variety of chronic diseases.[29] These diets could be customized to take into consideration not only individual genetic variation but also life stage, dietary preferences, and other aspects of health status. Some propose that, if followed, these personalized dietary recommendations may supplement and even replace prescription drugs.

Would the benefit of these individualized diet plans justify the expense of the genetic analyses? Some argue that this approach is unlikely to improve individual or public health.[29] Many people fail to follow current population-wide guidelines for a healthy diet, not because they lack the knowledge, money, or motivation to do so but simply because they choose not to. Therefore, it is unlikely that individuals will follow personalized guidelines any better or that genetic test results will motivate them to eat a healthier diet.[30] The priority for public health dollars should not be to fine-tune diet prescriptions but to find out what will make people change their diets and live healthier lives.

Other concerns with nutritional genomics are the ethics of widespread genetic testing and the possibility that commercial interests rather than any benefits to public health will drive nutritional genomics.[31] People who strictly adhere to their diet prescriptions will certainly benefit, but the big beneficiaries of personalized diet prescriptions could be biotech companies, which would profit from the genetic testing needed to establish disease profiles, and the food industry, which would benefit from the creation and sale of functional foods to prevent disease.

In the future, will we select breakfast cereals and dietary supplements based on our genes? Will these choices, which target the prevention of some potential chronic diseases, increase the risk of others? Will following personalized dietary prescriptions make us healthier than just choosing an overall healthy diet? In choosing foods based on nutritional genomics, will we lose track of the pleasure we get from food and the cultural and social roles that food plays in our lives?

Think Critically

If genetic testing determines that you are unlikely to become obese, does this mean that you can eat as much as you want? Why or why not?

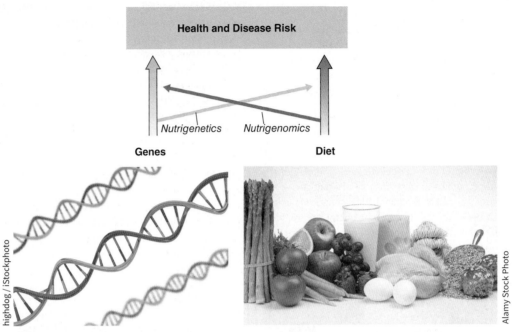

FIGURE 1.8 **Nutritional genomics** Both the genes you inherit and your dietary choices directly affect your health and disease risk. There is also interplay between these factors such that your genes, through nutrigenetics, influence how your diet affects your health and your diet, through nutrigenomics, affects how your genes impact your health.

The genes you inherit affect your tendency to develop nutrition-related chronic diseases such as heart disease and diabetes.

The diet you consume affects your risk of developing nutrition-related chronic diseases.

Think Critically

Can you become obese even if both of your parents are thin?

genes that allow them to consume more sodium without much of a rise in blood pressure. Those whose genes dictate a significant rise in blood pressure with a high-sodium diet can reduce their blood pressure, and the complications associated with high blood pressure, by eating a diet that is low in sodium.

Our increasing understanding of human genetics has given rise to the discipline of **nutritional genomics**, which explores the interaction between human genes, nutrition, and health.[27] It encompasses both the effects that the genes people inherit have on how their diet affects health (**nutrigenetics**) and the effects the nutrients and other food components they consume have on gene activity (**nutrigenomics**) (**Figure 1.8**). Research in these areas has led to the development of the concept of "personalized nutrition." The goal of personalized nutrition is to prescribe a diet based on the genes an individual has inherited in order to prevent, moderate, or cure chronic disease. Although today we do not have the tools to take a sample of everyone's DNA and use it to tell them what to eat to optimize their health, we do know that certain dietary patterns can reduce the risk of many chronic diseases.

Concept Check

1. **What** causes malnutrition?
2. **How** can your diet today affect your health 20 years from now?
3. **Why** might the diet that optimizes health be different for different people?

nutritional genomics The study of how our genes affect the impact of nutrients or other food components on health (nutrigenetics) and how nutrients affect the activity of our genes (nutrigenomics).

1.4 Choosing a Healthy Diet

LEARNING OBJECTIVES

1. **List** three reasons it is important to eat a variety of foods.
2. **Explain** why you can sometimes eat foods that are low in nutrient density and still have a healthy diet.
3. **Discuss** how dietary moderation can reduce the risk of chronic disease.

A healthy diet is one that provides the right number of calories to keep your weight in the desirable range; the proper balance of carbohydrates, proteins, and fat; plenty of water; and sufficient but not excessive amounts of vitamins and minerals. This healthy diet is rich in whole grains, fruits, and vegetables; high in fiber; and low in added sugars, sodium, and unhealthy fats (saturated and *trans* fats). In short, a healthy diet is based on variety, balance, and moderation (see *What Should I Eat?*).

Eat a Variety of Foods

In nutrition, choosing a variety of foods is important because no single food can provide all the nutrients the body needs for optimal health. *Variety* means choosing foods from different food groups—vegetables, grains, fruits, dairy products, and high-protein foods. Some of these foods are rich in vitamins and phytochemicals, others are rich in protein and minerals, and all are important.

Variety also means choosing diverse foods from within each food group. Different vegetables provide different nutrients. Potatoes, for example, are the only vegetable in many Americans' diets. Potatoes provide vitamin C but are low in vitamin A. If potatoes are your only vegetable, it is unlikely that you will meet your nutrient needs. If instead you have a salad, potatoes, and broccoli, you will be getting plenty of vitamins C and A, as well as many other vitamins and minerals. Choosing from all the food groups and making varied choices from within each food group is also important because nutrients and other food components interact. Such interactions may be positive, enhancing nutrient utilization, or negative, inhibiting nutrient availability. Variety averages out these interactions. Some foods may also contain toxic substances. Eating a variety of foods reduces the risk that you will consume enough of any one toxin to be harmful. For example, tuna may contain traces of mercury, but as long as you don't eat tuna too often, you are unlikely to consume a toxic amount.

Variety involves choosing different foods not only each day but also each week and throughout the year. If you had apples and grapes today, for example, have blueberries and cantaloupe tomorrow. If you can't find tasty tomatoes in December, replace them with a winter vegetable such as squash.

Balance Your Choices

Choosing a healthy diet is a balancing act. Healthy eating doesn't mean giving up your favorite foods. There is no such thing as a good food or a bad food—only healthy diets and unhealthy diets. Any food can be part of a healthy diet, as long as your diet throughout the day or week provides enough of all the nutrients you need without excesses of

---

What Should I Eat?

A Healthy Diet

Eat a variety of foods

- Mix up your snacks. For example, have salsa and chips one day and fruit, yogurt, or nuts another day.
- Add almonds and diced apples to your salad.
- Try a new vegetable or fruit each week. Tired of carrots? Try jicama.
- Vary your protein sources. Eat fish one day and beef the next—or skip the meat and have beans.

Balance your choices

- Going out to dinner? Have a salad for lunch.
- If you eat some extra fries, take some extra steps.
- When you have cookies for a snack, have fruit for dessert.
- Had ice cream with lunch? Have low-fat milk with dinner.

Practice moderation

- Push back from the table before you are stuffed and go for a walk.
- Reduce your portions by using a smaller bowl.
- Fill your plate using MyPlate and skip the seconds.
- Split your restaurant meal with a friend.

 Use iProfile to calculate the calories in your favorite fast-food meal.

any. When you choose a food that is lacking in fiber, for example, balance it with one that provides lots of fiber. When you choose a food that is high in salt, balance that choice with a low-salt one.

A balanced diet also balances the calories you take in with the calories you burn in your daily activities so that your body weight stays in the healthy range (**Figure 1.9**).

Practice Moderation

Moderation means not overdoing it—not having too many calories, too much saturated fat, too much sugar, too much salt, or too much alcohol. Choosing moderately will help you maintain a healthy weight and prevent some of the chronic diseases, such as heart disease and cancer, that are on the rise in the U.S. population.

The fact that more than 70% of adult Americans are overweight or obese demonstrates that we have not been practicing moderation when it comes to calorie intake.[22] One of the main culprits is likely the size of our food portions. The sandwiches, soft drinks, and French fry orders served in fast-food restaurants today are two to five times larger than what they were 50 years ago. The sizes of the snacks and meals we eat at home have also increased. As these portion sizes have grown, so has the amount we eat—and so has our weight.[32] Moderation

FIGURE 1.9 **Balance calories in with calories out** To keep your weight stable, you need to burn the same number of calories as you consume. Consuming extra calories during the day can be balanced by increasing the calories you burn in physical activity.

Choosing a Big Mac over a smaller burger means you will need to increase your energy expenditure by 300 Calories to maintain your weight.

You could do this by playing golf for about an hour, carrying your own clubs.

Choosing a grande Mocha Frappuccino over a regular iced coffee means you will need to increase your energy expenditure by 370 Calories to maintain your weight.

You could do this by jogging for about 30 minutes.

Ask Yourself

If you add a daily grande Mocha Frappuccino to your usual diet and do not increase your activity, what will happen to your weight?

Thinking It Through

A Case Study on Choosing a Healthy Diet

For many college students, their freshman year is the first time they are making all their own food choices, and they don't always make the best ones. Learning to apply the principles of variety, balance, and moderation can help improve these choices.

Helen doesn't really know how to choose a healthy diet, so she picks what she knows. Every day she eats cereal for breakfast, a peanut butter sandwich for lunch, and chicken, broccoli, and rice for dinner.

What's wrong with Helen's diet?
Your answer:

Amad's favorite breakfast is doughnuts, and he always has fast food for lunch—usually a burger and fries. He knows these are not the most nutrient-dense choices, but he is always in a hurry, rushing between school and work.

Suggest some more nutrient-dense foods that would be just as easy to grab as Amad's current breakfast and lunch choices.

Answer: He could grab yogurt for breakfast just as easily as a doughnut. If he has a doughnut for breakfast, he could balance this with a sandwich on whole-grain bread with turkey, lettuce, tomatoes, and peppers for lunch. He can get this meal just as quickly and easily as a burger and fries.

Suggest a nutrient-dense dinner that would help balance Amad's poor breakfast and lunch choices.
Your answer:

Sam has gained a few pounds and is worried that he will become a victim of the "freshman 15"—the 15 or so pounds gained by college students during the first year away from home. He thinks his weight gain is due to the dinner he eats on his all-you-can-eat college meal plan. He piles food on his plate, thinking he will only eat what he likes best but then ends up eating it all while he sits and relaxes with his friends.

Suggest two changes Sam could make to practice moderation with his dinner choices.
Your answer:

Marty knows that beans are a healthy choice, so she decides to try some Mexican food. She orders the beef and bean burrito platter shown here.

foodfolio / Alamy Stock Photo

How does Marty's meal stack up in terms of variety, balance, and moderation?
Your answer:

(Check your answers in Appendix L.)

---

makes it easier to balance your diet and allows you to enjoy a greater variety of foods (see *Thinking It Through: A Case Study on Choosing a Healthy Diet*).

Concept Check

1. **Why** is variety in a diet important?
2. **What** could you have for lunch to balance a breakfast that doesn't include any fruits or vegetables?
3. **How** are obesity and dietary moderation related?

1.5 Evaluating Nutrition Information

LEARNING OBJECTIVES

1. **List** the steps of the scientific method and give an example of how it is used in nutrition.
2. **Discuss** three different types of experiments used to study nutrition.
3. **Describe** the components of a sound scientific experiment.
4. **Distinguish** between reliable and unreliable nutrition information.

We are bombarded with nutrition information almost every day. The evening news, the morning papers, and the World Wide Web continually offer us tantalizing tidbits of nutrition advice. Food and nutrition information that used to take professionals

FIGURE 1.10 The scientific method The scientific method is a process used to ask and answer scientific questions through observation and experimentation.

Courtesy Lori Smolin

① The first step of the scientific method is to make an observation and ask questions about that observation.

Observation
More people get colon cancer in the United States than in Japan.

② The next step is to propose an explanation for this observation. This proposed explanation is called a hypothesis.

Hypothesis
The lower incidence of colon cancer in Japan than in the United States is due to differences in the diet.

③ Once a hypothesis has been proposed, experiments like this one are designed to test it. To generate reliable theories, the experiments done to test hypotheses must produce consistent, quantifiable results and must be interpreted accurately.

Experiment
Compare the incidence of colon cancer of Japanese people who move to the United States and consume a typical U.S. diet with Caucasian Americans who eat the same diet. **Result:** The Japanese people who eat the U.S. diet have the same higher incidence of colon cancer as Caucasian Americans.

④ If the results from repeated experiments support the hypothesis, a scientific theory can be developed. A single experiment is not enough to develop a theory; rather, repeated experiments showing the same conclusion are needed to develop a sound theory.

⑤ If experimental results do not support the hypothesis, a new hypothesis can be formulated.

Theory
The U.S. diet contributes to the development of colon cancer.

⑥ As new information becomes available, even a theory that has been accepted by the scientific community for years can be proved wrong.

Think Critically
A scientist has hypothesized that the difference in the incidence of colon cancer in Japan and the United States is due to differences in the genetic makeup of the populations. Based on the results of the experiment described in this illustration, explain why this hypothesis is not supported.

---

years to disseminate now travels with lightning speed, reaching millions of people within hours or days. Much of this information is reliable, but some can be misleading. In order to choose a healthy diet, we need to be able to sort out the useful material in this flood of information.

The Science Behind Nutrition

Like all other sciences, the science of nutrition is constantly evolving. As new discoveries provide clues to the right combination of nutrients needed for optimal health, new nutritional principles and recommendations are developed. Sometimes established beliefs and concepts give way to new information. Understanding the process of science can help consumers understand the nutrition information they encounter.

The systematic, unbiased approach that allows any science to acquire new knowledge and correct and update previous knowledge is the **scientific method**. The scientific method involves making observations of natural events, formulating **hypotheses** to explain these events, designing and performing experiments to test these hypotheses, and developing **theories** that explain the observed phenomenon based on the results of many studies (**Figure 1.10**). In nutrition, the scientific method is used to develop nutrient recommendations, understand the functions of nutrients, and learn about the role of nutrition in promoting health and preventing disease.

How Scientists Study Nutrition

Many different types of experiments are used to expand our knowledge of nutrition. Some make observations about

---

hypothesis A proposed explanation for an observation or a scientific problem that can be tested through experimentation.

theory A formal explanation of an observed phenomenon made after a hypothesis has been tested and supported through extensive experimentation.

relationships between diet and health; these are based on the science of **epidemiology**. Other types of experiments evaluate the effect of a particular dietary change on health. Some of these experiments study humans, others use animals; some look at whole populations, others study just a few individuals; and some use just cells or molecules (**Figure 1.11**).

A sound nutrition experiment studies the right experimental population, collects quantifiable data, includes proper experimental controls, and interprets the data accurately. The experimental population must be chosen to answer a specific question. For example, if a dietary supplement claims to increase bone strength in older women, a study to test this should use older women as subjects. For an experiment to determine whether a treatment does or does not have an effect, it must include enough subjects to demonstrate that the treatment causes the effect to occur more frequently than it would by chance. The number of subjects needed depends on how likely an effect is to occur without the treatment. For example, if weight training without a muscle-building supplement causes an increase in muscle mass, a large number of experimental subjects may be needed to demonstrate that there is a greater increase in muscle mass with the treatment—in this case, the muscle-building supplement. Results from studies with only a few subjects may not be able to distinguish effects that occur due to chance and should therefore be interpreted with caution.

Data collected in experiments must be quantifiable—that is, data must include parameters that can be measured reliably and repeatedly, such as body weight or blood pressure. Individual testimonies or opinions alone are not quantifiable, objective measures.

In order to know whether what is being tested has an effect, one must compare it with something. A **control group** acts as a standard of comparison for the factor, or **variable**, being studied. A control group is treated in the same way as the **experimental group** except that the control group does not receive the treatment being tested. For example, in a study examining the effect of a dietary supplement on muscle strength, the control group would consist of individuals of similar age, gender, and ability, eating similar diets and following similar workout regimens as individuals in the experimental group. While the experimental group would consume the supplement, the control subjects would consume a **placebo**, a harmless, inactive substance that is identical in appearance to the dietary supplement.

When an experiment has been completed, the results must be interpreted. Accurately interpreting results is just as important as conducting a study carefully. If a study conducted on a large group of young women indicates that a change in diet reduces breast cancer risk later in life, the results of that study cannot be used to claim that the same effect will occur if older women make a similar dietary change. Likewise, if the study looks only at the connection between a change in diet and breast cancer, the findings can't be used to claim a reduced risk for other cancers.

One way to ensure that the results of experiments are interpreted correctly is to have them reviewed by experts in the field who did not take part in the study being evaluated. Such a **peer-review process** is used in determining whether experimental results should be published in scientific journals. The reviewing scientists must agree that the experiments were conducted properly and that the results were interpreted appropriately. Nutrition articles that have undergone peer review can be found in many journals, including *The American Journal of Clinical Nutrition*, *The Journal of Nutrition*, *The Journal of the Academy of Nutrition and Dietetics*, *The New England Journal of Medicine*, and *The International Journal of Sport Nutrition*. Newsletters from reputable institutions, such as the *Tufts Health and Nutrition Letter,* the *Harvard Health Letter*, and *Nutrition Action Healthletter* are also reliable sources of nutrition and health information. The information in these newsletters comes from peer-reviewed articles but is written for a consumer audience.

Recommendations and policies regarding nutrition and healthcare practices are made by compiling the evidence from the wealth of well-controlled, peer-reviewed studies that are available. This is referred to as **evidence-based practice**.

Judging for Yourself

Not everything you hear is accurate. Because much of the nutrition information we encounter is intended to sell products, that information may be embellished to make it more appealing. Understanding the principles scientists use to perform nutrition studies can help consumers judge the nutrition information they encounter in their daily lives (see *What a Scientist Sees: Behind the Claims*). Some things that may tip you off to misinformation are claims that sound too good to be true, information from unreliable sources, information intended to sell a product, and information that is new or untested. The following questions can help you evaluate the validity of any nutrition information you encounter.

Does it make sense? Some claims are too outrageous to be true. For example, if a product claims to increase your muscle size without any exercise or decrease your weight without a change in diet, common sense should tell you that the claim is too good to be true. In contrast, an article that tells you that adding exercise to your daily routine will help you lose weight and increase your stamina is not so outrageous.

---

epidemiology The branch of science that studies health and disease trends and patterns in populations.

control group In a scientific experiment, the group of participants used as a basis of comparison. They are similar to the participants in the experimental group but do not receive the treatment being tested.

experimental group In a scientific experiment, the group of participants who undergo the treatment being tested.

FIGURE 1.11 **Types of nutrition studies** Scientists use a variety of methods to expand our understanding of nutrition.

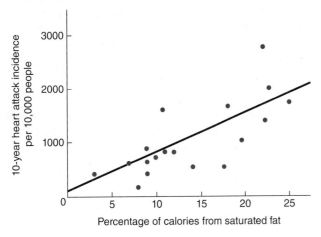

a. Epidemiological studies Epidemiological studies of populations around the world explore the impact of nutrition on health. If you were to measure saturated fat intake and the incidence of heart attacks in different populations, you might get a graph that looks like this one. It indicates that diets with a high percentage of calories from saturated fat are associated with an increased incidence of heart attacks. However, epidemiology does not determine cause-and-effect relationships; it just identifies patterns. Therefore, it cannot determine whether the higher incidence of heart attacks is caused by the high intake of saturated fat.

b. Clinical trials The observations and hypotheses that arise from epidemiology can be tested using clinical trials. In nutrition, clinical trials explore the health effects of altering people's diets—for instance, the possible effects of reducing saturated fat intake on blood cholesterol levels.

c. Animal studies Ideally, studies of human nutrition should be done with human subjects. However, because studying humans is costly, time-consuming, inconvenient for the subjects, and in some cases impossible for ethical reasons, many studies are done using animals. Guinea pigs provide a good model for studying heart disease, but even the best animal model is not the same as a human, and care must be taken when extrapolating animal results to humans.

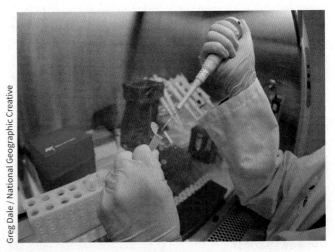

d. Biochemistry and molecular biology Laboratory-based techniques can be used to study nutrient functions in the body. For example, biochemistry can be used to study the chemical reactions that provide energy or synthesize molecules, such as cholesterol, and molecular biology can be used to study how nutrients regulate our genes.

What's the source? If a claim seems reasonable, find out where it came from. Personal testimonies are not a reliable source (**Figure 1.12**), but government recommendations regarding healthy dietary practices and information disseminated by universities generally are. Government recommendations are developed by committees of scientists who interpret the latest well-conducted research studies and use their conclusions to develop recommendations for the population as a whole. The information is designed to improve the health of the population. Information that comes from universities is supported by research studies that are well scrutinized and published in peer-reviewed journals. Many universities also provide

What a Scientist Sees

Behind the Claims

This product must be amazing! It will increase your muscle strength, decrease your body fat, and boost your drive and motivation. This is what consumers see. The claims sound great, but a scientist looking at the same ad may have some concerns.

First of all, the claims about muscle strength and motivation are testimonials based on individuals' feelings and impressions, and these are not objective measures.

A scientist would also question whether the research evidence supports the claim that the product increases muscle mass and decreases body fat. The study measured the amount of lean tissue (muscle, bone, and other nonfat body tissues) and fat tissue in weightlifters before and four weeks after they began consuming the POWER BOOST drink. The measures used provide quantifiable, repeatable data. The results report a gain of 5.2 lb of lean tissue and a loss of 4.5 lb of fat tissue in weightlifters taking POWER BOOST. This looks convincing, but the results for the control group are not reported in the ad. When the results for the experimental group are compared to those for the control group, a different picture emerges. This comparison (see graph) shows that the control group gained almost as much lean mass and lost slightly more fat mass than the group taking POWER BOOST.

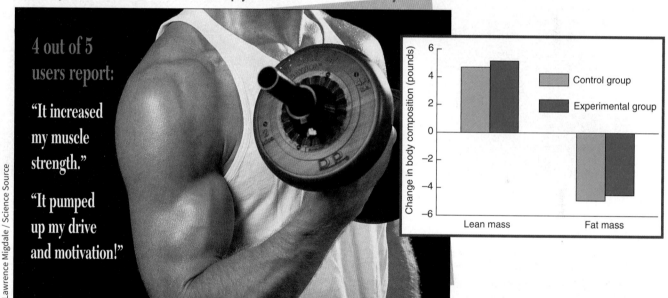

POWER BOOST

Years of research were needed to develop this special nutritional formulation. Just mix with water and drink one shake after your workout and with every meal or snack.

In a university study, 25 experienced weightlifters consumed one POWER BOOST shake 5 times a day for 4 weeks.

Lean body mass and fat mass were measured by underwater weighing before the study began and after 4 weeks of training while taking POWER BOOST.

RESULTS
The weightlifters gained an average of 5.2 lb of lean muscle and lost 4.5 lb of unwanted fat.

Think Critically

Based on the information in the graph, explain why you would or would not recommend this product.

FIGURE 1.12 **Individual testimonies are not proof** Weight-loss product advertisements commonly show before-and-after photos of people who have successfully lost weight using the product. These individuals' success stories are not a guarantee that the product will produce the same results for you or anyone else. These individuals' results are not compared to those for a control group or subjected to scientific evaluation. Therefore, it cannot be assumed that similar results will occur in other people.

Blend Images / Getty Images

Gazimal / Getty Images, Inc.

Ask Yourself

If an ad for a weight-loss product showed you these before-and-after photos with a quote saying "In just 3 weeks I went from a size 14 to a size 8 and looked fabulous on my vacation," what questions should you ask to determine whether this is valid information?

information that targets the general public. Not-for-profit organizations such as the Academy of Nutrition and Dietetics and the American Medical Association are also reliable sources of nutrition information.

If you are looking at an article in print or posted on a Web site, checking the author's credentials can help you evaluate the credibility of the information. Where does the author work? Does this person have a degree in nutrition or medicine? Although "nutrition counselors" may provide accurate information, this term is not legally defined and can be used by individuals with no formal nutrition or medical training.

One reliable source of nutrition information is a registered dietitian (RD), also called a registered dietitian nutritionist (RDN). RD/RDNs are nutrition professionals who are certified to provide nutrition education and counseling. To obtain certification, an RD/RDN must earn a four-year college degree that includes coursework approved by the Academy of Nutrition and Dietetics, complete a supervised internship, and pass a national exam.

Is it selling something? If a person or company will profit from the information presented, be wary. Advertisements are designed to increase product sales, and the company stands to profit if you believe the claims that are made. Information presented online, in newspapers and magazines, and on television may also be biased or exaggerated because it is designed to help boost sales, not necessarily to promote health and well-being. Even a well-designed, carefully executed study published in a peer-reviewed journal can be a source of misinformation if its results have been interpreted incorrectly or exaggerated (**Figure 1.13**).

Has it stood the test of time? Often the results of new scientific studies are on the news the same day they are

FIGURE 1.13 **Results may be misinterpreted in order to sell products** These rats, which were given large doses of vitamin E, lived longer than rats that consumed less vitamin E. Does this mean that dietary supplements of vitamin E will increase longevity in people? Not necessarily. The results of animal studies can't always be extrapolated to humans, but they are often the basis of claims in ads for dietary supplements.

presented at a meeting or published in a peer-reviewed journal. However, a single study cannot serve as a basis for a reliable theory. Results need to be reproduced and supported numerous times before they can be used as a foundation for nutrition recommendations.

Headlines based on a single study should therefore be viewed skeptically. The information may be accurate, but there is no way to know because there has not been enough time to repeat the work and reaffirm the conclusions. If, for example, someone has found the secret to easy weight loss, you will undoubtedly encounter this information again at some later time if the finding is valid. If the finding is not valid, it will fade away with all the other weight-loss concoctions that have come and gone.

Concept Check

1. **What** is the difference between a hypothesis and a theory?
2. **How** is an epidemiologic study different from a clinical trial?
3. **Why** are control groups important in any scientific experiment?
4. **Why** are personal testimonies not a source of reliable nutrition information?

Summary

1 Food Choices and Nutrient Intake 2

- The foods you choose determine which **nutrients** you consume. Choosing foods that are high in **nutrient density** allows you to obtain more nutrients in fewer calories, as shown in this graph. Fortified foods, or foods to which nutrients have been added, and **dietary supplements** can also contribute nutrients to the diet.

Figure 1.1a Nutrient density

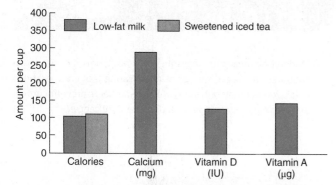

- Food contains not only nutrients but also nonnutritive substances, such as **phytochemicals**, that may provide additional health benefits. Foods that provide health benefits beyond basic nutrition are called **functional foods**. Some foods are naturally functional, and others are made functional through **fortification**.

- The food choices we make are affected not just by personal taste but by our **food environment**, which includes what foods are available in our community, what foods are advertised, how much food costs, and what we learn to eat from family, culture, and traditions.

2 Nutrients and Their Functions 6

- Nutrients are grouped into six classes. **Carbohydrates**, **lipids**, **proteins**, and water are referred to as **macronutrients** because they are needed in large amounts. **Vitamins** and **minerals** are **micronutrients** because they are needed in small amounts to maintain health.

- Carbohydrates, lipids, and proteins are nutrients that provide energy, typically measured in **calories**. Lipids, proteins, carbohydrates, minerals, and water perform structural roles, as shown in the following illustration, forming and maintaining the structure of our bodies. All six classes of nutrients help regulate body processes. The energy, structure, and regulation provided by nutrients are needed for growth, maintenance and repair of the body, and reproduction.

Figure 1.5b Nutrient functions

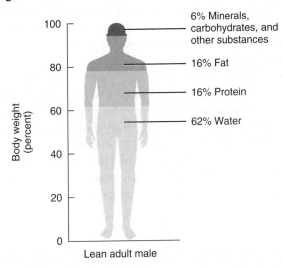

3 Nutrition in Health and Disease 9

- Your diet affects your health. The foods you choose contain the nutrients needed to keep you alive and healthy and prevent **malnutrition**. **Undernutrition** results from consuming too few calories and/or too few nutrients. **Overnutrition** can result from a toxic dose of a nutrient or from a chronic excess of nutrients or calories, which over time contributes to chronic diseases, such as those shown in this graph.

Figure 1.7 Overnutrition

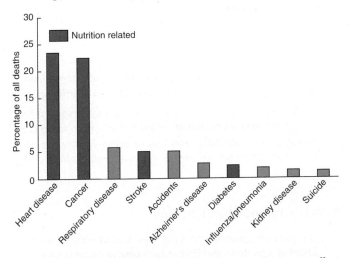

- Your genetic makeup and the diet you consume interact to affect your health risks. **Nutritional genomics** studies how the genes you inherit affect the impact of diet on health and how the diet you choose affects the activity of your genes.

4 Choosing a Healthy Diet 13

- A healthy diet includes a variety of nutrient-dense foods from the different food groups as well as a variety of foods from within each group. Variety is important to ensure that nutrient needs are met because different foods provide different nutrients and health-promoting substances as well as a variety of tastes.

- Balance means mixing and matching foods and meals in order to obtain enough of the nutrients you need and not too much of the ones that can potentially harm your health. Extra calories you consume during the day can be balanced by increasing the calories you burn in physical activity, as shown.

Figure 1.9 Balance calories in with calories out

- Moderation means not ingesting too many calories or too much saturated fat, sugar, salt, or alcohol. Eating moderate portions helps you maintain a healthy weight and helps prevent chronic diseases such as heart disease, diabetes, and cancer.

5 Evaluating Nutrition Information 15

- Nutrition uses the **scientific method** to study the relationships among food, nutrients, and health. The scientific method, illustrated below, involves observing and questioning natural events, formulating **hypotheses** to explain these events, designing and performing experiments to test the hypotheses, and developing **theories** that explain the observed phenomena based on the experimental results.

- To be valid, a nutrition experiment must provide quantifiable measurements, study the right type and number of subjects, and use appropriate **control groups**. When a study has been completed, the results must be interpreted fairly and accurately. The **peer-review process** ensures that studies published in professional journals adhere to a high standard of experimental design and interpretation of results.

- Not all the nutrition information we encounter is accurate. The first step in deciding whether a nutritional claim is valid is to ask whether the claim makes sense. If it sounds too good to be true, it probably is. It is also important to determine whether the information came from a reliable source, whether it is trying to sell a product, and whether it has been confirmed by multiple studies.

Figure 1.10 The scientific method

① The first step of the scientific method is to make an observation and ask questions about that observation.

Observation
More people get colon cancer in the United States than in Japan.

② The next step is to propose an explanation for this observation. This proposed explanation is called a hypothesis.

Hypothesis
The lower incidence of colon cancer in Japan than in the United States is due to differences in the diet.

③ Once a hypothesis has been proposed, experiments like this one are designed to test it. To generate reliable theories, the experiments done to test hypotheses must produce consistent, quantifiable results and must be interpreted accurately.

Experiment
Compare the incidence of colon cancer of Japanese people who move to the United States and consume a typical U.S. diet with Caucasian Americans who eat the same diet. **Result:** The Japanese people who eat the U.S. diet have the same higher incidence of colon cancer as Caucasian Americans.

④ If the results from repeated experiments support the hypothesis, a scientific theory can be developed. A single experiment is not enough to develop a theory; rather, repeated experiments showing the same conclusion are needed to develop a sound theory.

Courtesy Lori Smolin

⑤ If experimental results do not support the hypothesis, a new hypothesis can be formulated.

Theory
The U.S. diet contributes to the development of colon cancer.

⑥ As new information becomes available, even a theory that has been accepted by the scientific community for years can be proved wrong.

Key Terms

- amino acid 6
- calorie 2
- carbohydrates 6
- cholesterol 6
- control group 17
- designer food 3
- dietary supplement 3
- element 6
- energy-yielding nutrient 7
- empty calories 2
- epidemiology 17
- essential nutrient 2
- evidence-based practice 17
- experimental group 17
- fiber 6
- food environment 3
- fortification 3
- functional food 3
- genes 10
- hormone 9
- hypothesis 16
- kilocalorie 7
- lipids 6
- macronutrient 6
- malnutrition 9
- micronutrient 6
- mineral 6
- nutraceutical 3
- nutrient 2
- nutrient density 2
- nutrigenetics 12
- nutrigenomics 12
- nutritional genomics 12
- organic compound 6
- osteoporosis 10
- overnutrition 10
- peer-review process 17
- phytochemical 3
- placebo 17
- protein 6
- saturated fat 6
- scientific method 16
- theory 16
- undernutrition 9
- unsaturated fat 6
- variable 17
- vitamin 6
- zoochemical 3

What is happening in this picture?

An appealing display of candy, such as this one, is located near the cash register in almost every grocery store, drug store, and even hardware store across the country so you will see it while waiting in the checkout line. Food manufacturers pay placement fees to put their products in this and other high-profile spots within the store.

Think Critically

1. Why do candy companies pay to have their products placed here?
2. Would the healthfulness of your diet be affected if the display in the grocery store were apples instead of candy bars?
3. How do arrangements such as this between stores and food manufactures affect our food environment?

CHAPTER 2

Rex A. Stucky / National Geographic Creative

Guidelines for a Healthy Diet

"What you don't know could kill you" may have been the first nutrition recommendation. To swallow the wrong berry or gulp down water from a suspect source could have been fatal to early humans. Such lessons served as anecdotal guideposts to survival. As societies developed, dietary cautions turned into taboos, sometimes laws, and, ultimately, nutrition recommendations.

Governments have been providing what we would call modern nutrition information for the past 150 years. As the Industrial Revolution swept through Great Britain, urban populations—and poverty and hunger—swelled. To ensure a healthy workforce, the British government developed minimum dietary guidelines utilizing the cheapest foods. It wasn't until World War I that the British Royal Society determined that a healthy workforce required a healthy diet—not necessarily the cheapest. So fruits, vegetables, and milk became elements of nutritional guidance. Since then, virtually every nation has sought to establish dietary standards for its citizens.

Today, modern public health agencies provide valuable information regarding healthy food choices. However, this information isn't always understood or used properly. As portion sizes grow, so do waistlines—and the attendant health concerns. "What you don't know could kill you" remains as vital an admonition today as it was 40,000 years ago.

CHAPTER OUTLINE

Nutrition Recommendations 26
- Past and Present U.S. Recommendations

Debate: How Involved Should the Government Be in Your Food Choices?
- How We Use Nutrition Recommendations

What a Scientist Sees: Trends in Milk Consumption

Dietary Reference Intakes (DRIs) 31
- Recommendations for Nutrient Intake
- Recommendations for Energy Intake

The Dietary Guidelines for Americans 33
- Follow a Healthy Eating Pattern
- Shift to Healthier Choices

What Should I Eat? To Follow the Dietary Guidelines
- Support from All Segments of Society

MyPlate: Putting the Guidelines into Practice 38
- Choose the Right Amount from Each Food Group
- Choose a Variety of Nutrient-Dense Foods
- Limit Saturated Fat, Sodium, and Added Sugars
- MyPlate Compared with Choice (Exchange) Lists

Food and Supplement Labels 42
- Food Labels

Thinking It Through: A Case Study on Using Food Labels to Guide Food Choices
- Dietary Supplement Labels

2.1 Nutrition Recommendations

LEARNING OBJECTIVES

1. **Explain** the purpose of government nutrition recommendations.
2. **Discuss** how the focus of U.S. nutrition recommendations has changed over the past 100 years.
3. **Describe** the information needed to assess an individual's nutritional status.

What should we be eating if we want to satisfy our nutrient needs? Our personal preferences and our food environment influence the foods we choose; however, these choices may not always be healthy ones. Our taste buds respond to flavor rather than sensible nutrition, and restaurants and food manufacturers want to sell their products—not necessarily promote a healthy diet. Government recommendations, on the other hand, are designed with individual health as well as public health in mind. They can be used to plan diets and to evaluate what we are eating, both as individuals and as a nation.

Past and Present U.S. Recommendations

The U.S. federal government has been making nutritional recommendations for over 100 years. These recommendations have changed over time as our food intake patterns have changed and our knowledge of what constitutes a healthy diet has evolved.

The first dietary recommendations in the United States, published in 1894 by the U.S. Department of Agriculture (USDA), suggested amounts of protein, carbohydrate, fat, and "mineral matter" needed to keep Americans healthy.[1] At the time, specific vitamins and minerals essential for health had not been identified; nevertheless, this work set the stage for the development of the first **food guides**. Food guides are used to translate nutrient-intake recommendations into food choices (**Figure 2.1**). The food guide *How to Select Foods*, released in 1917, made recommendations based on five food groups: meat and milk, cereals, vegetables and fruit, fats and fatty foods, and sugars and sugary foods.

In the early 1940s, as the United States entered World War II, the Food and Nutrition Board was established to advise the Army and other federal agencies regarding problems related to food and the nutritional health of the armed forces and the general population. The Food and Nutrition Board developed the first set of recommendations for specific amounts of nutrients. These came to be known as the Recommended Dietary Allowances (RDAs). The original RDAs made recommendations on amounts of energy and on specific nutrients that were most likely to be deficient in people's diets—protein, iron, calcium, vitamins A and D, thiamin, riboflavin, niacin, and vitamin C. Recommended intakes were based on amounts that would prevent nutrient deficiencies.

Over the years since those first standards were developed, dietary habits and disease patterns have changed, and dietary recommendations have had to change along with them. Overt nutrient deficiencies are now rare in the United States, but the incidence of nutrition-related chronic diseases, such as heart disease, diabetes, osteoporosis, and obesity, has increased. To combat these more recent health concerns, recommendations are now intended to promote health as well as prevent deficiencies (see *Debate: How Involved Should the Government Be in Your Food Choices?*). The original RDAs have been expanded into the Dietary Reference Intakes, which address problems of excess as well as deficiency. The *Dietary Guidelines for Americans*, which make diet and lifestyle recommendations that promote health and reduce the risks of obesity and chronic disease, were introduced in 1980 and have been revised every five years.[2] Early food guides have evolved into *MyPlate*, an extensive online educational tool built around the Dietary Guidelines recommendations for amounts and types of food from five food groups (see Figure 2.1). In addition, standardized food labels have been developed to help consumers choose foods that meet these recommendations.

FIGURE 2.1 Today's Food Guide How food guides present recommendations has changed over the years, but the basic message has stayed the same: Choose the right combinations of foods to promote health. The MyPlate icon, shown here, represents the latest food guide; it shows the proportions recommended from each of the different food groups.

Debate

How Involved Should the Government Be in Your Food Choices?

The Issue

Poor dietary habits in the United States have resulted in a high incidence of nutrition-related chronic disease. Should the government intervene to change the types of food available to us?

The typical U.S. diet is not as healthy as it could be. Our lack of dietary discretion has contributed to our high rates of obesity, diabetes, high blood pressure, and heart disease.[2] This is not only the concern of the individuals whose lives are affected by these conditions but also the government. The dollar cost to our health care system is huge; half of the $150 billion per year the United States spends on obesity comes from government-funded Medicare and Medicaid.[3] Government concern is not just financial. The fact that almost one in four applicants to the military is rejected for being overweight is suggested to be a threat to national security and military readiness.[4]

So, who is responsible for our unhealthy diet, and who should be responsible for changing what we eat? Proponents of more government involvement in our food choices suggest that our food environment is the cause of our unhealthy eating habits. Obesity expert Kelly Brownell believes that environment plays a more powerful role in determining food choices than does personal irresponsibility.[5] Brownell and other proponents of government intervention argue that the government should treat our noxious food environment like any other public health threat and develop programs to keep us safe and healthy. Just as government regulations help to ensure that our food is not contaminated with harmful bacteria, laws could ensure that what you order at a restaurant will not contribute to heart disease or cancer. Unfortunately, unlike bacteria, individual foods are difficult to classify as healthy or unhealthy. Almost all food has some nutritional benefits, and the arguments as to what is a "junk food" and what we should add or subtract from our diets are ongoing. However, many people believe there are things that could be done to ensure healthier choices.

One option to encourage healthier choices suggested by proponents of government intervention is to tax junk food, making it more expensive, and to increase subsidies for fruits and vegetables, making them less expensive. Other suggestions include implementing zoning restrictions to keep fast-food restaurants away from schools and child-care facilities, placing limitations on the types of foods that can be advertised on children's television, and restricting the portion sizes that can be sold (see figure). All these ideas have pros and cons, and none will absolve individuals of the responsibility for getting more exercise and making healthier food choices.

Opponents of government involvement believe it is an infringement on personal freedom and suggest that individuals need to take responsibility for their actions. They propose

In 2013, New York City's Board of Health voted to prohibit the sale of sugar-sweetened drinks in cups larger than 16 ounces. This so-called "soda ban" was struck down by New York State's court of appeals. Other municipalities are trying different approaches. For example, Chicago and Philadelphia have approved a tax on sugary drinks.

that the food industry work with the public to make healthier food more available and affordable. Many food companies have already responded to the need for a better diet; General Mills and Kellogg's offer whole-grain cereals. And the giant food retailer Wal-Mart is working with suppliers to reduce the amount of sodium and added sugar and eliminate *trans* fat in packaged foods.

Our current food environment makes unhealthy eating easy. Fatty, salty, and sugary foods are available 24/7, and the portions offered are often massive. To preserve our public health, the United States needs to change the way it eats. This change could be driven by government regulations and taxes, it could come from changes in the food industry, or it could come from individuals taking more responsibility for their choices and their health. A synergy of policy intervention, industry cooperation, and personal efforts is likely needed to solve the crisis.

Think Critically

Do you think government restrictions on the size of beverages that can be sold will help curb obesity? Why or why not?

How We Use Nutrition Recommendations

Nutrition recommendations are developed to address the nutritional concerns of the population and help individuals meet their nutrient needs. These recommendations can also be used to evaluate the nutrient intake of populations and of individuals within populations. Determining what people eat and how their nutrient intake compares to nutrition recommendations is important for assessing their **nutritional status**.

When evaluating the nutritional status of a population, food intake can be assessed by having individuals record or recall their food intake or by using information about the amounts and types of food available to the population to identify trends in the diet (see *What a Scientist Sees: Trends in Milk Consumption*).

When an individual's food intake is evaluated in conjunction with information about his or her health and medical history, the person's nutritional status can be assessed (**Figure 2.2**). When this information is surveyed within populations, relationships between dietary intake and health

What a Scientist Sees

Trends in Milk Consumption

The graph below shows estimates of milk consumption in the United States from 1970 to 2010.[6] Anyone looking at the graph can tell that consumption of total milk as well as whole milk declined, while consumption of lower-fat milks increased. A nutrition scientist looking at this graph, however, would see not just changes in the amount of milk Americans drank but the nutritional and public health implications of these changes as well.

Because whole milk is high in saturated fat, replacing whole milk with lower-fat milk decreases saturated fat intake, potentially reducing the risk of heart disease. However, milk is the most important source of calcium in the North American diet, and calcium is needed for bone health, so a reduction in total milk consumption could cause an increase in the incidence of fractures due to low bone density, a condition called osteoporosis.[7]

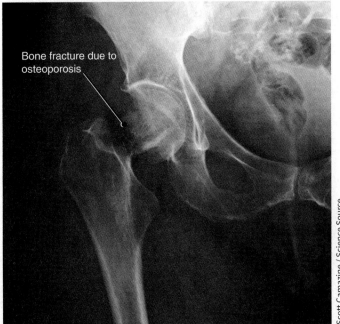

Bone fracture due to osteoporosis

Scott Camazine / Science Source

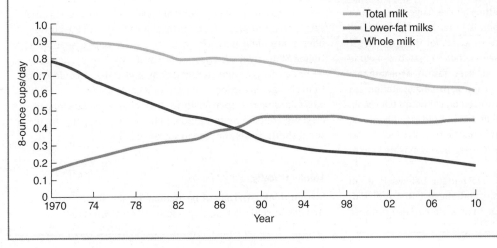

Interpret the Data

If 1 cup of milk has 300 milligrams of calcium, how much less calcium per person was consumed from milk in 2010 compared with 1970?

---

nutritional status An individual's health, as it is influenced by the intake and utilization of nutrients.

and disease can be identified. This is important for developing public health policies and programs that address nutritional problems. For example, population surveys such as the National Health and Nutrition Examination Survey (NHANES) helped public health officials recognize that low iron levels are a problem for many people, including young women, preschool children, and older adults. This information led to the fortification of grain products with iron beginning in the 1940s. Recent NHANES data have also shown that the number of calories Americans consume per day has increased over the past few decades and that the incidence of obesity has increased dramatically during the same period. This has led public health experts to develop programs to improve both the diet and the fitness of Americans.

PROCESS DIAGRAM

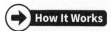

FIGURE 2.2 **Assessing nutritional status** A complete assessment of an individual's nutritional status includes a diet analysis, a physical exam, a medical history, and an evaluation of nutrient levels in the body. An interpretation of this information can determine whether an individual is well nourished, malnourished, or at risk of malnutrition.

1. Determine typical food intake. Typical food intake can be estimated by having people recall what they have eaten with the help of a trained interviewer or record all the food and beverages they consume as they consume them. Often a cell phone app is the most convenient way to do this. Because food intake varies from day to day, to obtain a realistic picture, intake should be monitored for more than one day.

Nutrient	My DRI	My Intake	Percent of My DRI (0% – 50% – 100%)
Vitamin A (RAE)	700 µg	525	75%
Vitamin C	75 mg	86	115%
Iron	18 mg	9.7	54%
Calcium	1000 mg	750	75%
Saturated fat	< 23.8 g	31.9	Above recommended range

Copyright © 2013 John Wiley & Sons, Inc. All rights reserved.

2. Compare intake to recommendations. A quick diet assessment can be done by comparing an individual's food intake to the food group recommendations of MyPlate. A more thorough analysis compares intake of individual nutrients to recommendations. In this example, which shows only a few nutrients, intakes of vitamin A, iron, and calcium are below the recommended amounts, and intakes of vitamin C and saturated fat are above the recommended amounts.

3. Evaluate physical health. A physical examination can detect signs of nutrient deficiencies or excesses. Measures of height and weight can be monitored over time or compared with standards for a given population. Drastic changes in measurements or measurements that are significantly above or below the standards could indicate nutritional deficiency or excess.

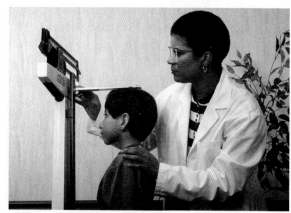

Blair Seitz / Science Source

(continues)

FIGURE 2.2 *(continued)*

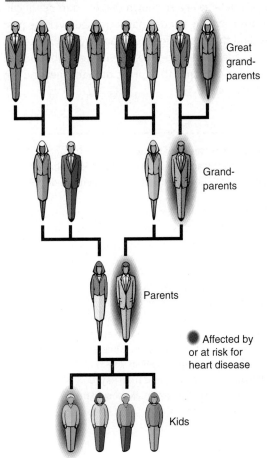

Affected by or at risk for heart disease

4 **Consider medical history and lifestyle.** Personal and family medical histories are important because genetic risk factors affect an individual's risk of developing a nutrition-related disease. For example, if you have high blood cholesterol and your father died of a heart attack at age 50, you have a higher-than-average risk of developing heart disease. Lifestyle factors such as physical activity level and eating habits can add to or reduce your inherited risk.

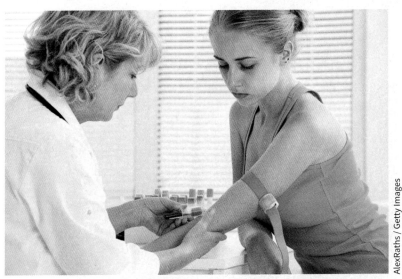

5 **Assess with laboratory tests.** Measures of nutrients, their by-products, or their functions in the blood, urine, and body cells can help detect nutrient deficiencies and excesses or the risk of nutrition-related chronic diseases (see Appendix F). For instance, an individual's iron status can be assessed by drawing a blood sample, measuring hemoglobin (an iron-carrying protein) and hematocrit (the proportion of blood volume that is red blood cells), and then comparing the values to the healthy range.

Think Critically

If your food record shows that your saturated fat intake is above recommendations, what other components of nutrition assessment would help you determine your overall risk for heart disease? Why?

The information obtained from population health and nutrition surveys is also used to determine whether the nation is meeting health and nutrition goals, such as those established by **Healthy People** (see Appendix C). This set of health-promotion and disease-prevention objectives is revised every 10 years, with the goal of increasing the quality and length of healthy lives for the population as a whole and eliminating health disparities among different segments of the population. The latest version of these objectives has been released as *Healthy People 2020*. The long-term goal is to create a physical and social environment in which everyone has a chance to live long, healthy lives.[8]

Concept Check

1. **How** do nutrition recommendations benefit individual and public health?
2. **Why** do the current DRIs focus on preventing chronic disease?
3. **How** is an individual's medical history used in assessing their nutritional status?

2.2 Dietary Reference Intakes (DRIs)

LEARNING OBJECTIVES

1. **Summarize** the purpose of the DRIs.
2. **Describe** the four sets of DRI values used in recommending nutrient intake.
3. **List** the factors that are considered when estimating an individual's energy needs (EERs).
4. **Explain** the concept of the Acceptable Macronutrient Distribution Ranges (AMDRs).

The **Dietary Reference Intakes (DRIs)** are recommendations for the amounts of energy, nutrients, and other food components that healthy people should consume in order to stay healthy, reduce the risk of chronic disease, and prevent deficiencies.[9] The DRIs can be used to evaluate whether a person's diet provides all the essential nutrients in adequate amounts. They include several types of recommendations that address both nutrient intake and energy intake and include values that are appropriate for people of different genders and stages of life (**Figure 2.3**).

Recommendations for Nutrient Intake

The DRI recommendations for nutrient intake include four sets of values. The **Estimated Average Requirements (EARs)** are average amounts of nutrients or other dietary components required by healthy individuals in a population (**Figure 2.4**). They are used to assess the adequacy of a population's food supply or typical nutrient intake and are not appropriate for evaluating an individual's intake. The **Recommended Dietary Allowances (RDAs)** are set higher than the EARs and represent amounts of nutrients and other dietary components that meet the needs of most healthy people (see Figure 2.4). When there aren't enough data about nutrient requirements to establish RDAs, **Adequate Intakes (AIs)** are set, based on what healthy people typically eat. RDA or AI values can be used as goals for individual intake and to plan and evaluate individual diets (see Appendix A). They are meant to represent the amounts that most healthy people should consume, on average, over several days or even weeks, not each and every day. Because they are set high enough to meet the needs of almost all healthy people, intake below the RDA or AI does not necessarily mean that an individual is deficient, but the risk of deficiency is greater than if the individual consumed the recommended amount.

FIGURE 2.3 **DRIs for all population groups** The DRIs include four types of nutrient intake recommendations and two types of recommendations related to energy intake. Because gender and life stage affect nutrient needs, recommendations have been set for each gender and for various life-stage groups. These values take into account the physiological differences that affect the nutrient needs of men and women, infants, children, adolescents, adults, older adults, and pregnant and lactating women.

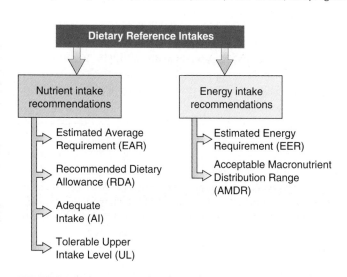

Estimated Average Requirements (EARs) Nutrient intakes estimated to meet the needs of 50% of the healthy individuals in a given gender and life-stage group.

Recommended Dietary Allowances (RDAs) Nutrient intakes that are sufficient to meet the needs of almost all healthy people in a specific gender and life-stage group.

Adequate Intakes (AIs) Nutrient intakes that should be used as a goal when no RDA exists. AI values are an approximation of the nutrient intake that sustains health.

FIGURE 2.4 **Understanding EARs, RDAs, and ULs** The EAR and RDA for a nutrient are determined by measuring the amount of the nutrient required by different individuals in a population group and plotting all the values. The resulting plot is a bell-shaped curve; a few individuals in the group need only a small amount of the nutrient, a few need a large amount, and the majority need an amount that falls between the extremes.

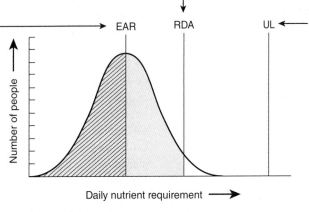

An EAR is the average amount of a nutrient required for good health. If everyone in the population consumed this amount, only 50% would obtain enough of the nutrient to meet their requirements (shown as diagonal lines).

The RDA is set by adding a safety factor to the EAR. About 97% of the population meets its needs by consuming this amount (shown as yellow shading). If nutrient intake meets the RDA, the risk of deficiency is very low. As intake falls, the risk of a deficiency increases.

The UL is set well above the needs of everyone in the population and represents the highest amount of the nutrient that will not cause toxicity symptoms in the majority of healthy people. As intake rises above the UL, the likelihood of toxicity increases.

Ask Yourself
1. Which DRI value(s) is/are set at a level that will meet the needs of most healthy people in the population?
2. Which DRI value represents the amount above which toxicity becomes more likely?

The fourth set of values, **Tolerable Upper Intake Levels (ULs)**, specifies the maximum amount of a nutrient that most people can consume on a daily basis without some adverse effect (see Figure 2.4). For most nutrients, it is difficult to exceed the UL by consuming food. Most foods do not contain enough of any one nutrient to cause toxicity; however, some dietary supplements and fortified foods may. For some nutrients, the UL is set for total intake from all sources, including food, fortified foods, and dietary supplements. For other nutrients, the UL refers to intake from supplements alone or from supplements and fortified foods. For many nutrients, there is no UL because not enough information is available to determine it.

Recommendations for Energy Intake

The DRIs make two types of recommendations about energy intake. The first, called **Estimated Energy Requirements (EERs)** (**Figure 2.5a**), provides an estimate of how many

FIGURE 2.5 **Meeting energy needs** The DRIs recommend amounts of energy (calories) and proportions of carbohydrate, fat, and protein that provide a healthy diet.

a. The EER represents the amount of energy required to maintain weight. Based on the EER calculation, a 19-year-old girl who is 5'4" tall, weighs 127 pounds, and gets no exercise needs about 1940 Calories a day. If she adds an hour of tennis or other moderate activity to her daily routine, her EER will increase to about 2400 Calories; she would need to eat an additional 460 Calories more per day to maintain her current weight.[9]

Think Critically

Why would gaining 20 pounds affect this 19-year-old's EER?

(continues)

Tolerable Upper Intake Levels (ULs) Maximum daily intake levels that are unlikely to pose risks of adverse health effects to almost all individuals in a given gender and life-stage group.

Estimated Energy Requirements (EERs) Energy intakes that are predicted to maintain body weight in healthy individuals.

FIGURE 2.5 *(continued)*

This meal contains approximately 480 Calories, of which about 55% is from carbohydrate, 20% is from protein, and 25% is from fat.

This meal contains approximately 740 Calories, of which about 30% is from carbohydrate, 35% is from protein, and 35% is from fat.

b. The AMDRs are a guide for selecting healthy proportions of carbohydrate, protein, and fat. As shown by these two meals, many different food combinations can provide a healthy diet. Although only the meal on the left provides proportions of carbohydrate, protein, and fat that fall within the AMDRs, the meal on the right can still be part of a diet that meets the AMDRs if other meals that day are lower in protein and fat and higher in carbohydrate.

calories are needed to keep body weight stable. EER calculations take into account a person's age, gender, weight, height, and level of physical activity (see Appendix A). A change in any of these variables changes the person's energy needs.

The second type of energy recommendation, called **Acceptable Macronutrient Distribution Ranges (AMDRs)**, makes recommendations about the proportions of calories that should come from carbohydrate, fat, and protein in a healthy diet. AMDRs are ranges—10 to 35% of calories from protein, 45 to 65% of calories from carbohydrate, and 20 to 35% of calories from fat—not exact values. This is because a wide range of macronutrient distributions is associated with health. AMDRs are intended to promote diets that minimize disease risk and allow flexibility in food intake patterns (**Figure 2.5b**).

Concept Check

1. **What** are RDAs and AIs used for?
2. **How** might you use ULs?
3. **What** happens to a person's EER as they increase their activity level?
4. **Why** are AMDR values given as ranges rather than as single numbers?

2.3 The Dietary Guidelines for Americans

LEARNING OBJECTIVES

1. **Explain** the purpose of the Dietary Guidelines.
2. **List** four characteristics of a healthy eating pattern.
3. **Name** three foods or food components that the Dietary Guidelines recommend we increase and three that we should decrease in our diets.
4. **Describe** how different segments of society can be involved in promoting a healthy lifestyle.

The DRIs tell you how much of each nutrient you need, but they do not help you choose foods that will meet these needs.

To help consumers choose diets that will meet their needs, the U.S. government has developed the ***Dietary Guidelines for Americans***. The *2015–2020 Dietary Guidelines for Americans* include five overarching evidence-based guidelines that encourage healthy eating patterns, recognize that most Americans need to make shifts in their food and beverage choices to achieve a healthy pattern, and urge all segments of society to play a role in supporting healthy eating and activity choices.[2] **Table 2.1** lists the five guidelines and the key recommendations of the 2015–2020 Dietary Guidelines. Adopting these recommendations will help support a healthy body weight, meet nutrient needs, and reduce the risk of chronic disease throughout all stages of life. The guidelines target Americans 2 years and older.

Acceptable Macronutrient Distribution Ranges (AMDRs) Healthy ranges of intake for carbohydrate, fat, and protein, expressed as percentages of total energy intake.

TABLE 2.1 Dietary Guidelines for Americans, 2015–2020[2]

The Guidelines

1. **Follow a healthy eating pattern across the life span.** All food and beverage choices matter. Choose a healthy eating pattern at an appropriate calorie level to help achieve and maintain a healthy body weight, support nutrient adequacy, and reduce the risk of chronic disease.

2. **Focus on variety, nutrient density, and amount.** To meet nutrient needs within calorie limits, choose a variety of nutrient-dense foods across and within all food groups in recommended amounts.

3. **Limit calories from added sugars and saturated fats and reduce sodium intake.** Consume an eating pattern low in added sugars, saturated fats, and sodium. Cut back on foods and beverages higher in these components to amounts that fit within healthy eating patterns.

4. **Shift to healthier food and beverage choices.** Choose nutrient-dense foods and beverages across and within all food groups in place of less healthy choices. Consider cultural and personal preferences to make these shifts easier to accomplish and maintain.

5. **Support healthy eating patterns for all.** Everyone has a role in helping to create and support healthy eating patterns in multiple settings nationwide, from home to school to work to communities.

Key Recommendations

The *Dietary Guidelines'* Key Recommendations for healthy eating patterns should be applied in their entirety, given the interconnected relationship that each dietary component can have with others.

Consume a healthy eating pattern that accounts for all foods and beverages within an appropriate calorie level.

A healthy eating pattern includes:

- A variety of vegetables from all of the subgroups—dark green, red and orange, legumes (beans and peas), starchy, and other
- Fruits, especially whole fruits
- Grains, at least half of which are whole grains
- Fat-free or low-fat dairy, including milk, yogurt, cheese, and/or fortified soy beverages
- A variety of protein foods, including seafood, lean meats and poultry, eggs, legumes (beans and peas), and nuts, seeds, and soy products
- Oils

A healthy eating pattern limits:

- Saturated fats and *trans* fats, added sugars, and sodium

Key Recommendations that are quantitative are provided for several components of the diet that should be limited. These components are of particular public health concern in the United States, and the specified limits can help individuals achieve healthy eating patterns within calorie limits:

- Consume less than 10% of calories per day from added sugars.
- Consume less than 10% of calories per day from saturated fats.
- Consume less than 2300 milligrams (mg) per day of sodium.
- If alcohol is consumed, it should be consumed in moderation—up to one drink per day for women and up to two drinks per day for men—and only by adults of legal drinking age.

In tandem with the recommendations above, Americans of all ages—children, adolescents, adults, and older adults—should meet the *Physical Activity Guidelines for Americans* to help promote health and reduce the risk of chronic disease. Americans should aim to achieve and maintain a healthy body weight. The relationship between diet and physical activity contributes to calorie balance and managing body weight. As such, the *Dietary Guidelines* includes a Key Recommendation to

- Meet the *Physical Activity Guidelines for Americans*.[a]

[a] U.S. Department of Health and Human Services. *2008 Physical Activity Guidelines for Americans*. ODPHP Publication No. U0036. Washington, DC, 2008. Available at www.health.gov/paguidelines. Accessed October 15, 2016.

Follow a Healthy Eating Pattern

An eating pattern, or *dietary pattern*, is the combination of foods and beverages that make up an individual's intake over time. In recommending patterns rather than specific foods, the Dietary Guidelines recognize that the components of an eating pattern have interactive and potentially cumulative effects on health that are greater than that of any individual food component in isolation. Many different dietary patterns can promote health; those recommended by the guidelines include the USDA Food Patterns, which reflect the U.S.-style eating pattern, as well as a Mediterranean-Style Eating Pattern, Vegetarian Patterns, and the DASH Eating Plan (**Figure 2.6**).[2] Healthy eating patterns provide an adaptable framework within which individuals can

NUTRITION INSIGHT

FIGURE 2.6 **Healthy eating patterns** The *Dietary Guidelines for Americans* suggest that there are many ways to choose a healthy diet (see Appendix B and C). Healthy dietary patterns are high in fruits, vegetables, and whole grains and include low-fat dairy products; varied protein sources such as fish, legumes, and nuts and seeds; and healthy oils. They are low in added sugars, saturated fat, and sodium.

The DASH Eating Plan

Food group	Servings
Grains	6–8/day
Vegetables	4–5/day
Fruits	4–5/day
Fat-free or low-fat milk and milk products	2–3/day
Lean meats, poultry, and fish	6 or less/day
Nuts, seeds, and legumes	4–5/week
Fat and oils	2–3/day
Sweets and added sugars	5 or less/week

The DASH Eating Plan focuses on increasing foods rich in potassium, calcium, magnesium, and fiber. It is plentiful in fruits and vegetables, whole grains, low-fat dairy, fish, poultry, beans, nuts, and seeds. It was first developed for lowering blood pressure and is discussed further in Chapter 8.

Mediterranean Eating Pattern

Foods	How often
Fruits, vegetables, grains (mostly whole), olive oil, nuts, legumes and seeds, herbs and spices	Every meal
Fish and seafood	At least twice a week
Cheese and yogurt	Moderate portions daily or weekly
Poultry and eggs	Moderate portions every 2 days or weekly
Meats and sweets	Less often

The traditional Mediterranean-Style Eating Pattern is based on fruits, vegetables, grains, olive oil, legumes, nuts, and seeds. It includes moderate portions of cheese and yogurt. Fish and seafood are consumed at least twice a week; poultry and eggs every few days; and red meat and sweets less often. The incidence of chronic diseases such as heart disease is low in populations consuming this diet (see Chapter 5).

USDA Food Patterns

Food group	Amount/day
Vegetables	2.5 cups
Fruit and juices	2.0 cups
Grains	6.0 ounces
Dairy products	3.0 cups
Protein foods	5.5 ounces
Oils	27 grams
Solid fats	16 grams
Added sugars	32 grams

USDA Vegetarian Adaptations

Food group	Amount/day	
	Lacto-Ovo	Vegan
Dairy products or dairy substitutes	3 cups milk, calcium fortified soymilk, or yogurt, 4½ oz. hard cheese	3 cups calcium fortified soymilk or other plant-based milks or yogurts
Protein foods		
Eggs	4 per week	
Beans and Peas	⅓ cup	½ cup
Soy products	½ cup tofu, 2 oz. tempeh, ½ cup soybeans	½ cup tofu, 2 oz. tempeh, ½ cup soybeans
Nuts and Seeds	¼ cup	¼ cup

The USDA Food Patterns and their vegetarian adaptations suggest amounts of foods from different food groups and subgroups for different calorie levels (2000 Calories shown here) (see Chapter 6). These were developed to help individuals follow the Dietary Guidelines recommendations and are the basis for the MyPlate recommendations.

36 CHAPTER 2 Guidelines for a Healthy Diet

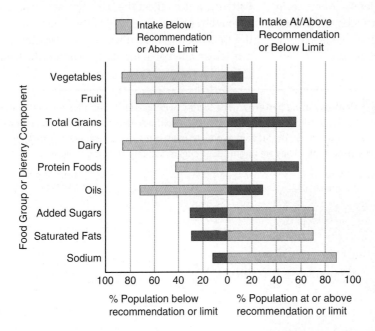

FIGURE 2.7 **How healthy is the American diet?** The current U.S diet needs improvement. The graph shown here illustrates how the typical American diet compares with recommendations for various food groups and dietary components.

Interpret the Data

What percentage of the population consumes fewer vegetables than recommended? What percentage consumes more sodium than recommended?

enjoy foods that meet their personal and sociocultural preferences, fit within their budgets, and meet their nutritional and medical needs.

Shift to Healthier Choices

The Dietary Guidelines suggest that most Americans do not consume a healthy eating pattern (**Figure 2.7**). To maintain a healthy weight and meet current recommendations for a nutrient-dense diet, most people need to shift food choices both within and across food groups. These shifts can occur slowly, beginning with small changes over the course of a week, a day, or even an individual meal (see What Should I Eat?).

Shift to balance calories with activity

The Dietary Guidelines emphasize balancing the calories consumed in foods and beverages with the calories expended through physical activity in order to achieve and maintain a healthy weight. Weight maintenance requires consuming the

---

What Should I Eat?

To Follow the Dietary Guidelines

Shift to Balance Calories with Activity

- If you stop for an ice cream cone go for a walk while eating it.
- Walk an extra 1000 steps; the more you exercise, the easier it is to keep your weight at a healthy level.
- Ride your bike to work or when running errands.
- Measure out your portions rather than snacking right out of the bag.

Shift to Meet Food Group Recommendations

- Have berries for dessert.
- Make sure your breakfast cereal is a whole-grain cereal.
- Be colorful by adding some red and orange vegetables to your salad.
- Toss some salmon on the grill to increase your seafood intake.

Shift to Limit Added Sugars, Saturated Fat, and Sodium

- Choose lean meat, fish, and low-fat dairy products in order to limit saturated fat.
- Have water and skip sugary soft drinks.
- Pass on the salt; instead, try lemon juice or some basil and oregano.
- Snack on baby carrots and apple slices; they are low in added sugars, saturated fat, and sodium.

 Use iProfile to look up the sugar and calorie content of your favorite dessert.

FIGURE 2.8 **Healthy weight and exercise recommendations** The Dietary Guidelines suggest that most Americans consume more calories than they expend. To achieve calorie balance, adults should decrease calorie intake and increase physical activity gradually over time to achieve a healthy weight.

a. Limiting portion sizes and reducing consumption of added sugars, solid fats, and alcohol, which provide calories but few essential nutrients, can help promote a healthy weight. There is no optimal proportion of macronutrients that can facilitate weight management; the critical issue is the right number of calories needed to maintain or lose weight over time.

b. To promote health and reduce disease risk, a minimum of 150 minutes of moderate-intensity aerobic exercise is recommended each week. Some adults need a higher level of physical activity than others to achieve and maintain a healthy body weight.

same number of calories as you burn; this means that if you eat more, you need to be more active (see Chapter 9). Most Americans would benefit from increasing their physical activity to a minimum of 150 minutes of moderate-intensity activity each week and from limiting their screen time and sedentary leisure activity. This, along with attention to portion sizes and a shift to more nutrient-dense choices from both foods and beverages, will promote a healthy weight (**Figure 2.8**).

Shift to meet food group recommendations

Most Americans do not meet recommendations for the amounts and variety of food from the various food groups (see Figure 2.7). Selecting more whole fruits and eating a variety of vegetables, especially dark-green and red and orange vegetables and beans and peas, and eating more whole grains will help align intake with recommendations. The guidelines also suggest that Americans increase their intake of fat-free or low-fat milk and milk products while limiting consumption of high-fat dairy products such as cheese.

This pattern of fruit, vegetable, grain, and dairy consumption will increase intakes of potassium, dietary fiber, calcium, and vitamin D, which are nutrients of concern in American diets. Most Americans consume enough protein foods but need to shift their choices to include more seafood along with a variety of other protein choices, such as lean meat, poultry, eggs, beans and peas, soy products, and unsalted nuts and seeds.

Shift to limit added sugars, saturated fat, and sodium

The Dietary Guidelines set specific limits on the amounts of added sugars, saturated fat, and sodium that can be included in a healthy dietary pattern (see Figure 2.7). Added sugars are sugars and syrups that are added to foods or beverages during processing or preparation. The Dietary Guidelines recommend that we reduce added sugars intake to less than 10% of calories (see Chapter 4). Currently, added sugars account for more than 13% of the calories Americans consume; most of this comes from sweetened beverages such as soft drinks, fruit drinks, sweetened coffee and tea, and sport and energy drinks (**Figure 2.9**).[2] Too much saturated fat increases the risk of heart disease (see Chapter 5), yet only about one-third of Americans meet the recommendation to limit saturated fat to less than 10% of calories. The Dietary Guidelines recommend shifting to oils in place of solid fats to reduce saturated fat intake. The American diet is also too high in sodium. The Dietary Guidelines recommend limiting sodium intake to less than 2300 mg/day (see Chapter 8).[2] This can be done by choosing whole foods instead of processed, packaged foods, which are the major source of sodium in the American diet.

Support from All Segments of Society

Americans consume too many calories, don't meet food group and nutrient recommendations, and don't get enough physical activity. Changing this pattern requires a new paradigm in which healthy lifestyle choices at home, at school, at work, and in the community are easy, accessible, and affordable. The Dietary Guidelines suggest that changes at all levels of society can help individuals embrace and maintain healthy eating and physical activity patterns. Policymakers can implement strategies and programs that provide nutrition education and access to affordable, healthy food. Business and industry can modify products and menus to align with the Dietary Guidelines. Communities can increase opportunities for physical activity by designing safe and accessible parks, recreation facilities, and sports programs. Individuals are encouraged to find their healthy eating and activity pattern by starting with small changes.

FIGURE 2.9 Beverages count Beverages contribute an average of about 400 Calories per day to the diets of American adults. Most of these calories come from sugar-sweetened sodas, fruit drinks, and alcoholic beverages, which are high in empty calories. The Dietary Guidelines recommend replacing sugar-sweetened drinks with water.[2]

Concept Check

1. **How** do the Dietary Guidelines recommendations differ from the DRIs?
2. **What** do healthy eating patterns have in common?
3. **Why** do the Dietary Guidelines recommend we decrease the amount of added sugars in our diet?
4. **What** could your community do to promote an increase in physical activity?

2.4 MyPlate: Putting the Guidelines into Practice

LEARNING OBJECTIVES

1. **Explain** the purpose of MyPlate.
2. **Describe** the food group recommendations of MyPlate.
3. **Use** the MyPlate recommendations to plan your healthy eating style.
4. **Distinguish** MyPlate food groups from the Choice Lists.

To help individuals apply the recommendations of the Dietary Guidelines to their own food choices, the USDA has developed **MyPlate**. This educational tool translates the recommendation of the Dietary Guidelines into the food choices that make up a healthy eating pattern. It divides foods into groups, based on the nutrients they supply most abundantly, and illustrates the appropriate proportions of foods from each food group that make up a healthy diet.

FIGURE 2.10 Building a healthy eating style The MyPlate Daily Checklist includes serving recommendations for each food group, along with activity recommendations and specific limits on the amounts of added sugars, saturated fat, and sodium in a healthy eating pattern. The serving recommendations shown are for someone who needs 2000 Calories per day.

The plate icon illustrates the proportions of food recommended from each of five food groups. Half of your plate should be fruits and vegetables, about a quarter grains, and about a quarter protein foods. Dairy should accompany meals, as shown by the small circle to the side.

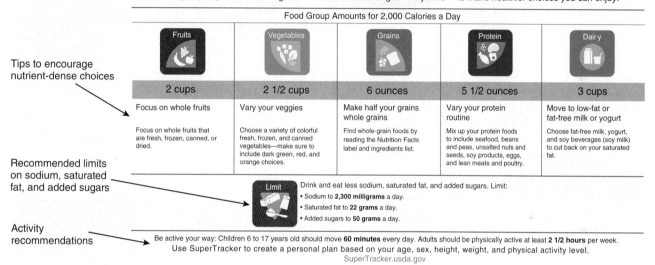

Tips to encourage nutrient-dense choices

Recommended limits on sodium, saturated fat, and added sugars

Activity recommendations

The recommendations of MyPlate are designed to help consumers build a healthy eating style. They focus on the amount, variety, and nutrient density of food and beverage choices and aim to meet but not exceed calorie needs. This healthy eating style also limits added sugars, saturated fat, and sodium and encourages physical activity. A healthy eating style can be achieved by making small changes over time that reflect personal preferences, culture, and tradition. You can obtain a personalized food plan, called a MyPlate Daily Checklist, by using the ChooseMyPlate Web site, www.ChooseMyPlate.gov (**Figure 2.10**).

Choose the Right Amount from Each Food Group

The MyPlate recommendations tell you how much to choose from each of five food groups: Fruits, vegetables, grains, protein foods, and dairy. The amounts recommended for fruits, vegetables, and dairy are given in cups. The amounts from the grains group are expressed in ounces. An ounce of grains is 1 cup of cold cereal, ½ cup of cooked cereal or grains, or a slice of bread. So if you have 2 cups of cereal and 2 slices of toast at breakfast, you have already consumed 4 ounces of grains for the day (two-thirds of the total for a 2000-Calorie diet). The amounts recommended for protein foods are also expressed in ounces. One ounce is equivalent to 1 ounce of cooked meat, poultry, or fish; one egg; 1 tablespoon of peanut butter; ¼ cup of cooked dry beans; or ¼ cup of nuts or seeds. The amounts recommended per day from each food group depend on your energy needs (see Figure 2.10). The amount of food chosen is important because how much you eat and drink affects your weight and your risk of chronic disease. Tracking how much you eat and drink can help you be aware of the calories in the foods you choose. Watching portion sizes can help meet amount recommendation, reduce calories, and balance calories consumed with calories burned.

It is easy to see where some foods in your diet fit on MyPlate. For example, a chicken breast is 3 ounces from the protein group; a scoop of rice is 2 ounces from the grains group. It is more difficult to see how much mixed foods such as pizza, stews, and casseroles contribute to each food group. To fit these on your plate, consider the individual ingredients. For example, a slice of pizza provides 1 ounce of grains, ⅛ cup of vegetables, and ½ cup of dairy. Having meat on your pizza adds about ¼ ounce from the protein group. Another example of how to break down meals into food groups is shown in **Figure 2.11**.

FIGURE 2.11 **How meals fit** The lunch that is part of this 2000-Calorie menu includes about a third of the amounts of grains, protein foods, and dairy recommended for the day, a quarter of the recommended fruit, and half the recommended oils but only a small proportion of the vegetables recommended. The lettuce and celery fit into the "other vegetables" category, so choices at other meals should come from dark-green, red and orange, and starchy vegetables as well as beans and peas. To find out what counts as an ounce or a cup, go to each food group at www.ChooseMyPlate.gov.

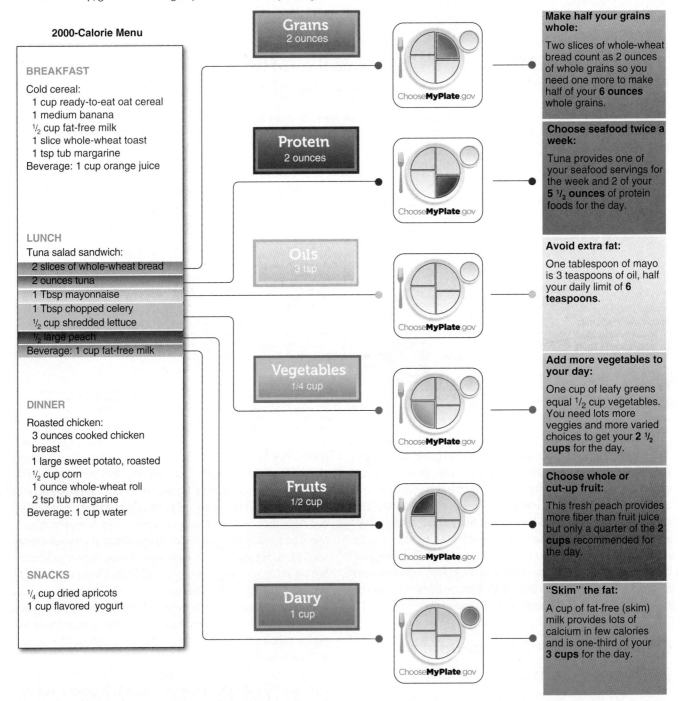

Choose a Variety of Nutrient-Dense Foods

Variety is important for a healthy diet because no one food or food group provides all the nutrients and food components the body needs. Choosing foods from all the food groups provides variety, but a variety of foods should also be selected from within each food group. The vegetables food group includes choices from five subgroups: dark-green vegetables such as broccoli, collard greens, and kale; red and orange vegetables such as carrots, sweet potatoes, and red peppers; starchy vegetables such as corn, green peas, and potatoes; other vegetables such as cabbage, asparagus, and artichokes; and beans and peas such as lentils, chickpeas, and black beans. Beans and peas are good sources of the nutrients found in both vegetables and protein foods, so they can be counted in either food group. Protein foods include meat, poultry, seafood, beans and peas, eggs, processed soy products, nuts, and seeds. Grains include whole grains such as whole-wheat bread, oatmeal, and brown rice as well as refined grains such as white bread, white rice, and white pasta (see Chapter 4). Fruits include fresh, canned, or dried fruit and 100% fruit juice. Dairy includes all fluid milk products and many foods made from milk, such as cheese, yogurt, and pudding, as well as calcium-fortified soy products.

The MyPlate Daily Checklist provides tips to help you make nutrient-dense choices within each food group (see Figure 2.10). Choosing nutrient-dense foods allows you to maximize the vitamins, minerals, fiber, and other healthful nutrients you consume while minimizing empty calories. This pattern helps manage weight and reduce the risk of chronic disease. Many of the foods Americans currently choose are high in empty calories. Some, such as butter, table sugar, soft drinks, and candy, don't belong in any food group because all their calories are empty. Other foods high in empty calories do belong to a food group. For example, cookies are in the grains group, but about half of their calories are empty calories (**Table 2.2**).

Limit Saturated Fat, Sodium, and Added Sugars

MyPlate also reinforces the recommendations of the Dietary Guidelines to keep saturated fat to less than 10% of calories, sodium to 2300 mg or less, and added sugars to 50 g/day or less. Specific recommendations based on calorie level are given at the bottom on the Daily Checklist (see Figure 2.10). Choosing nutrient-dense foods helps you meet these goals. Saturated fat can be limited by consuming less butter and lard and high-fat meats and dairy products and more liquid oils such as olive and corn oil, which are low in saturated fat. Sodium and added sugars can be limited by choosing fewer processed foods such as snack foods, baked goods, and sweetened beverages (see Table 2.2).

MyPlate Compared with Choice (Exchange) Lists

The **Choice Lists** are a set of food-group recommendations developed as the **Exchange Lists** in the 1950s to plan diets for people with diabetes. Since then, their use has been expanded to planning diets for anyone monitoring calorie intake. The most recent version, published as *Choose Your Foods*, has replaced the term "exchange" with "choice." The Choice Lists group foods based on their energy and macronutrient composition. Foods in the same list each contain approximately the same amounts of calories, carbohydrate, protein, and fat. The food groupings of the Choice Lists differ from the MyPlate food groups because the lists are designed to meet energy and macronutrient criteria, whereas the MyPlate groups are designed to

TABLE 2.2 Food group choices and empty calories

Food group	Foods with few or no empty calories	Foods with about half of their energy from empty calories
Grains	Whole-grain bagel	Donut
	Oatmeal	Oatmeal chocolate chip cookies
	Brown rice	Packaged flavored rice
Vegetables	Steamed broccoli	Broccoli with cheese sauce
	Sweet potatoes	Candied sweet potatoes
	Green beans	Green bean casserole
Fruits	Fresh cherries	Cherry pie
	Fresh strawberries	Frozen sweetened strawberries
	Fresh peach	Peach canned in heavy syrup
Dairy	Low-fat milk	Whole milk
	Vanilla nonfat yogurt	Fruit on the bottom low-fat yogurt
Protein	Extra lean ground beef	Pork sausage
	Roasted skinless chicken breast	Fried chicken breast with skin
	Grilled salmon	Fried breaded cod
	Black beans	Refried beans with cheese

be good sources of nutrients regardless of their energy content. For example, a potato is included in the starch food list because it contains about the same amount of energy, carbohydrate, protein, and fat as breads and grains, but in MyPlate a potato is in the vegetable group because it is a starchy vegetable that is a good source of vitamins, minerals, and fiber. The Choice Lists are a useful tool whether you are controlling calorie intake for purposes of weight loss or carbohydrate intake for purposes of diabetes management (see Appendix G).

Concept Check

1. **What** does the MyPlate graphic show you about a healthy diet?
2. **How** many ounces from the grains group does MyPlate recommend for you?
3. **Why** is variety important in a healthy eating style?
4. **Where** does a baked potato fit on MyPlate? In Choice Lists?

2.5 Food and Supplement Labels

LEARNING OBJECTIVES

1. **Discuss** how the information on food labels can help you choose a healthy diet.
2. **Determine** whether a food is high or low in fiber, sodium, or saturated fat.
3. **Explain** how the order of ingredients on a food label is determined.
4. **Explain** the types of claims that are common on food and dietary supplement labels.

The Dietary Guidelines and MyPlate recommend appropriate amounts of nutritious foods, but sometimes it is difficult to tell how nutritious a particular food is. How do you know whether your frozen entrée is a good source of iron, how much fiber your breakfast cereal provides, or how much calcium is in your daily vitamin/mineral supplement? You can find this information on food and supplement labels.

Food Labels

Standardized food labels are designed to help consumers make informed food choices by providing information about the nutrient composition of a food and how that food fits into the overall diet.[10] They are required on all packaged foods, except those produced by small businesses and those in packages too small to accommodate the information. Raw fruits, vegetables, and seafood are not required to carry individual food labels. However, grocery stores are asked to provide nutrition information voluntarily for the raw fruits, vegetables, and seafood most frequently purchased in the United States. The information can appear on large placards or in consumer pamphlets or brochures. Most fresh meat and poultry (whole cuts and ground and chopped products) are required to have a Nutrition Facts label on the package. Those that are not packaged must have the equivalent information displayed nearby.[11] Food served in restaurants, delicatessens, and bakeries is not required to carry labels unless the food is from an establishment that has 20 or more locations. These food chains must list calorie content information for standard menu items and provide other nutrient information upon request.[12]

Since 1990, food labels have included a **Nutrition Facts** panel and an ingredient list. In 2016, the Food and Drug Administration (FDA) announced an updated Nutrition Facts panel. This revision is designed to reflect new scientific information and make it easier for consumers to make informed food choices (**Figure 2.12**).[13]

Nutrition Facts All food labels must contain a Nutrition Facts panel that lists the number of servings per container and the serving size. The number of Calories per serving is shown on both the current and planned label. To help consumers be more aware of their calorie intake, the planned label uses a larger, bolder type for this information (see Figure 2.12b).[10, 14] No matter the font size, if a person eats twice the standard serving, he or she is consuming twice the number of calories listed (*See Thinking It Through*).

Both the current and planned Nutrition Facts panel list the amounts of nutrients contained in a serving by weight and as a percentage of the **Daily Value**. The % Daily Value is the amount of a nutrient in a food as a percentage of the amount recommended for a 2000-Calorie diet. For example, if a food provides 10% of the Daily Value for calcium, it provides 10% of the recommended daily intake for calcium in a 2000-Calorie diet (see Appendix H). These percentages help consumers see how a given food fits into their overall diet. Because a Daily Value is a single standard for all consumers, it may overestimate the amount of a nutrient needed for some population groups, but it does not underestimate the requirement for any group except pregnant and lactating women. Daily Values have not been established for *trans* fat or total sugars. The Daily Values used to calculate the % Daily Values on the planned label have been updated based on current science and disease risk.[15]

Ingredient list Do you want to know exactly what is in your food? The ingredient list is the place to look (see Figure 2.12c). The ingredient list presents the contents of the product, in order of their prominence by weight. An ingredient list is required on all products containing more than one ingredient and optional on products that contain a single ingredient. Food additives, including food colors and flavorings, must be listed among the ingredients.

Nutrient content and health claims Looking for low-fat or high-fiber foods? You may not even need to look at the Nutrition Facts panel. Food labels often contain **nutrient content claims**. These are statements that highlight specific characteristics of a product that might be of interest

---

Daily Value A reference value for the intake of nutrients used on food labels to help consumers see how a given food fits into their overall diet.

Food and Supplement Labels 43

NUTRITION INSIGHT | **FIGURE 2.12** **Food labels** Knowing how to interpret the information on food labels can help you choose foods that fit into a healthy eating pattern.

a. The current Nutrition Facts panel, developed in the 1990s, reflects the scientific understanding at that time regarding the roles of dietary fats and carbohydrate in health and disease as well as the nutrients considered at risk in the American diet.

Standard serving sizes are required to allow consumers to compare products. These are based on the portions people typically ate in the 1990s. They are smaller than people typically consume today.

The "% Daily Value" for certain macro and micronutrients is included. A % Daily Value of 5% or less is considered low, and a value of 20% or more is considered high.

The label provides information about the amounts of nutrients that should be limited—total fat, saturated fat, *trans* fat, cholesterol, and sodium. Sugars are listed, but it is not possible to tell how much of this is from added sugars.

The label provides information about the amounts of nutrients that tended to be low in the American diet in the 1990s—fiber, vitamins A and C, calcium, and iron.

The footnote gives the Daily Values for 2000- and 2500-Calorie diets to illustrate that for some nutrients the Daily Value increases with increasing caloric intake.

Think Critically

Would the more prominent presentation of the calories per serving seen in the planned label affect whether you choose this food?

Current Label

Nutrition Facts
Serving Size 2/3 cup (55g)
Servings Per Container About 8

Amount per servings
Calories 230   Calories from Fat 72

	% Daily Value*
Total Fat 8g	12%
Saturated Fat 1g	5%
Trans Fat 0g	
Cholesterol 0mg	0%
Sodium 160mg	7%
Total Carbohydrate 37g	12%
Dietary Fiber 4g	16%
Total Sugars 1g	
Protein 3g	

| Vitamin A 10% | • | Calcium 20% |
| Vitamin C 8% | • | Iron 45% |

* Percent Daily Values are based on a 2,000 calorie diet. Your daily value may be higher or lower depending on your calorie needs.

	Calories:	2,000	2,500
Total Fat	Less than	65g	80g
Sat Fat	Less than	20g	25g
Cholesterol	Less than	300mg	300mg
Sodium	Less than	2,400mg	2,400mg
Total Carbohydrate		300g	375g
Dietary Fiber		25g	30g

Planned Label

Nutrition Facts
8 servings per container
Serving size   2/3 cup (55g)

Amount per servings
Calories   **230**

	% Daily Value*
Total Fat 8g	10%
Saturated Fat 1g	5%
Trans Fat 0g	
Cholesterol 0mg	0%
Sodium 160mg	7%
Total Carbohydrate 37g	13%
Dietary Fiber 4g	14%
Total Sugars 12g	
Includes 10g Added Sugars	20%
Protein 3g	
Vitamin D 2mcg	10%
Calcium 260mg	20%
Iron 8mg	45%
Potassium 235mg	6%

* The % Daily Value (DV) tells you how much a nutrient in a serving of food contributes to a daily diet. 2,000 calories a day is used for general nutrition advice.

Standard serving sizes better reflect what people today actually eat. The serving size information is more prominent.

The amount per serving and Calories per serving are in large, bold type to highlight this information, which is important to address current public health concerns such as obesity, diabetes, and heart disease.

The Daily Values used to calculate % Daily Values have been updated and a new Daily Value established for added sugars.

Information about amounts of nutrients that should be limited includes the amount of added sugars in grams and as percentage of the Daily Value.

The nutrients considered at risk in the American diet have been updated to include vitamin D, potassium, calcium, and iron as well as fiber. The % Daily Values and amount by weight of each are provided.

A footnote clearly explains the meaning of % Daily Values.

b. The planned Nutrition Facts panel puts more emphasis on calories and serving size, less focus on total fat, and more focus on added sugars. It also includes amounts and % Daily Values for nutrients currently of concern in the American diet.

(continues)

FIGURE 2.12 (continued)

c. The ingredient list shows the exact contents of a food. This information can be useful for those who are curious about what is in their food, those who have food allergies, and those who restrict certain ingredients, such as vegans, who do not eat animal products.

Ingredients:
Enriched macaroni product (wheat flour, niacin, ferrous sulfate [iron], thiamine mononitrate, riboflavin, folic acid); cheese sauce mix (whey, modified food starch, milk fat, salt, milk protein concentrate, contains less than 2% of sodium tripolyphosphate, cellulose gel, cellulose gum, citric acid, sodium phosphate, lactic acid, calcium phosphate, milk, yellow 5, yellow 6, enzymes, cheese culture)

Labels must contain basic product information, such as the name of the product, the weight or volume of the contents, and the name and place of business of the manufacturer, packager, or distributor.

The ingredients are listed in descending order by weight, from the most abundant to the least abundant. The wheat flour in the macaroni is the most abundant ingredient in this product.

---

to consumers, such as "high fiber" or "low sodium." Standard definitions for these descriptors have been established by the FDA (see Appendix H). For example, foods that claim to be "sugar free" or "fat free" must contain less than 0.5 grams per serving of sugar or fat, respectively, and a food that is "high" in a particular nutrient must contain 20% or more of the Daily Value for that nutrient. These definitions also apply to food sold in restaurants. When a claim is made about a menu item's nutritional content or health benefits, such as "low-fat" or "heart healthy," nutrition information must be available upon request."[16]

Some of these standard definitions are outdated, and some of the descriptors we see on labels, such as "natural," have not been defined (**Figure 2.13**). The term "healthy" can currently be used on products that are low in fat, saturated fat, cholesterol, and sodium and also contain 10% of the Daily Value for vitamins A or C, iron, calcium, protein, or fiber. Based on this definition, almonds could not be labeled "healthy" because of their fat content, but sugary breakfast cereals could be labeled "healthy" despite their high added sugar content. The FDA is currently in the process of redefining the term "healthy" so it is more consistent with current dietary recommendations. Keeping labeling regulations consistent with current scientific knowledge and dietary recommendations is challenging because food label updates impact the packaging on all foods that carry a label, so changes can't be made often.

Because of the importance of certain types of foods and dietary components in disease prevention, food labels are permitted to include a number of **health claims**. Health claims refer to a relationship between a nutrient, food, food component, or dietary supplement and reduced risk of a disease or health-related condition. All health claims are reviewed by the FDA. To carry a health claim, a food must be a naturally good source of one of six nutrients (vitamin A, vitamin C, protein, calcium, iron, or fiber) and must not contain more than 20% of the Daily Value for fat, saturated fat, cholesterol, or sodium. Authorized health claims are supported by strong scientific evidence (**Figure 2.14** and Appendix H) Health claims for which there is emerging but not well-established evidence are called **qualified health claims**; such a claim must be accompanied by an explanatory statement to avoid misleading consumers.

Thinking It Through

A Case Study on Using Food Labels to Guide Food Choices

Frieda is trying to improve her diet. She needs to eat about 2000 Calories per day to maintain her weight. She has looked up her MyPlate recommendations and knows that she needs to make some changes.

Based on the MyPlate recommendations in Figure 2.10, how many cups of fruits and of vegetables should Frieda eat each day?

Your answer:

Frieda eats lots of fruits but eats only about a cup of vegetables each day. While shopping she notices vegetable chips. The bag says she will get a serving (½ cup) of vegetables in every ounce.

Vegetable Chips

Nutrition Facts	
Serving Size 1 oz (28g/about 14 chips)	
Servings Per Container About 7	
Amount Per Serving	
Calories 150	Calories from Fat 80
	% Daily Value*
Total Fat 9g	14%
Saturated Fat 1g	5%
Trans Fat 0g	
Cholesterol 0mg	0%
Sodium 150mg	2%
Total Carbohydrate 16g	5%
Dietary Fiber 3g	12%
Sugars 3g	
Protein 1g	

How many servings of chips would she have to eat to get the 1½ cups of vegetables she is lacking?

Your answer:

Based on the label shown here, how many Calories and how much sodium would this contribute? What percentage of her recommended intake would this be?

Your answer:

Use iProfile to compare the calories and fiber in the chips to the amounts in 18 baby carrots.

Your answer:

Another easy choice for Frieda would be a 100% fruit and vegetable juice blend. The bottle says that it provides ½ cup of fruit and ½ cup of vegetables in 8 ounces.

Juice Blend

Nutrition Facts	
Serving Size 8 fl oz (240 ml)	
Servings Per Container 1	
Amount Per Serving	
Calories 110	Calories from Fat 0
	% Daily Value*
Total Fat 0g	0%
Saturated Fat 0g	0%
Trans Fat 0g	
Cholesterol 0mg	0%
Sodium 70mg	3%
Total Carbohydrate 28g	9%
Dietary Fiber 0g	0%
Sugars 24g	
Protein 1g	

If Frieda wants to get all her fruits and vegetables from the juice, how much would she have to drink?

Your answer:

Based on the label shown here, how many Calories and how much fiber would this provide?

Your answer:

What are the advantages and disadvantages of the juice versus whole fruits and vegetables?

Your answer:

(Check your answers in Appendix L.)

Dietary Supplement Labels

Products ranging from multivitamin pills to protein powders and herbal elixirs can be defined as **dietary supplements**. These products are considered foods, not drugs, and therefore are regulated by the laws that govern food safety and labeling. To help consumers understand what they are choosing when they purchase these products, dietary supplements are required to carry a **Supplement Facts** panel similar to the Nutrition Facts panel found on food labels (**Figure 2.15**).[17] Along with changes in the Nutrition Facts label, the FDA has announced changes to the Supplement Facts label that will update the Daily Values and units of measure to current standards.[15]

---

dietary supplement A product sold to supplement the diet; may include nutrients, enzymes, herbs, or other substances.

FIGURE 2.13 Consider the entire label People often choose foods labeled "natural" because they think these products are good for them. It turns out that the term "natural" was not defined by the FDA. It currently appears on a wide variety of foods, some of which should be limited in a healthy diet. The FDA is in the process of developing a definition for "natural."

FIGURE 2.14 Health claims Oatmeal contains enough soluble fiber to be permitted to include this health claim about the relationship between soluble fiber and the risk of heart disease. Other health claims you may see on food labels are listed below and included in Appendix H[14]

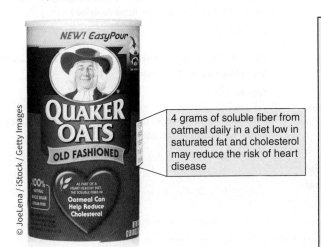

4 grams of soluble fiber from oatmeal daily in a diet low in saturated fat and cholesterol may reduce the risk of heart disease

- Calcium intake and calcium and vitamin D intake and the risk of osteoporosis
- Sodium intake and the risk of high blood pressure
- Saturated fat and cholesterol intake and the risk of heart disease
- Fiber-containing fruit, vegetable, and grain intake and the risk of heart disease and cancer
- Fruit and vegetable intake and the risk of cancer
- Dietary fat and the risk of cancer
- Whole-grain foods and the risk of heart disease and certain cancers

Labels on dietary supplements may also include nutrient content claims and FDA-approved health claims similar to those on food labels. For example, a product can claim to be an excellent source of a particular nutrient. To make this claim, one serving of the product must contain at least 20% of the Daily Value for that nutrient. A label may say "high potency" if one serving provides 100% or more of the Daily Value for the nutrient it contains. For multinutrient products, "high potency" means that a serving provides more than 100% of the Daily Value for two-thirds of the vitamins and minerals present.

Dietary supplement labels may also carry **structure/function claims**, which describe the role of a dietary ingredient

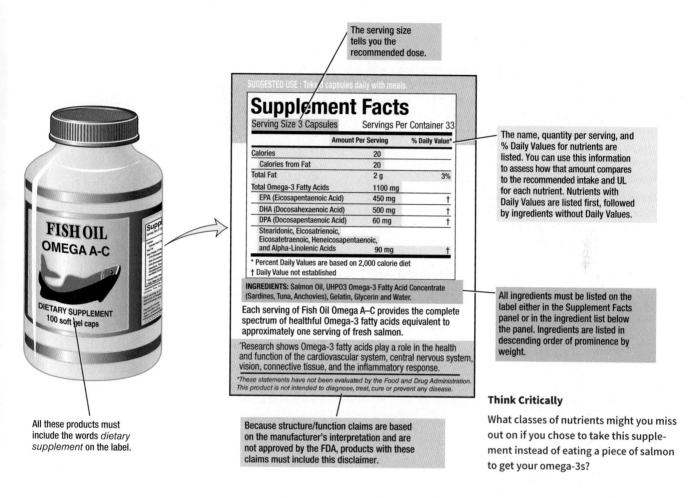

FIGURE 2.15 Current dietary supplement label Unlike food labels, dietary supplement labels must include directions for use and must provide information about ingredients that are not nutrients and for which Daily Values have not been established. For such ingredients, it is difficult to tell from the label whether the amount included in a serving is helpful, is harmful, or has no effect at all.

Think Critically

What classes of nutrients might you miss out on if you chose to take this supplement instead of eating a piece of salmon to get your omega-3s?

in maintaining normal structure, function, or general well-being. For example, a structure/function claim about calcium may state that "calcium builds strong bones"; one about fiber may say "fiber maintains bowel regularity." These statements can be misleading. For example, the health claim "lowers cholesterol" requires FDA approval, but the structure/function claim "helps maintain normal cholesterol levels" does not. It would not be unreasonable for consumers with high cholesterol to conclude that a product that "helps maintain normal cholesterol levels" would help lower their elevated blood cholesterol level to within the normal range.

Manufacturers must notify the FDA when including a structure/function claim on a dietary supplement label and are responsible for ensuring the accuracy and truthfulness of these claims. Structure/function claims are not approved by the FDA. For this reason, the law says that if a dietary supplement label includes such a claim, it must state in a disclaimer that the FDA has not evaluated the claim. The disclaimer must also state that

the dietary supplement product is not intended to "diagnose, treat, cure, or prevent any disease" because only a drug can legally make such a claim (see Figure 2.15). Structure/function claims may also appear on food labels, but the FDA does not require conventional food manufacturers to notify the FDA about their structure/function claims, and disclaimers are not required.

Concept Check

1. **Why** are serving sizes standardized on food labels?
2. **What** food label information helps you find foods that are low in saturated fat?
3. **Where** should you look to see if a food contains sesame seeds?
4. **How** do structure/function claims differ from health claims?

Summary

1 Nutrition Recommendations 26

- Nutrition recommendations are designed to encourage consumption of a diet that promotes health and prevents disease. Some of the earliest nutrition recommendations in the United States were in the form of **food guides**, which translate nutrient intake recommendations into food intake recommendations. The first set of Recommended Dietary Allowances, developed during World War II, focused on energy and the nutrients most likely to be deficient in a typical diet. Current recommendations focus on promoting health and preventing chronic disease as well as nutrient deficiencies.

- Dietary recommendations can be used as a standard for assessing the **nutritional status** of individuals and of populations. Records of dietary intake, such as the one shown here, along with information obtained from a physical examination, a medical history, and laboratory tests, can be used to assess an individual's nutritional status. Collecting information about the food intake and health of individuals in the population or surveying the foods available can help identify potential and actual nutrient deficiencies and excesses in a population and help policymakers improve nutrition recommendations.

Figure 2.2 Assessing nutritional status: Determine typical food intake

2 Dietary Reference Intakes (DRIs) 31

- **Dietary Reference Intakes (DRIs)** are recommendations for the amounts of energy, nutrients, and other food components that should be consumed by healthy people to promote health, reduce the incidence of chronic disease, and prevent deficiencies. **Estimated Average Requirements (EARs)** are average requirements, as seen in the illustration, and can be used to evaluate the adequacy of a population's nutrient intake. **Recommended Dietary Allowances (RDAs)** and **Adequate Intakes (AIs)** can be used by individuals as goals for nutrient intake, and **Tolerable Upper Intake Levels (ULs)** indicate safe upper intake limits.

Figure 2.4 Understanding EARs, RDAs, and ULs

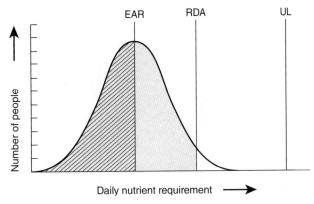

- The DRIs make two types of energy-intake recommendations. **Estimated Energy Requirements (EERs)** provide an estimate of how many calories are needed to maintain body weight. **Acceptable Macronutrient Distribution Ranges (AMDRs)** make recommendations about the proportion of energy that should come from carbohydrate, fat, and protein in a healthy diet.

3 The Dietary Guidelines for Americans 33

- The *Dietary Guidelines for Americans* are a set of diet and lifestyle recommendations designed to promote health and reduce the risk of overweight and obesity and chronic disease in the U.S. population.

- The Dietary Guidelines emphasize choosing a healthy eating pattern. Healthy eating patterns are higher in fruits, vegetables, whole grains, low-fat dairy products, and seafood than the current American diet as seen in the figure. The guidelines recommend that Americans shift to healthier choices to balance calorie intake with activity, meet food group recommendations, and reduce intake of saturated fat, sodium, and added sugars. They also suggest that all segments of society be involved in the process of developing a healthy lifestyle.

Figure 2.7 How healthy is the American diet?

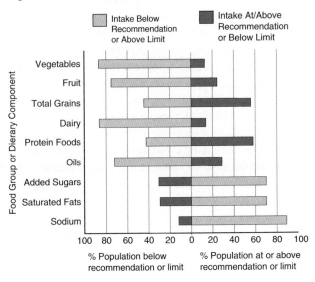

4 MyPlate: Putting the Guidelines into Practice 38

- **MyPlate** is the USDA's current food guide. It shows the proportions of foods from five food groups that make up a healthy diet. The MyPlate Daily Checklist, shown here, recommends amounts of food from each group based on individual energy needs.

Figure 2.10 Building a Healthy Eating Style

- MyPlate recommendations promote a healthy eating pattern that includes a variety of choices between and within food groups, stresses nutrient-dense choices, and limits the amounts of sodium, saturated fat, and added sugars. MyPlate also promotes the physical activity recommendations included in the Dietary Guidelines.
- **Choice (Exchange) Lists** are a food group system used to plan individual diets that provide specific amounts of energy, carbohydrate, protein, and fat.

5 Food and Supplement Labels 42

- Standardized food labels are designed to help consumers make healthy food choices by providing information about the nutrient composition of foods and about how a food fits into the overall diet. The current **Nutrition Facts** panel, illustrated here, as well as the planned label present the amounts of nutrients by weight and as a percentage of the **Daily Value**. A food label's ingredient list states the contents of the product, in order of prominence by weight. Food labels often include FDA-defined **nutrient content claims**, such as "low sodium" or "high fiber," and **health claims**, which refer to a relationship between a nutrient, food, food component, or dietary supplement and the risk of a particular disease or health-related condition. All health claims are reviewed by the FDA and permitted only when they are supported by scientific evidence, but the level of scientific support for such claims varies.

Figure 2.12a Food labels

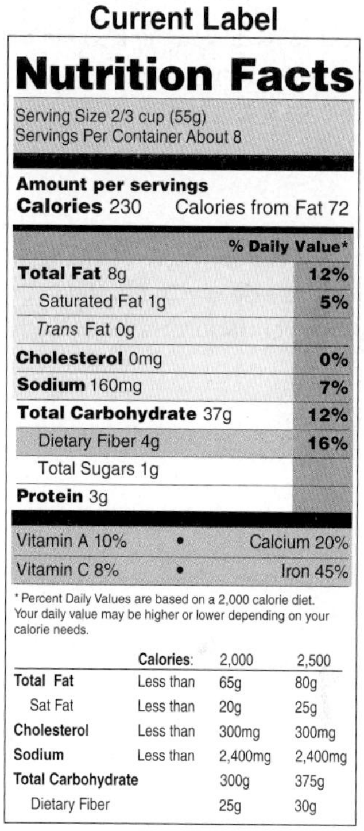

- A **Supplement Facts** panel appears on the label of every dietary supplement. Because **structure/function claims** are not FDA approved, when they appear on supplement labels, they must be accompanied by a disclaimer.

Key Terms

- Acceptable Macronutrient Distribution Ranges (AMDRs) 33
- Adequate Intakes (AIs) 31
- Choice Lists 41
- Daily Value 42
- dietary supplement 45
- *Dietary Guidelines for Americans* 33
- Dietary Reference Intakes (DRIs) 31
- Estimated Average Requirements (EARs) 31
- Estimated Energy Requirements (EERs) 32
- Exchange Lists 41
- food guide 26
- health claim 44
- *Healthy People* 30
- MyPlate 38
- nutrient content claim 42
- Nutrition Facts 42
- nutritional status 28
- qualified health claim 44
- Recommended Dietary Allowances (RDAs) 31
- structure/function claim 46
- Supplement Facts 45
- Tolerable Upper Intake Levels (ULs) 32

What is happening in this picture?

The Healthy Eating Plate is a variation of MyPlate. It emphasizes that a healthy diet is based on whole grains, fruits, vegetables, healthy protein sources, and oils. This dietary pattern recommends limiting red meat, refined grains, and dairy and avoiding sugary soft drinks, *trans* fat, and processed meats.

Think Critically

1. How does the Healthy Eating Plate differ from MyPlate?
2. Which of these tools do you think would better help Americans improve their diets? Why?

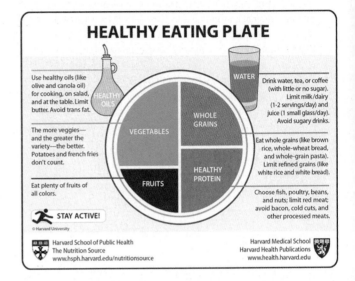

Copyright © 2011 Harvard University. For more information about The Healthy Eating Plate, please see The Nutrition Source, Department of Nutrition, Harvard School of Public Health, www.thenutritionsource.org.

CHAPTER 3

Digestion: From Meals to Molecules

The human body has been compared to a car: We fill the tank of our car with gasoline to get down the highway; we fill our body with food to get on with life. In both "machines," combustion with oxygen releases energy.

Our bodies are machine-like in another way as well: They are virtually identical to one another. Like cars, we look different on the outside but are basically the same on the inside. The processes that drive us—like the internal combustion engine, no matter where it's manufactured—are more similar than different because they are based on the same fundamental chemical reactions.

Despite the similarities, there are differences between human bodies and machines. An automobile cannot use gasoline to heal itself or to grow, as we do with nutrients. Although a gas tank and a stomach both store fuel, gasoline travels unchanged through the fuel line to the engine, whereas in humans the digestive system must break down the fuel into smaller units *before* it can be used by the body. Gas-powered cars are fueled only by gasoline, but the human digestive system must process fuel from many sources for use by the "high-performance machine" that is the human body.

CHAPTER OUTLINE

The Organization of Life 52

The Digestive System 55
- Organs of the Digestive System
- Digestive System Secretions

Digestion and Absorption of Nutrients 57
- The Mouth
- The Pharynx
- The Esophagus
- The Stomach
- The Small Intestine
- The Large Intestine

The Digestive System in Health and Disease 65
- The Intestinal Microbiota
- Gut Immune Function in Health and Disease

What a Scientist Sees: Bacteria on the Menu

Debate: Should You Be Gluten Free?
- Other Digestive System Problems and Discomforts

What Should I Eat? For Digestive Health

Thinking It Through: A Case Study on How Changes in the Digestive System Affect Nutrition

Delivering Nutrients and Eliminating Wastes 73
- The Cardiovascular System
- The Lymphatic System
- Elimination of Wastes

An Overview of Metabolism 77
- Releasing Energy
- Synthesizing New Molecules

3.1 The Organization of Life

LEARNING OBJECTIVES

1. **Describe** the organization of living things, from atoms to organisms.
2. **Name** the organ systems that work with the digestive system to deliver nutrients and eliminate wastes.

Matter, be it a meal you are about to eat or the plate you are about to eat it from, is made up of **atoms** (**Figure 3.1**). Atoms combine to form **molecules**, which can have different properties from those of the atoms they contain. In any living system, the molecules are organized into **cells**, the smallest units of life. Cells that are similar in structure and function form **tissues**. The human body contains four types of tissue: muscle, nerve, epithelial, and connective. These tissues are organized in varying combinations to form **organs**. In most cases, an organ does not function alone but is part of an **organ system**. Moreover, an organ may be part of more than one organ system. For example, the pancreas is part of the endocrine system and also part of the digestive system.

The body's 11 organ systems interact to perform all the functions necessary for life (**Table 3.1**). For example, the digestive system, which is the primary organ system responsible for moving nutrients into the body, is assisted by the endocrine system, which secretes **hormones** that help regulate how much we eat and how quickly food and nutrients travel through the digestive system. The digestive system is also aided by the nervous system, which sends nerve signals that help control the passage of food through the digestive tract;

PROCESS DIAGRAM | **FIGURE 3.1** **From atoms to organisms** The organization of life begins with atoms that form molecules, which are then organized into cells to form tissues, organs, organ systems, and whole organisms.

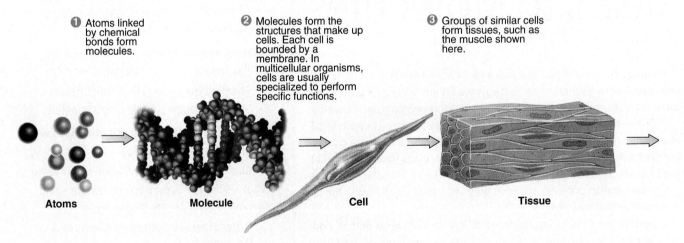

❶ Atoms linked by chemical bonds form molecules.

❷ Molecules form the structures that make up cells. Each cell is bounded by a membrane. In multicellular organisms, cells are usually specialized to perform specific functions.

❸ Groups of similar cells form tissues, such as the muscle shown here.

Atoms   Molecule   Cell   Tissue

(continues)

atom The smallest unit of an element that retains the properties of the element.

molecule A group of two or more atoms of the same or different elements bonded together.

cell The basic structural and functional unit of living things.

organ A discrete structure composed of more than one tissue that performs a specialized function.

hormone A chemical messenger that is produced in one location in the body, is released into the blood, and travels to other locations, where it elicits responses.

FIGURE 3.1 (continued)

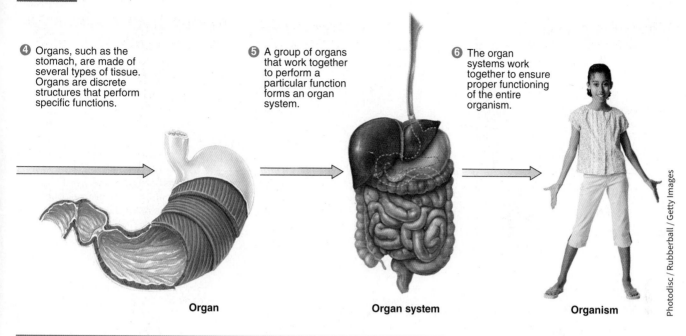

④ Organs, such as the stomach, are made of several types of tissue. Organs are discrete structures that perform specific functions.

⑤ A group of organs that work together to perform a particular function forms an organ system.

⑥ The organ systems work together to ensure proper functioning of the entire organism.

Organ   Organ system   Organism

Photodisc / Rubberball / Getty Images

TABLE 3.1 The major organ systems of the human body

Organ system	What it includes	What it does
Nervous	Nerves, sense organs, brain, and spinal cord	Responds to stimuli from external and internal environments; conducts impulses to activate muscles and glands; integrates activities of other systems.
Respiratory	Lungs, trachea, and air passageways	Supplies the blood with oxygen and removes carbon dioxide.
Urinary	Kidneys, bladder, and associated structures	Eliminates wastes and regulates the balance of water, electrolytes, and acid in the blood.
Reproductive	Testes, ovaries, and associated structures	Produces offspring.

(continues)

TABLE 3.1 The major organ systems of the human body *(continued)*

Organ system	What it includes	What it does
Cardiovascular/circulatory	Heart and blood vessels	Transports blood, which carries oxygen, nutrients, and wastes.
Lymphatic/immune	Lymph, lymph structures, and white blood cells	Defends against foreign invaders; picks up fluid leaked from blood vessels; transports fat-soluble nutrients.
Muscular	Skeletal muscles	Provides movement and structure.
Skeletal	Bones and joints	Protects and supports the body, provides a framework for the muscles to use for movement.
Endocrine	Pituitary, adrenal, thyroid, pancreas, and other ductless glands	Secretes hormones that regulate processes such as growth, reproduction, and nutrient use.
Integumentary	Skin, hair, nails, and sweat glands	Covers and protects the body; helps control body temperature.
Digestive	Mouth, esophagus, stomach, intestines, pancreas, liver, and gallbladder	Ingests and digests food; absorbs nutrients into the blood; eliminates nonabsorbed food wastes.

by the cardiovascular system, which transports nutrients to individual cells in the body; and by the urinary, respiratory, and integumentary systems, which eliminate wastes generated in the body.

Concept Check

1. **How** are atoms, molecules, and cells related to one another?
2. **How** do the endocrine and nervous systems interact with the digestive system?

3.2 The Digestive System

LEARNING OBJECTIVES

1. **Define** digestion and absorption.
2. **List** the organs that make up the digestive system.
3. **Describe** the tissue layers that make up the wall of the gastrointestinal tract.
4. **Explain** the roles of mucus, enzymes, nerves, and hormones in digestion.

The digestive system is the organ system that is primarily responsible for **digestion** and for the **absorption** of nutrients into the body. When you eat a taco, for example, the tortilla, meat, cheese, lettuce, and tomato are broken apart, releasing the nutrients and other food components they contain. Water, vitamins, and minerals are taken into the body without being broken into smaller units, but proteins, carbohydrates, and fats must be digested further. Proteins are broken down into amino acids, most of the carbohydrate is broken down into sugars, and most fats are digested to produce molecules with long carbon chains called **fatty acids**. The sugars, amino acids, and fatty acids can then be absorbed into the body. The fiber in whole grains, fruits, and vegetables cannot be digested and therefore is not absorbed into the body. It and other unabsorbed substances pass through the digestive tract and are eliminated in **feces**.

Organs of the Digestive System

The digestive system is composed of the **gastrointestinal tract** and accessory organs (**Figure 3.2a**). The gastrointestinal tract is a hollow tube, about 30 feet long, that runs from the mouth to the anus. It is also called the gut, GI tract, alimentary canal, or digestive tract. The inside of the tube is the **lumen** (**Figure 3.2b**). Food in the lumen is not technically inside the body because it has not been absorbed. When you swallow something that cannot be digested, such as a whole sesame seed or an unpopped kernel of popcorn, it passes through your digestive tract and exits in the feces, without ever entering your blood or cells. Only after substances have been absorbed into the cells that line the gastrointestinal tract can they be said to be inside the body.

The lumen is lined with a type of epithelial cells called **mucosal cells** that together with underlying tissues make up the **mucosa**. Because mucosal cells are in direct contact with churning food and harsh digestive secretions, they live only about two to five days. The dead cells are sloughed off into the lumen, where some components are digested and absorbed and the rest are eliminated in feces. New mucosal cells are formed continuously to replace those that die. To allow for this rapid replacement, the mucosa has high nutrient requirements and is one of the first parts of the body to be affected by nutrient deficiencies.

The time it takes food to travel the length of the GI tract from mouth to anus is called the **transit time**. The shorter the transit time, the more rapidly material is passing through the digestive tract. In a healthy adult, transit time is 24 to 72 hours, depending on the composition of the individual's diet and his or her level of physical activity, emotional state, health status, and use of medications.

Digestive System Secretions

Digestion is aided by substances secreted into the digestive tract from cells in the mucosa and from a number of accessory organs. One of these substances is **mucus**, which moistens, lubricates, and protects the digestive tract. **Enzymes** are

---

digestion The process by which food is broken down into components small enough to be absorbed into the body.

absorption The process of taking substances from the gastrointestinal tract into the interior of the body.

feces Body waste, including unabsorbed food residue, bacteria, mucus, and dead cells, which is eliminated from the gastrointestinal tract by way of the anus.

mucus A viscous fluid secreted by glands in the digestive tract and other parts of the body. It lubricates, moistens, and protects cells from harsh environments.

enzyme A protein molecule that accelerates the rate of a chemical reaction without itself being changed.

FIGURE 3.2 **Structure of the digestive system** This overview of the digestive system illustrates structure and summarizes organ functions.

a. The digestive system consists of the organs of the digestive tract—mouth, pharynx, esophagus, stomach, small intestine, and large intestine—plus four accessory organs—salivary glands, liver, gallbladder, and pancreas.

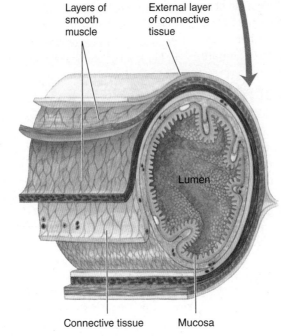

Organs of the gastrointestinal tract

Mouth: Chews food and mixes it with saliva

Pharynx: Swallows chewed food mixed with saliva

Esophagus: Moves food to the stomach

Stomach: Churns and mixes food; secretes acid and a protein-digesting enzyme

Small intestine: Completes digestion; absorbs nutrients into blood or lymph

Large intestine: Absorbs water and some vitamins and minerals; heavily populated by bacteria; passes waste material
{ Colon
  Rectum

Anus: Opens to allow waste to leave the body

Accessory organs

Salivary glands: Produce saliva, which contains a starch-digesting enzyme

Liver: Makes bile, which aids in digestion and absorption of fat

Pancreas: Releases bicarbonate to neutralize intestinal contents; produces enzymes that digest carbohydrate, protein, and fat

Gallbladder: Stores bile and releases it into the small intestine when needed

Ask Yourself

Bile is made in the _____ and stored in the _____. It is released into the _____, where it is important for the digestion and absorption of _____.

b. This cross section through the wall of the small intestine reveals the four tissue layers that make up the wall of the gastrointestinal tract.

Layers of smooth muscle

External layer of connective tissue

Lumen

Connective tissue

Mucosa

Think Critically

Why is food in the lumen still outside the body?

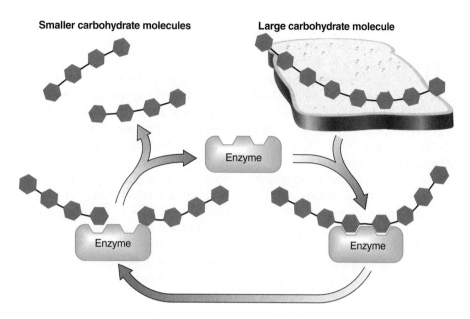

FIGURE 3.3 **Enzyme activity** Enzymes are needed to break down different food components. The enzyme shown here, called an *amylase*, breaks large carbohydrate molecules, such as those in bread, into smaller ones. Amylases have no effect on fat, whereas enzymes called *lipases* digest fat and have no effect on carbohydrate.

also present in digestive system secretions. They accelerate the chemical reactions that break down food into units small enough to be absorbed (**Figure 3.3**).

The gastrointestinal tract is part of the endocrine system as well as the digestive system. It releases hormones that help prepare different parts of the gut for the arrival of food and thus regulate digestion and the rate at which food moves through the digestive tract. Some hormonal signals slow digestion, whereas others facilitate it. For example, when the nutrients from your lunch reach your small intestine, they trigger the release of hormones that signal the pancreas and gallbladder to secrete digestive substances into the small intestine.

Concept Check

1. **What** happens to food during digestion?
2. **Which** organs make up the gastrointestinal tract?
3. **What** are mucosal cells?
4. **How** are enzymes important for digestion and absorption?

3.3 Digestion and Absorption of Nutrients

LEARNING OBJECTIVES

1. **Describe** what happens in each of the organs of the gastrointestinal tract.
2. **Discuss** how food moves through the GI tract.
3. **Explain** how the structure of the small intestine aids in its function.
4. **Distinguish** passive diffusion from active transport.

Imagine warm slices of freshly baked bread smeared with melting butter. Is your mouth watering? You don't even need to put food in your mouth for activity to begin in the digestive tract. Sensory input alone—the sight of the bread being lifted out of the oven, the smell of the bread, the clatter of the butter knife—may make your mouth water and your stomach begin to secrete digestive substances. This response occurs when the nervous system signals the digestive system to ready itself for a meal. In order for food to be used by the body, however, you need to do more than smell your meal. The food must be ingested and digested, and the nutrients must be absorbed and transported to the body's cells. This involves the combined functions of all the organs of the digestive system, as well as the help of some other organ systems.

The Mouth

Digestion involves chemical and mechanical processes, both of which begin in the mouth. The presence of food in the mouth stimulates the flow of **saliva** from the salivary glands. Saliva moistens the food and carries dissolved food molecules to the taste buds, most of which are located on the tongue. Signals from the taste buds, along with the aroma of food, allow us to enjoy the taste of the food we eat. Saliva contains the enzyme **salivary amylase**, which begins the chemical digestion of food by breaking starch molecules into shorter sugar chains (see Figure 3.3). Saliva also helps protect against tooth decay because it washes away food particles and contains substances that inhibit the growth of bacteria that cause tooth decay.

Chewing food begins the mechanical aspect of digestion. Adult humans have 32 teeth, which are specialized for biting, tearing, grinding, and crushing foods. Chewing breaks food into small pieces. This makes the food easier to swallow and increases the surface area in contact with digestive juices. The tongue helps mix food with saliva and aids chewing by constantly repositioning food between the teeth. Chewing also breaks up fiber, which traps nutrients. If the fiber is not broken up, some of the nutrients in the food cannot be absorbed. For example, if the fibrous skin of a raisin is not broken open by the teeth, the nutrients inside the raisin remain inaccessible, and the raisin travels, undigested, through the intestines for elimination in the feces.

The Pharynx

The **pharynx**, the part of the gastrointestinal tract that is responsible for swallowing, is also part of the respiratory tract. Food passes through the pharynx on its way to the stomach, and air passes through the pharynx on its way to and from the lungs. As we prepare to swallow, the tongue moves the bolus of chewed food mixed with saliva to the back of the mouth. During swallowing, the air passages are blocked by a valvelike flap of tissue called the **epiglottis** so that food goes to the esophagus and not to the lungs (**Figure 3.4a**). Sometimes eating too quickly or talking while eating interferes with the movement of the epiglottis, and food passes into an upper air passageway. This food can usually be dislodged with a cough, but if it becomes stuck and causes choking, it may need to be forced out by means of the Heimlich maneuver (**Figure 3.4b**).

FIGURE 3.4 **The role of the epiglottis** The epiglottis prevents food from entering the airway, and the Heimlich maneuver can be used to expel food that becomes lodged in the airway.

a. When a bolus of food is swallowed, it normally pushes the epiglottis down over the opening to the passageway that leads to the lungs.

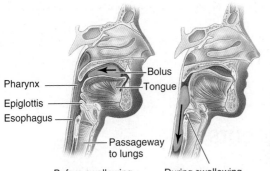

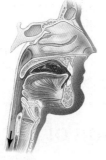

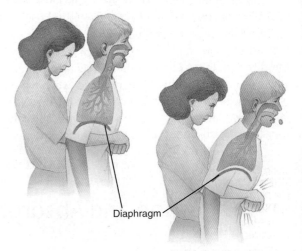

b. If food becomes lodged in the passageway leading to the lungs, it can block the flow of air. The Heimlich maneuver, which involves a series of thrusts directed upward from under the diaphragm (the muscle separating the chest and abdominal cavities), forces air out of the lungs, blowing the lodged food out of the air passageway.

---

saliva A watery fluid that is produced and secreted into the mouth by the salivary glands. It contains lubricants, enzymes, and other substances.

epiglottis A piece of elastic connective tissue that covers the opening to the lungs during swallowing.

The Esophagus

The esophagus connects the pharynx with the stomach. In the esophagus, the bolus of food is moved along by rhythmic contractions of the smooth muscles, an action called **peristalsis** (**Figure 3.5**). The contractions of peristalsis are strong enough so that even if you ate while standing on your head, food would reach your stomach. This contractile movement, which is controlled automatically by the nervous system, occurs throughout the gastrointestinal tract, pushing the bolus along from the pharynx through the large intestine.

To leave the esophagus and enter the stomach, food must pass through a **sphincter**, a muscle that encircles the tube of the digestive tract and acts as a valve. When the sphincter contracts, the valve is closed; when it relaxes, the valve is open (see Figure 3.5). The sphincter, located between the esophagus and the stomach, prevents food from moving from the stomach back into the esophagus, but occasionally stomach contents do move in this direction. This is what occurs with heartburn (as discussed later in this chapter): Some of the acidic stomach contents leak up through this sphincter into the esophagus, causing a burning sensation.

Food also moves from the stomach into the esophagus during vomiting. Vomiting is initiated by a complex series of signals from the brain that cause the sphincter to relax and the muscles to contract, forcing the stomach contents upward, out of the stomach and toward the mouth.

The Stomach

The stomach is an expanded portion of the gastrointestinal tract that serves as a temporary storage place for food. Here the bolus is mashed and mixed with highly acidic stomach secretions to form a semiliquid food mass called **chyme**. The mixing of food in the stomach is aided by an extra layer of smooth muscle in the stomach wall (**Figure 3.6a**). Some digestion takes place in the stomach, but, with the exception of some water, alcohol, and a few drugs, such as aspirin and acetaminophen (Tylenol), very little absorption occurs here.

Gastric juice

Gastric juice, which is produced by gastric glands in pits that dot the stomach lining, promotes chemical digestion in the stomach (**Figure 3.6b**). Gastric juice is a mixture of water, mucus, hydrochloric acid, and an inactive form of the protein-digesting enzyme **pepsin**. This enzyme is secreted in an inactive form so that it will not damage the gastric glands that produce it. The hydrochloric acid in gastric juice kills most of the bacteria present in food. It also stops the activity of the

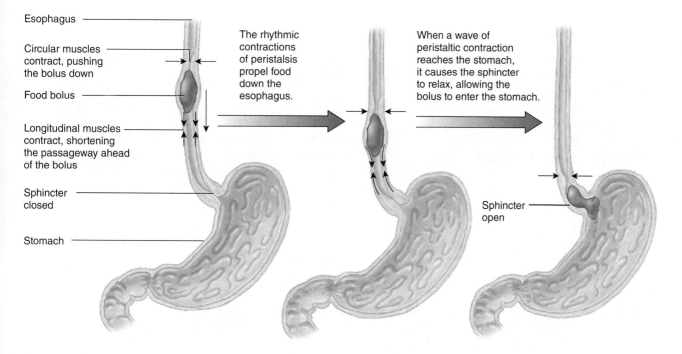

FIGURE 3.5 Moving food through the GI tract The food we swallow doesn't just fall down the esophagus and into the stomach. It is pushed along by muscular contractions and enters the stomach in response to the opening and closing of the sphincter, located where the esophagus meets the stomach.

---

peristalsis Coordinated muscular contractions that move material through the GI tract.

FIGURE 3.6 Stomach structure and function The stomach contributes to both the mechanical and chemical breakdown of food.

a. Most of the gastrointestinal tract is surrounded by two layers of smooth muscle, one that is longitudinal and one that is circular, but the stomach contains a third smooth muscle layer running diagonally. The presence of this diagonal layer allows for the powerful contractions that churn and mix the stomach contents. The flow of chyme from the stomach into the small intestine is regulated by a sphincter at the lower end of the stomach.

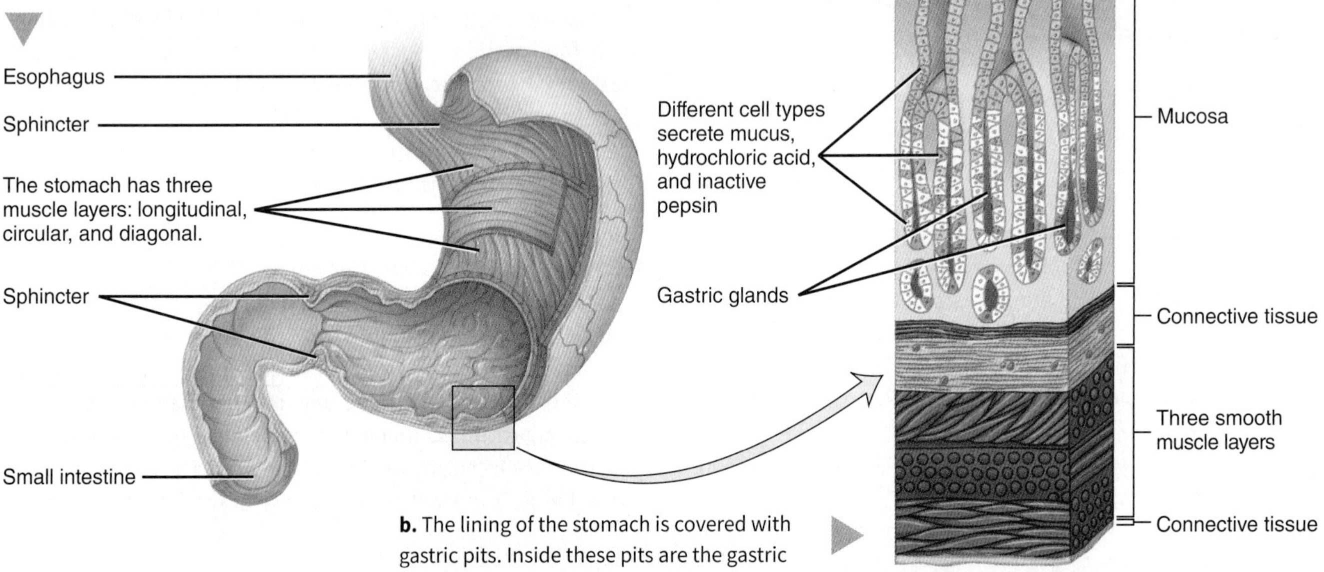

- Esophagus
- Sphincter
- The stomach has three muscle layers: longitudinal, circular, and diagonal.
- Sphincter
- Small intestine

- Gastric pits
- Surface epithelium
- Different cell types secrete mucus, hydrochloric acid, and inactive pepsin
- Gastric glands
- Mucosa
- Connective tissue
- Three smooth muscle layers
- Connective tissue

b. The lining of the stomach is covered with gastric pits. Inside these pits are the gastric glands, made up of different types of cells that produce the mucus, hydrochloric acid, and the inactive form of pepsin contained in gastric juice.

Think Critically
Why is the protein-digesting enzyme pepsin produced in an inactive form?

carbohydrate-digesting enzyme salivary amylase and helps begin the digestion of protein by activating pepsin and unfolding proteins. A thick layer of mucus prevents the protein that makes up the stomach wall from being damaged by the hydrochloric acid and pepsin in gastric juice.

Regulation of stomach activity How much your stomach churns and how much gastric juice is released are regulated by signals from both nerves and hormones. These signals originate from three sites—the brain, the stomach, and the small intestine (**Figure 3.7**).

As chyme moves out of the stomach, signals sent by the small intestine help regulate the rate at which the stomach empties. The small intestine stretches as it fills with chyme; this distension inhibits the stomach from emptying. Chyme normally empties from the stomach within two to six hours, but this rate varies with the size and composition of the meal that has been consumed. A large meal takes longer to leave the stomach than does a small meal. Liquids empty quickly, but solids linger until they are well mixed with gastric juice and are liquefied; hence, solids leave the stomach more slowly than liquids.

The nutritional composition of a meal also affects how long it stays in the stomach. A meal that consists primarily of starch or sugar leaves quickly, but a meal that is high in fiber or protein takes longer to leave the stomach. A high-fat meal stays in the stomach the longest. Because the nutrient composition of a meal affects how quickly it leaves your stomach, it affects how soon after eating you feel hungry again (**Figure 3.8**).

The Small Intestine

The small intestine is a narrow tube about 20 feet long. Here the chyme is propelled along by peristalsis and mixed by rhythmic constrictions called **segmentation** that slosh the material back and forth. The small intestine is the main site for the chemical digestion of food, completing the process that the mouth and stomach have started. It is also the primary site for the

Digestion and Absorption of Nutrients **61**

PROCESS DIAGRAM — **How It Works**

FIGURE 3.7 **The regulation of stomach motility and secretion** Stomach activity is affected by food that has not yet reached the stomach, by food that is in the stomach, and by food that has left the stomach.

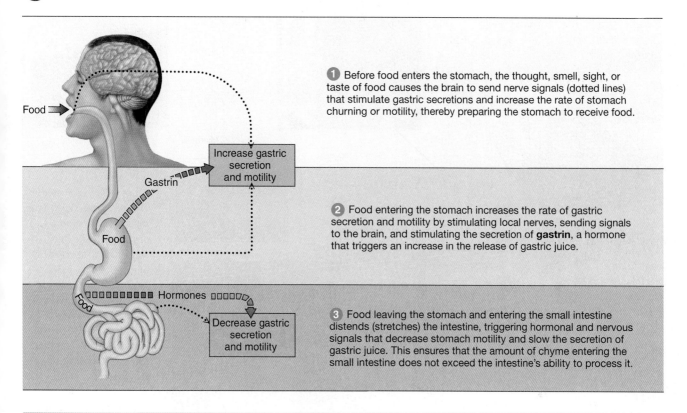

① Before food enters the stomach, the thought, smell, sight, or taste of food causes the brain to send nerve signals (dotted lines) that stimulate gastric secretions and increase the rate of stomach churning or motility, thereby preparing the stomach to receive food.

② Food entering the stomach increases the rate of gastric secretion and motility by stimulating local nerves, sending signals to the brain, and stimulating the secretion of **gastrin**, a hormone that triggers an increase in the release of gastric juice.

③ Food leaving the stomach and entering the small intestine distends (stretches) the intestine, triggering hormonal and nervous signals that decrease stomach motility and slow the secretion of gastric juice. This ensures that the amount of chyme entering the small intestine does not exceed the intestine's ability to process it.

FIGURE 3.8 **Hunger and meal composition** What you choose for breakfast can affect how soon you become hungry for lunch. A small, carbohydrate-rich meal of dry toast and coffee will leave your stomach far more quickly than a larger meal containing more protein, fiber, and fat, such as this vegetable-and-cheese omelet with whole-wheat toast and butter.

dirkr / iStockphoto

absorption of water, vitamins, minerals, and the products of carbohydrate, fat, and protein digestion.

A number of unique structural features contribute to the small intestine's digestive function and enhance the amount of surface area available for absorption (**Figure 3.9**). Together these features provide a surface area that is about the size of a tennis court (about 2700 ft^2).

Secretions that aid digestion In the small intestine, secretions from the pancreas, the gallbladder, and the small intestine itself aid digestion. The pancreas secretes **pancreatic juice**, which contains **bicarbonate**, and digestive enzymes. Bicarbonate, which is a base, neutralizes the acid in the chyme, making the environment in the small intestine neutral or slightly basic rather than acidic, as in the stomach. This

NUTRITION INSIGHT

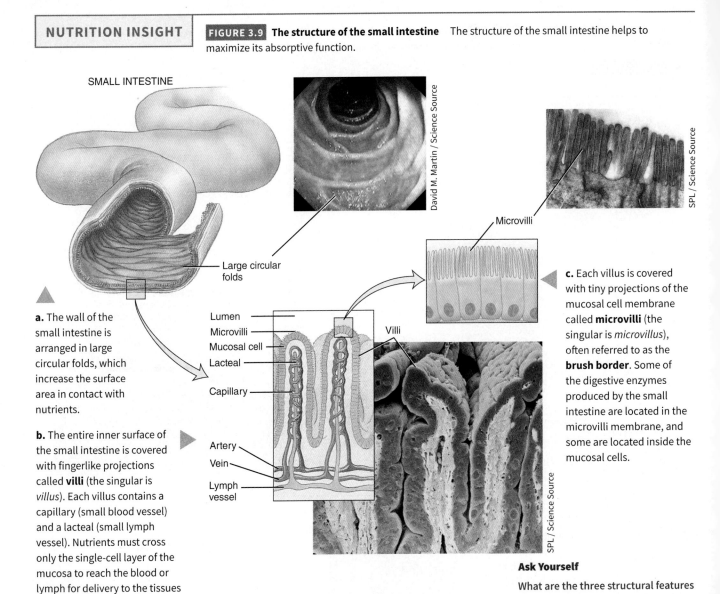

FIGURE 3.9 The structure of the small intestine The structure of the small intestine helps to maximize its absorptive function.

a. The wall of the small intestine is arranged in large circular folds, which increase the surface area in contact with nutrients.

b. The entire inner surface of the small intestine is covered with fingerlike projections called **villi** (the singular is *villus*). Each villus contains a capillary (small blood vessel) and a lacteal (small lymph vessel). Nutrients must cross only the single-cell layer of the mucosa to reach the blood or lymph for delivery to the tissues of the body.

c. Each villus is covered with tiny projections of the mucosal cell membrane called **microvilli** (the singular is *microvillus*), often referred to as the **brush border**. Some of the digestive enzymes produced by the small intestine are located in the microvilli membrane, and some are located inside the mucosal cells.

Ask Yourself

What are the three structural features of the small intestine that increase its surface area?

---

neutrality allows enzymes from the pancreas and small intestine to function.

Pancreatic amylase is an enzyme that continues the job of breaking down starches into sugars that was started in the mouth by salivary amylase. Pancreatic **proteases** (protein-digesting enzymes), such as trypsin and chymotrypsin, break protein into shorter and shorter chains of amino acids, and fat-digesting enzymes called **lipases** break down fats into fatty acids. The pancreatic proteases, like the pepsin produced by the stomach, are released in an inactive form so that they will not digest the glands that produce them. Intestinal digestive enzymes, found in the cell membranes or inside the cells lining the small intestine, aid the digestion of double sugars (those that contain two sugar units) into single sugar units and the digestion of short amino acid chains into single amino acids. The sugars from carbohydrate digestion and the amino acids from protein digestion pass into the blood and are delivered to the liver (**Figure 3.10**).

The gallbladder stores and secretes **bile**, a fluid containing bile acids and cholesterol, which is produced in the liver and

---

bile A digestive fluid made in the liver and stored in the gallbladder that is released into the small intestine, where it aids in fat digestion and absorption.

PROCESS DIAGRAM **FIGURE 3.10** **Digestion and absorption in the small intestine** Most digestion and absorption occurs in the small intestine.

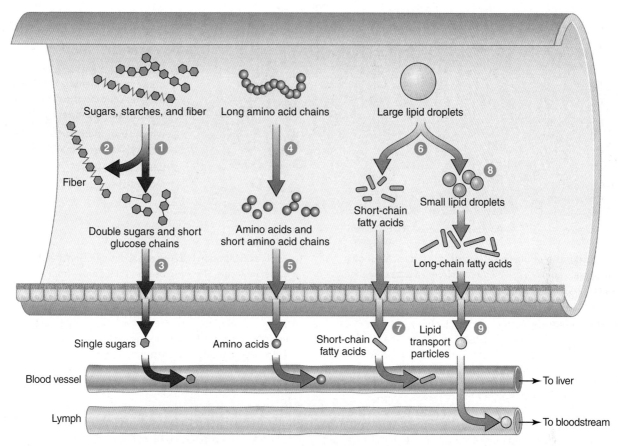

1. Pancreatic amylase digests starch to double sugars and short glucose chains.
2. Fiber, which cannot be digested by human enzymes, passes to the large intestine.
3. Enzymes in the microvilli digest double sugars into single sugars, which are absorbed into the blood.
4. Pancreatic proteases, along with proteases in the microvilli, digest long amino acid chains into amino acids and short amino acid chains.
5. Amino acids and short amino acid chains are absorbed into the mucosal cells, where they are digested into single amino acids, which pass into the blood.
6. Bile helps divide large fat globules. Pancreatic lipases digest fat molecules into fatty acids.
7. Short-chain fatty acids are absorbed into the mucosal cells and then pass directly into the blood.
8. Long-chain fatty acids and other lipids combine with bile to form small droplets that aid the absorption of fatty acids and other fat-soluble substances into the mucosal cell.
9. Absorbed lipids are incorporated into transport particles that pass into the lymph. They enter the blood without first passing through the liver.

is necessary for the digestion and absorption of fat. Bile that is secreted into the small intestine mixes with fat. Bile acids help divide the large lipid droplets into small globules, allowing lipases to access and digest the fat molecules more efficiently. The bile acids and digested fats then form small droplets that facilitate the absorption of fat into the mucosal cells. Inside the mucosal cells, the products of fat digestion are incorporated into transport particles. These are absorbed into the lymph before passing into the blood (see Figure 3.10).

Absorption The small intestine is the main site for the absorption of nutrients. To be absorbed, nutrients must pass from the lumen of the GI tract into the mucosal cells lining the tract and then into either the blood or the lymph. Several different mechanisms are involved (**Figure 3.11**). Some rely on **diffusion**, which is the net movement of substances from an area of higher concentration to an area of lower concentration. **Simple diffusion**, in which material moves freely across a cell membrane; **osmosis**, which is the diffusion of water; and **facilitated diffusion**,

simple diffusion The unassisted diffusion of a substance across a cell membrane.

osmosis The unassisted diffusion of water across a cell membrane.

facilitated diffusion Assisted diffusion of a substance across a cell membrane.

FIGURE 3.11 **Absorption mechanisms** A variety of mechanisms are involved in transporting nutrients from the lumen of the small intestine into the mucosal cells.

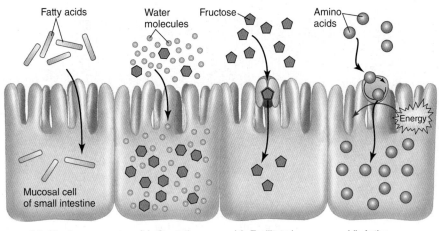

Ask Yourself
1. Which absorption mechanism(s) can only move nutrients from an area with a higher concentration of that nutrient to an area with a lower concentration?
2. Which absorption mechanism(s) require(s) a carrier molecule?
3. Which absorption mechanism(s) require(s) energy?

a. In simple diffusion, substances such as the fatty acids shown here pass freely across a cell membrane from an area of higher concentration to an area of lower concentration, and no energy is required.

b. Osmosis is the passage of water molecules from an area with a lower concentration of dissolved substances, such as the glucose shown here (red hexagons), to an area with a higher concentration of dissolved substances. Water can move both into and out of the lumen of the GI tract by osmosis.

c. Facilitated diffusion is a type of passive diffusion that requires a carrier molecule. Here fructose molecules move from an area of higher concentration to an area of lower concentration, with the help of a carrier molecule.

d. Active transport requires energy and a carrier molecule and can transport substances from an area of lower concentration to an area of higher concentration. Active transport allows nutrients, such as the amino acids shown here, to be absorbed even when they are present in higher concentrations in the mucosal cell than in the lumen.

in which a carrier molecule is needed for the substance to cross a cell membrane, depend on diffusion and are passive, requiring no energy. **Active transport** requires energy and a carrier molecule. This process can transport material from an area of lower concentration to one of higher concentration.

The Large Intestine

Materials not absorbed in the small intestine pass through a sphincter between the small intestine and the large intestine. This sphincter prevents material from the large intestine from reentering the small intestine.

The large intestine is about 5 feet long and is divided into the colon, which makes up the majority of the large intestine, and the rectum, the last 8 inches. The large intestine opens to the exterior of the body at the anus. Although most nutrient absorption occurs in the small intestine, water and some vitamins and minerals are also absorbed in the colon.

Peristalsis occurs more slowly in the large intestine than in the small intestine. Water, nutrients, and fecal matter may spend 24 hours in the large intestine, in contrast to the 3 to 5 hours it takes these materials to move through the small intestine. This slow movement favors the growth of microorganisms. These microorganisms, collectively referred to as the **intestinal microbiota**, produce nutrients and have other effects that impact the health of the host body, as discussed below.[1,2] Another bacterial by-product is gas, which causes flatulence. In normal adult humans between 200 and 2000 mL of intestinal gas is produced per day.

Material that is not absorbed in the colon passes into the rectum, where it is stored temporarily and then evacuated through the anus as feces. The feces are a mixture of undigested, unabsorbed matter, dead cells, secretions from the GI tract, water, and bacteria. The amount of bacteria varies but can make up more than half the weight of the feces. The amount of water in the feces is affected by fiber and fluid intake. Because fiber retains water, when adequate fiber and fluids are consumed, feces have a higher water content and are more easily passed.

Concept Check

1. **What** are the functions of the stomach?
2. **What** does peristalsis do?
3. **How** do the villi and microvilli aid absorption?
4. **Why** must some nutrients be absorbed by active transport rather than passive diffusion?

active transport The transport of substances across a cell membrane with the aid of a carrier molecule and the expenditure of energy.

3.4 The Digestive System in Health and Disease

LEARNING OBJECTIVES

1. **Explain** how the intestinal microbiota affects health.
2. **Discuss** the role of the gastrointestinal tract in protecting us from infection.
3. **Describe** how the symptoms of food allergies and celiac disease are triggered.
4. **Discuss** the potential effect of dental problems, ulcers, GERD, and gallstones on nutritional status.

The health of the GI tract is essential to our overall health. The gut not only takes nutrients into the body, but also interacts with the intestinal microbiota and serves as a barrier between the body and hazardous substances in the outside world. Food allergies and sensitivities have their origins in the GI tract, but most common gastrointestinal problems are minor and do not affect long-term health.

The Intestinal Microbiota

The human gut is home to about 100 trillion (10^{14}) microorganisms, 10 times the number of cells that make up the human body itself. These microorganisms are present throughout the GI tract but are more numerous and diverse in the colon than in other parts.[3,4] The right mix is important for optimal gastrointestinal function, maintenance of immune function, and overall health.[1]

Functions of the intestinal microbiota The intestinal microbiota acts on unabsorbed portions of food and on substances secreted by the GI tract, such as mucus, to produce nutrients that are used by the bacteria and can affect human health. For example, bacterial breakdown of undigested carbohydrates produces small fatty acids referred to as *short-chain fatty acids*, that provide energy for colonic cells and are important for regulating metabolism in the gut and other parts of the body (**Figure 3.12a**).[5] The presence of the microbiota and the substances they produce helps maintain the mucosal layer that lines the intestine and serves as a barrier, modulates the amount of inflammation in the gut, and prevents the growth of disease-causing bacteria. The gut microbiota is also important for the maturation of the immune system because it allows the immune system to tolerate harmless bacteria and respond to harmful ones.[6] In addition, there is evidence that the gut microbiota affects metabolism throughout the body by regulating energy intake, expenditure, and storage, as well as glucose homeostasis.[2,5,7]

A healthy microbiota A healthy microbiota is made up of many different types of microorganisms; it resists change under physiological stress and supports health. An unhealthy microbiota is less diverse and is unable to maintain a good balance between the number of protective versus harmful bacteria (**Figure 3.12b**).[1] An unhealthy microbiota has been implicated in the development of intestinal diseases, such as Crohn's disease, as well as systemic chronic diseases such as obesity, type 2 diabetes, colon cancer, and heart disease.[8]

FIGURE 3.12 Gut microbiota, nutritional status, and health Our diet affects the makeup of the intestinal microbiota, which impacts our overall health.

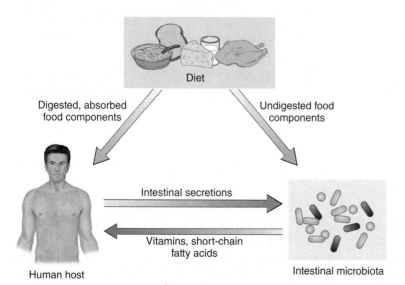

a. What we eat affects the gut microbiota, and the gut microbiota in turn enables the digestion of substances that can't be broken down by human enzymes.

(continues)

FIGURE 3.12 *(continued)*

b. A healthy microbiota helps prevent the growth of disease-causing organisms and moderate inflammation. Antibiotic treatment or other environmental factors that disrupt the microbial community may allow disease-causing bacteria to grow. This leads to inflammation and damage to the mucosal barrier and, possibly, systemic infection.

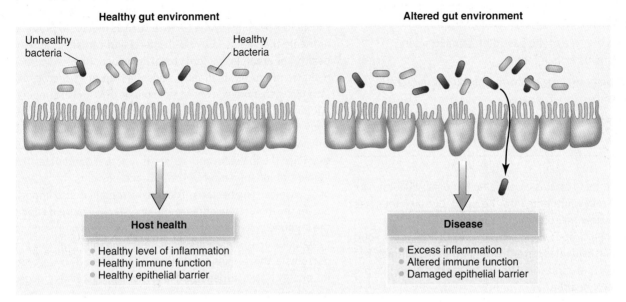

The intestinal microbiota is established in infancy and early childhood and is affected throughout life by diet and antibiotic use.[7] Diet affects the microbiota because it determines the food available to the microorganisms in the gut and therefore what bacteria grow there (see Figure 3.12a). A diet high in **prebiotics** promotes the maintenance of a healthy microbiota. Prebiotics are found naturally in certain foods and are sold as dietary supplements. Another option for maintaining a healthy intestinal microbiota is the consumption of live beneficial bacteria, referred to as **probiotics** (see *What a Scientist Sees: Bacteria on the Menu*).[1]

Gut Immune Function in Health and Disease

Food almost always contains bacteria and other contaminants, but it rarely makes us sick. This is because acid in the stomach kills most bacteria, a healthy microbiota prevents harmful bacteria from growing, and the mucosa serves as a barrier, preventing the absorption of toxins and disease-causing organisms. The gastrointestinal tract is also an important site of immune system activity. If an invading substance, or **antigen**, enters the lumen or is absorbed into the mucosa, the immune system can use a number of weapons to destroy it. These include various types of white blood cells, which circulate in the blood and reside in the mucosa of the gastrointestinal tract.

When an antigen is present, **phagocytes** are the first type of white blood cell to come to the body's defense. If the phagocytes do not eliminate the invader, white blood cells called **lymphocytes**, which target specific antigens, join the battle. Some lymphocytes destroy specific antigens by binding to them. This type of lymphocyte helps eliminate cancer cells, foreign tissue, and cells that have been infected by viruses and bacteria. Other lymphocytes produce and secrete protein molecules called **antibodies**. Antibodies bind to antigens and help destroy them. Each antibody is able to fight off only one type of antigen. Once antibodies to a specific antigen have been made, the immune system remembers and is ready to fight that antigen any time it enters the body again.

If harmful organisms infect the GI tract, the body may help out the immune system by using diarrhea or vomiting to flush them out.

Food allergies Our immune system protects us from many invaders without our being aware of it. Unfortunately,

---

prebiotic A substance that passes undigested into the colon and stimulates the growth and/or activity of certain types of bacteria.

probiotic Live bacteria that, when consumed, live temporarily in the colon and confer health benefits on the host.

antigen A foreign substance that, when introduced into the body, stimulates an immune response.

antibody A protein, released by a type of lymphocyte, that interacts with and neutralizes specific antigens.

What a Scientist Sees

Bacteria on the Menu

Ads claim that eating specialized yogurts such as those shown in the photo will help regulate the digestive system. Consumers see these products as a tasty way to help prevent and cure certain digestive problems. Scientists recognize that these products as well as most other yogurts contain active cultures of beneficial bacteria, including *Lactobacillus* and *Bifidobacterium*. When these living bacteria are consumed in adequate amounts, they live temporarily in the colon (see figure), where they inhibit the growth of harmful bacteria. The bacteria must be consumed frequently because they are flushed out in the feces. The best evidence for the benefits of

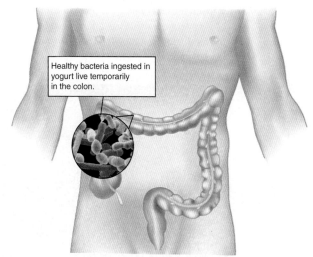

Healthy bacteria ingested in yogurt live temporarily in the colon.

SCIMAT / Getty Images

Andy Washnik

probiotics is for the prevention and treatment of antibiotic-induced diarrhea, acute infectious diarrhea, and persistent diarrhea in children.[4]

Think Critically

Why might your doctor recommend consuming yogurt during and after a course of antibiotics?

---

the response of the immune system to a foreign substance is also to blame for allergic reactions. An allergic reaction occurs when the immune system produces antibodies to a substance, called an **allergen**, that is present in our diet or environment. **Food allergies** occur when the body identifies proteins present in food as foreign substances and initiates an immune response (see Chapter 6). The immune response causes symptoms that range from hives to life-threatening anaphylactic reactions such as breathing difficulties or a drop in blood pressure.

The first time a food is consumed, it does not trigger an allergic reaction, but in a susceptible person, this first exposure begins the process. As the food is digested, tiny fragments of undigested protein trigger the production of antibodies. When the food protein is eaten again, it binds to the antibodies, signaling the release of chemicals that cause redness, swelling, and other allergy symptoms. The best way to avoid allergy symptoms is to avoid foods to which you are allergic (**Figure 3.13**).

Celiac disease and gluten-related disorders

Celiac disease is a condition in which the protein gluten, found in wheat, barley, and rye, triggers an immune system response that damages or destroys the villi of the small intestine. For most of us, the gluten in our foods is digested and absorbed like other proteins. However, for people with celiac disease, also called celiac sprue, nontropical sprue, and gluten-sensitive enteropathy, consuming even a tiny amount of gluten can cause abdominal pain, diarrhea, and fatigue. Eventually this damage can lead to malnutrition, weight loss, anemia, osteoporosis, intestinal cancer, and other chronic illnesses.[11,12] Celiac disease is an inherited condition that affects about 1% of the population.[13] It can be diagnosed only by a blood test or an intestinal biopsy. There are also several other **gluten-related disorders**, which cause reactions to gluten-containing foods. These include wheat allergy and non-celiac gluten sensitivity (NCGS). Wheat allergy is a food allergy; it is diagnosed by measuring blood levels of antibodies specifically to wheat. NCGS is not an allergy but consumption of gluten-containing foods causes symptoms such as bloating, gas, diarrhea, headache, joint and muscle pain, chronic fatigue, weight loss, and anemia. These symptoms disappear when gluten is excluded from the diet and may reappear when gluten is reintroduced.[14,15] It is unclear whether gluten is the culprit in NCGS; it is possible that wheat proteins other than

---

allergen   A substance that causes an allergic reaction.

FIGURE 3.13 **Allergy information on food labels** Food allergies affect about 5% of adults and 8% of children in the United States.[9] When allergic individuals are exposed, severe reactions, referred to as anaphylaxis, may occur and can be fatal. To protect consumers, food manufacturers are required to clearly state on the label whether a product contains any of the eight major ingredients that are most likely to cause allergic reactions: peanuts, tree nuts, milk, eggs, fish, shellfish, soy, and wheat. Sesame seeds are the ninth most common food allergen, but current rules do not require that they be clearly labeled in the ingredient list.[10]

gluten or small indigestible carbohydrates found in grains are responsible for the symptoms.[16, 17]

For people with gluten-related disorders, consuming a diet that eliminates gluten provides relief from symptoms and long-term complications. This means eliminating all products made from wheat, barley, or rye, including most breads, crackers, pastas, cereals, cakes, and cookies. It also requires eliminating foods ranging from packaged gravies to soy sauce that contain trace amounts of these grains (**Figure 3.14**). Whether a gluten-free diet has health benefits for people without these conditions is open to debate (*see Debate: Should You Be Gluten Free?*).

FIGURE 3.14 **Identifying gluten-free foods** The ingredient list can help people with gluten-related disorders eliminate gluten-containing foods, but consumers need to know what to look for. Products containing any type of wheat, including wheat berries, durum, semolina, spelt, farina, faro, and graham; rye; barley; or triticale should be avoided.

Think Critically

Neither potatoes nor onions contain gluten. If you had celiac disease, based on the ingredients shown here, which would be a safer choice: the French fries or the onion rings?

Debate

Should You Be Gluten Free?

The Issue

Gluten-free diets are essential for people with gluten-related disorders, but a gluten-free diet has also been promoted for weight loss and to treat a host of other ailments. Should you be gluten free?

You see the term *gluten free* on breakfast cereals, cake mixes, pastas, soups, and a host of other products. General Mills, the cereal giant, even has a Web site dedicated to helping people eat gluten free. Celebrities are touting the benefits of going gluten free. Gwyneth Paltrow and Miley Cyrus swear by it for weight loss. Rachael Ray devoted a show to gluten-free cooking. Gluten-free diets are essential for people who have gluten-related disorders; however, they have also been promoted as a healthier way of eating for everyone. Advocates claim that eliminating gluten promotes weight loss and helps those suffering with joint pain, rheumatoid arthritis, osteoporosis, anemia, and diabetes. They contend that individuals with these conditions have undiagnosed celiac disease or non-celiac gluten sensitivity (NCGS). This may be the case for some people since only about 10 to 15% of patients with celiac disease are believed to have been diagnosed, and it is estimated that as many as 18 million Americans have NCGS.[18] For both these groups eating a gluten-free diet is likely to be of benefit. However, there is little evidence that switching to gluten free will help relieve these conditions in those who do not have a gluten-related disease.[19]

What about going gluten free for weight loss or to improve overall health? Marketing has made following a gluten-free diet appear to be a healthier way of eating. However, gluten-free foods are not necessarily healthier or lower in calories than other foods (see photo). Gluten is found in breads, pastas, and breakfast cereals—all good sources of B vitamins, iron, and fiber—so if you just eliminate these foods, you will increase the risk of nutrient deficiencies. If you replace gluten-containing processed foods such as cereals, pasta, and muffins with processed gluten-free versions of these foods, you probably won't decrease your calories or increase the nutrient density of your diet. If you skip high-carbohydrate, gluten-containing snacks like crackers and cookies but then replace them with potato

Gluten-free bread is made by using gluten-free flours such as rice flour, garbanzo bean flour, tapioca flour, and millet flour instead of wheat flour and including xanthan gum to give it elasticity.

chips and ice cream, your diet could be higher in calories than your original diet. But if you replace gluten-containing refined grain products with gluten-free whole grains such as amaranth, millet, and quinoa and other whole foods such as fresh fruits and vegetables, the switch may help you lose weight and improve your overall diet.

Is the gluten-free fad good or bad? Proponents think a gluten-free diet will improve everyone's health. And, in general, any change that makes you carefully plan your diet is a good thing. Individuals with gluten-related disorders are benefitting from the craze because it has increased the availability and quality of gluten-free foods, improved the labeling of gluten-free products, and heightened awareness of gluten-sensitive diseases. However, for those without gluten-related disorders, it unnecessarily eliminates a large number of nutrient-dense food options.

Think Critically

Is gluten-free bread more or less healthy than whole-wheat bread for someone without a gluten-related disorder? Why?

Other Digestive System Problems and Discomforts

Almost everyone experiences digestive system problems from time to time. These problems often cause discomfort and frequently limit the types of foods a person can consume (**Figure 3.15a**). They also can interfere with nutrient digestion and absorption. Problems may occur anywhere in the digestive tract, from the mouth to the anus, and can affect the accessory organs that provide the secretions that are essential for proper GI function.

Heartburn and GERD Heartburn occurs when the acidic contents of the stomach leak back into the esophagus (**Figure 3.15b**). The medical term for the leakage of stomach contents into the esophagus is *gastroesophageal reflux*. Occasional heartburn is common, but if it occurs more than twice a week, it may indicate a condition called **gastroesophageal reflux disease (GERD)**. If left untreated, GERD can eventually lead to more serious health problems, such as esophageal bleeding, ulcers, and cancer.

The discomforts of heartburn and GERD can be avoided by limiting the amounts and types of foods consumed. Eating small

---

heartburn A burning sensation in the chest or throat caused when acidic stomach contents leak back into the esophagus.

gastroesophageal reflux disease (GERD) A chronic condition in which acidic stomach contents leak into the esophagus, causing pain and damaging the esophagus.

meals and consuming beverages between rather than with meals prevents heartburn by reducing the volume of material in the stomach. Avoiding fatty and fried foods, chocolate, peppermint, and caffeinated beverages, which increase stomach acidity or slow stomach emptying, can help minimize symptoms. Remaining upright after eating, wearing loose clothing, avoiding smoking and alcohol, and losing weight may also help prevent heartburn (see *What Should I Eat?*). For many people, medications that neutralize acid or reduce acid secretion are needed to manage symptoms.

Peptic ulcers Peptic ulcers occur when the mucus barrier protecting the stomach, esophagus, or upper small intestine is penetrated and the acid and pepsin in digestive secretions damage the gastrointestinal lining (**Figure 3.15c**). Mild ulcers cause abdominal pain; more severe ulcers can cause life-threatening bleeding.

Peptic ulcers can result from GERD or from misuse of medications such as aspirin or nonsteroidal anti-inflammatory drugs (such as Motrin and Advil) but are more often caused by infection with the bacterium *Helicobacter pylori* (*H. pylori*). These bacteria burrow into the mucus and destroy the protective mucosal layer.[20] About 30 to 40% of people in the United States become infected with *H. pylori*, but not everyone who is infected develops ulcers.[21] *H. pylori* infection can be treated using antibiotics.

Gallstones Clumps of solid material that accumulate in either the gallbladder or the bile duct are referred to as **gallstones** (**Figure 3.15d**). They can cause pain when the gallbladder contracts in response to fat in the intestine. Gallstones can interfere with bile secretion and reduce fat absorption. They are usually treated by removing the gallbladder. After the gallbladder has been removed, bile, which is produced in the liver, drips directly into the intestine as it is produced rather than being stored and squeezed out in larger amounts when fat enters the intestine.

Diarrhea and constipation Diarrhea and constipation are common discomforts that are related to problems in the intestines. **Diarrhea** refers to frequent, watery stools. It occurs when material moves through the colon too quickly for sufficient water to be absorbed or when water is drawn into the lumen from cells lining the intestinal tract.

NUTRITION INSIGHT | **FIGURE 3.15 Digestive disorders** Abnormalities in any of the organs of the digestive system can affect nutritional status and overall health.

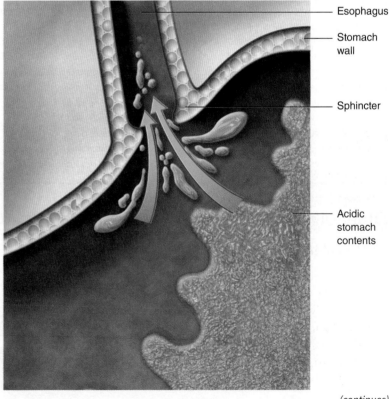

a. Tooth loss and dental pain can make chewing difficult. This may limit the intake of certain foods and reduce nutrient absorption because poorly chewed food may not be completely digested. Tooth decay and gum disease are more likely when saliva production is reduced. Reduced saliva production, which is a side effect of many medications, can also cause changes in taste and difficulty swallowing.

b. Heartburn and GERD occur when stomach acid leaks back through the sphincter and irritates the lining of the esophagus. Stomach contents also pass through this sphincter during vomiting. Vomiting may be caused by an illness, a food allergy, medication, an eating disorder, or pregnancy.

(continues)

peptic ulcer An open sore in the lining of the stomach, esophagus, or upper small intestine.

FIGURE 3.15 *(continued)*

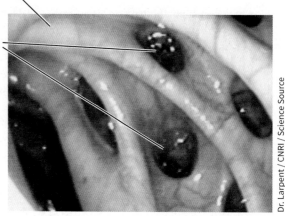

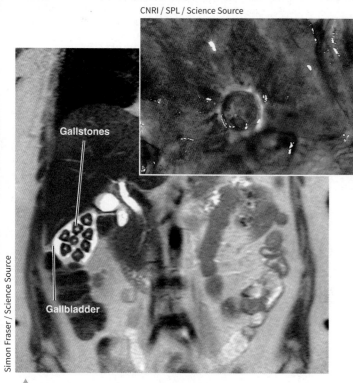

c. Peptic ulcers occur when the mucosa is destroyed, exposing underlying tissues to gastric juices. Damage that reaches the nerve layer causes pain, and bleeding can occur if blood vessels are damaged. If the wall of the stomach or esophagus is perforated because of an ulcer, a serious abdominal infection can occur.

d. Gallstones, visible in this image of the abdomen, are deposits of cholesterol, bile pigments, and calcium that can be in the gallbladder or the bile duct. They can block bile from entering the small intestine, causing pain when the gallbladder contracts and reducing fat digestion and absorption.

e. Constipation increases pressure in the colon and can lead to outpouches in the colon wall, shown here, called diverticula (discussed further in Chapter 4).

What Should I Eat?

For Digestive Health

Reduce your risk of adverse reactions

- Avoid foods that you are allergic to or can't tolerate by reading food labels carefully.
- Chew each bite thoroughly to maximize digestion and avoid choking.
- Don't talk with food in your mouth.
- Learn the Heimlich maneuver: You could save a life.

Reduce your chances of heartburn

- Eat enough to satisfy your hunger but not so much that you are stuffed.
- Wait 10 minutes between your first and second courses to see how full you feel.
- Stay upright after you eat; don't flop on the couch in front of the television.

Avoid constipation by consuming enough fiber and fluid

- Choose whole-grain cereals such as oatmeal or raisin bran.
- Double your servings of vegetables at dinner.
- Eat two pieces of fruit with your lunch.
- Choose whole-grain bread.
- Have one or two unsweetened beverages with or before each meal.

 Use iProfile to find the fiber content of your favorite fruits and vegetables.

Thinking It Through

A Case Study on How Changes in the Digestive System Affect Nutrition

Changes in the digestive system affect how our bodies process the food we eat. For each patient described here, think about how digestion and absorption are affected and the consequences for the patient's nutritional health.

A 50-year-old man is taking medication that reduces the amount of saliva he produces.

What effect might this have on his nutrition and dental health?
Your answer:

An 80-year-old woman with poorly-fitting dentures likes raw carrots. She still eats them but can't chew them thoroughly.

How might this affect the digestion and absorption of nutrients contained in the carrots?
Your answer:

After reading about the benefits of dietary fiber, a 25-year old woman dramatically increases the amount of fiber she consumes.

How might increasing dietary fiber intake affect bowel movements? The amount of intestinal gas?
Your answer:

A 47-year-old woman undergoes treatment for colon cancer, which requires that most of her large intestine be surgically removed.

How does this change affect the amount of fluid she needs to consume?
Your answer:

A 56-year-old man has gallstones, which cause pain when his gallbladder contracts.

What types of food should he avoid, and why?
Your answer:

A 50-year-old man has a deficiency of pancreatic enzymes.

How would this affect nutrient digestion?
Your answer:

A 40-year-old woman weighing 300 pounds has undergone a surgical procedure called gastric banding to help her lose weight. The diagram shows how her stomach was altered.

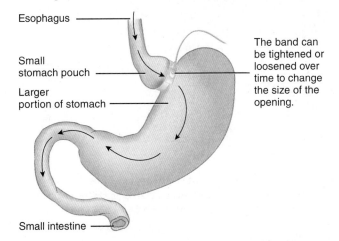

Why can't she eat as much food as before? Will the procedure affect nutrient absorption? Why or why not?
Your answer:

(Check your answers in Appendix L.)

Diarrhea can be caused by bacterial or viral infections, irritants that inflame the lining of the GI tract, the passage of undigested food into the large intestine, medications, and chronic intestinal diseases. Diarrhea causes loss of fluids and minerals. Severe diarrhea lasting more than a day or two can be life threatening.

Constipation refers to hard, dry stools that are difficult to pass (**Figure 3.15e**). Constipation can be caused by a diet containing insufficient fluid or fiber, lack of exercise, a weakening of the muscles of the large intestine, and a variety of medications. It can be prevented by drinking plenty of liquids, consuming a high-fiber diet, and getting enough exercise (see *Thinking It Through*).

Concept Check

1. **Why** do the types of microorganisms in your GI tract affect your risk of infection?
2. **How** does the GI tract prevent harmful substances from entering the body?
3. **How** can the symptoms of gluten-related disorders be prevented?
4. **What** diet and lifestyle changes can reduce symptoms of GERD?

3.5 Delivering Nutrients and Eliminating Wastes

LEARNING OBJECTIVES

1. **Trace** the path of blood circulation.
2. **Discuss** how blood flow is affected by eating and by activity.
3. **Explain** the functions of the lymphatic system.
4. **List** four ways in which waste products are eliminated from the body.

After food has been digested and the nutrients have been absorbed, the nutrients must be delivered to the cells. This delivery is handled by the **cardiovascular system**, which consists of the heart and blood vessels. Amino acids from protein, single sugars from carbohydrate, and the water-soluble products of fat digestion are absorbed into **capillaries** in the villi of the small intestine and transported via the blood to the liver (see Figure 3.9b). The products of digestion that are not water soluble, such as cholesterol and large fatty acids, are absorbed into **lacteals**, which are part of the **lymphatic system**, before entering the blood.

The Cardiovascular System

The cardiovascular system circulates blood throughout the body. Blood carries nutrients and oxygen to the cells of all the organs and tissues of the body and removes carbon dioxide and other waste products from these cells (**Figure 3.16**). Blood also carries other substances, such as hormones, from one part of the body to another.

PROCESS DIAGRAM

FIGURE 3.16 **Blood circulation** Blood pumped to the lungs picks up oxygen and delivers nutrients. Blood pumped to the rest of the body delivers oxygen and nutrients.

① Oxygen-poor blood that reaches the heart from the rest of the body is pumped through the arteries to the capillaries of the lungs.

② In the capillaries of the lungs, oxygen from inhaled air is picked up by the blood, and carbon dioxide is released into the lungs and exhaled.

③ Oxygen-rich blood returns to the heart from the lungs via veins.

④ Oxygen-rich blood is pumped out of the heart into the arteries that lead to the rest of the body.

⑤ In the capillaries of the body, nutrients and oxygen move from the blood to the body's tissues, and carbon dioxide and other waste products move from the tissues to the blood, to be carried away.

⑥ Oxygen-poor blood returns to the heart via veins.

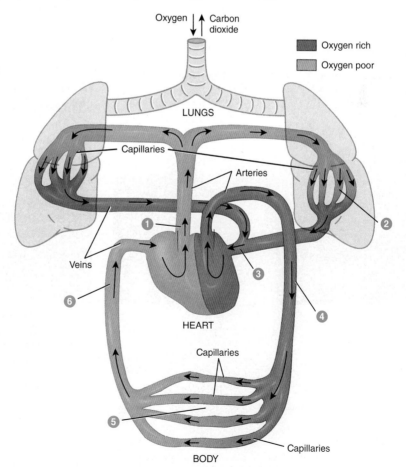

Think Critically

Why is the cardiovascular system important in nutrition?

---

capillary A small, thin-walled blood vessel through which blood and the body's cells exchange gases and nutrients.

lacteal A lymph vessel in the villi of the small intestine that picks up particles containing the products of fat digestion.

The heart and blood vessels The heart is the workhorse of the cardiovascular system. It is a muscular pump with two circulatory loops—one that carries blood to and from the lungs and one that carries blood to and from the rest of the body (see Figure 3.16).

The blood vessels that transport blood and dissolved substances toward the heart are called **veins**, and those that transport blood and dissolved substances away from the heart are called **arteries**. As arteries carry blood away from the heart, they branch many times to form smaller and smaller blood vessels. The smallest arteries are called **arterioles**. Arterioles branch to form capillaries. Blood from capillaries flows into the smallest veins, the **venules**, which converge to form larger and larger veins for returning blood to the heart.

The exchange of nutrients and gases occurs across the thin walls of the capillaries. In most body tissues, oxygen and nutrients carried by the blood pass from the capillaries into the cells, and carbon dioxide and other waste products pass from the cells into the capillaries. In the capillaries of the lungs, blood releases carbon dioxide to be exhaled and picks up oxygen to be delivered to the cells. In the capillaries of the GI tract, blood delivers oxygen and picks up water-soluble nutrients absorbed from the diet.

The amount of blood, and hence the amounts of nutrients and oxygen, delivered to a specific organ or tissue depends on the need. When you are resting, about 25% of your blood goes to your digestive system, about 20% to your skeletal muscles, and the rest to the heart, kidneys, brain, skin, and other organs.[22] This distribution changes when you eat or exercise. When you have eaten a large meal, a greater proportion of your blood goes to your digestive system to provide the oxygen and nutrients needed by the GI muscles and glands for digestion of the meal and absorption of nutrients. When you are exercising strenuously, about 70% of your blood is directed to your skeletal muscles to deliver nutrients and oxygen and remove carbon dioxide and other waste products (**Figure 3.17**).

FIGURE 3.17 **Blood flow at rest and during exercise** Blood flow changes depending on the needs of various organs and tissues.

a. At rest between meals, the amount of blood directed to the abdomen, which includes the organs, muscles, and glands of the digestive system, is similar to the amount that goes to the skeletal muscles.[22]

b. During exercise, blood flow increases to the muscles so that more oxygen and nutrients can be delivered. As a result, only a small proportion of the blood is directed to the abdomen.[22] You may get cramps if you exercise right after eating a big meal because your body cannot direct enough blood to the intestines and the muscles at the same time. The muscles win out, and food remains in your intestines, often causing cramps.

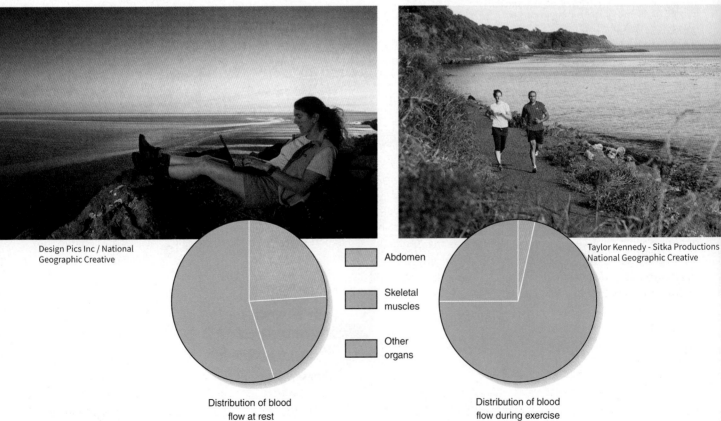

Interpret the Data

Based on these charts, which area receives the least blood flow during exercise?
a. skeletal muscle
b. abdomen
c. other organs
d. blood flow is equal to all areas.

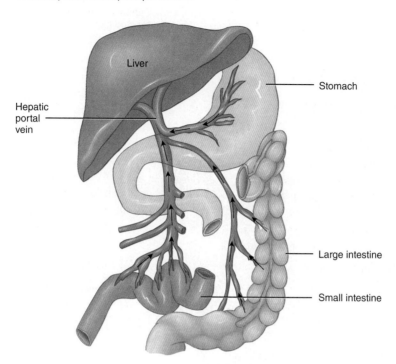

FIGURE 3.18 Hepatic portal circulation The hepatic portal circulation delivers nutrients to the liver. Water-soluble substances absorbed into the capillaries of the villi move into venules, which merge to form larger veins that eventually form the hepatic portal vein.

Delivering nutrients to the liver Water-soluble molecules in the small intestine, including amino acids, sugars, water-soluble vitamins, and the water-soluble products of fat digestion, cross the mucosal cells of the villi and enter the capillaries (see Figures 3.9 and 3.10). Once in the capillaries, these molecules are carried to the liver via the **hepatic portal vein** (**Figure 3.18**).

The liver acts as a gatekeeper between the body and substances absorbed from the intestine. Some nutrients are stored in the liver, some are changed into different forms, and others are allowed to pass through unchanged. The liver determines whether individual nutrients are stored or delivered immediately to the cells, depending on the body's needs. The liver is also important in the synthesis and breakdown of amino acids, proteins, and lipids. It modifies the products of protein breakdown to form molecules that can be safely transported to the kidney for excretion. The liver also contains enzyme systems that protect the body from toxins absorbed by the gastrointestinal tract.

The Lymphatic System

The lymphatic system consists of a network of tubules (lymph vessels) and lymph organs that contain infection-fighting cells. Fluid that collects in tissues and between cells drains into the lymphatic system. This prevents the fluid from accumulating and causing swelling.

The lymphatic system is an important part of the immune system. Fluid that drains into lymph vessels is filtered past a collection of infection-fighting cells before being returned to the blood. If the fluid contains antigen, an immune response is triggered. White blood cells and antibodies produced as a result of this response enter the blood and help destroy the foreign substance.

In the small intestine, the lymph vessels aid in the absorption and transport of fat-soluble substances such as cholesterol, fatty acids, and fat-soluble vitamins. These nutrients pass from the intestinal mucosa into the lacteals located in the villi (see Figure 3.9b). The lacteals drain into larger lymph vessels. Lymph vessels from the intestine and most other organs drain into the thoracic duct, which empties into the blood near the neck. Therefore, substances absorbed into the lymphatic system do not pass through the liver before entering the general blood circulation.

Elimination of Wastes

Material that is not absorbed from the gut into the body is eliminated from the gastrointestinal tract in the feces. Wastes that are generated in the body, such as carbon dioxide, minerals, and nitrogen-containing wastes, must also be eliminated. The same highway of blood vessels that picks up absorbed nutrients and oxygen helps remove wastes from the body (**Figure 3.19**). Carbon dioxide and some water are lost via the

FIGURE 3.19 **Organ systems involved in elimination of wastes** Substances in food that cannot be absorbed are eliminated in the feces. Wastes generated from nutrient metabolism, called metabolic wastes, are eliminated from the body by the skin and the urinary and respiratory systems.

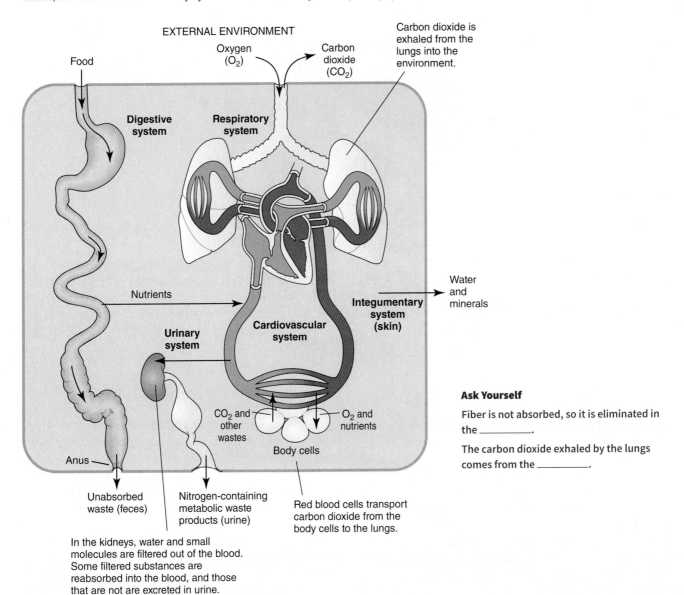

Ask Yourself

Fiber is not absorbed, so it is eliminated in the _____.

The carbon dioxide exhaled by the lungs comes from the _____.

lungs, and some water, minerals, and nitrogen-containing wastes are lost through the skin, but the kidney is the primary site for the excretion of metabolic wastes. Water, minerals, and the nitrogen-containing by-products of protein breakdown are filtered out of the blood by the kidneys and excreted in the urine.

Concept Check

1. **Where** does blood go after it leaves the lungs?
2. **Why** is it not a good idea to exercise after eating a large meal?
3. **How** does the lymphatic system aid fat absorption?
4. **What** wastes are excreted by the kidneys? By the lungs?

3.6 An Overview of Metabolism

LEARNING OBJECTIVES

1. **Discuss** two ways in which nutrients are used after they have been absorbed.
2. **Describe** what happens in cellular respiration.
3. **List** the types of molecules that can be made from glucose, from fatty acids, and from amino acids.

Once nutrients are inside the body's cells, they are used either for energy or to synthesize all the structural and regulatory molecules needed for growth and maintenance. Together, the chemical reactions that break down molecules to provide energy and those that synthesize larger molecules are referred to as **metabolism**. Many of the reactions of metabolism occur in series known as **metabolic pathways**. Molecules that enter these pathways are modified at each step, with the help of enzymes. Some of the pathways use energy to build body structures, and others break large molecules into smaller ones, releasing energy. Reactions that synthesize molecules occur in different cellular compartments from those that break down molecules for energy. For example, ribosomes are cellular structures that specialize in the synthesis of proteins, and **mitochondria** are cellular organs that are responsible for breaking down molecules to release energy. Metabolism is discussed in more detail in appropriate individual chapters and reviewed in the online minichapter *Metabolism: Energy for Life*.

Releasing Energy

In the mitochondria, glucose, fatty acids, and amino acids derived from carbohydrates, fats, and proteins, respectively, are broken down in the presence of oxygen to produce carbon dioxide and water and release energy. This process, called **cellular respiration**, is like cell breathing: Oxygen goes into the cell, and carbon dioxide comes out. The energy released by cellular respiration is used to make a molecule called **adenosine triphosphate (ATP)** (Figure 3.20). ATP can be thought of as the

PROCESS DIAGRAM

FIGURE 3.20 Producing ATP Cellular respiration uses oxygen to convert glucose, fatty acids, and amino acids into carbon dioxide, water, and energy, in the form of ATP.

How It Works

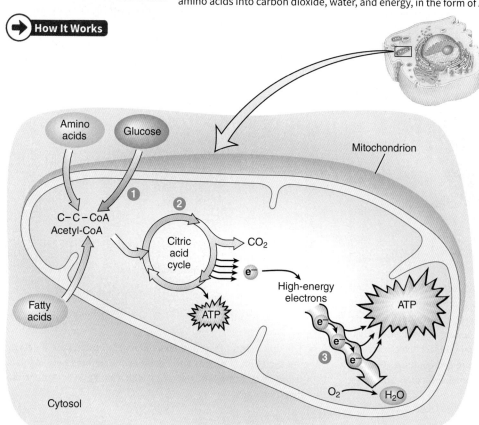

① In the presence of oxygen, glucose, fatty acids, and amino acids can be metabolized to produce a two-carbon molecule (acetyl-CoA).

② Each acetyl-CoA molecule enters a circular pathway, called the citric acid cycle, that produces two molecules of carbon dioxide (CO_2).

③ In the final step of this metabolic pathway, most of the energy released from the glucose, fatty acid, or amino acid molecules is used to produce ATP, and oxygen combines with hydrogen to form water.

adenosine triphosphate (ATP) A high-energy molecule that the body uses to power activities that require energy.

cell's energy currency. The chemical bonds of ATP are very high in energy, and when they break, the energy is released and can be used either to power body processes, such as muscle contraction or the transport of molecules across membranes, or to synthesize new molecules needed to maintain and repair body tissues.

Synthesizing New Molecules

Glucose, fatty acids, and amino acids that are not broken down for energy are used, with the input of energy from ATP, to synthesize structural, regulatory, or storage molecules. Glucose molecules are used to synthesize the glucose-storage molecule glycogen and, in some cases, fatty acids. Fatty acids are used to make body fat, cell membranes, and regulatory molecules. Amino acids are used to synthesize the various proteins that the body needs and, when necessary, to make glucose. Excess amino acids can also be converted into fatty acids and stored.

Concept Check

1. **How** are absorbed nutrients used to provide energy?
2. **Why** can cellular respiration be thought of as cell breathing?
3. **What** types of molecules can be made from amino acids?

Summary

1 The Organization of Life 52

- Our bodies and the foods we eat are all made from the same building blocks—**atoms**. Atoms are linked together by chemical bonds to form **molecules**. Molecules can form **cells**, and cells with similar structures and functions are organized into **tissues**. Tissues are organized into **organs**, such as the stomach shown here, and the **organ systems** that make up an organism. The body organ systems work together; for example, the passage of food through the digestive system and the secretion of digestive substances are regulated by the nervous and endocrine systems.

Figure 3.1 From atoms to organisms: Organ

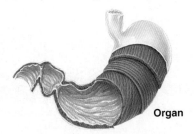

2 The Digestive System 55

- The digestive system has two major functions: **digestion** and **absorption**. Digestion breaks down food and nutrients into units that are small enough to be absorbed. Absorption transports nutrients into the body. The main component of the digestive system, illustrated here, is the **gastrointestinal tract**, which consists of a hollow tube that begins at the mouth and continues through the pharynx, esophagus, stomach, small intestine, and large intestine, ending at the anus.

Figure 3.2a Structure of the digestive system

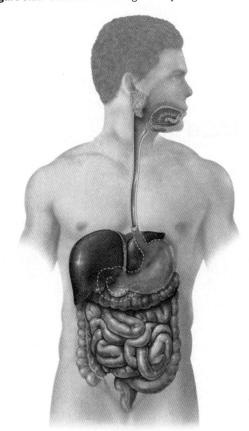

- The digestion of food and absorption of nutrients are aided by the secretion of **mucus** and **enzymes**. **Hormones** help regulate digestive processes and the rate at which food moves through the digestive tract.

Summary

3 Digestion and Absorption of Nutrients   57

- The processes involved in digestion begin in response to the smell or sight of food and continue as food enters the digestive tract at the mouth, where it is broken down into smaller pieces by the teeth and mixed with **saliva** to form a bolus. Carbohydrate digestion is begun in the mouth by **salivary amylase**. From the mouth, the bolus passes through the **pharynx** and into the esophagus. The rhythmic contractions of **peristalsis** propel it down the esophagus to the stomach.

- The stomach is a temporary storage site for food. The muscles of the stomach mix the food into a semiliquid mass called **chyme**, and **gastric juice**, which contains hydrochloric acid and **pepsin**, begins the digestion of protein. The rate at which the stomach empties varies with the amount and composition of food consumed and is regulated by hormones and signals from nerves.

- The small intestine is the primary site for nutrient digestion and absorption. The circular folds, **villi**, shown here, and **microvilli** of the small intestine, ensure a large absorptive surface area. In the small intestine, **bicarbonate** from the pancreas neutralizes stomach acid, and pancreatic and intestinal enzymes digest carbohydrate, fat, and protein. The digestion and absorption of fat in the small intestine are aided by **bile** from the gallbladder.

Figure 3.9b The structure of the small intestine

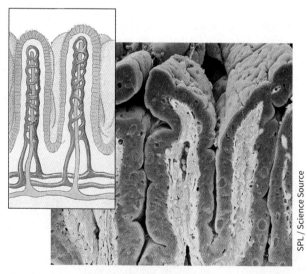

- The absorption of nutrients across the intestinal **mucosa** occurs by means of several different transport mechanisms. **Simple diffusion**, **osmosis**, and **facilitated diffusion** do not require energy, but **active transport** does.

- Components of **chyme** that are not absorbed in the small intestine pass on to the large intestine, where some water and other nutrients are absorbed. The large intestine is populated by a large number and variety of microorganisms referred to as the **intestinal microbiota**. The remaining unabsorbed materials are eliminated in the **feces**.

4 The Digestive System in Health and Disease   65

- The intestinal microbiota breaks down undigested carbohydrates, producing gas, some vitamins, and short chain fatty acids, which fuel colonic cells. A healthy microbiota includes the right balance of bacteria and supports digestive and immune health. It is believed to protect against intestinal disorders as well as chronic diseases such as obesity, heart disease, and diabetes.

- Immune system cells and tissues located in the gastrointestinal tract help prevent disease-causing organisms or chemicals from entering the body. An **antigen** entering the digestive tract is attacked first by **phagocytes**. If it is not eliminated by the phagocytes, **lymphocytes** respond specifically to the antigen by producing **antibodies**.

- The symptoms of **food allergies** and **celiac disease** are due to abnormal reactions of the immune system. Individuals with food allergies must avoid foods that cause a reaction. Eliminating gluten can prevent the symptoms of celiac disease and other gluten-related disorders.

- Diseases or discomforts that affect any part of the digestive system can interfere with food intake, digestion, or nutrient absorption. Common difficulties include dental problems, reduced saliva production, **heartburn**, GERD, **peptic ulcers** (such as the one shown here), **gallstones**, vomiting, **diarrhea**, and **constipation**.

Figure 3.15c Digestive disorders

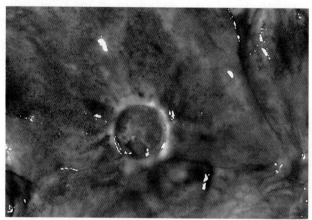

5 Delivering Nutrients and Eliminating Wastes   73

- Absorbed nutrients are delivered to the cells by the **cardiovascular system**. The heart pumps blood to the lungs to pick up oxygen and release carbon dioxide. From the lungs, blood returns to the heart and is then pumped to the rest of the body to deliver oxygen and nutrients and remove carbon dioxide and other wastes before returning to the heart. Exchange of nutrients and gases occurs at the **capillaries**.

- The products of carbohydrate and protein digestion and the water-soluble products of fat digestion enter capillaries in the intestinal villi and are transported to the liver via the **hepatic portal vein**, as illustrated here. The liver removes the absorbed substances for storage, converts them into other forms, or allows them to pass unaltered. The liver also protects the body from toxic substances that may have been absorbed.

Figure 3.18 Hepatic portal circulation

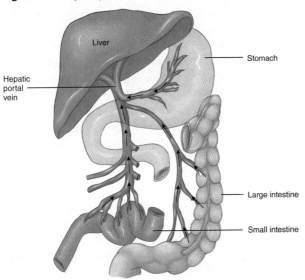

- The fat-soluble products of digestion enter **lacteals** in the intestinal villi. Lacteals join larger lymph vessels. The nutrients absorbed via the **lymphatic system** enter the blood without first passing through the liver.

- Unabsorbed materials are eliminated in the feces. The waste products of metabolism are excreted by the lungs, skin, and kidneys.

6 An Overview of Metabolism 77

- In the cells, glucose, **fatty acids**, and amino acids absorbed from the diet can be broken down by means of **cellular respiration**, as shown here, to provide energy in the form of **ATP**.

Figure 3.20 Producing ATP

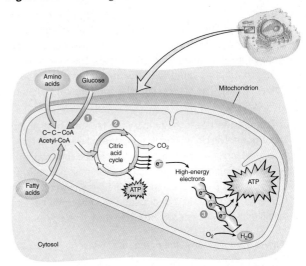

- In the presence of ATP, glucose, fatty acids, and amino acids can be used either to synthesize structural or regulatory molecules or to synthesize energy-storage molecules.

Key Terms

- absorption 55
- active transport 64
- adenosine triphosphate (ATP) 77
- allergen 67
- antibody 66
- antigen 66
- arteriole 74
- artery 74
- atom 52
- bicarbonate 61
- bile 62
- brush border 62
- capillary 73
- cardiovascular system 73
- celiac disease 67
- cell 52
- cellular respiration 77
- chyme 59
- constipation 72
- diarrhea 70
- diffusion 63
- digestion 55
- enzyme 55
- epiglottis 58
- facilitated diffusion 63

- fatty acid 55
- feces 55
- food allergy 67
- gallstone 70
- gastric juice 59
- gastrin 61
- gastroesophageal reflux disease (GERD) 69
- gastrointestinal tract 55
- gluten-related disorders 67
- heartburn 69
- hepatic portal vein 75
- hormone 52
- intestinal microbiota 64
- lacteal 73
- lipase 62
- lumen 55
- lymphatic system 73
- lymphocyte 66
- metabolic pathway 77
- metabolism 77
- microvillus 62
- mitochondrion 77
- molecule 52
- mucosa 55
- mucosal cells 55

- mucus 55
- organ 52
- organ system 52
- osmosis 63
- pancreatic amylase 62
- pancreatic juice 61
- pepsin 59
- peptic ulcer 70
- peristalsis 59
- phagocyte 66
- pharynx 58
- prebiotic 66
- probiotic 66
- protease 62
- saliva 58
- salivary amylase 58
- segmentation 60
- simple diffusion 63
- sphincter 59
- tissue 52
- transit time 55
- vein 74
- venule 74
- villus 62

What is happening in this picture?

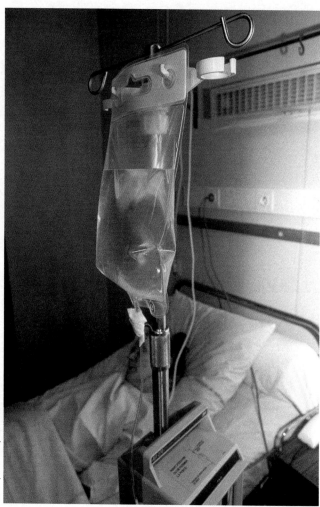

This patient has Crohn's disease, an inflammatory disease of the intestine that is interfering with his ability to absorb nutrients. Doctors are ensuring that he is being nourished by infusing a nutrient solution into his blood through a process called total parenteral nutrition (TPN).

Think Critically

1. Would the TPN solution delivered meet his needs if it contained undigested starches and proteins? Why or why not?
2. What would happen to the patient if an essential nutrient were missing from the TPN solution?
3. How would waste products from the TPN be eliminated?

CHAPTER 4

Carbohydrates: Sugars, Starches, and Fibers

The grains that humans consumed 10,000 years ago were soaked and sprouted and then toasted or boiled into gruels and porridges. Dark in color and course in texture, these crude mixtures contained all the components of the grains from which they were made. With the introduction of grinding, grains could be made into different types of products. The smoother, purer products were reserved for higher social classes. From the Egyptian dynasties into the 19th century, coarse-grained bread was distributed to slaves and the lower classes, while finer-grained varieties were reserved for athletes and the upper classes. The less refined bread was actually more nutritious, containing more germ and bran, but the "upper crust" disdained all but the finest—and unbeknownst to them—less nutritious—flours.

The great steel roller mills of the Industrial Revolution increased our ability to refine grains on a mass scale, making refined flour a commonplace staple. Industrial grinding also eliminated more and more nutrients from the grain itself by removing even more of the coarser parts of the plant. Eventually, the almost exclusive use of refined grains prompted government mandates that nutrients be added back to prevent nutrient deficiencies. Today recommendations suggest that we move back toward a pattern based on less refined grains.

CHAPTER OUTLINE

Carbohydrates in Our Food 83
- What Is a Whole Grain?
- What Is Added Sugar?

Types of Carbohydrates 85
- Simple Carbohydrates
- Complex Carbohydrates

Carbohydrate Digestion and Absorption 88
- Carbohydrate Digestion
- Carbohydrate Absorption

What a Scientist Sees: Glycemic Response

Carbohydrate Functions 92
- Getting Enough Glucose to Cells
- Glucose as a Source of Energy

Carbohydrates in Health and Disease 96
- Diabetes

Debate: Is Sugar Making Us Sick?
- Hypoglycemia
- Dental Caries
- Weight Management
- Bowel Health
- Heart Disease

Meeting Carbohydrate Needs 102
- Carbohydrate Recommendations
- Choosing Carbohydrates Wisely

What Should I Eat? Carbohydrates

Thinking It Through: A Case Study on Healthy Carbohydrate Choices

4.1 Carbohydrates in Our Food

LEARNING OBJECTIVES

1. **Identify** sources of carbohydrate in the diet.
2. **Distinguish** refined carbohydrates from unrefined carbohydrates.
3. **Compare** the nutrients in whole grains to those in enriched grains.
4. **Describe** how foods high in added sugars differ from foods that are naturally high in sugars.

High-carbohydrate foods, including rice, cereal, bread, pasta, and starchy vegetables such as potatoes and beans, are the basis of our diets (**Figure 4.1a**). These inexpensive staples make up the bulk of most meals—oatmeal for breakfast, a sandwich for lunch, and potatoes or rice and beans at dinner. The diet of early humans consisted almost entirely of **unrefined foods**—foods eaten either just as they are found in nature or with only minimal processing, such as cooking. Today we still consume unrefined sources of carbohydrate, such as quinoa, lentils, bananas, and brown rice, but many of the foods we consume are made with **refined** grains like white flour and white rice and contain **added sugars** (**Figure 4.1b**).

The increased consumption of refined carbohydrates that has occurred around the world over the past few decades has been implicated as one of the causes of the current obesity epidemic and the rising incidence of chronic diseases. Recommendations for a healthy diet suggest that we select more unrefined sources of carbohydrates, including whole grains, vegetables, and fruits, and that we limit foods high in refined carbohydrates, such as candies, cookies, and sweetened beverages.

What Is a Whole Grain?

When you eat a bowl of oatmeal or a slice of whole-wheat toast, you are consuming a **whole-grain product**. Whole-grain

FIGURE 4.1 **Sources of carbohydrate** Carbohydrate comes from whole food and refined sources.

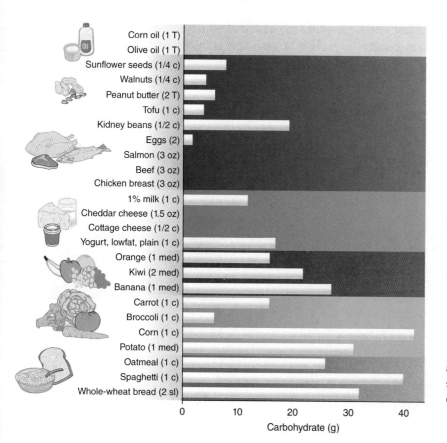

a. Grains, vegetables, legumes, fruits, nuts, seeds, and milk are all whole sources of carbohydrate.

(continues)

---

refined Refers to foods that have undergone processing that changes or removes various components of the original food.

added sugars Refined sugars and syrups that have been added to foods during processing or preparation.

FIGURE 4.1 (continued)

b. Corn is an unrefined source of carbohydrate. The cornflakes in your cereal bowl are a refined carbohydrate produced from corn by grinding, cooking, extruding, and drying. The sugar you add to your cornflakes is also a refined carbohydrate; it has been refined from sugar cane or sugar beets.

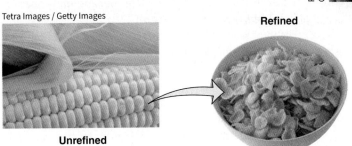

Ask Yourself

Which is less refined: canned peaches or fresh peaches? Whole-wheat bread or white bread? Granola cereal or oatmeal?

products include the entire kernel of the grain: the **germ**, the **bran**, and the **endosperm** (**Figure 4.2a**). Refined grain products, such as white bread, include just the endosperm. The bran and germ are discarded during refining, and along with them the fiber and some vitamins and minerals are lost. To make up some of these losses, refined grains sold in the United States are required to be enriched. **Enrichment**, adds back some, but not all, of the nutrients lost in processing (**Figure 4.2b**). For example, the thiamin, niacin, riboflavin, and iron that are lost when grains are milled are later added back to levels that are equal to or higher than originally present. Since 1998, folic acid has also been added to enriched grains. Other nutrients, including vitamin E and vitamin B_6, are also removed by milling, but they are not added back. Therefore, foods made with enriched refined grains contain more of some nutrients and less of others than foods made from whole grains.

What Is Added Sugar?

Refined sugars added to food during processing or at the table account for over 13% of the calories consumed in the typical American diet.[1] Refined sugars are nutritionally and chemically

FIGURE 4.2 **Whole grains** Whole-grain products provide greater amounts of many nutrients than refined grains but lesser amounts of a few nutrients that are added in the enrichment process.

a. A kernel of grain is made up of the endosperm, the bran, and the germ. Together they provide all the nutrients present in whole grain.

The **endosperm** is the largest part of the kernel. It is made up of primarily starch, but it also contains most of the kernel's protein, along with some vitamins and minerals.

The outermost **bran** layers contain most of the fiber and are a good source of many vitamins and minerals.

The **germ**, located at the base of the kernel, is the embryo where sprouting occurs. It is a source of oil and is rich in vitamin E.

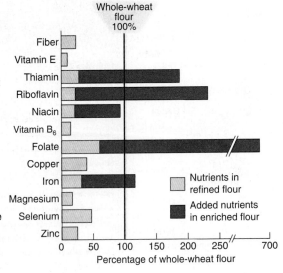

b. The amounts of many nutrients in refined flour (yellow bars) are much lower than the amounts originally present in the whole grain (100% line). In enriched flour, thiamin, riboflavin, niacin, iron, and folate have been added back in amounts that are equal to or exceed the original levels (red bars).

---

enrichment The addition to food of specific amounts of nutrients to replace those lost during processing.

FIGURE 4.3 **Added versus naturally occurring sugars** Choosing three kiwis rather than four pieces of red licorice is a more nutrient-dense choice. The kiwis are an unrefined source of sugar that also provides fiber and vitamin C, folate, and calcium. Most of the calories in the licorice are from added sugars; licorice is lower in nutrient density because it provides almost no other nutrients other than sugar.

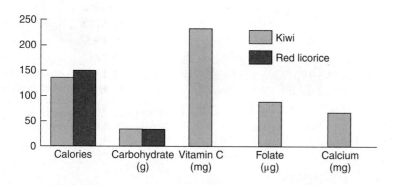

identical to sugars that occur naturally in foods. When separated from their plant sources, however, refined sugars no longer come with the fiber, vitamins, minerals, and other substances found in the original plant. Therefore, added refined sugars contribute empty calories to the diet. Foods that naturally contain sugars, such as fruits and milk, provide vitamins, minerals, and phytochemicals, along with the calories from the sugar, making them higher in nutrient density (**Figure 4.3**). To help consumers identify packaged foods that are high in added sugars, the planned Nutrition Facts label includes the amount of added sugars per serving of a product both in grams and as a percentage of the Daily Value.

Concept Check

1. **Which** foods in your diet contribute the most carbohydrate?
2. **What** is the difference between brown rice and white rice?
3. **Why** is there more vitamin B_6 and less thiamin in a slice of whole-wheat bread than in a slice of white bread?
4. **Why** are foods high in added sugars said to contribute empty calories?

4.2 Types of Carbohydrates

LEARNING OBJECTIVES

1. **Name** the basic unit of carbohydrate.
2. **Classify** carbohydrates as simple or complex.
3. **Describe** the types of complex carbohydrates.
4. **Distinguish** soluble fiber from insoluble fiber.

Chemically, carbohydrates are a group of compounds made up of one or more **sugar units** that contain carbon (*carbo*) as well as hydrogen and oxygen in the same two-to-one proportion found in water (*hydrate*, H_2O). Carbohydrates made up of only one sugar unit are called **monosaccharides**, those made up of two sugar units are called **disaccharides**, and those made up of more than two sugar units are called **polysaccharides**.

Simple Carbohydrates

Monosaccharides and disaccharides are classified as **simple carbohydrates**. The three most common monosaccharides in the diet are **glucose**, **fructose**, and **galactose**. Each contains 6 carbon, 12 hydrogen, and 6 oxygen atoms ($C_6H_{12}O_6$), but these three sugars differ in the arrangement of these atoms (**Figure 4.4a**). Glucose, the sugar referred to as *blood sugar*, is the most important carbohydrate fuel for the human body.

The most common disaccharides in our diet are **maltose**, **sucrose**, and **lactose** (**Figure 4.4b**).

---

sugar unit A sugar molecule that cannot be broken down to yield other sugars.

monosaccharide A carbohydrate made up of a single sugar unit.

disaccharide A carbohydrate made up of two sugar units.

polysaccharide A carbohydrate made up of many sugar units linked together.

glucose A six-carbon monosaccharide that is the primary form of carbohydrate used to provide energy in the body.

NUTRITION INSIGHT | **FIGURE 4.4** **Carbohydrate structures and sources** Simple carbohydrates include monosaccharides and disaccharides. Complex carbohydrates include glycogen, starches, and fiber.

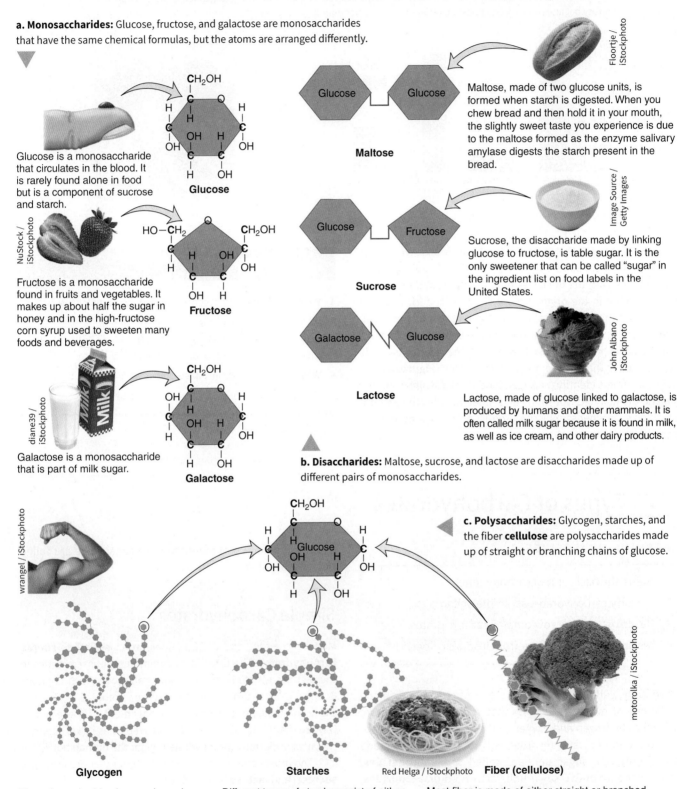

a. Monosaccharides: Glucose, fructose, and galactose are monosaccharides that have the same chemical formulas, but the atoms are arranged differently.

Glucose is a monosaccharide that circulates in the blood. It is rarely found alone in food but is a component of sucrose and starch.

Fructose is a monosaccharide found in fruits and vegetables. It makes up about half the sugar in honey and in the high-fructose corn syrup used to sweeten many foods and beverages.

Galactose is a monosaccharide that is part of milk sugar.

Maltose, made of two glucose units, is formed when starch is digested. When you chew bread and then hold it in your mouth, the slightly sweet taste you experience is due to the maltose formed as the enzyme salivary amylase digests the starch present in the bread.

Sucrose, the disaccharide made by linking glucose to fructose, is table sugar. It is the only sweetener that can be called "sugar" in the ingredient list on food labels in the United States.

Lactose, made of glucose linked to galactose, is produced by humans and other mammals. It is often called milk sugar because it is found in milk, as well as ice cream, and other dairy products.

b. Disaccharides: Maltose, sucrose, and lactose are disaccharides made up of different pairs of monosaccharides.

c. Polysaccharides: Glycogen, starches, and the fiber **cellulose** are polysaccharides made up of straight or branching chains of glucose.

Glycogen
The polysaccharide glycogen is made of highly branched chains of glucose. This branched structure allows glycogen, which is found in muscle and liver, to be broken down quickly when the body needs glucose.

Starches
Different types of starch consist of either straight chains or branched chains of glucose. We consume a mixture of starches in grain products, legumes, and other starchy vegetables.

Fiber (cellulose)
Most fiber is made of either straight or branched chains of monosaccharides, but the bonds that link the sugar units cannot be broken by human digestive enzymes. For example, cellulose, shown here, is a fiber made up of straight chains of glucose molecules. Sources include wheat bran and broccoli.

Complex Carbohydrates

Complex carbohydrates are polysaccharides; they are generally not sweet tasting the way simple carbohydrates are. They include **glycogen** in animals and starches and fibers in plants (**Figure 4.4c**). Glycogen, sometimes called *animal starch*, is the storage form of glucose in humans and other animals. It is found in the liver and muscles, but we don't consume it in our diet because the glycogen in animal muscles is broken down soon after the animal is slaughtered.

Starch is made up of glucose molecules linked together in either straight or branched chains (see Figure 4.4c). It is the storage form of carbohydrate in plants and provides energy for plant growth and reproduction. When we eat plants, we consume the energy stored in the starch (**Figure 4.5**).

Fiber includes complex carbohydrates that cannot be broken down by human digestive enzymes. Thus fiber cannot be absorbed in the human small intestine, and it passes into the large intestine. A number of chemical substances are classified as fiber. **Soluble fiber** dissolves in water to form viscous solutions. Soluble fiber is found around and inside plant cells and includes pectins, gums, some hemicelluloses, indigestible **oligosaccharides**, and **resistant starch**. Although human enzymes can't digest soluble fiber, the intestinal microbiota can break it down, creating gas and other by-products. Beans, oats, apples, and seaweed are food sources of soluble fiber (**Figure 4.6**).

Insoluble fiber does not dissolve in water. It comes primarily from the structural parts of plants, such as cell walls,

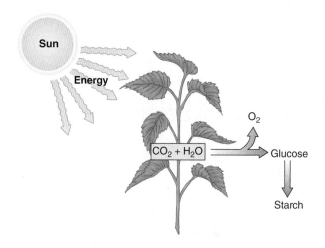

FIGURE 4.5 **Photosynthesis** Glucose is produced in plants through the process of **photosynthesis**, which uses energy from the sun to convert carbon dioxide and water to glucose. Plants most often convert glucose to starch. When a human eats plants, digestion converts the starch back to glucose.

FIGURE 4.6 **Sources of fiber** Fiber is found naturally in foods and added to food during processing

a. Whole grains, legumes, fruits, and vegetables are good sources of soluble and/or insoluble fiber.

Good sources of soluble fiber:
Legumes
Prunes
Apricots
Raisins
Bananas
Oats
Apples
Flaxseed

Good sources of insoluble fiber:
Wheat bran
Whole-wheat bread
Broccoli
Corn
Eggplant
Apple skins
Nuts and seeds

b. Fiber is added to change the physical properties or nutritional profile of foods. The soluble fiber pectin, found in fruits and vegetables, is added to jams and jellies as a thickener. Gums such as locust bean gum, agar, and carrageenan are also used as thickeners because they combine with water to keep solutions from separating. Wheat bran and oat bran are added to breads and muffins to boost their fiber content.

---

glycogen The storage form of carbohydrate in animals, made up of many glucose molecules linked together in a highly branched structure.

starch A carbohydrate found in plants, made up of many glucose molecules linked in straight or branched chains.

soluble fiber Fiber that dissolves in water or absorbs water and is readily broken down by the intestinal microbiota.

oligosaccharide A carbohydrate made up of 3 to 10 sugar units.

resistant starch Starch that escapes digestion in the small intestine of healthy people.

insoluble fiber Fiber that does not dissolve in water and is less readily broken down by bacteria in the large intestine.

and includes cellulose and some hemicelluloses as well as lignin, which is a non-carbohydrate component of the cell wall. This type of fiber adds bulk to fecal matter because it passes, relatively unchanged, through the gastrointestinal tract. Food sources of insoluble fiber include wheat and rye bran, broccoli, and celery (see Figure 4.6).

Concept Check

1. **What** molecules make up starch?
2. **Why** is sucrose classified as a simple carbohydrate?
3. **What** is glycogen?
4. **Which** type of fiber is plentiful in beans?

4.3 Carbohydrate Digestion and Absorption

LEARNING OBJECTIVES

1. **Describe** the steps of carbohydrate digestion.
2. **Explain** what is meant by lactose intolerance.
3. **Discuss** how fiber affects the colon and feces.
4. **Draw** a graph that compares blood glucose levels after drinking soda and after eating beans.

Disaccharides and complex carbohydrates must be digested to monosaccharides before they can be absorbed into the body. When the disaccharide lactose is not completely digested, it causes digestive problems. In contrast, fiber, which is not completely digested or absorbed, has a positive impact on the gastrointestinal tract and overall health. Carbohydrates that are absorbed travel in the blood to the liver.

Carbohydrate Digestion

Carbohydrate digestion begins in the mouth, but most starch digestion and the breakdown of disaccharides occur in the small intestine (**Figure 4.7**). Fiber passes into the colon, where some is broken down by bacteria. Material that cannot be absorbed is excreted in the feces.

PROCESS DIAGRAM

FIGURE 4.7 Carbohydrate digestion During digestion, enzymes break starches and sugars into monosaccharides, which are absorbed. Most of the fiber is excreted in the feces.

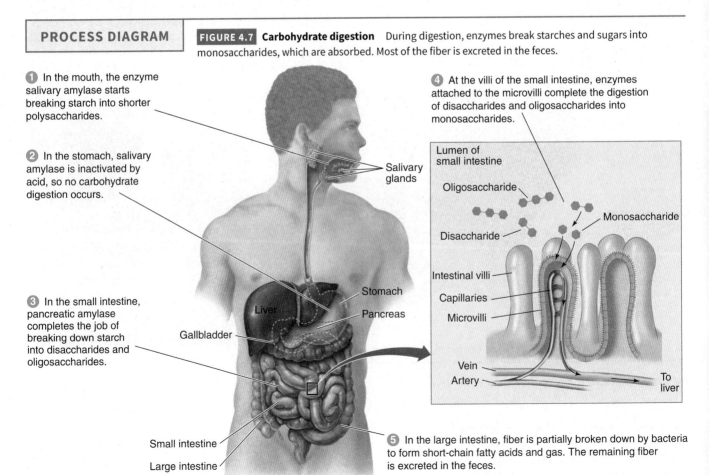

1. In the mouth, the enzyme salivary amylase starts breaking starch into shorter polysaccharides.

2. In the stomach, salivary amylase is inactivated by acid, so no carbohydrate digestion occurs.

3. In the small intestine, pancreatic amylase completes the job of breaking down starch into disaccharides and oligosaccharides.

4. At the villi of the small intestine, enzymes attached to the microvilli complete the digestion of disaccharides and oligosaccharides into monosaccharides.

5. In the large intestine, fiber is partially broken down by bacteria to form short-chain fatty acids and gas. The remaining fiber is excreted in the feces.

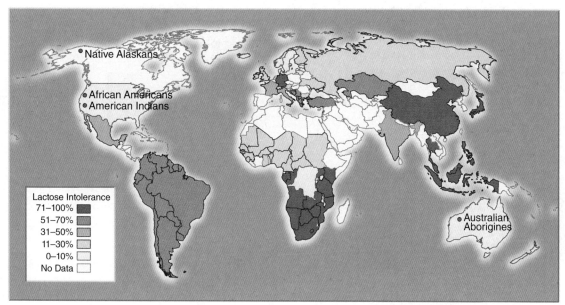

FIGURE 4.8 **Lactose intolerance** This map illustrates the dramatic variation in the incidence of lactose intolerance around the world. In the United States, between 30 and 50 million people are lactose intolerant; it is more common in some ethnic and racial populations than in others. Up to 80% of African Americans, 80 to 100% of Native Americans, and 90 to 100% of Asian Americans are lactose intolerant, but only about 15% of Caucasian Americans are.[2]

Interpret the Data

Based on this graph, Americans whose ancestors came from which parts of the world are most likely to have a low calcium intake because they can't drink milk?

Lactose intolerance The disaccharide lactose is broken down by the enzyme lactase in the small intestine. **Lactose intolerance** occurs when there is insufficient lactase to digest the lactose. When lactose intolerant individuals consume milk or other dairy products, the lactose passes into the large intestine, where it draws in water and is metabolized by bacteria, producing gas and causing abdominal distension, cramping, and diarrhea. On rare occasions infants are lactose intolerant at birth but most people are born with adequate levels of lactase. In many people lactase levels decline so much with age that they become lactose intolerant. The incidence of lactose intolerance varies among populations (**Figure 4.8**).

Because milk is the primary source of calcium in the U.S. diet, lactose-intolerant individuals may have difficulty meeting calcium needs. Many people who are lactose intolerant can handle small amounts of lactose and therefore can meet their calcium needs by consuming small portions of milk throughout the day and eating cheese and yogurt, which contain less lactose than milk. Those who cannot tolerate any lactose can get their calcium from nondairy sources, such as tofu, legumes, dark-green vegetables, and canned salmon and sardines, which are consumed with the bones, as well as from calcium-fortified foods, calcium supplements, and lactase-treated milk (such as Lactaid). Another option is to take lactase tablets with or before consuming milk products to digest the lactose before it passes into the large intestine.

Fiber Fiber is not digested and is therefore not readily absorbed. Some fibers such as cellulose, found in celery and fruit peels, and oligosaccharides, found in legumes, are not digested because human enzymes cannot break the bonds that hold together their subunits. Resistant starch is found in legumes, whole grains, nuts, unripe bananas, and cold cooked potatoes, rice, and pasta. It is not digested for a number of reasons. For example, in whole grains and legumes, it is not digested because the natural structure of the plant protects the starch molecules; in cold cooked potatoes and pasta the digestibility of the starch is reduced by cooling.

As fiber passes through the gastrointestinal tract, it slows the rate at which nutrients, such as glucose, are absorbed (**Figure 4.9a**). Fiber can also bind to certain minerals, preventing their absorption. For instance, wheat bran fiber binds zinc, calcium, magnesium, and iron. Fiber also speeds transit through the intestine by increasing the amount of water and the volume of material in the intestine (**Figure 4.9b**). This stimulates peristalsis, causing the muscles of the large intestine to work more and function better, helping to prevent constipation.

Some fibers are digested by intestinal bacteria when they reach the large intestine, producing short-chain fatty acids and gas (**Figure 4.9c**). The fatty acids can be used as a fuel source for cells in the colon and other body tissues; they help ensure a healthy microbiota and may play a role in regulating cellular processes and preventing disease (see Chapter 3).[3-5]

Carbohydrate Absorption

After a meal, the monosaccharides from carbohydrate digestion enter the portal circulation and travel to the liver. Galactose and fructose can be converted to glucose. Glucose can be used to provide energy, stored as liver glycogen, or delivered via the general blood circulation to other body tissues, causing

lactose intolerance The inability to completely digest lactose due to a reduction in the levels of the enzyme lactase.

NUTRITION INSIGHT | **FIGURE 4.9** **The effects of fiber** Fiber promotes health by slowing digestion and absorption, reducing transit time, increasing stool weight, and promoting the growth of a healthy intestinal microbiota.

a. As shown on the left, the bulk and volume of a high-fiber meal dilute the gastrointestinal contents. This dilution slows the digestion of food and absorption of nutrients (shown as green dots moving slowly out of the intestine), causing a delay and a blunting of the rise in blood glucose that occurs after a meal (see graph). With a low-fiber meal, as shown on the right, nutrients are more concentrated; digestion and absorption occur more rapidly (shown as green dots moving quickly out of the intestine), causing a quicker, sharper rise in blood glucose (see graph).

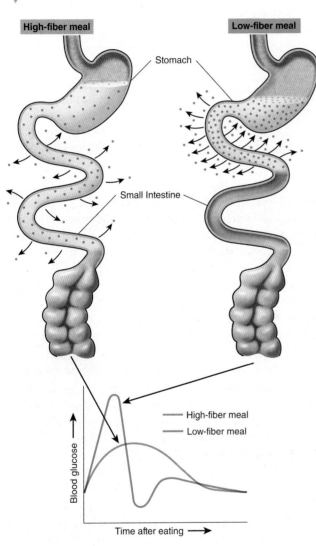

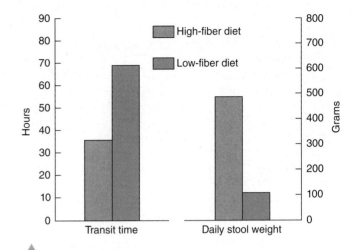

b. Stool weights are greater and transit times shorter for people consuming a diet high in fiber than for those who consume a more refined, low-fiber diet.[6]

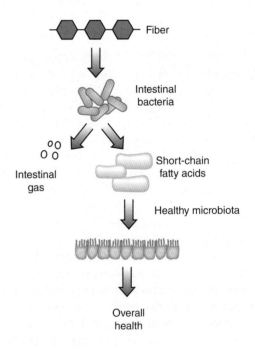

c. When bacteria break down fiber, gas and short-chain fatty acids are formed. The acidic conditions that result inhibit the growth of undesirable bacteria and favor the growth of healthy ones.

Ask Yourself

Why does blood glucose rise more slowly after a high-fiber meal than after a low-fiber meal?

What a Scientist Sees

Glycemic Response

Potatoes and beans are both sources of unrefined carbohydrate, but scientists know that the effect potatoes have on blood glucose is very different from the effect beans have. Beans are much higher in fiber and protein, both of which slow digestion and absorption and therefore reduce the glycemic response.

The glycemic response of potatoes versus kidney beans is shown here graphically, but it can also be expressed using the **glycemic index**, which is a ranking of how a food affects blood glucose relative to the effect of an equivalent amount of carbohydrate from a reference food, such as white bread or pure glucose. For example, on a glycemic index scale on which white bread is 100, potatoes are 90 and kidney beans are about 25. This means that blood glucose levels do not increase as much after eating kidney beans as they do after eating white bread or potatoes.

A shortcoming of the glycemic index is that it is measured using a set amount of carbohydrate in a food (usually 50 grams), not the typical serving of food that we eat. For example, it takes over 4 cups of strawberries to supply 50 g of carbohydrate, but people typically eat only about 1 cup. **Glycemic load** compares the effect of typical portions of food on blood glucose, so it is a more practical way to assess the effect of a food on blood glucose levels.

A shortcoming of both the glycemic index and glycemic load is that they are determined for individual foods rather than for meals, which contain mixtures of foods. We typically eat meals, so knowing the glycemic index or glycemic load of a single food doesn't tell us much about the rise in blood glucose that will occur after eating a meal.

Think Critically

How would the graph of blood glucose levels after eating a meal of meat and potatoes differ from the graph shown here for just the potatoes?

blood glucose levels to rise. **Glycemic response** is a measure of the impact a food has on blood glucose levels. How quickly and how high blood glucose levels rise are affected by how long it takes a food to leave the stomach and by how fast the food is digested and the glucose absorbed.

Foods high in refined sugars and starches generally cause a greater glycemic response than unrefined carbohydrates because sugars and starches consumed alone leave the stomach quickly and are rapidly digested and absorbed. For example, when you drink a bottle of sugary soda, your blood glucose increases within minutes. Because fiber takes longer to leave the stomach and slows absorption in the small intestine, a fiber-containing food such as oatmeal would take longer to leave your stomach and would therefore cause a lower glycemic response (see *What a Scientist Sees: Glycemic Response*). When carbohydrate, fat, and protein are consumed together, stomach emptying is slowed, delaying both digestion and absorption of carbohydrate, so blood glucose rises more slowly than when carbohydrate is consumed alone. For instance, after a meal of chicken, brown rice, and green beans, which contains carbohydrate, fat, protein, and fiber, blood glucose doesn't begin to rise for 30 to 60 minutes.

Concept Check

1. **Where** does most starch digestion occur?
2. **Why** does lactose in the colon cause gas and diarrhea?
3. **How** does fiber affect the type of bacteria in the colon?
4. **How** does fiber affect the rate at which blood glucose rises after a meal?

---

glycemic response The rate, magnitude, and duration of the rise in blood glucose that occurs after food is consumed.

4.4 Carbohydrate Functions

LEARNING OBJECTIVES

1. **Describe** the functions of carbohydrate in the body.
2. **Contrast** the roles of insulin and glucagon in blood glucose regulation.
3. **Compare** anaerobic and aerobic metabolism.
4. **Discuss** what happens to protein and fat metabolism when dietary carbohydrate is insufficient.

The main function of carbohydrates is to provide energy, but carbohydrates also play other roles in the body. For example, nerve tissue needs the sugar galactose, and in breast-feeding women, galactose combines with glucose to produce the milk sugar lactose. The monosaccharides ribose and deoxyribose play nonenergy roles as components of RNA and DNA, respectively, the two molecules that contain a cell's genetic information. Ribose is also a component of the B vitamin riboflavin. Oligosaccharides are associated with cell membranes, where they help signal information about cells, and large polysaccharides found in connective tissue provide cushioning and lubrication.

Getting Enough Glucose to Cells

Glucose is an important fuel for body cells. Many body cells can use energy sources other than glucose, but brain cells, red blood cells, and a few others must have glucose to stay alive. In order to provide a steady supply of glucose, the concentration of glucose in the blood is regulated by the liver and by hormones secreted by the pancreas. The rise in blood glucose levels after eating stimulates the pancreas to secrete the hormone **insulin**, which allows glucose to enter muscle and fat cells, thereby lowering the level of glucose in the blood. In muscle insulin stimulates the synthesis of glycogen from glucose. In fat-storing cells, it promotes fat synthesis. In the liver, insulin promotes the storage of glucose as glycogen and, to a lesser extent, fat. Insulin also stimulates protein synthesis. The overall effect of insulin is to remove glucose from the blood and promote energy storage (**Figure 4.10**).

A few hours after eating, blood glucose levels—and consequently the amount of glucose available to the cells—have decreased enough to trigger the pancreas to secrete the hormone **glucagon** (see Figure 4.10). Glucagon raises blood glucose by signaling liver cells to break down glycogen into glucose, which is released into the blood. At the same time, glucagon signals the liver to synthesize new glucose molecules, which are also released into the blood, bringing blood glucose levels back to normal.

Glucose as a Source of Energy

Cells use glucose to provide energy via cellular respiration (see Chapter 3). Cellular respiration uses oxygen to convert glucose to carbon dioxide and water and provide energy in the form of ATP (**Figure 4.11**).

The first step in cellular respiration is **glycolysis** (*glyco* = glucose, *lysis* = to break down). Glycolysis can rapidly produce two molecules of ATP from each glucose molecule. Because oxygen is not needed for this step, glycolysis is sometimes called anaerobic glycolysis, or **anaerobic metabolism**. When oxygen is available, the complete breakdown of glucose can proceed. This **aerobic metabolism** produces about 36 molecules of ATP for each glucose molecule, 18 times more ATP than is generated by anaerobic metabolism.

Limited carbohydrate increases protein breakdown Glucose is an essential fuel for brain cells and red blood cells. If adequate amounts of glucose are not available, it can be synthesized from three-carbon pyruvate molecules (see Figure 4.11, step 1). Fatty acids cannot be used to synthesize glucose because the reactions that break them

---

insulin A hormone made in the pancreas that allows glucose to enter cells and stimulates the synthesis of protein, fat, and liver and muscle glycogen.

glucagon A hormone made in the pancreas that raises blood glucose levels by stimulating the breakdown of liver glycogen and the synthesis of glucose.

glycolysis An anaerobic metabolic pathway that splits glucose into two three-carbon pyruvate molecules; the energy released from one glucose molecule is used to make two molecules of ATP.

anaerobic metabolism Metabolism in the absence of oxygen.

aerobic metabolism Metabolism in the presence of oxygen. It can completely break down glucose to yield carbon dioxide, water, and energy in the form of ATP.

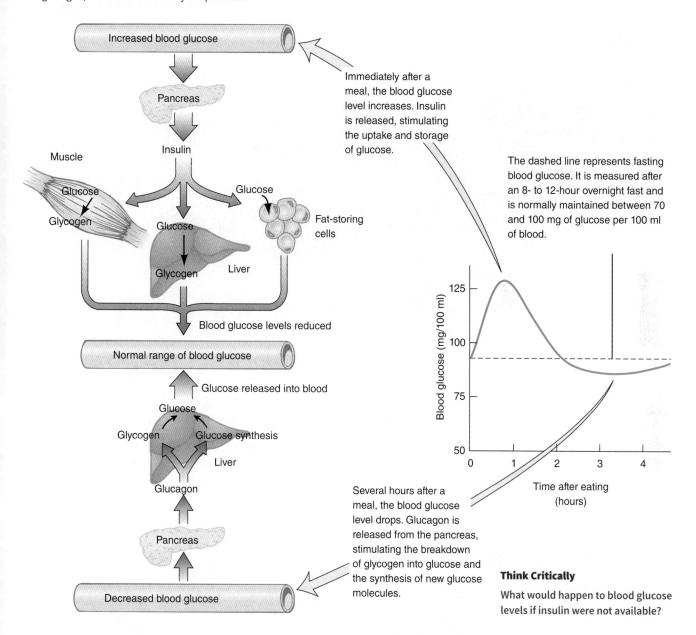

FIGURE 4.10 Blood glucose regulation Blood glucose levels are regulated by the hormones insulin and glucagon, which are secreted by the pancreas.

Immediately after a meal, the blood glucose level increases. Insulin is released, stimulating the uptake and storage of glucose.

The dashed line represents fasting blood glucose. It is measured after an 8- to 12-hour overnight fast and is normally maintained between 70 and 100 mg of glucose per 100 ml of blood.

Several hours after a meal, the blood glucose level drops. Glucagon is released from the pancreas, stimulating the breakdown of glycogen into glucose and the synthesis of new glucose molecules.

Think Critically
What would happen to blood glucose levels if insulin were not available?

down produce two-carbon, rather than three-carbon, molecules. Some of the amino acids from protein breakdown can supply the three-carbon molecules needed for glucose synthesis (**Figure 4.12**). However, this use of amino acids takes them away from body proteins. Body proteins that are broken down to make glucose are no longer available to do their job, whether that job is to speed up a chemical reaction or contract a muscle. Sufficient dietary carbohydrate ensures that protein is not utilized in this way; carbohydrate is therefore said to *spare* protein.

Limited carbohydrate interferes with fat breakdown
Most of the energy stored in the body is stored as fat. Fatty acids are broken down into two-carbon units that form acetyl-CoA. To proceed through aerobic metabolism, acetyl-CoA must combine with a molecule derived primarily from carbohydrate. When carbohydrate is in short supply, such as during starvation or when the diet is very low in carbohydrate, acetyl-CoA molecules cannot proceed through aerobic metabolism and instead react with each other to form

PROCESS DIAGRAM **FIGURE 4.11 Cellular respiration** Inside body cells, the reactions of cellular respiration split the bonds between carbon atoms in glucose, releasing energy that is used to synthesize ATP. ATP is used to power the energy-requiring processes in the body.

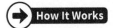

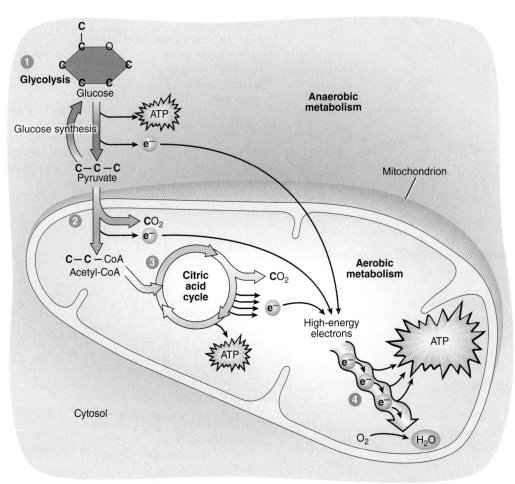

1. Glycolysis, which takes place in the cytosol, splits glucose, a six-carbon molecule, into two three-carbon molecules (pyruvate). This step releases high-energy electrons (purple balls) and produces a small amount of ATP. Pyruvate is then either broken down to produce more ATP or is used to remake glucose.

2. Pyruvate can be used to produce more ATP when oxygen is available. In the mitochondria, pyruvate is broken down, releasing carbon dioxide (CO_2) and high-energy electrons and forming acetyl-CoA (2 carbons), which continues through aerobic metabolism.

3. Acetyl-CoA enters the citric acid cycle, where CO_2 and high-energy electrons are released and where a small amount of ATP is produced.

4. Most ATP is produced in the final step of aerobic metabolism. Here the energy in the high-energy electrons released in previous steps is transferred to ATP, and the electrons are combined with oxygen and hydrogen to form water.

molecules called **ketones**, or **ketone bodies** (see Figure 4.12). The heart, muscles, and kidneys can use ketones for energy. After about three days of fasting, even the brain adapts and can obtain about half of its energy from ketones. The use of ketones for energy helps spare glucose and decreases the amount of protein that must be broken down to synthesize glucose.

Ketones not used for energy can be excreted in the urine. However, when ketone production is high, they build up in the blood, a condition known as **ketosis**. Mild ketosis can occur

ketone or **ketone body** An acidic molecule formed when there is not sufficient carbohydrate to break down acetyl-CoA.

ketosis High levels of ketones in the blood.

FIGURE 4.12 What happens when carbohydrate is limited? The availability of carbohydrate affects the metabolism of both protein and fat. When carbohydrate is limited, protein is broken down to supply amino acids that can be used to make glucose. Because the complete breakdown of fat requires some carbohydrate, when carbohydrate is limited, ketones are formed. Ketones can be used as a source of energy, but high levels can accumulate in the blood and are excreted in the urine.

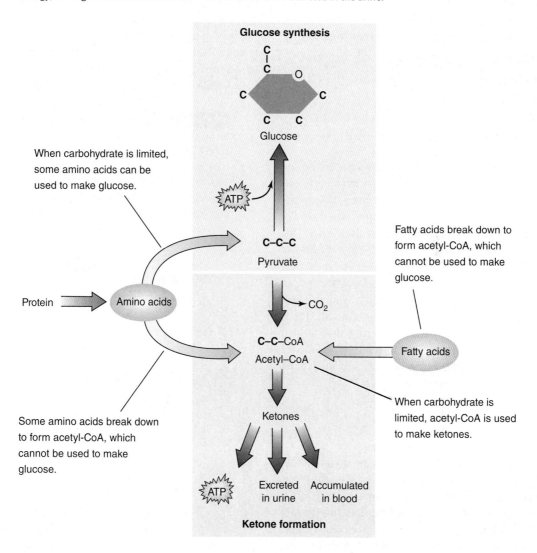

during starvation or when consuming a low-carbohydrate weight-loss diet and can cause symptoms such as reduced appetite, headaches, dry mouth, and odd-smelling breath. Severe ketosis can occur with untreated diabetes and can increase the blood's acidity so much that normal body processes are disrupted, resulting in coma and even death.

Concept Check

1. **What** is the main function of glucose in the body?
2. **How** does insulin affect blood glucose levels?
3. **What** process breaks down glucose in the presence of oxygen to yield ATP?
4. **Why** is carbohydrate said to *spare* protein?

4.5 Carbohydrates in Health and Disease

LEARNING OBJECTIVES

1. **Define** diabetes and explain its health consequences.
2. **Discuss** the role of carbohydrates in weight control.
3. **Describe** how carbohydrates contribute to the development of dental caries.
4. **Explain** how a diet high in fiber promotes health.

Are carbohydrates good for you or bad for you? On the one hand, they have been blamed for everything from obesity and heart disease to dental caries. On the other hand, U.S. guidelines for a healthy diet recommend that people base their diet on carbohydrate-rich foods in order to reduce disease risk. This incongruity relates to the health effects of different types of dietary carbohydrates: Diets high in unrefined carbohydrates from whole grains, fruits, and vegetables are associated with a lower incidence of a variety of chronic diseases, whereas diets high in refined carbohydrates, such as refined grains and foods high in added sugars, increase chronic disease risk.[7] Added sugars are currently being implicated in the rising incidence of diabetes, obesity, cardiovascular disease, and **metabolic syndrome** (see *Debate: Is Sugar Making Us Sick?*).

Diabetes

Diabetes mellitus, commonly referred to simply as diabetes, is a disease characterized by high blood glucose levels (**Figure 4.13**). Uncontrolled diabetes damages the heart, blood vessels, kidneys, eyes, and nerves. It is the leading cause of adult blindness and accounts for over 40% of new cases of kidney failure and more than 60% of nontraumatic lower-limb amputations. In the United States, 29.1 million people have diabetes, and 8.1 million of these people have not been diagnosed.[18]

Types of diabetes
Type 1 diabetes is an **autoimmune disease** in which the insulin-secreting pancreatic cells are destroyed by the body's immune system. This form of diabetes accounts for only 5 to 10% of diagnosed cases and usually develops before age 30. Because no insulin is produced, people with type 1 diabetes must inject insulin in order to keep blood glucose levels in the normal range. When insulin levels are low, the lack of glucose inside cells leads to ketone formation. In uncontrolled type 1 diabetes, ketone levels can get high enough to increase the acidity of the blood. This condition, called diabetic **ketoacidosis**, can lead to coma and death.

FIGURE 4.13 Blood glucose levels in diabetes Normal blood glucose is less than 100 mg/100 ml blood after an eight-hour fast; a fasting blood level from 100 to 125 mg/100 ml is defined as prediabetes; a fasting level of 126 mg/100 ml or above is defined as diabetes. Two hours after consuming 75 g of glucose, normal blood levels are less than 140 mg/100 ml; prediabetes levels are from 140 to 199 mg/100 ml; diabetes levels are 200 mg/100 ml or greater.

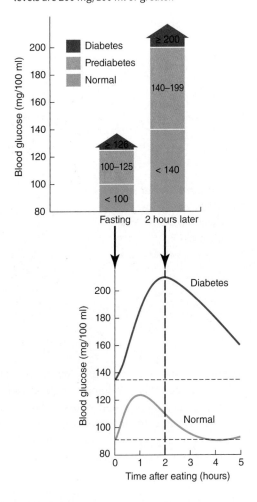

metabolic syndrome A collection of health risks, including high blood pressure, altered blood lipids, high blood glucose, and a large waist circumference, that increase the chance of developing heart disease, stroke, and diabetes.

diabetes mellitus A disease characterized by elevated blood glucose due to either insufficient production of insulin or decreased sensitivity of cells to insulin.

type 1 diabetes The form of diabetes caused by autoimmune destruction of insulin-producing cells in the pancreas, usually leading to absolute insulin deficiency.

autoimmune disease A disease that results from immune reactions that destroy normal body cells.

Debate

Is Sugar Making Us Sick?

The Issue

Few components of our diet have been more scrutinized than sugar. While the Sugar Association calls it "nature's sweetener," the popular press and some scientists have called it "toxic" or even the "new tobacco." They claim it not only rots your teeth but also causes obesity, diabetes, heart disease, and even cancer.[8] Few would argue that the calories we consume as sugar would be better spent on healthier foods, but whether current levels of sugar consumption pose a serious health risk is hotly debated.

Most of the sugar in the modern diet is from added sugars, such as sucrose (a disaccharide formed from glucose and fructose) and high-fructose corn syrup (a syrup that is approximately half glucose and half fructose). The connection between added sugars, obesity, and other chronic diseases relates to both the calories sugars contribute and the fructose component of these sweeteners.[9] Added sugar intake rose drastically during the last part of the 20th century, and as the amount rose, so did the incidence of obesity (see graph). As obesity rates increase, so does the incidence of heart disease, diabetes, and other chronic diseases.

Added sugars provide empty calories and, when consumed in excess, contribute to weight gain. Much of the increase in added sugars since 1980 has been from sugar-sweetened beverages; intake of these has been correlated with obesity.[10, 11] In addition to the calories they provide, these beverages contribute to weight gain because liquids are not satiating and therefore do not suppress intake of other foods. However, the sugars in solid food are also problematic. A meta-analysis of sugar intake and body weight showed that intake of sugars in either food or beverages increases energy intake and thus body weight.[12]

The fructose component of these sweeteners is believed to contribute to the development of obesity and other chronic diseases because it is metabolized differently than glucose. Fructose is broken down in the liver, producing by-products that are used to make fat. It has been shown to increase fat in the liver, alter blood lipids to increase the risk of heart disease, and increase insulin resistance.[9, 11] We have been consuming fructose in fruits and vegetables since the dawn of civilization, so why is it a problem now? The reason is that we now consume more fructose than ever before, and even moderate amounts can cause these metabolic changes.[13]

Despite the correlation between sugar intake and obesity, many argue that sugar per se is not the cause of obesity or other health problems. Just because two events occur at the same time does not mean one causes the other. Obesity rates continued to rise even after sugar intake began decreasing at the turn of the century. Sugar adds calories, but there is no convincing evidence that sugar has a unique or harmful effect relative to any other source of calories on the development of obesity or diabetes.[14] Sugar and the fructose it contains cause weight gain when caloric intake exceeds needs, but the same weight gain would occur with excess calorie intake from other sources, such as starch or fat. Those who do not believe sugar is to blame for our health problems also contend that the metabolic changes caused by a high-fructose diet do not necessarily lead to diabetes, heart disease, and metabolic syndrome in free-living populations. Long-term studies showing that these changes result in metabolic disease have not been done.[15–17]

In the 1990s fat was stigmatized as the cause of all health problems, so people stocked up on low-fat foods. Then carbohydrates were blamed, so everyone went on low-carb diets. Most would agree that cutting down on added sugars is a good thing, but we need to be careful not to blame all our health problems on one dietary component. If we cut down on sugars, we need to be mindful of which nutrients take their place.

Think Critically

Suggest some factors other than sugar intake that might help explain the rise in obesity that has occurred over the past few decades.

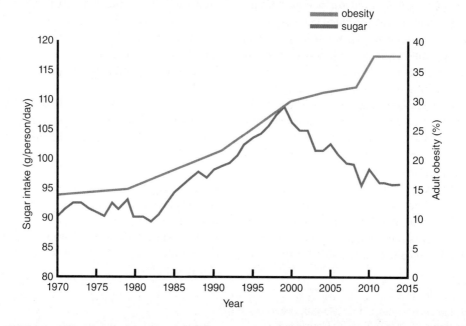

FIGURE 4.14 **Risk factors for type 2 diabetes** Some of the risk factors for type 2 diabetes can't be changed, but dietary and lifestyle risk factors can be modified in order to reduce the risk of developing diabetes.

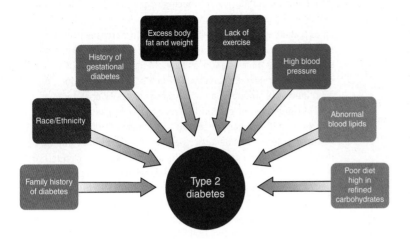

Think Critically

Are high blood pressure and abnormal blood lipids modifiable risk factors? Why or why not?

The more common form of diabetes is **type 2 diabetes**, which accounts for 90 to 95% of all cases. It occurs when the amount of insulin produced by the body is not sufficient to keep blood glucose in the normal range. This can occur because body cells lose their sensitivity to insulin, a condition called **insulin resistance**, or because the amount of insulin secreted is reduced. Type 2 diabetes is believed to be due to both genetic and lifestyle factors (**Figure 4.14**). Incidence in the United States is higher among African Americans, Hispanics, Asian Americans, and Native Americans than Caucasians.[18] Evidence is accumulating that the risk of developing type 2 diabetes is increased by poor dietary choices. Diets high in refined carbohydrates, particularly sugar-sweetened beverages, are associated with an increased risk of developing diabetes.[19, 20] Risk is lower in populations with diets rich in whole grains, fruits, vegetables, legumes, and nuts and low in refined grains, red or processed meats, and sugar-sweetened beverages.[21–23]

Type 2 diabetes is more commonly diagnosed in adulthood, but we now know that people can develop this disease at any age. Twenty years ago, type 2 diabetes was rare in adolescents (ages 10 to 19), but as in adults, the incidence is rising, especially among minorities.[18] A progressive disease, it usually begins with **prediabetes**, a condition in which glucose levels are above normal but not high enough to be diagnosed as diabetes (see Figure 4.13). In many cases, adjustments in diet and lifestyle can prevent prediabetes from progressing to type 2 diabetes.

Gestational diabetes is diabetes diagnosed during pregnancy that is not clearly either type 1 or type 2 diabetes.[24] The high levels of glucose in the mother's blood are passed to the fetus, frequently resulting in a baby that is large for gestational age and at increased risk of complications. Women who have had gestational diabetes are at increased risk of developing type 2 diabetes later in life.

Symptoms and complications of diabetes The symptoms and complications of all types of diabetes result from the inability to use glucose normally and from high glucose levels in the blood. Cells that require insulin in order to take up glucose are starved for glucose, and cells that can use glucose without insulin are exposed to damaging high levels.

Early symptoms of diabetes include frequent urination, excessive thirst, blurred vision, and weight loss. Frequent urination and excessive thirst occur because as blood glucose levels rise, the kidneys excrete the extra glucose and as a result must also excrete extra water, increasing the volume of urine. The additional loss of water from the body makes the individual thirsty. Blurred vision occurs when excess glucose enters the lens of the eye, drawing in water and causing the lens to swell. Weight loss occurs because cells are unable to take up blood glucose and use it as an energy source. As a result, the body must break down its energy stores to meet needs.

The long-term complications of diabetes include damage to the heart, blood vessels, kidneys, eyes, and nerves (**Figure 4.15**). These complications are believed to be due to prolonged exposure to high glucose levels.

Managing blood glucose The goal in treating diabetes is to maintain blood glucose levels within the normal range. This requires an individualized program of diet, exercise, and, in many cases, medication, along with frequent monitoring of blood glucose levels.[24] Carbohydrate intake should be coordinated with

---

type 2 diabetes The form of diabetes characterized by insulin resistance and relative (rather than absolute) insulin deficiency.

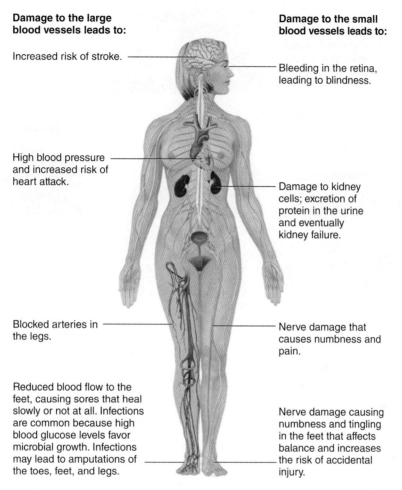

FIGURE 4.15 Diabetes complications The long-term complications of diabetes result from damage to the large blood vessels, which leads to an increased risk of heart attack and stroke, and changes in small blood vessels, which can cause blindness, kidney failure, and nerve dysfunction.

Damage to the large blood vessels leads to:
- Increased risk of stroke.
- High blood pressure and increased risk of heart attack.
- Blocked arteries in the legs.
- Reduced blood flow to the feet, causing sores that heal slowly or not at all. Infections are common because high blood glucose levels favor microbial growth. Infections may lead to amputations of the toes, feet, and legs.

Damage to the small blood vessels leads to:
- Bleeding in the retina, leading to blindness.
- Damage to kidney cells; excretion of protein in the urine and eventually kidney failure.
- Nerve damage that causes numbness and pain.
- Nerve damage causing numbness and tingling in the feet that affects balance and increases the risk of accidental injury.

exercise and medication schedules so that glucose and insulin are available in the right proportions (**Figure 4.16**).

Dietary management of diabetes involves limiting the amount of carbohydrate consumed at each meal to prevent a rapid or prolonged rise in blood glucose.[24] A diet providing unrefined carbohydrates is recommended because these carbohydrate sources cause a slower rise in blood glucose than refined carbohydrates. Diets should be limited in saturated fat and *trans* fat in order to reduce the risk for cardiovascular disease. Weight management is an important component of diabetes care because excess body fat increases the resistance of body cells to insulin. Exercise is important not only because it helps to achieve and maintain a healthy body weight but also because it increases the sensitivity of body cells to insulin.[24]

Individuals with type 1 diabetes require insulin injections because they no longer make insulin. Insulin must be given by injection because it is a protein and would therefore be digested in the gastrointestinal tract if taken orally. Individuals with type 2 and gestational diabetes are often able to manage blood glucose levels with diet and exercise but may also require oral medications and/or insulin injections.

Hypoglycemia

Another condition that involves abnormal blood glucose levels is **hypoglycemia**. Symptoms of hypoglycemia include low blood sugar (below 70 mg glucose/100 ml blood), irritability, sweating, shakiness, anxiety, rapid heartbeat, headache, hunger, weakness, and sometimes seizures and coma. Hypoglycemia occurs most frequently in people who have diabetes as a result of overmedication. It can also occur in people without diabetes; it is caused by abnormalities in insulin production or by abnormalities in the way the body responds to insulin or to other hormones.

---

hypoglycemia Abnormally low blood glucose levels.

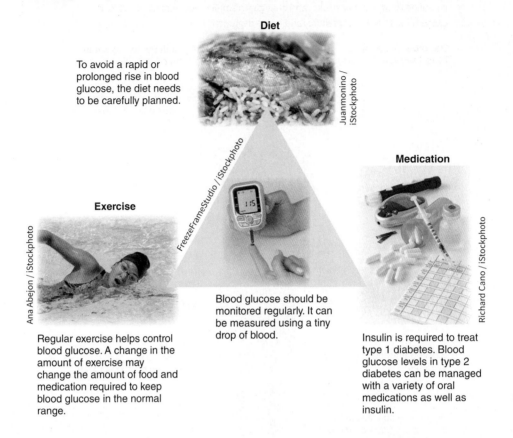

FIGURE 4.16 Managing diabetes Managing diabetes requires monitoring blood glucose levels and controlling blood glucose with diet and exercise and, in some cases, medication.

Fasting hypoglycemia, which occurs when an individual has not eaten, is often related to some underlying condition, such as excess alcohol consumption, hormonal deficiencies, or tumors. Treatment involves identifying and treating the underlying disease. **Reactive hypoglycemia** occurs in response to the consumption of high-carbohydrate foods. The rise in blood glucose from the carbohydrate stimulates insulin release. However, too much insulin is secreted, resulting in a rapid fall in blood glucose to abnormally low levels. To prevent the rapid changes in blood glucose that occur with reactive hypoglycemia, the diet should consist of small, frequent meals that contain protein and fiber and are low in simple carbohydrates.

Weight Management

As low-carbohydrate diets have gained popularity, carbohydrates have gotten a reputation for being fattening. In reality, carbohydrates are no more fattening than other nutrients. Weight gain is caused by excess intake of calories, no matter whether the excess is from carbohydrate, fat, or protein. Carbohydrates provide only 4 Calories/gram, less than half the 9 Calories/gram provided by fat.

The type and amount of carbohydrates you consume can affect how hungry you feel and whether you lose or gain weight. High intakes of refined carbohydrates, particularly of added sugars, increase the risk of excess body weight and obesity (see *Debate: Is Sugar Making Us Sick?*).[25] Foods high in refined carbohydrates stimulate the release of insulin, and insulin promotes fat storage. Therefore, a diet high in refined carbohydrates may shift metabolism toward fat storage.[26] In contrast, a low-carbohydrate diet causes less insulin release and hence does not promote fat storage.[27] Weight loss while consuming a very low-carbohydrate diet may also be aided by the accumulation of ketones, which help suppress appetite, and the high-protein content of these diets, which increases satiety. In addition, very low-carbohydrate diets limit food choices to such an extent that the monotony of the diet may cause the dieter to eat less. Diets that rely on unrefined carbohydrates are high in fiber and aid weight loss because they increase the sense of fullness by adding bulk and slowing digestion, allowing you to feel satisfied with less food.[28]

Weight loss is caused by a reduction in energy intake relative to expenditure. Whether a low-carbohydrate diet or a diet based on unrefined carbohydrates or some other approach to weight loss is most effective depends on the individual.[27]

Heart Disease

The effect carbohydrate intake has on heart disease risk depends on the type of carbohydrate. There is evidence that

FIGURE 4.17 Cholesterol and soluble fiber Soluble fiber helps lower blood cholesterol by increasing excretion of bile acids and cholesterol.

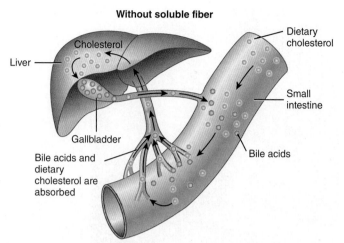
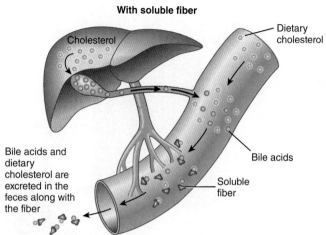

a. In the absence of soluble fiber, dietary cholesterol and bile, which contains cholesterol and bile acids made from cholesterol, are absorbed into the blood and transported to the liver for use in the body.

b. When soluble fiber is present in the digestive tract, the fiber binds cholesterol and bile acids so that they are excreted rather than absorbed. This helps reduce the amount of cholesterol in the body.

diets high in sugar can raise blood lipid levels and thereby increase the risk of heart disease,[29] whereas dietary patterns that are high in fiber from grains, vegetables, and fruits reduce the risk of heart disease.[28, 30, 31]

Dietary patterns that are high in fiber lower heart disease risk by helping lower blood cholesterol, reduce blood pressure, normalize blood glucose levels, and prevent obesity, as well as by affecting a number of other parameters that impact heart disease risk.[30] Soluble fiber lowers blood cholesterol levels by reducing the intestinal absorption of dietary cholesterol and bile acids, which are made from cholesterol (**Figure 4.17**).[32, 33] Soluble fiber may also enhance health by affecting the types of bile acids produced by the intestinal microbiota.[32] Insoluble fibers are also beneficial for heart health but have less effect on blood cholesterol.

Dental Caries

Dental caries, or cavities, are the most well-documented health problem associated with carbohydrate intake. Ninety-one percent of adults aged 20 to 64 have had caries.[34] They occur when bacteria that live in the mouth form colonies, known as plaque, on the tooth surface. If the plaque is not brushed, flossed, or scraped away, the bacteria metabolize carbohydrate from the food we eat, producing acids. These acids can dissolve tooth enamel and the underlying tooth structure, forming dental caries.

Although all carbohydrates can contribute to dental caries, sucrose is the most cariogenic because it is easily metabolized to acid by bacteria, and it is needed for the synthesis of materials that help bacteria stick to the teeth and form plaque.[35] The longer teeth are exposed to carbohydrates—for example, through frequent snacking, slowly sipping juice and other beverages that contain sugar, consuming foods that stick to the teeth, and sucking hard candy—the greater the risk of caries. Limiting intake of sweet or sticky foods and proper dental hygiene can help prevent dental caries.

Bowel Health

Fiber adds bulk and absorbs water in the gastrointestinal tract, making the feces larger and softer and reducing the pressure needed for defecation. This helps reduce the incidence of constipation and **hemorrhoids**, the swelling of veins in the rectal or anal area. It also reduces the risk of developing outpouches in the wall of the colon called **diverticula** (the singular is *diverticulum*) (**Figure 4.18**). Fecal matter can accumulate in these pouches, causing irritation, pain, and inflammation—a condition known as **diverticulitis**. Diverticulitis may lead to infection. Treatment usually includes antibiotics to eliminate the infection and a low-fiber diet to prevent irritation of inflamed tissues. Once the inflammation is resolved, a high-fiber diet is recommended to ease stool elimination and reduce future attacks of diverticulitis.

Although fiber speeds movement of the intestinal contents, when the diet is low in fluid, fiber can contribute to constipation. The more fiber in the diet, the more water is needed to keep the stool soft. When too little fluid is consumed, the stool becomes hard and difficult to eliminate. In severe cases of excessive fiber intake and low fluid intake, intestinal blockage can occur.

A diet high in fiber, particularly from whole grains, may reduce the risk of colon cancer, although not all studies support

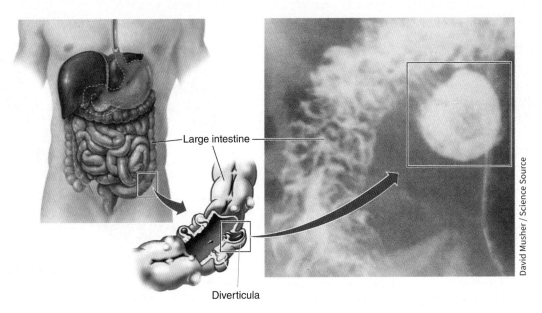

FIGURE 4.18 Diverticulosis Diverticulosis is a condition in which outpouches form in the wall of the colon. These diverticula form at weak points due to pressure exerted when the colon contracts.

this finding.[4, 36, 37] By diluting the colon contents and reducing transit time, fiber decreases the amount of contact between the cells lining the colon and potentially cancer-causing substances in the feces.[38] Fiber in the colon also affects the intestinal microbiota and its by-products. These by-products may directly affect colon cells or may change the environment of the colon in a way that can affect the development of colon cancer. Some of the protective effect may also be due to antioxidant vitamins and phytochemicals present in fiber-rich whole grains, fruits, and vegetables.

Concept Check

1. **Why** is it important to keep blood glucose levels in the normal range?
2. **When** does a low-carbohydrate diet promote weight loss?
3. **Why** does frequent snacking on high-carbohydrate foods promote dental caries?
4. **How** does fiber benefit colon health?

4.6 Meeting Carbohydrate Needs

LEARNING OBJECTIVES

1. **Compare** the carbohydrate intake of Americans with recommendations for a healthy diet.
2. **Calculate** the percentage of calories from carbohydrate in your diet.
3. **Use** food labels to identify foods that are high in fiber and low in added sugar.

Recommendations for carbohydrate intake focus on two main points: getting enough carbohydrate to meet the need for glucose and choosing the types that promote health and prevent disease.

Carbohydrate Recommendations

The RDA for carbohydrate is 130 g per day, based on the average minimum amount of glucose used by the brain.[39] In a diet that meets energy needs, this amount provides adequate glucose and prevents ketosis. Additional carbohydrate provides an important source of energy in the diet, and carbohydrate-containing foods can add vitamins, minerals, fiber, and phytochemicals. Therefore, the Acceptable Macronutrient Distribution Range for carbohydrate is 45 to 65% of total calorie intake. A diet in this range meets energy needs without excessive amounts of protein or fat (**Figure 4.19**).

The typical U.S. diet meets the recommendation for the amount of carbohydrate, but most of this comes from refined

FIGURE 4.19 How much carbohydrate do you eat? To calculate the percentage of calories from carbohydrate in a diet, first determine the number of grams of carbohydrate and multiply this value by 4 Calories/gram. For example, the vegetarian food shown here, which represents a day's intake, provides about 300 g of carbohydrate:

300 g × 4 Calories/g = 1200 Calories from carbohydrate

Next divide the number of Calories from carbohydrate by the total number of Calories in the diet and multiply by 100 to convert it to a percentage. In this example, the diet contains 2000 total Calories, and so it provides:

(1200 Calories from carbohydrate/2000 Calories total) × 100 = 60% of Calories from carbohydrate

Ask Yourself

What is the percentage of calories from carbohydrate in a diet that provides 240 g of carbohydrate and 2400 Calories?

a. 10
b. 40
c. 50
d. 60

sources, making the diet lower in fiber and higher in added sugars than recommended. The Adequate Intake for fiber is 38 g per day for men and 25 g per day for women; the typical intake is only about 16 g per day.[40]

Americans currently consume over 13% of their calories as added sugars.[1] Although there is no RDA for added sugars, a Daily Value of 50 g per day for a 2000-Calorie diet has been established. This is consistent with the recommendations of the 2015–2020 Dietary Guidelines to limit added sugar intake to less than 10% of calories. For a 2000-Calorie diet, this would mean limiting added sugar intake to less than 200 Calories per day; this is 50 grams or about 12 teaspoons—the amount in one 20-ounce bottle of sugar-sweetened beverage.

Because no specific toxicity is associated with high intake of any type of carbohydrate, no UL has been established for total carbohydrate intake, for fiber intake, or for added sugar intake.

Choosing Carbohydrates Wisely

To promote a healthy, balanced diet, the 2015–2020 Dietary Guidelines and MyPlate recommend increasing consumption of whole grains, fruits and vegetables, and low-fat dairy products while limiting foods high in refined grains and added sugars, such as soft drinks and other sweetened beverages, sweet bakery products, and candy (see *What Should I Eat?*). Because

---

What Should I Eat?

Carbohydrates

Make half your grains whole

- Have your sandwich on whole-wheat, oat bran, rye, or pumpernickel bread.
- Switch to whole-wheat pasta and brown rice.
- Fill your cereal bowl with plain oatmeal and add a few raisins for sweetness.
- Check the ingredient list for the words *whole* or *whole grain* before the grain ingredient's name.

Increase your fruits and veggies

- Don't forget beans. Kidney beans, chickpeas, black beans, and others are a good source of fiber.
- Add berries and bananas to your cereal or dessert.
- Pile the veggies on your sandwich.
- Have more than one vegetable at dinner.

Limit added sugars

- Switch from a 20-oz bottle of sugar-sweetened soda to water or low-fat milk.
- Use one-quarter less sugar in your recipe next time you bake.
- Snack on a piece of fruit instead of a candy bar.
- Swap your sugary breakfast cereal for an unsweetened whole-grain variety.

 Use iProfile to look up the fiber content of some of your favorite foods.

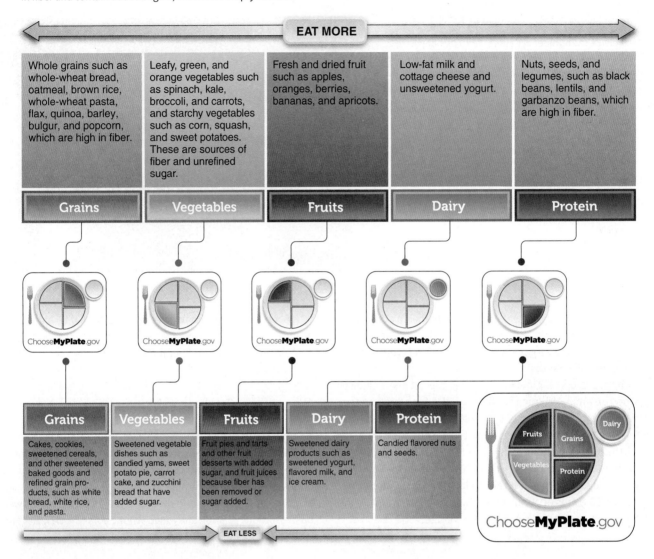

FIGURE 4.20 **Healthy MyPlate carbohydrate choices** The healthiest carbohydrate choices are whole grains, legumes, and fresh fruits and vegetables, which are low in added sugars and often are good sources of fiber. Foods containing refined carbohydrates should be limited because they are typically low in fiber and contain added sugars, which add empty calories.

almost half of the added sugars Americans consume come from beverages, the Dietary Guidelines specifically recommend reducing intake of sugar-sweetened beverages such as soda, energy drinks, sports drinks, and sugar-sweetened fruit drinks.[1]

Using MyPlate to make healthy choices For a 2000-Calorie diet, MyPlate recommends 6 oz of grains (half of which should be whole grains), 2 cups of fruit, and 2½ cups of vegetables. As **Figure 4.20** suggests, refined carbohydrates can be replaced with unrefined ones to make the diet healthier. For example, an apple provides about 80 Calories and 3.7 g of fiber, making it a better choice than 1 cup of apple juice, which has the same amount of energy but almost no fiber (0.2 g).

Interpreting food labels Food labels can help in choosing the right mix of carbohydrates (**Figure 4.21**). The current and planned Nutrition Facts labels can help consumers find foods that are good sources of fiber. The current label provides information on sugars but does not distinguish between naturally occurring and added sugars. The planned Nutrition Facts label includes the amount of added sugars per serving in grams and as a percentage of the Daily Value. The ingredient list helps identify whole-grain products and the sources of added sugars (see *Thinking It Through*). Nutrient content claims such as "high in fiber" or "no sugar added" and health claims such as those highlighting the relationship between fiber intake and the risk of heart disease and cancer help identify foods that meet the recommendations for fiber and added sugar intake.

Meeting Carbohydrate Needs 105

FIGURE 4.21 Choosing carbohydrates from the label The planned Nutrition Facts labels shown here along with the ingredient lists provide information about the types and amounts of carbohydrates in packaged foods.

Whole-Wheat Bread

Nutrition Facts	Amount/serving	% Daily Value*	Amount/serving	% Daily Value*
16 servings per container	Total Fat 1.5g	2%	Total Carbohydrate 18g	7%
Serving size 1	Saturated Fat 0g	0%	Dietary Fiber 3g	11%
1 slice (43g)	Trans Fat 0g		Total Sugars 1g	
Calories per serving 100	Cholesterol 0mg	0%	Includes 1g of Added Sugars	2%
	Sodium 210mg	9%	Protein 4g	
	Vitamin D 0mcg 0% • Calcium 50mg 4% • Iron 1mg 6% • Potassium 470mg 10%			
	Thiamin 8% • Riboflavin 2% • Niacin 6%			

*The % Daily Value (DV) tells you how much a nutrient in a serving of food contributes to a daily diet. 2,000 calories a day is used for general nutrition advice.

INGREDIENTS: WHOLE WHEAT FLOUR, WATER, SWEETENERS (HIGH FRUCTOSE CORN SYRUP, MOLASSES), WHEAT GLUTEN, SOYBEAN OIL, CONTAINS 2% OR LESS OF THE FOLLOWING: YEAST, DOUGH CONDITIONERS (MONO & DIGLYCERIDES, ETHOXYLATED MONO & DIGLYCERIDES, CALCIUM STEAROYL-2-LACTYLATE), YEAST NUTRIENTS (CALCIUM SULFATE, MONO-CALCIUM PHOSPHATE), CALCIUM PROPIONATE (A PRESERVATIVE).

The Nutrition Facts panel of a food label lists the number of grams of total carbohydrate and fiber and gives these amounts as a percentage of the Daily Value.

The Daily Value for fiber used on the planned Nutrition Facts label has been increased from 25 grams to 28 grams because of the importance of fiber for health.

To identify products made mostly from *whole* grains, look for the word "whole" before the name of the grain. If this is the first ingredient listed, the product is made from mostly whole grain. "Wheat flour" simply means it was made with wheat, not whole wheat. Note that foods labeled with the words "multigrain," "stone-ground," "100% wheat," "cracked wheat," "seven-grain," or "bran" are not necessarily 100% whole-grain products and may not contain any whole grains.

Foods labeled "high fiber" contain 20% or more of the Daily Value per serving.

Foods labeled "good source of fiber" contain between 10 and 19% of the Daily Value per serving.

Ask Yourself

Is a product that lists wheat flour as the first ingredient and whole-wheat flour as the second made mostly from whole grains? Why or why not?

Products labeled "reduced sugar" contain 25% less sugar than the regular, or reference, product.

The ingredient list helps identify the sources of added sugars. They, like other ingredients, are listed in order of prominence by weight.

INGREDIENTS: CULTURED PASTEURIZED GRADE A REDUCED FAT MILK, SUGAR, NONFAT MILK, HIGH FRUCTOSE CORN SYRUP, STRAWBERRY PUREE, MODIFIED CORN STARCH, KOSHER GELATIN, TRI-CALCIUM PHOSPHATE, NATURAL FLAVOR, COLORED WITH CARMINE, VITAMIN A ACETATE, VITAMIN D₃

On the ingredient list, all these are added sugar: Brown sugar, corn sweetener, corn syrup, dextrose, fructose, fruit juice, glucose, high fructose corn syrup, honey, invert sugar, lactose, maltose, malt syrup, molasses, raw sugar, sucrose, and sugar syrup concentrates.

Nutrition Facts

1 servings per container
Serving size 1 container (170g)

Amount per serving
Calories 190

	% Daily Value*
Total Fat 3.5g	4%
Saturated Fat 2g	10%
Trans Fat 0g	
Cholesterol 15mg	5%
Sodium 100mg	4%
Total Carbohydrate 32g	12%
Dietary Fiber 0g	0%
Total Sugars 28g	
Includes 21g Added Sugars	42%
Protein 7g	
Vitamin D 0mcg	0%
Calcium 300mg	23%
Iron 0mg	0%
Potassium 367mg	8%

* The % Daily Value (DV) tells you how much a nutrient in a serving of food contributes to a daily diet. 2,000 calories a day is used for general nutrition advice.

Foods labeled "sugar free" contain less than 0.5g of sugar per serving.

The number of grams of total sugars on both the current and planned Nutrition Facts labels includes naturally occurring sugars from the fruit and yogurt and added sugars. The planned label also includes the amount that is from added sugars in grams and as a percentage of the Daily Value.

Thinking It Through

A Case Study on Healthy Carbohydrate Choices

Venicia is busy and tends to grab whatever is quick and easy to eat. She knows that she isn't choosing well and is concerned about her intake of added sugars. Venicia analyzes a typical day's diet using iProfile. For breakfast she has a bowl of frosted corn flakes cereal with milk and orange drink, lunch is chips and a sugar-sweetened soda, and dinner is a burrito. She always drinks sweetened iced tea with dinner. She has another soda at night while studying. Her iProfile analysis shows that she eats 2319 Calories, 70 g protein, 71 g fat, 350 g carbohydrate, and 12 g fiber per day.

How does her intake compare with the DRI recommendations for total carbohydrate and fiber?

Answer: Venicia calculates the percentage of calories from carbohydrate (350 g carbohydrate × 4 Calories/gram ÷ 2319 Calories × 100 = 60%), and finds it to be in the recommended range of 45 to 65% of calories. However, she consumes only 12 g of fiber, 13 g less than the 25 g recommended for a woman of her age.

Venicia is surprised to see that despite her poor choices, her total carbohydrate intake meets recommendations, but she recognizes that her intake of added sugars is high.

What foods are sources of added sugars in her diet?
Your answer:

Use iProfile to determine how many Calories and how many grams of added sugar Venicia would eliminate from her diet if she replaced the two 20-oz sodas with water.
Your answer:

Venicia decides that she could cut out more added sugar and add some fiber by switching her breakfast cereal to a whole-grain cereal with fruit and nuts that isn't coated with sugar.

Using the Nutrition Facts labels, calculate how much Veronica's fiber and sugar intake would change if she switched from a serving of frosted corn flakes to a serving of whole-grain flakes.
Your answer:

Use the ingredient lists to identify the sources of the whole grains and added sugars in these two products.
Your answer:

(Check your answers in Appendix L.)

Frosted Corn Flakes

Nutrition Facts

8 servings per container
Serving size   1 cup (41g)

Amount per serving
Calories   **160**

	% Daily Value
Total Fat 0g	0%
Saturated Fat 0g	0%
Trans Fat 0g	
Cholesterol 0mg	0%
Sodium 200mg	9%
Total Carbohydrate 37g	13%
Dietary Fiber 1g	4%
Total Sugars 16g	
Includes 16g Added Sugars	32%
Protein 1g	

INGREDIENTS: Milled corn, sugar, malt flavoring, high fructose corn syrup, salt, sodium ascorbate and ascorbic acid (Vitamin C), niacinamide, iron, pyridoxine hydrochloride (vitamin B_6) riboflavin (vitamin B_2), thiamin hydrochloride (vitamin B_1,) vitamin A palmitate, folic acid, BHT (preservative), vitamin B_{12} and vitamin D.

Whole-Grain Flakes

Nutrition Facts

8 servings per container
Serving size   1 cup (55g)

Amount per serving
Calories   **220**

	% Daily Value
Total Fat 0g	3%
Saturated Fat 1g	4%
Trans Fat 0g	
Cholesterol 0mg	0%
Sodium 280mg	12%
Total Carbohydrate 43g	16%
Dietary Fiber 5g	18%
Total Sugars 12g	
Includes 9g Added Sugars	18%
Protein 4g	

INGREDIENTS: Whole Grain Wheat, Corn Meal, Sugar, Raisins, Rice Flour, Almond Pieces, Brown Sugar Syrup, Whole Grain Oats, Salt, Glycerin, Dried Cranberries, Palm Kernel Oil, Corn Syrup, Brown Sugar, Barley Malt Extract, Dextrose, Honey, Color (caramel color and annatto extract), Cinnamon, Soy Lecithin, Baking soda, Natural Flavor, Vitamin E (mixed tocopherols) Added to Preserve Freshness.

Pros and cons of nonnutritive sweeteners

One way to reduce the amount of refined sugar in the diet is to replace sugar with **nonnutritive sweeteners** (also called **artificial sweeteners**). The products listed in **Table 4.1** have been approved for use as nonnutritive sweeteners in food.[41] The FDA has defined **acceptable daily intakes (ADIs)**—levels that should not be exceeded when using these products.

When nonnutritive sweeteners are used to replace added sugars in the diet, they can help reduce the incidence of dental caries and manage blood sugar levels. Their effectiveness for weight loss is more controversial. The majority of studies on the effects of nonnutritive sweeteners on body weight suggest that obesity rates are lower when artificially sweetened beverages replace sugar-sweetened beverages, and there is no evidence that the use of artificial sweeteners causes higher body weight in adults.[42, 43] Animal studies suggest that artificial sweeteners may stimulate appetite, leading to weight gain, but this hypothesis has not been supported by studies done in humans.[42]

TABLE 4.1 Nonnutritive sweeteners

Sweetener	Brand names	What is it?	ADI
Acesulfame K (Ace-K)	Sunett, Sweet One	A heat-stable sweetener that is often used in combination with other sweeteners. It is 200 times sweeter than sucrose.	15 mg/kg of body weight/day (23 sweetener packets*)
Advantame		A heat-stable sweetener made from aspartame and vanillin that can be used in baked goods. It is 2000 times sweeter than sucrose.	32.8 mg/kg of body weight/day (4920 sweetener packets*)
Aspartame	Nutrasweet® Equal® Sugar Twin®	Made of two amino acids (phenylalanine and aspartic acid; see What a Scientist Sees: Phenylketonuria, in Chapter 6). Because it breaks down when heated, it is typically used in cold products or added after cooking. It is 200 times sweeter than sucrose.	50 mg/kg of body weight/day (75 packets*); should be limited in the diets of people with phenylketonuria
Luo Han Guo (*Siraitia grosvenorii*)	Monk fruit, Nectresse, Pure Lo, Sweet Sensation	A sweetener containing *Siraitia grosvenorii* fruit extract (SGFE) primarily from a plant native to Southern China. It is 100 to 250 times sweeter than sucrose.	None specified
Neotame	Newtame®	Made from the same two amino acids as aspartame, but because the bond between them is harder to break. It is heat stable and can be used in baking. It is used in soft drinks, dairy products, and gum. It is 7000 to 13,000 times sweeter than sucrose.	0.3mg/kg of body weight/day (23 sweetener packets*)

(continues)

TABLE 4.1 Nonnutritive sweeteners *(continued)*

Sweetener	Brand names	What is it?	ADI
Saccharin	Sweet'N Low, SweetTwin, NectaSweet	The oldest of the nonnutritive sweeteners, developed in 1879. It was once considered a carcinogen but was taken off the government's list of cancer-causing substances in 2000. It is 200 to 700 times sweeter than sucrose and has a bitter aftertaste.	15 mg/kg of body weight/day (45 packets*)
Stevia, steviol glycosides	Truvia, Pure Via, Enliten	A sweetener made from the leaf of the stevia plant.[41] It is about 200 to 400 times sweeter than sucrose.	4 mg/kg of body weight/day (9 packets*)
Sucralose	Splenda	Made from sucrose molecules that have been modified so that they cannot be digested or absorbed. It is heat stable, so it can be used in cooking. It is 600 times sweeter than sucrose.	5 mg/kg of body weight/day (23 packets*)

*Number of tabletop sweetener packets a 60-kg (132-pound) person would need to consume to reach the ADI.

Weight gain seen in some studies of artificial sweetener users is more likely to be due to the fact that individuals at higher risk of obesity are more likely to use artificial sweeteners to try to control weight.

If you think switching to nonnutritive sweeteners will make your diet healthier, think again. Foods that are high in added sugar tend to be nutrient poor. Replacing them with artificially sweetened alternatives does not necessarily increase the nutrient density of the diet or improve overall diet quality.

Concept Check

1. **How** does the typical U.S. diet compare with recommendations for fiber and added sugar?
2. **What** is the percentage of calories from carbohydrate in a serving of breakfast cereal that has 170 Calories and 44 grams of carbohydrate?
3. **Where** on a food label can you find information about whole grains? Added sugars?

Summary

1 Carbohydrates in Our Food 83

- High-carbohydrate foods are the basis of our diets. Unrefined whole grains, fruits, and vegetables are good sources of fiber and micronutrients. When these foods are **refined**, micronutrients and fiber are lost. Whole grains contain the entire kernel, as shown here, which includes the **endosperm, bran**, and **germ**; refined grains include only the endosperm. Refined grains are enriched with some of the B vitamins and iron, but not all the nutrients lost in refining are added back. Recommendations for a healthy diet suggest that we select more unrefined sources of carbohydrates and limit foods that are high in refined carbohydrates.

Figure 4.2 Whole grains

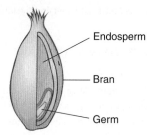

- Refined sugars contain calories but few nutrients; for this reason, foods high in **added sugars** are low in nutrient density.

2 Types of Carbohydrates 85

- Carbohydrates contain carbon as well as hydrogen and oxygen, in the same proportion as water. **Simple carbohydrates** include **monosaccharides** and **disaccharides** and are found in foods such as table sugar, honey, milk, and fruit. **Complex carbohydrates** are **polysaccharides**; they include **glycogen** in animals and **starches** (illustrated here) and **fiber** in plants.

Figure 4.4c Structures and sources of carbohydrates: Complex carbohydrates

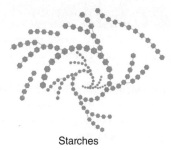

- Fiber cannot be digested in the stomach or small intestine and therefore is not absorbed into the body. **Soluble fiber** dissolves in water to form a viscous solution and is digested by bacteria in the colon; **insoluble fiber** is not readily digested by bacteria and adds bulk to fecal matter.

3 Carbohydrate Digestion and Absorption 88

- Disaccharides and starches must be digested to monosaccharides, as shown here, before they can be absorbed. In individuals with **lactose intolerance**, lactose passes into the colon undigested, causing cramps, gas, and diarrhea. Fiber, including some **oligosaccharides**, and **resistant starch**, can increase intestinal gas, but benefits overall health by increasing bulk in the stool, slowing nutrient absorption, and leading to the production of short-chain fatty acids that promote a healthy intestinal microbiota.

Figure 4.7 Carbohydrate digestion

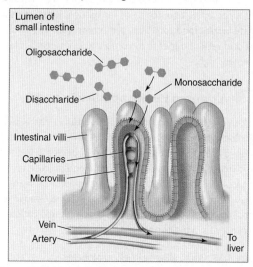

- After a meal, blood **glucose** levels rise. The rate, magnitude, and duration of this rise are referred to as the **glycemic response**. Glycemic response is affected by the amount and type of carbohydrate consumed and by other nutrients ingested with the carbohydrate.

4 Carbohydrate Functions 92

- Carbohydrate, primarily as glucose, provides energy to the body. Blood glucose levels are maintained by the hormones **insulin** and **glucagon**. As depicted here, when blood glucose levels rise, insulin from the pancreas allows cells to take up glucose from the blood and promotes the synthesis of glycogen, fat, and protein. When blood glucose levels fall, glucagon increases them by causing glycogen breakdown and glucose synthesis.

Figure 4.10 Blood glucose regulation

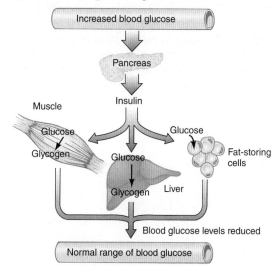

- Glucose is metabolized through cellular respiration. It begins with **glycolysis**, which breaks each six-carbon glucose molecule into two three-carbon pyruvate molecules, producing ATP even when oxygen is unavailable. The complete breakdown of glucose through **aerobic metabolism** requires oxygen and produces carbon dioxide, water, and more ATP than glycolysis.
- When carbohydrate intake is limited, amino acids from the breakdown of body proteins can be used to synthesize glucose. Therefore, an adequate carbohydrate intake is said to spare protein. Limited carbohydrate intake also results in the formation of **ketones** (**ketone bodies**) by the liver. These can be used as an energy source by other tissues. Ketones that accumulate in the blood can cause symptoms that range from headache and lack of appetite to coma and even death if levels are extremely high.

5 Carbohydrates in Health and Disease   96

- As shown in the graph, **diabetes mellitus** is characterized by high blood glucose levels, which occur either because insufficient insulin is produced and/or because of a decrease in the body's sensitivity to insulin. Over time, high blood glucose levels damage tissues and contribute to the development of heart disease, kidney failure, blindness, and infections that may lead to amputations. Treatment includes diet, exercise, and medication to keep glucose levels in the normal range.

Figure 4.13 Blood glucose levels in diabetes

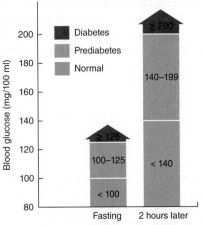

- **Hypoglycemia**, or low blood glucose, causes symptoms such as sweating, headaches, and rapid heartbeat.
- Weight gain is caused by excess intake of calories, no matter whether the excess is from carbohydrate, fat, or protein. Gram for gram, carbohydrates provide fewer calories than fat; however, diets high in refined carbohydrates are associated with higher body weights. Diets high in fiber aid weight loss because they increase the sense of fullness, allowing you feel satisfied with less food. Very low-carbohydrate diets limit food choices to such an extent that dieters eat less.
- High-sugar diets can increase heart disease risk by raising blood lipids. Dietary patterns that are high in fiber lower heart disease risk by helping lower blood cholesterol, reduce blood pressure, normalize blood glucose levels, and prevent obesity. Soluble fiber in particular helps lower blood cholesterol.
- Diets high in carbohydrate, particularly sucrose, increase the risk of dental caries. Sucrose helps bacteria stick to the teeth, and the bacteria then use sucrose and other carbohydrates as a food supply, producing acids that damage the teeth.
- Fiber makes the stool larger and softer. This reduces the pressure needed to move material through the colon, lowering the risk of constipation, **hemorrhoids**, and **diverticula**. A dietary pattern high in fiber may reduce the risk of colon cancer by diluting the colon contents, reducing transit time, and affecting the intestinal microbiota.

6 Meeting Carbohydrate Needs   102

- Guidelines for a healthy diet recommend consuming 45 to 65% of energy from carbohydrates. Most of this should come from whole grains, legumes, fruits, and vegetables, such as those in this photo. Foods high in added sugar should be consumed in moderation.

Figure 4.19 How much carbohydrate do you eat?

- The recommendations of MyPlate and the information on food labels can be used to select healthy amounts and sources of carbohydrate.
- **Nonnutritive sweeteners** can reduce added sugar intake. They do not raise blood glucose levels or contribute to tooth decay and can aid weight loss if the sugar calories they replace are not added back from other food sources.

Key Terms

- acceptable daily intake (ADI)   107
- added sugars   83
- aerobic metabolism   92
- anaerobic metabolism   92
- autoimmune disease   96
- bran   84
- cellulose   86
- complex carbohydrate   87
- diabetes mellitus   96
- disaccharide   85
- diverticula   101
- diverticulitis   101
- diverticulosis   102
- endosperm   84
- enrichment   84
- fasting hypoglycemia   100
- fiber   87
- fructose   85
- galactose   85
- germ   84
- gestational diabetes   98
- glucagon   92
- glucose   85
- glycemic index   91
- glycemic load   91
- glycemic response   91
- glycogen   87
- glycolysis   92
- hemorrhoid   101
- hypoglycemia   99
- insoluble fiber   87
- insulin   92
- insulin resistance   98
- ketoacidosis   96
- ketone or ketone body   94
- ketosis   94
- lactose   85
- lactose intolerance   89
- maltose   85
- metabolic syndrome   96
- monosaccharide   85
- nonnutritive sweetener or artificial sweetener   107
- oligosaccharide   87
- photosynthesis   87
- polysaccharide   85
- prediabetes   98
- reactive hypoglycemia   100
- refined   83
- resistant starch   87
- simple carbohydrate   85
- soluble fiber   87
- starch   87
- sucrose   85
- sugar unit   85
- type 1 diabetes   96
- type 2 diabetes   98
- unrefined food   83
- whole-grain product   83

What is happening in this picture?

These students are choosing fruit drinks, sports drinks, and sweetened iced tea because they believe these beverages are healthier choices than soda.

Think Critically

1. How does the amount of added sugar in these products compare to the amount in soft drinks?
2. Do these beverages provide significant amounts of any essential nutrients?
3. Suggest beverage alternatives that would be lower in added sugar.

CHAPTER 5

Lipids: Fats, Phospholipids, and Sterols

Lard—an animal fat—was until relatively recently the principal fat that Americans used to prepare food, and the source of that lard—swine—was an American staple. In the 1800s salt-cured pork "hogged" the menu, and the enormous quantity of fat generated in its production was used in everything from soap to candles.

The Long Depression of the 1870s impelled William Procter and James Gamble to seek cheaper sources for the now-expensive pork fat needed to make soap. They settled on a combination of palm and coconut oil, which produced a product so high in fat that it floated in water: Ivory soap.

In their quest for additional sources of fat, Procter and Gamble turned their attention to a waste product of textile manufacturing: cottonseed oil. But how to make a solid fat from a liquid oil? The answer came from a German chemist in 1907: hydrogenation. This process ultimately led to a product that would replace virtually every tub of lard in the land: Crisco.

The dangers associated with hydrogenated fats would not become widely understood until almost a century later, when it was recognized that increased consumption of the *trans* fats present in hydrogenated oils increased the risk of heart disease. This finding prompted first New York City and then the FDA to initiate efforts to remove this material from foods and encourage healthier alternatives.

CHAPTER OUTLINE

Fats in Our Food 114

Types of Lipids 115
- Triglycerides and Fatty Acids
- Phospholipids

Debate: Coconut Oil: Does a Tablespoon a Day Keep the Doctor Away?
- Sterols

Absorbing and Transporting Lipids 121
- Digestion and Absorption of Lipids
- Transporting Lipids in the Blood

Lipid Functions 125
- Essential Fatty Acids
- Fat as a Source of Energy

Lipids in Health and Disease 128
- Heart Disease
- Diet and Cancer
- Dietary Fat and Obesity

Meeting Lipid Needs 133
- Fat and Cholesterol Recommendations
- Choosing Fats Wisely

What a Scientist Sees: Are Eggs OK?

What Should I Eat? Healthy Fats

Thinking It Through: A Case Study on Meeting Fat Recommendations

5.1 Fats in Our Food

LEARNING OBJECTIVES

1. **Describe** the roles of fat in our food.
2. **Identify** some obvious and some hidden sources of fat in the diet.
3. **Name** sources of saturated and unsaturated fats in your diet.

Fat is the term we commonly use for lipids. The fats in our food contribute to their texture, flavor, and aroma. It is the fat that gives ice cream its smooth texture and rich taste. Olive oil imparts a unique flavor to salads and many traditional Italian and Greek dishes. Sesame oil gives egg rolls and other Chinese foods their distinctive aroma. But while the fats in our foods contribute to their appeal, they also add more calories than other nutrients (9 Calories/gram compared to 4 Calories/gram for carbohydrate and protein), so consuming too much can contribute to weight gain. The types of fats we eat can also affect our health; too much of the wrong types can increase the risk of heart disease and cancer.

The typical American diet contains about 34% of calories from fat.[1] Some of this fat is naturally present in foods such as meats, dairy products, nuts, seeds, and vegetable oils, but much of it is added to our food by manufacturers in processing and by consumers when they add butter, margarine, salad dressings, and sauces (**Figure 5.1**).

Sometimes the fat in our food is obvious. You can see the butter you spread on a slice of toast, the layer of fat around the outside of your steak, and the oil on your pepperoni pizza. When you choose these items, you know you are eating a high-fat food. Not all sources of dietary fat are obvious. Cheese, ice cream, and whole milk are high in fat, and foods that we think of as sources of carbohydrate, such as baked goods, noodle and rice dishes, and snack foods, make up much of the fat in our diets (**Figure 5.2**). We also add invisible fat when we fry foods: French fries start as potatoes, which are low in fat, but when

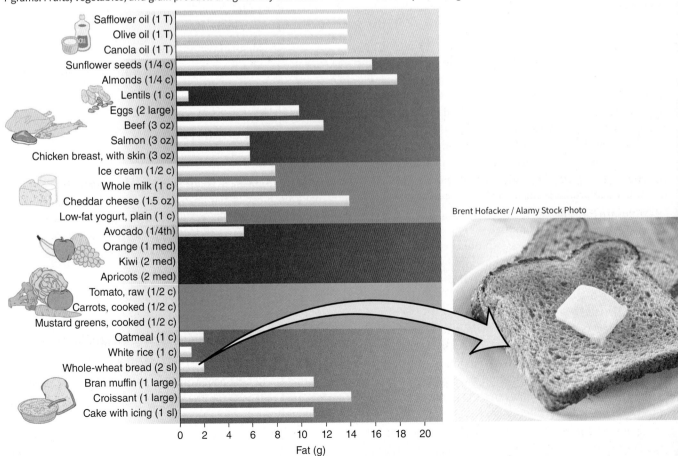

FIGURE 5.1 Sources of fat Fat in our diet comes from fats naturally found in foods like vegetable oils, meats, dairy products, and nuts and those we add to foods in processing, preparation, and at the table. For example, a slice of bread has only 2 grams of fat, but adding a teaspoon of butter increases this to 7 grams. Fruits, vegetables, and grain products are generally low in fat unless it is added in processing.

Brent Hofacker / Alamy Stock Photo

FIGURE 5.2 Hidden fats The fat in food is not always obvious. The two strips of bacon in this breakfast provide a total of 8 grams of fat, and the muffin provides twice this much.

Muffin: 16 g fat
2 strips Bacon: 8 g fat
2 Eggs: 14 g fat
½ cup Hash brown potatoes: 10 g fat

they are immersed in hot oil for frying, they soak up fat, which increases their calorie content.

The type of fat as well as the overall dietary pattern can affect health risk. High intakes of **saturated fat** from meat and dairy products, and **trans fat** used in shortening and margarine and added to processed foods are associated with an increased incidence of heart disease and certain types of cancer. Diets high in **unsaturated fats** from fish, nuts, and vegetable oils seem to protect against chronic disease. A healthy diet includes the right kinds of fats along with plenty of whole grains, fruits, and vegetables.[2]

Concept Check

1. **How** does adding fat affect the calorie content of a food?
2. **What** are some hidden sources of fat in the diet?
3. **Which** types of foods are sources of saturated fat?

5.2 Types of Lipids

LEARNING OBJECTIVES

1. **Explain** the relationship between triglycerides and fatty acids.
2. **Compare** the structures of saturated, monounsaturated, polyunsaturated, omega-6, omega-3, and *trans* fatty acids.
3. **Describe** how phospholipids and cholesterol are used in the body.
4. **Name** foods that are sources of cholesterol and saturated, monounsaturated, polyunsaturated, omega-6, omega-3, and *trans* fatty acids.

Lipids are substances that do not dissolve in water. Although we tend to use the term *fat* to refer to lipids, we are usually referring to types of lipids called **triglycerides**. Triglycerides make up most of the lipids in our food and in our bodies. The structure of triglycerides includes lipid molecules called **fatty acids**. Two other types of lipids that are important in nutrition but are present in the body in smaller amounts are **phospholipids** and **sterols**.

triglyceride The major type of lipid in food and the body, consisting of three fatty acids attached to a glycerol molecule.

fatty acid A molecule made up of a chain of carbons linked to hydrogens, with an acid group at one end of the chain.

phospholipid A type of lipid whose structure includes a phosphorus atom.

sterol A type of lipid with a structure composed of multiple chemical rings.

Triglycerides and Fatty Acids

A triglyceride consists of the three-carbon molecule glycerol with three fatty acids attached to it (**Figure 5.3**). Triglycerides may contain any combination of fatty acids. Fatty acids consist of chains of carbon atoms that vary in length from short-chain fatty acids (4 to 7 carbons) to medium-chain (8 to 12 carbons) and long-chain fatty acids (more than 12 carbons). Most fatty acids in plants and animals contain between 14 and 22 carbons. The fatty acids in a triglyceride determine its function in the body and the properties it gives to food. It is the fatty acids in triglycerides that we are really talking about when we refer to *trans* fat or saturated fat—these terms really mean **trans fatty acids** and **saturated fatty acids**, respectively.

Saturated and unsaturated fatty acids

Fatty acids are classified as saturated fatty acids or **unsaturated fatty acids**, depending on whether they contain carbon–carbon double bonds. A double bond forms between adjacent carbons in the carbon chain of a fatty acid when they each have only one hydrogen atom attached (**Figure 5.4**). The number and location of these double bonds affect the characteristics that fatty acids give to food and the health effects they have in the body. Saturated fatty acids have straight carbon chains that pack tightly together. Therefore, triglycerides that are high in saturated fatty acids, such as those found in beef, butter, and lard, tend to be solid at room temperature. Diets high in saturated fatty acids have been shown to increase the risk of heart disease. Unsaturated fatty acids have bent chains, so they don't pack together. This makes triglycerides that are higher in unsaturated fatty acids, such as those found in corn, safflower, and sunflower oils, liquid at room temperature. Unsaturated fats are susceptible to spoilage or *rancidity* because the unsaturated bonds in fatty acids are easily damaged by oxygen. When fats go rancid, they give food an "off" flavor. Diets high in unsaturated fatty acids are associated with a lower risk of heart disease.

The body is capable of synthesizing most of the fatty acids it needs from glucose or other sources of carbon, hydrogen, and oxygen, but the body cannot make some of the fatty acids it needs. These must be consumed in the diet and are referred to as **essential fatty acids**.

Saturated fatty acids are more plentiful in foods from animal sources, such as meat and dairy products, than in those from plant sources. Plant oils are generally low in saturated fatty acids (**Figure 5.5**). However, **tropical oils**, which include palm oil, palm kernel oil, and coconut oil, are exceptions. These saturated plant oils are called tropical oils because they are found in plants that are common in tropical climates. Saturated oils are useful in food processing because they are less susceptible to spoilage than are more unsaturated oils. Even though tropical oils are high in saturated fats, the types of saturated fats they contain do not increase the risk of heart disease and

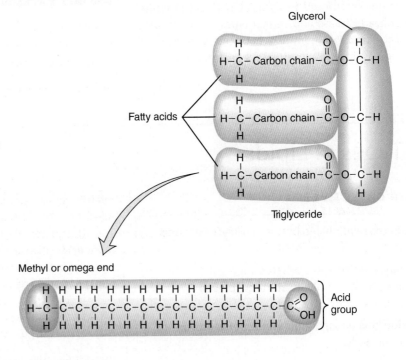

FIGURE 5.3 **Triglycerides**
A triglyceride contains glycerol and three fatty acids. A fatty acid is a chain of carbon atoms with an acid group (COOH) at one end of the chain. The other end of the chain is called the omega, or methyl, end and consists of a carbon attached to three hydrogen atoms. Each of the carbon atoms in between is attached to two carbon atoms and up to two hydrogen atoms.

---

saturated fatty acid A fatty acid in which the carbon atoms are bonded to as many hydrogen atoms as possible; it therefore contains no carbon–carbon double bonds.

unsaturated fatty acid A fatty acid that contains one or more carbon–carbon double bonds; may be either monounsaturated or polyunsaturated.

essential fatty acid A fatty acid that must be consumed in the diet because it cannot be made by the body or cannot be made in sufficient quantities to meet the body's needs.

NUTRITION INSIGHT

FIGURE 5.4 Fatty acids Fatty acids are categorized as saturated or unsaturated based on the types of bonds between the carbons. Unsaturated fatty acids are further subdivided based on the number and locations of the double bonds.

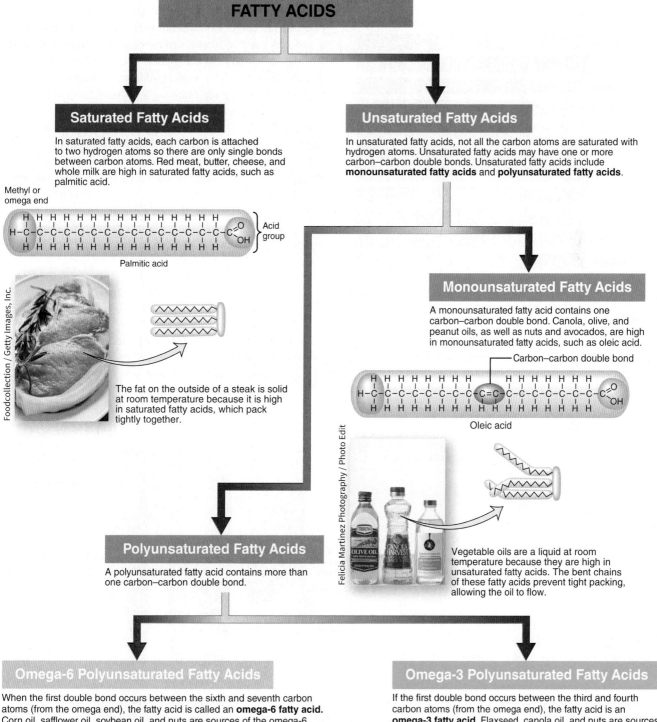

FATTY ACIDS

Saturated Fatty Acids

In saturated fatty acids, each carbon is attached to two hydrogen atoms so there are only single bonds between carbon atoms. Red meat, butter, cheese, and whole milk are high in saturated fatty acids, such as palmitic acid.

Methyl or omega end — Palmitic acid — Acid group

The fat on the outside of a steak is solid at room temperature because it is high in saturated fatty acids, which pack tightly together.

Unsaturated Fatty Acids

In unsaturated fatty acids, not all the carbon atoms are saturated with hydrogen atoms. Unsaturated fatty acids may have one or more carbon–carbon double bonds. Unsaturated fatty acids include **monounsaturated fatty acids** and **polyunsaturated fatty acids**.

Monounsaturated Fatty Acids

A monounsaturated fatty acid contains one carbon–carbon double bond. Canola, olive, and peanut oils, as well as nuts and avocados, are high in monounsaturated fatty acids, such as oleic acid.

Carbon–carbon double bond — Oleic acid

Vegetable oils are a liquid at room temperature because they are high in unsaturated fatty acids. The bent chains of these fatty acids prevent tight packing, allowing the oil to flow.

Polyunsaturated Fatty Acids

A polyunsaturated fatty acid contains more than one carbon–carbon double bond.

Omega-6 Polyunsaturated Fatty Acids

When the first double bond occurs between the sixth and seventh carbon atoms (from the omega end), the fatty acid is called an **omega-6 fatty acid**. Corn oil, safflower oil, soybean oil, and nuts are sources of the omega-6 polyunsaturated fatty acid linoleic acid.

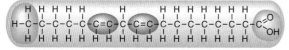

Linoleic acid

Omega-3 Polyunsaturated Fatty Acids

If the first double bond occurs between the third and fourth carbon atoms (from the omega end), the fatty acid is an **omega-3 fatty acid**. Flaxseed, canola oil, and nuts are sources of the omega-3 polyunsaturated fatty acid alpha-linolenic acid, and fish oils are high in longer-chain omega-3 fatty acids.

Alpha-linolenic acid

118 CHAPTER 5 Lipids: Fats, Phospholipids, and Sterols

FIGURE 5.5 Fatty acids in fats and oils The fats and oils in our diets contain combinations of saturated, monounsaturated, and polyunsaturated fatty acids. These combinations determine the texture, taste, and physical characteristics of the triglycerides.

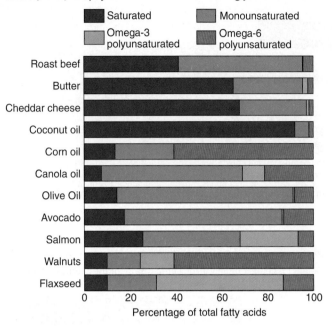

Ask Yourself

Based on the information in the graph:

1. Which two foods are highest in omega-3 fatty acids?
2. Which two foods are highest in saturated fatty acids?

may have some health benefits (see *Debate: Coconut Oil: Does a Tablespoon a Day Keep the Doctor Away?*).[3,4]

***Trans* fatty acids** The bonds in unsaturated fatty acids can occur in the *cis* or *trans* configuration (**Figure 5.6**). Small amounts of *trans* fatty acids occur naturally in meat, dairy products and oils. Larger amounts are generated by **hydrogenation**, a process used to make unsaturated oils more saturated. Hydrogenation solidifies vegetable oils into hard margarine and shortening. Food manufacturers add hydrogenated oils to foods because they are more solid at room temperature and can be stored longer without becoming rancid. A disadvantage of hydrogenation is that in addition to converting some double bonds into saturated bonds, it transforms some double bonds from the *cis* to the *trans* configuration. The consumption of these synthetic *trans* fats increases the risk of developing heart disease. Because of this, the FDA recently determined that **partially hydrogenated oils** are not generally recognized as safe for use in our food (see Chapter 13). Therefore, food manufacturers must petition the FDA before they can include them in their products.[11]

Phospholipids

Phospholipids, though present in small amounts, are important in food and in the body because they allow water and fat to mix. They can do this because one side of the molecule

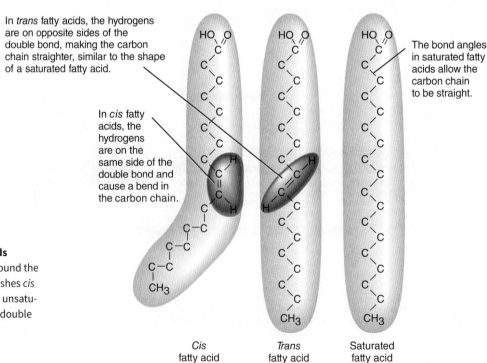

FIGURE 5.6 *Cis* and *trans* **fatty acids**
The orientation of hydrogen atoms around the carbon–carbon double bond distinguishes *cis* fatty acids from *trans* fatty acids. Most unsaturated fatty acids found in nature have double bonds in the *cis* configuration.

---

hydrogenation A process used to make partially hydrogenated oils in which hydrogen atoms are added to the carbon–carbon double bonds of unsaturated fatty acids, making them more saturated. *Trans* fatty acids are formed during the process.

Debate

Coconut Oil: Does a Tablespoon a Day Keep the Doctor Away?

The Issue

Claims abound that coconut oil can protect you from heart disease and help you lose weight. You can add it to your coffee, put it in your smoothie, bake it into goodies, or use it for frying. But adding fat adds calories, and coconut oil, one of the tropical oils, is high in saturated fat. So, why is it being marketed to improve your health?

Coconut oil, like any other oil, provides 9 Calories per gram, or about 120 Calories in a tablespoon. But coconut oil is unusual because it contains a high percentage of triglycerides made up of medium-chain fatty acids, called *medium-chain triglycerides* (*MCTs*). Our bodies metabolize MCTs differently than the long-chain triglycerides that make up most oils. Instead of being absorbed into the lymph, the fatty acids from MCTs are absorbed into the blood and transported directly from the intestines to the liver, where they can be broken down to provide energy. Coconut oil is also high in lauric acid, a saturated fatty acid that raises blood levels of total cholesterol; however, unlike other saturated fatty acids, lauric acid also raises levels of healthy cholesterol.[5]

So, does this mean that a diet high in coconut oil is good for your heart? Unfortunately, data on the effect of coconut oil on heart disease are scarce. We know that the incidence of heart disease is not elevated in Polynesian island populations where coconut oil intake is high, but this does not mean that coconut oil is the reason. The data on coconut oil and blood lipid levels is mixed. Filipino women with the highest coconut oil intake were found to have higher blood levels of good cholesterol.[4] A study that compared low-calorie diets supplemented with coconut oil versus soybean oil found an increase in blood levels of good cholesterol in the coconut oil group.[6] But another study showed that coconut oil caused an increase in blood levels of bad cholesterol as well as good cholesterol compared to olive oil.[7]

What about weight loss? MCT oil is less likely to be stored in adipose tissue. Does this mean that adding coconut oil will make weight loss easier? A study of overweight subjects compared those who consumed 18 grams of MCT oil as part of their daily diet with those who consumed 18 grams of a long-chain triglyceride oil found that the MCT oil group lost more weight and decreased their waist circumference.[8] A study that looked at MCT oil versus olive oil as part of a low-calorie diet showed greater weight loss in the MCT oil group, but the difference was small.[9] In another study, which compared food intake after a premeal dose of MCTs versus long-chain triglycerides, food and calorie intake in the MCT group was reduced.[10] These studies suggest that MCT oil may promote weight loss, but they were done using 100% MCT. Coconut oil is only 60% MCT. It is unclear whether coconut oil would have a similar effect. One study that compared coconut oil and soy oil in addition to a low-calorie diet showed similar weight loss in both groups but a greater reduction in waist circumference in the coconut oil group.[6]

Coconut oil is gaining favor. Vegans are using it as a plant-based alternative to butter on their toast, and chefs are discovering its unique properties in food. But is adding coconut oil likely to improve your health? Adding it to your diet may give you a slight weight loss edge but only if you compensate for the calories it adds by eating less of something else.

Coconut oil contains an unusual blend of fatty acids that may have health benefits.

Think Critically

Will adding coconut oil to a dietary pattern that doesn't meet health recommendations make it a healthy diet?

---

dissolves in water, and the other side dissolves in fat (**Figure 5.7a**). In foods, substances that allow fat and water to mix are referred to as **emulsifiers**. For example, the phospholipids in egg yolks allow the oil and water in cake batter to mix; phospholipids in salad dressings prevent the oil and vinegar in the dressing from separating.

One of the best-known phospholipids is **lecithin** (shown in Figure 5.7a). Eggs and soybeans are natural sources of lecithin. The food industry uses lecithin as an emulsifier in margarine, salad dressings, chocolate, frozen desserts, and baked goods to prevent oil from separating from the other ingredients (**Figure 5.7b**). In the body, lecithin is a major constituent of cell membranes (**Figure 5.7c**). It is also used to synthesize the neurotransmitter acetylcholine, which activates muscles and plays an important role in memory.

Sterols

The best-known sterol is **cholesterol**. It is needed in the body, but because the liver manufactures it, it is not essential in the

---

cholesterol   A sterol, produced by the liver and consumed in the diet, which is needed to build cell membranes and make hormones and other essential molecules.

NUTRITION INSIGHT

FIGURE 5.7 Phospholipids Phospholipid structure supports its function because it allows one end of the molecule to be soluble in fat, while the other end is soluble in water.

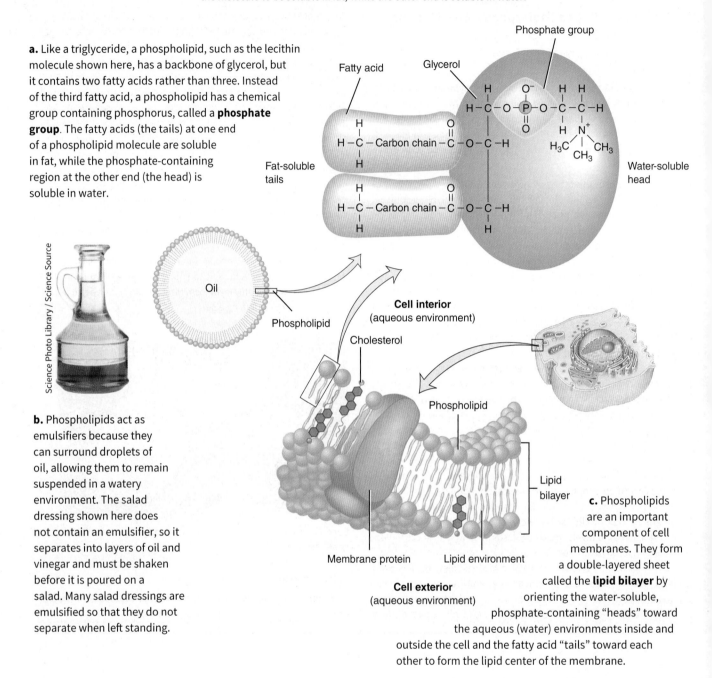

a. Like a triglyceride, a phospholipid, such as the lecithin molecule shown here, has a backbone of glycerol, but it contains two fatty acids rather than three. Instead of the third fatty acid, a phospholipid has a chemical group containing phosphorus, called a **phosphate group**. The fatty acids (the tails) at one end of a phospholipid molecule are soluble in fat, while the phosphate-containing region at the other end (the head) is soluble in water.

b. Phospholipids act as emulsifiers because they can surround droplets of oil, allowing them to remain suspended in a watery environment. The salad dressing shown here does not contain an emulsifier, so it separates into layers of oil and vinegar and must be shaken before it is poured on a salad. Many salad dressings are emulsified so that they do not separate when left standing.

c. Phospholipids are an important component of cell membranes. They form a double-layered sheet called the **lipid bilayer** by orienting the water-soluble, phosphate-containing "heads" toward the aqueous (water) environments inside and outside the cell and the fatty acid "tails" toward each other to form the lipid center of the membrane.

Think Critically

How does the structure of a phospholipid facilitate its function as an emulsifier?

FIGURE 5.8 **Cholesterol** Cholesterol is a sterol that is found only in animal products.

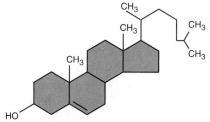

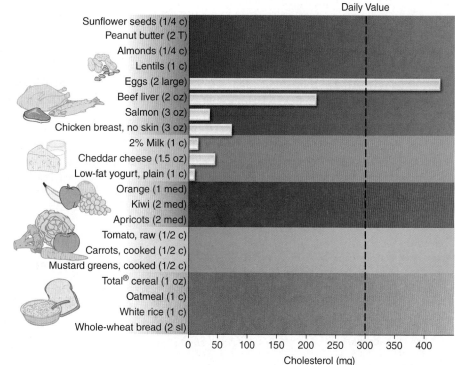

a. The cholesterol structure shown here illustrates the four interconnected rings of carbon atoms that form the backbone structure that is common to all sterols.

b. Egg yolks and organ meats such as liver and kidney are high in cholesterol. Lean red meats and skinless chicken are low in total fat but still contain some cholesterol. Cholesterol is not found in plant cell membranes, so even high-fat plant foods, such as nuts, peanut butter, and vegetable oils, do not contain cholesterol.

diet. More than 90% of the cholesterol in the body is found in cell membranes (see Figure 5.7c). It is also part of myelin, the insulating coating on many nerve cells. Cholesterol is needed to synthesize other sterols, including vitamin D; bile acids, which are emulsifiers in bile; cortisol, which is a hormone that regulates our physiological response to stress; and testosterone and estrogen, which are hormones necessary for reproduction.

In the diet, cholesterol is found only in foods from animal sources (**Figure 5.8**). Plant foods do not contain cholesterol unless it has been added in the course of cooking or processing. Plants do contain other sterols, however, and these **plant sterols** have a role similar to that of cholesterol in animals: They help form plant cell membranes. Plant sterols are found in small quantities in most plant foods; when consumed in the diet, they can help reduce cholesterol levels in the body.

Concept Check

1. **How** are triglycerides and fatty acids related?
2. **What** is the structural difference between saturated and unsaturated fatty acids?
3. **Why** are phospholipids good emulsifiers?
4. **Which** food groups contain the most saturated fat and cholesterol?

5.3 Absorbing and Transporting Lipids

LEARNING OBJECTIVES

1. **Discuss** the steps involved in the digestion and absorption of lipids.
2. **Describe** how lipids are transported in the blood and delivered to cells.
3. **Compare** the functions of LDLs and HDLs.

The fact that oil and water do not mix poses a problem for the digestion and absorption of lipids in the watery environment of the small intestine and the transport of lipids in the blood, which is mostly water. Therefore, the body has special mechanisms that allow it to digest, absorb, and transport lipids.

Digestion and Absorption of Lipids

In healthy adults, most fat digestion and absorption occurs in the small intestine (**Figure 5.9**). Here, the bile acids in bile act as emulsifiers, breaking down large lipid droplets into small globules. The triglycerides in the globules can then be digested by enzymes from the pancreas. The resulting mixture of fatty acids, **monoglycerides**, cholesterol, and bile acids forms smaller droplets called **micelles**, which facilitate absorption (see Figure 5.9). Once inside the mucosal cells of the intestine, the fatty acids, cholesterol, and other fat-soluble substances must be processed further before they can be transported in the blood. The bile acids absorbed from the micelles are returned to the liver to be reused.

The fat-soluble vitamins (A, D, E, and K) are absorbed through the same process as other lipids. These vitamins are not digested but must be incorporated into micelles to be absorbed. The amounts absorbed can be reduced if dietary fat

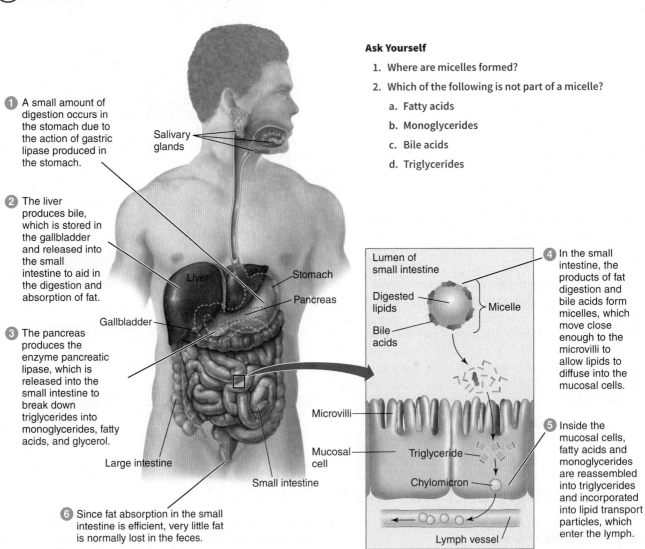

PROCESS DIAGRAM

FIGURE 5.9 Lipid digestion and absorption The bulk of our dietary lipids are triglycerides, which need to be digested before they can be absorbed. The diet also contains smaller amounts of phospholipids, which are partially digested, and cholesterol and fat-soluble vitamins, which are absorbed without digestion.

Ask Yourself

1. Where are micelles formed?
2. Which of the following is not part of a micelle?
   a. Fatty acids
   b. Monoglycerides
   c. Bile acids
   d. Triglycerides

① A small amount of digestion occurs in the stomach due to the action of gastric lipase produced in the stomach.

② The liver produces bile, which is stored in the gallbladder and released into the small intestine to aid in the digestion and absorption of fat.

③ The pancreas produces the enzyme pancreatic lipase, which is released into the small intestine to break down triglycerides into monoglycerides, fatty acids, and glycerol.

④ In the small intestine, the products of fat digestion and bile acids form micelles, which move close enough to the microvilli to allow lipids to diffuse into the mucosal cells.

⑤ Inside the mucosal cells, fatty acids and monoglycerides are reassembled into triglycerides and incorporated into lipid transport particles, which enter the lymph.

⑥ Since fat absorption in the small intestine is efficient, very little fat is normally lost in the feces.

monoglyceride A glycerol molecule with one fatty acid attached.

micelle A particle that is formed in the small intestine when the products of fat digestion are surrounded by bile acids. It facilitates the absorption of lipids.

Absorbing and Transporting Lipids   **123**

is very low or if disease, other dietary components, or medications such as the diet drug Alli, interfere with fat absorption (see Figure 9.20 in Chapter 9).

Transporting Lipids in the Blood

Lipids that are consumed in the diet are absorbed into the intestinal mucosal cells. From here, small fatty acids, which are soluble in water, are absorbed into the blood and travel to the liver for further processing. Long-chain fatty acids, cholesterol, and fat-soluble vitamins, which are not soluble in water, are not absorbed directly into the blood and must be packaged for transport. They are covered with a water-soluble envelope of protein, phospholipids, and cholesterol to form particles called **lipoproteins** (**Figure 5.10a**). Different types of lipoproteins transport dietary lipids from the small intestine to body cells, from the liver to body cells, and from body cells back to the liver for disposal (**Figure 5.10b**).

Transport from the small intestine   After long-chain fatty acids (from the digestion of triglycerides) have been absorbed into the mucosal cells, they are reassembled into triglycerides. These triglycerides, along with cholesterol and fat-soluble vitamins, are packaged with phospholipids and protein to form lipoproteins called **chylomicrons**. Chylomicrons are too large to enter the capillaries in the small intestine, so

FIGURE 5.10 Lipoprotein structure   Lipoproteins allow water insoluble lipids to be transported in the blood.

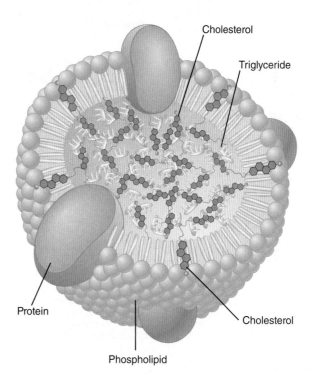

a. A lipoprotein consists of a core of triglycerides and cholesterol surrounded by a shell of protein, phospholipids, and cholesterol. Phospholipids orient with their fat-soluble "tails" toward the interior and their water-soluble "heads" toward the outside. This allows the fat-soluble substances in the interior to travel through the aqueous blood.

Ask Yourself

1. Which lipoprotein contains the highest proportion of cholesterol?
2. Which lipoprotein has the highest proportion of triglyceride?

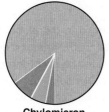

Chylomicron
Chylomicrons are the largest lipoproteins and contain the greatest proportion of triglycerides.

VLDL
VLDLs are smaller than chylomicrons but still contain a high proportion of triglycerides.

LDL
LDLs contain a higher proportion of cholesterol than do other lipoproteins.

HDL
HDLs are high in cholesterol and are the densest lipoproteins due to their high protein content.

b. Lipoproteins vary in size and in the proportions of triglycerides, cholesterol, phospholipid, and protein. Denser lipoproteins have less triglyceride and more protein.

---

lipoprotein   A particle that transports lipids in the blood.

chylomicron   A lipoprotein that transports lipids from the mucosal cells of the small intestine and delivers triglycerides to other body cells.

124 CHAPTER 5 Lipids: Fats, Phospholipids, and Sterols

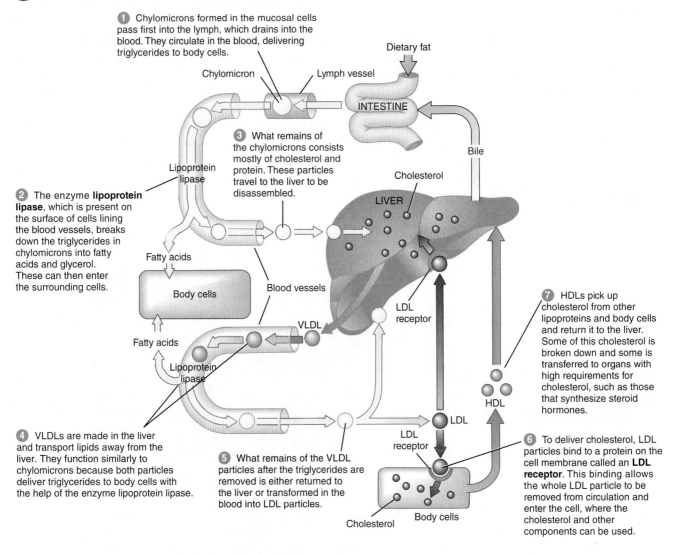

PROCESS DIAGRAM **FIGURE 5.11** **Lipid transport and delivery** Chylomicrons and very-low-density lipoproteins transport triglycerides and deliver them to body cells. Low-density lipoproteins transport and deliver cholesterol, and high-density lipoproteins help return cholesterol to the liver for reuse or elimination.

① Chylomicrons formed in the mucosal cells pass first into the lymph, which drains into the blood. They circulate in the blood, delivering triglycerides to body cells.

② The enzyme **lipoprotein lipase**, which is present on the surface of cells lining the blood vessels, breaks down the triglycerides in chylomicrons into fatty acids and glycerol. These can then enter the surrounding cells.

③ What remains of the chylomicrons consists mostly of cholesterol and protein. These particles travel to the liver to be disassembled.

④ VLDLs are made in the liver and transport lipids away from the liver. They function similarly to chylomicrons because both particles deliver triglycerides to body cells with the help of the enzyme lipoprotein lipase.

⑤ What remains of the VLDL particles after the triglycerides are removed is either returned to the liver or transformed in the blood into LDL particles.

⑥ To deliver cholesterol, LDL particles bind to a protein on the cell membrane called an **LDL receptor**. This binding allows the whole LDL particle to be removed from circulation and enter the cell, where the cholesterol and other components can be used.

⑦ HDLs pick up cholesterol from other lipoproteins and body cells and return it to the liver. Some of this cholesterol is broken down and some is transferred to organs with high requirements for cholesterol, such as those that synthesize steroid hormones.

---

they pass from the intestinal mucosa into the lymph, which then delivers them to the blood. They circulate in the blood, delivering triglycerides to body cells (**Figure 5.11**). To enter the cells, the triglycerides must first be broken down into fatty acids and glycerol, which can diffuse across the cell membrane. Once inside the cells, fatty acids can either be used to provide energy or reassembled into triglycerides for storage.

Transport from the liver The liver can synthesize lipids. Lipids are transported from the liver in **very-low-density lipoproteins (VLDLs)**. Like chylomicrons, VLDLs are lipoproteins that circulate in the blood, delivering triglycerides to body cells (see Figure 5.11). When the triglycerides have been removed from the VLDLs, a denser, smaller particle remains. About two-thirds of these particles are returned to the liver, and the rest are transformed in the blood into **low-density lipoproteins (LDLs)**. LDLs are the primary cholesterol delivery system for cells. They contain a higher proportion of cholesterol than do chylomicrons or VLDLs (see Figure 5.10b). High levels of LDLs in the blood have been associated with an increased risk for heart disease. For this reason, they are sometimes referred to as "bad cholesterol."

Eliminating cholesterol Because most body cells have no system for breaking down cholesterol, cholesterol must be returned to the liver to be eliminated from the body.

---

low-density lipoprotein (LDL) A lipoprotein that transports cholesterol to cells.

This reverse cholesterol transport is accomplished by **high-density lipoproteins (HDLs)** (see Figure 5.11). HDL cholesterol is often called "good cholesterol" because high levels of HDL in the blood are associated with a reduction in the risk of heart disease.

> **Concept Check**
>
> 1. **How** does bile help in the digestion and absorption of lipids?
> 2. **Why** are lipoproteins needed to transport lipids?
> 3. **What** is the primary function of LDLs?

5.4 Lipid Functions

LEARNING OBJECTIVES

1. **List** the functions of lipids in the body.
2. **Explain** why the ratio of omega-3 and omega-6 fatty acids is important in the diet.
3. **Summarize** how and when fatty acids are used to provide energy.
4. **Describe** how fat is stored and how it is retrieved from storage.

Lipids are necessary to maintain health. In our diet, fat is needed to absorb fat-soluble vitamins and is a source of essential fatty acids and energy. In our bodies, lipids form structural and regulatory molecules and are broken down to provide ATP. As discussed earlier, cholesterol plays both regulatory and structural roles: It is used to make steroid hormones, and it is an important component of cell membranes and the myelin coating that is necessary for brain and nerve function.

Most lipids in the body are triglycerides stored in **adipose tissue**, which is body fat that lies under the skin and around internal organs (**Figure 5.12**). The triglycerides in our adipose tissue provide a lightweight energy storage molecule, help cushion our internal organs, and insulate us from changes in temperature. Triglycerides are also found in oils that lubricate body surfaces, keeping the skin soft and supple.

Essential Fatty Acids

Humans are not able to synthesize fatty acids that have double bonds in the omega-6 and omega-3 positions (see Figure 5.4). Therefore, the fatty acids **linoleic acid** (omega-6) and

FIGURE 5.12 **Adipose tissue** The cells that make up our adipose tissue enlarge when we gain weight (store more triglyceride) and shrink when the amount of stored triglyceride decreases.

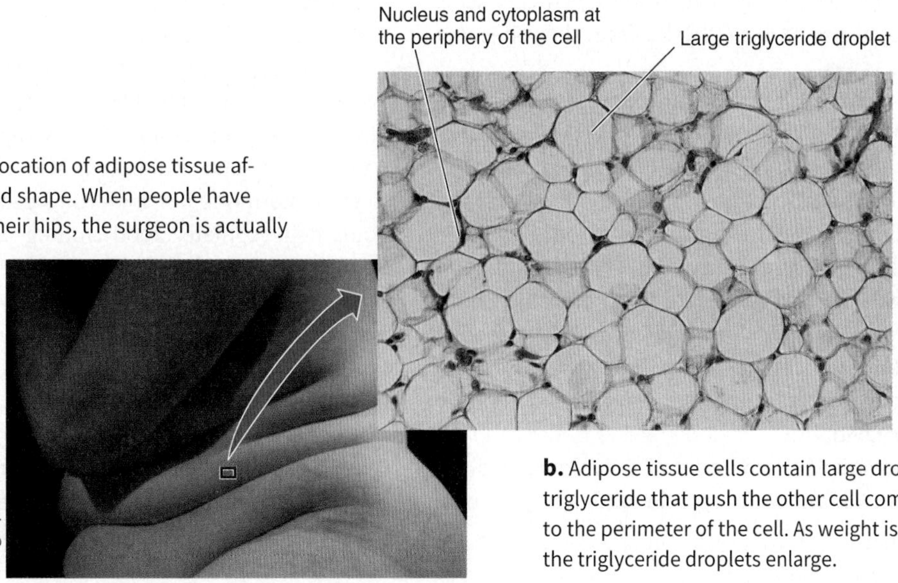

a. The amount and location of adipose tissue affect our body size and shape. When people have liposuction to slim their hips, the surgeon is actually vacuuming out fat cells from the adipose tissue in the region.

b. Adipose tissue cells contain large droplets of triglyceride that push the other cell components to the perimeter of the cell. As weight is gained, the triglyceride droplets enlarge.

---

high-density lipoprotein (HDL) A lipoprotein that picks up cholesterol from cells and transports it to the liver so that it can be eliminated from the body.

FIGURE 5.13 Essential fatty acids and health The omega-6 fatty acid linoleic acid can be used to synthesize **arachidonic acid**. The omega-3 fatty acid α-linolenic acid can be used to synthesize **eicosapentaenoic acid (EPA)** and **docosahexaenoic acid (DHA)**. Arachidonic acid and EPA compete for the enzymes that synthesize eicosanoids so the proportions of essential fatty acids in the diet affect the types of eicosanoids synthesized. These impact overall health through the regulation of factors like blood pressure, blood clotting, and inflammation.

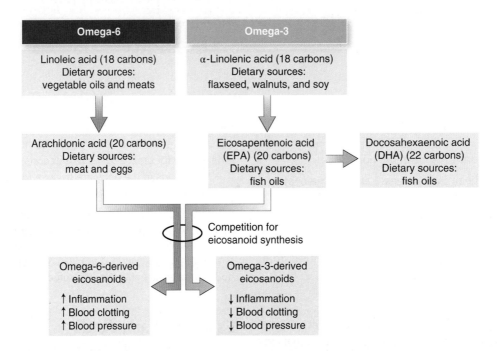

alpha-linolenic acid (α-linolenic acid) (omega-3) are considered essential fatty acids. They must be consumed in the diet because they cannot be made in the body. As illustrated in the top portion of **Figure 5.13**, they can be used to synthesize longer-chain omega-6 and omega-3 fatty acids. If the diet is low in linoleic acid and/or α-linolenic acid, longer-chain fatty acids that the body would normally synthesize from them become dietary essentials as well.

Omega-6 and omega-3 fatty acids are important for health. They are needed for the formation of the phospholipids that give cell membranes their structure and functional properties. Therefore, they are essential for growth, development, and fertility, as well as for maintaining the structure of red blood cells and cells in the skin and nervous system. The omega-3 fatty acid DHA is particularly important in the retina of the eye. Both DHA and the omega-6 fatty acid arachidonic acid are needed to synthesize cell membranes in the central nervous system and are therefore important for normal brain development in infants and young children.

If adequate amounts of linoleic and α-linolenic acid are not consumed, an **essential fatty acid deficiency** results. Symptoms include scaly, dry skin, liver abnormalities, poor wound healing, impaired vision and hearing, and failure to grow in infants. Because the requirement for essential fatty acids is well below the typical intake in the United States, essential fatty acid deficiencies are rare in this country. However, deficiencies have occurred in infants and young children consuming low-fat diets and in individuals unable to absorb lipids.

Getting enough essential fatty acids in your diet prevents deficiency. The ratio of dietary omega-6 to omega-3 fatty acids also affects your health. This is because the omega-6 and omega-3 polyunsaturated fatty acids made from them are used to make hormone-like molecules called **eicosanoids**. Eicosanoids help regulate blood clotting, blood pressure, immune function, and other body processes. The effect of eicosanoids on these functions depends on the fatty acid from which they are made. For example, when the omega-6 fatty acid arachidonic acid is the starting material, the eicosanoid synthesized increases blood clotting; when the omega-3 fatty acid EPA is the starting material, the eicosanoid made decreases blood clotting. The ratio of dietary omega-6 to omega-3 fatty acids affects the balance of the omega-6 and omega-3 fatty acids in the body

---

essential fatty acid deficiency A condition characterized by dry, scaly skin and poor growth that results when the diet does not supply sufficient amounts of linoleic acid and α-linolenic acid.

eicosanoids Regulatory molecules that can be synthesized from omega-3 and omega-6 fatty acids.

and, therefore, the balance of the omega-6- and omega-3-derived eicosanoids produced (see Figure 5.13).

The U.S. diet contains a higher ratio of omega-6 to omega-3 fatty acids than is optimal for health. Increasing consumption of foods that are rich in omega-3 fatty acids increases the proportion of omega-3 eicosanoids. This reduces the risk of heart disease by decreasing inflammation, lowering blood pressure, and reducing blood clotting (see Figure 5.13).[12] The American Heart Association recommends eating two or more servings per week of fish, which is a good source of EPA and DHA, along with plant sources of omega-3 fatty acids, such as walnuts, canola oil, and flaxseed.[13]

Fat as a Source of Energy

Fat is an important source of energy in the body (**Figure 5.14**). Triglycerides that are consumed in the diet can be either used immediately to fuel the body or stored in adipose tissue. Depositing fat in adipose tissue is an efficient way to store energy because each gram of fat provides 9 Calories, compared with only 4 Calories per gram from carbohydrate or protein. This allows a large amount of energy to be stored in the body without a great increase in body size or weight. For example, even a lean man stores over 50,000 Calories in his adipose tissue.

Throughout the day, as we eat and then go for hours without eating, triglycerides are stored and then retrieved from storage, depending on the body's immediate energy needs. For example, after we have feasted on a meal, some triglycerides are stored; then, in the small fasts between meals, some of the stored triglycerides are broken down to provide energy. When the energy consumed in the diet equals the body's energy requirements, the net amount of body fat does not change.

Feasting When we consume more calories than we need, the excess is stored primarily as fat. Excess fat from our diet is packaged in chylomicrons and transported directly from the intestines to the adipose tissue. Because the fatty acids in our body fat come from the fatty acids we eat, what we eat affects the fatty acid composition of our adipose tissue; therefore, if you eat more saturated fat, there will be more saturated fat in your adipose tissue. Excess calories that are consumed as carbohydrate or protein can also be stored as fat, but they are stored less efficiently. This is because the absorbed glucose

PROCESS DIAGRAM

FIGURE 5.14 **Metabolism of fat** Triglycerides are broken down to fatty acids and a small amount of glycerol. The fatty acids provide most of the energy stored in a triglyceride molecule. Fatty acids are transported into the mitochondria, where, in the presence of oxygen, they are broken down to form acetyl-CoA, which can be further metabolized to generate ATP. The glycerol molecules, which contain three carbon atoms, can also be used to generate ATP or small amounts of glucose.

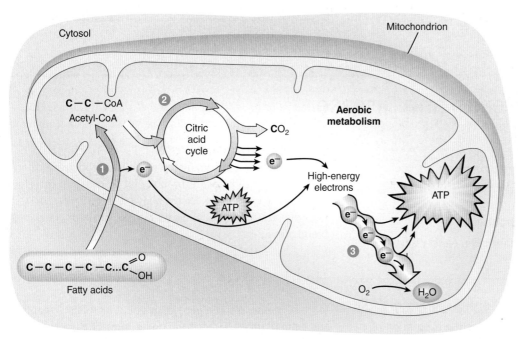

FIGURE 5.15 Feasting and fasting
When we eat more than we need, excess energy is stored as triglycerides. When we don't eat enough, triglycerides in adipose tissue are broken down, releasing fatty acids, which can be used to provide energy.

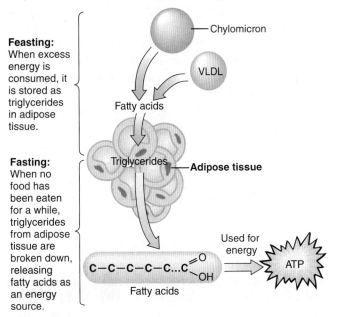

Feasting: When excess energy is consumed, it is stored as triglycerides in adipose tissue.

Fasting: When no food has been eaten for a while, triglycerides from adipose tissue are broken down, releasing fatty acids as an energy source.

Ask Yourself

It is 1 PM, and you have not eaten since 7:30 AM. Where is your body getting energy?

and amino acids must first go to the liver, where they can be used to synthesize fatty acids, which are then assembled into triglycerides, packaged in VLDLs, and transported in the blood to adipose tissue (**Figure 5.15**).

The ability of the body to store excess triglycerides is theoretically limitless. Cells in your adipose tissue can increase in weight by about 50 times, and new fat cells can be made when existing cells reach their maximum size.

Fasting When you eat fewer calories than you need, your body takes energy from its fat stores. In this situation, an enzyme inside the fat cells receives a signal to break down stored triglycerides. The fatty acids and glycerol that result are released directly into the blood and circulate throughout the body. They are taken up by cells and used to produce ATP (see Figures 5.14 and 5.15).

To be used for energy, fatty acids are broken into two-carbon units that form acetyl-CoA. When oxygen and carbohydrate are available, acetyl-CoA can be used to generate ATP (see Figure 5.14). If there is not enough carbohydrate available in cells to allow the acetyl-CoA to enter the citric acid cycle, it is used to make *ketones* (see Chapter 4). Many tissues in the body can use ketones as an energy source. During prolonged fasting, even the brain can adapt itself to use ketones to meet about half of its energy needs. For the other half, the brain continues to require glucose. Fatty acids cannot be used to make glucose, and only a small amount of glucose can be made from the glycerol released from triglyceride breakdown.

Concept Check

1. **Why** would eliminating all fat from your diet cause health problems?
2. **How** does eating fish regularly affect the types of eicosanoids produced in the body?
3. **How** are fatty acids used to produce ATP?
4. **What** happens to excess dietary fat after it has been absorbed?

5.5 Lipids in Health and Disease

LEARNING OBJECTIVES

1. **Describe** the events that lead to the development of atherosclerosis.
2. **Evaluate** your risk of heart disease.
3. **Discuss** the roles of dietary fat in cancer and obesity.

The amount and types of fat you eat can affect your health. A diet that is too low in fat can reduce the absorption of fat-soluble vitamins, slow growth, and impair the functioning of the skin, eyes, liver, and other body organs. Eating the wrong types of fat can contribute to chronic diseases such as heart disease and cancer. Consuming too much fat can increase calorie intake and contribute to extra body fat storage and therefore weight gain and obesity. Excess body fat, in turn, is associated with an increased risk of heart disease, diabetes, and high blood pressure.

Heart Disease

More than one in three people, or about 85 million people in the United States, suffer from some form of **cardiovascular disease**, which is any disease that affects the heart and blood vessels. It is the number-one cause of death for both men and women in the United States.[14] **Atherosclerosis** is a type of cardiovascular disease in which cholesterol is deposited in the artery walls, reducing their elasticity and eventually blocking the flow of blood. The development of atherosclerosis has been linked to dietary patterns that tend to be higher in cholesterol, saturated fat, and *trans* fat and lower in unsaturated fats and fiber.[2]

How atherosclerosis develops

Inflammation, the process whereby the body responds to injury, drives the formation of **atherosclerotic plaque**. For example, cutting yourself triggers an inflammatory response. White blood cells, which are part of the immune system, rush to the injured area, blood clots form, and soon new tissue grows to heal the wound. Similar inflammatory responses occur when an artery is injured, but instead of resulting in healing, they lead to the development of atherosclerotic plaque (**Figure 5.16**). Therefore, the atherosclerotic process begins with an injury, and the response to this injury causes changes in the lining of the artery wall.

The exact cause of the injuries that initiate the development of atherosclerosis is not known but may be related to elevated blood levels of LDL cholesterol, glucose, or the amino acid homocysteine; high blood pressure; cigarette smoking; diabetes; genetic alterations; or infectious microorganisms. The specific cause may be different in different people.

Risk factors for heart disease

Diabetes, high blood pressure, obesity, and high blood levels of LDL cholesterol are considered primary risk factors for heart disease because they directly increase risk. Other factors that affect risk include age, gender, genetics, and lifestyle factors such as smoking, exercise, and diet (**Table 5.1**).

TABLE 5.1 What affects the risk of heart disease?

Risk factor	How it affects risk
Obesity	Obesity increases blood pressure, blood cholesterol levels, and the risk of developing diabetes. It also increases the amount of work the heart must do to pump blood throughout the body.
Diabetes	High blood glucose damages blood vessel walls, contributing to atherosclerosis.
High blood pressure	High blood pressure can damage blood vessel walls, contributing to atherosclerosis. It forces the heart to work harder, causing it to enlarge and weaken over time.
Blood lipid levels	High blood levels of LDL cholesterol and triglycerides, and low levels of HDL cholesterol increase risk (see Appendix F). An individual's specific goal for blood levels of LDL cholesterol depends on his or her other risk factors for heart disease. Cholesterol-lowering medications, called statins, are recommended for individuals with a high overall risk.[16]
Gender	Men and women are both at risk for heart disease, but men are generally affected a decade earlier than are women. This difference is due in part to the protective effect of the hormone estrogen in women. As women age, the effects of menopause—including a decline in estrogen level and a gain in weight—increase heart disease risk.
Age	The risk of heart disease is increased in men age 45 and older and in women age 55 and older.
Family history	Individuals with a male family member who exhibited heart disease before age 55 or a female family member who exhibited heart disease before age 65 are considered to be at increased risk. African Americans have a higher risk of heart disease than the general population, in part due to the high incidence of high blood pressure among African Americans.[14]
Smoking	Tobacco use increases the risk of heart disease by itself. It also increases blood pressure, decreases exercise tolerance, and increases the tendency for blood to clot.
Physical activity	Regular physical activity decreases risk by reducing blood pressure, lowering LDL cholesterol, increasing healthy HDL cholesterol levels, reducing the risk of diabetes, and promoting a healthy weight.
Diet	The risk of heart disease is decreased by a dietary pattern that emphasizes fruits, vegetables, whole grains, low-fat dairy products, poultry, fish, and nuts and that limits red meat and sugary foods and beverages. This pattern is low in saturated fat and *trans* fat and high in fiber.

---

atherosclerosis A type of cardiovascular disease that involves the buildup of fatty material in the artery walls.

atherosclerotic plaque Cholesterol-rich material that is deposited in the arteries of individuals with atherosclerosis.

130 CHAPTER 5 Lipids: Fats, Phospholipids, and Sterols

PROCESS DIAGRAM **FIGURE 5.16** **Development of atherosclerosis** The inflammation that occurs in response to an injury to the artery wall precipitates the development of atherosclerotic plaque. The buildup of plaque can eventually lead to a heart attack or stroke.

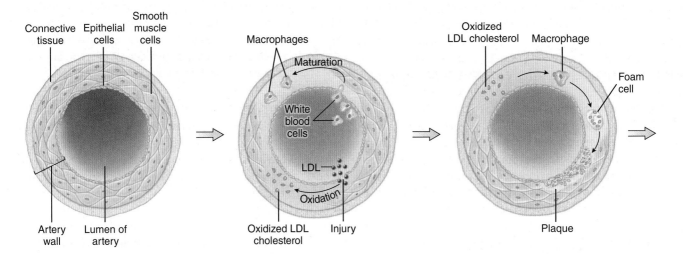

1 Normal Artery

The wall of a normal, healthy artery is lined with a layer of epithelial cells that are surrounded by smooth muscle.

2 Damaged Artery

Damage to the lining of an artery causes inflammation and begins the process of plaque formation. The injury attracts white blood cells, which mature into macrophages, and makes the lining more permeable to LDL particles. Inside the artery wall, LDL is oxidized to form **oxidized LDL cholesterol**.

3 Plaque Formation

Inside the artery wall, **macrophages** fill up with oxidized LDL cholesterol and are transformed into **foam cells**. These cells get so full that they burst, depositing cholesterol.[15] The accumulation of cholesterol and proteins forms a plaque.

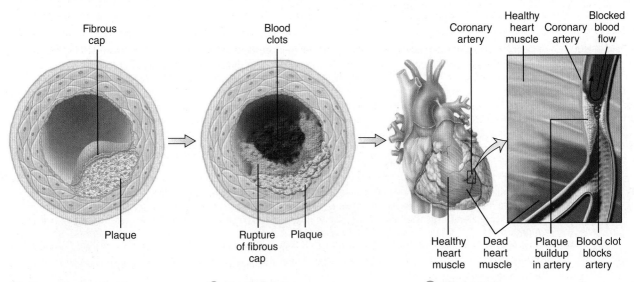

4 Plaque Enlargement

As inflammation continues, plaque builds up, causing the artery to narrow and lose its elasticity. A cap of smooth muscle cells and fibrous proteins forms over the plaque, walling it off from the lumen of the artery.

5 Plaque Rupture

If the inflammation continues, the fibrous cap covering the plaque degrades. If the cap ruptures or erodes, blood clots can rapidly form around it. The blood clots can completely block the artery at that spot or break loose and block an artery elsewhere.

6 Heart Attack

If blood flow in a coronary artery is blocked, a heart attack results. Heart muscle cells are cut off from their blood supply and die, causing pain and reducing the heart's ability to pump blood. If blood flow to the brain is interrupted, a stroke results. Brain cells are cut off from their blood supply and die.

Think Critically

How is blood LDL cholesterol related to the formation of atherosclerotic plaque?

Dietary patterns and heart disease risk Evidence now suggests that a healthy dietary pattern can reduce the risk of heart disease. Heart-healthy dietary patterns are rich in whole grains, fruits, vegetables, fish, and nuts, which contain nutrients and other dietary components that protect heart health (**Figure 5.17**).[17] The heart-protective effect of healthy dietary patterns has prompted nutrition experts to promote the USDA Food Patterns and their vegetarian adaptations, healthy Mediterranean-style eating patterns, and the DASH dietary pattern (see Chapters 2 and 8 and Appendix B) to reduce the risk of heart disease in the United States.[2, 18]

Much of the impact that a dietary pattern has on heart disease risk depends on the abundance of nutrients and other dietary components that affect blood cholesterol levels (see Figure 5.17). For example, saturated fatty acids from high-fat meats and whole-fat dairy products can cause an increase in blood levels of total and LDL cholesterol and hence increase heart disease risk. Soluble fibers and plant sterols reduce risk by lowering blood cholesterol. Replacing foods that are high in cholesterol-raising fats with foods that provide omega-3 fatty acids, monounsaturated fat, soluble fiber, and plant sterols, which have been shown to lower total and LDL cholesterol, can reduce the risk of heart disease.[17]

Nutrients and other dietary components can also affect heart disease risk through mechanisms unrelated to blood cholesterol level. For example, adequate intakes of vitamin B_6, vitamin B_{12}, and folate help protect against heart disease because they help maintain low blood levels of the amino acid homocysteine (discussed further in Chapter 7). Elevated homocysteine levels are associated with a higher incidence of heart disease.[26] Much of the heart-protective effect of omega-3 fatty acids, such as those found in fish, is due to the eicosanoids made from omega-3 fatty acids; these eicosanoids prevent the growth of atherosclerotic plaque, reduce blood clotting and blood pressure, and decrease inflammation.[19] The heart protective effect of plant foods is due not only to soluble fiber, which can reduce blood cholesterol, but also to the antioxidant vitamins, minerals, and phytochemicals they provide. **Antioxidants** decrease the oxidation of LDL cholesterol and, therefore, may prevent the development of plaque in artery walls (see Figure 5.16).[27]

Diet and Cancer

Cancer is the second-leading cause of death in the United States. As with heart disease, there is evidence that the risk of cancer can be reduced with changes in diet and activity patterns.[28] Populations consuming diets that are high in fruits and vegetables tend to have a lower risk of cancer than populations with lower intakes. These foods are rich in antioxidants such as vitamin C, vitamin E, and β-carotene. In contrast, in populations that consume diets that are high in animal fats and red meat, the incidence of certain cancers is higher.[29]

The good news is that the same dietary pattern that protects you from cardiovascular disease may also reduce the risk of certain forms of cancer. For example, the Mediterranean diet, which is high in monounsaturated fat from olive oil and omega-3 fatty acids from fish, is associated with a low risk of cancers of the breast, colon, prostate, and upper digestive and respiratory tracts.[30, 31] *Trans* fatty acids, on the other hand, not only increase the risk of heart disease but are also believed to increase mortality from breast cancer.[32]

NUTRITION INSIGHT

FIGURE 5.17 Diet and heart disease risk The combination of foods we consume and the nutrients they provide make up our dietary pattern. The overall pattern of food intake, often referred to as the total diet, has a greater impact on the risk of heart disease than individual nutrients and foods.

Individual nutrients and food components

Factors That Increase Risk	Factors That Reduce Risk
Saturated fat	Polyunsaturated fat
Trans fat	Monounsaturated fat
Sodium	Fiber
Excess sugar	B vitamins
Excess energy	Antioxidants
Excess alcohol	Moderate alcohol

Nutrients are found in foods.

Whole foods

Fish is high in omega-3 fatty acids, which protect against heart disease by lowering blood triglyceride levels, reducing blood pressure, decreasing lipid deposition in the artery wall, modulating heartbeats, and reducing inflammation.[19]

Whole foods

Nuts are a good source of monounsaturated fat, which lowers LDL cholesterol and makes it less susceptible to oxidation. Nuts are also high in omega-3 fatty acids, fiber, vegetable protein, antioxidants, and plant sterols. Diets containing nuts may improve blood lipids, reduce inflammation, and have beneficial effects on the cells lining the artery wall.[20]

(continues)

---

antioxidant A substance that decreases the adverse effects of reactive molecules.

FIGURE 5.17 *(continued)*

Whole foods

Oatmeal and brown rice are good sources of soluble fiber, which has been shown to reduce blood cholesterol levels. Whole grains also provide omega-3 fatty acids, B vitamins, and antioxidants, as well as phytochemicals that may protect against heart disease.[21]

Whole foods

Plant sterols resemble cholesterol chemically, making it difficult for the digestive tract to distinguish them from cholesterol. Consuming plant sterols reduces cholesterol absorption, lowering total and LDL cholesterol levels.[22] Plant sterols are found in vegetable oils, nuts, seeds, cereals, legumes, and many fruits and vegetables and are added to specialized products such as margarines, salad dressings, and orange juice.

Whole foods

Modest consumption of dark chocolate is associated with reduced risk of heart disease. This effect is attributed to the phytochemicals in dark chocolate.[23] In addition, most of the fat in chocolate is from stearic acid, a saturated fatty acid that does not cause an increase in blood levels of LDL cholesterol.

Whole foods

Moderate alcohol consumption—that is, one drink a day for women and two a day for men (one drink is equivalent to 5 ounces of wine, 12 ounces of beer, or 1.5 ounces of distilled spirits)—reduces blood clotting and increases HDL cholesterol.[24] Higher alcohol intake increases the risk of heart disease and causes other health and societal problems.

Whole foods

Soy-based diets are associated with a reduced risk of heart disease. Soy has a small LDL cholesterol-lowering effect but may have anti-inflammatory properties that help lower the risk of heart disease.[25]

Foods make up dietary patterns →

Dietary patterns

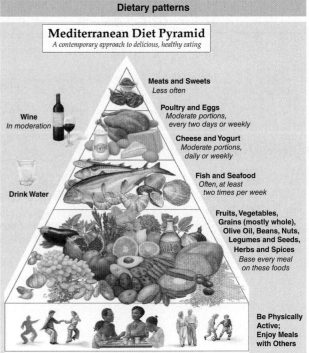

The Mediterranean-style eating pattern, illustrated here as a pyramid, is associated with low rates of heart disease. This pattern is based on fruits, vegetables, legumes, and olive oil and limits meats and sweets. It is one of several dietary patterns that promote heart health.

Dietary Fat and Obesity

Excess dietary fat consumption can contribute to weight gain and obesity. One reason is that fat contains 9 Calories/gram, more than twice the calorie content of carbohydrate or protein. Therefore, a high-fat meal contains more calories in the same volume than does a lower-fat meal. Because people have a tendency to eat a certain weight or volume of food, consuming meals that are high in fat leads to more calories being consumed.[33, 34] Dietary fat may also contribute to weight gain because it is stored very efficiently as body fat.

Despite the fact that fat is fattening, the fat content of the U.S. diet is unlikely to be the sole reason for the high rate of obesity in the United States.[35] Weight gain occurs when energy intake exceeds energy expenditure, regardless of whether the extra energy comes from fat, carbohydrate, or protein. The increasing prevalence of overweight and obesity in the United States and worldwide is likely due to a general increase in calorie intake combined with a decrease in energy expenditure.[2]

Concept Check

1. **How** does atherosclerotic plaque lead to heart attack?
2. **What** are three dietary factors that increase the risk of heart disease?
3. **Why** might eating a high-fat diet contribute to weight gain?

5.6 Meeting Lipid Needs

LEARNING OBJECTIVES

1. **Discuss** the recommendations for fat and cholesterol intake.
2. **Choose** heart-healthy foods from each section of MyPlate.
3. **Use** food labels to choose foods containing healthy fats.

The amount of fat the body requires from the diet is small, but a diet that provides only the minimum amount would be very high in carbohydrate, would not be very palatable, and would not necessarily be any healthier than diets with more fat. Therefore, the recommendations for fat intake focus on meeting essential fatty acid needs and choosing the amounts and types of fat that promote health and prevent disease.

Fat and Cholesterol Recommendations

The DRIs recommend a total fat intake of 20 to 35% of calories for adults. The percentage of calories from fat in a diet can be calculated by multiplying the number of grams of fat by 9 Calories/gram to determine the Calories from fat and then dividing this by the total Calories in the diet and multiplying by 100. For example, a diet that provides 65 grams of fat has 585 Calories from fat (65 g × 9 Calories/g = 585 Calories from fat). In a 2000-Calorie diet, this is equal to 29% of calories from fat (585 Calories from fat ÷ 2000 total Calories × 100 = 29% of calories from fat). A small proportion of the fat in the diet needs to come from the essential fatty acids. The AI for linoleic acid is 12 g/day for women and 17 g/day for men. You can meet your requirement by consuming a half-cup of almonds or 2 tablespoons of corn oil. For α-linolenic acid, the AI is 1.1 g/day for women and 1.6 g/day for men. Your requirement can be met by eating a quarter-cup of walnuts or 1 tablespoon of ground flaxseed. Consuming these amounts provides the recommended ratio of linoleic to α-linolenic acid of between 5:1 and 10:1.[36]

To reduce the risk of heart disease, the 2015–2020 Dietary Guidelines recommend limiting saturated fat intake to less than 10% of total calories and minimizing *trans* fat intake. The Guidelines do not place a limit on dietary cholesterol intake but recommend keeping intake as low as possible (see *What a Scientist Sees: Are Eggs OK?*).[2] The American Heart Association recommends that adults with elevated cholesterol consume only 5 to 6% of calories from saturated fat and as little *trans* fat as possible.[18]

Children need more fat than adults to allow for growth and development, so their acceptable ranges of fat intake are higher: 30 to 40% of calories for ages 1 to 3 and 25 to 35% of calories for ages 3 to 18, compared to 20 to 35% for adults. Like adults, adolescents and children over age 2 should consume a diet that is low in saturated and *trans* fat.[2]

Choosing Fats Wisely

The typical U.S. diet falls within the recommended 20 to 35% of calories from fat. However, many Americans exceed the recommended limit on saturated fat and don't get enough omega-3 polyunsaturated fatty acids.[2] Shifting the sources of dietary fat can improve the proportion of healthy fats in your diet (see *What Should I Eat?*). Limiting sources of solid fats in the diet helps reduce intake of saturated fat, cholesterol, and *trans* fat. Saturated fat can be removed by trimming the fat from meat and removing the skin from poultry. Replacing

What a Scientist Sees

Are Eggs OK?

Eggs contain more cholesterol than any other individual food, about 200 mg per egg. Because elevated cholesterol in the blood increases the risk of heart disease, it was thought that a diet high in cholesterol would also increase risk. Therefore, since the 1960s dietary recommendations have told us to limit our egg consumption. However, most studies do not find a relationship between dietary cholesterol or egg consumption and heart disease.[37] As a result, the 2015–2020 Dietary Guidelines did away with the recommendation to limit dietary cholesterol to 300 mg/day.[2] Consumers saw this as an opportunity to enjoy a delicious plate of scrambled eggs without cholesterol anxiety (see breakfast A).

an egg, though high in cholesterol, contains less than 2 g of saturated fat and only about 75 calories, and it provides high-quality protein, vitamin D, vitamin A, calcium, and the carotenoids lutein and zeaxanthin (antioxidants that protect your eyes). The bacon, sausage, gravy, and even the biscuit are high in saturated fat, which does increase blood cholesterol and the risk of heart disease. Meals like this contribute to a dietary pattern that is high in saturated fat and calories and low in plant-based foods—a pattern that increases the risk of heart disease. A meal that contains eggs but also limits saturated fat and includes whole grains, fruits, and vegetables would meet the recommendations for a healthy eating pattern (see breakfast B).

Breakfast A

This breakfast is high in saturated fat as well as cholesterol and lacks whole grains, fruits, and vegetables.

When a nutrition scientist looks at this meal, she sees not just a plate of eggs but also bacon, sausage, gravy, and biscuits, without any whole grains, fruits, or vegetables. The scientist recognizes that

Breakfast B

This breakfast is high in cholesterol but low in saturated fat and includes whole grains, fruits, and vegetables.

Think Critically

Why do you think the Dietary Guidelines removed the limit on dietary cholesterol but still recommend that people eat as little cholesterol as possible?

What Should I Eat?

Healthy Fats

Limit your intake of cholesterol, *trans* fat, and saturated fat.

- Choose low-fat milk and yogurt.
- Trim the fat from your meat and serve chicken and fish but don't eat the skin.
- Cut in half your usual amount of butter and use soft rather than stick margarine.
- Check the ingredient list to avoid hydrogenated fats.

Select sources of polyunsaturated and monounsaturated fats.

- Snack on nuts and seeds.
- Add olives and avocados to your salads.
- Choose olive, peanut, or canola oil for cooking and salad dressing.
- Use corn, sunflower, or safflower oil for baking.

Up your omega-3 intake

- Sprinkle ground flaxseeds on your cereal or yogurt.
- Have a serving of mackerel, lake trout, sardines, tuna, or salmon.
- Pick a leafy green vegetable with dinner.
- Add walnuts to your salad or cereal.

 Use iProfile to find the varieties of nuts and fish that are highest in omega-3 fatty acids.

full-fat dairy products with low-fat ones and fatty cuts of meat and high-fat processed meats with lean meats, seafood, and beans also helps reduce saturated fat and cholesterol intake. Eating more nuts and avocados and cooking with canola and olive oils boost monounsaturated fat intake, and eating more fish, nuts, and flaxseed increases intake of omega-3 fatty acids.

Most of the *trans* fat in the American diet comes from foods containing hydrogenated oils. As food manufacturers replace partially hydrogenated oils with fats that do not contain synthetic *trans* fat, the amount of *trans* fat in our diet has been declining. This decline will continue as the FDA restriction on the use of hydrogenated oils is implemented.[11] Avoiding foods such as hard margarines and baked goods that contain partially hydrogenated fats can ensure a low *trans* fat intake.

Making wise MyPlate choices Your choices from each food group can have a significant impact on the amounts and types of fats in your diet (**Figure 5.18**). Generally, grains, fruits, and vegetables are low in total fat and saturated fat. However, choices from these groups need to be made with care to avoid fats that are added in processing or preparation. Smart choices from the protein group and the dairy group can reduce your intake of unhealthy fats (see *Thinking it Through*).

Oils are not a food group but are a part of a healthy eating pattern. Therefore, MyPlate recommends that we consume fats from food such as olives, nuts, fish, and vegetable oils, which provide unsaturated fats, rather than butter, lard, and high-fat dairy products and meats, which are high in unhealthy saturated fat. MyPlate suggests an oil allowance that meets recommendations for healthy fats without exceeding calorie needs. For example a 20-year-old women could consume the equivalent of 5 teaspoons oil by including foods such as peanut butter, soft margarine, and salad dressings with her meals.

Looking at food labels Food labels are an accessible source of information about the fat content of packaged foods. Both the current and planned Nutrition Facts panels

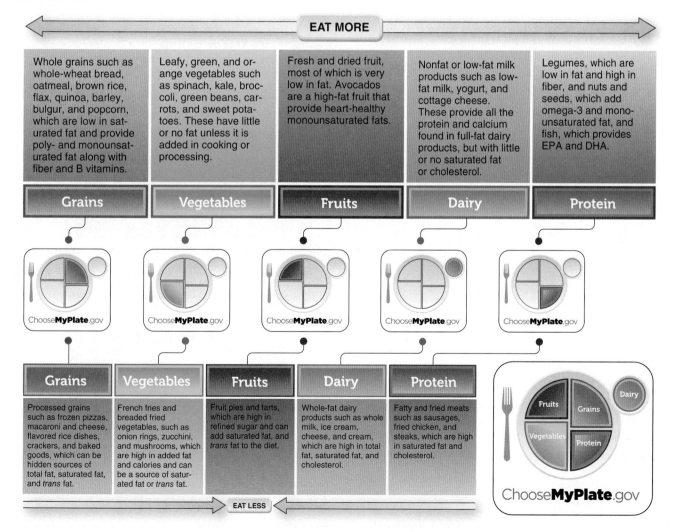

FIGURE 5.18 Healthy MyPlate choices Making healthy choices from each of the MyPlate food groups will provide a diet that is high in mono- and polyunsaturated fats and limited in saturated fat, *trans* fat, and cholesterol.

Thinking It Through

A Case Study on Meeting Fat Recommendations

Isabel has a busy schedule—working full time and going to college. She picks up breakfast and lunch on the run and makes a quick dinner when she gets home at night. She would like to improve her diet by choosing healthier fats. An analysis of Isabel's total fat and saturated fat intake is shown in the table.

Food	Serving	Fat (g)	Saturated Fat (g)
Breakfast			
Bran muffin	1 large	6	2.6
Margarine	2 tsp	7.6	1.4
Coffee	1 cup	0	0
Whole milk	1 cup	8	5
Lunch			
Big burger	1 (2 oz bun, 5 oz beef, 1 oz cheese, 2 T dressing)	30	12.5
French fries	small order	11	1.5
Diet Coke	16 oz	0	0
Snack			
Apple	1 med	0	0
Dinner			
Chicken Nuggets	4 oz	19.8	9
Rice a roni	¾ cup	9	2
Hot tea	8 oz	0	0
Apple pie	1 slice	19	4.7
TOTAL		110.4	38.7

If Isabel's diet provides 2200 Calories, what is her percentage of calories from fat? Saturated fat? Are these within the recommended range?

Your answer:

Suggest two foods you could substitute for her current choices at breakfast and two at lunch that would reduce Isabel's total fat and saturated fat intake.

Your answer:

Suggest a dinner that would be lower in saturated fat than Isabel's current dinner and would provide omega-3 fatty acids.

Your answer:

Fat is not the only factor in a healthy diet. Isabel compares her current diet to her MyPlate recommendations.

FOOD GROUP	MyPlate Recommendations	ISABEL'S DIET
Grains	7 oz (3.5 whole grains)	8 oz (3 whole grains)
Vegetables	3 cups	½ cup
Fruits	2 cups	1 cup
Dairy	3 cups	2 cups
Protein	6 oz	7 oz

Isabel's weight is stable on this diet. How can she be short on foods from three food groups (fruits, vegetables, and dairy) and still be getting the right number of calories?

Your answer:

Which of the snacks below is a better choice for Isabel based on the MyPlate recommendations and the type of fat they provide? Why?

Carrots served with hummus, a traditional Middle Eastern food made from chickpeas, olive oil, and spices.

A whole-grain bagel served with cream cheese.

Your answer:

(Check your answers in Appendix L)

provide information about amounts and types of fat, and the ingredient list indicates the sources of fat—for example, whether a food contains corn oil, soybean oil, coconut oil, or partially hydrogenated vegetable oil (**Figure 5.19**). Nutrient content claims such as "low fat," "fat free," and "low saturated fat" on food labels can also be used to identify foods that help you meet the recommendations for fat intake (**Figure 5.20a** and Appendix H). The terms *lean* and *extra lean* are used to describe the fat content of fresh meats, such as pork, beef, and poultry, as well as packaged meats, such as hot dogs and lunch meat. "Lean" means that the meat contains less than 10% fat by weight, and "extra lean" means that it contains less than 5% fat by weight. Ground beef often includes a claim that it is a certain "percent lean," but percent lean is not the same as lean, as shown in **Figure 5.20b**. Health claims can also help you choose foods that meet your nutritional goals. For example, foods that are low in saturated fat and cholesterol may contain a health claim stating that they help reduce the risk of heart disease.

The role of fat replacers
People often choose low-fat and reduced-fat products in order to reduce the total fat and calories in their diets. Some of these foods, such as low-fat and nonfat milk and yogurt, are made by simply removing the fat, but in other products, the fat is replaced with ingredients that mimic the taste and texture of the fat. Some reduced-fat foods contain added sugars to improve the taste and texture. Some contain soluble fiber or modified proteins that simulate fat, and others contain fats that have been altered to have fewer calories or reduce or prevent absorption (**Figure 5.21**).[38] A problem with nonabsorbable fats is that they reduce the absorption of the fat-soluble substances in the diet, including the fat-soluble vitamins, A, D, E, and K. To avoid depleting these vitamins, products made with the nonabsorbable fat substitute Olestra

FIGURE 5.19 Lipids on food labels Both the current and planned Nutrition Facts panels, along with the ingredient list, help identify the types and sources of lipids in packaged foods. The planned label, on the right, has less emphasis on the amount of total fat; the *Calories from Fat* is not included and the Daily Value has been increased from 65 grams in a 2000-Calorie diet to 78 grams.

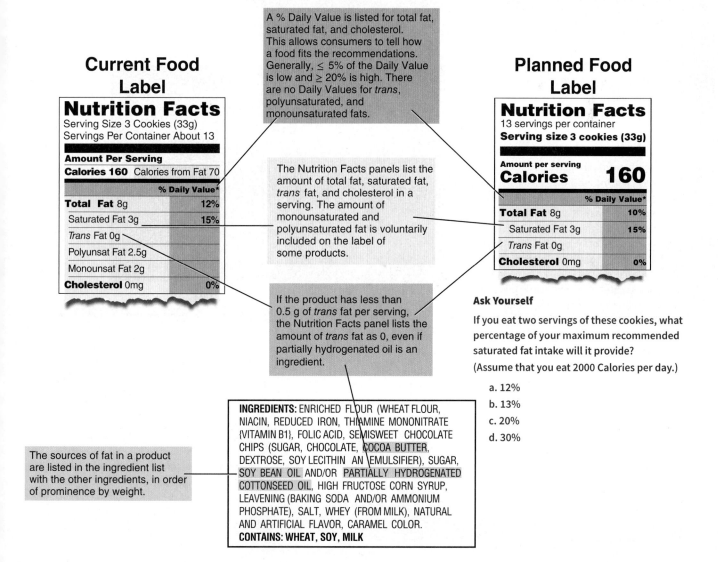

Ask Yourself

If you eat two servings of these cookies, what percentage of your maximum recommended saturated fat intake will it provide?
(Assume that you eat 2000 Calories per day.)

a. 12%
b. 13%
c. 20%
d. 30%

FIGURE 5.20 **Understanding Fat Descriptors** Food labeling regulations have developed standard definitions for descriptors used on food labels, but some of these terms can still be misleading (see Appendix H).

Fat free: Contains < 0.5 g fat per serving

Reduced fat: Contains at least 25% less fat per serving than the regular or reference product

Low fat: Contains ≤ 3 g fat per serving

a. Descriptors such as those shown here help consumers identify products that meet fat recommendation for a healthy diet. However, even foods low in fat or saturated fat can be high in sodium or added sugars.

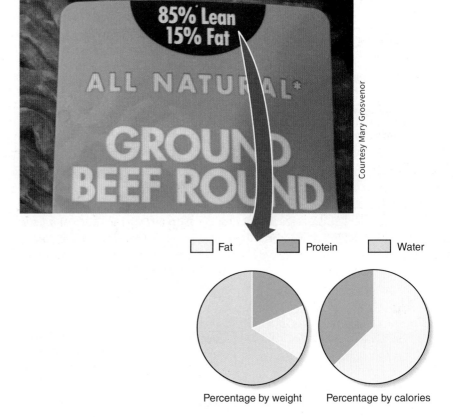

b. Ground beef that is 85% lean looks like a good choice. However, "percent lean" refers to the weight of the meat that is lean, not the percentage of calories provided by lean meat. So when the label says it is 85% lean, it means that 15% of the weight of the meat is fat, or that there are 15 g of fat in 100 g (3.5 oz) of raw meat. Because fat contains 9 Calories/gram, a relatively small percentage of fat by weight can make a large calorie contribution. In this example, the 15% fat by weight in the ground beef contributes 63% of the calories in the meat.

FIGURE 5.21 Fat replacers Carbohydrates and proteins added to replace fat add calories to foods. In some cases, so much is added that the low-fat food is not much lower in calories than the original product. Some fat replacers are made from fats that have been modified to reduce how well they can be digested and absorbed. The number of calories they provide depends on how much is absorbed.

The artificial fat Olestra (sucrose polyester) is made from sucrose with fatty acids attached. Olestra cannot be digested by either human enzymes or bacterial enzymes in the gastrointestinal tract. Therefore, it is excreted in the feces without being absorbed.

Soluble fibers such as pectins and gums are often used in baked goods, as well as salad dressings, sauces, and ice cream, to mimic the texture that fat provides. They reduce the amount of fat in a product and at the same time add soluble fiber.

The sugar sucrose is usually added to low-fat and nonfat baked goods to improve flavor and add volume. Sucrose adds 4 Calories per gram.

Protein-based fat replacers are made from whey protein or milk and egg proteins processed to mimic the creamy texture of fat.[38] They are used in frozen desserts, cheese foods, and other products but cannot be used for frying because they break down at high temperatures.

have been fortified with them. However, these products are not fortified with β-carotene and other fat-soluble substances that may be important for health. Another potential problem with Olestra is that it can cause abdominal cramping and loose stools in some individuals because it passes into the colon without being digested.

Will using low-fat and reduced-fat products improve your diet? Some low-fat foods make an important contribution to a healthy diet. Low-fat dairy products are recommended because they provide all the essential nutrients contained in the full-fat versions but have fewer calories and less saturated fat and cholesterol. Using these products increases the nutrient density of the diet as a whole. However, not all reduced-fat foods are nutrient dense. Low-fat baked goods often have more sugar than the full-fat versions because extra sugar is needed to add volume and make up for the flavor that is lost when the fat is removed. Some are just lower-fat versions of nutrient-poor choices, such as cookies and chips. If these reduced-fat desserts and snack foods replace whole grains, fruits, and vegetables, the resulting diet could be low in fat but also low in fiber, vitamins, minerals, and phytochemicals.

Using low-fat foods does not necessarily transform a poor diet into a healthy one or improve overall diet quality, but if used appropriately, fat-modified foods can be part of a healthy diet. For example, if a low-fat salad dressing replaces a full-fat version, it allows you to enhance the appeal of a nutrient-rich salad without as much added fat and calories from the dressing. Low-fat products can also be used in conjunction with weight-loss diets because they are often lower in calories. But check the label. Although most are lower in calories, they are by no means calorie free and cannot be consumed liberally without adding calories to the diet and possibly contributing to weight gain.

Concept Check

1. **How** much fat is recommended in a healthy diet?
2. **Which** foods are highest in saturated fat?
3. **How** can labels help you identify foods that are low in saturated and *trans* fat?

Summary

1 Fats in Our Food   114

- Fat adds calories, texture, and flavor to foods such as those in the photo. Some of the fats we eat are visible, but others are hidden. The type of fat as well as the overall dietary pattern can affect health risk.

Figure 5.2 Hidden fat

2 Types of Lipids   115

- **Lipids** are a diverse group of organic compounds, most of which do not dissolve in water. **Triglycerides**, commonly referred to as fat, are the type of lipid that is most abundant in our food and in our **adipose tissue**. As shown here, a triglyceride contains three **fatty acids** attached to a molecule of glycerol.

Figure 5.3 Triglycerides

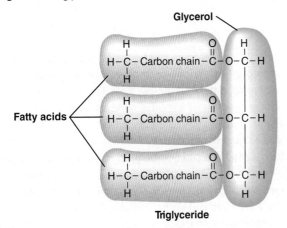

- The structure of fatty acids affects their chemical properties and functions in the body. Each carbon atom in the carbon chain of a **saturated fatty acid** is attached to as many hydrogen atoms as possible, so no carbon–carbon double bonds form. Saturated fatty acids are most abundant in animal products. Exceptions include saturated plant oils often called **tropical oils**. A **monounsaturated fatty acid** contains one carbon–carbon double bond. A **polyunsaturated fatty acid** contains more than one carbon–carbon double bond. The location of the first double bond determines whether it is an **omega-3** or **omega-6 fatty acid**. The orientation of hydrogen atoms around a carbon–carbon double bond distinguishes *cis* fatty acids from ***trans* fatty acids**. **Hydrogenation** transforms some carbon–carbon double bonds to the *trans* configuration.

- A **phospholipid** contains a **phosphate group** and two fatty acids attached to a backbone of glycerol. One end of the molecule is water soluble, and one end is fat soluble. Phospholipids therefore make good **emulsifiers**. In the human body, they are an important structural component of cell membranes and **lipoproteins**.

- **Sterols**, of which **cholesterol** is the best known, are made up of multiple chemical rings. Cholesterol is made by the body and consumed in animal foods in the diet. In the body, it is a component of cell membranes and is used to synthesize vitamin D, bile acids, and some hormones.

3 Absorbing and Transporting Lipids   121

- In the small intestine, muscular churning mixes chyme with bile from the gallbladder to break fat into small globules. This allows pancreatic lipase to access triglycerides for digestion. The products of triglyceride digestion, cholesterol, phospholipids, and other fat-soluble substances combine with bile to form **micelles**, as depicted here, which facilitate the absorption of these materials.

Figure 5.9 Lipid digestion and absorption

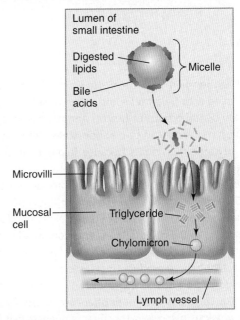

- Lipids absorbed from the intestine are packaged with protein to form **chylomicrons**. The triglycerides in chylomicrons are broken down by **lipoprotein lipase** on the surface of cells lining the blood vessels. The fatty acids released are taken up by surrounding cells, and what remains of the chylomicrons is taken up by the liver.

- **Very-low-density lipoproteins (VLDLs)** are synthesized by the liver. With the help of lipoprotein lipase, they deliver triglycerides to body cells. **Low-density lipoproteins (LDLs)** deliver cholesterol to tissues by binding to **LDL receptors** on the cell surface. **High-density lipoproteins (HDLs)** help remove cholesterol from cells and transport it to the liver for disposal.

4 Lipid Functions   125

- Dietary fat is needed for the absorption of fat-soluble vitamins and to provide essential fatty acids. In the body, triglycerides in adipose tissue provide a concentrated source of energy and insulate the body against shock and temperature changes. **Essential fatty acids** are needed for normal structure and function of cell membranes, particularly those in the retina and central nervous system. Omega-6 and omega-3 polyunsaturated fatty acids are used to synthesize **eicosanoids**, which help regulate blood clotting, blood pressure, immune function, and other body processes. The ratio of dietary omega-6 to omega-3 fatty acids affects the balance of omega-6 and omega-3 eicosanoids made and hence their overall physiological effects.

- Throughout the day, triglycerides are continuously stored in adipose tissue and then broken down to release fatty acids, as shown in the illustration, depending on the immediate energy needs of the body. To generate ATP from fatty acids, the carbon chain is broken into two carbon units that form acetyl-CoA, which can then be metabolized in the presence of oxygen.

Figure 5.15 Feasting and fasting

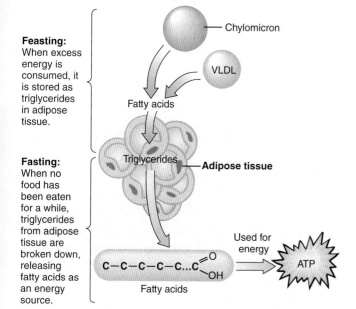

5 Lipids in Health and Disease   128

- **Atherosclerosis** is a disease characterized by the formation of **atherosclerotic plaque** in the artery wall. It starts with an injury to the artery wall that triggers **inflammation**, leading to plaque formation. A key event in the process is the oxidation of LDL cholesterol in the artery wall. **Oxidized LDL cholesterol** promotes inflammation and is taken up by macrophages, forming **foam cells** that deposit in the artery wall, as depicted here. High blood levels of LDL cholesterol are a risk factor for heart disease, and high blood HDL cholesterol levels protect against heart disease. The risk of atherosclerosis is also increased by diabetes, high blood pressure, and obesity.

Figure 5.16 Development of atherosclerosis

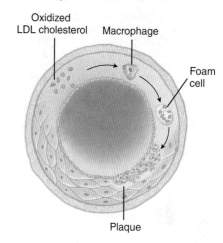

- Diets high in saturated and *trans* fat increase the risk of heart disease. Diets high in omega-6 and omega-3 polyunsaturated fatty acids, monounsaturated fatty acids, certain B vitamins, and plant foods containing fiber, antioxidants, and phytochemicals reduce the risk of heart disease. The total dietary and lifestyle pattern is more important than any individual dietary factor in reducing heart disease risk.

- Diets high in fat correlate with an increased incidence of certain types of cancer. In general, the same types of lipids and other dietary components that protect you from heart disease will also protect you from certain forms of cancer.

- Fat contains 9 Calories per gram. A high-fat diet may therefore increase energy intake and promote weight gain, but it is not the primary cause of obesity. Consuming more energy than expended leads to weight gain, regardless of whether the energy is from fat, carbohydrate, or protein.

6 Meeting Lipid Needs   133

- The DRIs recommend that adults consume a diet that provides 20 to 35% of energy from fat and is low in cholesterol, saturated fat, and *trans* fat. The Dietary Guidelines recommend limiting saturated fatty acid intake to less than 10% of total calories and avoiding *trans* fat. It is recommended that individuals with heart disease or a high risk of heart disease limit their intake even further.

- The U.S. diet is not too high in fat, but it often does not contain the healthiest types of fats. To reduce saturated fat intake, limit solid fats and choose liquid oils, fish, and nuts and seeds often. Use food labels like the planned Nutrition Facts panel shown to avoid processed foods that are high in saturated and *trans* fat. A diet that is based on whole grains, fruits, vegetables, and includes lean meats and low-fat dairy products meets the recommendations for fat intake.

Figure 5.19 Lipids on food labels

- Fat replacers are used to create reduced-fat products with taste and texture similar to the original. Some low-fat products are made by using mixtures of carbohydrates or proteins to simulate the properties of fat, and some use lipids that are modified to reduce their digestion and absorption. Products containing fat replacers can help reduce fat and energy intake when used in moderation as part of a balanced diet.

Key Terms

- adipose tissue 125
- alpha-linolenic acid (α-linolenic acid) 126
- antioxidant 131
- arachidonic acid 126
- atherosclerosis 129
- atherosclerotic plaque 129
- cardiovascular disease 128
- cholesterol 119
- chylomicron 123
- docosahexaenoic acid (DHA) 126
- eicosanoids 126
- eicosapentaenoic acid (EPA) 126
- emulsifier 119
- essential fatty acid 116
- essential fatty acid deficiency 126
- fatty acid 115
- foam cell 130

- high-density lipoprotein (HDL) 125
- hydrogenation 118
- inflammation 129
- LDL receptor 124
- lecithin 119
- linoleic acid 125
- lipid 115
- lipid bilayer 120
- lipoprotein 123
- lipoprotein lipase 124
- low-density lipoprotein (LDL) 124
- macrophage 130
- micelle 122
- monoglyceride 122
- monounsaturated fatty acid 117
- omega-3 fatty acid 117
- omega-6 fatty acid 117

- oxidized LDL cholesterol 130
- partially hydrogenated oil 118
- phosphate group 120
- phospholipid 115
- plant sterol 121
- polyunsaturated fatty acid 117
- saturated fat 115
- saturated fatty acid 116
- sterol 115
- *trans* fat 115
- *trans* fatty acid 116
- triglyceride 115
- tropical oil 116
- unsaturated fat 115
- unsaturated fatty acid 116
- very-low-density lipoprotein (VLDL) 124

What is happening in this picture?

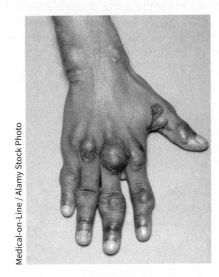

This hand belongs to an individual with familial hypercholesterolemia, a rare genetic disease in which the LDL receptors on cells do not function properly. It causes cholesterol levels so high that the cholesterol deposits in body tissues, seen here as raised lumps.

Think Critically

1. Why would this condition cause elevated blood cholesterol?
2. How would this condition affect the risk of developing heart disease?

CHAPTER 6

Proteins and Amino Acids

The word *protein* comes from the Greek *proteios*, meaning "of primary importance." Proteins are essential to every cellular function. Proteins catalyze chemical reactions, are fundamental components of skeletal and muscular tissue, and are vital to immune response. The cultivation and ingestion of adequate amounts of protein are essential for human survival.

Although protein is found in plants and animals, developed Western nations fulfill most of their protein needs from animal products, while less affluent societies rely on plant proteins. As affluence increases in non-Western societies, they adopt Western habits, including increasing the amount of beef in their diets. This places greater demands on agriculture to provide grain for cattle, which will convert it to the preferred beef protein.

The current dietary recommendations, encouraging a plant-based diet, have shifted some of the focus of meeting protein needs in developed nations. Foods that have been staples in developing countries for centuries are now trendy in Western agriculture and Western diets. Soy, a traditional Asian staple, is the second-most-planted crop in the United States, and the production of quinoa, the high-protein grain shown in the photo that was a staple for the ancient Inca civilization, has increased fourfold over the past decade. Obtaining enough protein and maintaining an adequate balance of plant and animal protein for the world's population, like protein itself, are of primary importance.

CHAPTER OUTLINE

Proteins in Our Food 144

The Structure of Amino Acids and Proteins 145
- Amino Acid Structure
- What a Scientist Sees: Phenylketonuria
- Protein Structure

Protein Digestion and Absorption 147

Protein Synthesis and Functions 149
- Synthesizing Proteins
- Proteins Provide Structure and Regulation
- Protein as a Source of Energy

Protein in Health and Disease 153
- Protein Deficiency
- High-Protein Diets and Health
- Proteins and Food Allergies and Intolerances

Debate: Is a High-Protein Diet Safe and Effective for Weight Loss?

Meeting Protein Needs 157
- Balancing Protein Intake and Losses
- Recommended Protein Intake
- Choosing Protein Wisely

What Should I Eat? Protein Sources

Vegetarian Diets 163
- Benefits of Vegetarian Diets

Thinking It Through: A Case Study on Choosing a Healthy Vegetarian Diet

- Risks of Vegetarian Diets
- Planning Vegetarian Diets

6.1 Proteins in Our Food

LEARNING OBJECTIVES

1. **List** some plant and animal sources of protein.
2. **Compare** the nutrients in plant sources of protein with those in animal sources of protein.

When we think of protein, we usually think of a steak, a plate of scrambled eggs, or a glass of milk. These animal foods provide the most concentrated sources of protein in our diet, but plant foods such as grains, nuts, and **legumes** are also important sources of dietary protein. The proteins found in plants are made up of different combinations of **amino acids** than proteins found in animals. Because of this difference, most plant proteins are not used as efficiently as animal proteins to build proteins in the human body. Nevertheless, a diet that includes a variety of plant proteins can easily meet most people's protein needs.

The sources of protein in your diet have an impact not only on the amount of protein and variety of amino acids available to your body but also on what other nutrients you are consuming (**Figure 6.1**). Animal products provide B vitamins and readily absorbable sources of minerals, such as iron, zinc, and calcium. They are low in fiber, however, and are often high in saturated fat—a nutrient mix that increases the risk of heart disease.

Plant sources of protein are generally excellent sources of fiber, phytochemicals, and unsaturated fats—dietary substances that promote health. They provide B vitamins except vitamin B_{12} and supply iron, zinc, and calcium, but often these minerals are less absorbable than they are from animal products. Current recommendations promote dietary patterns that are based on whole grains, vegetables, and fruits and include smaller amounts of meats and dairy products. Following these guidelines will provide plenty of protein from a mixture of plant and animal sources. A vegetarian diet that excludes some or all animal products is one of the dietary patterns recommended by the 2015–2020 Dietary Guidelines.

Concept Check

1. **Which** is higher in protein: an egg or a cup of rice?
2. **How** do the nutrients provided by meat and milk differ from those in grains and legumes?

FIGURE 6.1 Animal and plant proteins Animal foods, such as those on the left, are concentrated sources of protein but also add saturated fat and cholesterol to the diet. Plant sources of protein, such as those on the right, are rich in fiber, phytochemicals, and monounsaturated and polyunsaturated fats.

1 cup milk: 8 grams protein
One egg: 7 grams protein
3 ounces meat: over 20 grams protein
½ cup legumes: 6–10 grams protein
1 slice bread: about 2 grams protein
¼ cup nuts or seeds: 5–10 grams protein
½ cup rice, pasta, or cereal: 2–3 grams protein

Ask Yourself
Why might a diet high in animal protein increase the risk of heart disease?

legume The starchy seed of a plant that produces bean pods; includes peas, peanuts, beans, soybeans, and lentils.

amino acids The building blocks of proteins. Each amino acid contains an amino group, an acid group, and a unique side chain.

6.2 The Structure of Amino Acids and Proteins

LEARNING OBJECTIVES

1. **Describe** the general structure of an amino acid.
2. **Distinguish** between essential and nonessential amino acids.
3. **Discuss** how the order of amino acids in a polypeptide chain affects a protein's shape and function.
4. **Explain** how a protein's structure is related to its function.

What do the proteins in a lamb chop, a kidney bean, and your thigh muscle have in common? They are all constructed of amino acids linked together to form one or more folded, chainlike strands. Twenty amino acids are commonly found in proteins. Each kind of protein contains a different number, combination, and sequence of amino acids. These differences give proteins their specific functions in living organisms and their unique characteristics in foods.

Amino Acid Structure

Each amino acid consists of a central carbon atom that is bound to a hydrogen atom; an amino group, which contains nitrogen; an acid group; and a side chain (**Figure 6.2**). The nitrogen in amino acids distinguishes protein from carbohydrate and fat; all three contain carbon, hydrogen, and oxygen, but only protein contains nitrogen. The side chains of amino acids vary in size and structure; they give different amino acids their unique properties.

Nine of the amino acids needed by the adult human body must be consumed in the diet because they cannot be made in the body (see Figure 6.2). If the diet is deficient in one or more of these **essential amino acids** (also called **indispensable amino acids**), the body cannot make new proteins without breaking down existing proteins to provide the needed amino acids. The other 11 amino acids that are commonly found in protein are **nonessential**, or **dispensable**, **amino acids** because they can be made in the body.

Under certain conditions, some of the nonessential amino acids cannot be synthesized in sufficient amounts to meet the body's needs. These are therefore referred to as **conditionally essential amino acids**. For example, the amino acid tyrosine can be made in the body from the essential amino acid phenylalanine. In individuals who have the inherited disease **phenylketonuria (PKU)**, phenylalanine cannot be converted into tyrosine, so tyrosine is an essential amino acid for these individuals (see *What a Scientist Sees: Phenylketonuria*).

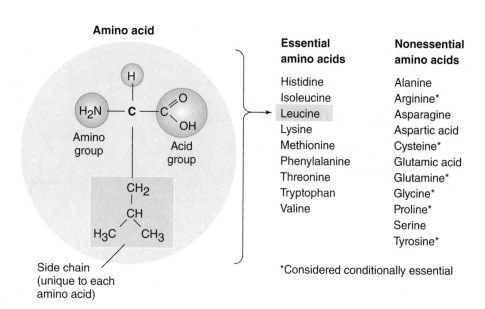

FIGURE 6.2 **Amino acids** All "amino" "acids" have a similar structure that includes an amino group and an acid group, giving them their name, but each has a different side chain. Of the 20 amino acids in proteins, 9 are considered dietary essentials because they cannot be made in the body.

Essential amino acids: Histidine, Isoleucine, Leucine, Lysine, Methionine, Phenylalanine, Threonine, Tryptophan, Valine

Nonessential amino acids: Alanine, Arginine*, Asparagine, Aspartic acid, Cysteine*, Glutamic acid, Glutamine*, Glycine*, Proline*, Serine, Tyrosine*

*Considered conditionally essential

---

essential, or **indispensable**, **amino acid** An amino acid that cannot be synthesized by the body in sufficient amounts to meet its needs and therefore must be included in the diet.

phenylketonuria (PKU) A genetic disease in which the amino acid phenylalanine cannot be metabolized normally, causing it to build up in the blood. If untreated, the condition results in brain damage.

What a Scientist Sees

Phenylketonuria

The warning on this can of diet soda probably doesn't mean much unless you have the genetic disease phenylketonuria (PKU). Individuals with PKU must limit their intake of the amino acid phenylalanine. Usually this means limiting their consumption of high-protein foods. When a scientist looks at this label, she recognizes that the artificial sweetener aspartame in this soda is the source of the phenylalanine. The breakdown of aspartame in the digestive tract releases phenylalanine, which cannot be properly metabolized by individuals with PKU. If they consume large amounts of this amino acid, compounds called phenylketones build up in their blood. In infants and young children, phenylketones interfere with brain development, and in pregnant women, they cause birth defects in the baby. To prevent these effects, individuals with PKU must consume a diet that provides just enough phenylalanine to meet the body's needs but not so much that phenylketones build up in their blood.

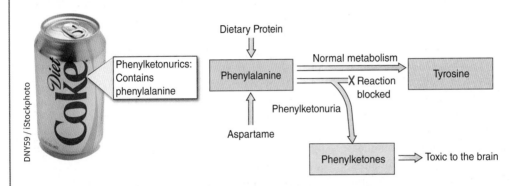

Think Critically

Why do you think this warning appears on diet soda labels but not on labels for high-protein foods such as meat and milk?

Protein Structure

To form proteins, amino acids are linked together by **peptide bonds**, which join the acid group of one amino acid to the amino group of another amino acid (**Figure 6.3**). Many amino acids bonded together constitute a **polypeptide**. A protein is made up of one or more polypeptide chains that are folded into three-dimensional shapes. The order and chemical properties of the amino acids in a polypeptide determine its final shape because the folding of the chain occurs in response to forces that attract or repel amino acids from one another or from water. The folded polypeptide chain may constitute the final protein, or it may join with several other folded polypeptide chains to form the final structure of the protein (see Figure 6.3).

The shape of a protein is essential to its function. For example, the elongated shape of the protein collagen, found in connective tissue, helps it give strength to tendons and ligaments. The spherical shape of the protein hemoglobin contributes to the proper functioning of red blood cells, and the linear shape of muscle proteins allows them to overlap and shorten muscles during contraction.

When the shape of a protein is altered, the protein no longer functions normally. For example, when the enzyme salivary amylase, which is a protein, enters the stomach, the acid

FIGURE 6.3 **Proteins are made from amino acids** Amino acids linked by peptide bonds are called **peptides**. When two amino acids are linked, they form a **dipeptide**; three form a **tripeptide**. Many amino acids bonded together constitute a polypeptide. Polypeptide chains may contain hundreds of amino acids. The chemical properties of the amino acids in the polypeptide chain cause it to fold, contributing to the three-dimensional structure of the protein. The final protein may consist of one or more folded polypeptide chains.

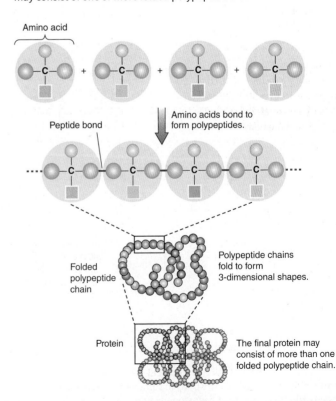

---

polypeptide A chain of amino acids linked by peptide bonds that is part of the structure of a protein.

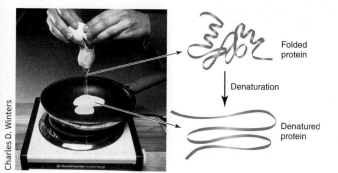

FIGURE 6.4 **Protein denaturation** When an egg is cooked, the heat denatures the protein, causing the polypeptide chains to unfold. The protein in a raw egg white forms a clear, viscous liquid, but when heat denatures it, the cooked egg white becomes white and firm and cannot be restored to its original form. The denaturation of proteins in our food also creates other characteristics we desire. For example, whipped cream is made when mechanical agitation denatures the protein in cream.

causes the structure of the enzyme protein to change, and it no longer functions in the digestion of starch. This change in structure is called **denaturation**, referring to a change from the natural. Proteins in food are often denatured during processing and cooking (**Figure 6.4**).

Concept Check

1. **Which** chemical elements are found in all amino acids?
2. **What** determines whether an amino acid is essential in the diet?
3. **How** does the order of amino acids in a polypeptide affect the function of a protein?
4. **How** does denaturation affect the function of proteins?

6.3 Protein Digestion and Absorption

LEARNING OBJECTIVES

1. **Describe** the process of protein digestion.
2. **Discuss** how amino acids are absorbed.

Proteins must be digested before their amino acids can be absorbed into the body. The chemical digestion of protein begins in the acid environment of the stomach. Here, hydrochloric acid denatures proteins, opening up their folded structure to make the polypeptide chains more accessible for breakdown by enzymes. Stomach acid also activates the protein-digesting enzyme *pepsin*, which breaks some of the peptide bonds in the polypeptide chains, leaving shorter polypeptides. Most protein digestion occurs in the small intestine, where polypeptides are broken into even smaller peptides and amino acids by protein-digesting enzymes produced in the pancreas and small intestine. Single amino acids, dipeptides, and tripeptides are absorbed into the mucosal cells of the small Intestine (**Figure 6.5**).

Because protein must be broken down in order to be absorbed, proteins consumed in the diet enter the body as a collection of individual amino acids, not as functioning proteins. Supplements of specific proteins therefore do not retain their function in the body. For example, an enzyme supplement advertised to destroy free radicals, eliminate toxins, or enhance muscle growth will not provide these functions in the body because the amino acids that make up the enzyme, not the enzyme itself, are what enters the body. Supplements of enzymes that function in the GI tract, such as lactase, taken to prevent the symptoms of lactose intolerance, retain enzyme activity long enough to act on materials in the intestine but are also eventually digested to amino acids, which are then absorbed.

Amino acids from protein digestion enter your body by crossing from the lumen of the small intestine into the mucosal cells and then into the blood. This process involves one of several energy-requiring amino acid transport systems. Amino acids with similar structures use the same transport system (see Figure 6.5). As a result, amino acids may compete with one another for absorption. If there is an excess of any one of the amino acids sharing a transport system, more of it will be absorbed, slowing the absorption of competing amino acids. This competition for absorption is usually not a problem

---

denaturation Alteration of a protein's three-dimensional structure.

PROCESS DIAGRAM — FIGURE 6.5 Protein digestion and absorption
Protein must be broken down into small peptides and amino acids to be absorbed into the mucosal cells.

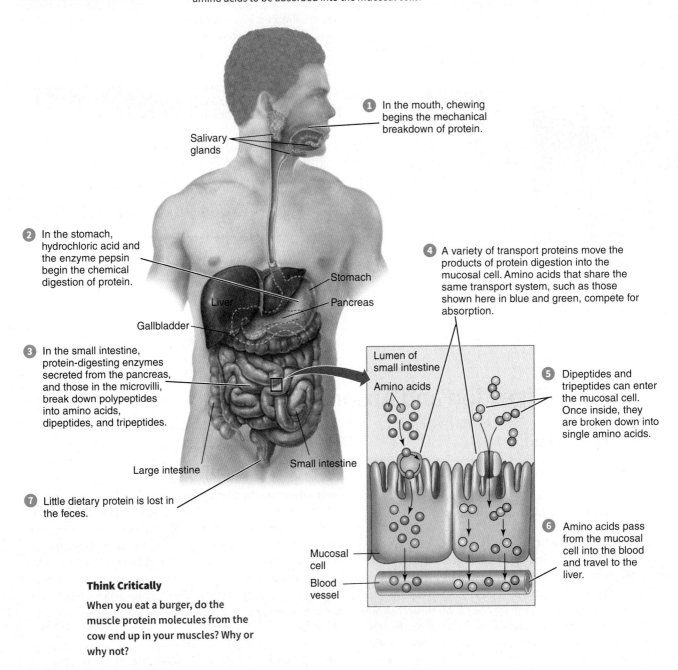

1. In the mouth, chewing begins the mechanical breakdown of protein.
2. In the stomach, hydrochloric acid and the enzyme pepsin begin the chemical digestion of protein.
3. In the small intestine, protein-digesting enzymes secreted from the pancreas, and those in the microvilli, break down polypeptides into amino acids, dipeptides, and tripeptides.
4. A variety of transport proteins move the products of protein digestion into the mucosal cell. Amino acids that share the same transport system, such as those shown here in blue and green, compete for absorption.
5. Dipeptides and tripeptides can enter the mucosal cell. Once inside, they are broken down into single amino acids.
6. Amino acids pass from the mucosal cell into the blood and travel to the liver.
7. Little dietary protein is lost in the feces.

Think Critically

When you eat a burger, do the muscle protein molecules from the cow end up in your muscles? Why or why not?

---

because foods contain a variety of amino acids, none of which are present in excessive amounts. However, when people consume amino acid supplements, the supplemented amino acid may overwhelm the transport system, reducing the absorption of other amino acids that share the same transport system. For example, weight lifters often take supplements of the amino acid arginine. Because arginine shares a transport system with lysine, large doses of arginine can inhibit the absorption of lysine, upsetting the balance of amino acids in the body.

Concept Check

1. **Where** does the chemical digestion of protein begin?
2. **Why** might supplementing one amino acid reduce the absorption of other amino acids?

6.4 Protein Synthesis and Functions

LEARNING OBJECTIVES

1. **Discuss** the steps involved in synthesizing proteins.
2. **Explain** the term *limiting amino acid*.
3. **Name** four general functions provided by body proteins.
4. **Describe** the conditions under which the body uses protein to provide energy.

As discussed earlier, proteins are made from amino acids. Amino acids are also used to make other nitrogen-containing molecules, including neurotransmitters; the units that make up DNA and RNA; the skin pigment melanin; the vitamin niacin; creatine phosphate, which is used to fuel muscle contraction; and histamine, which causes blood vessels to dilate. In some situations, amino acids from proteins are also used to provide energy or synthesize glucose or fatty acids.

The amino acids available for these functions come from the proteins consumed in the diet and from the breakdown of body proteins. These amino acids are referred to collectively as the **amino acid pool** (Figure 6.6). There is not actually a "pool" in the body containing a collection of amino acids, but these molecules are available in body fluids and cells to provide the raw materials needed to synthesize proteins and other molecules.

Synthesizing Proteins

The instructions for making proteins are contained in the nucleus of the cell in stretches of DNA called **genes**. When a protein is needed, the process of protein synthesis begins, and the information contained in the gene is used to make the necessary protein.

Regulating protein synthesis Both the types of proteins made and when they are made are carefully regulated by increasing or decreasing **gene expression**. When a gene is expressed, the protein it codes for is made. Not all genes are expressed in all cells or at all times; only the proteins that are needed are made at any given time. Which genes are expressed is also affected by genetic background and by the nutrients and other food components we consume.

The regulation of gene expression allows the body to save energy and resources. For example, when your diet is high in iron, expression of the gene that codes for the protein ferritin, which stores iron, is increased. This causes more ferritin to be synthesized and allows the body to store extra iron in this protein. When the diet is low in iron, the production of ferritin is suppressed so that the body doesn't waste amino acids and energy making large amounts of a protein that it doesn't need.

Limiting amino acids During the synthesis of a protein, a shortage of one amino acid can stop the process. This is similar to an assembly line, where if one part is missing, the line stops; a different part cannot be substituted. If the missing amino acid is a nonessential amino acid, it can be made in the body, and protein synthesis can continue. Most nonessential amino acids are made through a process called **transamination**, which involves transferring the amino group

FIGURE 6.6 Amino acid pool Amino acids enter the available pool from the diet and from the breakdown of body proteins. Of the approximately 300 g of protein synthesized by the body each day, only about 100 g are made from amino acids consumed in the diet. The other 200 g are produced by the recycling of amino acids from protein broken down in the body. Amino acids in the pool can be used to synthesize body proteins and other nitrogen-containing molecules, to provide energy, or to synthesize glucose or fatty acids.

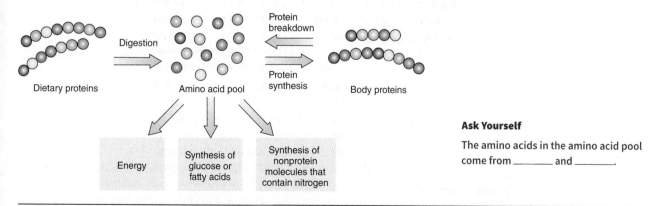

Ask Yourself

The amino acids in the amino acid pool come from _____ and _____.

---

amino acid pool All the amino acids in body tissues and fluids that are available for use by the body.

gene A length of DNA that contains the information needed to synthesize a polypeptide chain or a molecule of RNA.

transamination The process by which an amino group from one amino acid is transferred to a carbon compound to form a new amino acid.

NUTRITION INSIGHT

FIGURE 6.7 Protein synthesis Protein is synthesized from amino acids. All of the amino acids in the protein must be available for protein synthesis to proceed.

a. Amino acids come from protein in the diet and from the breakdown of body proteins. Some can be made in the body by transamination.

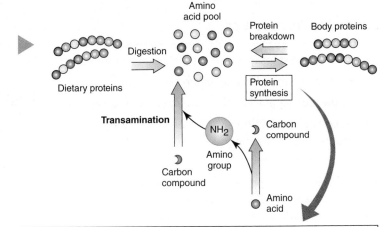

b. The instructions for protein synthesis come from genes. The process of protein synthesis involves **transcription** and **translation**.

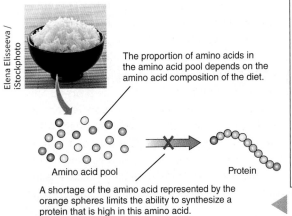

The proportion of amino acids in the amino acid pool depends on the amino acid composition of the diet.

A shortage of the amino acid represented by the orange spheres limits the ability to synthesize a protein that is high in this amino acid.

① In the nucleus, the code for a protein is copied from the DNA into a molecule of messenger RNA (mRNA).

② The mRNA takes the genetic information from the nucleus to ribosomes in the cytosol, where proteins are made.

③ At the ribosomes, transfer RNA (tRNA) reads the genetic code and delivers the needed amino acids to the ribosome to form a polypeptide chain.

c. The amino acids needed for protein synthesis come from the amino acid pool. If the protein to be made requires more of a particular amino acid than is available, that amino acid limits protein synthesis and is referred to as the *limiting amino acid*.

---

from one amino acid to a carbon-containing molecule to form the needed amino acid (**Figure 6.7a**). If the missing amino acid is an essential amino acid, the body cannot make the amino acid, but it can break down its own protein to obtain it, allowing protein synthesis to proceed (**Figure 6.7b**). If an amino acid cannot be supplied, protein synthesis stops.

The essential amino acid that is present in shortest supply relative to the body's need for it is called the **limiting amino acid**; the lack of this amino acid limits the body's ability to synthesize a protein that contains it (**Figure 6.7c**). Different food sources of protein provide different combinations of amino acids. The limiting amino acid in a food is the one supplied in the lowest amount relative to the body's need. For example, lysine is the limiting amino acid in wheat, whereas methionine is the limiting amino acid in beans. When the diet provides adequate amounts of all the essential amino acids needed to synthesize a specific protein, synthesis of the polypeptide chains that make up the protein can be completed (see Figure 6.7b).

Proteins Provide Structure and Regulation

When you think of the protein in your body, you probably think of muscle, but muscle contains only a few of the many types of proteins found in your body. There are more than 500,000 proteins in the human body, each with a specific function. Some perform important structural roles, and others help regulate specific body processes.

Structural proteins are found in skin, hair, ligaments, tendons, and bones (**Figure 6.8a**). Proteins also provide structure to individual cells, where they are an integral part of the cell membrane, cytosol, and organelles. Proteins such as enzymes,

---

limiting amino acid The essential amino acid that is available in the lowest concentration relative to the body's need.

NUTRITION INSIGHT

FIGURE 6.8 Protein functions The proteins that are part of the human body perform a myriad of structural and regulatory roles.

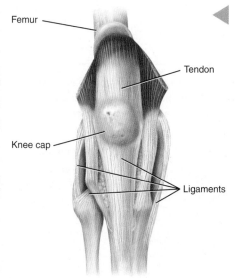

a. Collagen is the most abundant protein in the body. It is the major protein in ligaments, which attach bones to other bones, and in tendons, which attach muscles to bones, and it forms the protein framework of bones and teeth.

b. Enzymes, such as this one, are protein molecules. Almost all the chemical reactions occurring within the body require the help of enzymes. Each enzyme has a structure or shape that allows it to interact with the specific molecules in the reaction it accelerates. Without enzymes, metabolic reactions would occur too slowly to support life.

c. Proteins help transport materials throughout the body and into and out of cells. The protein hemoglobin, which gives these red blood cells their color, shuttles oxygen to body cells and carries away carbon dioxide.

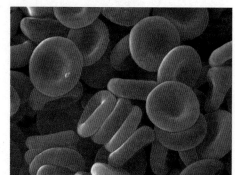

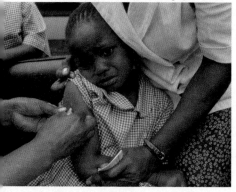

e. Proteins help us move. The proteins actin and myosin in the arm and leg muscles of this rock climber are able to slide past each other to contract the muscles. A similar process causes contraction in the heart muscle and in the muscles of the digestive tract, blood vessels, and body glands.

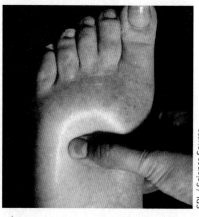

d. Proteins help protect us from disease. This child is being immunized against measles. The vaccine contains a small amount of dead or inactivated measles virus. It does not make the child sick, but it does stimulate her immune system to make proteins called *antibodies*, which help destroy the measles virus and prevent the child from contracting the disease.

f. Proteins help regulate fluid balance through their effect on *osmosis*. Blood proteins help hold fluid in the blood because they contribute to the number of dissolved particles in the blood. If protein levels in the blood fall too low, water leaks out of the blood vessels and accumulates in the tissues, causing swelling known as **edema**, shown here. Proteins also regulate fluid balance because some are membrane transporters, which pump dissolved substances from one side of a membrane to the other.

which speed up biochemical reactions (**Figure 6.8b**), and transport proteins that travel in the blood or help materials cross membranes, regulate processes throughout the body (**Figure 6.8c**).

Proteins are an important part of the body's defense mechanisms. Skin, which is made up primarily of protein, is the first barrier against infection and injury. Foreign particles such as dirt or bacteria that are on the skin cannot enter the body and can be washed away. If the skin is broken and blood vessels are injured, blood-clotting proteins help prevent too much blood from being lost. If a foreign material does get into the body, *antibodies*, which are immune system proteins, help destroy it (**Figure 6.8d**).

PROCESS DIAGRAM **FIGURE 6.9 Producing ATP from amino acids** Amino acids from dietary or body protein can be used to produce ATP. First, the amino group (NH_2) must be removed through a process called **deamination**. The remaining compound, composed of carbon, hydrogen, and oxygen, can then be broken down to produce ATP or used to make glucose or fatty acids.

Ask Yourself

If you are not eating anything, how can the body supply itself with glucose?

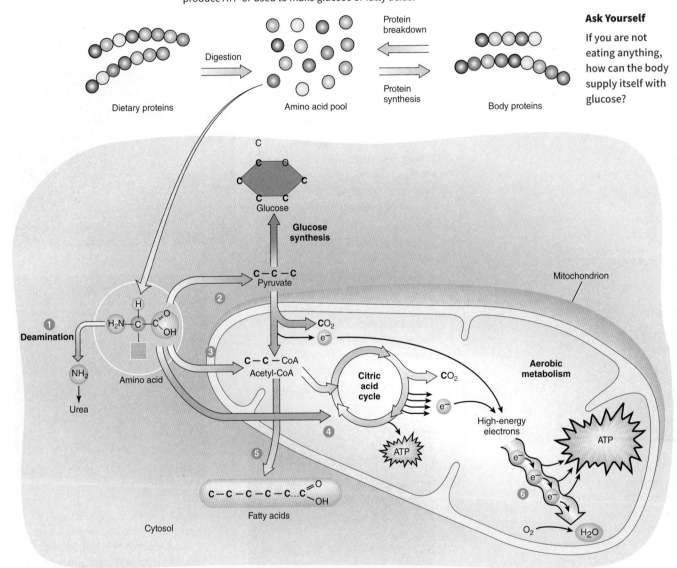

① The amino group is removed by deamination and converted into the waste product **urea**. Urea is removed from the blood by the kidneys and excreted in the urine.

② Deamination of some amino acids results in three-carbon molecules that can be used to synthesize glucose.

③ Deamination of some amino acids results in acetyl-CoA that enters the citric acid cycle.

④ Deamination of some amino acids forms molecules that enter the citric acid cycle directly.

⑤ The acetyl-CoA derived from the breakdown of amino acids can be used to synthesize fatty acids. This occurs when calories and protein are consumed in excess of needs.

⑥ In the final step of aerobic metabolism, the energy released from the amino acid molecules is transferred to ATP.

Some proteins have contractile properties, which allow muscles to move various parts of the body (**Figure 6.8e**). Others are hormones, which regulate biological processes. The hormones insulin, growth hormone, and glucagon are made from amino acids. These protein hormones act rapidly because they affect the activity of proteins that are already present in the cell.

Proteins also help regulate fluid balance (**Figure 6.8f**) and prevent the level of acidity in body fluids from deviating from the normal range. The chemical reactions of metabolism require a specific level of acidity, or pH, to function properly. Inside the body, pH must be maintained at a relatively neutral level in order to allow metabolic reactions to proceed normally. If the pH changes, these reactions slow or stop. Proteins both within cells and in the blood help prevent large changes in pH.

Protein as a Source of Energy

In addition to all the essential functions performed by body proteins, under some circumstances, proteins can be broken down and their amino acids used to provide energy or synthesize glucose or fatty acids (**Figure 6.9**). When the diet does not provide enough energy to meet the body's needs, such as during starvation or when consuming a low-calorie weight-loss diet, body protein is used to provide energy. Because our bodies do not store protein, functional body proteins, such as enzymes and muscle proteins, must be broken down to yield amino acids, which can then be used as fuel or to make glucose. This ensures that cells have a constant energy supply but also robs the body of the functions performed by these proteins.

Amino acids are also used for energy when the amount of protein consumed in the diet is greater than that needed to make body proteins and other molecules. This occurs in most Americans every day because our typical diet contains more protein than we need. The body first uses amino acids from the diet to make body proteins and other nitrogen-containing molecules. Then, because extra amino acids can't be stored, they are metabolized to provide energy. When your diet includes more energy and protein than you need, amino acids can be converted into fatty acids, which are stored as triglycerides, thus contributing to weight gain.

Concept Check

1. **How** does the body know in what order to assemble the amino acids when making a protein?
2. **Why** does protein synthesis stop when the supply of an amino acid is limited?
3. **What** type of protein molecule speeds up chemical reactions?
4. **Why** is much of the protein consumed in a typical American diet used as an energy source?

6.5 Protein in Health and Disease

LEARNING OBJECTIVES

1. **Distinguish** kwashiorkor from marasmus.
2. **Explain** why protein-energy malnutrition is more common in children than in adults.
3. **Discuss** the potential risks associated with high-protein diets.
4. **Explain** how a protein in the diet can trigger an allergic reaction.

We need to eat protein to stay healthy. If we don't eat enough of it, less-essential body proteins are broken down, and their amino acids are used to synthesize proteins that are critical for survival. For example, when the diet is deficient in protein, muscle protein is broken down to provide amino acids to make hormones and enzymes for which there is an immediate need. If protein deficiency continues, eventually so much body protein is lost that all life-sustaining functions cannot be supported. In some cases, too much protein or the wrong proteins can also contribute to health problems.

Protein Deficiency

Protein deficiency is a great concern in the developing world but generally not a problem in economically developed societies, where plant and animal sources of protein are abundant (**Figure 6.10a**). Usually, protein deficiency occurs along with a general lack of food and other nutrients. The term **protein-energy malnutrition (PEM)** is used to refer to a

---

protein-energy malnutrition (PEM) A nutritional deficiency resulting from inadequate intake of energy and/or protein. It includes marasmus and kwashiorkor.

FIGURE 6.10 Protein-energy malnutrition
Protein-energy malnutrition affects the health of children in many parts of the world.

a. Protein-energy malnutrition is uncommon in the United States and other developed nations, but in developing countries, it is a serious public health problem that results in high infant and child mortality. The highest prevalence is in sub-Saharan Africa and South Asia.

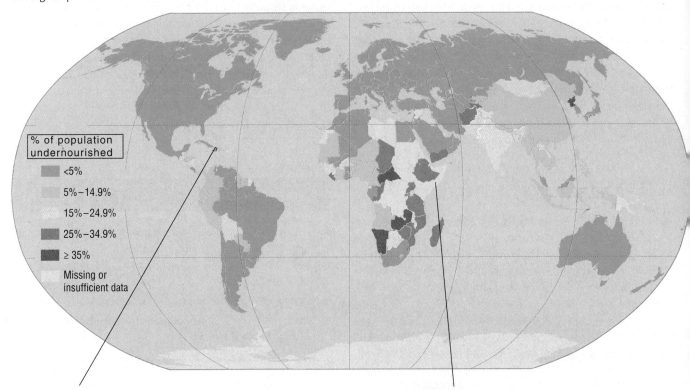

b. Kwashiorkor, seen in this child from Haiti, is characterized by a swollen belly, which results from fluid accumulating in the abdomen and fat accumulating in the liver. Growth is impaired, but because energy intake is not necessarily low, the child may not appear extremely thin. The lack of protein also causes poor immune function and an increase in infections, changes in hair color, and impaired nutrient absorption.

c. Marasmus, seen in this Somalian boy, is characterized by depletion of fat stores and wasting of muscle. It has devastating effects on infants because most brain growth takes place in the first year of life; malnutrition in the first year causes decreases in intelligence and learning ability that persist throughout the individual's life.

continuum of conditions ranging from pure protein deficiency, called **kwashiorkor**, to an overall energy deficiency, called **marasmus**.

Protein deficiency occurs when the diet is very low in protein or when protein needs are high, as they are in young children. Hence, kwashiorkor is typically a disease found in children (**Figure 6.10b**). The word *kwashiorkor* comes from the Ga language of coastal Ghana. It means "the disease that the first child gets when a second child is born."[1] When the new baby is born, the older child is no longer breast-fed. Rather than receive protein-rich breast milk, the young child is fed a watered-down version of the diet eaten by the rest of the family. This diet is low in protein and often high in fiber and difficult to digest. The child, even if he or she is able to obtain adequate calories from the diet, may not be able to eat a large enough quantity to get adequate protein. Because children are growing, their protein needs per unit of body weight are higher than those of adults, and a deficiency occurs more quickly. Although kwashiorkor only occurs in those whose diet is deficient in protein, it is associated with, or even triggered by, infectious diseases.

---

kwashiorkor A form of protein-energy malnutrition characterized by edema that is usually the result of a severe restriction in protein intake.

marasmus A form of protein-energy malnutrition characterized by wasting of both muscle and fat that is usually the result of a severe energy restriction.

At the other end of the continuum of protein-energy malnutrition is marasmus, meaning "to waste away" (**Figure 6.10c**). Marasmus is caused by starvation; the diet doesn't supply enough calories or nutrients to meet the body's needs. Marasmus may have some of the same symptoms as kwashiorkor, but there are differences. In kwashiorkor, some fat stores are retained because energy intake is adequate. In marasmus, individuals appear emaciated because their stores of body fat have been depleted to provide energy. Although they are most common in children, both marasmus and kwashiorkor can occur in individuals of all ages.

High-Protein Diets and Health

The recent popularity of high-protein, low-carbohydrate diets for weight loss (see Chapter 9) has raised questions about whether consuming too much protein can be harmful (see *Debate: Is a High-Protein Diet Safe and Effective for Weight Loss?*). As protein intake increases, so does the production of protein-breakdown products, such as urea (see Figure 6.9), which must be eliminated from the body by the kidneys. This increases the workload of the kidneys and may contribute to the development of kidney disease and speed the progression of renal failure in people who already have kidney disease.[2, 3] Because high-protein diets increase the amount of wastes that must be excreted in urine, they also increase water loss. Although not a concern for most people, this can be a problem if the kidneys are not able to concentrate urine, as is the case for infants. Feeding a newborn an infant formula that is too high in protein increases the amount of water lost in the urine and can lead to dehydration.

A diet high in protein increases calcium excretion in the urine. Because of this, there has been concern that high-protein diets would negatively affect bone health.[4] However, when calcium intake is adequate, high-protein diets are associated with greater bone mass.[5] This seems to be a contradiction, but it is now recognized that the higher calcium excretion with a high-protein diet is accompanied by an increase in calcium absorption, so there is no net loss of calcium.[6, 7]

It has also been suggested that the increase in urinary calcium excretion with a high-protein diet increases the risk of kidney stones. Kidney stones are deposits of calcium and other substances in the kidneys and urinary tract. In individuals prone to kidney stones, diets that are rich in animal protein, high in sodium, and low in fluid can contribute to kidney stone formation.[8]

The best-documented concern with high-protein diets is related more to the rest of the diet than to the amount of protein consumed. Typically, high-protein diets are also high in animal products; this dietary pattern is high in saturated fat and low in fiber, and it therefore increases the risk of heart disease. These diets are also typically low in grains, vegetables, and fruits, and they may be high in red meat and processed meats, a pattern associated with an increased risk of certain types of cancer.[9, 10]

Proteins and Food Allergies and Intolerances

The proteins in food can trigger a **food allergy**. The first time the protein is consumed by a susceptible individual, fragments of undigested protein stimulate the immune system. When the same protein is consumed again, the immune system sees it as a foreign substance and mounts an attack, causing an allergic reaction (see Chapter 3). Allergic reactions cause symptoms throughout the body and can be life threatening. The proteins from milk, eggs, peanuts, tree nuts, wheat, soy, fish, and shellfish are the eight most common food allergens and must be listed on food labels (**Figure 6.11**). Sesame seeds are also a common cause of food allergies but are not highlighted on food labels. Individuals with allergies to food components other than the eight required on food labels may have a more challenging time determining whether a food is safe for them to eat.

Not all adverse reactions to proteins and amino acids are due to allergies; some are due to **food intolerances**, also called **food sensitivities**. These reactions do not involve the immune system. The symptoms of a food intolerance can range from minor discomfort, such as the abdominal distress some people feel after eating raw onions, to more severe reactions. Some people report having a reaction after consuming **monosodium glutamate (MSG)**. MSG is a flavor enhancer made up of the amino acid glutamic acid bound to sodium. It is used in meat tenderizers and commonly added to Chinese food. Although research has been unable to confirm that MSG ingestion causes any adverse reactions, some people report experiencing a collection of symptoms such as flushed face, tingling or burning sensations, headache, rapid heartbeat, chest pain, and general weakness that are collectively referred to as **MSG symptom complex**, commonly called **Chinese restaurant syndrome**.[25] Sensitive individuals should ask for food to be prepared without added MSG and should check ingredient lists for monosodium glutamate or potassium glutamate before consuming packaged foods.

Celiac disease is another type of adverse reaction to food. Celiac disease is an autoimmune disease in which gluten, a protein found in wheat, rye, and barley, causes the body's immune

---

food allergy An adverse immune response to a specific food protein.

food intolerance or **food sensitivity** An adverse reaction to a food that does not involve the production of antibodies by the immune system.

celiac disease A disorder that causes damage to the intestines when the protein gluten is eaten.

Debate

Is a High-Protein Diet Safe and Effective for Weight Loss?

The Issue

High-protein diets are promoted for weight loss. Some of these diets allow all the steak, bacon, and eggs you can eat; others suggest leaner protein choices. Is all that meat good for you? Are high-protein diets effective for weight loss? And, are they healthy?

When talking about a high-protein diet, it's necessary to define what counts as high protein. Most Americans consume more than the RDA for protein.[11] This gives the impression that we are all eating high-protein diets, but at 16% of calories from protein the typical American diet is at the low end of the AMDR of 10 to 35% of calories from protein.[12] Most studies on the effects of high-protein consider a high-protein diet to be one with 25% of calories or more from protein. For the average person, this is about 134 grams per day, which is equivalent to eating a dozen eggs and a 12-ounce steak every day. Is such a diet the best way to lose weight?

Weight loss requires consuming fewer calories than you expend. So, any diet that limits calorie intake will promote weight loss. Protein is the most satiating of the macronutrients, which is why high-protein diets promote weight loss; the protein keeps you feeling satisfied, so you eat less.[13] Protein also affects energy expenditure. When you eat less to lose weight, your metabolism slows to conserve energy. High-protein diets have been found to cause less of a decrease in energy expenditure than low-fat or high-carbohydrate diets.[14, 15] Some of this may be because the body uses more calories to process protein than it does for carbohydrate or fat, so eating a high-protein diet increases calorie needs. A high protein intake also helps maintain the body's muscle mass, and the greater one's muscle mass, the greater one's calorie needs. So, even people who aren't counting calories may eat less and expend more energy when consuming a high-protein diet than they would with less protein. So, a high-protein diet may make weight loss easier. But high-protein diets can vary in the amounts of other nutrients. Do the other components of the diet matter?

The amount of carbohydrate in the diet may affect appetite. Some high-protein diets severely restrict carbohydrate. Diets very low in carbohydrate lead to the production of ketones, which further suppress appetite. However, the beneficial effect that a high-protein diet has on weight management has been found to depend on the high-protein component of the diet, not on the low-carb component.[16] Also, severely restricting carbohydrate intake means all but eliminating entire food groups and may not promote weight loss more than would a less restrictive diet.

Does the type of protein you choose matter? A high-protein diet could be high in red meat and high-fat dairy products (see figure), or it could be high in legumes, nuts, fish, and poultry. The source of the protein in the diet, whether plant or animal, does not seem to affect satiety. Plant proteins suppress appetite as well as animal proteins.[17] But the source of the protein may be important for the overall health of the diet.

Many high-protein diets are high in animal products and low in whole grains, fruits, and vegetables; this dietary pattern has been shown to promote heart disease.[18] Despite this, improvements in risk factors for heart disease have been seen in the short term with these diets. An analysis of studies of high-protein diets lasting about six months found small improvements in body fat, blood pressure, and triglyceride levels.[19] Even after a year or more, high-protein diets exerted neither specific beneficial nor detrimental effects on risk factors for heart disease.[20] However, when the effect of high-protein diets is followed in the long term (an average of about 16 years), some studies find an increase in heart disease risk and overall mortality as well.[21, 22] Some of the reason for the discrepancy in these results may be the overall dietary pattern; protein sources such as red meat and high-fat dairy products are associated with an elevated risk of coronary heart disease, while higher intakes of poultry, fish, and nuts correlate with a lower risk.[23] Processed meat is also a carcinogen. A dietary pattern that is high in red and processed meat is associated with an increased risk of cancer, especially colorectal cancer. A study that monitored people for 18 years found that those 50 to 65 years of age who had a high protein intake had a 75% increase in overall mortality and a fourfold increase in cancer death risk.[24] These cancer and mortality associations were either abolished or attenuated if the proteins were plant derived.

High-protein diets promote weight loss and weight maintenance and help maintain lean body mass. This is good for your overall health, but if the same diet increases the risk of heart disease and cancer over the long term, is it the best choice?

A high-protein diet may include meals such as this one that are based on a large serving of red meat, or they could include fish, poultry, or plant proteins such as legumes.

Think Critically

Assume that you want to lose weight on a high-protein diet. Plan a meal that provides 40 grams of plant-based protein.

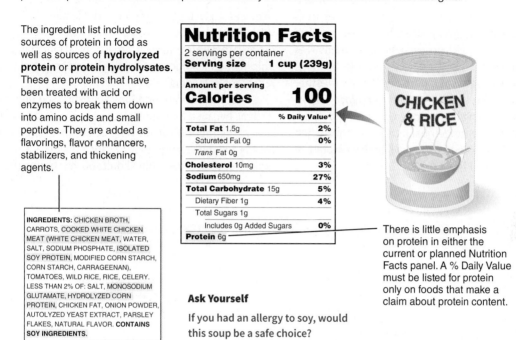

FIGURE 6.11 Protein and allergen information on food labels The Nutrition Facts panel on food labels provides information on the amount of protein in a food. The ingredient list specifies the source of the protein; this information can be life-saving for individuals with food allergies. The eight major food allergens must be included in the ingredient list, but often, as on this label, they are also highlighted at the end of the list, in a statement such as "Contains soy ingredients." Warnings such as "manufactured in a facility that processes peanuts" are included on products that may be cross-contaminated with these allergens.

The ingredient list includes sources of protein in food as well as sources of **hydrolyzed protein** or **protein hydrolysates**. These are proteins that have been treated with acid or enzymes to break them down into amino acids and small peptides. They are added as flavorings, flavor enhancers, stabilizers, and thickening agents.

There is little emphasis on protein in either the current or planned Nutrition Facts panel. A % Daily Value must be listed for protein only on foods that make a claim about protein content.

Ask Yourself

If you had an allergy to soy, would this soup be a safe choice?

system to attack the villi in the small intestine. When individuals with this disease consume foods containing gluten, they develop symptoms ranging from diarrhea, abdominal bloating, and cramps to weight loss, anemia, and malnutrition. Celiac disease affects about 3 million people in the United States. A more common sensitivity to gluten, non-celiac gluten sensitivity, affects about 18 million Americans.[26, 27] The only treatment for either is to eliminate gluten from the diet (see *Debate: Should You Be Gluten Free?* in Chapter 3).

Concept Check

1. **Why** do children with marasmus appear more emaciated than those with kwashiorkor?
2. **Why** might kwashiorkor occur in a young child after his mother has another baby?
3. **Who** should be concerned about excessive protein intake?
4. **How** can someone with a food allergy avoid an allergic reaction?

6.6 Meeting Protein Needs

LEARNING OBJECTIVES

1. **Describe** how protein needs are determined.
2. **Explain** what is meant by protein quality.
3. **Review** a diet and replace the animal proteins with complementary plant proteins.

In order to stay healthy, you have to eat enough protein to replace the amount you lose every day. Most Americans get plenty of protein, and healthy diets can contain a wide range of intakes from both plant and animal sources. An individual's protein needs may be increased by growth, injury, and illness, as well as by some types of physical activity.

Balancing Protein Intake and Losses

Current protein intake recommendations are based on **nitrogen balance** studies. These studies compare the amount of nitrogen consumed with the amount excreted. Studying nitrogen balance allows researchers to evaluate protein balance because most of the nitrogen we consume comes from dietary protein. Most of the nitrogen we lose is excreted in urine. Smaller amounts are lost in feces, skin, sweat, menstrual fluids, hair, and nails. When your body is in nitrogen balance, your nitrogen intake equals your nitrogen losses; in other words, you are consuming enough protein to replace losses. You are not gaining or losing body protein; you are maintaining it at a constant level. Nitrogen balance is negative when you're losing body protein and positive when the amount of body protein is increasing (**Figure 6.12**).

Recommended Protein Intake

The RDA for protein for adults is 0.8 g/kg of body weight. For the average American adult male who weighs 89 kg (196 pounds), the RDA is about 71 g of protein/day. Most of us eat more protein than we need: Adult males in the United States consume almost 100 g of protein/day and adult females about 73 grams.[11] RDAs have also been developed for each of the essential amino acids;[28] these are not a concern in typical diet planning, but this information is important for the development of solutions for intravenous feeding.

Protein recommendations are expressed per unit of body weight because protein is needed to maintain and repair the body. The more a person weighs, the more protein he or she needs for those purposes. Because children are small, they need less total protein than adults do, but because new protein

FIGURE 6.12 Nitrogen balance A comparison of nitrogen intake to nitrogen output can be used to determine whether the body is gaining or losing protein.

a. Nitrogen balance
Nitrogen intake = Nitrogen output. This indicates that the amount of protein being synthesized is equal to the amount being broken down, so the total amount of protein in the body is not changing. Healthy adults who consume adequate amounts of protein and are maintaining a constant body weight are in nitrogen balance.

b. Negative nitrogen balance
Nitrogen intake < Nitrogen output. This indicates that more protein is being broken down than is being synthesized, so body protein is decreasing. Negative nitrogen balance occurs due to injury or illness as well as when the diet is too low in protein or calories.

c. Positive nitrogen balance
Nitrogen intake > Nitrogen output. This indicates that there is more protein synthesis than degradation, so the body is gaining protein. This occurs in individuals who are pregnant, growing, gaining weight, or increasing their muscle mass by lifting weights.

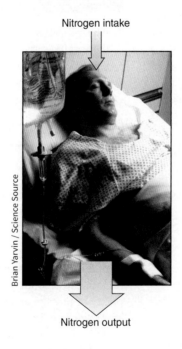

Think Critically
In a healthy adult, what will happen to nitrogen excretion if more protein is consumed than is needed by the body?

nitrogen balance The amount of nitrogen consumed in the diet compared with the amount excreted over a given period.

FIGURE 6.13 **Protein is needed for growth** Protein is needed for growth. During the first year of life, growth is rapid, so a large amount of protein is required per unit of body weight. As growth rate slows during childhood and adolescence, requirements per unit of body weight decrease, but continue to be greater than adult requirements, until age 19.

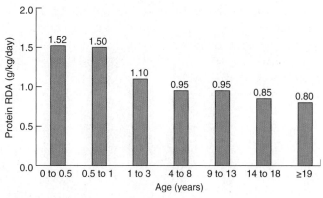

Interpret the Data

What is the RDA for a 2-year-old child who weighs 14 kg?

a. 14 g/day
b. 15.4 g/day
c. 11.2 g/day
d. 21 g/day

must be synthesized for growth to occur, protein requirements per unit of body weight are much greater for infants and children than for adults (**Figure 6.13**). To calculate protein needs per day, multiply weight in kilograms (which equals weight in pounds multiplied by 0.45) by the recommended amount for the individual's age.

Protein needs are increased during pregnancy and lactation. Additional protein is needed during pregnancy to support the expansion of maternal blood volume, the growth of the uterus and breasts, the formation of the placenta, and the growth and development of the fetus. The RDA for pregnant women is 25 g of protein/day higher than the recommendation for nonpregnant women. An extra 25 g/day is also needed during lactation to provide protein for the production of breast milk.

The RDA for protein is not different for older adults, but because there is a decrease in energy needs with aging, the diet must be higher in protein relative to calories than in younger adults. Declining muscle mass is a problem that affects health and the ability to maintain independence in the elderly. Protein intake that meets or, many believe, exceeds the current RDA, along with strength training exercise, may help delay loss of muscle mass and maintain physical function and optimal health. There is increasing evidence that to maintain muscle mass, older adults may benefit from a diet with 1.2 to 2.0 g protein/kg body weight.[29]

Extreme stresses on the body, such as infections, fevers, burns, or surgery, increase the amount of protein that is broken down. For the body to heal and rebuild, the amount of protein lost must be replaced. The extra amount needed for healing depends on the injury. A severe infection may increase the body's protein needs by about 30%; a serious burn can increase protein requirements by 200 to 400%.

Recommendations for protein intake are higher for athletes than for the general population, ranging from 1.2 to 2.0 g protein/kg of body weight (**Figure 6.14a**).[30] Athletes often think they need supplements to meet their higher protein needs, but because they have high energy needs, most athletes can easily meet their protein needs with food sources of high-quality protein. Taking protein supplements can however be a convenient way to meet protein needs and consume protein at times found to optimize muscle protein synthesis (**Figure 6.14b**).

In addition to the RDA, the DRIs include a recommendation for protein intake as a percentage of calories: The AMDR for protein is 10 to 35% of calories. This range allows for different food preferences and eating patterns. A protein intake in this range will meet protein needs and allow sufficient intakes of other nutrients to promote health. A diet that provides 10% of calories from protein will meet the RDA but is a relatively low-protein diet compared with typical eating patterns in the United States. The upper end of this healthy range—35% of calories—is a high-protein diet, about twice as much protein as the average American eats. This amount of protein is not harmful, but if the diet is this high in protein, it is probably high in animal products, which tend to be high in saturated fat and cholesterol. Therefore, unless protein sources are chosen carefully, a diet that contains 35% protein would tend to include more saturated fat than would a diet with the same number of calories that contains only 10% protein. For example, a 2500-Calorie diet that provides 35% of calories from protein would include the animal protein equivalent of one pound of turkey and 3 quarts of milk daily.

FIGURE 6.14 **Protein needs of athletes** Physical activity increases protein needs.

a. Athletes need extra protein because some protein is used to provide energy and to maintain blood glucose during activity and because protein provides the raw materials needed for muscle growth and repair.

b. Many athletes choose supplements to boost total protein intake and to add individual amino acids.

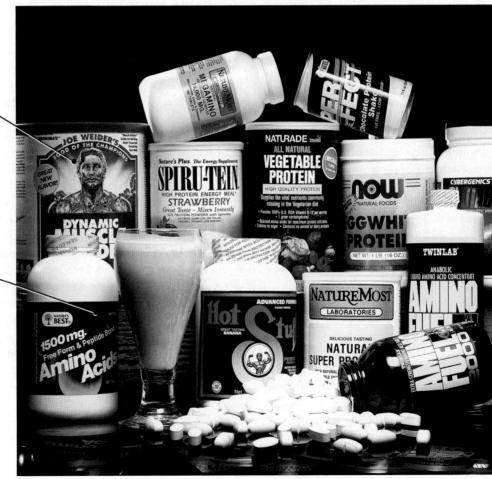

Protein powders can be an easy way to consume high-quality protein immediately after an exercise session when protein is most effective for repairing and building muscle (see Chapter 10).

Many promises are made about amino acid supplements, from aiding sleep to enhancing athletic performance. There is weak evidence to support some of these claims, but consuming large amounts of one amino acid may interfere with the absorption of others. Due to insufficient research, no ULs have been set for amino acids.

Choosing Protein Wisely

To evaluate protein intake, it is important to consider both the amount and the quality of protein in the diet. **Protein quality** is a measure of how good the protein in a food is at providing the essential amino acids the body needs to synthesize proteins (see Appendix J for methods used to measure protein quality). Because animal amino acid patterns are similar to those of humans, the animal proteins in our diet generally provide a mixture of amino acids that better matches our needs than the amino acid mixtures provided by plant proteins. Animal proteins also tend to be digested more easily than plant proteins; only protein that is digested can contribute amino acids to meet the body's requirements.[31] Because they are easily digested and supply essential amino acids in the proper proportions for human use, foods of animal origin are generally sources of **high-quality protein**, or **complete dietary protein**. When your diet contains high-quality protein, you don't have to eat as much total protein to meet your needs.

Compared to animal proteins, plant proteins are usually more difficult to digest and are lower in one or more of the essential amino acids. They are therefore generally referred to as **incomplete dietary protein**. Exceptions include quinoa and soy protein, which are both high-quality plant proteins.

Complementary proteins If you get your protein from a single source and that source is an incomplete protein, it will be difficult to meet your body's protein needs. However, combining proteins that are limited in different amino acids can supply a complete mixture of essential amino acids. For example, legumes are limited in methionine but high in lysine. When legumes are consumed with grains, which are high in methionine and low in lysine, the combination provides all the needed amino acids (**Figure 6.15a**). Vegetarian diets rely

FIGURE 6.15 **Protein complementation** The amino acids in grains, nuts, or seeds complement the amino acids in legumes to provide complete proteins.

a. The amino acids that are most often limited in plant proteins are lysine, methionine, and cysteine. As a general rule, legumes are deficient in methionine and cysteine but high in lysine. Grains, nuts, and seeds are deficient in lysine but high in methionine and cysteine.

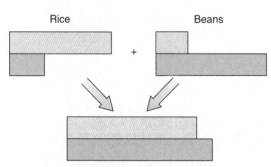

b. Many of the food combinations consumed in traditional diets take advantage of complementary plant proteins, such as lentils and rice or chickpeas and rice in India, rice and pinto or black beans in Mexico and South America, hummus (chickpeas and sesame seeds) in the Middle East, and bread and peanut butter (peanuts are a legume) in the United States.

Grains, Nuts, and Seeds

Rice and	Beans
Rice and	Lentils
Bread and	Peanut butter
Cashew and	Tofu stirfry
Corn tortilla and	Beans
Sesame seeds and	Chick peas
Corn bread and	Black-eyed peas
Sesame seeds and	Peanut sauce
Nuts and	Soy beans
Rice and	Tofu

Legumes

on this technique, called **protein complementation**, to meet protein needs. By eating plant proteins that have complementary amino acid patterns, a person can meet his or her essential amino acid requirements without consuming any animal proteins. Many vegetarian meals include complementary proteins (**Figure 6.15b**). However, complementary proteins do not have to be consumed in the same meal; eating an assortment of plant foods throughout the day can provide enough of all the essential amino acids.[32]

MyPlate and Dietary Guidelines recommendations

MyPlate and the Dietary Guidelines include recommendations regarding both animal and plant sources of protein to meet your need for protein and essential amino acids (**Figure 6.16**). The MyPlate food groups that provide the most protein per serving are the dairy and protein groups; 1 cup of milk provides about 8 g of protein, 1 ounce of meat or fish provides about 7 g, and ½ cup of beans or ¼ cup of nuts or

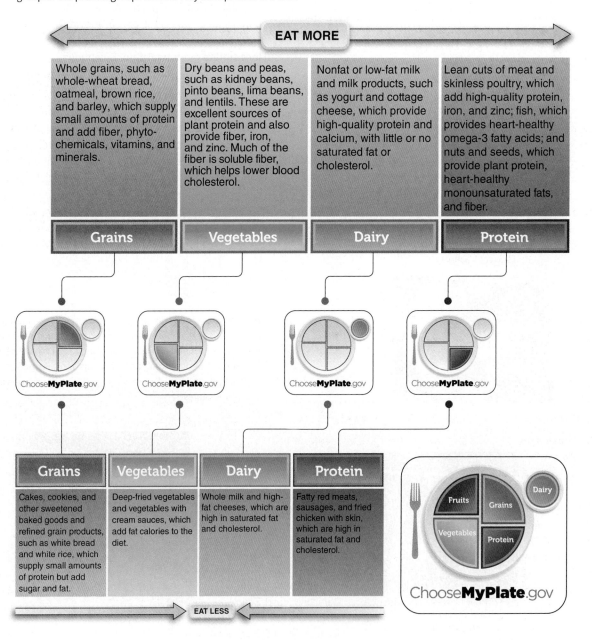

FIGURE 6.16 Choosing healthy protein sources The protein group and the dairy group of MyPlate provide the most concentrated sources of protein. Nuts, dry beans, and peas are the most concentrated sources of plant protein. When you eat dry beans or peas, you can count them in either the vegetables group or the protein group. There is very little protein in fruits.

---

protein complementation The process of combining proteins from different sources so that they collectively provide the proportions of amino acids required to meet the body's needs.

> ### What Should I Eat?
>
> **Protein Sources**
>
> **Increase your plant protein**
> - Add nuts and seeds to salads.
> - Snack on almonds or edamame.
> - Use quinoa with your stir-fry.
> - Try a veggie burger.
>
> **Get protein without too much saturated fat**
> - Skip the sausage and grill some fish.
> - Have a chicken sandwich rather than a burger.
> - Go for round steaks, loin chops, or other lean cuts of meat.
> - Choose low-fat milk and yogurt.
>
> **Reduce your portions of animal proteins**
> - Half your serving of beef or chicken and stir-fry it with lots of vegetables.
> - Serve a small portion of meat over noodles.
> - Have your crackers with hummus instead of cheddar.
> - Add veggies to your omelets rather than ham.
>
> Use iProfile to look up the protein content of your favorite vegetarian entrée.

seeds provides 6 to 10 g. Each serving from the grains group and the vegetables group provides 2 to 4 g, so choosing the recommended number of servings from these groups will provide a significant proportion of your protein needs. To assure an overall healthy diet, the Dietary Guidelines recommend that we choose a variety of protein foods, including seafood, lean meat and poultry, eggs, beans, soy products, and unsalted nuts and seeds. They recommend increasing consumption of seafood and low-fat or fat-free dairy products and replacing protein foods that are high in solid fats with those that are lower in solid fats and calories (see *What Should I Eat?*).

> ### Concept Check
>
> 1. **What** circumstances result in a positive nitrogen balance?
> 2. **Why** is the protein in beef considered to be higher quality than the protein in wheat?
> 3. **What** plant-based food could you serve with rice to increase the overall protein quality of the meal?

6.7 Vegetarian Diets

LEARNING OBJECTIVES

1. **Compare** the different types of vegetarian diets.
2. **Discuss** the benefits and nutritional risks of vegetarian diets.

In many parts of the world, **vegetarian diets** have evolved mostly out of necessity because animal sources of protein are limited, either physically or economically. Because animals require more land and resources to raise and are more expensive to purchase than are plants, much of the developing world relies primarily on plant foods to meet protein needs. For example, in rural Mexico, most of the protein in the diet comes from beans, rice, and tortillas (corn), and in India, protein comes from lentils and rice. As a population's economic prosperity rises, the proportion of animal foods in its diet typically increases, but in developed countries, many people eat vegetarian diets for reasons other than economics, such as health, religion, personal ethics, or environmental awareness. **Vegan diets** eliminate all animal products, but there are other types of vegetarian diets that are less restrictive (**Table 6.1**).

Benefits of Vegetarian Diets

A vegetarian diet can be a healthy alternative to the traditional American meat-and-potatoes diet. Vegetarians have been shown to have lower body weight relative to height and

---

vegetarian diet A diet that includes plant-based foods and eliminates some or all foods of animal origin.

vegan diet A plant-based diet that eliminates all animal products.

Dietary pattern	Description	Red meat	Poultry/fish	Dairy/eggs
Nonvegetarian	Includes all types of animal products.	✓	✓	✓
Semivegetarian	Occasional red meat, fish, and poultry, as well as dairy products and eggs.	✓	✓	✓
Pescetarian	Excludes all animal flesh except fish. Usually includes dairy and eggs.	⊘	✓ (fish only)	✓
Lacto-ovo vegetarian	Excludes all animal flesh but does include eggs and dairy products such as milk and cheese.	⊘	⊘	✓
Lacto-vegetarian	Excludes animal flesh and eggs but does include dairy products.	⊘	⊘	✓ (dairy only)
Vegan	Excludes all food of animal origin.	⊘	⊘	⊘

TABLE 6.1 Types of vegetarian diets

a reduced incidence of obesity and of other chronic diseases, such as diabetes, cardiovascular disease, high blood pressure, and some types of cancer. The lower body weight of vegetarians is a result of the low energy density of vegetarian diets, which makes these diets more filling. The reductions in the risk of other chronic diseases may be due to lower body weight and to the fact that these diets are lower in saturated fat, which increases disease risk. Or it could be that vegetarian diets are higher in whole grains, legumes, nuts, vegetables, and fruits, which add fiber, vitamins, minerals, antioxidants, and phytochemicals—substances that have been shown to lower disease risk. It is likely that the total dietary pattern, rather than a single factor, is responsible for the health-promoting effects of vegetarian diets.

In addition to reducing disease risks, diets that rely more heavily on plant protein than on animal protein are more economical. For example, a vegetarian stir-fry over rice costs about half as much as a meal of steak and potatoes. Yet both meals provide a significant portion of the day's protein requirement. A small steak, a baked potato with sour cream, and a tossed salad provides about 50 g of protein, whereas a dish of rice with tofu and vegetables provides about 25 g.

Thinking It Through

A Case Study on Choosing a Healthy Vegetarian Diet

Simon is 26 years old and weighs 154 pounds. A year ago, he decided to stop eating meat because he thought it would make his diet healthier. Now that he is studying nutrition, he has become concerned that his vegetarian diet may not be as healthy as he thought. Simon records his food intake for one day and then uses iProfile to assess his nutrient intake. His analysis reveals that his diet provides 2900 Calories, 78 g of protein, and 43 g of saturated fat.

Does Simon's diet provide enough protein to meet his RDA of 0.8 g of protein per kilogram of body?

Your answer:

How does the percentage of calories from saturated fat in Simon's diet compare with recommendations?

Your answer:

This is a photo of Simon's typical lunch. Why is it high in saturated fat?

Your answer:

Simon typically has cereal with whole milk for breakfast, this sandwich for lunch, and cheese pizza or cheese lasagna for dinner. He snacks on chips and ice cream.

Vegetarian diets are often deficient in calcium, vitamin D, zinc, and iron. Based on his meal pattern, which of these are likely low in Simon's diet?

Your answer:

Simon's choices are lacking in several MyPlate food groups. What does he need to add to his diet to meet these recommendations?

Your answer:

To reduce his saturated fat intake, Simon wants to try a vegan lunch. Suggest a vegan sandwich Simon could have that makes use of complementary plant proteins.

Your answer:

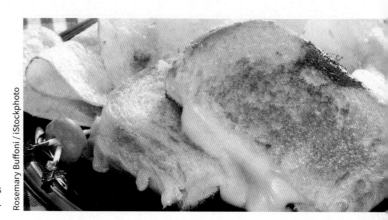

Rosemary Buffoni / iStockphoto

(Check your answers in Appendix L.)

Risks of Vegetarian Diets

Despite the health and economic benefits of vegetarian diets, a poorly planned vegetarian diet can result in nutrient deficiencies or excesses (see *Thinking It Through*). Protein deficiency is a risk when vegan diets that contain little high-quality protein are consumed by small children or by adults with increased protein needs, such as pregnant women and those recovering from illness or injury. Most people can easily meet their protein needs with lacto and lacto-ovo vegetarian diets. These diets contain high-quality animal proteins from eggs or milk, which complement the limiting amino acids in the plant proteins.

Vitamin and mineral deficiencies are a greater concern for vegetarians than is protein deficiency. Of primary concern to vegans is vitamin B_{12}. Because this B vitamin is found almost exclusively in animal products, vegans must take vitamin B_{12} supplements or consume foods fortified with vitamin B_{12} to meet their needs for this nutrient. Another nutrient of concern is calcium. Dairy products are the major source of calcium in the North American diet, so diets that eliminate these foods must include plant sources of calcium, such as greens and tofu. Likewise, because much of the dietary vitamin D comes from fortified milk, this vitamin must be made in the body from exposure to sunlight or consumed in other sources. Iron and zinc may be deficient in vegetarian diets because they exclude red meat, which is an excellent source of these minerals, and iron and zinc are poorly absorbed from plant sources. Because dairy products are low in iron and zinc, lacto-ovo and lacto vegetarians as well as vegans are at risk for deficiencies of these minerals. Vegan diets may also be low in iodine and the omega-3 fatty acids EPA and DHA (see Chapter 5). **Table 6.2** provides suggestions for how vegans can meet the needs for the nutrients just discussed.

Planning Vegetarian Diets

Well-planned vegetarian diets, including vegan diets, can meet nutrient needs at all stages of the life cycle, from infancy, childhood, and adolescence to early, middle, and late adulthood, and during pregnancy and lactation.

TABLE 6.2 Meeting nutrient needs with a vegan diet

Nutrient at risk	Sources in vegan diets
Protein	Soy-based products, legumes, seeds, nuts, grains, and vegetables
Vitamin B_{12}	Products fortified with vitamin B_{12}, such as soy beverages, rice milk,* almond milk,* and breakfast cereals; fortified nutritional yeast; dietary supplements
Calcium	Tofu processed with calcium; broccoli, kale, collard and mustard greens, bok choy, and legumes; products fortified with calcium, such as soy beverages, rice milk,* almond milk,* grain products, and orange juice
Vitamin D	Sunshine; products fortified with vitamin D, such as soy beverages, rice milk,* almond milk,* breakfast cereals, and margarine
Iron	Legumes, tofu, dark-green leafy vegetables, dried fruit, whole grains, iron-fortified grain products (absorption is improved when iron-containing foods are consumed with vitamin C found in citrus fruit, tomatoes, strawberries, and dark-green vegetables)
Zinc	Whole grains, wheat germ, legumes, nuts, tofu, and fortified breakfast cereals
Iodine	Iodized salt, sea vegetables (seaweed), and foods grown near the sea
Omega-3 fatty acids	Canola oil, flaxseed and flaxseed oil, soybean oil, walnuts, and sea vegetables (seaweed), which provide fatty acids that can be used to synthesize EPA and DHA; DHA-rich microalgae

*Most rice milk and almond milk products are low in protein (about 1 g per serving) but fortified with vitamin B_{12}, calcium, and vitamin D.

FIGURE 6.17 Using MyPlate to choose a vegetarian dietary pattern MyPlate can be used to choose a healthy vegetarian dietary pattern by making simple modifications to the choices from the protein and dairy groups.

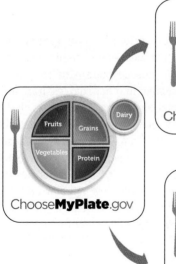
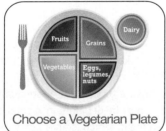
Choose a Vegetarian Plate
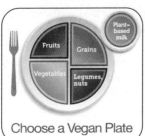
Choose a Vegan Plate

One way to plan a healthy vegetarian diet is to modify the selections from MyPlate (**Figure 6.17**). The food choices and recommended amounts from the grains, vegetables, and fruits groups should stay the same for vegetarians. Including 1 cup of dark-green and colorful vegetables daily will help meet iron and calcium needs. The dairy group and the protein group include foods of animal origin. Vegetarians who consume eggs and milk can still choose these foods. Those who avoid all animal foods can choose dry beans, nuts and seeds, and soy products from the protein group. Fortified soymilk and protein-enriched rice milk can be substituted for dairy foods. To obtain adequate vitamin B_{12}, vegans must take supplements or use products fortified with vitamin B_{12}. Obtaining plenty of omega-3 fatty acids from foods such as canola oil, nuts, and flaxseed ensures adequate synthesis of the long-chain omega-3 fatty acids DHA and EPA.

Concept Check

1. **What** is a lacto-vegetarian diet?
2. **Why** are vegans at risk for vitamin B_{12} deficiency?

Summary

1 Proteins in Our Food 144

- Dietary protein comes from both animal and plant sources. Animal sources of protein are generally good sources of iron, zinc, and calcium but are high in saturated fat and cholesterol. Plant sources of protein, such as the grains, nuts, and legumes shown in the photo, are higher in unsaturated fat and fiber and provide phytochemicals.

Figure 6.1b Plant proteins

2 The Structure of Amino Acids and Proteins 145

- **Amino acids** are the building blocks from which proteins are made. Each amino acid contains an amino group, an acid group, and a unique side chain. The amino acids that the body is unable to make in sufficient amounts are called **essential amino acids** and must be consumed in the diet.
- Proteins are made by linking amino acids by **peptide bonds**, as shown here, to form **polypeptides**. Polypeptide chains fold to create unique three-dimensional protein structures. The shape of a protein determines its function.

Figure 6.3 Proteins are made from amino acids

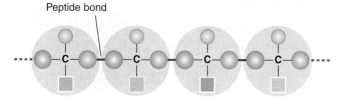

3 Protein Digestion and Absorption 147

- Digestion breaks dietary protein into small **peptides** and amino acids that can be absorbed. Because amino acids that share the same transport system, such as those pictured here in green and blue, compete for absorption, an excess of one can inhibit the absorption of another.

Figure 6.5 Protein digestion and absorption

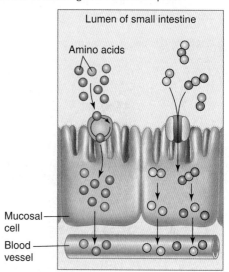

4 Protein Synthesis and Functions 149

- Amino acids are used to synthesize proteins and other nitrogen-containing molecules. **Genes** located in the nucleus of the cell code for the order of amino acids in the polypeptide chains that make up proteins. Regulatory mechanisms ensure that proteins are made only when they are needed. For a protein to be synthesized, all the amino acids it contains must be available. The essential amino acid present in shortest supply relative to need, depicted here as the orange spheres, is called the **limiting amino acid**.

Figure 6.7c Protein synthesis

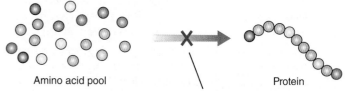

A shortage of the amino acid represented by the orange spheres limits the ability to synthesize a protein that is high in this amino acid.

- In the body, protein molecules form structures, regulate body functions, transport molecules through the blood and in and out of cells, function in the immune system, and aid in muscle contraction, fluid balance, and acid balance.
- When the diet is deficient in energy, amino acids from body proteins are used as an energy source and to synthesize glucose. When the diet contains more protein than needed, amino acids are used as an energy source and can be used to synthesize fatty acids. Before amino acids can be used for these purposes, the amino group must be removed via **deamination**.

5 Protein in Health and Disease 153

- **Protein-energy malnutrition (PEM)** is a health concern primarily in developing countries. **Kwashiorkor**, shown here, occurs when the protein content of the diet is deficient but energy is adequate. It is most common in children. **Marasmus** occurs when total energy intake is deficient.

Figure 6.10b Protein-energy malnutrition

- High-protein diets increase the production of urea and other waste products that must be excreted in the urine and therefore can increase water losses. High protein intakes increase urinary calcium losses, but when calcium intake is adequate, high-protein diets are associated with greater bone mass and fewer fractures. Diets high in animal proteins and low in fluid are associated with an increased risk of kidney stones. High-protein diets can be high in saturated fat and cholesterol.

- In sensitive individuals, protein fragments can trigger an immune system reaction, resulting in a **food allergy**. Some amino acids and proteins can also cause **food intolerances**.

6 Meeting Protein Needs 157

- Protein requirements are determined by looking at **nitrogen balance**, the amount of nitrogen consumed as dietary protein compared with the amount excreted as protein waste products.

- For healthy adults, the RDA for protein is 0.8 g/kg of body weight. Older adults and athletes may benefit from higher protein intakes, and growth, pregnancy, lactation, illness, and injury increase requirements. Recommendations for a healthy diet are to ingest 10 to 35% of calories from protein.

- Animal proteins are considered **high-quality proteins** because their amino acid composition matches that needed to synthesize body proteins. Most plant proteins are limited in one or more of the essential amino acids needed to make body protein; therefore, they are considered **incomplete proteins**. The **protein quality** of plant sources can be increased through **protein complementation**. As illustrated here, it combines proteins with different limiting amino acids to supply enough of all the essential amino acids.

Figure 6.15b Protein complementation

7 Vegetarian Diets 163

- Vegetarian diets eliminate some or all animal products. Compared with meat-based diets, **vegetarian diets** are lower in saturated fat and cholesterol and higher in fiber, certain vitamins and minerals, antioxidants, and phytochemicals. People consuming **vegan diets** must plan their diets carefully to meet their needs for vitamin B_{12}, calcium, vitamin D, iron, zinc, iodine, and omega-3 fatty acids.

- Modifications to MyPlate, as shown here, can be used to meet needs for vegetarian and vegan diets.

Figure 6.17 Using MyPlate to choose a vegetarian dietary pattern

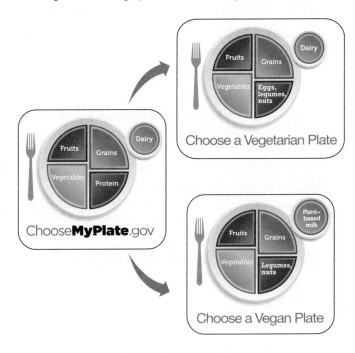

Key Terms

- amino acid 144
- amino acid pool 149
- celiac disease 155
- conditionally essential amino acid 145
- deamination 152
- denaturation 147
- dipeptide 146
- edema 151
- essential, or indispensable, amino acid 145
- food allergy 155
- food intolerance or food sensitivity 155
- gene 149
- gene expression 149
- high-quality protein or complete dietary protein 161
- hydrolyzed protein or protein hydrolysate 157
- incomplete dietary protein 161
- kwashiorkor 154
- legume 144
- limiting amino acid 150
- marasmus 154
- monosodium glutamate (MSG) 155
- MSG symptom complex or Chinese restaurant syndrome 155
- nitrogen balance 158
- nonessential, or dispensable, amino acid 145
- peptide bond 146
- peptide 146
- phenylketonuria (PKU) 145
- polypeptide 146
- protein complementation 162
- protein quality 161
- protein-energy malnutrition (PEM) 153
- transamination 149
- transcription 150
- translation 150
- tripeptide 146
- urea 152
- vegan diet 163
- vegetarian diet 163

What is happening in this picture?

ChiccoDodiFC / Shutterstock

Cheese is made by heating milk and then adding bacterial cultures to ferment the lactose in the milk, converting it to lactic acid. Then a milk-clotting enzyme called rennet is added to cause the protein casein, which is normally dissolved, to precipitate and clump together, forming curds. The curds are then separated from the liquid and further processed to create a final product that contains the milk fat and some of the milk protein.

Think Critically

How is the milk protein casein changed during cheese production?

What protein function is needed for cheese production?

How is cheese different from milk nutritionally?

CHAPTER 7

Vitamins

Multiple sclerosis (MS) is a degenerative neurological disease. Although this disease occurs around the world, more cases are reported in northern and southern latitudes, prompting investigators to wonder what factors might contribute to such a distribution.

At the extreme northern edge of the Scottish mainland lies an archipelago called the Orkney Islands. These islands have the largest concentration of MS in the world. One in every 170 women suffers from the condition. The reasons proposed for this remarkable number of cases range from a genetic predisposition to a deficiency of vitamin D. This vitamin can be made in the body when exposure to sunlight is plentiful, but in northern Scotland, sunlight is limited by short winter days and frequent cloud cover. Interestingly, rickets, also caused by a lack of vitamin D, is making a comeback in Scotland and northern parts of England.

Populations enduring sun-starved northern winters have traditionally acquired vitamin D from a diet rich in fish and eggs. But as consumption of these foods has declined in favor of a more modern diet high in processed foods, vitamin D deficiency has increased. Health authorities in the United Kingdom are proposing fortifying milk with vitamin D, a common practice in the United States, to ensure that consumers are getting enough to promote better health and perhaps prevent disease.

CHAPTER OUTLINE

A Vitamin Primer 171
- Vitamins in Our Food

Debate: Discretionary Fortification: Is It Beneficial?
- Vitamin Bioavailability
- Vitamin Functions
- Understanding Vitamin Needs

Vitamins and Energy Metabolism 179
- Thiamin
- Riboflavin
- Niacin
- Biotin
- Pantothenic Acid
- Choline and Other Vitamin-like Compounds

Vitamins and Healthy Blood 186
- Vitamin B_6
- Folate (Folic Acid)
- Vitamin B_{12}
- Vitamin K

What a Scientist Sees: Anticoagulants Take Lives and Save Them

Antioxidant Vitamins 196
- Vitamin C
- Vitamin E

Vitamins in Gene Expression 200
- Vitamin A

Thinking It Through: A Case Study on Vitamins in the Modern Diet
- Vitamin D

What Should I Eat? Vitamins

Meeting Needs with Dietary Supplements 211
- Who Needs Vitamin/Mineral Supplements?
- Herbal Supplements
- Choosing Supplements with Care

7.1 A Vitamin Primer

LEARNING OBJECTIVES

1. **Discuss** the dietary sources of vitamins.
2. **Describe** the information about vitamins presented on food labels.
3. **Explain** how bioavailability affects vitamin requirements.
4. **Name** the DRI values that recommend enough and not too much of each vitamin.

Vitamins are organic compounds that are essential in small amounts in the diet to promote and regulate body processes necessary for growth, reproduction, and the maintenance of health. When a vitamin is lacking in the diet, deficiency symptoms occur. When the vitamin is restored to the diet, the symptoms resolve.

Vitamins have traditionally been assigned to two groups, based on their solubility in water or fat. The **water-soluble vitamins** include the B vitamins and vitamin C. The **fat-soluble vitamins** include vitamins A, D, E, and K (**Figure 7.1**). This classification allows generalizations to be made about how the vitamins are absorbed, transported, excreted, and stored.

Vitamins in Our Food

Almost all foods contain some vitamins, and all the food groups contain foods that are good sources of a variety of vitamins (**Figure 7.2**). The amount of a vitamin in a food depends on the

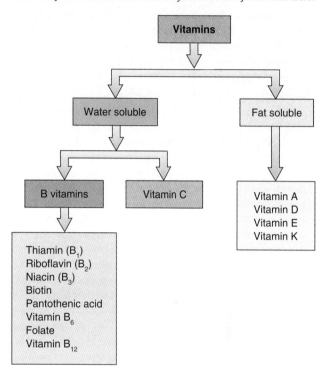

FIGURE 7.1 **The vitamins** The vitamins were initially named alphabetically, in approximately the order in which they were identified. The B vitamins were first thought to be a single chemical substance but were later found to be many different vitamins, and for this reason they were distinguished by numbers. Vitamins B_6 and B_{12} are the only ones that are still routinely referred to by their numbers.

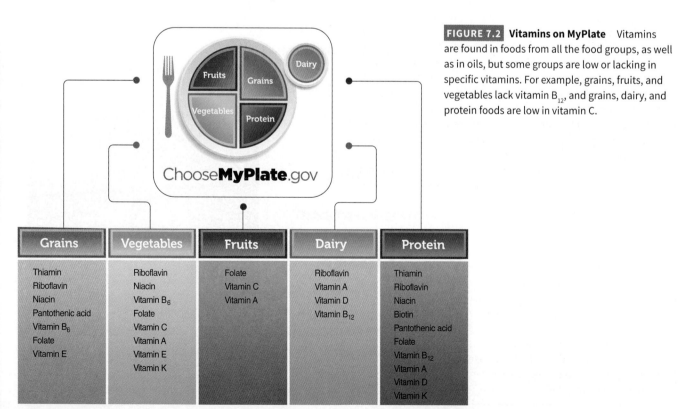

FIGURE 7.2 **Vitamins on MyPlate** Vitamins are found in foods from all the food groups, as well as in oils, but some groups are low or lacking in specific vitamins. For example, grains, fruits, and vegetables lack vitamin B_{12}, and grains, dairy, and protein foods are low in vitamin C.

amount that is naturally present in the food, what is added to it, and how the food is processed, prepared, and stored.

Nutrients are added to foods for a variety of reasons (see Debate: Discretionary Fortification: Is It Beneficial?). Sometimes they are added to comply with government fortification programs that mandate such additions in order to prevent vitamin or mineral deficiencies and promote health in the population. For example, milk is fortified with vitamin D

Debate

Discretionary Fortification: Is It Beneficial?

The Issue

Some foods, such as protein bars, breakfast cereals, flavored waters, and energy drinks, are fortified with nutrients at the discretion of manufacturers to enhance and sell their products. Do we need these extra nutrients? Can they pose a risk of toxicity if eaten in large quantities?

Historically certain foods were fortified with specific nutrients to improve public health. For example, the enrichment of grains, begun in the 1940s, helped prevent deficiencies of B vitamins and iron. Today many foods continue to be fortified with small amounts of specific nutrients according to government guidelines to address nutrient deficiencies; these programs have helped prevent birth defects and deficiencies of vitamin A, vitamin D, and iodine. However, not all fortification is mandated or regulated by the government. Discretionary fortification by food manufacturers can add nutrients that don't necessarily address public health needs.

Advocates of discretionary fortification point out that fortified foods provide consumers with a greater variety of nutrient sources and help people from all segments of the population meet their needs.[1] For example, someone who cannot tolerate dairy products can meet his or her calcium needs with calcium-fortified orange juice. Analyses of sources of nutrients in the American diet show that without fortified foods, a high percentage of children and adolescents would have inadequate intakes of many micronutrients, including thiamin, folate, and iron.[2] However, these analyses did not distinguish mandated from discretionary fortification.

Despite the benefits associated with some nutrient fortification, many of the nutrients manufacturers add to foods to help market their products are ones that are not even low in the population's diet.[3] For example, vitamins B_6 and B_{12} are commonly added to energy drinks and sports beverages, yet there is little evidence that these nutrients are lacking in the population of teens and young adults that typically consumes them. These nutrients are more likely to be deficient in the diets of older adults, who are typically not major consumers of energy drinks.

Those concerned about discretionary fortification point out the risk of nutrient toxicity. Because the nutrients naturally found in foods like an orange, a tomato, a slice of whole-grain bread, or a piece of salmon are present in relatively small amounts, the risk of consuming a toxic amount of a nutrient is very low. In contrast, it is not difficult to swallow a very high dose of one or more nutrients from a few servings of a highly fortified food. For example, if you drank the recommended 2 to 3 liters of fluid as water fortified with vitamin C, niacin, vitamin E, and vitamin B_6, you could exceed the UL for these vitamins. Then if you also consumed 2 cups of fortified breakfast cereal and two protein bars during the day, your risk of toxicity would increase even more. While the risk of toxicity is low in adults, an analysis of nutrient intakes in toddlers and preschoolers showed that a significant percentage had intakes of preformed vitamin A, folate, and zinc that exceeded the ULs (see graph). Much of the excess zinc is likely to be from fortified breakfast cereals.[4] The government labels these fortified products as foods, and we eat them like foods, but they may have the same toxicity risks as supplements.

There is also concern that the appeal of added nutrients may encourage the consumption of processed foods high in added sugars or salt. For example, vitamin C–fortified fruit snacks may be consumed as a source of vitamin C. While this would help children meet their vitamin C requirement, it would also increase their intake of added sugars and reduce the likelihood that they will meet their needs for other nutrients.

So, are these products helpful or harmful? It depends on what is in them, how much you consume, and who you are. Should the government get involved in regulating the amounts of nutrients that can be added to all foods? The answer depends on your view of the government's role in food regulation. Should we be gobbling down super-fortified foods without a thought? Probably not.

Think Critically

Should the amounts and types of nutrients added to foods be regulated by the government?

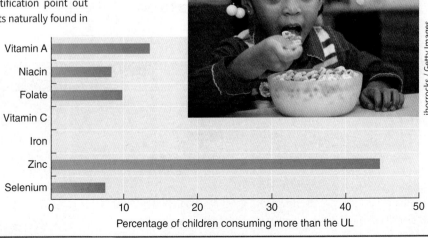

This graph shows the percentage of children 2 to 8 years of age whose usual intakes of these nutrients are equal to or greater than the UL. These percentages consider the nutrients intrinsic in foods and those added in fortification but do not include nutrients from dietary supplements. The percentage exceeding the UL would be even greater if supplements were included.

FIGURE 7.3 Vitamins on food labels The planned Nutrition Facts label—this one from canned salmon—is required to include the amount of vitamin D. The information on vitamin A and vitamin B_{12} is listed voluntarily. As a general guideline, if the % Daily Value is 20% or more, the food is an excellent source of that nutrient; if it is 10 to 19%, the food is a good source; and if it is 5% or less, the food is a poor source of that nutrient.

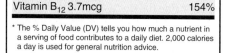

Interpret the Data

How much salmon would you need to eat to get 100% of the Daily Value for vitamin D?

to prevent vitamin D deficiency. Refined grains are fortified with thiamin, riboflavin, niacin, and iron to prevent deficiencies of these nutrients and with folic acid to reduce the risk of birth defects. In other cases, manufacturers add nutrients for marketing and sales purposes rather than to target nutrition problems.

Food labels can help identify packaged foods that are good sources of vitamins. Some vitamins are required on labels, and others can be listed voluntarily (**Figure 7.3**). The nutrients that must be listed are based on which nutrients are concerns for deficiency or excess in the diet at the time the labeling regulations are developed. Vitamin D is the only vitamin required on the planned Nutrition Facts labels; the current labels developed in 1990 require vitamins A and C and list amounts only as a percentage of the Daily Value. The planned labels list the amounts of required and voluntary nutrients both by weight and as a percentage of the Daily Value. Fresh fruits, vegetables, and fish, which are excellent sources of many vitamins, do not carry food labels. The FDA has therefore asked that grocery stores voluntarily provide nutrition information for the raw fruits, vegetables, and fish that are most frequently purchased, and about 75% of stores comply.[5]

The vitamins in foods can be damaged by exposure to light or oxygen, washed away during preparation, and destroyed by cooking. Thus, processing steps used by food producers can cause nutrient losses, as can cooking and storage methods used at home (**Figure 7.4**). Vitamin losses can be minimized through food preparation methods that reduce exposure to heat and light, which destroy some vitamins, and to water, which washes away water-soluble vitamins (**Table 7.1**).

FIGURE 7.4 **Which choice is highest in vitamins?** Because heat, light, air, and the passage of time all cause foods to lose nutrients, most of us try to purchase fresh produce, but is fresh always best?

The high temperatures used in canning reduce nutrient content. However, because canned foods keep for a long time, do not require refrigeration, and are often less expensive than fresh or frozen foods, they provide an available, affordable source of nutrients that may be the best choice in some situations.

Sometimes "fresh" produce is lower in nutrients than you would expect because it has spent a week in a truck, traveling to your store, several days on a shelf, and maybe another week in your refrigerator.

Frozen foods are often frozen in the field in order to minimize nutrient losses. Thus, frozen fruits and vegetables may supply more vitamins than "fresh" ones.

TABLE 7.1 Tips for preserving the vitamins in your food
• Store food away from heat and light and eat it soon after purchasing it.
• Cut fruits and vegetables as close as possible to the time when they will be cooked or served.
• Don't soak vegetables.
• Cook vegetables with as little water as possible by microwaving, steaming, pressure-cooking, roasting, grilling, stir-frying, or baking rather than boiling them.
• If foods are cooked in water, use the cooking water to make soups and sauces so that you can retrieve some of the nutrients.
• In order to avoid washing away water-soluble vitamins, don't rinse rice before cooking it.

Vitamin Bioavailability

About 40 to 90% of the vitamins in food are absorbed, primarily from the small intestine (**Figure 7.5**). The composition of the diet and conditions in the digestive tract and the rest of the body influence vitamin **bioavailability**. For example, fat-soluble vitamins are absorbed along with dietary fat (see Chapter 5). If the diet is very low in fat, absorption of these vitamins is impaired.

Once they have been absorbed into the blood, vitamins must be transported to the cells. Most of the water-soluble vitamins are bound to blood proteins for transport. Fat-soluble vitamins are incorporated into chylomicrons for transport from the intestine. The bioavailability of a vitamin depends on the availability of these transport systems.

Some vitamins are absorbed in an inactive form called a **provitamin**, or **vitamin precursor**. To perform vitamin functions, provitamins must be converted into active vitamin forms once they are inside the body. How much of each provitamin can be converted into the active vitamin and the rate at which this process occurs affect the amount of a vitamin available to function inside the body.

Vitamin Functions

Vitamins promote and regulate the body's activities. Each vitamin has one or more important functions (**Figure 7.6**). For example, vitamin A is needed for vision as well as normal growth and development. Vitamin K is needed for blood clotting and bone health. Often more than one vitamin is needed to ensure the health of a particular organ or system. Some vitamins act in a similar manner to do their jobs. For example, all the B vitamins are needed for the production of ATP from the energy-yielding nutrients, and vitamins C and E both protect us from damage caused by oxygen.

Understanding Vitamin Needs

The RDAs and AIs of the DRIs recommend amounts for each of the vitamins that will prevent deficiency and promote

---

bioavailability The extent to which the body can absorb and use a nutrient.

provitamin or **vitamin precursor** A compound that can be converted into the active form of a vitamin in the body.

PROCESS DIAGRAM — **FIGURE 7.5** **Vitamin absorption** Most vitamin absorption takes place in the small intestine. The mechanism by which vitamins are absorbed and transported affects their bioavailability.

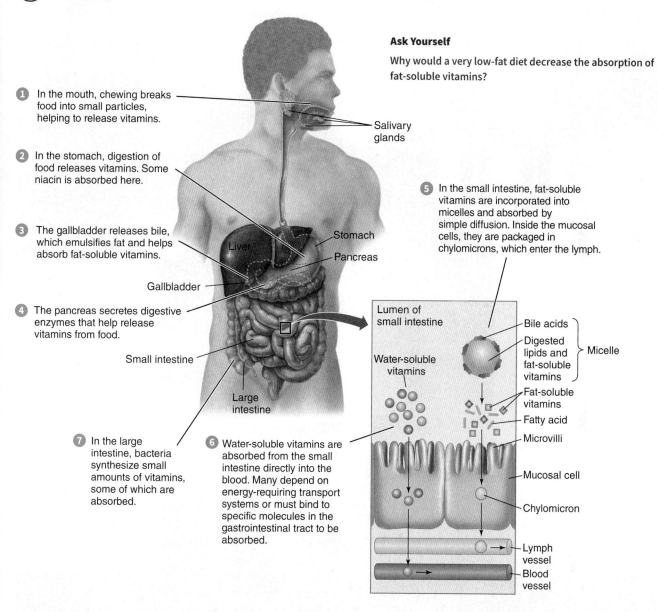

health (see Chapter 2 and Appendix A). Meeting vitamin needs requires a steady supply of vitamin-rich foods or supplements. Because most water-soluble vitamins are not stored to any great extent, deficiencies can develop more quickly than with the fat-soluble vitamins, which can be stored in the liver and fatty tissues.

Because more is not always better when it comes to nutrient intake, the DRIs have also established ULs as a guide to amounts that avoid the risk of toxicity. Fat-soluble vitamins carry a greater risk of toxicity because their solubility in fat limits the routes of excretion. Although excesses of most water-soluble vitamins are excreted in the urine, toxicity is still a concern for some of these vitamins.

Despite our knowledge of what vitamins do and how much of each we need, not everyone consumes the right amounts. In developing countries, vitamin deficiencies remain a major public health problem. In developed countries, thanks to the more varied food supply, along with fortification, vitamin-deficiency diseases have been almost eliminated. In these countries, concern now focuses on meeting the needs of high-risk groups,

NUTRITION INSIGHT

FIGURE 7.6 Functions of vitamins Each vitamin has one or more functions, and many have complementary roles in supporting health.

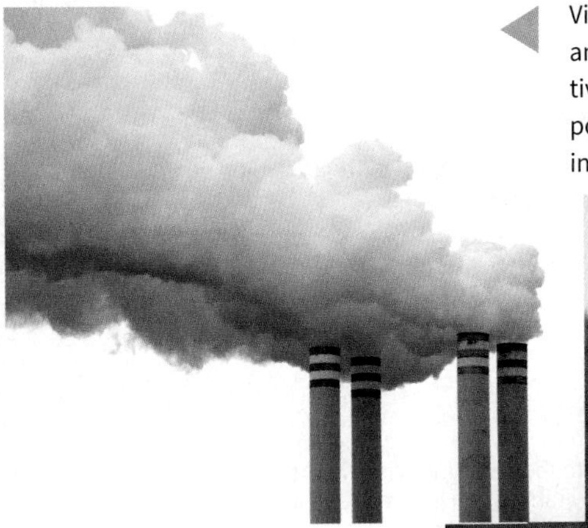

Thaddeus Robertson / Getty Images

Vitamin C, vitamin E, and provitamin A are antioxidants, which help protect us from oxidative damage that can be caused by industrial pollutants or reactive molecules generated inside the body.

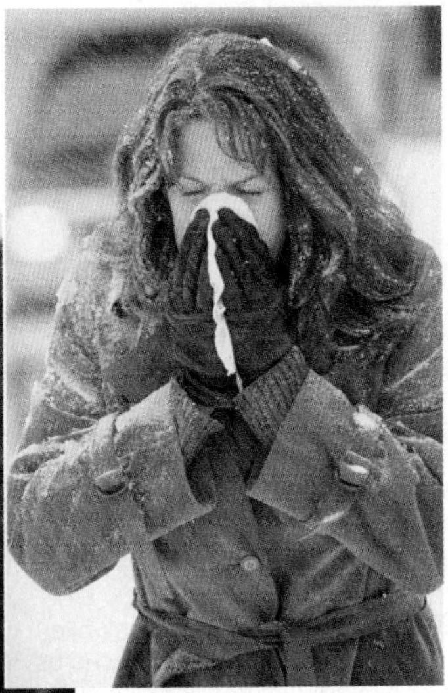

Design Pics Inc / National Geographic Creative

Vitamins A, B_6, C, and D, as well as folate, are needed for healthy immune function, and thus help protect us from infection.

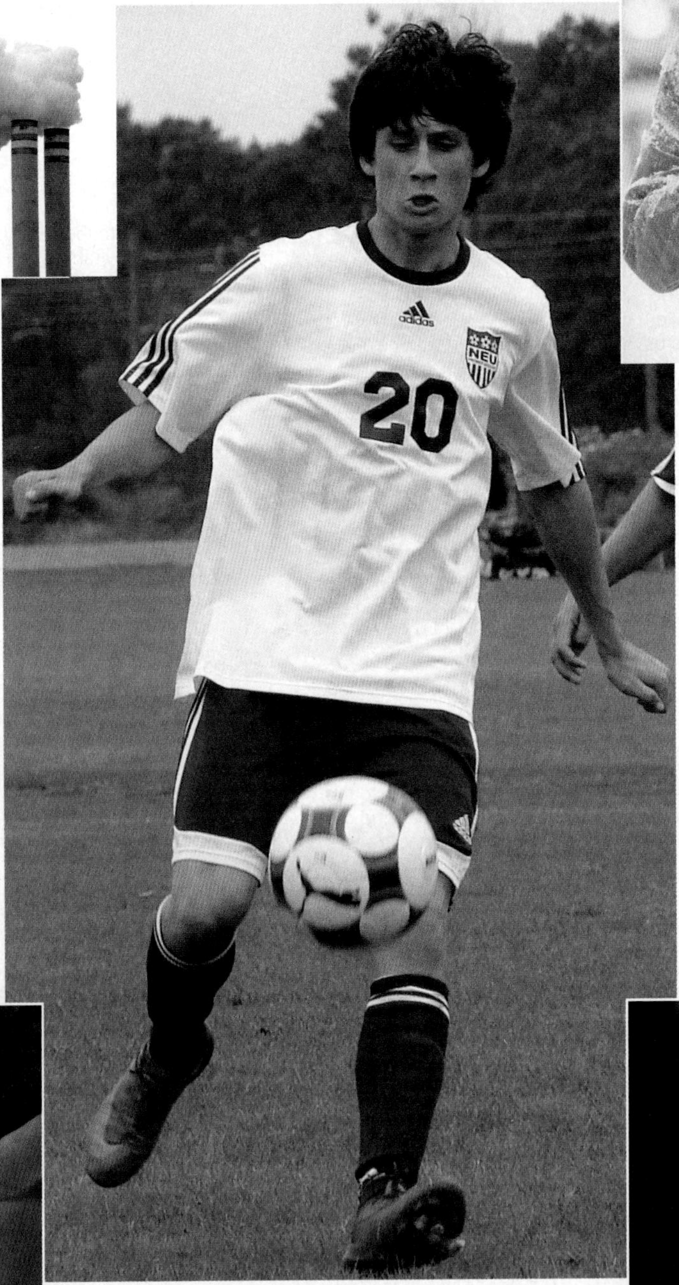

Courtesy Lori Smolin

Vitamin A and vitamin D are needed for normal growth and development.

The B vitamins thiamin, riboflavin, niacin, biotin, pantothenic acid, and vitamin B_6 are needed to produce ATP from carbohydrate, fat, and protein.

Folate, vitamin B_6, and vitamin B_{12} are important for protein and amino acid metabolism.

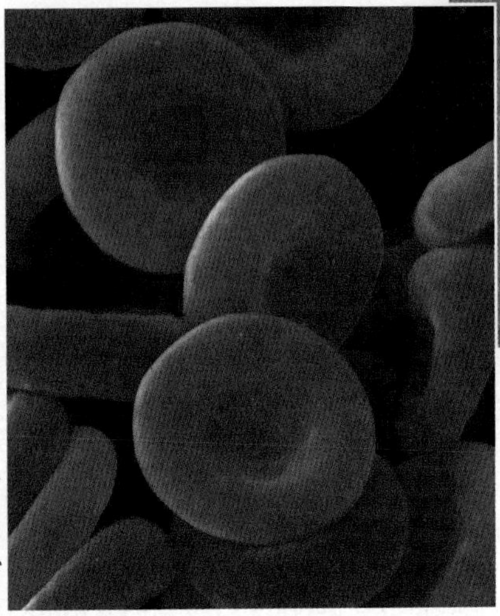

Cheryl Power / Science Source

Folate, vitamin B_6, vitamin B_{12}, and vitamin K are needed to keep blood healthy.

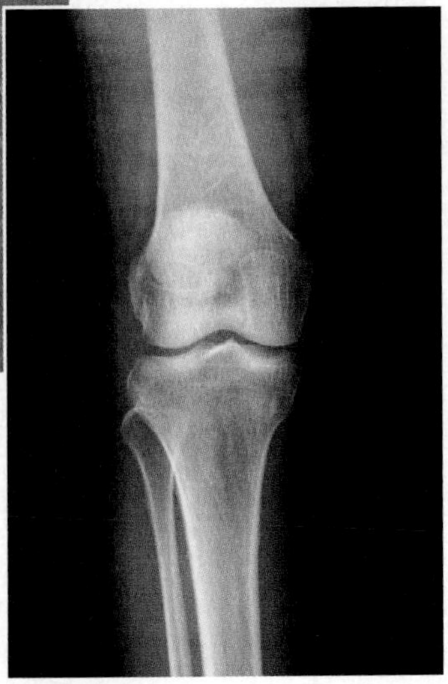

© stockdevil / iStockphoto

Vitamins A, D, K, and C are needed for bone health.

such as children and pregnant women; determining the consequences of marginal deficiencies, such as the effect of low vitamin D on the risk of multiple sclerosis; and evaluating the risks of consuming toxic amounts of certain vitamins.

Table 7.2 provides a summary of the recommended intakes of the water-soluble and fat-soluble vitamins and choline, along with their sources, functions, and deficiency and toxicity symptoms.

TABLE 7.2 A summary of the vitamins and choline

			Water-soluble vitamins				
Vitamin	Sources	Recommended intake for adults	Major functions	Deficiency diseases and symptoms	Groups at risk for deficiency	Toxicity	UL
Thiamin (vitamin B₁, thiamin mononitrate)	Pork, whole and enriched grains, seeds, nuts, legumes	1.1–1.2 mg/day	Coenzyme in glucose and energy metabolism; needed for neurotransmitter synthesis and normal nerve function	Beriberi: weakness, apathy, irritability, nerve tingling, poor coordination, paralysis, heart changes	Alcoholics, those living in poverty	None reported	ND
Riboflavin (vitamin B₂)	Dairy products, whole and enriched grains, dark-green vegetables, meats	1.1–1.3 mg/day	Coenzyme in energy and lipid metabolism	Inflammation of the mouth and tongue, cracks at corners of the mouth	None	None reported	ND
Niacin (nicotinamide, nicotinic acid, vitamin B₃)	Beef, chicken, fish, peanuts, legumes, whole and enriched grains; can be made from tryptophan	14–16 mg NE/day	Coenzyme in energy metabolism and lipid synthesis and breakdown	Pellagra: diarrhea, dermatitis on areas exposed to sun, dementia	Those consuming a limited diet based on corn; alcoholics	Flushing, nausea, rash, tingling extremities	35 mg/day from fortified foods and supplements
Biotin	Liver, egg yolks; synthesized in the gut	30 µg/day	Coenzyme in glucose synthesis and energy and fatty acid metabolism	Dermatitis, nausea, depression, hallucinations	Those consuming large amounts of raw egg whites; alcoholics	None reported	ND
Pantothenic acid (calcium pantothenate)	Meat, legumes, whole grains; widespread in foods	5 mg/day	Coenzyme in energy metabolism and lipid synthesis and breakdown	Fatigue, rash	Alcoholics	None reported	ND
Vitamin B₆ (pyridoxine, pyridoxal phosphate, pyridoxamine)	Meat, fish, poultry, legumes, whole grains, nuts and seeds	1.3–1.7 mg/day	Hemoglobin and neurotransmitter synthesis, coenzyme in protein and amino acid metabolism, and many other reactions	Headache, convulsions, other neurological symptoms, nausea, poor growth, anemia	Alcoholics	Numbness, nerve damage	100 mg/day
Folate (folic acid, folacin, pteroylglutamic acid)	Leafy green vegetables, legumes, seeds, enriched grains, orange juice	400 µg DFE/day	Coenzyme in DNA synthesis and amino acid metabolism, particularly important for rapidly dividing cells, including red blood cells	Macrocytic anemia, inflammation of tongue, diarrhea, poor growth, neural tube defects	Pregnant women, alcoholics	Masks B₁₂ deficiency	1000 µg/day from fortified food and supplements

(continues)

TABLE 7.2 A summary of the vitamins and choline *(continued)*

		Water-soluble vitamins					
Vitamin	Sources	Recommended intake for adults	Major functions	Deficiency diseases and symptoms	Groups at risk for deficiency	Toxicity	UL
Vitamin B_{12} (cobalamin, cyanocobalamin)	Animal products	2.4 µg/day	Coenzyme in folate and homocysteine metabolism; needed for normal nerve function and maintenance of myelin	Pernicious anemia, macrocytic anemia, nerve damage	Vegans, those over age 50, people with stomach or intestinal disease	None reported	ND
Vitamin C (ascorbic acid, ascorbate)	Citrus fruit, broccoli, strawberries, greens, peppers	75–90 mg/day	Antioxidant; coenzyme in collagen synthesis and in hormone and neurotransmitter synthesis	Scurvy: poor wound healing, bleeding gums, loose teeth, bone fragility, joint pain, pinpoint hemorrhages	Alcoholics, older adults	GI distress, diarrhea	2000 mg/day
Choline*	Egg yolks, organ meats, wheat germ, meat, fish, nuts, synthesis in the body	425–550 mg/day	Synthesis of cell membranes and neurotransmitters	Fatty liver, muscle damage, abnormal prenatal development	None	Sweating, low blood pressure, liver damage	3500 mg/day
		Fat-soluble vitamins					
Vitamin A (retinol, retinal, retinoic acid, vitamin A acetate, vitamin A palmitate, retinyl palmitate, provitamin A, carotene, β-carotene, carotenoids)	Retinol: liver, fish, fortified milk and margarine, butter, eggs; carotenoids: carrots, leafy greens, sweet potatoes, broccoli, apricots, cantaloupe	700–900 µg/day	Gene expression, vision, health of cornea and other epithelial tissue, cell differentiation, reproduction, immune function	Xerophthalmia: night blindness, dry cornea, eye infections; poor growth, dry skin, impaired immune function	People living in poverty (particularly children and pregnant women), people who consume very low-fat or low-protein diets	Headache, vomiting, hair loss, liver damage, skin changes, bone pain, bone fractures, birth defects	3000 µg/day of preformed vitamin A
Vitamin D (calciferol, cholecalciferol, calcitriol, ergocalciferol, dihydroxy vitamin D)	Egg yolk, liver, fish oils, tuna, salmon, fortified milk, synthesis from sunlight	600–800 IU/day (15–20 µg/day)	Gene expression, absorption of calcium and phosphorus, maintenance of bone	Rickets in children: abnormal growth, misshapen bones, bowed legs, soft bones; osteomalacia in adults: weak bones and bone and muscle pain	Some breast-fed infants; children and older adults (especially those with dark skin and little exposure to sunlight); people with kidney disease	Calcium deposits in soft tissues, growth retardation, kidney damage	4000 IU/day (100 µg/day)
Vitamin E (tocopherol, alpha-tocopherol)	Vegetable oils, leafy greens, seeds, nuts, peanuts	15 mg/day	Antioxidant, protects cell membranes	Broken red blood cells, nerve damage	People with poor fat absorption, premature infants	Inhibition of vitamin K activity	1000 mg/day from supplemental sources
Vitamin K (phylloquinones, menaquinone)	Vegetable oils, leafy greens, synthesis by intestinal bacteria	90–120 µg/day	Synthesis of blood-clotting proteins and proteins needed for bone health and cell growth	Hemorrhage	Newborns (especially premature newborns), people on long-term antibiotics	Anemia, brain damage	ND

* Choline is technically not a vitamin, but recommendations have been made for its intake.

Note: UL, Tolerable Upper Intake Level; NE, niacin equivalent; DFE, dietary folate equivalent; ND, not determined due to insufficient data.

Concept Check

1. **What** food groups contain the greatest variety of vitamins?
2. **How** can the % Daily Value on food labels be used to identify foods that are good sources of vitamins?
3. **Why** might a low-fat diet affect the bioavailability of fat-soluble vitamins?
4. **Which** DRI value is designed to caution against toxicity?

7.2 Vitamins and Energy Metabolism

LEARNING OBJECTIVES

1. **Discuss** the role of thiamin, riboflavin, niacin, biotin and pantothenic acid in ATP production.
2. **List** food sources of thiamin, riboflavin, and niacin.
3. **Relate** the functions of thiamin, riboflavin, and niacin to their deficiency symptoms.
4. **Discuss** factors that cause losses of thiamin, riboflavin, and niacin during food processing, cooking, and storage.

Vitamins do not provide energy, but many are necessary for the production of ATP from the energy-yielding nutrients carbohydrate, fat, and protein. The B vitamins thiamin, riboflavin, niacin, biotin, and pantothenic acid function as **coenzymes** involved in transferring the energy in carbohydrate, fat, and protein to ATP, the high-energy molecule that is used to fuel the body (**Figure 7.7**). Without these coenzymes, the reactions that produce ATP cannot proceed. Each of these vitamins also has other individual functions. Deficiency symptoms can reflect their complementary roles in energy metabolism and their other functions.

Thiamin

Thiamin, the first of the B vitamins to be discovered, is sometimes called vitamin B_1. **Beriberi**, the disease that results from a deficiency of this vitamin, flourished in East Asian countries for over 1000 years. It came to the attention of Western medicine in colonial Asia in the 19th century. The disease became such a problem that the Dutch East India Company sent a team of scientists to determine its cause. A young physician named Christiaan Eijkman worked on this problem for over 10 years. His success came as a result of a twist of fate. He ran out of food for his experimental chickens and, instead of the usual brown rice, fed them white rice. Shortly thereafter, the chickens displayed beriberi-like symptoms. When he fed them brown rice again, their health was restored. These events provided evidence that beriberi was not caused by a poison or a microorganism, as had previously been thought, but rather by something that was missing from the diet.

Knowledge gained from Eijkman's studies made it possible to prevent and cure beriberi by feeding people a diet adequate in thiamin; however, the vitamin itself was not isolated until 1926. We now know that polishing the bran layer off rice kernels to make white rice removes the thiamin-rich portion of the grain. Therefore, in populations where white rice was the staple of the diet, beriberi became a common health problem. The incidence of beriberi in East Asia increased dramatically in the late 1800s due to the rising popularity of polished rice.

Thiamin is a water-soluble vitamin that functions as a coenzyme in the breakdown of glucose to provide energy. It is particularly important for nerve function because glucose is the energy source for nerve cells. In addition to its role in energy metabolism, thiamin is needed for the synthesis of **neurotransmitters**, the metabolism of other sugars and certain amino acids, and the synthesis of ribose and deoxyribose, sugars that are part of the structure of RNA (ribonucleic acid) and DNA, respectively.

---

coenzyme An organic nonprotein substance that binds to an enzyme to promote its activity.

beriberi A thiamin deficiency disease that may manifest in one of two forms: dry beriberi, which causes weakness and nerve degeneration, or wet beriberi, which causes heart changes.

neurotransmitter A chemical substance produced by a nerve cell that can stimulate or inhibit another cell.

PROCESS DIAGRAM **FIGURE 7.7 Coenzymes** Coenzymes are needed for enzyme activity. They act as carriers of electrons, atoms, or chemical groups that participate in the reaction. All the B vitamins as well as vitamins C and K are coenzymes; there are also coenzymes that are not dietary essentials and therefore are not vitamins.

How It Works

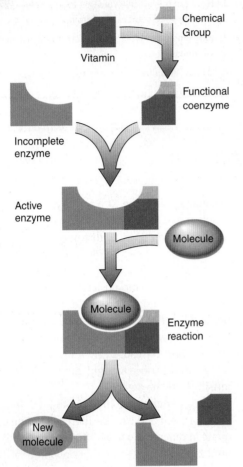

1. The vitamin combines with a chemical group to form the functional coenzyme (active vitamin).

2. The functional coenzyme combines with the incomplete enzyme to form the active enzyme.

3. The active enzyme binds to one or more molecules and accelerates the chemical reaction to form one or more new molecules.

4. The new molecules are released, and the enzyme and coenzyme (vitamin) can be reused or separated.

When thiamin intake is deficient, neurological symptoms such as mental confusion, weakness, impaired tendon reflexes, and tingling and lack of feeling in the hands and feet (dry beriberi), or cardiovascular symptoms such as rapid heart rate, enlargement of the heart, and accumulation of fluid in the tissues (wet beriberi) occur. The neurological symptoms of dry beriberi can be related to the functions of thiamin (**Figure 7.8**), but it is not clear why thiamin deficiency causes the cardiovascular symptoms seen with wet beriberi.[6]

Thiamin deficiency can also result in **Wernicke–Korsakoff syndrome**, which involves confusion, loss of coordination, vision changes, and hallucinations and can progress to coma and death. It occurs most often in alcoholics because alcohol decreases thiamin absorption, and diets high in alcohol are typically low in micronutrients. It has also been identified in those with eating disorders, in those on long-term intravenous nutrition, and in individuals who have undergone gastric bypass surgery.[7]

Thiamin is abundant in pork, legumes, and seeds. In addition to being found in the bran layer of brown rice and other whole grains, thiamin is added to enriched grains (**Figure 7.9**).

Although thiamin is needed to provide energy, unless the diet is deficient in thiamin, increasing thiamin intake does not increase the ability to produce ATP. There is no UL for thiamin because no toxicity has been reported when an excess of this vitamin is consumed from either food or supplements.

Riboflavin

Riboflavin is a B vitamin that provides a visible indicator of excessive consumption; the excess is excreted in the urine, turning it a bright fluorescent yellow. The color may surprise you, but it is harmless. No adverse effects of high doses of riboflavin from either foods or supplements have been reported.

Riboflavin is a water-soluble vitamin that forms two active coenzymes. They serve as electron carriers that

FIGURE 7.8 Thiamin functions and deficiency The role of thiamin in producing ATP from glucose and synthesizing neurotransmitters has been linked to the neurological symptoms of beriberi.

a. Thiamin is needed to convert pyruvate into acetyl-CoA. Acetyl-CoA can continue through cellular respiration to produce ATP. Acetyl-CoA is also needed to synthesize the neurotransmitter acetylcholine.

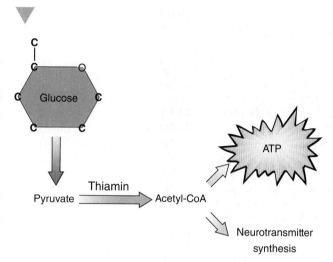

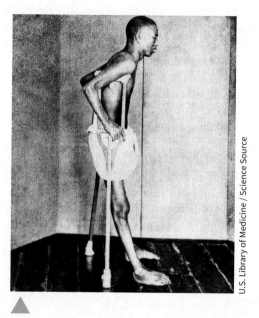

b. In Sri Lanka, the word *beriberi* means "I cannot," referring to the extreme weakness that is one of the earliest symptoms of the disease. Weakness and other neurological symptoms such as poor coordination, tingling sensations, and paralysis may result from the inability of nerve cells to produce ATP from glucose, or the body's inability to synthesize certain neurotransmitters when thiamin is deficient.

Ask Yourself

How would a deficiency of thiamin affect the amounts of ATP and neurotransmitters available?

FIGURE 7.9 Meeting thiamin needs A large proportion of the thiamin consumed in the United States comes from enriched grain products that we eat in abundance, such as pasta, rice dishes, baked goods, and breakfast cereals. The dashed lines indicate the RDAs for adult men and women, which are 1.2 mg/day and 1.1 mg/day, respectively.

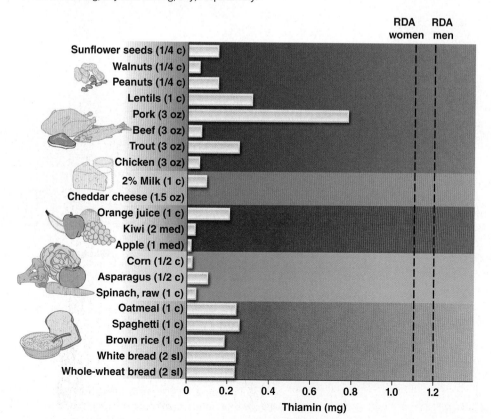

function in the reactions needed to produce ATP from carbohydrate, fat, and protein. Riboflavin is also involved directly or indirectly in converting a number of other vitamins, including folate, niacin, vitamin B_6, and vitamin K, into their active forms.

One of the best sources of riboflavin in the diet is milk (**Figure 7.10**). When riboflavin is deficient, injuries heal poorly because new cells cannot grow to replace the damaged ones. The tissues that grow most rapidly, such as the skin and the linings of the eyes, mouth, and tongue, are the first to be affected. Deficiency causes symptoms such as cracking of the lips and at the corners of the mouth; increased sensitivity to light; burning, tearing, and itching of the eyes; and flaking of the skin around the nose, eyebrows, and earlobes.

A deficiency of riboflavin usually occurs in conjunction with deficiencies of other B vitamins because the same foods are also sources of those vitamins and because riboflavin is needed to convert other vitamins into their active forms. Some of the symptoms seen in cases of riboflavin deficiency therefore reflect deficiencies of these other nutrients as well.

Niacin

In the early 1900s, psychiatric hospitals in the southeastern United States were filled with patients with the niacin-deficiency disease **pellagra** (**Figure 7.11**). **Niacin** is a B vitamin that forms coenzymes essential for glucose metabolism and the synthesis of fatty acids and cholesterol. The need for this water-soluble vitamin is so widespread in metabolism that a deficiency causes major changes throughout the body. The early symptoms of pellagra include fatigue, decreased appetite, and indigestion. These are followed by symptoms that can be remembered as the three Ds: dermatitis, diarrhea,

FIGURE 7.10 **Meeting riboflavin needs** Riboflavin is found in a variety of foods but is susceptible to losses when exposed to light.

a. Ever wonder why milk doesn't come in glass bottles anymore? The reason is that riboflavin is destroyed by light. Cloudy plastic containers block some of the light, but opaque ones, such as cardboard containers, are the most effective. Exposure to light can also cause an "off" flavor and losses of vitamins A and D.[8]

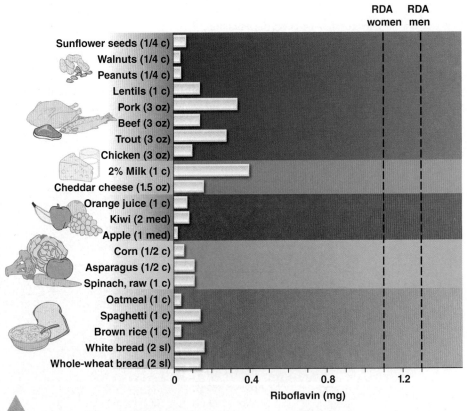

b. Major dietary sources of riboflavin include dairy products, red meat, poultry, fish, whole grains, and enriched breads and cereals. Vegetable sources include asparagus, broccoli, mushrooms, and leafy green vegetables such as spinach. The dashed lines indicate the RDAs for adult men and women, which are 1.3 mg/day and 1.1 mg/day, respectively.

pellagra A disease resulting from niacin deficiency, which causes dermatitis, diarrhea, dementia, and, if not treated, death.

FIGURE 7.11 Tracking down the cause of pellagra In 1914, Dr. Joseph Goldberger was appointed by the U.S. Public Health Service to investigate the pellagra epidemic in the South. He unraveled the mystery of its cause by using the scientific method.

Observation
Goldberger observed that individuals in institutions such as hospitals, orphanages, and prisons suffered from pellagra, but the staff did not. If pellagra were an infectious disease, both populations would be equally affected.

Hypothesis
Goldberger hypothesized that pellagra was due to a deficiency in the diet.

Experiments
Experimental design: Goldberger and coworkers added nutritious foods, including meat, milk, and vegetables, to the diets of children in two orphanages.
Results: Those consuming the healthier diets recovered from pellagra. Those without the disease who ate the new diet did not contract pellagra, supporting the hypothesis that it was caused by a dietary deficiency.
Experimental design: Goldberger and colleagues fed 11 volunteers a diet believed to be lacking in the dietary substance that prevents pellagra.
Results: Six of the eleven developed symptoms of pellagra after 5 months of consuming the experimental diet, supporting the hypothesis that it was caused by a dietary deficiency.
Continued experiments: Human and animal studies by a number of scientists lead to the identification of nicotinic acid, better known as the water-soluable B vitamin niacin, in 1937, as the dietary component that cures and prevents pellagra.

Theory
Pellagra is caused by a deficiency of the B vitamin niacin.

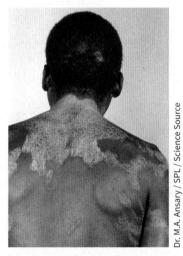

This photograph illustrates the cracked, inflamed skin that is characteristic of pellagra. The rash most commonly appears on areas of the skin that are exposed to sunlight or other stresses.

and dementia. If left untreated, niacin deficiency results in a fourth D—death.

Meats and grains are good sources of niacin. Niacin can also be synthesized in the body from the essential amino acid tryptophan (**Figure 7.12**). Tryptophan, however, is used to make niacin only if enough of it is available to first meet the needs of protein synthesis. When the diet is low in tryptophan, it is not used to synthesize niacin. Because some of the requirement for niacin can be met through the synthesis of niacin from tryptophan, the RDA is expressed as **niacin equivalents (NEs)**.

One NE is equal to 1 mg of niacin or 60 mg of tryptophan, the amount needed to make 1 mg of niacin.

Niacin deficiency was rampant in the South in the early 1900s because the local diet among the poor was based on corn. Corn is low in tryptophan, and the niacin found naturally in corn is bound to other molecules and therefore is not well absorbed. Today, as a result of the enrichment of grains with an available form of niacin, pellagra is rare in the United States, but it remains a problem in areas of Africa where the diet is based on corn.[9] Despite the corn-based diet in Mexico and Central

FIGURE 7.12 Meeting niacin needs Niacin needs can be met by consuming foods that are sources of niacin and foods that contain tryptophan, which can be converted to niacin.

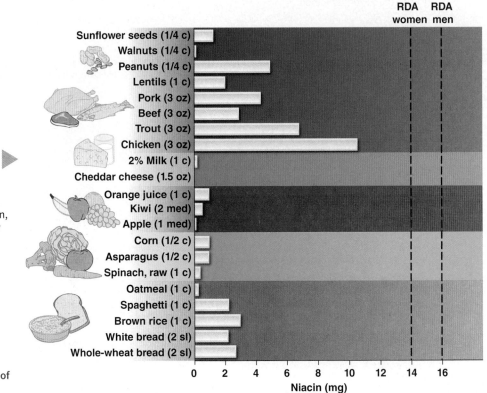

a. Meat, fish, peanuts, and whole and enriched grains are the best sources of niacin. Other sources include legumes and wheat bran. The dashed lines indicate the RDAs for adult men and women, which are 16 mg NE/day and 14 mg NE/day, respectively.

Think Critically

Using the values in this graph, plan a dinner that provides 100% of the RDA of niacin for women.

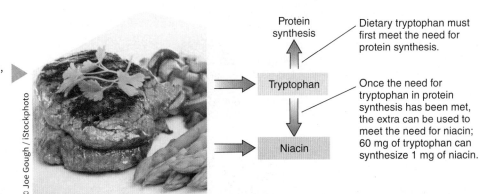

b. The essential amino acid tryptophan, found in dietary protein, can be used to synthesize niacin but only after the body's need for tryptophan in protein synthesis has been met. Values given in food composition tables and databases do not include the amount of niacin that could be made from the tryptophan in a food.

The diet provides food sources of both niacin and tryptophan.

Dietary tryptophan must first meet the need for protein synthesis.

Once the need for tryptophan in protein synthesis has been met, the extra can be used to meet the need for niacin; 60 mg of tryptophan can synthesize 1 mg of niacin.

American countries, pellagra is uncommon in those countries, in part because the treatment of corn with limewater, as is done during the making of tortillas, enhances the bioavailability of niacin. The diet in these regions also includes legumes, which provide both niacin and a source of tryptophan for the synthesis of niacin.

There is no evidence of any adverse effects from consuming the niacin that occurs naturally in foods, but niacin supplements can be toxic. Excess niacin supplementation can cause flushing of the skin, a tingling sensation in the hands and feet, a red skin rash, nausea, vomiting, diarrhea, high blood sugar levels, abnormalities in liver function, and blurred vision. The UL for adults is 35 mg/day from fortified foods and supplements.

Doses of 50 mg/day or greater of one form of niacin are used as a drug to treat elevated blood cholesterol; this amount should be consumed only when prescribed by a physician.

Biotin

The B vitamin **biotin** is a coenzyme that functions in energy metabolism and glucose synthesis. This water-soluble vitamin is also important in the metabolism of fatty acids and amino acids. Good sources of biotin in the diet include cooked eggs, liver, yogurt, and nuts. Fruit and meat are poor sources. Bacteria in the gastrointestinal tract synthesize

FIGURE 7.13 **Raw eggs and biotin bioavailability** Raw egg whites contain a protein called avidin that tightly binds biotin and prevents its absorption, potentially leading to a deficiency. Even if you were not concerned with biotin deficiency, eggs should never be eaten raw because they can contain harmful bacteria. Thoroughly cooking eggs kills bacteria and denatures avidin so that it cannot bind biotin.

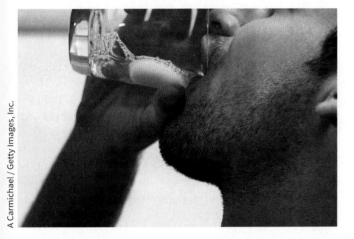

biotin, and some of this is absorbed into the body and helps meet our biotin needs. An AI of 30 μg/day has been established for adults.[6]

Although biotin deficiency is uncommon, it has been observed in people with malabsorption and those taking certain medications for long periods. Eating raw eggs can also cause biotin deficiency (**Figure 7.13**). Biotin deficiency in humans causes nausea, thinning hair, loss of hair color, a red skin rash, depression, lethargy, hallucinations, and tingling of the extremities. High doses of biotin have not resulted in toxicity symptoms; there is no UL for biotin.

Pantothenic Acid

Pantothenic acid, which gets its name from the Greek word *pantothen*, meaning "from everywhere," is a B vitamin that is widely distributed in foods. It is particularly abundant in meat, eggs, whole grains, and legumes, and it is found in lesser amounts in milk, vegetables, and fruits.

In addition to being "from everywhere" in the diet, pantothenic acid seems to be needed everywhere in the body. It is part of coenzyme A (CoA), which is needed for the breakdown of carbohydrates, fatty acids, and amino acids, as well as for the modification of proteins and the synthesis of neurotransmitters, steroid hormones, and **hemoglobin**. Pantothenic acid is also needed to form a molecule that is essential for the synthesis of cholesterol and fatty acids.

The wide distribution of this water-soluble vitamin in foods makes deficiency rare in humans. The AI is 5 mg/day for adults. Pantothenic acid is relatively nontoxic, and there are insufficient data to establish a UL.

Choline and Other Vitamin-like Compounds

A number of substances perform vitamin-like functions in the body but are not classified as vitamins because adequate amounts can be synthesized in the body so that they are not required in the diet. For example, carnitine is needed to transport fatty acids into the mitochondria, where they are broken down to produce ATP. Lipoic acid functions as a coenzyme, and inositol is important for membrane function. **Choline** is discussed here because a dietary recommendation has been established for it.

Choline is a water-soluble substance that you may see included in supplements called "vitamin B complex." It is needed for the synthesis of the neurotransmitter acetylcholine, the structure and function of cell membranes, lipid transport, and metabolism of the amino acid homocysteine. It can be synthesized to a limited extent by humans. Although it is not currently classified as a vitamin, it is recognized as an essential nutrient. Deficiency during pregnancy can interfere with brain development in the fetus, and deficiency in adults causes fatty liver and muscle damage.[10] The DRIs have set AIs for this compound: 550 mg/day for men and 425 mg/day for women.

Choline is found in many foods, with large amounts in egg yolks, liver, meat and fish, wheat germ, and nuts.[11] Because the average daily choline intake in the United States exceeds the recommended intake, a deficiency is unlikely in healthy humans in this country.

Excess choline intake can cause a fishy body odor, sweating, reduced growth rate in children, low blood pressure, and liver damage. The amounts needed to cause these symptoms are much higher than can be obtained from foods. The UL for choline for adults is 3.5 g/day.

Concept Check

1. **What** common role do thiamin, riboflavin, niacin, biotin, and pantothenic acid play in generating ATP?
2. **What** food groups are the best sources of niacin?
3. **Why** does a thiamin deficiency cause neurological symptoms?
4. **Why** is milk packaged in opaque containers?

---

hemoglobin An iron-containing protein in red blood cells that binds and transports oxygen through the bloodstream to cells.

7.3 Vitamins and Healthy Blood

LEARNING OBJECTIVES

1. **Discuss** the functions of vitamin B_6.
2. **Explain** the relationship between folate and vitamin B_{12}.
3. **List** some food sources of vitamin B_6, of folate, and of vitamin B_{12}.
4. **Describe** the role of vitamin K in blood clotting.

Blood transports oxygen, nutrients, and hormones to cells throughout the body and carries waste products away from cells for elimination. Healthy blood contains the right ratio of blood cells to blood plasma (the liquid portion of blood) and the right amounts of a variety of blood proteins. The vitamins discussed here are needed to support the synthesis of red blood cells and blood-clotting proteins. Minerals such as iron and copper, also essential for healthy blood, are discussed in Chapter 8.

Vitamin B_6

Vitamin B_6 is a water-soluble vitamin that is particularly important for amino acid and protein metabolism. There are three forms of vitamin B_6: pyridoxal, pyridoxine, and pyridoxamine. These can be converted into the active coenzyme **pyridoxal phosphate**, which is needed for the activity of more than 100 enzymes. For example, the coenzyme form of vitamin B_6 is needed to produce hemoglobin for red blood cell synthesis, to convert tryptophan to niacin, for the metabolism of glucose and amino acids, and for the synthesis of lipids needed for nerve function (**Figure 7.14**).

NUTRITION INSIGHT

FIGURE 7.14 Vitamin B_6 functions and deficiency The functions of vitamin B_6 in amino acid metabolism, red and white blood cell synthesis, and nerve function help explain the symptoms that occur when this vitamin is deficient.

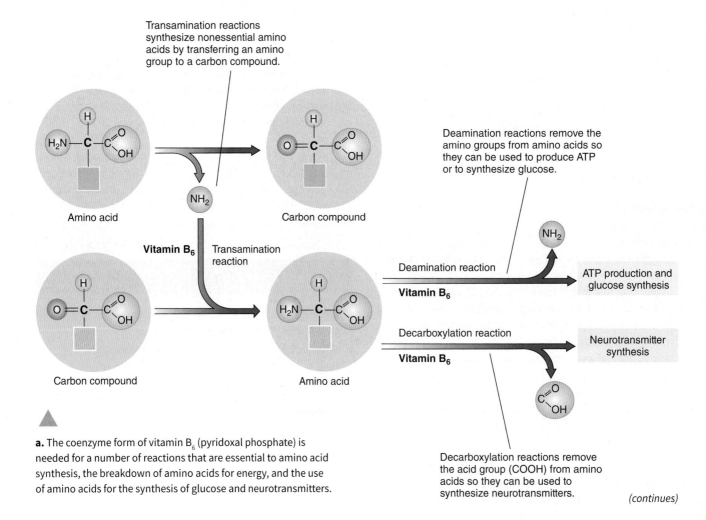

a. The coenzyme form of vitamin B_6 (pyridoxal phosphate) is needed for a number of reactions that are essential to amino acid synthesis, the breakdown of amino acids for energy, and the use of amino acids for the synthesis of glucose and neurotransmitters.

(continues)

FIGURE 7.14 *(continued)*

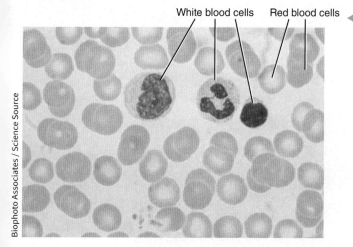

b. When vitamin B_6 is deficient, hemoglobin cannot be made; the result is a type of anemia characterized by small, pale red blood cells. Vitamin B_6 is also needed to form white blood cells, which are part of the immune system, so deficiency reduces immune function.

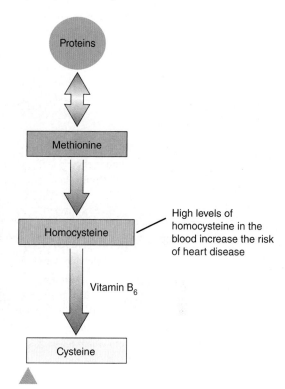

c. Vitamin B_6 is needed for the synthesis of lipids that are part of the **myelin** coating on nerves. Myelin is essential for nerve transmission. The role of vitamin B_6 in myelin formation and neurotransmitter synthesis may explain the neurological symptoms that occur with deficiency, such as numbness and tingling in the hands and feet, depression, headaches, confusion, and seizures.

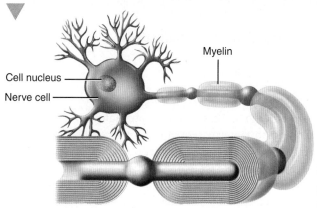

d. If vitamin B_6 status is low, homocysteine, which is formed from the amino acid methionine, cannot be converted to the amino acid cysteine, so levels rise. Even a mild elevation in blood homocysteine levels has been shown to increase the risk of heart disease.[12]

Ask Yourself
What functions of vitamin B_6 might explain why a deficiency of this vitamin interferes with nerve function?

Vitamin B_6 deficiency leads to poor growth, **anemia**, decreased immune function, neurological symptoms, and skin lesions. Because vitamin B_6 is needed for amino acid metabolism, the onset of a deficiency can be hastened by a diet that is low in vitamin B_6 but high in protein. Many of the symptoms can be linked to the biochemical reactions that depend on this vitamin coenzyme (see Figure 7.14). For example, poor growth, skin lesions, and decreased antibody formation may occur with a diet that is low in vitamin B_6 because of the central role of vitamin B_6 in protein and energy metabolism. Anemia can be linked to the role of vitamin B_6 in hemoglobin synthesis and neurological symptoms can be linked to the importance of vitamin B_6 in myelin formation and neurotransmitter synthesis.

Meat and fish are excellent animal sources of vitamin B_6, and whole grains and legumes are good plant sources (**Figure 7.15**). Refined grain products such as white rice and white bread are not good sources of vitamin B_6 because the vitamin is lost during refining and is not added back through enrichment. This vitamin is, however, added to many fortified breakfast cereals, making these good sources. Vitamin B_6 is destroyed by heat and light, so it can easily be lost during processing.

---

anemia A condition in which the oxygen-carrying capacity of the blood is decreased by a reduced number of red blood cells or a reduced amount of hemoglobin in the cells.

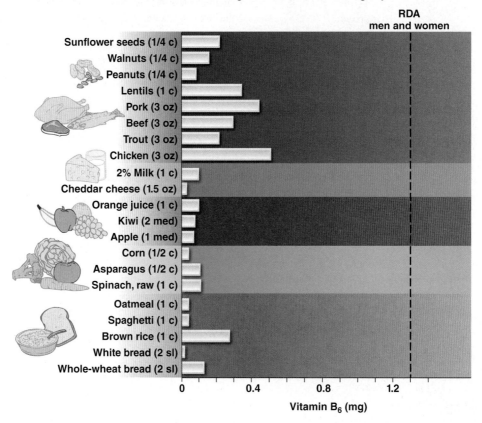

FIGURE 7.15 Meeting vitamin B₆ needs Animal sources of vitamin B₆ include chicken, fish, pork, and organ meats. Good plant sources include whole-wheat products, brown rice, soybeans, sunflower seeds, and some fruits and vegetables, such as bananas, broccoli, and spinach. The dashed line indicates the RDA for adult men and women ages 19 to 50, which is 1.3 mg/day.

No adverse effects have been associated with high intake of vitamin B₆ from foods, but the large doses in some supplements can cause severe nerve impairment. To prevent nerve damage, the UL for adults is set at 100 mg/day from food and supplements. Despite the potential for toxicity, supplements of vitamin B₆ containing 100 mg or more per dose (5000% of the Daily Value) are available over the counter, making it easy exceed the UL.

People take vitamin B₆ supplements to reduce the symptoms of premenstrual syndrome (PMS), treat carpal tunnel syndrome, and strengthen immune function. There is some evidence that supplements may reduce the symptoms of carpal tunnel syndrome and PMS.[13,14] Vitamin B₆ supplements also improve immune function, but only in individuals with vitamin B₆ deficiency. There is no evidence that supplements will enhance immune function in those with adequate vitamin B₆ status.

Folate (Folic Acid)

Folate is a B vitamin that is needed for the synthesis of DNA and the metabolism of some amino acids. Cells must synthesize DNA in order to replicate, so this water-soluble vitamin is particularly important in tissues in which cells are dividing rapidly, such as the intestines, skin, embryonic and fetal tissues, and bone marrow, where red blood cells are made. When folate is deficient, cells cannot divide normally. This leads to one of the most notable symptoms of folate deficiency: a type of anemia called **macrocytic anemia** (**Figure 7.16**). Other symptoms of folate deficiency include poor growth, problems with nerve development and function, diarrhea, and inflammation of the tongue.

Folate is also important during pregnancy for the development of the embryo. Low folate intake increases the risk of birth defects called **neural tube defects** (**Figure 7.17**). Neural tube defects are not a true folate-deficiency symptom because not every pregnant woman with inadequate folate levels will bear a child with a neural tube defect. Instead, neural tube defects are probably due to a combination of factors that include low folate levels and a genetic predisposition. The formation of the neural tube, which later develops into the brain and spinal cord, occurs very early during pregnancy. Therefore, to reduce the risk of neural tube

---

macrocytic anemia A reduction in the blood's hemoglobin content and hence capacity to carry oxygen that is characterized by abnormally large red blood cells.

neural tube defect An abnormality in the brain or spinal cord that results from errors that occur during prenatal development.

FIGURE 7.16 Folate deficiency and macrocytic anemia Folate is needed for DNA replication. Without folate, developing red blood cells cannot divide. Instead, they just grow bigger. The abnormally large immature red blood cells, called *megaloblasts*, then mature into abnormally large red blood cells called *macrocytes*. Fewer red blood cells are produced, and they often contain less hemoglobin, resulting in macrocytic anemia.

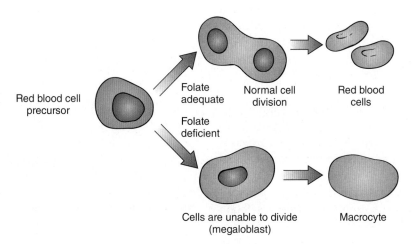

FIGURE 7.17 Neural tube defects Abnormalities in the development of the neural tube, which forms the brain and spinal cord, result in neural tube defects, such as spina bifida.

a. During embryonic development, a groove forms in the plate of neural tissue; then the edges fold up to form the neural tube, which will become the brain and spinal cord. If this sequence does not occur normally, the baby is born with a neural tube defect.

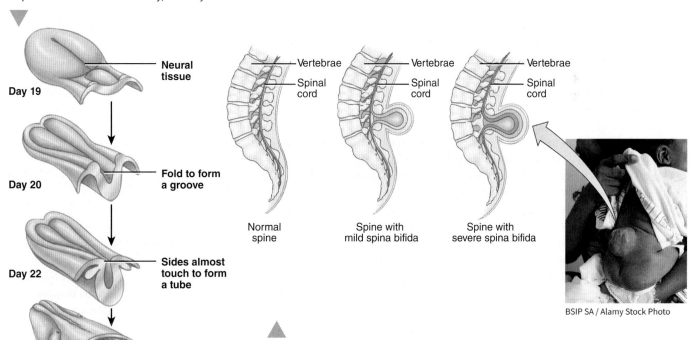

b. If a lower portion of the neural tube does not close normally, the result is **spina bifida** (shown here), a condition in which the spinal cord forms abnormally. Many babies with spina bifida have learning disabilities and nerve damage that causes varying degrees of paralysis of the lower limbs. If the head end of the neural tube does not close properly, the brain doesn't form completely; the result is anencephaly (partial or total absence of the brain). Babies with anencephaly are usually blind, deaf, and unconscious, and they die soon after birth.

defects, a woman's folate status must be adequate before she becomes pregnant and during the critical early days of pregnancy.

To help ensure adequate folate intake in women of childbearing age, in 1998 the FDA mandated that the **folic acid** form of the vitamin be added to all enriched grains and cereal products. Since then, the incidence of neural tube defects in the United States has decreased by about 30%, and reductions ranging from 19 to 55% have been observed in other countries where folic acid fortification has been introduced (see *What a Scientist Sees: Folate Fortification and Neural Tube Defects* in Chapter 11).[15,16]

Folate and *folacin* are general terms for compounds whose chemical structures and nutritional properties are similar to those of folic acid. The folic acid form, which is added to enriched grains and other fortified products and used in dietary supplements, is more easily absorbed, so its bioavailability is about twice that of folate found naturally in foods. The RDA for folate is expressed in **dietary folate equivalents (DFEs)**. DFEs correct for differences in the bioavailability of different forms of folate. One DFE is equal to 1 microgram (μg) of food folate, 0.6 μg of synthetic folic acid from fortified food or supplements consumed with food, or 0.5 μg of synthetic folic acid consumed on an empty stomach (see Appendix J).

Low folate status may also increase the risk of developing heart disease. Folate's connection with heart disease has to do with the metabolism of the amino acid homocysteine. Elevated levels of homocysteine have been associated with increased risk of heart disease. Folate and vitamins B_{12} and B_6 are all needed to prevent homocysteine levels from rising (**Figure 7.18**). However, supplementation with these vitamins has failed to exert significant effects on cardiovascular risk.

Population groups most at risk of folate deficiency include pregnant women and premature infants (because of their rapid rates of cell division and growth), older adults (because of their limited intake of foods high in folate), alcoholics (because alcohol inhibits the absorption of folate), and tobacco smokers (because smoke inactivates folate in the cells lining the lungs).

Asparagus, oranges, legumes, liver, and yeast are excellent food sources of folate (**Figure 7.19**). Whole grains are a fair source, and, as discussed earlier, folic acid is added to enriched grain products, including enriched breads, flours, corn meal, pasta, grits, and rice. Because supplementing folic acid early in pregnancy has been shown to reduce neural tube defects in the fetus, it is recommended that women

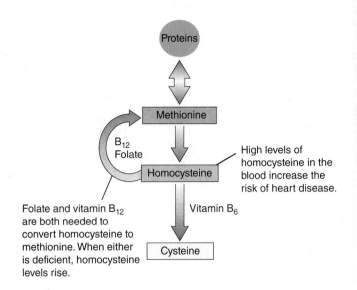

FIGURE 7.18 **B vitamins, homocysteine, and the risk of heart disease** Folate and vitamin B_{12} are both needed to convert homocysteine to methionine. When either folate, vitamin B_{12}, or, as discussed earlier, vitamin B_6 is deficient, homocysteine levels rise, increasing the risk of developing heart disease.

who are capable of becoming pregnant consume 400 μg of synthetic folic acid from fortified foods and/or supplements in addition to the naturally occurring folate consumed in a varied diet. To get 400 μg of folic acid, women of childbearing age would need to eat four to six servings of fortified grain products each day or take a supplement containing folic acid.

Although there is no known folate toxicity, a high intake from fortified foods or supplements could potentially mask the early symptoms of vitamin B_{12} deficiency, allowing it to go untreated. Based on this concern, a UL for adults has been set at 1000 μg/day of folic acid from supplements and/or fortified foods.

Vitamin B_{12}

In the early 1900s, **pernicious anemia** amounted to a death sentence. There was no cure. In the 1920s, researchers George Minot and William Murphy pursued their belief that pernicious anemia could be cured by something in the diet. They discovered that they could restore patients' health by feeding them about 4 to 8 ounces of slightly cooked liver at every meal.

---

folic acid An easily absorbed form of the vitamin folate that is used in dietary supplements and fortified foods.

pernicious anemia A macrocytic anemia resulting from vitamin B_{12} deficiency that occurs when dietary vitamin B_{12} cannot be absorbed due to a lack of intrinsic factor.

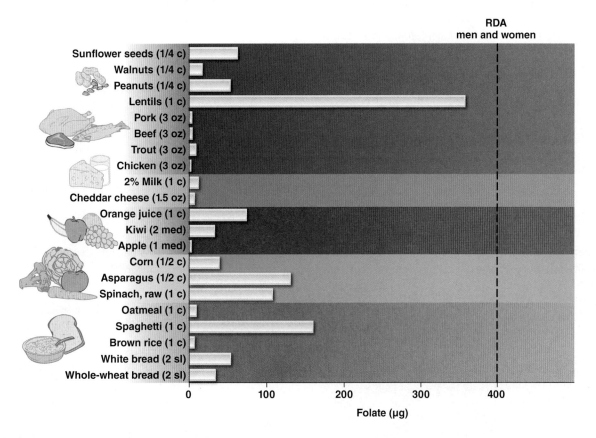

FIGURE 7.19 Meeting folate needs The word *folate* comes from the Latin for *foliage* because leafy greens, such as spinach, are good sources of this vitamin. Legumes, nuts, enriched grains, and orange juice are also good sources. Whole grains and many vegetables are fair sources. Only small amounts of folate are found in meats, cheese, milk, and most fruits. The dashed line indicates the RDA for adult men and women, which is 400 µg/day.

Interpret the Data

Which three of the foods in the graph provide the most folate? Are any of these foods fortified?

Today we know that eating liver cured pernicious anemia because liver is a concentrated source of **vitamin B$_{12}$**. Individuals with pernicious anemia lack a protein produced in the stomach, called **intrinsic factor**, that enhances vitamin B$_{12}$ absorption (**Figure 7.20**). Eating large amounts of liver was an effective treatment because a small proportion of dietary B$_{12}$ can be absorbed by passive diffusion even when intrinsic factor is absent. Today, pernicious anemia is treated with injections or mega-doses of vitamin B$_{12}$ rather than with plates full of liver.

Vitamin B$_{12}$, also known as **cobalamin**, is a water-soluble vitamin necessary for the production of ATP from certain fatty acids, to maintain the myelin coating on nerves (see Figure 7.14c), and for a reaction that converts homocysteine to methionine and converts folate to the form that is active for DNA synthesis (**Figure 7.21**). When vitamin B$_{12}$ is deficient, homocysteine levels rise, and folate is trapped in an inactive form. Without adequate active folate, DNA synthesis slows, red blood cells do not divide normally, and macrocytic anemia develops. Lack of vitamin B$_{12}$ also leads to degeneration of the myelin that coats nerves including those in the spinal cord and brain, resulting in symptoms such as numbness and tingling, abnormalities in gait, memory loss, and disorientation. If not treated, vitamin B$_{12}$ deficiency can cause irreversible nerve damage and eventually death.

---

intrinsic factor A protein produced in the stomach that aids in the absorption of vitamin B$_{12}$.

192 CHAPTER 7 Vitamins

PROCESS DIAGRAM **FIGURE 7.20** **Vitamin B_{12} absorption** The body stores and reuses vitamin B_{12} more efficiently than it does most other water-soluble vitamins, so deficiency is typically caused by poor absorption rather than by low intake alone. Absorption of adequate amounts of vitamin B_{12} from food depends on the presence of stomach acid, protein-digesting enzymes, and intrinsic factor.

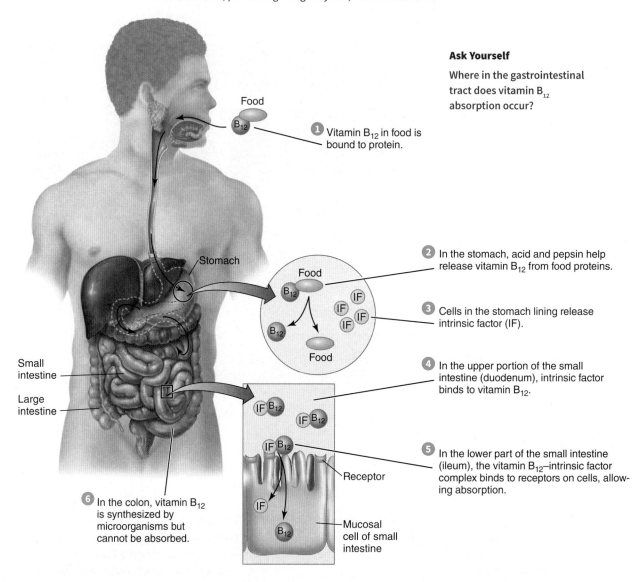

Ask Yourself

Where in the gastrointestinal tract does vitamin B_{12} absorption occur?

① Vitamin B_{12} in food is bound to protein.

② In the stomach, acid and pepsin help release vitamin B_{12} from food proteins.

③ Cells in the stomach lining release intrinsic factor (IF).

④ In the upper portion of the small intestine (duodenum), intrinsic factor binds to vitamin B_{12}.

⑤ In the lower part of the small intestine (ileum), the vitamin B_{12}–intrinsic factor complex binds to receptors on cells, allowing absorption.

⑥ In the colon, vitamin B_{12} is synthesized by microorganisms but cannot be absorbed.

Vitamin B_{12} is found naturally only in animal products (**Figure 7.22**). Therefore, meeting vitamin B_{12} needs is a concern among vegans—those who consume no animal products. Vegans must consume supplements or foods fortified with vitamin B_{12} in order to meet their needs for this vitamin.[17] Vitamin B_{12} deficiency is also a concern in adults over 50 years of age because of a condition called **atrophic gastritis**, which reduces the secretion of stomach acid. Without sufficient stomach acid, the enzymes that release the vitamin B_{12} bound to proteins in food cannot function properly, so vitamin B_{12} remains bound to the food proteins and cannot be absorbed (see Figure 7.20). In addition, lack of stomach acid allows large numbers of microbes to grow in the gut and compete for available vitamin B_{12}, reducing the amount absorbed. Atrophic gastritis affects 10 to 30% of adults over age 50. To ensure adequate vitamin B_{12} absorption, it is recommended that individuals over age 50 meet their RDA by consuming foods fortified with vitamin B_{12} or taking vitamin B_{12} supplements. The vitamin B_{12} in these products is not bound to proteins, so it is absorbed even when stomach acid levels are low.

atrophic gastritis An inflammation of the stomach lining that results in reduced secretion of stomach acid, microbial overgrowth, and, in severe cases, a reduction in the production of intrinsic factor.

FIGURE 7.21 **The relationship between folate and B_{12}** When vitamin B_{12} is deficient, it causes a secondary folate deficiency, which means folate is available, but the B_{12} deficiency prevents it from being activated. The folic acid provided by supplements and fortified foods does not need to be activated. This has raised concerns that our folic acid–fortified food supply will prevent folate-deficiency symptoms, such as macrocytic anemia, from occurring when B_{12} is deficient and allow the B_{12} deficiency to go unnoticed until irreversible neurological symptoms occur.

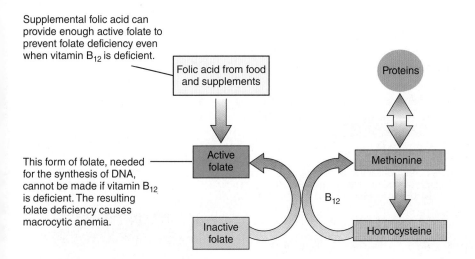

Supplemental folic acid can provide enough active folate to prevent folate deficiency even when vitamin B_{12} is deficient.

This form of folate, needed for the synthesis of DNA, cannot be made if vitamin B_{12} is deficient. The resulting folate deficiency causes macrocytic anemia.

Think Critically

If you were deficient in vitamin B_{12} but took large amounts of folic acid from supplements, would you develop macrocytic anemia? Why or why not?

FIGURE 7.22 **Meeting vitamin B_{12} needs** Animal foods provide vitamin B_{12}, but plant foods do not unless they have been fortified with it or have been contaminated by bacteria, soil, insects, or other sources of B_{12}. The dashed line indicates the RDA for adult men and women of all ages, which is 2.4 µg/day. No toxic effects have been reported for vitamin B_{12} intakes of up to 100 µg/day from food or supplements. Insufficient data are available to establish a UL for vitamin B_{12}.

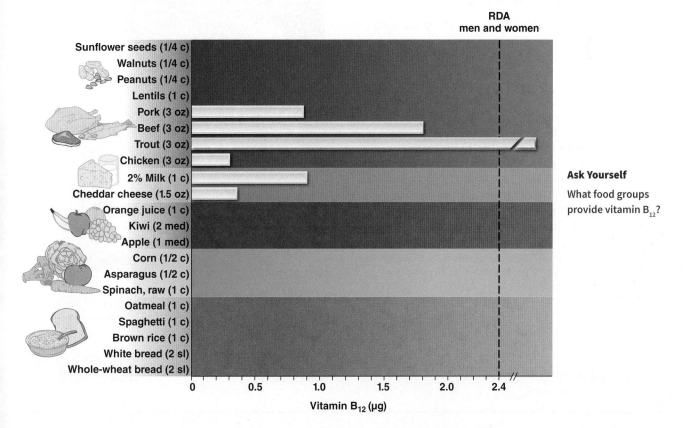

Ask Yourself

What food groups provide vitamin B_{12}?

Vitamin K

Blood is a fluid that flows easily through your blood vessels, but when you cut yourself, blood must solidify or clot to stop your bleeding. **Vitamin K** is a fat-soluble vitamin that is essential for the production of several blood proteins, called clotting factors, that are needed for blood to clot (**Figure 7.23**). The *K* in vitamin K comes from the Danish word for coagulation, *koagulation*, which means "blood clotting." Abnormal blood coagulation is the major symptom of vitamin K deficiency. Without vitamin K, even a bruise or small scratch could cause you to bleed to death (see *What a Scientist Sees: Anticoagulants Take Lives and Save Them*).

FIGURE 7.23 **Vitamin K and blood clotting** Blood clotting requires a series of reactions that result in the formation of a fibrous protein called **fibrin**. Fibrin fibers form a net that traps platelets and blood cells and forms the structure of a blood clot. Several of the clotting factors, including **prothrombin**, require vitamin K for synthesis. If vitamin K is deficient, they are not made correctly, and the blood will not clot.

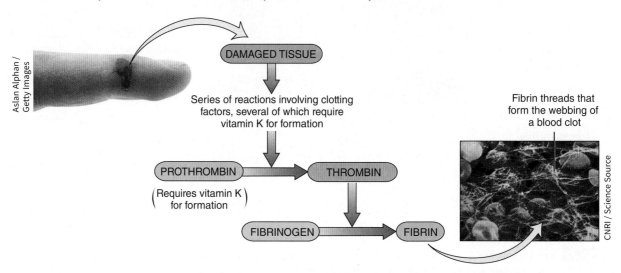

What a Scientist Sees

Anticoagulants Take Lives and Save Them

Consumers see warfarin as a means of eliminating some unwanted houseguests. A scientist sees that this rat poison has saved millions of lives. Warfarin is an anticoagulant, a substance that prevents blood from clotting. It acts by blocking the activation of vitamin K (see figure). When rats eat warfarin, minor bumps and scrapes cause them to bleed to death. Dicoumarol, a derivative of warfarin, was first isolated from moldy clover hay in 1940. At that time, cows across the Midwest were bleeding to death because they were being fed this moldy hay during the winter months.

Blood clot formation is essential to survival, but blood clots in the arteries cause heart attacks and strokes and are responsible for killing about half a million Americans annually. Scientists took advantage of what they learned about dicoumarol and vitamin K to save human lives. Dicoumarol was the first anticoagulant used to treat humans that could be taken orally rather than by injection. In the 1950s, the more potent anticoagulant warfarin, also known by the brand name Coumadin, was developed. It was used to treat President Eisenhower when he had a heart attack in 1955. Today this "rat poison" is being replaced by newer anticoagulant drugs, but since its development, it has been used to treat millions of heart attack patients.

Think Critically

Why might patients taking warfarin need to avoid vitamin K supplements?

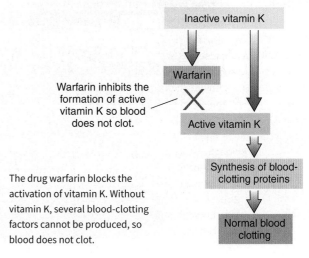

The drug warfarin blocks the activation of vitamin K. Without vitamin K, several blood-clotting factors cannot be produced, so blood does not clot.

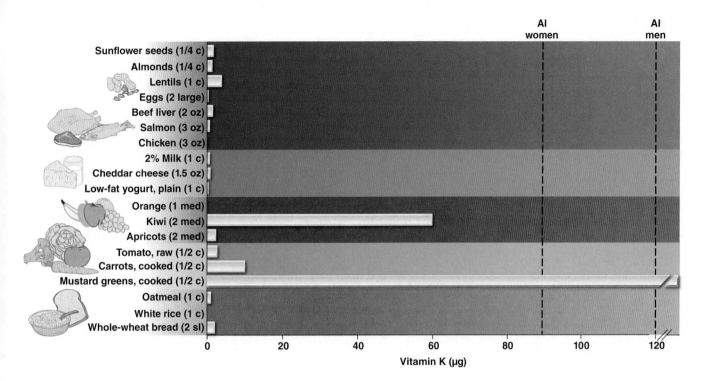

FIGURE 7.24 **Meeting vitamin K needs** Leafy green vegetables, such as spinach, broccoli, Brussels sprouts, kale, and turnip greens, and some vegetable oils are good sources of vitamin K. The dashed lines indicate the AIs for adult men and women, which are 120 μg/day and 90 μg/day, respectively.

Vitamin K is also needed for the synthesis of several proteins involved in bone formation and breakdown, inhibition of blood vessel calcification, and regulation of cell growth.[18] Because of the roles of these vitamin K–dependent proteins, mild vitamin K deficiency is believed to be responsible for a reduction in bone mineral density accompanied by an increased risk of fractures due to osteoporosis and an increased risk of atherosclerosis and cancer.[19]

Vitamin K is used more rapidly than other fat-soluble vitamins, so a constant supply is necessary. Only a small number of foods provide a significant amount of vitamin K (**Figure 7.24**). There is concern that typical intakes in North America provide enough vitamin K for normal blood coagulation but not enough for adequate synthesis of other vitamin K–dependent proteins. Evidence is accumulating that the current recommendations may not be high enough to provide for all the functions of vitamin K.[20] The Daily Value for vitamin K used on the planned Nutrition Facts label has been increased from 80 μg/day to 120 μg/day.[21]

A frank deficiency is rare in adults because vitamin K is synthesized by bacteria in the large intestine. Deficiency can be precipitated by a poor diet; Crohn's disease, which damages the colon, limiting absorption; and long-term antibiotic use, which kills the bacteria in the gastrointestinal tract that synthesize the vitamin. Newborns are at risk of deficiency because when a baby is born, no bacteria are present in the GI tract to synthesize vitamin K. In addition, newborns are at risk because little vitamin K is transferred to the baby from the mother before birth, and breast milk is a poor source of this vitamin. To ensure normal blood clotting, infants are typically given a vitamin K injection within six hours of birth.

Concept Check

1. **What** is the role of vitamin B_6 in amino acid metabolism?
2. **How** can folate and vitamin B_{12} deficiency both cause macrocytic anemia?
3. **Why** is vitamin B_{12} deficiency a concern in vegans?
4. **Why** does a vitamin K deficiency cause bleeding?

7.4 Antioxidant Vitamins

LEARNING OBJECTIVES

1. **Discuss** the relationship between oxidative stress and antioxidants.
2. **Relate** the role of vitamin C in the body to the symptoms of scurvy.
3. **Discuss** how vitamin E protects cell membranes.
4. **List** food sources of vitamin C and vitamin E.

Some vitamins function as **antioxidants**, substances that protect against **oxidative stress**. Damage caused by oxidative stress has been related to everything from aging to heart disease, cancer, and Alzheimer's disease.

Oxidative stress allows reactive oxygen molecules to steal electrons from other compounds, such as DNA, proteins, carbohydrates, or unsaturated fatty acids, resulting in changes in their structure and function. Reactive oxygen molecules are generated by normal oxygen-requiring reactions inside the body, such as cellular respiration, and come from environmental sources such as air pollution or cigarette smoke. Antioxidants act by reducing the formation of or destroying **free radicals** and other reactive oxygen molecules before they can cause damage (**Figure 7.25**). Some antioxidants are produced in the body; others, such as vitamin C, vitamin E, and the mineral selenium (discussed in Chapter 8), are consumed in the diet.[22]

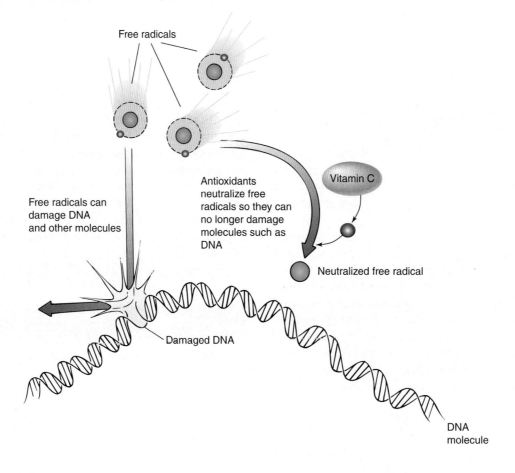

FIGURE 7.25 **How antioxidants work** Many antioxidants, including vitamin C, function by donating electrons to free radicals. A donated electron stabilizes the free radical so that it is no longer reactive and cannot steal electrons from important molecules in and around cells.

antioxidant A substance that decreases the adverse effects of reactive molecules on normal physiological function.

oxidative stress An imbalance between reactive oxygen molecules and antioxidant defenses that results in damage.

free radical A type of highly reactive atom or molecule that causes oxidative damage.

Vitamin C

Vitamin C, also called **ascorbic acid**, is a water-soluble vitamin that functions as an antioxidant in the blood and other body fluids (see Figure 7.25). Vitamin C's antioxidant action regenerates the active antioxidant form of vitamin E. The ability of vitamin C to donate electrons also enhances iron absorption in the small intestine by keeping iron in its more readily absorbed form.

Another important role for vitamin C is in the synthesis and maintenance of **collagen** (**Figure 7.26**). Collagen, the most abundant protein in the body, can be thought of as the glue that holds the body together. It forms the base of all connective tissue. It is the framework for bones and teeth; it is the main component of ligaments, tendons, and the scars that bind a wound together; and it gives structure to the walls of blood vessels. When vitamin C is lacking, collagen cannot be formed and maintained, and the symptoms of **scurvy** appear. In the 17th and 18th centuries, sailors were far more likely to die of scurvy than to be killed in shipwrecks or battles.

Vitamin C is also necessary for reactions that synthesize neurotransmitters, hormones, bile acids, and carnitine, which is needed for the breakdown of fatty acids.

Citrus fruits are an excellent source of vitamin C. A large orange contains enough vitamin C to meet the RDA of 90 mg/day for men and 75 mg/day for women. Other fruits and vegetables are also good sources of this vitamin (**Figure 7.27**).

Vitamin C is destroyed by oxygen, light, and heat, so it is readily lost in cooking. This loss is accelerated in low-acid foods and by the use of copper or iron cooking utensils. Although most Americans consume enough vitamin C to prevent severe deficiency, marginal vitamin C deficiency is a concern for individuals who consume few fruits and vegetables. Cigarette smoking increases the requirement for vitamin C because the vitamin is used to break down compounds in cigarette smoke. It is recommended that cigarette smokers consume an extra 35 mg of vitamin C daily—an amount that can easily be supplied by a half-cup of broccoli.

One-third of the population of the United States takes vitamin C supplements—usually in the hope that they will prevent the common cold. Although vitamin C does not prevent colds or reduce their severity, regular vitamin C supplementation may help reduce the duration of cold symptoms.[23] It has also been suggested that vitamin C supplements reduce the risk of cardiovascular disease and cancer, but there is insufficient evidence to support this claim.

Taking high doses of supplemental vitamin C can cause diarrhea, nausea, and abdominal cramps. In susceptible individuals, high-dose supplements may increase the risk of kidney stone formation because vitamin C can be metabolized to a compound found in some types of kidney stones.[24] In individuals who are unable to regulate iron absorption, taking vitamin C supplements, which increase iron absorption, increases the risk that toxic amounts of iron will accumulate in the body. For those

FIGURE 7.26 **Vitamin C function and deficiency** A deficiency of vitamin C results in the inability to form healthy collagen, which leads to the symptoms of scurvy.

a. A reaction requiring vitamin C is essential for the formation of bonds that hold together adjacent collagen strands and give the protein strength. Like all other body proteins, collagen is continuously being broken down and re-formed. Without vitamin C, the bonds holding together adjacent collagen molecules cannot be formed, so the collagen that is broken down is replaced with abnormal collagen.

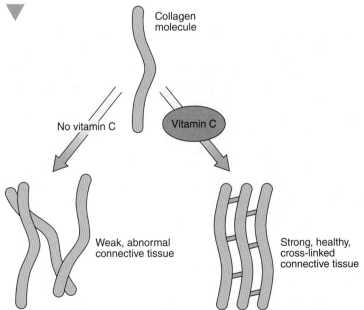

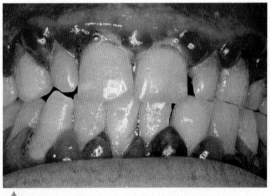

b. When vitamin C intake is below 10 mg/day, the symptoms of scurvy begin to appear. The gums become inflamed, swell, and bleed. The teeth loosen and eventually fall out. The capillary walls weaken and rupture, causing bleeding under the skin and into the joints. This causes raised red spots on the skin, joint pain and weakness, and easy bruising. Wounds do not heal, old wounds may reopen, and bones fracture. People with scurvy become tired and depressed, and they suffer from hysteria.

scurvy A vitamin C deficiency disease characterized by bleeding gums, tooth loss, joint pain, bleeding into the skin and mucous membranes, and fatigue.

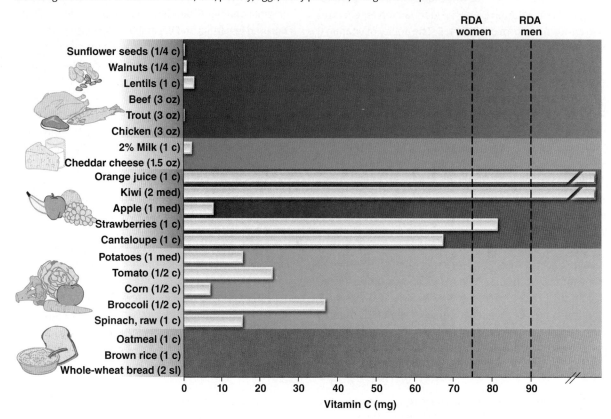

FIGURE 7.27 **Meeting vitamin C needs** Fruits that are high in vitamin C include citrus fruits, strawberries, kiwis, and cantaloupe. Vegetables in the cabbage family, such as broccoli, cauliflower, bok choy, and Brussels sprouts, as well as dark-green vegetables, green and red peppers, okra, tomatoes, and potatoes, are also good sources of vitamin C. Meat, fish, poultry, eggs, dairy products, and grains are poor sources.

with sickle cell anemia, excess vitamin C can worsen symptoms. In those taking medication to reduce blood clotting, taking more than 3000 mg/day of vitamin C can interfere with the effectiveness of the medication. In large doses, chewable vitamin C supplements can dissolve tooth enamel. The UL for vitamin C has been set at 2000 mg/day from food and supplements.

Vitamin E

Vitamin E, or **tocopherol**, is a fat-soluble vitamin that functions primarily as an antioxidant. It protects lipids throughout the body by neutralizing reactive oxygen compounds before they can cause damage (**Figure 7.28**). Vitamin E protects the membranes of red blood cells, white blood cells, nerve cells, and lung cells, as well as the lipids in lipoproteins (see Chapter 5). Research has also identified roles for vitamin E in modulating immune response, reducing inflammation, allowing cells to communicate with each other, regulating genes, and inhibiting an early step in blood clot formation.[25, 26]

Because vitamin E is needed to protect cell membranes, a deficiency causes those membranes to break down. Red blood cells and nerve tissue are particularly susceptible. With a vitamin E deficiency, red blood cell membranes may rupture, causing a type of anemia called **hemolytic anemia**. This is most common in premature infants. All newborn infants have low blood vitamin E levels because there is little transfer of this vitamin from mother to fetus until the last weeks of pregnancy. The levels are lower in premature infants because they are born before much vitamin E has been transferred from the mother. To prevent vitamin E deficiency in premature infants, special formulas for these infants contain higher amounts of vitamin E.

Vitamin E deficiency is rare in adults, occurring only when other health problems interfere with fat absorption, which reduces vitamin E absorption. In such cases, the vitamin E deficiency is usually characterized by symptoms associated with nerve degeneration, such as poor muscle coordination, weakness, and impaired vision.

The RDA for vitamin E is expressed as mg **alpha-tocopherol (α-tocopherol)**; older information may list the recommendation as International Units (IUs), which is a measure based on vitamin activity (see Appendix J). Although there are several forms of vitamin E that occur naturally in food, only the α-tocopherol form can meet vitamin E requirements in humans. Synthetic α-tocopherol used in supplements and fortified foods provides only half as much vitamin E activity as the natural form.

Most Americans do not consume enough vitamin E to meet the RDA.[27] Nuts, seeds, and plant oils are the best sources of vitamin E; fortified products such as breakfast cereals also make a significant contribution to our vitamin E intake (**Figure 7.29**). The need for vitamin E increases as polyunsaturated fat intake increases because polyunsaturated fats are particularly susceptible to oxidative damage; fortunately, polyunsaturated oils are

FIGURE 7.28 **The antioxidant role of vitamin E** Vitamin E neutralizes free radicals in cell membranes, as shown here. After vitamin E is used to eliminate free radicals, its antioxidant function can be restored by vitamin C.

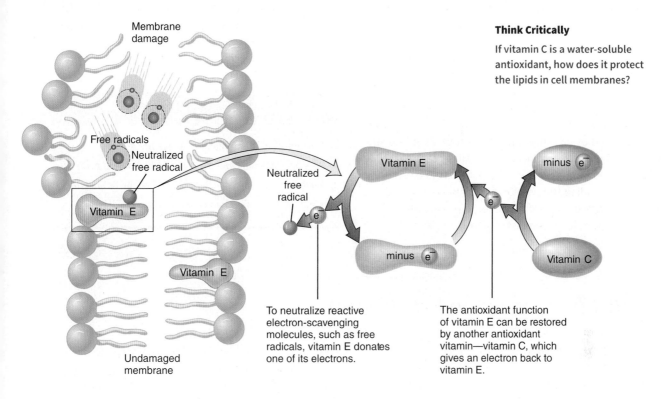

Think Critically

If vitamin C is a water-soluble antioxidant, how does it protect the lipids in cell membranes?

To neutralize reactive electron-scavenging molecules, such as free radicals, vitamin E donates one of its electrons.

The antioxidant function of vitamin E can be restored by another antioxidant vitamin—vitamin C, which gives an electron back to vitamin E.

FIGURE 7.29 **Meeting vitamin E needs** Dietary sources of vitamin E include sunflower seeds, nuts, peanuts, and refined plant oils such as canola, safflower, and sunflower oils. Vitamin E is also found in leafy green vegetables, such as spinach and mustard greens, and in wheat germ and fortified breakfast cereals. The dashed line represents the RDA for adults of 15 mg α-tocopherol/day.

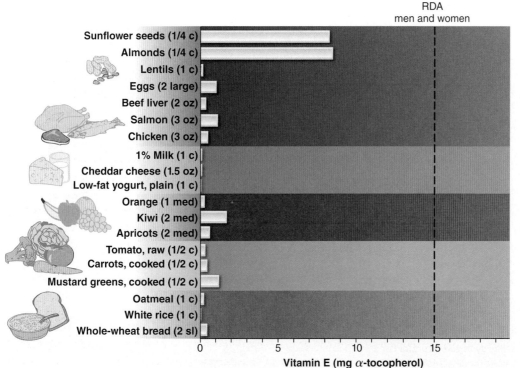

Interpret the Data

Snacking on a quarter cup of almonds will provide over half of the RDA for vitamin E.

a. true
b. false

one of the best sources of dietary vitamin E. Because vitamin E is sensitive to destruction by oxygen, metals, light, and heat, when vegetable oils are repeatedly used for deep-fat frying, most of the vitamin E in them is destroyed.

Vitamin E supplements are popular primarily because people believe that they will boost antioxidant function in the body. As an antioxidant, vitamin E may help reduce the risk of heart disease and other chronic diseases associated with oxidative stress. Studies done in cells and animals as well as epidemiological studies have reported that vitamin E may help reduce the incidence of cardiovascular disease.[28] However, intervention studies investigating the effect of vitamin E supplements have failed to provide evidence of any cardiovascular benefits.[29,30] There are a number of hypotheses to explain this discrepancy, including the possibility that forms of vitamin E found in food, not the α-tocopherol form in the supplements used, are important for preventing atherosclerosis.

There is no evidence of adverse effects from consuming large amounts of vitamin E naturally present in foods. The amounts typically contained in supplements are also safe for most people; however, large doses can interfere with blood clotting, so individuals taking blood-thinning medications should not take vitamin E supplements. The UL is 1000 mg/day from supplemental sources.

Concept Check

1. **How** do antioxidants stabilize free radicals?
2. **Why** does vitamin C deficiency cause bleeding gums?
3. **How** does vitamin E deficiency cause hemolytic anemia?
4. **What** food groups provide the most vitamin C?

7.5 Vitamins in Gene Expression

LEARNING OBJECTIVES

1. **Explain** how changes in gene expression affect body function.
2. **Describe** two vitamin A roles that keep eyes healthy.
3. **Discuss** vitamin A deficiency as a world health issue.
4. **Relate** the functions of vitamin D to the symptoms that occur when it is deficient in the body.

Vitamins A and D have functions that depend on their ability to regulate **gene expression**. When a specific gene is expressed, it instructs the cell to make a particular protein. These proteins have structural and regulatory functions within cells and throughout the body. Therefore, by regulating the expression of certain genes, these vitamins can increase or decrease levels of specific proteins and alter various cellular and body functions (**Figure 7.30**). By affecting gene expression, vitamin A helps ensure healthy eyes, and vitamin D promotes strong bones.

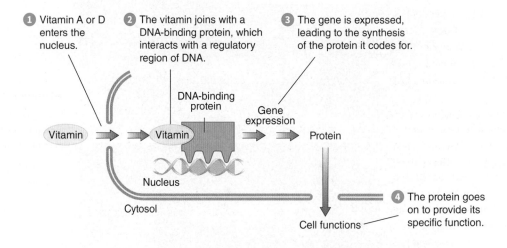

FIGURE 7.30 Turning on genes Vitamins A and D can turn on genes, thereby increasing the amounts of the particular proteins for which they code and their specific functions.

1. Vitamin A or D enters the nucleus.
2. The vitamin joins with a DNA-binding protein, which interacts with a regulatory region of DNA.
3. The gene is expressed, leading to the synthesis of the protein it codes for.
4. The protein goes on to provide its specific function.

gene expression The events of protein synthesis in which the information coded in a gene is used to synthesize a protein or a molecule of RNA.

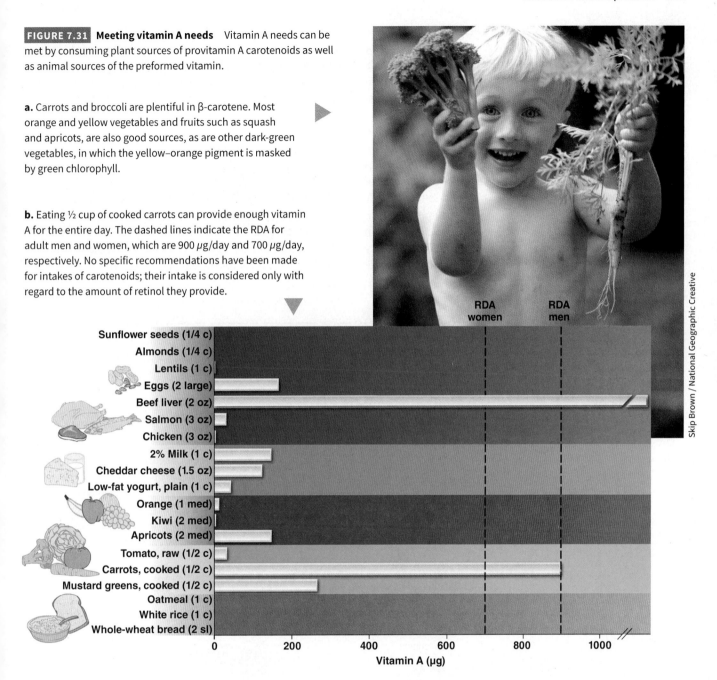

FIGURE 7.31 Meeting vitamin A needs Vitamin A needs can be met by consuming plant sources of provitamin A carotenoids as well as animal sources of the preformed vitamin.

a. Carrots and broccoli are plentiful in β-carotene. Most orange and yellow vegetables and fruits such as squash and apricots, are also good sources, as are other dark-green vegetables, in which the yellow–orange pigment is masked by green chlorophyll.

b. Eating ½ cup of cooked carrots can provide enough vitamin A for the entire day. The dashed lines indicate the RDA for adult men and women, which are 900 μg/day and 700 μg/day, respectively. No specific recommendations have been made for intakes of carotenoids; their intake is considered only with regard to the amount of retinol they provide.

Vitamin A

Have you ever heard that eating carrots will help you see in the dark? It turns out to be true. Carrots are a good source of **beta-carotene (β-carotene)**, a provitamin that can be converted into **vitamin A** in your body. Vitamin A is needed for vision and healthy eyes.

Vitamin A in the diet
Vitamin A is a fat-soluble vitamin found both preformed and in provitamin form in our diet. Preformed vitamin A compounds are known as **retinoids**. Three retinoids are active in the body: retinal, retinol, and retinoic acid. Retinal and retinoic acid are formed in the body from retinol consumed in the diet. **Carotenoids** are yellow–orange pigments found in plants, some of which are vitamin A precursors; once inside the body, they can be converted into retinoids (**Figure 7.31a**). Beta-carotene is the most potent vitamin A precursor. **Alpha-carotene (α-carotene)**, found in dark-green vegetables, carrots, and squash, and **beta-cryptoxanthin**

---

retinoids A group of compounds with vitamin A activity. There are three retinoids: retinol, retinal, and retinoic acid.

carotenoids Yellow, orange, and red pigments synthesized by plants and many microorganisms. Some carotenoids can be converted to vitamin A.

(β-cryptoxanthin), found in papaya, sweet red peppers, and winter squash, are also provitamin A carotenoids, but they are not converted into retinoids as efficiently as β-carotene. Carotenoids that are not converted into retinoids may function as antioxidants and thus may play a role in protecting against cancer and heart disease.

You can meet your needs for vitamin A by eating animal products, such as dairy products, eggs, liver, and fish that are sources of retinol, and by eating fruits and vegetables such as carrots, broccoli, cantaloupe, and squash that are sources of provitamin A carotenoids (**Figure 7.31b**). Because carotenoids are not absorbed as well as retinol and are not completely converted into vitamin A in the body, you get less functional vitamin A from this form. To account for this difference, **retinol activity equivalents (RAEs)** are used to express the amount of usable vitamin A in foods; 1 RAE is the amount of retinol, β-carotene, α-carotene, or β-cryptoxanthin that provides vitamin A activity equal to 1 μg of retinol (see Appendix J).[31]

Both carotenoids and retinol are bound to proteins in food. To be absorbed, they must be released from protein by pepsin and other protein-digesting enzymes. In the small intestine, they combine with bile acids and other dietary fats in order to be absorbed. The fat content of the diet can affect the amount of vitamin A absorbed. When dietary fat intake is very low (less than 10 g/day), vitamin A absorption is impaired. This is rarely a problem in the United States and other industrialized countries, where typical fat intake is greater than 50 g/day. However, in developing countries, vitamin A deficiency may occur not only because the diet is low in vitamin A but also because the diet is too low in fat for the vitamin to be absorbed efficiently.

PROCESS DIAGRAM

FIGURE 7.32 **The visual cycle** Looking into the bright headlights of an approaching car at night is temporarily blinding for all of us, but for someone with vitamin A deficiency, the blindness lasts a lot longer. This occurs because of the role of vitamin A in the visual cycle.

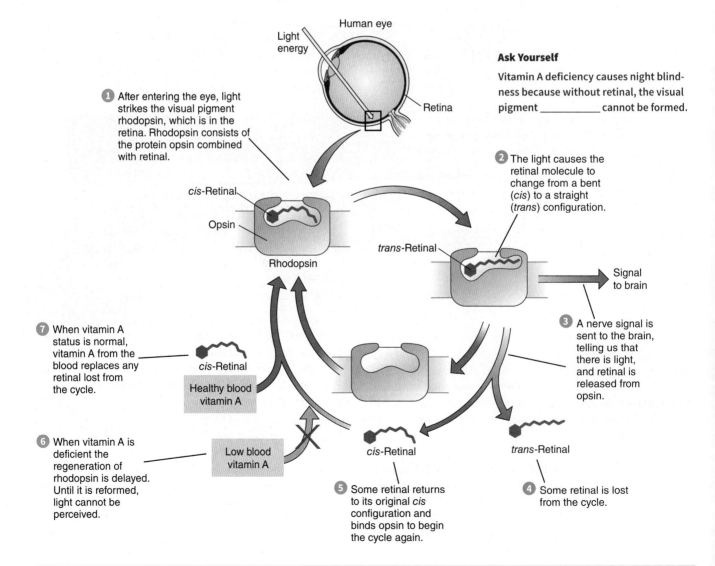

Ask Yourself

Vitamin A deficiency causes night blindness because without retinal, the visual pigment _____ cannot be formed.

Diseases that cause fat malabsorption can also interfere with vitamin A absorption and cause deficiency.

Protein and zinc status are also important for healthy vitamin A status. To move from liver stores to other body tissues, vitamin A must be bound to a protein called **retinol-binding protein**. When protein is deficient, the amount of retinol-binding protein made is inadequate, so vitamin A cannot be transported to the tissues where it is needed. Likewise, when zinc is deficient, a vitamin A deficiency may occur because zinc is needed to make proteins involved in vitamin A transport and metabolism.

Vitamin A functions and deficiency

Vitamin A is needed for vision and eye health because it is involved in the perception of light and because it is needed for normal **cell differentiation**, the process whereby immature cells change in structure and function to become specialized.

Vitamin A helps us see light because retinal is part of **rhodopsin**, a visual pigment in the eye. When light strikes rhodopsin, it initiates a series of events that break apart rhodopsin and cause a nerve signal to be sent to the brain, which allows us to perceive the light (**Figure 7.32**). After the light stimulus has passed, rhodopsin is re-formed. Because some retinal is lost in these reactions, it must be replaced by vitamin A from the blood. If blood levels of vitamin A are low, as they are in someone who is vitamin A deficient, there is a delay in the regeneration of rhodopsin. This delay causes difficulty seeing in dim light, a condition called **night blindness**. Night blindness is one of the first and most easily reversible symptoms of vitamin A deficiency. If the deficiency progresses, more serious and less reversible symptoms can occur.

The role of vitamin A in gene expression makes it essential for normal differentiation of cells in epithelial tissue. Epithelial tissue is the type of tissue that covers internal and external body surfaces. It includes the skin and the linings of the eyes, intestines, lungs, vagina, and bladder. When vitamin A is deficient, epithelial tissue cannot be maintained; all epithelial tissue is affected by vitamin A deficiency, but that in the eye is particularly susceptible (**Figure 7.33**). The eye disorders associated with vitamin A deficiency are collectively known as

---

NUTRITION INSIGHT

FIGURE 7.33 **Vitamin A deficiency** Vitamin A deficiency causes night blindness and, if left untreated, leads to blindness by interfering with the maintenance of epithelial tissue. This deficiency is a threat to the sight, health, and lives of millions of children in the developing world.

The lining of the eye normally contains cells that secrete mucus, which lubricates the eye. When these cells die, immature cells differentiate to become new mucus-secreting cells that replace the dead ones. Without vitamin A, the immature cells can't differentiate normally, and instead of mucus-secreting cells, they become cells that produce a hard protein called keratin.

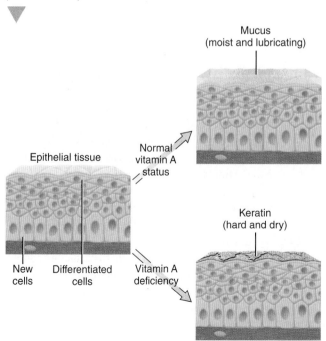

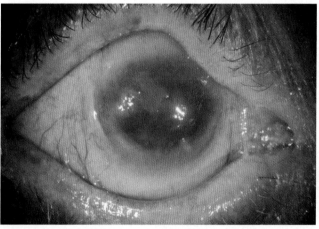

When mucus-secreting cells are replaced by keratin-producing cells, the surface of the eye becomes dry and cloudy. As xerophthalmia progresses, the drying of the cornea results in ulceration and infection. If left untreated, the damage is irreversible and causes permanent blindness.

(continues)

FIGURE 7.33 *(continued)*

Interpret the Data

On which continents is clinical vitamin A deficiency most prevalent?

It is estimated that more than 250 million preschool children worldwide are vitamin A deficient and that 250,000 to 500,000 children go blind annually due to vitamin A deficiency. Children with clinical vitamin A deficiency have poor appetites, are anemic, are more susceptible to infections, and are more likely to die in childhood.[33]

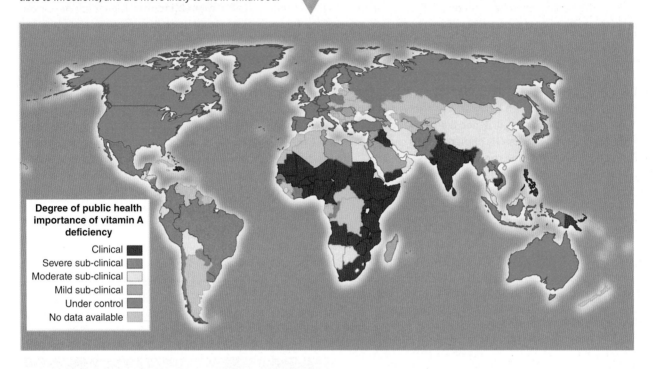

xerophthalmia. Night blindness is an early stage of xeropthalmia and can be treated by increasing vitamin A intake. If left untreated, xerophthalmia affects the epithelial lining of the eye and can result in permanent blindness.

The ability of vitamin A to regulate the growth and differentiation of cells makes it essential throughout life for normal reproduction, growth, and immune function. In a developing embryo, vitamin A is needed to direct cells to differentiate and to form the shapes and patterns needed for the development of a complete organism. In growing children, vitamin A affects the activity of cells that form and break down bone; a deficiency early in life can cause abnormal jawbone growth, resulting in crooked teeth and poor dental health. In the immune system, vitamin A is needed for the differentiation that produces the different types of immune cells. When vitamin A is deficient, the activity of specific immune cells cannot be stimulated; the result is increased susceptibility to infections.

Vitamin A deficiency is uncommon in developed countries, but low intake can still occur due to poor food choices.

In the United States, vitamin A intake may be below the RDA if consumption of dairy products, which provide preformed vitamin A, and of fruits and vegetables, many of which provide provitamin A, do not meet recommendations (see *Thinking It Through*).[32]

Vitamin A toxicity Preformed vitamin A is toxic in large doses, causing symptoms such as nausea, vomiting, headache, dizziness, blurred vision, and lack of muscle coordination (**Figure 7.34a**). Excess vitamin A is a particular concern for pregnant women because it may contribute to birth defects. Derivatives of vitamin A that are used to treat acne (Retin-A and Accutane) should never be used by pregnant women because they cause birth defects. High intakes of vitamin A have also been found to cause liver damage.[34] The UL is set at 2800 μg/day of preformed vitamin A for 14- to 18-year-olds and 3000 μg/day for adults.

Because preformed vitamin A can be toxic, dietary supplements typically contain β-carotene. Carotenoids are not toxic

---

xerophthalmia A spectrum of eye conditions resulting from vitamin A deficiency that may lead to blindness.

Thinking It Through

A Case Study on Vitamins in the Modern Diet

John lives on his own, works, and goes to school. He doesn't have much time to cook so eats the same foods most days. He has cereal for breakfast at home and usually takes a turkey sandwich and an apple for lunch. Five nights a week he has dinner at the McDonald's fast-food restaurant where he works. He is worried that his boring diet isn't providing everything he needs. He looks up his MyPlate recommendations: 9 oz. of grains, 3½ cups of vegetables, 2½ cups of fruit, 3 cups of dairy, and 6½ oz of protein foods. Use the table below to compare his diet to these food group recommendations.

Food	Myplate Servings
BREAKFAST	
Cheerios, 2 cups	2 grains (whole)
2% milk, 1 cup	1 dairy
Coffee, black	—
LUNCH	
Turkey sandwich	
Whole-wheat bread, 2 slices	2 grains (whole)
Turkey, 2 oz	2 protein
Mayonnaise, 1 tsp	1 tsp oils
Tomato, ½ medium	¼ c vegetable
Lettuce, 2 leaves	¼ c vegetable
Apple, 1 small	1 c fruit
Soda, 20 oz	—
SNACK	
Snickers bar	—
DINNER	
McDonald's Quarter Pounder, 2 each	5 grains, 5 ½ protein, ¼ c vegetable
McDonalds fries, 1 medium order	¾ cup vegetable
Cola, 1 medium	—

Which food groups does John need to increase? How much more of each of these does he need?

Your answer:

What could John have for lunch instead of the 250-Calorie soda that would help to meet his dairy recommendations without increasing his calorie intake?

Your answer:

John's diet is likely low in vitamins because he does not meet all of his food group recommendations. He looks at Figure 7.2 to see which vitamins are found in these food groups

Which vitamin is plentiful in all the food groups that are lacking in John's diet (see Figure 7.2)?

Your answer:

John recognizes that he might improve the quality and variety of his diet by getting dinner from one of the other take-out restaurants near where he works. The photos show two meal options that are nearby.

What components of the meals shown here provide vitamin A?

Your answer:

Suggest some other take-out meals that would increase John's intake of fruits, vegetables, and/or dairy.

Your answer:

(Check your answers in Appendix L.)

FIGURE 7.34 Vitamin A toxicity Only preformed vitamin A is toxic.

a. Although foods generally do not naturally contain large enough amounts of nutrients to be toxic, polar bear liver contains about 100,000 μg of vitamin A in just 1 ounce—enough to have caused vitamin A toxicity in Arctic explorers. Polar bear liver is not a common dish at most dinner tables, but supplements of preformed vitamin A also have the potential to deliver a toxic dose.

b. β-carotene supplements or regular consumption of large amounts of carrot juice can cause hypercarotenemia. The hand on the right illustrates this harmless buildup of carotenoids in the adipose tissue, which makes the skin look yellow–orange, particularly on the palms of the hands and the soles of the feet.

because when they are consumed in high doses, their absorption from the diet decreases, and their conversion to active vitamin A is limited. However, large daily intakes of carotenoids from supplements or the diet can lead to a harmless condition known as **hypercarotenemia** (Figure 7.34b). β-carotene supplements have also been associated with an increase in lung cancer in cigarette smokers.[35] Therefore, smokers are advised to avoid β-carotene supplements. There is no UL for carotenoids, and the small amounts found in standard-strength multivitamin supplements are not likely to be harmful for any group.

Vitamin D

Vitamin D is a fat-soluble vitamin known as the sunshine vitamin because it can be made in the skin with exposure to ultraviolet (UV) light. Because vitamin D can be made in the body, it is essential in the diet only when exposure to sunlight is limited or the body's ability to synthesize it is reduced.

Vitamin D, whether from the diet or from synthesis in the skin, is inactive until it is modified by biochemical reactions in both the liver and the kidney (**Figure 7.35**). Active vitamin D is needed to maintain normal levels of the minerals calcium and phosphorus in the blood. Calcium is important for bone health, but it is also needed for proper functioning of nerves, muscles, glands, and other tissues. Blood levels of calcium are regulated so that a steady supply of the mineral is available when and where it is needed.

When calcium levels in the blood drop too low, the body responds immediately to correct the problem. The response starts with the release of **parathyroid hormone (PTH)**, which stimulates the activation of vitamin D by the kidneys. Active vitamin D enters the blood and travels to its major target tissues—intestine, bone, and kidneys—where it acts to increase calcium and phosphorus levels in the blood (see Figure 7.35). The functions of vitamin D, like vitamin A, are due to its role in gene expression. In the intestine, it increases the production of proteins needed for the absorption of calcium. In the bone, it increases the production of proteins that are needed for the differentiation of cells that break down bone.

It is also now recognized that in addition to its role in bone health, vitamin D affects cells in the pancreas, heart, skeletal muscles, and immune system.[36] Vitamin D is hypothesized to protect against cancer by decreasing the proliferation of cancer cells, promoting cancer cell death, and inhibiting the spread of cancerous cells. There is also evidence that vitamin D protects against a variety of chronic and autoimmune diseases.[37]

Vitamin D deficiency Low blood levels of vitamin D are now believed to play an important role in the development of multiple sclerosis, type 1 diabetes, inflammatory bowel disease, infections, cardiovascular disease, cancer, and even neurological disorders such as Alzheimer's. More severe vitamin D deficiency interferes with calcium absorption; when vitamin D

---

hypercarotenemia A condition caused by the accumulation of carotenoids in the adipose tissue, causing the skin to appear yellow–orange.

parathyroid hormone (PTH) A hormone released by the parathyroid gland that acts to increase blood calcium levels.

PROCESS DIAGRAM

FIGURE 7.35 Vitamin D activation and function In order to function, vitamin D from food and from synthesis in the skin must be activated.

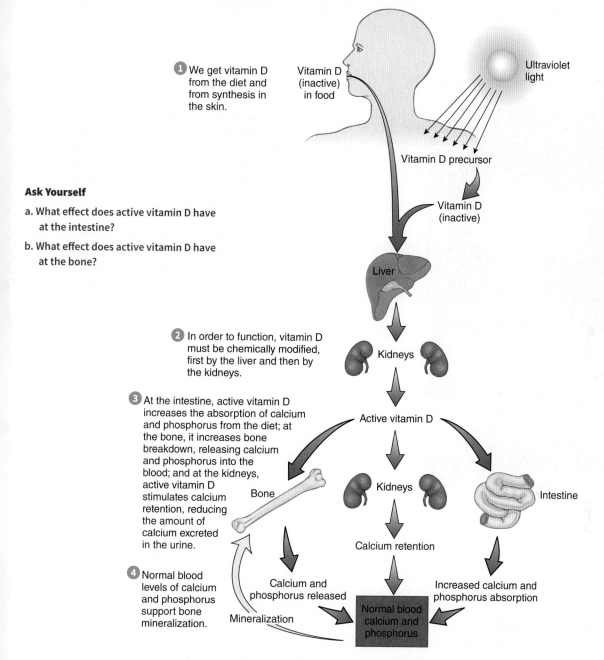

1. We get vitamin D from the diet and from synthesis in the skin.

2. In order to function, vitamin D must be chemically modified, first by the liver and then by the kidneys.

3. At the intestine, active vitamin D increases the absorption of calcium and phosphorus from the diet; at the bone, it increases bone breakdown, releasing calcium and phosphorus into the blood; and at the kidneys, active vitamin D stimulates calcium retention, reducing the amount of calcium excreted in the urine.

4. Normal blood levels of calcium and phosphorus support bone mineralization.

Ask Yourself

a. What effect does active vitamin D have at the intestine?

b. What effect does active vitamin D have at the bone?

---

is deficient, only about 10 to 15% of the calcium in the diet can be absorbed, increasing the risk of abnormalities in bone structure. In children, vitamin D deficiency causes **rickets**; it is characterized by narrow rib cages known as pigeon breasts and by bowed legs (**Figure 7.36**). In adults, vitamin D deficiency causes a condition called **osteomalacia**. Bone deformities do not occur with osteomalacia because adults are no longer growing, but the bones are weakened because not enough calcium is available to form the mineral deposits needed to maintain healthy bone. Insufficient bone mineralization leads to fractures of the weight-bearing bones, such as those in the hips and spine. This lack of calcium in bones can precipitate or

---

rickets A vitamin D deficiency disease in children, characterized by poor bone development due to inadequate calcium absorption.

osteomalacia A vitamin D deficiency disease in adults, characterized by loss of minerals from bone, bone pain, muscle aches, and an increase in bone fractures.

FIGURE 7.36 Rickets The vitamin D deficiency disease rickets causes short stature and bone deformities. The characteristic bowed legs occur because the bones are too weak to support the body. It is most common in children with poor diets and little exposure to sunlight, in those with disorders that affect fat absorption, and in vegan children who do not receive adequate exposure to sunlight.

exacerbate **osteoporosis**, which is a loss of total bone mass, not just minerals (discussed further in Chapter 8). Osteomalacia is common in adults with kidney failure because the conversion of vitamin D to the active form is reduced in these patients.

Meeting vitamin D needs Vitamin D is not very widespread in the diet. It is found naturally in liver, egg yolks, and oily fish such as salmon (**Figure 7.37a**). Foods fortified with vitamin D include milk, milk substitutes, margarine, and some yogurts, cheeses, and breakfast cereals (see *What Should I Eat?*). National surveys indicate that average vitamin D intake is below recommendations, but average blood levels of vitamin D are above the level needed for good bone health. This dichotomy suggests that vitamin D synthesis from sun exposure contributes enough vitamin D to allow the majority of the population to meet vitamin D needs even when intake is below recommendations.[38] Anything that interferes with the transmission of UV radiation to Earth's surface or its penetration into the skin will affect the synthesis of vitamin D. Therefore, living at higher latitudes, staying indoors, and keeping the skin covered when outdoors increase the risk of vitamin D deficiency (**Figure 7.37b–e**).

The RDA for vitamin D is expressed both in International Units (IUs) and in μg. One IU is equal to 0.025 μg of vitamin D (40 IU = 1 μg of vitamin D; see Appendix J). The RDA for vitamin D for children and adults 70 and under is set at 600 IU (15 μg)/day—an amount that ensures that vitamin D levels in the blood are high enough to support bone health even when sun exposure is minimal. Due to physiological changes that occur with aging, such as less efficient vitamin D synthesis in the skin, the amount of vitamin D needed to reduce the risk of fractures is higher in older adults. The RDA for adults older than 70 is 800 IU (20 μg)/day.

What Should I Eat?

Vitamins

Focus on folate, vitamin A, and vitamin K

- Snack on an orange—you'll get your folate as well as vitamin C.
- Add beans—such as lentils and kidneys—to soups and tacos.
- Add color and carotenoids to your meal with peppers, carrots, or kale.
- Have a salad with a heaping helping of leafy greens to add folate, vitamin K, and β-carotene.

B (vitamin) sure

- Don't forget the whole grains for vitamin B_6 as well as fiber.
- Enrich your diet with some enriched grains.
- Have a bowl of fortified breakfast cereal for B_{12} insurance.
- Enjoy lean meats, poultry, and fish to get a B_6 and B_{12} boost.

Get your antioxidants

- Snack on nuts and seeds and cook with canola oil to increase your vitamin E.
- Try for five different colors of fruits and veggies each day; different colors indicate different antioxidants.
- Add carrot sticks to your lunch or snack to increase your β-carotene intake.
- Savor some strawberries and kiwis for dessert—they are loaded with vitamin C.

Soak up some D

- Exercise outside to stay fit and make some vitamin D.
- Have three servings of milk per day to boost your intake of vitamin D.

 Use iProfile to calculate your vitamin C intake for a day.

NUTRITION INSIGHT

FIGURE 7.37 Barriers to meeting vitamin D needs Limited food sources as well as factors that reduce the amount of sunlight that reaches the skin make it difficult for many people to meet their vitamin D needs.

a. Only a few foods are natural sources of vitamin D. Without adequate sun exposure, supplements, or fortified foods, it is difficult to meet the body's needs for this vitamin, particularly for vegans. The dashed line indicates the RDA for ages 1 to 70, which is 600 IU (15 μg)/day. It would take about 5 cups of vitamin D–fortified milk to provide this much vitamin D.

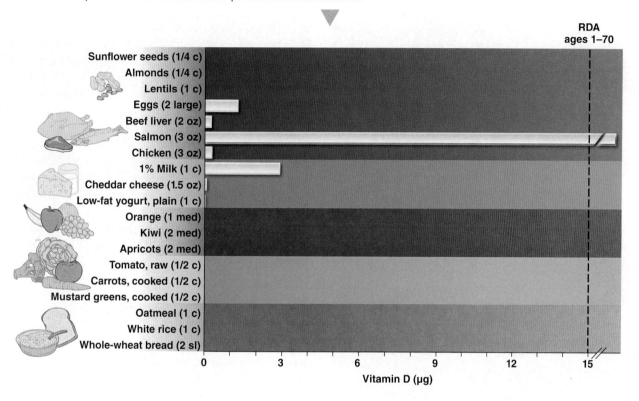

b. The angle at which the sun strikes the Earth affects the body's ability to synthesize vitamin D in the skin. During the winter at latitudes greater than about 40 degrees north or south, there is not enough UV radiation to synthesize adequate amounts. However, during the spring, summer, and fall at 42 degrees latitude, as little as 5 to 10 minutes of midday sun exposure three times weekly can provide a light-skinned individual with adequate vitamin D.[39]

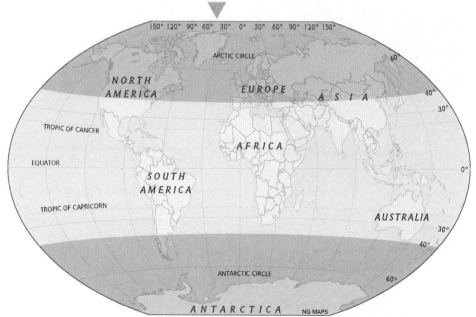

(continues)

FIGURE 7.37 *(continued)*

c. Sunscreen with an SPF of 15 decreases vitamin D synthesis by 99%.[40] Sunscreen is important for reducing the risk of skin cancer, but some time in the sun without sunscreen may be needed to meet vitamin D needs. In the summer, children and active adults usually spend enough time outdoors without sunscreen to meet their vitamin D requirements.

d. Dark skin pigmentation prevents UV light rays from penetrating into the layers of the skin where vitamin D is formed and reduces the body's ability to make vitamin D in the skin by as much as 99%. Dark-skinned individuals living in temperate climates have a higher rate of vitamin D deficiency than do those living near the equator.

e. Concealing clothing worn by certain cultural and religious groups prevents sunlight from striking the skin. This explains why vitamin D deficiency occurs in women and children in some of the sunniest regions of the world. Older adults also often wear concealing clothing when they are outdoors. The risk of vitamin D deficiency in older adults is further compounded by a diet that is low in vitamin D and an age-related decline in the ability to synthesize vitamin D in the skin and to convert the vitamin to its active form.

Too much vitamin D in the body can cause high calcium concentrations in the blood and urine, deposition of calcium in soft tissues such as the blood vessels and kidneys, and cardiovascular damage. Synthesis of vitamin D from exposure to sunlight does not produce toxic amounts because the body regulates vitamin D formation. The UL for ages 9 and older for vitamin D is 4000 IU (100 μg)/day.

Concept Check

1. **What** happens when a gene is expressed?
2. **How** does vitamin A help us see in the dark?
3. **Where** is vitamin A deficiency most common?
4. **Why** do the leg bones bow in children with vitamin D deficiency?

7.6 Meeting Needs with Dietary Supplements

LEARNING OBJECTIVES

1. **List** population groups that may benefit from taking vitamin and/or mineral supplements.
2. **Explain** how the safety of dietary supplements is regulated.
3. **Evaluate** the nutrient content of a dietary supplement using a Supplement Facts panel.

Currently, more than 60% of Americans adults take dietary supplements.[41] We take supplements for a variety of reasons—to energize ourselves, to protect ourselves from disease, to cure illnesses, to lose weight, to enhance athletic performance, and simply to prevent deficiencies. These products may be beneficial and even necessary under some circumstances for some people, but they also have the potential to cause harm.

Some dietary supplements contain vitamins and minerals, some contain herbs and other plant-derived substances, and some contain compounds that are found in the body but are not essential in the diet. While supplements can help us obtain adequate amounts of specific nutrients, they do not provide all the benefits of foods. A pill that meets a person's vitamin needs does not provide the energy, protein, minerals, fiber, or phytochemicals supplied by food sources of these vitamins.

Who Needs Vitamin/Mineral Supplements?

Eating a variety of foods is the best way to meet nutrient needs, and most healthy adults who consume a reasonably good diet do not need supplements.[42] In fact, an argument against the use of supplements is that supplement use gives people a false sense of security, causing them to pay less attention to the nutrient content of the foods they choose. For some people, however, taking supplements may be the only way to meet certain nutrient needs because of low intakes, increased needs, or excess losses (**Table 7.3**).

TABLE 7.3 Groups for whom dietary supplements are recommended[43]

Group	Recommendation
Dieters	People who consume fewer than 1600 Calories/day should take a multivitamin/multimineral supplement.
Vegans and those who eliminate all dairy foods	To obtain adequate vitamin B_{12}, people who do not eat animal products need to take supplements or consume vitamin B_{12}–fortified foods. Because dairy products are an important source of calcium and vitamin D, those who do not consume dairy products may benefit from taking supplements that provide calcium and vitamin D.
Infants and children	Supplemental fluoride, vitamin D, and iron are recommended under certain circumstances.
Young women and pregnant women	Women of childbearing age should consume 400 μg of folic acid daily from either fortified foods or supplements. Supplements of iron and folic acid are recommended for pregnant women, and multivitamin/multimineral supplements are usually prescribed during pregnancy.
Older adults	Because of the high incidence of atrophic gastritis in adults over age 50, vitamin B_{12} supplements or fortified foods are recommended. It may also be difficult for older adults to meet the RDAs for vitamin D and calcium, so supplements of these nutrients are often recommended.
Individuals with dark skin pigmentation	People with dark skin may be unable to synthesize enough vitamin D to meet their needs for this vitamin and may therefore require supplements.
Individuals with restricted diets	Individuals with health conditions that affect what foods they eat or how nutrients are used may require vitamin and mineral supplements.
People taking medications	Medications may interfere with the body's use of certain nutrients.
Cigarette smokers and alcohol users	People who smoke heavily require more vitamin C and possibly vitamin E than do nonsmokers[22,44] Alcohol consumption inhibits the absorption of B vitamins and may interfere with B vitamin metabolism.

Herbal Supplements

Technically, an herb is a nonwoody, seed-producing plant that dies at the end of the growing season. However, the term *herb* is generally used to refer to any botanical or plant-derived substance. Throughout history, folk medicine has used herbs to prevent and treat disease. Today, herbs and herbal supplements are still popular (**Figure 7.38**). They are readily available and relatively inexpensive, and they can be purchased without a trip to the doctor or a prescription. Although these features are appealing to consumers who want to manage their own health, some herbs may be toxic, either alone or in combination with other herbs and drugs. They should not be taken to replace prescribed medication without the knowledge of your physician.[45] For more information on the risks and benefits of herbal supplements see Appendix K or go to the National Center for Complementary and Integrative Health website.[46]

FIGURE 7.38 Popular herbal supplements *Ginkgo biloba*, St. John's wort, ginseng, garlic, Echinacea, and saw palmetto are among the best-selling herbal supplements in the United States today.

Ginkgo biloba, also called "maidenhair," is marketed to enhance memory and to treat a variety of circulatory ailments. *Ginkgo biloba* extract does not appear to enhance cognitive function in healthy adults, but it can slow cognitive decline in patients with dementia.[47,48] Side effects include headaches and gastrointestinal symptoms.[49] *Ginkgo biloba* also interacts with a number of medications. It can cause bleeding when combined with warfarin or aspirin, elevated blood pressure when combined with a thiazide diuretic, and coma when combined with the antidepressant trazodone.[50]

St. John's wort, taken to promote mental well-being, contains low doses of the chemical found in the antidepressant drug fluoxetine (Prozac). The results of clinical trials suggest that it may be beneficial for the treatment of depression.[51] Side effects include nausea and sensitivity to sunlight. St John's wort should not be used in conjunction with prescription antidepressant drugs, and it has been found to interact with many different medications, including anticoagulants, heart medications, birth control pills, immunosuppressants, antibiotics, and medications used to treat HIV infection.[52]

Ginseng has been used in Asia for centuries for its energizing, stress-reducing, and aphrodisiac properties. Today it is popular because it is suggested to improve mental and physical performance, augment immune function, and enhance sexual function. Although ginseng contains substances that have antioxidant, anti-inflammatory, and immunostimulating effects, there is little evidence that it has any health benefits.[53] Adverse effects include diarrhea, headache, and insomnia.

Hippocrates recommended garlic for treating pneumonia and other infections, as well as cancer and digestive disorders. Although it is no longer recommended for those purposes, recent research has shown that garlic may cause a modest reduction in blood cholesterol and triglyceride levels.[54] Even though we often spice our food with garlic, garlic supplements are not safe for everyone. They interfere with drugs used to treat HIV infection and can increase the risk of bleeding, so they should be used with caution before having surgery or dental work.[55]

(continues)

FIGURE 7.38 *(continued)*

Native Americans used petals of the Echinacea plant as a treatment for colds, flu, and infections. Today, the plant's root is typically used, and it is a popular herbal cold remedy. Echinacea is believed to act as an immune system stimulant; studies are mixed as to whether it is beneficial in either preventing or treating the common cold.[56] Although side effects have not been reported, allergies are possible.

Saw palmetto, which comes from the berries of the American dwarf palm, is marketed to treat prostate enlargement and therefore improve urinary flow. A review of the effectiveness of this supplement found that it provides mild to moderate improvement in urinary symptoms and flow measures. Most side effects, including abdominal pain, decreased libido, headache, fatigue, nausea, and rhinitis, are mild.[57]

Choosing Supplements with Care

Using dietary supplements can be part of an effective strategy to promote good health, but supplements are not a substitute for a healthy diet, and they are not without risks. The Dietary Supplement Health and Education Act (DSHEA) of 1994 defined the term *dietary supplement* and created standards for labeling these products (see Chapter 2). However, it left most of the responsibility for manufacturing practices and safety in the hands of manufacturers.

To help ensure the safety of dietary supplements, the FDA has established "Current Good Manufacturing Practice" regulations for dietary supplements. These address the manufacturing, labeling, and storage of products to ensure purity, strength, and composition.[58] A more stringent set of manufacturing regulations has been established by the U.S. Pharmacopeia (USP). Manufacturers that choose to participate in this voluntary dietary supplement verification program can include the USP-verified mark on their product label (**Figure 7.39a**).

Supplement manufacturers are responsible for ensuring product safety, and most of these products do not require FDA approval before they are marketed. If a problem arises with a specific product, the FDA must prove that the supplement represents a risk before it can require the manufacturer to remove the supplement from the market. The exception to this rule is products that contain new ingredients. The manufacturer of any supplement containing an ingredient not sold in the United States before October 15, 1994, must provide the FDA with safety data prior to marketing the product. Ingredients that were sold prior to this date are presumed to be safe, based on their history of safe use by humans.

Because supplements are not regulated as strictly as drugs, consumers need to use care and caution if they choose to use them. A safe option is a multivitamin/multimineral supplement that does not exceed 100% of the Daily Values. Although there is little evidence that the average person benefits from such a supplement, there is also little evidence of harm. Here are a number of suggestions that will help you when choosing or using dietary supplements:

- **Consider why you want it.** If you are taking it to prevent a deficiency, does it provide the specific nutrients or other substances you need?
- **Compare product costs.** Just as more is not always better, more expensive is not always better either.
- **Make sure it is safe.** Does the supplement contain a potentially toxic level of any nutrient or other substance? Are you taking the dose recommended on the label? For any nutrients that exceed 100% of the Daily Value, check to see if they exceed the UL (see Appendix A). Does it contain any nonvitamin/nonmineral ingredients? If so, have any of them been shown to be toxic to someone like you (**Figure 7.39b**)?
- **Check the expiration date.** Some nutrients degrade over time, so expired products will have a lower nutrient content than is shown on the label. This is particularly true if the product has not been stored properly.
- **Consider your medical history.** Do you have a medical condition that recommends against certain nutrients or other ingredients? Are you taking prescription medication that an ingredient in the supplement may interact with?

FIGURE 7.39 Check the label Supplements must include a Supplement Facts panel and many also provide information about manufacturing practices.

a. The USP symbol indicates that the supplement has been verified as having been manufactured according to the quality, purity, and potency standards set by the USP convention.

b. The current Supplement Facts panel shown here is from a supplement marketed to reduce appetite and therefore promote weight loss. It contains vitamins, minerals, herbal products, and other ingredients. The planned Supplement Facts label includes updates to the Daily Values and units of measure.

Supplement Facts

Serving Size: 2 Capsules
Servings Per Container: 60

	Amount Per Serving	DV%
Vitamin C (as ascorbic acid)	60mg	100%
Pantothenic Acid (as calcium, pantothenate)	20mg	200%
Vitamin B-6 (as pyridoxine HCL)	8mg	400%
Niacin	5mg	25%
Folate (as folic acid)	100mcg	25%
Zinc (as zinc gluconate)	5mg	33%
Copper (as copper gluconate)	500mcg	25%
NADH (Nicotinamide Adenine Dinucleotide)	1000mcg	*
Hoodia Gordonii Extract (20:1 Extract- Equal to 2000 mg of whole plant)	100mg	*
5-Hydroxytryptophan (Griffonia Simplicifolia)	25mg	*
N, N Dimethylglycine	50mg	*
Trimethylglycine	75mg	*
L-Phenylalanine	600mg	*
Decaffinated Green Tea Extract (Total Catechins 130mg, Epigaliocatechin Galiate (EGCG) 70mg)	175mg	*
Salvia Scalarea Extract	50mg	*
Choline (as bitartrate)	75mg	*

*Daily Value (DV, not established)
Recommended Use: As a dietary supplement take two capsules before breakfast on a empty stomach (or before exercise) and two capsules at mid-afternoon preferably with 8 oz of water.

This supplement contains more than 100% of the Daily Value for vitamin B_6. Does this amount exceed the UL?

This supplement contains many ingredients that are not vitamins or minerals and therefore have no Daily Value or UL. The ones shown in blue are herbs. Are they safe when taken in these amounts?

Think Critically
Would you recommend this product? Why or why not?

Check with a physician, dietitian, or pharmacist to help identify these interactions.

- **Approach herbal supplements with caution.** If you are pregnant, ill, or taking medications, consult your physician before taking herbs. Do not give them to children. Do not take combinations of herbs. Do not use herbs for long periods. Stop taking any product that causes side effects.

- **Report harmful effects.** If you suffer a harmful effect or an illness that you think is related to the use of a supplement, seek medical attention and go to the FDA Reporting Web site, at www.fda.gov/safety/medwatch/howtoreport/, for information about how to proceed.

Concept Check

1. **Why** is it recommended that vegans and older adults take vitamin B_{12} supplements?
2. **Who** regulates the safety of dietary supplements?
3. **How** can the UL be used when evaluating a dietary supplement?

Summary

1 A Vitamin Primer 171

- **Vitamins** are essential organic nutrients that do not provide energy and are required in small quantities in the diet to maintain health. As shown in the illustration, vitamins are classified by their solubility in either water or fat.

Figure 7.1 The vitamins

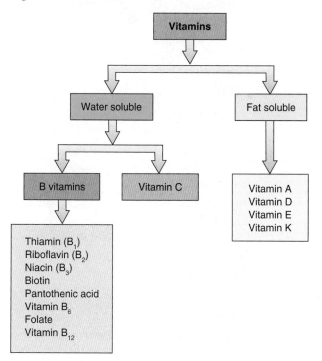

- Vitamins are present naturally in foods and are also added through fortification and enrichment. Food labels can help identify packaged foods that are good sources of certain vitamins.
- Vitamin **bioavailability** is affected by the composition of the diet, conditions in the digestive tract, and the ability to transport and activate the vitamin once it has been absorbed.
- Vitamins promote and regulate body activities. Each vitamin has one or more important functions.
- The RDAs and AIs recommend amounts for each of the vitamins that will prevent deficiency and promote health. In developing countries, vitamin deficiencies remain a major public health problem. They are less common in industrialized countries due to enrichment and fortification and a more varied food supply.

2 Vitamins and Energy Metabolism 179

- The B vitamins thiamin, riboflavin, niacin, biotin, and pantothenic acid, as well as a number of vitamins and vitamin-like compounds function as **coenzymes** involved in transferring the energy in carbohydrate, fat, and protein to ATP.

- **Thiamin** is a coenzyme that is particularly important for glucose metabolism and **neurotransmitter** synthesis, as shown; a deficiency of thiamin, called **beriberi**, causes nervous system abnormalities and cardiovascular changes. Thiamin is found in whole and enriched grains.

Figure 7.8a Thiamin functions and deficiency

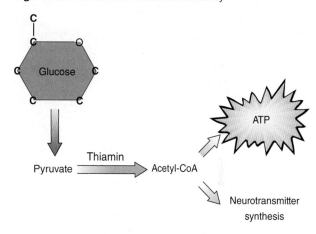

- **Riboflavin** coenzymes are needed for ATP production and for the utilization of several other vitamins. Milk is one of the best food sources of riboflavin.
- **Niacin** coenzymes are needed for the breakdown of carbohydrate, fat, and protein and for the synthesis of fatty acids and cholesterol. A deficiency results in **pellagra**, which is characterized by dermatitis, diarrhea, and dementia. The amino acid tryptophan can be converted into niacin. High doses lower blood cholesterol but can cause toxicity symptoms.
- **Biotin** is needed for the synthesis of glucose and for energy, fatty acid, and amino acid metabolism. Some of our biotin need is met by bacterial synthesis in the gastrointestinal tract.
- **Pantothenic acid** is part of coenzyme A. It is required for the production of ATP from carbohydrate, fat, and protein and for the synthesis of cholesterol and fat. It is widespread in the food supply.

3 Vitamins and Healthy Blood 186

- Blood transports oxygen, nutrients, and hormones to cells throughout our bodies and carries waste products away from cells for elimination. The right mix of nutrients in the diet is needed to keep blood healthy and functioning.
- **Vitamin B_6** is particularly important for amino acid and protein metabolism and is needed for **hemoglobin** synthesis. Deficiency causes **anemia**, numbness and tingling, and it may increase the risk of heart disease because vitamin B_6 is needed to keep levels of homocysteine low. Food sources include meats and whole grains. Large doses of vitamin B_6 can cause nervous system abnormalities.

- **Folate** is necessary for the synthesis of DNA, so it is especially important for rapidly dividing cells. Folate deficiency prevents red blood cell precursor cells from dividing, as shown here, resulting in **macrocytic anemia**. Low levels of folate before and during early pregnancy are associated with an increased incidence of **neural tube defects**. Food sources include liver, legumes, oranges, leafy green vegetables, and fortified grains. A high intake of folate can mask some of the symptoms of vitamin B_{12} deficiency.

Figure 7.16 Folate deficiency and macrocytic anemia

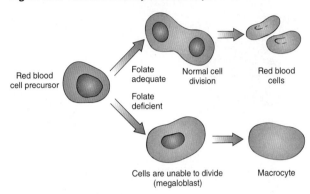

- **Vitamin B_{12}** is needed for the metabolism of folate and fatty acids and to maintain **myelin**. Absorption of vitamin B_{12} requires stomach acid and **intrinsic factor**. If intrinsic factor is not produced, only tiny amounts of vitamin B_{12} are absorbed, and **pernicious anemia** occurs. In addition to causing macrocytic anemia, deficiency results in nerve damage. Vitamin B_{12} is found almost exclusively in animal products. Deficiency is a concern in vegans and in older individuals with **atrophic gastritis**.

- **Vitamin K** is essential for blood clotting and is important for bone health. It is found in leafy greens and is synthesized by bacteria in the gastrointestinal tract. Because vitamin K deficiency is a problem in newborns, they are routinely given vitamin K injections at birth.

4 Antioxidant Vitamins   196

- Some vitamins function as **antioxidants**, substances that protect against **oxidative stress**. Damage caused by oxidative stress has been related to aging and a variety of chronic diseases.

- **Vitamin C** is an antioxidant that acts by donating electrons to neutralize **free radicals**. It is also necessary for the synthesis of **collagen**, hormones, and neurotransmitters. Vitamin C deficiency, called **scurvy**, is characterized by poor wound healing, bleeding, and other symptoms related to the improper formation and maintenance of collagen. The best food sources are citrus fruits.

- **Vitamin E** functions primarily as a fat-soluble antioxidant. It is necessary for reproduction and, as shown here, protects cell membranes from oxidative damage. It is found in nuts, plant oils, green vegetables, and fortified cereals.

Figure 7.28 The antioxidant **role of vitamin E**

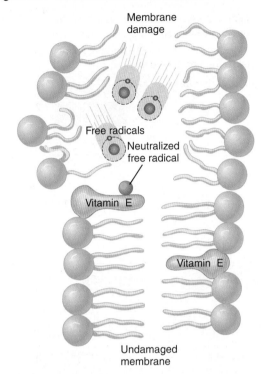

5 Vitamins in Gene Expression   200

- By regulating **gene expression**, some vitamins can increase or decrease the amount of protein coded for by the gene. This in turn affects cell function and body processes.

- **Vitamin A** is a fat-soluble vitamin needed in the visual cycle and for growth and **cell differentiation**. Its role in regulating gene expression makes it essential for maintenance of epithelial tissue, reproduction, and immune function. Vitamin A deficiency causes blindness and death. Preformed vitamin A **retinoids** are found in liver, eggs, fish, and fortified dairy products. High intakes are toxic and have been linked to birth defects and bone loss. Provitamin A **carotenoids**, such as **β-carotene**, are found in yellow–orange fruits and vegetables such as mangoes, carrots, and apricots, as well as leafy greens. Some carotenoids are antioxidants. Carotenoids are not toxic, but a high intake can give the skin a yellow–orange appearance.

- **Vitamin D** can be made in the skin with exposure to sunlight, so dietary needs vary depending on the amount synthesized. Vitamin D is found in fish oils and fortified milk. It helps maintain proper levels of calcium and phosphorus in the body by regulating gene expression. A deficiency in children results in **rickets**, shown here; in adults, vitamin D deficiency causes **osteomalacia**.

Figure 7.36 Rickets

- women of childbearing age, older adults, and other nutritionally vulnerable groups.
- Herbal supplements are currently popular. These products, which are made from plants (such as the St. John's wort shown here), may have beneficial physiological actions, but their dosage is not regulated, and they can be toxic either on their own or in combination with other herbs or medications or in people with certain medical conditions.

Figure 7.38 Popular herbal supplements: St. John's wort

- Manufacturers are responsible for the consistency and safety of supplements before they are marketed. The FDA regulates the labeling of dietary supplements and can monitor their safety once they are being sold. When choosing a dietary supplement, it is important to carefully consider the potential risks and benefits of the product.

6 Meeting Needs with Dietary Supplements 211

- Over 60% of the adult population in the United States takes some type of dietary supplement. Vitamin and mineral supplements are recommended for dieters, vegetarians, pregnant women and

Key Terms

- alpha-carotene (α-carotene) 201
- alpha-tocopherol (α-tocopherol) 198
- anemia 187
- antioxidant 196
- ascorbic acid 197
- atrophic gastritis 192
- beriberi 179
- beta-carotene (β-carotene) 201
- beta-cryptoxanthin (β-cryptoxanthin) 201
- bioavailability 174
- biotin 184
- carotenoids 201
- cell differentiation 203
- choline 185

- cobalamin 191
- coenzyme 179
- collagen 197
- dietary folate equivalent (DFE) 190
- fat-soluble vitamin 171
- fibrin 194
- folate 188
- folic acid 190
- free radical 196
- gene expression 200
- hemoglobin 185
- hemolytic anemia 198
- hypercarotenemia 206
- intrinsic factor 191

- macrocytic anemia 188
- myelin 187
- neural tube defect 188
- neurotransmitter 179
- niacin 182
- niacin equivalent (NE) 183
- night blindness 203
- osteomalacia 207
- osteoporosis 208
- oxidative stress 196
- pantothenic acid 185
- parathyroid hormone (PTH) 206
- pellagra 182
- pernicious anemia 190

218 CHAPTER 7 Vitamins

- prothrombin   194
- provitamin or vitamin precursor   174
- pyridoxal phosphate   186
- retinoids   201
- retinol activity equivalent (RAE)   202
- retinol-binding protein   203
- rhodopsin   203
- riboflavin   180
- rickets   207
- scurvy   197
- spina bifida   189
- thiamin   179
- tocopherol   198
- vitamin   171
- vitamin A   201
- vitamin B_6 186
- vitamin B_{12} 191
- vitamin C   197
- vitamin D   206
- vitamin E   198
- vitamin K   194
- water-soluble vitamin   171
- Wernicke-Korsakoff syndrome   180
- xerophthalmia   204

What is happening in this picture?

These children, who live in Russia, are being exposed to UV radiation to prevent vitamin D deficiency.

Russia Dean Conger / National Geographic Creative

Think Critically

1. Why does this treatment help them meet their need for vitamin D?
2. Why are children in Russia at risk for vitamin D deficiency?
3. What else could be done to ensure that they get adequate amounts of vitamin D?

CHAPTER 8

Water and Minerals

It is in the water of Earth's first seas that scientists propose life itself originally appeared. These primitive seas were not plain water but rather a complex mixture of minerals, organic compounds, and water. Over time, as simple organisms developed greater complexity, many left this mineral-rich external marine environment behind. To survive, they needed to bring with them a similar internal environment. Within our bodies today, minerals and water make up an "internal sea" that allows the chemistry of life to function.

Getting the right amounts of minerals and water remains essential to survival. Inadequate and excessive intakes of certain minerals are world health problems contributing to such conditions as high blood pressure, bone fractures, anemia, and increased risk of infection. Getting enough clean, fresh water may prove to be an even greater challenge to human survival than meeting mineral needs. Pollution and population growth are making fresh water an increasingly rare commodity: Approximately 1.8 billion people around the world lack access to a safe water supply.[1] In the United States clean water is accessible to most, but the huge aquifers beneath the surface are diminishing. The water crisis may make shrinking oil supplies seem like a minor inconvenience for the simple reason that without water, we die.

CHAPTER OUTLINE

Water 220
- Water in the Body
- Water Balance
- The Functions of Water
- Water in Health and Disease
- Meeting Water Needs

Debate: Is Bottled Water Better?

An Overview of Minerals 226
- Minerals in Our Food
- Mineral Bioavailability
- Mineral Functions
- Understanding Mineral Needs

Electrolytes: Sodium, Potassium, and Chloride 232
- Electrolytes in the Body
- Electrolyte Deficiency and Toxicity
- Hypertension
- Meeting Electrolyte Needs

What Should I Eat? Water and Electrolytes
What a Scientist Sees: Hidden Sodium

Minerals and Bone Health 240
- Osteoporosis
- Calcium
- Phosphorus
- Magnesium
- Fluoride

What Should I Eat? Calcium, Phosphorus, Magnesium, and Fluoride

Minerals and Healthy Blood 249
- Iron

Thinking It Through: A Case Study on Iron Deficiency
- Copper

Antioxidant Minerals 253
- Selenium
- Zinc
- Manganese
- Sulfur
- Molybdenum

Minerals and Energy Metabolism 257
- Iodine
- Chromium

What Should I Eat? Trace Minerals

8.1 Water

LEARNING OBJECTIVES

1. **Describe** factors that affect water distribution in the body.
2. **Explain** how water is regulated in the body.
3. **Describe** the functions of water in the body.
4. **Discuss** factors that increase water needs.

Water is an overlooked but essential nutrient. Lack of sufficient water in the body causes deficiency symptoms more rapidly than does a deficiency of any other nutrient. These symptoms can be alleviated almost as rapidly as they appeared by drinking enough water to restore the body's water balance.

Water in the Body

In adults, water accounts for about 60% of body weight. Water is found in different proportions in different tissues. For example, about 75% of muscle weight is water, whereas only about 25% of the weight of bone is due to water. Water is found both inside cells (intracellular) and outside cells (extracellular) in the blood, the lymph, and the spaces between cells (**Figure 8.1**). The cell membranes that separate the intracellular and extracellular spaces are not watertight; water can pass right through them.

The distribution of water between various intra- and extracellular spaces is affected by the concentrations of dissolved substances, or **solutes**, such as proteins, sodium, potassium, and other small molecules. The concentration differences of these substances drive *osmosis*, the diffusion of water in a direction that equalizes the concentration of dissolved substances on either side of a membrane (see Chapter 3). Water distribution is also affected by **blood pressure**, which forces water from the capillary blood vessels into the spaces between the cells of the surrounding tissues (see Figure 8.1). The body regulates the amount of water in cells and in different extracellular spaces by adjusting the concentration of dissolved particles and relying on osmosis to move the water.

FIGURE 8.1 Water distribution About two-thirds of the water in the body is intracellular (located inside the cells). The other one-third is extracellular (located outside the cells). The extracellular portion includes the water in the blood and lymph, the water between cells, and the water in the digestive tract, eyes, joints, and spinal cord. The distribution of water between intracellular and extracellular spaces is affected by blood pressure and the force generated by osmosis.

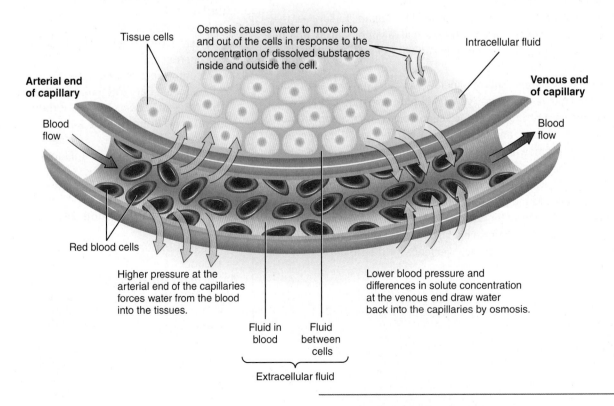

blood pressure The amount of force exerted by the blood against the walls of arteries.

Water Balance

The amount of water in the body remains relatively constant over time. Because water cannot be stored in the body, water intake and output must be balanced to maintain the right amount. Most of our water intake comes from water itself and other beverages, such as coffee, tea, milk, juice, and soft drinks. Solid foods also provide water; most fruits and vegetables are over 80% water, and even roast beef is about 50% water. A small amount of water is also produced in the body as a by-product of metabolic reactions. We lose water from our bodies in urine and feces, through evaporation from the lungs and skin, and in sweat (**Figure 8.2**).

The majority of water loss is usually through urinary excretion; the amount varies with water intake and the amount of waste that needs to be excreted in the urine. The amount of water lost in the feces of a healthy person is small—usually less than a cup per day. This is remarkable because every day about 9 L (38 cups) of fluid enters the gastrointestinal tract from liquids we consume and from secretions from the cells and organs of the gastrointestinal tract, but more than 95% of this is absorbed before the feces are eliminated. However, in cases of severe diarrhea, large amounts of water can be lost via this route, compromising health.

We are continuously losing water from our skin and respiratory tract due to evaporation. The amount of water lost through evaporation varies greatly, depending on body size, activity level, and environmental temperature and humidity. In a temperate climate, an inactive person loses about 1 L (4 cups) per day; the amount increases with increases in activity, environmental temperature, and body size, as well as when humidity is low.

In addition to being lost through evaporation, water is lost through sweat when you exercise or when the environment is hot. More sweat is produced as exercise intensity increases and as the environment becomes hotter and more humid. An individual

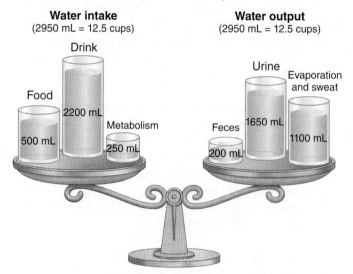

FIGURE 8.2 Water balance To maintain water balance, intake from food and drink and water produced by metabolism must equal water output from urine, feces, evaporation, and sweat. This figure illustrates approximate amounts of water that enter and leave the body daily in a typical woman who is in water balance. Increases in temperature or activity increase evaporative and sweat losses. If losses from evaporation and sweat remain constant, increasing water consumption proportionately increases urinary excretion.

doing light work at a temperature of 84°F (29°C) will lose about 2 to 3 L (8 to 12 cups) of sweat per day. Strenuous exercise in a hot environment can increase this to 2 to 4 L (8 to 16 cups) in an hour.[2] Clothing that permits the evaporation of sweat helps keep the body cool and therefore decreases the amount of sweat produced.

Regulating water intake When water losses increase, intake must increase to keep body water at a healthy level (**Figure 8.3**). The need to consume water is signaled by the

PROCESS DIAGRAM

FIGURE 8.3 Stimulating water intake The sensation of thirst motivates fluid intake in order to restore water balance.

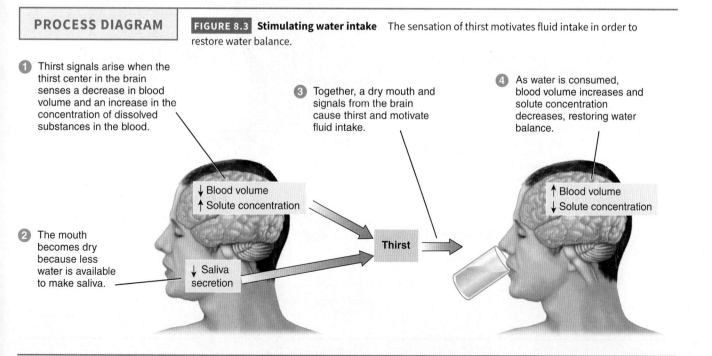

1. Thirst signals arise when the thirst center in the brain senses a decrease in blood volume and an increase in the concentration of dissolved substances in the blood.

2. The mouth becomes dry because less water is available to make saliva.

3. Together, a dry mouth and signals from the brain cause thirst and motivate fluid intake.

4. As water is consumed, blood volume increases and solute concentration decreases, restoring water balance.

sensation of **thirst**. Thirst is caused by dryness in the mouth as well as by signals from the brain. It is a powerful urge but often lags behind the need for water, and we don't or can't always drink when we are thirsty. Therefore, thirst alone cannot be relied on to maintain water balance.

Regulating water loss: The kidneys

To maintain water balance, the kidneys regulate water loss in urine. The kidneys typically produce about 1 to 2 L (4 to 8 cups) of urine per day, but urine production varies, depending on the amount of water consumed and the amount of waste that needs to be excreted. Wastes that must be eliminated in the urine include urea and other nitrogen-containing molecules (produced by protein breakdown), ketones (from incomplete fat breakdown), sodium, and other minerals.

The kidneys function like a filter. As blood flows through them, water molecules and other small molecules move through the filter and out of the blood, while blood cells and large molecules are retained in the blood. Some of the water and other molecules that are filtered out are reabsorbed into the blood, and the rest are excreted in the urine.

The amount of water that is reabsorbed into the blood rather than excreted in the urine depends on conditions in the body. When the solute concentration in the blood is high, as it would be in someone who has exercised strenuously and not consumed enough water, a hormone called **antidiuretic hormone (ADH)** signals the kidneys to reabsorb water, reducing the amount lost in the urine (**Figure 8.4**). This reabsorbed water is returned to the blood, maintaining body water and preventing the concentration of dissolved particles from increasing further. When the solute concentration in the blood is low, as it might be after someone has guzzled several glasses of water, ADH levels decrease. With less ADH, the kidneys reabsorb less water, and more water is excreted in the urine, allowing the solute concentration in the blood to increase to its normal level. The amount of sodium in the blood, blood volume, and blood pressure also play a role in regulating body water (as discussed later in the chapter).

Even though the kidneys work to control how much water is lost, their ability to concentrate urine is limited, so there is a minimum amount of water that must be lost to excrete dissolved wastes. If there are a lot of wastes to be excreted, more water must be lost.

The Functions of Water

Water doesn't provide energy, but it is essential to life. Water in the body serves as a medium for and participant in metabolic

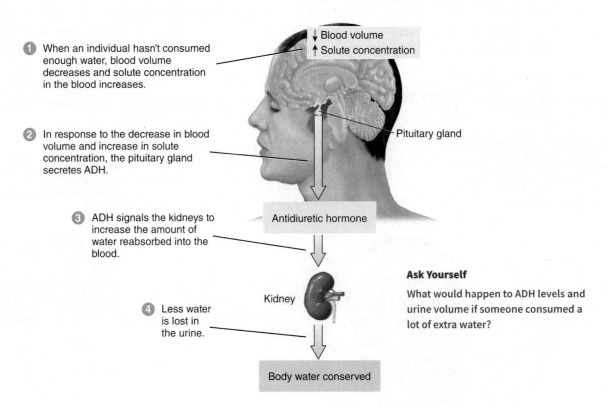

PROCESS DIAGRAM **FIGURE 8.4** **Regulating urinary water losses** The kidneys help regulate water balance by adjusting the amount of water lost in the urine in response to the release of antidiuretic hormone (ADH).

1. When an individual hasn't consumed enough water, blood volume decreases and solute concentration in the blood increases.
2. In response to the decrease in blood volume and increase in solute concentration, the pituitary gland secretes ADH.
3. ADH signals the kidneys to increase the amount of water reabsorbed into the blood.
4. Less water is lost in the urine.

Ask Yourself

What would happen to ADH levels and urine volume if someone consumed a lot of extra water?

reactions, helps regulate acid–base balance, transports nutrients and wastes, provides protection, and helps regulate body temperature.

Water in metabolism and transport

Water is an excellent **solvent**; glucose, amino acids, minerals, proteins, and many other molecules dissolve in water. The chemical reactions of metabolism that support life take place in water. Water also participates in a number of reactions that join small molecules together or break apart large ones. Some of the reactions in which water participates help maintain the proper level of acidity in the body. Water is the primary constituent of blood, which flows through our bodies, delivering oxygen and nutrients to cells and transporting waste products to the lungs and kidneys for excretion.

Water as protection

Water bathes the cells of the body and lubricates and cleanses internal and external body surfaces. Water in tears lubricates the eyes and washes away dirt; water in synovial fluid lubricates the joints; and water in saliva lubricates the mouth, helping us chew and swallow food. Because water resists compression, it cushions the joints and other parts of the body against shock. The cushioning effect of water in the amniotic sac protects a fetus as it grows inside the uterus.

Water and body temperature

The fact that water holds heat and changes temperature slowly helps keep body temperature constant, but water is also actively involved in temperature regulation (**Figure 8.5**). The water in blood helps regulate body temperature by increasing or decreasing the amount of heat lost at the surface of the body. When body temperature starts to rise, the blood vessels in the skin dilate, causing more blood to flow close to the surface, where it can release some of the heat to the surrounding air. Cooling is aided by the production of sweat. When body temperature increases, the brain triggers the sweat glands in the skin to produce sweat, which is mostly water. As the water in sweat evaporates from the skin, additional heat is lost, cooling the body.

Water in Health and Disease

Without food, you could probably survive for about eight weeks, but without water, you would last only a few days. Too

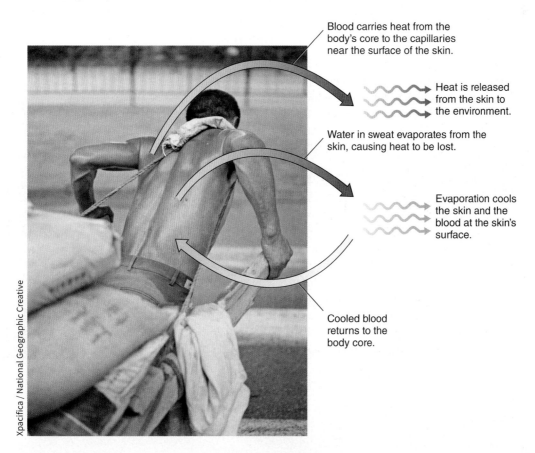

FIGURE 8.5 **Water helps cool the body** In hot weather and during strenuous activity blood flow increases at the surface of the body. Shunting blood to the surface warms the skin and allows heat to be transferred to the surroundings. Evaporation of sweat cools the skin and the blood near the surface of the skin.

FIGURE 8.6 **Are you at risk for dehydration?** Urine color is an indication of whether you are drinking enough. Pale yellow urine indicates that you are well hydrated. The darker the urine, the greater the level of dehydration.

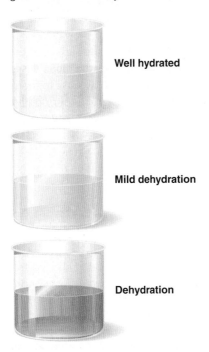

much water can also be a problem if it changes the osmotic balance, disrupting the distribution of water within the body.

Dehydration Dehydration occurs when water loss exceeds water intake. It causes a reduction in blood volume, which impairs the ability to deliver oxygen and nutrients to cells and remove waste products. Early symptoms of dehydration include thirst, headache, fatigue, loss of appetite, dry eyes and mouth, and dark-colored urine (**Figure 8.6**). Dehydration affects physical and cognitive performance. As dehydration worsens, it causes nausea, difficulty concentrating, confusion, and disorientation. The milder symptoms of dehydration disappear quickly after water or some other beverage is consumed, but if left untreated, dehydration can become severe enough to require medical attention (**Figure 8.7**). A water loss amounting to about 10 to 20% of body weight can be fatal. Athletes are at risk for dehydration because they may lose large amounts of water in sweat. Older adults are at risk because the thirst mechanism becomes less sensitive with age. Infants are at risk because their body surface area relative to their weight is much greater than that of adults, so they lose proportionately more water through evaporation and because their kidneys cannot concentrate urine efficiently, so they lose proportionately more water in urine. In addition, they cannot tell us they are thirsty.

Water intoxication It is difficult to consume too much water under normal circumstances. However, overhydration, or **water intoxication**, can occur under some conditions (see Chapter 10). When there is too much water relative to the amount of sodium in the body, the concentration of sodium in the blood is too low—a condition called **hyponatremia**. When this occurs, water moves out of the blood vessels and into the tissues by osmosis, causing them to swell. Swelling in the brain can cause disorientation, convulsions, coma, and death. The early symptoms of water intoxication may be similar to those of dehydration: nausea, muscle cramps, disorientation, slurred speech, and confusion. It is important to determine whether the symptoms are due to dehydration or water intoxication because while drinking water will alleviate dehydration, it will worsen the symptoms of water intoxication.

Meeting Water Needs

The AI for water is 3.7 L/day for men and 2.7 L/day for women. As discussed above, however, the amount of water you need each day varies, depending on your activity level,

FIGURE 8.7 **Rehydration saves lives** Dehydration due to diarrhea is a major cause of child death in the developing world. Replacing fluids and electrolytes in the right combinations can save lives. In some cases, oral rehydration therapy is sufficient. Drinking mixtures made by simply dissolving a large pinch of salt (½ tsp) and a fistful of sugar (2 Tbsp) in 1 L of clean water can restore the body's water balance by promoting the absorption of water and sodium. In severe cases of diarrhea, administration of intravenous fluids, as seen in this hospital in Bangladesh, is needed to restore hydration.

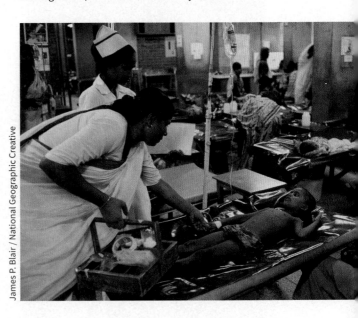

James P. Blair / National Geographic Creative

---

dehydration A condition that results when not enough water is present to meet the body's needs.

water intoxication A condition that occurs when a person drinks enough water to significantly lower the concentration of sodium in the blood.

Debate

Is Bottled Water Better?

The Issue

Americans consume over 36 gallons of bottled water per person per year.[5] We choose it because it is convenient and because we think it tastes better and is safer than tap water. But the cost to our pocketbooks and our environment is high. Should we be drinking from the tap instead?

It is easy to grab a bottle of water, and most bottled water has no chlorine or other unpleasant aftertaste. Words like *pure*, *crisp*, and *fresh tasting* on the label entice consumers to buy bottled water because they think it is better than water that comes from the tap (see graph). But much of the bottled water sold in the United States is tap water or tap water that has been filtered, disinfected, or otherwise treated. By definition, bottled water can be any water, as long as it has no added ingredients (except antimicrobial agents or fluoride). Labels may help you distinguish: Distilled water and purified water are treated tap water; artesian water, spring water, well water, and mineral water come from underground sources.

Is bottled water safer than tap water? Municipal (tap) water is regulated by the Environmental Protection Agency (EPA), and bottled water sold in interstate commerce is regulated by the Food and Drug Administration (FDA). The FDA uses most of the EPA's tap water standards, so it would make sense that tap water and bottled water would be equally safe. However, tap water advocates argue that tap water may actually be safer. A certified outside laboratory tests municipal water supplies every year, while bottled water companies are permitted to do their own tests for purity.[6] Tap water must also be filtered and disinfected; there are no federal filtration or disinfection requirements for bottled water.

Contamination is a safety concern for both bottled water and tap water. A study of contaminants in bottled water found that 10 popular brands contained a total of 38 chemical pollutants—everything from caffeine and Tylenol to bacteria, radioactive isotopes, and fertilizer residue.[7] Sounds scary, but bottled water advocates argue that public drinking water may also fall short of pollutant standards. Since 2004, testing by water utilities has found more than 300 pollutants in tap water. Some of these are among the 90 contaminants for which the EPA sets limits, but more than half of the chemicals detected are not subject to health or safety regulations and can legally be present in any amount.[8] The EPA requires that public water systems notify consumers in a timely manner when monitoring reveals a violation that may affect public health.[9] However, high lead levels in the water supply in Flint, Michigan, went unreported for months, and this case has fueled distrust of public water supplies.[10]

One of the strongest arguments against bottled water is the cost, both to the consumer and the environment. Bottled water typically costs about $7.50 per gallon—almost 2000 times the cost of public tap water.[5] Bottled water drinkers may feel that the added cost is worth it because of the advantages in terms of taste and convenience. Opponents argue that even if you can afford it, the planet can't. Americans consume about 50 billion bottles of water per year, but only about 23% of these bottles are recycled. Manufacturing the bottles consumes more than 17 million barrels of oil annually, and this does not include the gasoline and jet fuel used to transport them.[11] Bottled water proponents argue that despite the large amount of plastic waste, discarded water bottles still represent less than 1% of total municipal waste. And even though bottled water production is more energy intensive than the production of tap water, it comprises only a small share of total U.S. energy demand.[6]

So which is better? In the United States, bottled water and tap water are both generally safe. If you recycle your bottle, does it matter which you choose?

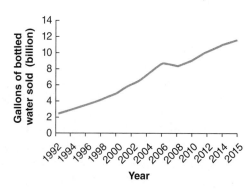

This graph shows bottled water sales in the United States from 1992 to 2015. A slight dip in consumption occurred during the 2008 recession as many Americans switched from bottled to tap water, but sales have soared since then.

Think Critically

Suggest some factors that might be fueling the rise in bottled water consumption since 2010.

the environmental temperature, and the humidity. Diet can also affect water needs because more water is lost when more waste must be excreted in the urine. A high-protein diet increases water needs because the urea produced from protein breakdown is excreted in the urine. A low-calorie diet increases water needs because as body fat and protein are broken down to fuel the body, ketones and urea are produced and must be excreted in the urine. A high-salt diet increases water needs because eliminating the excess sodium in the urine increases water losses. A high-fiber diet increases water needs because more water is held in the intestines and lost in the feces.

In the United States, beverages provide about 80% of the body's requirement for water—about 3 L (13 cups) for men and 2.2 L (9 cups) for women.[3] The rest comes from the water consumed in food. Most beverages, whether water, milk, juice, or soda, help meet the overall need for water (see *Debate: Is Bottled Water Better?*). Beverages containing caffeine, such as coffee, tea, and cola, increase water losses in the short term because caffeine is a **diuretic**. However, the increase in water loss is small, so the net amount of water that caffeinated beverages add to the body is similar to the amount contributed by non-caffeinated beverages. Alcohol is also a diuretic; the overall effect it has on water balance depends on the relative amounts of water and alcohol in the beverages being consumed.[3] Most adults in the United States consume adequate amounts of water in the their diet, from water and other beverages; however, low water intake is a concern for older adults: 83% of men and 95% of women 71 years and older have intakes below the AI.[4]

Concept Check

1. **How** does osmosis affect the distribution of water in the body?
2. **What** is the role of ADH?
3. **How** does water help cool the body?
4. **Why** do water needs increase when you exercise more?

8.2 An Overview of Minerals

LEARNING OBJECTIVES

1. **Define** minerals in terms of nutrition.
2. **Review** factors that affect mineral bioavailability.
3. **Discuss** the functions of minerals in the body.
4. **Describe** why it can be challenging to meet trace mineral needs.

Minerals are found in the ground on which we walk, the jewels we wear on our fingers, and even some of the makeup we wear on our faces. But perhaps the most significant impact of minerals on our lives comes from their importance in our nutritional health. You need to consume more than 20 different **minerals** in your food to stay healthy. Some of these make up a significant portion of your body weight; others are found in minute quantities. If more than 100 milligrams of a mineral is required in the diet each day, an amount equivalent in weight to about two drops of water, the mineral is considered a **major mineral**; these include sodium, potassium, chloride, calcium, phosphorus, magnesium, and sulfur. Minerals that are needed in smaller amounts are referred to as **trace minerals**. The small amounts of some of the trace minerals that are present in the body and in the diet have made it difficult to determine their functions and requirements. The DRI committee has determined that there are sufficient data to establish recommendations for the trace minerals iron, copper, zinc, selenium, iodine, chromium, fluoride, manganese, and molybdenum. There is evidence that arsenic, boron, nickel, silicon, and vanadium play a role in human health, but there have been insufficient data to establish recommended intakes for any of these. Other trace minerals that are believed to play a physiological role in human health include aluminum, bromine, cadmium, germanium, lead, lithium, rubidium, and tin. Their specific functions have not been defined, and they have not been evaluated by the DRI committee.

Just because you need more of the major minerals than of the trace minerals doesn't mean that one group is more important than the other. A deficiency of a trace mineral is just as damaging to your health as a deficiency of a major mineral. A summary of the sources, recommended intakes, functions, and deficiency and toxicity symptoms of the major and trace minerals for which there are DRI recommendations is provided in **Table 8.1**.

---

diuretic   A substance that increases the amount of urine passed from the body.

mineral   In nutrition, an element needed by the body to maintain structure and regulate chemical reactions and body processes.

major mineral   A mineral required in the diet in an amount greater than 100 mg/day or present in the body in an amount greater than 0.01% of body weight.

trace mineral   A mineral required in the diet in an amount of 100 mg or less per day or present in the body in an amount of 0.01% of body weight or less.

TABLE 8.1 A summary of the minerals

Mineral	Sources	Recommended intake for adults	Major functions	Deficiency diseases and symptoms	Groups at risk for deficiency	Toxicity	UL
Major minerals							
Sodium	Table salt, processed foods	< 2300 mg/day; ideally 1500 mg/day	Major positive extracellular ion, nerve transmission, muscle contraction, fluid balance	Muscle cramps, nausea, disorientation, confusion, excess fluid in tissues	People consuming too much water compared to sodium, those with vomiting or diarrhea or who sweat excessively	High blood pressure in sensitive people	2300 mg/day
Potassium	Fresh fruits and vegetables, legumes, whole grains, milk, meat	4700 mg/day or more	Major positive intracellular ion, nerve transmission, muscle contraction, fluid balance	Irregular heartbeat, fatigue, muscle cramps	People with vomiting or diarrhea, those taking thiazide diuretics	Abnormal heartbeat	ND
Chloride	Table salt, processed foods	< 3600 mg/day; ideally 2300 mg/day	Major negative extracellular ion, fluid balance	Unlikely	None	None likely	3600 mg/day
Calcium	Dairy products, fish consumed with bones, leafy green vegetables, fortified foods	1000–1200 mg/day	Bone and tooth structure, nerve transmission, muscle contraction, blood clotting, blood pressure regulation, hormone secretion	Increased risk of osteoporosis	Postmenopausal women; older adults; those who consume a vegan diet, are lactose intolerant, have low vitamin D status, or have kidney disease	Elevated blood calcium, kidney stones, and other problems in susceptible individuals	2000–2500 mg/day from food and supplements
Phosphorus	Meat, dairy, cereals, baked goods	700 mg/day	Structure of bones and teeth, cell membranes, ATP, and DNA; acid–base balance	Bone loss, weakness, lack of appetite	Premature infants, alcoholics, older adults	None likely	4000 mg/day
Magnesium	Greens, whole grains, legumes, nuts, seeds	310–420 mg/day	Bone structure, ATP stabilization, enzyme activity, nerve and muscle function	Nausea, vomiting, weakness, muscle pain, heart changes, associated with increased risk of type 2 diabetes and other chronic diseases	Alcoholics, individuals with kidney and gastrointestinal disease	Nausea, vomiting, low blood pressure	350 mg/day from nonfood sources
Sulfur	Protein foods, preservatives	None specified	Part of some amino acids and vitamins, acid–base balance	None when protein needs are met	None	None likely	ND
Trace minerals							
Iron	Red meats, leafy greens, dried fruit, legumes, whole and enriched grains	8–18 mg/day	Part of hemoglobin (which delivers oxygen to cells), myoglobin (which holds oxygen in muscle), and proteins needed for ATP production; needed for immune function	Iron deficiency anemia: fatigue; weakness; small, pale red blood cells; low hemoglobin levels; inability to maintain normal body temperature	Infants and preschool children, adolescents, women of childbearing age, pregnant women, athletes, vegetarians	Acute: Gastrointestinal upset, liver damage Chronic: fatigue, heart and liver damage, increased risk of diabetes and cancer	45 mg/day

(continues)

TABLE 8.1   A summary of the minerals *(continued)*

Mineral	Sources	Recommended intake for adults	Major functions	Deficiency diseases and symptoms	Groups at risk for deficiency	Toxicity	UL
Copper	Organ meats, nuts, seeds, whole grains, seafood, cocoa	900 µg/day	A component of proteins needed for iron transport, lipid metabolism, collagen synthesis, nerve and immune function, protection against oxidative damage	Anemia, poor growth, skeletal abnormalities	People who consume excessive amounts of zinc in supplements	Vomiting, abdominal pain, diarrhea, liver damage	10 mg/day
Zinc	Meat, seafood, whole grains, dairy products, legumes, nuts	8–11 mg/day	Regulates protein synthesis; functions in growth, development, wound healing, immunity, and antioxidant enzymes	Poor growth and development, skin rashes, decreased immune function	Vegetarians, low-income children, older adults	Decreased copper absorption, depressed immune function	40 mg/day
Selenium	Meats, seafood, eggs, whole grains, nuts, seeds	55 µg/day	Antioxidant as part of glutathione peroxidase, synthesis of thyroid hormones, spares vitamin E	Muscle pain, weakness, Keshan disease	Populations in areas where the soil is low in selenium	Nausea, diarrhea, vomiting, fatigue, changes in hair and nails	400 µg/day
Iodine	Iodized salt, seafood, seaweed, dairy products	150 µg/day	Needed for synthesis of thyroid hormones	Goiter, cretinism, impaired brain function, growth and developmental abnormalities	Populations in areas where the soil is low in iodine and iodized salt is not used	Enlarged thyroid	1110 µg/day
Chromium	Brewer's yeast, nuts, whole grains, meat, mushrooms	25–35 µg/day	Enhances insulin action	High blood glucose	Malnourished children	None reported	ND
Fluoride	Fluoridated water, tea, fish, toothpaste	3–4 mg/day	Strengthens tooth enamel, enhances remineralization of tooth enamel, reduces acid production by bacteria in the mouth	Increased risk of dental caries	Populations in areas with unfluoridated water, those who drink mostly bottled water	Fluorosis: mottled teeth, kidney damage, bone abnormalities	10 mg/day
Manganese	Nuts, legumes, whole grains, tea, leafy vegetables	1.8–2.3 mg/day	Functions in carbohydrate and cholesterol metabolism and antioxidant enzymes	Growth retardation	None	Nerve damage	11 mg/day
Molybdenum	Milk, organ meats, grains, legumes	45 µg/day	Cofactor for a number of enzymes	Unknown in humans	None	Arthritis, joint inflammation	2 mg/day

Note: UL, Tolerable Upper Intake Level; ND, not determined.

Minerals in Our Food

Minerals in our diet come from both plant and animal sources (**Figure 8.8**). Some minerals are present as functioning components of the plant or animal and are therefore present in consistent amounts. For instance, the iron content of beef is predictable because iron is part of the muscle protein that gives beef its red color. The mineral content of other foods depends on where the food is grown and how it is handled and processed. For example, plants grown in an area where the soil is high in selenium are higher in selenium than plants grown in other areas. In countries where the diet consists

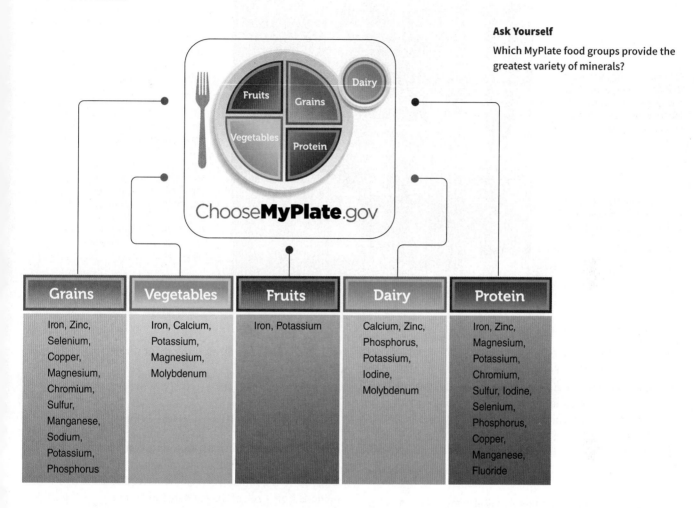

FIGURE 8.8 Minerals on MyPlate Minerals are found in all the MyPlate food groups, but some groups are particularly good sources of specific minerals. Eating a variety of foods, including fresh fruits, vegetables, nuts, legumes, whole grains and cereals, milk, seafood, and lean meats, can maximize your diet's mineral content.

Ask Yourself
Which MyPlate food groups provide the greatest variety of minerals?

predominantly of locally grown foods, trace mineral deficiencies and excesses are more likely to occur because all of the food supply is from the same geographic area. In areas where modern transportation systems provide access to food from many different locations, trace mineral deficiencies and toxicities are less likely.

Processing affects the mineral content of foods for a variety of reasons. Some minerals enter food as contaminants. For example, milk from dairies that use sterilizing solutions that contain iodine is higher in iodine. Some minerals are intentionally added to food during processing. Sodium is added to soups and crackers as a flavor enhancer; iron is added to refined grain products as part of the enrichment process; and calcium, iron, and other minerals are typically added to fortified breakfast cereals. Processing can also remove minerals from foods. For example, when vegetables are cooked, the cells are broken down, and potassium is lost in the cooking water. When the skins of fruits and vegetables or the bran and germ of grains are detached, magnesium, iron, selenium, zinc, and copper are lost.

Mineral Bioavailability

The bioavailability of the minerals that we consume in foods varies. For some minerals, such as sodium, we absorb almost all that is present in our food, but for others, we absorb only a small percentage. For instance, we typically absorb only about 25% of the calcium in our diet, and iron absorption may be as low as 5%. How much of a particular mineral is absorbed may vary from food to food, meal to meal, and person to person.

In general, the minerals in animal products are better absorbed than those in plant foods. The difference in absorption is due in part to the fact that plants contain substances such as phytates (also called phytic acid), tannins, oxalates, and fiber that bind minerals in the gastrointestinal tract and can reduce absorption (**Figure 8.9**). The North American diet generally does not contain enough of any of these components to cause a mineral deficiency, but diets in developing countries may. For example, in some populations, the phytate content of the diet is high enough to cause a zinc deficiency.

FIGURE 8.9 **Compounds that interfere with mineral absorption** Plant foods such as these contain substances that can reduce mineral absorption when consumed in large amounts.

Oxalates, found in spinach, rhubarb, beet greens, and chocolate, have been found to interfere with the absorption of calcium and iron.

Tannins, found in tea and some grains, can interfere with the absorption of iron.

Phytates, found in whole grains, bran, and soy products, bind calcium, zinc, iron, and magnesium, limiting the absorption of these minerals. Phytates can be broken down by yeast, so the bioavailability of minerals is higher in yeast-leavened foods such as breads.

The presence of one mineral can also interfere with the absorption of another. For example, mineral **ions** that carry the same charge compete for absorption in the gastrointestinal tract. Calcium, magnesium, zinc, copper, and iron all carry a 2+ charge, so a high intake of one may reduce the absorption of another. Although this is generally not a problem when whole foods are consumed, a large dose of one mineral from a dietary supplement may interfere with the absorption of other minerals.

The body's need for a mineral may also affect how much of that mineral is absorbed. For instance, if plenty of iron is stored in your body, you will absorb less of the iron you consume. Life stage can also affect absorption; for example, calcium absorption doubles during pregnancy, when the body's needs are high.

Mineral Functions

Minerals contribute to the body's structure and help regulate body processes. Many serve more than one function. For example, we need calcium to keep our bones strong as well as to maintain normal blood pressure, allow muscles to contract, and transmit nerve signals from cell to cell. Some minerals help regulate water balance, others help regulate energy metabolism, and some affect growth and development through their role in the expression of certain genes. Many minerals act as **cofactors** needed for enzyme activity (**Figure 8.10**). None of the minerals we require act in isolation. Instead, they interact with each other as well as with other nutrients and other substances in the body.

Understanding Mineral Needs

Minerals are needed in small amounts compared with the macronutrients. Nonetheless, meeting mineral needs can be challenging. To meet needs, consumers need to choose a variety of foods from each of the MyPlate food groups (see Figure 8.8). Some minerals are found in large amounts in a limited number of foods. For example, the best sources of calcium are dairy products and leafy greens. A diet that doesn't include these foods will likely be deficient in calcium unless fortified foods or supplements are consumed. Some minerals, such as selenium, are found in small amounts in many foods, depending on where they are grown or processed, so a diet that includes a variety of foods from different locations is needed to meet needs. Some minerals are found in such minute quantities that they may be present due to environmental exposure rather than physiological need.

Deficient intakes of minerals can have major health consequences. Insufficient intake of iron threatens the health of children and pregnant women throughout the world. Insufficient iodine causes brain damage in the developing world, and too little calcium increases the risk of bone fractures in both developed and developing countries. Marginal deficiencies are also a concern: Too little zinc can suppress immune function, and too little selenium can increase cancer risk.

---

ion An atom or a group of atoms that carries an electrical charge.

cofactor An inorganic ion or coenzyme that is required for enzyme activity.

An Overview of Minerals 231

NUTRITION INSIGHT

FIGURE 8.10 Mineral functions While each mineral has one or more unique functions, many also have complimentary roles in supporting health.

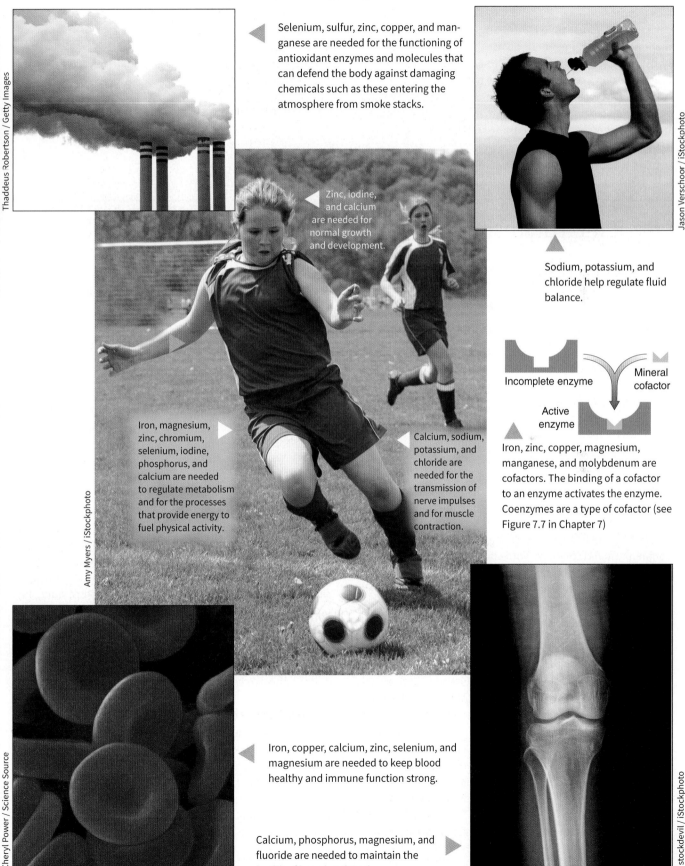

Selenium, sulfur, zinc, copper, and manganese are needed for the functioning of antioxidant enzymes and molecules that can defend the body against damaging chemicals such as these entering the atmosphere from smoke stacks.

Zinc, iodine, and calcium are needed for normal growth and development.

Sodium, potassium, and chloride help regulate fluid balance.

Iron, magnesium, zinc, chromium, selenium, iodine, phosphorus, and calcium are needed to regulate metabolism and for the processes that provide energy to fuel physical activity.

Calcium, sodium, potassium, and chloride are needed for the transmission of nerve impulses and for muscle contraction.

Iron, zinc, copper, magnesium, manganese, and molybdenum are cofactors. The binding of a cofactor to an enzyme activates the enzyme. Coenzymes are a type of cofactor (see Figure 7.7 in Chapter 7)

Iron, copper, calcium, zinc, selenium, and magnesium are needed to keep blood healthy and immune function strong.

Calcium, phosphorus, magnesium, and fluoride are needed to maintain the health of bones and teeth.

DRI recommendations have been established for the seven major minerals and for nine of the trace minerals. The current Nutrition Facts panel on food labels is required to provide information about the amounts of three minerals: sodium, calcium, and iron. Planned Nutrition Facts labels require information about the amount of potassium as well. The amounts of calcium, iron, and potassium are important because many Americans don't get enough of these minerals. Sodium is listed because we typically get too much. Having this information available on labels can help consumers choose foods that will meet but not exceed their needs.

> **Concept Check**
>
> 1. **How** do minerals differ from vitamins?
> 2. **How** do phytates, oxalates, and tannins decrease mineral bioavailability?
> 3. **What** is the function of a cofactor?
> 4. **Why** is a varied diet important for meeting trace mineral needs?

8.3 Electrolytes: Sodium, Potassium, and Chloride

LEARNING OBJECTIVES

1. **Explain** how electrolytes function in the body.
2. **Define** hypertension and describe its symptoms and consequences.
3. **Discuss** how diet affects blood pressure.
4. **Contrast** the dietary sources of sodium and potassium.

We think of **electrolytes** as the things we get by guzzling a sports drink. But what exactly are they, and why do we need them? In terms of nutrition, *electrolytes* typically refer to the three principal ions in body fluids: sodium, potassium, and chloride. Sodium and potassium carry a positive charge, and chloride carries a negative charge. In the diet, sodium is most commonly found combined with chloride as **sodium chloride**, what we call either "salt" or "table salt." In our bodies, these electrolytes are important in maintaining fluid balance and allowing nerve impulses to travel throughout our bodies, signaling the activities that are essential for life.

Electrolytes in the Body

The concentrations of sodium, potassium, and chloride inside cells differ dramatically from those outside cells. Potassium is the principal positively charged intracellular ion, sodium is the most abundant positively charged extracellular ion, and chloride is the principal negatively charged extracellular ion.

Functions of electrolytes Electrolytes help regulate fluid balance; the distribution of water throughout the body depends on the concentration of electrolytes and other solutes. Water moves by osmosis in response to differences in solute concentration. So, for example, if the concentration of sodium in the blood increases, water will move into the blood from intracellular and other extracellular spaces to equalize the concentration of sodium and other dissolved substances. A high sodium concentration in the blood also stimulates thirst; water intake helps dilute blood sodium (**Figure 8.11a**).

Sodium, potassium, and chloride are also essential for generating and conducting nerve impulses. Nerve impulses are created by the movement of sodium and potassium ions across the nerve cell membrane. When a nerve cell is at rest, potassium is concentrated inside the nerve cell, and sodium stays outside the cell. Sodium and potassium ions cannot pass freely across the cell membrane. But when a nerve is stimulated, the cell membrane becomes more permeable to sodium, allowing sodium ions to rush into the nerve cell, which initiates a nerve impulse (**Figure 8.11b**).

Regulating electrolyte balance Our bodies are efficient at regulating the concentration of electrolytes, even when dietary intake varies dramatically. Sodium and chloride

---

electrolyte A positively or negatively charged ion that conducts an electrical current in solution. Commonly refers to sodium, potassium, and chloride.

FIGURE 8.11 Electrolyte functions
Electrolytes have important roles in fluid balance and nerve conductivity.

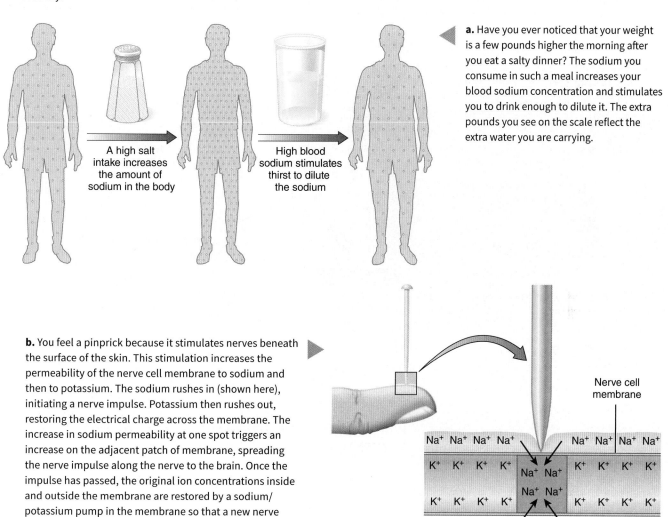

a. Have you ever noticed that your weight is a few pounds higher the morning after you eat a salty dinner? The sodium you consume in such a meal increases your blood sodium concentration and stimulates you to drink enough to dilute it. The extra pounds you see on the scale reflect the extra water you are carrying.

b. You feel a pinprick because it stimulates nerves beneath the surface of the skin. This stimulation increases the permeability of the nerve cell membrane to sodium and then to potassium. The sodium rushes in (shown here), initiating a nerve impulse. Potassium then rushes out, restoring the electrical charge across the membrane. The increase in sodium permeability at one spot triggers an increase on the adjacent patch of membrane, spreading the nerve impulse along the nerve to the brain. Once the impulse has passed, the original ion concentrations inside and outside the membrane are restored by a sodium/potassium pump in the membrane so that a new nerve signal can be triggered.

balance is regulated to some extent by the intake of both salt and water. When sodium chloride intake is high, thirst is stimulated in order to increase water intake. Very low salt intake stimulates a "salt appetite" that causes you to crave salt. The craving that triggers your desire to plunge into a bag of salty chips, however, is not due to this salt appetite. It is a learned preference, not a physiological drive. If you cut back on your salt intake, you will find that your taste buds become more sensitive to the presence of salt, and foods taste saltier.

Thirst and salt appetite help ensure that appropriate proportions of sodium chloride and water are taken in, but the kidneys are the primary regulator of sodium, potassium, and chloride concentrations in the body. Excretion of these electrolytes in the urine is decreased when intake is low and increased when intake is high.

Because water follows sodium by osmosis, the ability of the kidneys to conserve sodium provides a mechanism for conserving water in the body. This mechanism also helps regulate blood pressure. When the concentration of sodium in the blood increases, water follows, causing an increase in blood volume. An increase in blood volume can increase blood pressure. When blood pressure decreases, it triggers the production and release of proteins and hormones that stimulate thirst and cause the blood vessels to constrict and the kidneys to retain sodium and hence water (**Figure 8.12**).

Regulation of blood potassium levels is also important. Even a small increase can be dangerous. If blood potassium levels begin to rise, body cells are stimulated to take up potassium. This short-term regulation prevents the amount of potassium in the extracellular fluid from getting lethally high. Long-term regulation of potassium balance depends on the release of proteins that cause the kidney to excrete potassium and retain sodium.

Electrolyte Deficiency and Toxicity

Deficiencies and toxicities of electrolytes are uncommon in healthy people. Sodium, potassium, and chloride are plentiful

PROCESS DIAGRAM **FIGURE 8.12 Regulation of blood pressure** When blood pressure decreases, the kidneys help to raise it. As blood pressure increases, these events are inhibited so that blood pressure does not continue to rise.

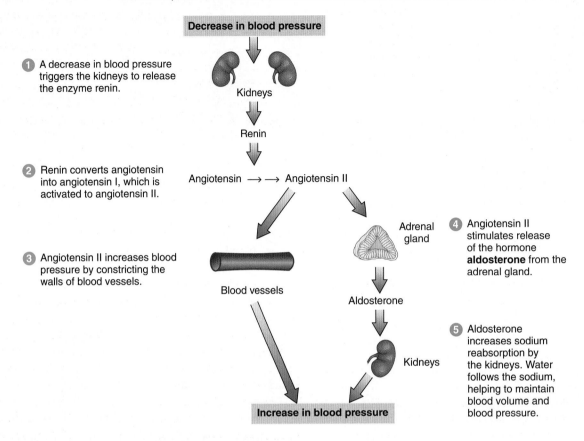

1. A decrease in blood pressure triggers the kidneys to release the enzyme renin.
2. Renin converts angiotensin into angiotensin I, which is activated to angiotensin II.
3. Angiotensin II increases blood pressure by constricting the walls of blood vessels.
4. Angiotensin II stimulates release of the hormone **aldosterone** from the adrenal gland.
5. Aldosterone increases sodium reabsorption by the kidneys. Water follows the sodium, helping to maintain blood volume and blood pressure.

in most diets, and the kidneys of healthy individuals are efficient at regulating the amounts of these electrolytes in the body. Acute deficiencies and excesses can occur due to illness or extreme conditions.

Low levels of sodium, chloride, and potassium can occur when losses of these electrolytes are increased by diarrhea, vomiting, heavy and persistent sweating, or kidney disorders that lead to excessive excretion. Medications can also cause electrolyte depletion. For example, certain diuretic medications used to treat high blood pressure cause potassium loss. Deficiencies of any of the electrolytes can cause poor appetite, muscle cramps, confusion, apathy, constipation, and disturbances in acid–base and fluid balance. An electrolyte imbalance due to excess losses or to reduced intake as a result of fasting, anorexia nervosa, or starvation can eventually cause an irregular heartbeat and lead to sudden death.

Excess potassium levels in the blood can potentially cause death due to an irregular heartbeat. It is not possible for healthy people to consume too much potassium from foods. If, however, supplements are consumed in excess or kidney function is compromised, blood levels of potassium can increase to dangerous levels.

High blood sodium is uncommon because a high sodium intake increases thirst, so we usually drink more water when we consume more sodium. Elevation of blood sodium can result from massive ingestion of salt, as may occur from drinking seawater or consuming salt tablets. The most common cause of high blood sodium is dehydration, and the symptoms of high blood sodium are similar to those of dehydration.

Hypertension

The health problem most commonly associated with electrolyte imbalance is **hypertension**, or high blood pressure (**Figure 8.13**). Hypertension is a complex disorder caused by a narrowing of the blood vessels and/or an increase in blood volume resulting from disturbances in one or more of the mechanisms that control body fluid and electrolyte balance. It has been called "the silent killer" because it has no outward symptoms but can lead to atherosclerosis, heart attack, stroke, kidney disease, and early death. Hypertension is a serious public health concern in the United States; about one-third of adult

---

hypertension Blood pressure that is consistently elevated to 140/90 millimeters (mm) mercury or greater.

Electrolytes: Sodium, Potassium, and Chloride   **235**

FIGURE 8.13 **Assessing blood pressure** Blood pressure is measured in millimeters of mercury (mm Hg) and consists of two readings: systolic over diastolic. Systolic pressure is the maximum pressure in the arteries during a heartbeat and diastolic pressure is the minimum, between heartbeats. A blood pressure assessment determines whether you have a healthy blood pressure, **prehypertension**, or hypertension (see Appendix F).[12]

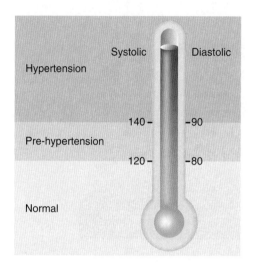

Americans ages 20 and older have hypertension. Only about 53% of those with diagnosed hypertension have their blood pressure under control.[12]

Risk factors for hypertension Some of the risk of developing hypertension is genetic; your risk is increased if you have a family history of the disease. It is more common in African Americans than in Mexican Americans or non-Hispanic whites.[12] Whether you are genetically predisposed to hypertension or not, your risk of developing high blood pressure increases as you grow older and is higher if you are overweight, particularly if the excess fat is in your abdominal region. Lack of physical activity, heavy alcohol consumption, and stress can also increase blood pressure.[13] Regular exercise can prevent or delay the onset of hypertension, and weight loss can help reduce blood pressure in obese individuals. Your risk of developing high blood pressure can also be increased or decreased by your dietary choices (**Figure 8.14**).

Preventing and treating hypertension Maintaining a healthy diet and lifestyle can help prevent hypertension. Dietary intake of sodium, chloride, potassium, calcium, and magnesium can affect your blood pressure and your risk of hypertension. In general, as the sodium content of the diet increases, so does blood pressure.[17] In contrast, diets that are high in potassium, calcium, and magnesium are associated with a lower average blood pressure.[18] Other components of the diet, such as the amount of fiber and the type and amount of fat, may also affect your risk of developing hypertension. While changes in each of these individual components have varying effects on blood pressure, a dietary pattern, such as the **DASH (Dietary Approaches to Stop Hypertension) eating plan**, that incorporates all these recommendations has been shown to cause a significant reduction in blood pressure.[16] This pattern, recommended by the Dietary Guidelines, provides plenty of fiber, potassium, magnesium, and calcium; is low in total fat, saturated fat, and cholesterol; and is lower in sodium than the typical American diet (see Appendix B). Consuming a diet that follows the DASH pattern lowers blood pressure in individuals with elevated blood

---

NUTRITION INSIGHT   **FIGURE 8.14** **The effect of diet on blood pressure** Epidemiology and clinical trials have helped identify dietary patterns that are associated with healthy blood pressure.

This Yanomami Indian boy lives in the rain forest of Brazil. The Yanomami diet consists of locally grown crops, nuts, insects, fish, and game. It contains less than 1 gram (g) of sodium chloride a day, the lowest salt intake recorded for any population. The Yanomami have very low average blood pressure and no hypertension. Studying the salt intake and blood pressure of the Yanomami and 51 other populations around the world helped to establish a relationship between high salt intake and hypertension.[14]

A diet that is high in fruits and vegetables, which are good sources of potassium, magnesium, and fiber, reduces blood pressure compared to a similar diet containing fewer fruits and vegetables.[15] The amounts in the measuring cups shown here, about 2 cups of fruit and 2½ cups of vegetables, represent the amount recommended for a 2000-Calorie diet.

(continues)

FIGURE 8.14 (continued)

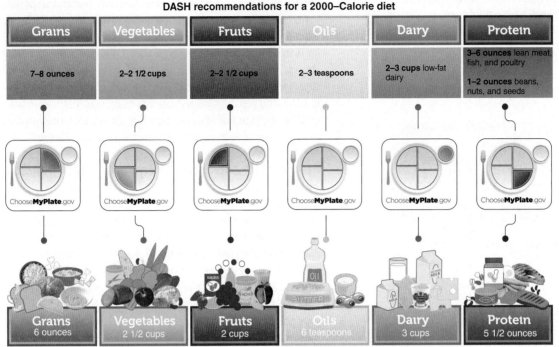

The recommendations of the DASH eating plan (DASH stands for Dietary Approaches to Stop Hypertension) (upper portion of the figure) are similar to those of MyPlate (lower portion). Both recommend eating plenty of fruits and vegetables; choosing whole grains; having beans, nuts, seeds, and fish more often; and choosing lean meats and low-fat dairy products.

Reductions in blood pressure are seen with the DASH diet pattern compared with a typical American eating pattern. Reducing the sodium content of both diets further reduces blood pressure.[16]

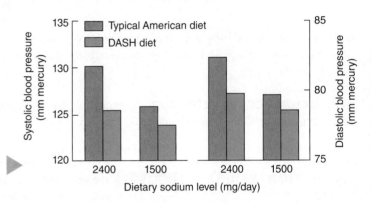

pressure even when sodium levels are not severely restricted. Reductions in blood pressure are greater when sodium intake is lower (see Figure 8.14).[16,19] Monitoring blood pressure regularly can allow early treatment and avoid the potentially lethal effects of elevated blood pressure. A number of medications are available to treat elevated blood pressure.

Meeting Electrolyte Needs

Most people in the United States need to reduce their salt intake and increase their potassium intake to meet recommendations for a healthy diet (see *What Should I Eat? Water and Electrolytes*). The 2015–2020 Dietary Guidelines recommend a dietary pattern that limits sodium to less than 2300 mg/day for adults.[20] This amount is the same as the UL for sodium, which is set to avoid the increase in blood pressure seen with higher sodium intakes. For individuals with prehypertension and hypertension, further reducing sodium intake to 1500 mg/day may result in greater blood pressure reduction. About 90% of people 1 year of age and older in the United States consume more than the UL for sodium; average sodium intake is 3440 mg/day.[20] Because salt is 40% sodium and 60% chloride by weight, 3440 mg of sodium represents 8600 mg (8.6 g) of salt per day.

The DRIs recommend a potassium intake of 4700 mg/day. Usual intake in the United States is about 2600 mg/day; fewer than 2% of adults meet recommendations.[21] No UL has been set for potassium.

One of the reasons our diet is high in salt (sodium chloride) and low in potassium is that we eat many processed foods, which are high in sodium and chloride, and too few fresh unprocessed foods, such as fruits, vegetables, and whole grains, which are high in potassium. Over three-quarters of the salt we eat comes from foods that have had salt added during processing and manufacturing (**Figure 8.15**). Some of this salt is obvious, but much of it is found in foods we may not think of as salty (see *What a Scientist Sees: Hidden Sodium*).

Electrolytes: Sodium, Potassium, and Chloride 237

What Should I Eat?

Water and Electrolytes

Stay hydrated
- Drink before, during, and after exercise.
- Guzzle two extra glasses of water when you are outside on a hot day.
- Bring a water bottle with you in your car.

Boost your potassium intake
- Double your vegetable serving at dinner.
- Take two pieces of fruit for lunch.
- Have orange juice or, better yet, an orange, instead of drinking soda or punch.
- Start with five and aim for nine servings of fruits and vegetables per day.

Reduce your sodium intake
- Have plain rice or noodles; prepackaged mixes have a lot of extra sodium.
- Do not add salt to the water when cooking rice, pasta, and cereals.
- Eat at home more frequently and prepare more meals from scratch. Flavor foods with salt-free herbs and spices rather than processed sauces.
- Limit salty snacks such as salted potato chips, nuts, popcorn, and crackers.
- Limit high-sodium condiments such as soy sauce, barbecue sauce, ketchup, and mustard.

 Use iProfile to compare the sodium content of fresh vegetables with that of canned vegetables.

FIGURE 8.15 Processing adds sodium Most of the sodium in the American diet comes from processed foods.

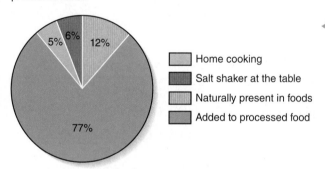

a. About 77% of the salt we eat comes from processed foods. Only about 12% comes from salt found naturally in food, while 11% comes from salt added in cooking and at the table.[22]

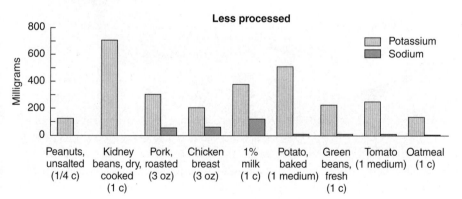

b. Processed foods are generally higher in sodium and may also be lower in potassium than unprocessed foods. Some of the sodium in processed foods comes from salt added for flavoring; some is added as a preservative to inhibit microbial growth. In addition to sodium chloride, sodium bicarbonate, sodium citrate, and sodium glutamate are added to preserve and flavor foods.

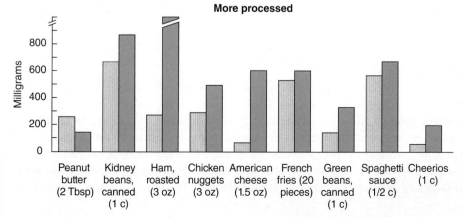

Ask Yourself

For a typical American, which of the following will result in the greatest reduction in sodium intake?

a. Taking away the salt shaker during meals
b. Reducing the amount of salt added during meal preparation
c. Consuming fewer processed foods

What a Scientist Sees

Hidden Sodium

A bowl of soup and a sandwich: What a consumer sees is an inexpensive healthy meal. It is quick, easy to prepare, and moderate in calories. If you add an apple and glass of milk, you have all your food groups.

What a scientist sees is a single meal that has more sodium than you should eat in the entire day (see figure). The Centers for Disease Control and Prevention includes both sandwiches and soups in their list of the top sources of sodium in the U.S. diet.[23] More than 40% of the sodium Americans consume each day comes from these sources: breads and rolls, cold cuts and cured meats, pizza, poultry, soups, sandwiches, cheese, pasta dishes, meat dishes, and snacks. Some of these, such as potato chips, which are sprinkled with salt, are obvious sources of sodium, but others, such as breads and soups, are less obvious sources. This hidden sodium is a concern because a diet high in sodium increases the risk of developing high blood pressure.[15]

Reducing sodium in the American diet requires limiting processed foods and choosing the ones we do include wisely. Food labels can be helpful in choosing packaged foods that are lower in sodium. Manufacturers are also working, with guidance from the FDA, to reduce the sodium content of processed foods. The FDA has suggested a stepwise reduction in sodium in processed foods that, if followed, would reduce the average amount of sodium consumed by Americans to 3000 mg/day within 2 years and to about 2300 mg/day within 10 years.[24]

Think Critically

What substitutions could you make to reduce the sodium content of this meal?

This bowl of soup and turkey and cheese sandwich provide over 2300 mg of sodium.

To reduce the sodium and increase the potassium in your diet, cut down on the amount of salt added to foods and replace many of your processed foods with fresh and minimally processed alternatives. Food labels can be helpful when choosing packaged foods (**Figure 8.16**). Some medications can also contribute a significant amount of sodium. Drug facts labels on over-the-counter medications can help identify those that contain large amounts of sodium. A dietary pattern that is high in fruits and vegetables will easily meet potassium intake recommendations; daily intakes of 8000 to 11,000 mg are not uncommon.[3]

Electrolytes: Sodium, Potassium, and Chloride 239

FIGURE 8.16 Sodium and potassium on food labels Both the current and planned Nutrition Facts panels list the sodium content of packaged foods in milligrams and as a percentage of the Daily Value. The planned food label also lists potassium and uses updated Daily Values for both sodium and potassium. A food that is low in sodium may include on the label the health claim that diets that are low in sodium may reduce the risk of high blood pressure. A food that is low in sodium and a good source of potassium may include the health claim that diets containing foods that are good sources of potassium and low in sodium may reduce the risk of high blood pressure and stroke.

Current food label

Nutrition Facts
Serving Size 1/2 cup (125g)
Servings Per Container about 3½

Amount Per Serving	
Calories 50	Calories from Fat 10
	%Daily Value**
Total Fat 1g	2%
Saturated Fat 0g	0%
Trans Fat 0g	
Cholesterol 0mg	0%
Sodium 250mg	10%
Potassium 530mg	15%
Total Carbohydrate 9g	3%
Dietary Fiber 1g	4%
Sugars 7g	
Protein 2g	
Vitamin A 10% • Vitamin C 25%	
Calcium 2% • Iron 10%	

*Percent Daily Values are based on a 2,000 calorie diet. Your daily values may be higher or lower depending on your calorie needs.

	Calories:	2,000	2,500
Total Fat	Less than	65g	80g
Sat Fat	Less than	20g	25g
Cholesterol	Less than	300mg	300mg
Sodium	Less than	2,400mg	2,400mg
Potassium		3,500mg	3,500mg
Total Carbohydrate		300g	375g
Dietary Fiber		25g	30g

The Daily Value for sodium for a 2000-Calorie diet is given at the bottom of the label.

Light Spaghetti Sauce, 250 milligrams (mg) per serving
Regular Spaghetti Sauce, 500 mg per serving

On the current label the amount of potassium may be listed voluntarily. The planned label requires that the amount of potassium be included along with the amounts of calcium, iron, and vitamin D.

Planned food label

Nutrition Facts
3½ servings per container
Serving size   1/2 cup (125g)

Amount per serving	
Calories	**50**
	% Daily Value*
Total Fat 1g	1%
Saturated Fat 0g	0%
Trans Fat 0g	
Cholesterol 0mg	0%
Sodium 250mg	11%
Total Carbohydrate 9g	3%
Dietary Fiber 1g	4%
Total Sugars 7g	
Includes 5g Added Sugars	10%
Protein 2g	
Vitamin D 0mcg	0%
Calcium 20mg	2%
Iron 2mg	10%
Potassium 530mg	11%

* The % Daily Value (DV) tells you how much a nutrient in a serving of food contributes to a daily diet. 2,000 calories a day is used for general nutrition advice.

"Light in sodium" means that this product contains 50% less sodium than regular spaghetti sauce. Other sodium descriptors seen on food labels include:
Sodium free—a food contains less than 5 mg of sodium per serving
Low sodium—a food contains 140 mg or less of sodium per serving (about 5% of the Daily Value)
Reduced sodium—a food contains at least 25% less sodium than a reference food.

Ask Yourself

If instead of a serving of this spaghetti sauce, which is light in sodium, you ate a serving of regular spaghetti sauce, how much more sodium would you consume?

Concept Check

1. **Why** does eating a salty meal cause your weight to increase temporarily?
2. **Why** is hypertension called "the silent killer"?
3. **What** is the DASH eating plan?
4. **Which** types of foods contribute the most sodium to the American diet?

8.4 Minerals and Bone Health

LEARNING OBJECTIVES

1. **Describe** factors that affect the risk of osteoporosis.
2. **List** foods that are good sources of calcium, of phosphorus, and of magnesium.
3. **Explain** how blood levels of calcium and phosphorus are regulated.
4. **Describe** functions of calcium, phosphorus, and magnesium that are unrelated to their role in bone.
5. **Discuss** the role of fluoride in maintaining dental health.

Bones are the hardest, strongest structures in the human body. Bone is strong because it is composed of a protein framework, or matrix, that is hardened by deposits of minerals. This matrix consists primarily of the protein collagen. The mineral portion of bone is composed mainly of calcium associated with phosphorus, but it also contains magnesium, sodium, fluoride, and a number of other minerals. Healthy bone requires adequate dietary protein and vitamin C to maintain the collagen. Adequate vitamin D is needed to maintain appropriate levels of calcium and phosphorus. There is also growing evidence of the importance of vitamin K for bone health.[25] Even when all these nutrients are consumed in adequate amounts, bone mass decreases as adults grow older. For many people, the loss of bone mass is so great that the risk of fractures is seriously increased, a condition called **osteoporosis**.

Osteoporosis

Like other tissues in the body, bone is alive, and it is constantly being broken down and re-formed through a process called **bone remodeling**. Most bone is formed early in life. During childhood, bones grow larger; even after growth stops, bone mass continues to increase into young adulthood (**Figure 8.17a**). The maximum amount of bone that you have in your lifetime, called **peak bone mass**, is achieved somewhere between ages 16 and 30. Up to this point, bone formation occurs more rapidly than breakdown, so the total amount of bone increases. After about age 35 to 45, the amount of bone that is broken down begins to exceed the amount that is formed, so total bone mass decreases. Over time, if enough bone is lost, osteoporosis results; the skeleton is weakened, and fractures occur more easily (**Figure 8.17b** and **c**).

Factors affecting the risk of osteoporosis The risk of developing osteoporosis depends on the level of peak bone mass and the rate at which bone is lost. These variables are affected by genetics, gender, age, hormone levels, body weight, and lifestyle factors such as smoking, alcohol consumption, exercise, and diet (**Table 8.2**).

Women have a higher risk of osteoporosis because they have lower bone density than men and lose bone faster as they age. **Age-related bone loss** occurs in both men and women, but women lose additional bone for a period of about 5 to 10 years surrounding **menopause**. This **postmenopausal bone loss** is related to the decline in estrogen level that occurs with menopause. Low estrogen increases calcium release from bone and decreases the amount of calcium absorbed in the intestines.

Body weight affects the risk of osteoporosis.[28] Being small and light increases risk whereas having a greater body weight, whether it is due to being tall, having a large muscle mass, or to excess body fat, decreases risk. Being heavier increases bone mass because it increases the amount of weight the bones must support. In other words, stressing the bones makes them grow stronger. A similar effect is seen in people of all body weights who engage in weight-bearing exercise. In postmenopausal women, excess body fat may also reduce risk because adipose tissue is an important source of estrogen, which helps maintain bone mass and enhances calcium absorption.

Dietary factors that impact calcium status may affect the risk of osteoporosis. A low calcium intake is the most significant dietary factor contributing to osteoporosis. High intakes of phytates, oxalates, and tannins can increase the risk for osteoporosis by reducing calcium absorption. High intakes of dietary sodium and protein have been found to increase calcium loss in the urine. However, when intakes of calcium and vitamin D are adequate, neither high protein nor high sodium intakes are believed to adversely affect bone health and the risk of osteoporosis.[29, 30]

Preventing and treating osteoporosis You can't feel your bones weakening, so people with osteoporosis may not know that their bone mass is dangerously low until they are in their 50s or 60s and experience a bone fracture. Once osteoporosis has developed, it is difficult to restore lost bone. Therefore, the best treatment for osteoporosis is to prevent it by achieving a high peak bone mass and slowing the rate of bone loss. During childhood, adolescence, and young adulthood, diet and exercise can help prevent osteoporosis by helping to

---

osteoporosis A bone disorder characterized by reduced bone mass, increased bone fragility, and increased risk of fractures.

bone remodeling The process whereby bone is continuously broken down and re-formed to allow for growth and maintenance.

peak bone mass The maximum bone density attained at any time in life, usually occurring in young adulthood.

age-related bone loss Bone loss that occurs in both men and women as they advance in age.

menopause The time in a woman's life when the menstrual cycle ends.

postmenopausal bone loss Accelerated bone loss that occurs in women for about 5 to 10 years surrounding menopause.

Minerals and Bone Health

NUTRITION INSIGHT

FIGURE 8.17 Bone mass and osteoporosis Osteoporosis is a major public health problem. In the United States, more than 10 million people over age 50 have osteoporosis, and another 43.4 million are at risk due to low bone mass, called **osteopenia**.[26] It is estimated that by 2025, osteoporosis will lead to 3 million fractures and account for $25.2 billion annually in direct medical costs.[27]

a. Changes in the balance between bone formation and bone breakdown cause bone mass to increase in children and adolescents and decrease in adults as they grow older.

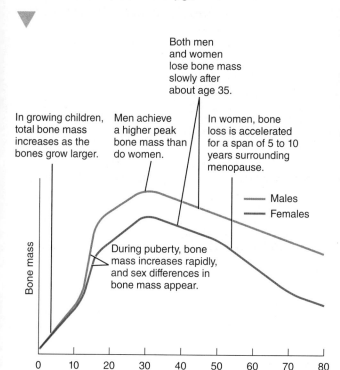

b. The decrease in bone mass and strength that occurs with osteoporosis is illustrated here by comparing osteoporotic bone with normal bone.

Interpret the Data

Based on the information in the graph, why are women over 50 at greater risk of osteoporosis than men over 50?

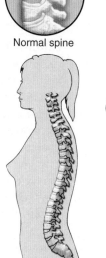

When weakened by osteoporosis, the front edge of the vertebrae collapses more than the back edge, so the spine bends forward.

c. Spinal compression fractures, shown here, are common and may result in loss of height and a stooped posture (called a "dowager's hump").

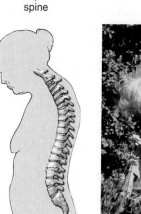

TABLE 8.2 Factors affecting the risk of osteoporosis

Risk factor	How it affects risk
Gender	Fractures due to osteoporosis are about twice as common in women as in men. Men are larger and heavier than women and therefore have a greater peak bone mass. Women lose more bone than men due to postmenopausal bone loss.
Age	Bone loss is a normal part of aging, so the risk of osteoporosis increases with age.
Race	African Americans have denser bones than do Caucasians and Southeast Asians, so their risk of osteoporosis is lower.[26]
Family history	Having a family member with osteoporosis increases risk.
Body weight	Individuals who are small and light have an increased risk because they have less bone mass.
Smoking	Tobacco use increases risk by reducing bone mass.
Exercise	Weight-bearing exercise, such as walking and jogging, throughout life strengthens bone, and increasing weight-bearing exercise at any age can increase your bone density and decrease risk.
Alcohol abuse	Long-term alcohol abuse increases risk by reducing bone formation and interfering with the body's ability to absorb calcium.
Diet	A diet that is lacking in calcium and vitamin D increases the risk of osteoporosis. Low calcium intake during the years of bone formation results in a lower peak bone mass, and low calcium intake in adulthood can accelerate bone loss.

maximize peak bone density. A diet that contains adequate amounts of calcium and vitamin D produces greater peak bone mass during the early years and slows bone loss as adults age. Supplements can help meet needs if the diet does not provide adequate calcium and vitamin D. Adequate intakes of zinc, magnesium, potassium, fiber, vitamin K, and vitamin C—nutrients that are plentiful in fruits and vegetables—are also important for bone health. Weight-bearing exercise before about age 35 helps increase peak bone mass, and maintaining an active lifestyle that includes weight-bearing exercise throughout life helps maintain bone density. Limiting smoking and alcohol consumption can also help increase and maintain bone density.

Osteoporosis is commonly treated with drugs that prevent bone breakdown. One class of drugs, called bisphosphonates, inhibits the activity of cells that break down bone. Bisphosphonates have been shown to prevent postmenopausal bone loss, increase bone mineral density, and reduce the risk of fractures.[31] Another class of drugs called selective estrogen receptor modulators (SERMS) provides the benefits of estrogen (reduced bone breakdown and increased calcium absorption), without the negative side effects, such as increasing the risk of heart disease and some types of cancer.

Calcium

In an average person, about 1.5% of body weight is due to calcium, and 99% of this is found in the bones and teeth. The remaining calcium is located in body cells and fluids, where it is needed for release of neurotransmitters, muscle contraction, blood pressure regulation, cell communication, blood clotting, and other essential functions. Neurotransmitter release is critical to nerve function because neurotransmitters relay nerve impulses from one nerve to another and from nerves to other cells. Calcium in muscle cells is essential for muscle contraction because it allows the muscle proteins actin and myosin to interact. Calcium may help regulate blood pressure by controlling the contraction of muscles in the blood vessel walls and signaling the secretion of substances that regulate blood pressure.

Calcium in health and disease The various roles of calcium are so vital to survival that powerful regulatory mechanisms maintain calcium concentrations both inside and outside cells. Slight changes in blood calcium levels trigger the release of hormones that work to keep calcium levels constant. When calcium levels drop, parathyroid hormone (PTH) is released (see Chapter 7). PTH acts in a number of tissues to increase blood calcium levels (**Figure 8.18**). If blood calcium levels become too high, PTH secretion stops, and **calcitonin** is released. Calcitonin is a hormone that acts primarily on bone to inhibit the release of calcium into the blood.

When too little calcium is consumed, the body maintains normal blood levels by breaking down bone to release calcium, a process called **bone resorption**. This process provides a steady supply of calcium and causes no short-term symptoms. Over time, however, inadequate calcium intake can reduce bone mass. Low calcium intake during the years of bone formation results in lower peak bone mass. If calcium

FIGURE 8.18 **Raising blood calcium levels** Low blood calcium triggers the secretion of PTH from the parathyroid gland. PTH stimulates the release of calcium from bone and causes the kidneys to reduce calcium loss in the urine and to activate vitamin D. Activated vitamin D increases the absorption of calcium from the gastrointestinal tract and acts with PTH to stimulate calcium release from the bone. The overall effect of PTH is to rapidly restore blood calcium levels to normal.

Think Critically

When blood calcium is normal, PTH levels are low. How does this affect the amount of calcium excreted in the urine and the amount absorbed from the gastrointestinal tract?

intake is low after peak bone mass has been achieved, the rate of bone loss may be increased and, along with it, the risk of osteoporosis.

Too much calcium can also affect health. Consuming too much calcium from foods is unlikely, but high intakes of calcium from supplements can interfere with the availability of iron, zinc, magnesium, and phosphorus; cause constipation; and contribute to elevated blood and urinary calcium levels. Because the level of calcium in the blood is finely regulated, elevated blood calcium is rare and is most often caused by cancer and by disorders that increase the secretion of PTH. It can also result from increases in intestinal calcium absorption due to excessive vitamin D intake or high intakes of calcium from supplements or antacids.[32] Elevated blood calcium levels can cause symptoms such as loss of appetite, abnormal heartbeat, weight loss, fatigue, frequent urination, and soft tissue calcification. High urinary calcium can damage the kidneys and increase the risk of kidney stones.[33]

The UL for calcium in young adults ages 19 to 50 is 2500 mg/day. In older adults the UL is lower, only 2000 mg/day, based on the occurrence of kidney stones in older age groups.[33] Some postmenopausal women taking supplements may be getting too much calcium and increasing their risk of kidney stones.

Meeting calcium needs For adults ages 19 to 50, 1000 mg/day of calcium is recommended to maintain bone health. Because bone loss is accelerated in women due to menopause, the RDA for women 51 to 70 is set at 1200 mg/day to slow bone loss. For men in this age group it remains at 1000 mg/day. Bone loss and resulting osteoporotic fractures are of concern for both men and women 70 and older, so the RDA for both genders in this age group is 1200 mg/day.[33]

Milk and other dairy products are the main sources of calcium in the U.S. diet, but milk consumption has been declining. Many people drink sugar-sweetened beverages instead of milk; almost one-third of Americans drink at least one sugar-sweetened beverage every day.[34] The shift from milk to sugar-sweetened beverages has reduced calcium intake (**Figure 8.19a**), which is often below the RDA, especially among adolescent girls and adults.[35]

FIGURE 8.19 **Food sources of calcium** Dairy products are an important source of calcium and other essential nutrients.

a. Choosing a glass of low-fat milk over a can of soda increases the amounts of protein, calcium, magnesium, vitamin A, vitamin D, riboflavin, and vitamin B_{12} in your diet and eliminates almost 10 teaspoons of added sugars.

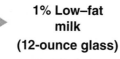

1% Low–fat milk (12-ounce glass)

Cola soft drink (12-ounce can)

1% Low-fat milk		Cola
154	Energy (Calories)	150
0	Added sugars (g)	39
12	Protein (g)	0
458	Calcium (mg)	0
348	Phosphorus (mg)	45
40	Magnesium (mg)	0
0.7	Riboflavin (mg)	0
1.7	Vitamin B_{12} (µg)	0
212	Vitamin A (µg)	0
4	Vitamin D (µg)	0

(continues)

FIGURE 8.19 *(continued)*

b. Foods in the dairy group are high in calcium. Other good sources of calcium include fish, such as sardines, that are consumed with the bones, legumes, almonds, and some dark-green vegetables, such as kale and broccoli. Grains are a moderate source, but because we consume them in large quantities, they make a significant contribution to our calcium intake.

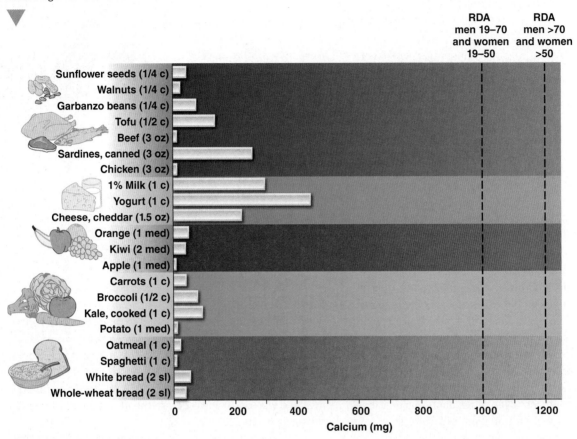

Interpret the data

If bioavailability were the same, how much cooked kale would you need to eat to get as much calcium as you do from a cup of milk?

Those who do not or cannot consume dairy products can meet their calcium needs by consuming dark-green vegetables, fish consumed with bones, foods processed with calcium, and foods fortified with calcium, such as milk substitutes, juices, and breakfast cereals (**Figure 8.19b**). Although calcium-rich foods and beverages are the preferred source of calcium, individuals who do not meet their calcium needs through their diet alone can benefit from calcium supplements (**Figure 8.20**). In young individuals, supplemental calcium can increase peak bone mass. In postmenopausal women, calcium supplements are not effective at increasing bone mass, but in individuals with low dietary calcium intake, supplements can help reduce the rate of bone loss.[36]

Whether your calcium comes from foods or from supplements, bioavailability must be considered. Vitamin D is the nutrient that has the most significant impact on calcium absorption. When calcium intake is high, calcium is absorbed by diffusion, but when intake is low to moderate, as it typically is, absorption depends on the active form of vitamin D. When vitamin D is deficient, less than 10% of dietary calcium may be absorbed, compared to the typical 25% when it is present. Other dietary components that affect calcium absorption include acidic foods, lactose, and fat, which increase calcium absorption, and oxalates, phytates, tannins, and fiber, which inhibit calcium absorption. For example, spinach is a high-calcium vegetable, but only about 5% of its calcium is absorbed; the rest is bound by oxalates and excreted in the feces.[37] Vegetables such as broccoli, kale, collard greens, turnip greens, mustard greens, and Chinese cabbage are low in oxalates, so their calcium is more readily absorbed. Chocolate also contains oxalates, but chocolate milk is still a good source of calcium because the amount of oxalates from the chocolate added to a glass of milk is small.

FIGURE 8.20 Calcium supplements If you are not getting enough calcium from foods, a supplement that contains a calcium compound alone or calcium with vitamin D can help you meet your calcium needs. A multivitamin/multimineral supplement will provide only a small amount of the calcium you need. Use the Supplement Facts label to choose an appropriate supplement.

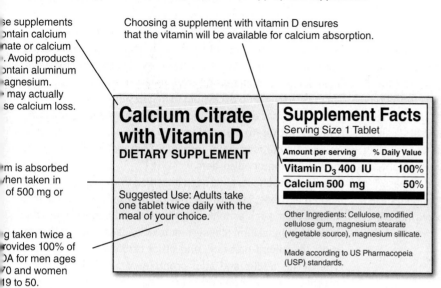
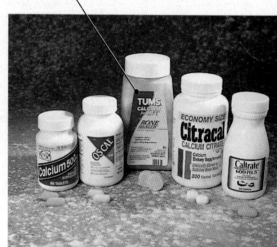

se supplements ontain calcium nate or calcium . Avoid products ontain aluminum agnesium. may actually se calcium loss.

Choosing a supplement with vitamin D ensures that the vitamin will be available for calcium absorption.

Some antacids are sources of calcium. These are over-the-counter medications, so they carry a Drug Facts panel rather than a Supplement Facts panel.

m is absorbed hen taken in of 500 mg or

g taken twice a rovides 100% of DA for men ages 70 and women 19 to 50.

George Semple

Phosphorus

The phosphorus in our bodies is present as part of a chemical group called a phosphate group (see Chapter 5). Most is associated with calcium as part of the hard mineral crystals in bones and teeth. The smaller amount of phosphorus in soft tissues performs an essential role as a structural component of phospholipids, DNA and RNA, and ATP. It is also important in regulating the activity of enzymes and maintaining the proper level of acidity in cells (**Figure 8.21**).

Phosphorus in health and disease
Blood levels of phosphorus are not controlled as strictly as calcium levels, but the kidneys help maintain phosphorus levels in a ratio with calcium that allows minerals to be deposited into bone.

FIGURE 8.21 Nonskeletal functions of phosphorus Most of the phosphorus in the body helps form the structure of bones and teeth, but phosphorus also plays an important role in a host of cellular activities.

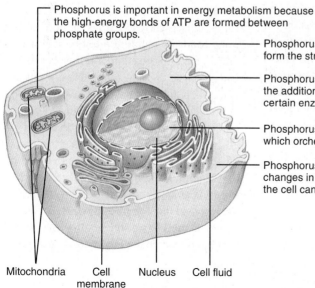

Phosphorus is important in energy metabolism because the high-energy bonds of ATP are formed between phosphate groups.

Phosphorus is a component of phospholipids, which form the structure of cell membranes.

Phosphorus is involved in regulating enzyme activity; the addition of a phosphorus-containing group to certain enzymes can activate or inactivate them.

Phosphorus is a major constituent of DNA and RNA, which orchestrate the synthesis of proteins.

Phosphorus is part of a compound that can prevent changes in acidity so that chemical reactions inside the cell can proceed normally.

Mitochondria   Cell membrane   Nucleus   Cell fluid

A deficiency of phosphorus can lead to bone loss, weakness, and loss of appetite. Inadequate phosphorus intake is rare because phosphorus is widely distributed in food. Marginal phosphorus status may be caused by losses due to chronic diarrhea or poor absorption due to overuse of aluminum-containing antacids.

An excess of dietary phosphorus is a growing concern. Phosphorus is found in many processed foods; an increase in the use of phosphorus-containing food additives has caused phosphorus intakes to rise. Too much phosphorus in the diet disrupts the hormonal regulation of phosphate, calcium, and vitamin D, which contributes to decreased bone mass, calcification of blood vessels, and impaired kidney function.[38] High intakes of dietary phosphorus and elevated serum phosphorus levels are both associated with an increase in heart disease.[39,40]

The UL for phosphorus is 4000 mg/day for adults, based on an amount associated with the upper limits of normal blood phosphorus levels.[41]

Meeting phosphorus needs The RDA for phosphorus is 700 mg/day for adults; and the typical intake is well above this.[20,41] Dairy products such as milk, yogurt, and cheese, as well as meat, cereals, bran, eggs, nuts, and fish, are good sources of phosphorus. Many food additives used in baked goods, cheese, processed meats, carbonated beverages, and convenience foods contain phosphorus. These additives are a major contributor to our increasing phosphorus intake.[38]

Magnesium

Magnesium is far less abundant in the body than are calcium and phosphorus, but it is still essential for healthy bones. About 50 to 60% of the magnesium in the body is in bone, where it helps maintain bone structure. The rest of the magnesium is found in cells and fluids throughout the body. Magnesium is involved in regulating calcium homeostasis and is needed for the action of vitamin D and many hormones, including PTH. Magnesium is important for the regulation of blood pressure and may play a role in maintaining cardiovascular health. In addition, magnesium forms a complex with ATP that stabilizes ATP's structure. It is therefore needed in every metabolic reaction that generates or uses ATP. This includes reactions needed for the release of energy from carbohydrate, fat, and protein; the functioning of the nerves and muscles; and the synthesis of DNA, RNA, and protein, making it particularly important for dividing, growing cells.

Magnesium in health and disease Severe magnesium deficiency can cause nausea, muscle weakness and cramping, irritability, mental derangement, and changes in blood pressure and heartbeat. Although overt magnesium deficiency is rare, the typical intake of magnesium in the United States is below the RDA.[42] Low intake of magnesium has been associated with a number of chronic diseases, including type 2 diabetes, hypertension, atherosclerosis, and osteoporosis.[42] As discussed earlier, dietary patterns that are high in magnesium are associated with lower blood pressure, and the risk of other types of cardiovascular disease is lower for people with adequate magnesium intake than for those with less magnesium in their diet.[43] Over the past 30 years nutritional messages have focused on increasing calcium intake. This has caused an increase in the ratio of calcium to magnesium in the diet. There is concern that either our low magnesium intake or the rising ratio of calcium to magnesium may play a role in the development of metabolic syndrome, type 2 diabetes, osteoporosis, and inflammatory disorders.[42,44]

No toxic effects have been observed from magnesium consumed in foods, but toxicity may occur from drugs containing magnesium, such as milk of magnesia, and supplements that include magnesium. Magnesium toxicity causes nausea, vomiting, low blood pressure, and other cardiovascular changes. The UL for adults and adolescents over age 9 is 350 mg of magnesium from nonfood sources such as supplements and medications.[41]

Meeting magnesium needs Magnesium is found in many foods but in small amounts, so you can't get all you need from a single food (**Figure 8.22**). Enriched grain products are poor sources because magnesium is lost in processing, and it is not added back by enrichment. For example, removing the bran and germ from wheat kernels reduces the magnesium content from 166 mg in 1 cup of whole-wheat flour to only 28 mg in 1 cup of white flour. Magnesium absorption is enhanced by the active form of vitamin D and decreased by the presence of phytates and calcium.

Fluoride

Fluoride is a trace mineral that is incorporated into the mineral crystals in bones and teeth. It can increase bone strength but is best known for its role in strengthening tooth enamel and preventing **dental caries** (cavities) in both children and adults.[45] During tooth formation (up until about age 13), ingested fluoride is incorporated into the crystals that make up tooth enamel. These fluoride-containing crystals are more resistant to acid than are crystals formed when fluoride is not present. Ingested fluoride, whether it is from the diet, the water supply, or supplements, is also secreted in saliva, which continually bathes the teeth. Fluoride in saliva prevents cavities in children and adults by reducing the amount of acid produced by bacteria, inhibiting the dissolution of tooth enamel by acid, and increasing enamel remineralization after acid exposure. Therefore, ingested fluoride benefits dental health throughout life.[46] Fluoride in toothpaste and mouth rinses have the same topical effect but remain in the mouth for only a few hours after use.

FIGURE 8.22 Sources of magnesium Magnesium is a component of the green pigment chlorophyll, so leafy greens such as spinach and kale are good sources of this mineral. Nuts, seeds, legumes, bananas, and the germ and bran of whole grains are also good sources of magnesium. The dashed lines represent the RDAs of 420 and 320 mg/day for adult men and women over age 30, respectively.[41]

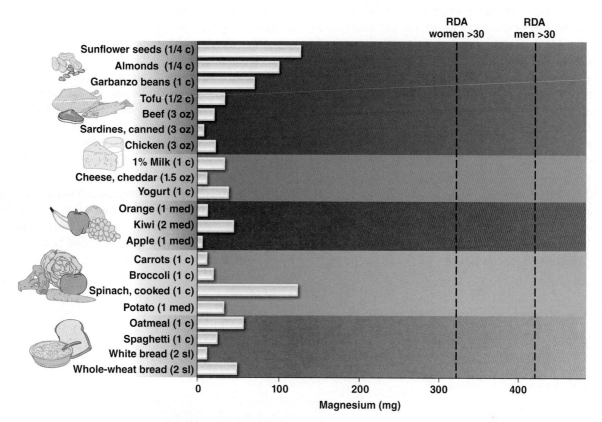

Meeting fluoride needs

Fluoride is present in small amounts in almost all soil, water, plants, and animals. The richest dietary sources of fluoride are toothpaste, tea, marine fish consumed with their bones, and fluoridated water (see *What Should I Eat? Calcium, Phosphorus, Magnesium, and Fluoride*). Because food readily absorbs the fluoride in cooking water, the fluoride content of food can be significantly increased when it is handled and prepared using water that contains fluoride. Cooking utensils also affect the fluoride content of foods. Foods cooked with Teflon utensils can pick up fluoride from the Teflon, whereas aluminum cookware can decrease the fluoride content of foods. Bottled water usually does not contain fluoride, so people who habitually drink bottled water need to obtain fluoride from other sources.

What Should I Eat?

Calcium, Phosphorus, Magnesium, and Fluoride

Get calcium into your body and your bones
- Have three servings of dairy a day: milk, yogurt, cheese.
- Bone up on calcium by eating sardines or canned salmon, which are eaten with the bones.
- Choose leafy greens for a crisp source of calcium.
- Walk, jog, or jump up and down; weight-bearing exercises build up bone.

Don't fret about phosphorus—it's in almost everything you eat

Maximize your magnesium
- Choose whole grains.
- Sprinkle nuts and seeds on your salad, cereal, and stir-fry.
- Go for the green: Whenever you eat green, you are eating magnesium; most greens contain calcium, too.

Find your fluoride
- Check to see if your water is fluoridated.
- Have a cup of tea to boost your fluoride intake.

 Use iProfile to find a nondairy source of calcium.

The recommended fluoride intake for people 6 months of age and older is 0.05 mg/kg/day. This is equivalent to about 3.8 mg/day for a 76-kg man and 3.1 mg/day for a 61-kg woman. Fluoridated water provides about 0.7 mg/L of fluoride.[47] The American Academy of Pediatrics suggests a fluoride supplement of 0.25 mg/day for children 6 months to 3 years of age, 0.5 mg/day for children ages 3 to 6 years, and 1.0 mg/day for those ages 6 to 16 years who are receiving less than 0.3 mg/L of fluoride in the water supply. These supplements are available by prescription for children living in areas with low fluoride concentrations in the water supply. In adults, there is evidence that fluoride supplements stimulate bone formation and increase bone density, but there is no clear evidence that fluoride therapy reduces fractures.[46]

Fluoridation of water To promote dental health, fluoride is added to public water supplies in many communities. Currently, about 75% of people served by public water systems receive fluoridated water.[48] When fluoride intake is low, tooth decay is more frequent (**Figure 8.23**). Water fluoridation is a safe, inexpensive way to prevent dental caries, but some people still believe that the added fluoride increases the risk of cancer and other diseases. These beliefs are not supported by scientific facts; the small amounts of fluoride consumed in drinking water promote dental health and do not pose a risk for health problems such as cancer, kidney failure, or bone disease.[46] The amount of fluoride added to drinking water does not affect bone health.[49]

Although the levels of fluoride included in public water supplies are safe, too much fluoride can be toxic. In children, too much fluoride causes **fluorosis** (see Figure 8.23). Recently, there has been an increase in fluorosis in the United States due to chronic ingestion of fluoride-containing toothpaste.[50] The fluoride in toothpaste is good for your teeth, but swallowing it can increase fluoride intake to dangerous levels. Swallowed toothpaste is estimated to contribute about 0.6 mg/day of fluoride in young children. Due to concern about excess fluoride intake, the following warning is now required on all fluoride-containing toothpastes: "If you accidentally swallow more than used for brushing, seek professional help or contact a poison control center immediately."

FIGURE 8.23 **Just the right amount of fluoride** This graph illustrates that the incidence of dental caries in children increases when the concentration of fluoride in the water supply is lower. If the fluoride concentration of the water is too high, it increases the risk of fluorosis.

Too little fluoride makes teeth more susceptible to dental caries.

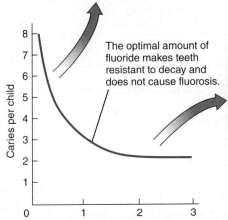

The optimal amount of fluoride makes teeth resistant to decay and does not cause fluorosis.

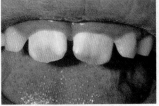

Too much fluoride (intakes of 2 to 8 mg/day or greater in children) causes teeth to appear mottled, a condition called fluorosis.

Interpret the Data

Based on this graph, what concentration of water fluoride will protect against dental caries but not cause fluorosis?

a. 2.5 mg/L
b. 2 mg/L
c. 1 mg/L
d. 0.5 mg/L

fluorosis A condition caused by chronic overconsumption of fluoride, characterized by black and brown stains and cracking and pitting of the teeth.

The UL for fluoride is set at 0.1 mg/kg/day for infants and children younger than 9 years and at 10 mg/day for those 9 years and older.[41] In adults, doses of 20 to 80 mg/day of fluoride can result in changes in bone health that may be crippling, as well as changes in kidney function and possibly nerve and muscle function. Death has been reported in cases involving intake of 5 to 10 g/day.

> **Concept Check**
>
> 1. **Why** are women at greater risk for osteoporosis than men?
> 2. **How** can vegans meet their calcium needs?
> 3. **What** happens when blood calcium levels decrease?
> 4. **What** is the relationship between phosphorus, magnesium, and ATP?
> 5. **How** does fluoride protect against dental caries?

8.5 Minerals and Healthy Blood

LEARNING OBJECTIVES

1. **Describe** the primary function of iron in the body.
2. **Discuss** factors that affect iron absorption.
3. **Explain** how copper is related to iron status.

Iron and copper are trace minerals essential for the synthesis of adequate amounts of the protein **hemoglobin**, which gives blood its red color and transports oxygen throughout the body. Deficiencies of either of these minerals will result in red blood cells that are small and pale and unable to transport adequate amounts of oxygen to body cells.

Iron

In the 1800s, iron tablets were used to treat young women whose blood lacked "coloring matter." Today we know that the "coloring matter" is the iron-containing protein hemoglobin. The hemoglobin in red blood cells transports oxygen to body cells and carries carbon dioxide away from them for elimination by the lungs. Most of the iron in the body is part of hemoglobin, but iron is also needed for the production of other iron-containing proteins. It is part of **myoglobin**, a protein found in muscle that enhances the amount of oxygen available for use in muscle contraction. Iron is a part of several proteins needed in aerobic metabolism, so it is essential for ATP production. Iron-containing proteins are also involved in drug metabolism and immune function.

Iron absorption and transport The amount of iron absorbed from the intestine depends on the form of the iron and on the dietary components consumed along with it. Much of the iron in meats is **heme iron**—iron that is part of a chemical complex found in proteins such as hemoglobin and myoglobin. Heme iron is absorbed more than twice as efficiently as nonheme iron, which is the only form found in plant sources such as leafy green vegetables, legumes, and grains. The amount of nonheme iron absorbed can be enhanced or reduced by the foods and nutrients consumed in the same meal.

Once iron is absorbed, the amount that is delivered to the cells of the body depends to some extent on the body's needs. When the body's iron status is high, less iron is delivered to body cells and more is trapped in the mucosal cells of the small intestine and lost when the cells die and are sloughed into the intestinal lumen. When iron status is low, more iron is transported out of the mucosal cells and delivered to body cells (**Figure 8.24**).[51] This regulation is important because once iron has entered the blood and other tissues, it is not easily eliminated. Even when red blood cells die, the iron in their hemoglobin is not lost from the body; instead, it is recycled and can be incorporated into new red blood cells. Even in healthy individuals, most iron loss occurs through blood loss, including blood lost during menstruation and the small amounts lost from the gastrointestinal tract. Some iron is also lost through the shedding of cells from the intestines, skin, and urinary tract.

Meeting iron needs Iron in the diet comes from both plant and animal sources (**Figure 8.25**). Animal products provide both heme and nonheme iron, but only the less readily absorbed nonheme iron is found in plants. Nonheme iron absorption can be enhanced as much as sixfold if it is consumed along with foods that are rich in vitamin C. Consuming beef, fish, or poultry in the same meal as nonheme iron also increases absorption. For example, a small amount of hamburger in a pot of chili will enhance the body's absorption of iron from the beans. If the pot is made of iron, it increases iron intake because the iron leaches into food, particularly if the food is acidic. Iron absorption is decreased by fiber, phytates, tannins, and oxalates, which bind iron in the gastrointestinal tract. The presence of other minerals with the same charge, such as calcium, may also decrease iron absorption.

---

heme iron A readily absorbable form of iron found in meat, fish, and poultry that is chemically associated with certain proteins.

FIGURE 8.24 **Iron absorption, uses, and loss** The amount of dietary iron that reaches body cells depends on both the amount absorbed into the mucosal cells of the small intestine and the amount transported from the mucosal cells to the rest of the body. The iron that is transported may be used to synthesize iron-containing proteins, such as hemoglobin needed for red blood cell formation, or increase iron stores in the liver or spleen.

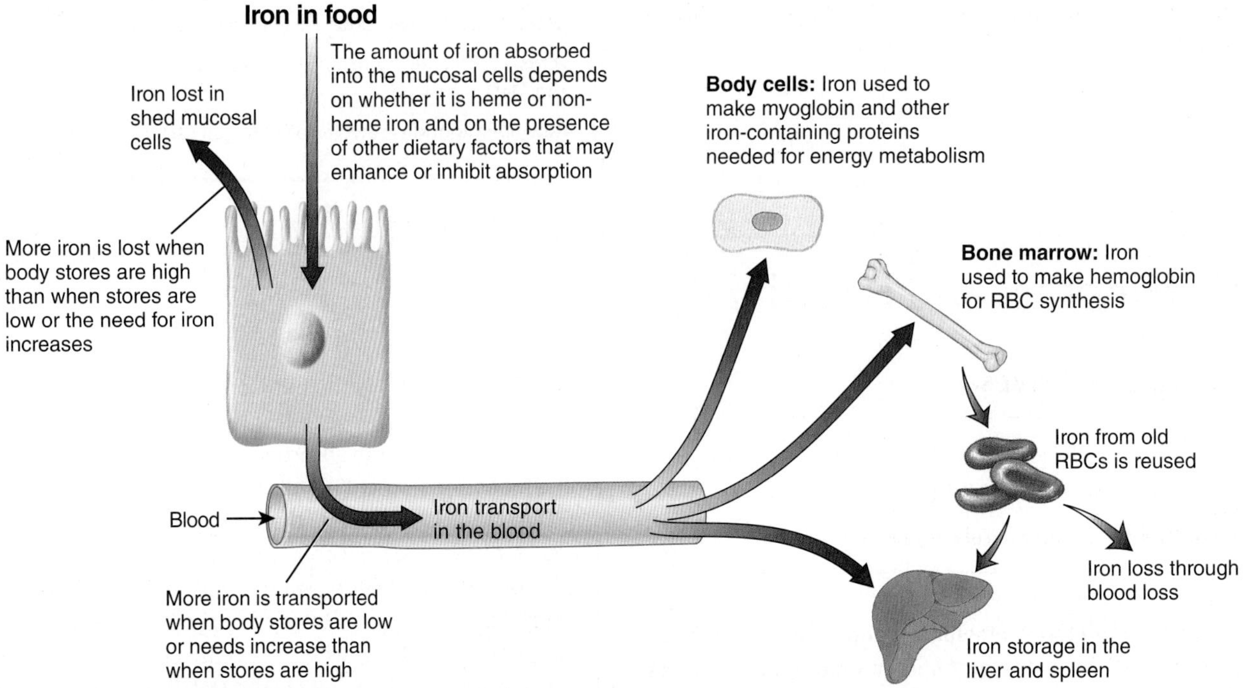

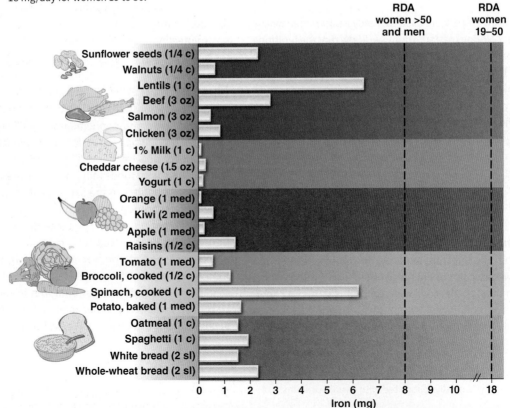

FIGURE 8.25 **Sources of iron** The best sources of highly absorbable heme iron are red meats and organ meats such as liver and kidney. Legumes, leafy greens, and whole grains are good sources of non-heme iron; refined grains are also good sources of nonheme iron because it is added during enrichment. The RDA for adult men and postmenopausal women is 8 mg/day of iron. Due to menstrual losses, the RDA for women of childbearing age is set much higher, 15 mg/day for young women 14 to 18 years and 18 mg/day for women 19 to 50.

Minerals and Healthy Blood 251

The RDA for iron assumes that the diet contains both plant and animal sources of iron.[52] A separate RDA category has been created for vegetarians. These recommendations are higher to take into account lower iron absorption from plant sources. People who have difficulty consuming enough iron can increase their iron intake by choosing foods fortified with iron.

Iron in health and disease When there is too little iron in the body, hemoglobin cannot be produced. When sufficient hemoglobin is not available, the red blood cells that are formed are small and pale, and they are unable to deliver adequate oxygen to the tissues. This condition is known as **iron deficiency anemia (Figure 8.26)**. Symptoms of iron deficiency

NUTRITION INSIGHT

FIGURE 8.26 Iron deficiency Iron deficiency anemia is the final stage of iron deficiency. It results when there is too little iron to synthesize adequate amounts of hemoglobin. Children and young women are at the greatest risk.

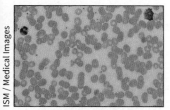
Normal red blood cells

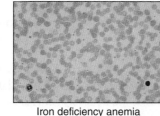
Iron deficiency anemia

Iron deficiency anemia results in red blood cells that are small and pale and unable to transport as much oxygen as red blood cells containing normal amounts of hemoglobin.

Inadequate iron intake first causes a decrease in the amount of stored iron, followed by low iron levels in the blood plasma. It is only after plasma levels are depleted that there is no longer enough iron to maintain hemoglobin in red blood cells and the symptoms of iron deficiency anemia appear.

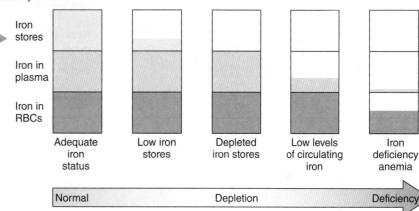

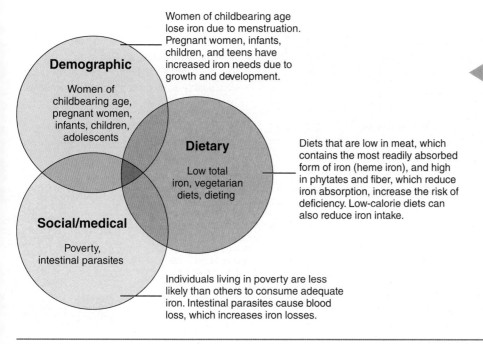

The risk of iron deficiency is highest among individuals with greater iron losses, those with greater needs due to growth and development, and those who are unable to obtain adequate dietary iron. In the United States, about 10 million people are iron deficient, including 5 million who have iron deficiency anemia.[53] The incidence is greatest among low-income and minority women and children.

---

iron deficiency anemia An iron deficiency disease that occurs when the oxygen-carrying capacity of the blood is decreased because there is insufficient iron to make hemoglobin.

Thinking It Through

A Case Study on Iron Deficiency

Hanna is a 23-year-old graduate student from South Carolina. She has been working long hours and is always tired. She tries to eat a healthy diet; she has been a lacto-ovo vegetarian for the past six months.

What factors increase Hanna's risk for iron deficiency anemia?

Your answer:

Hanna goes to the health center to see if there is a medical reason for her exhaustion. A nurse draws her blood to check her for anemia. The results of Hanna's blood work are summarized in the figure below.

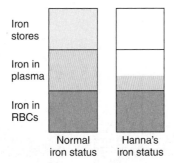

Does Hanna have iron deficiency anemia? Why or why not?

Your answer:

Hanna meets with a dietitian. A diet analysis shows that because she consumes no meat, much of her protein comes from dairy products. She consumes about four servings of dairy, eight servings of whole grains, three fresh fruits, and about a cup of cooked vegetables daily. She drinks four glasses of iced tea daily. Her average iron intake is about 12 mg/day.

Name three dietary factors that put Hanna at risk for iron deficiency.

Your answer:

Suggest two changes Hanna could make to increase her iron intake while sticking with her lacto-ovo vegetarian diet.

Your answer:

What could Hanna add to her diet to increase the absorption of the nonheme iron she consumes?

Your answer:

(Check your answers in Appendix L.)

---

anemia include fatigue, weakness, headache, decreased work capacity, inability to maintain body temperature in a cold environment, changes in behavior, decreased resistance to infection, impaired development in infants, and increased risk of lead poisoning in young children. Anemia is the last stage of iron deficiency. Earlier stages have no symptoms because they do not affect the amount of hemoglobin in red blood cells, but levels of iron in the blood plasma and in body stores are low (see *Thinking It Through*). Iron deficiency anemia is the most common nutritional deficiency; more than 2 billion people, or over 30% of the world's population, suffer from anemia, many due to iron deficiency.[54]

Iron also causes health problems when levels in the GI tract and in the body are too high. Acute iron toxicity caused by excessive consumption of iron-containing supplements is one of the most common forms of poisoning among children under age 6. Iron poisoning may cause damage to the lining of the intestine, abnormalities in body acidity, shock, and liver failure. Even a single large dose can be fatal (**Figure 8.27**). A UL has been set at 45 mg/day from all sources.[52]

Accumulation of iron in the body over time, referred to as **iron overload**, is most commonly due to an inherited condition called **hemochromatosis**. Hemochromatosis occurs when genes involved in regulating iron uptake are defective, so excess iron is allowed to enter the circulation.[56] It is the most common genetic disorder in the Caucasian population,

affecting more than 1 million people in the United States.[57] It has no symptoms early in life, but in middle age, nonspecific symptoms such as weight loss, fatigue, weakness, and abdominal pain develop. If allowed to progress, the accumulation of excess iron can damage the heart and liver and increase the

FIGURE 8.27 **Iron toxicity** To protect children, the labels on iron-containing drugs and supplements are required to display this warning.[55] Iron-containing products should be stored out of the reach of children and other individuals who might consume them in excess.

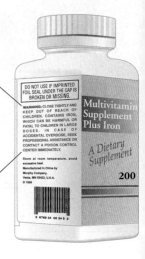

---

hemochromatosis An inherited disorder that results in increased iron absorption.

individual's risks for diabetes and cancer. The treatment for hemochromatosis is simple: regular blood withdrawal. Iron loss through blood withdrawal will prevent the complications of iron overload, but to be effective, the treatment must be initiated before organs have been damaged. Therefore, genetic screening is essential to identify and treat individuals before any damage occurs.

Copper

It is logical that consuming too little iron will cause iron deficiency anemia, but consuming too little copper can also cause this problem. Iron status and copper status are interrelated because a copper-containing protein is needed for iron to be transported from the intestinal cells. Even if iron intake is adequate, iron can't get to cells throughout the body if copper is not present. Thus copper deficiency results in a secondary iron deficiency that may lead to anemia. Copper also functions as a component of a number of important proteins and enzymes that are involved in connective tissue synthesis, antioxidant protection, lipid metabolism, maintenance of heart muscle, and immune and central nervous system function.[52]

We consume copper in seafood, nuts and seeds, whole-grain breads and cereals, and chocolate; the richest dietary sources of copper are organ meats such as liver and kidney. As with many other trace minerals, soil content affects the amount of copper in plant foods. The RDA for copper for adults is 900 micrograms (μg)/day.[52]

When there is too little copper in the body, the protein collagen does not form normally, resulting in skeletal changes similar to those seen in vitamin C deficiency. Copper deficiency also causes elevated blood cholesterol, reflecting copper's role in cholesterol metabolism. Copper deficiency has been associated with impaired growth, degeneration of the heart muscle and the nervous system, and changes in hair color and structure. Because copper is needed to maintain the immune system, a diet that is low in copper increases the incidence of infections. Also, because copper is an essential component of one form of the antioxidant enzyme *superoxide dismutase*, a copper deficiency will weaken antioxidant defenses.

Severe copper deficiency is relatively rare, although it may occur in premature infants. It can also occur if zinc intake is high because high dietary zinc interferes with the absorption of copper. Copper toxicity from dietary sources is also rare but has occurred as a result of drinking from contaminated water supplies or consuming acidic foods or beverages that have been stored in copper containers. Toxicity is more likely to occur from supplements containing copper. Excessive copper intake causes abdominal pain, vomiting, and diarrhea. The UL has been set at 10 mg/day of copper.[52]

Concept Check

1. **Why** does iron deficiency reduce the amount of oxygen delivered to cells?
2. **What** foods increase iron absorption?
3. **Why** can a copper deficiency lead to iron deficiency anemia?

8.6 Antioxidant Minerals

LEARNING OBJECTIVES

1. **Discuss** the relationship between selenium and vitamin E.
2. **Explain** the role of zinc in gene expression.
3. **Discuss** the antioxidant functions of zinc, copper, and manganese.
4. **Explain** how a vegetarian diet might impact selenium and zinc status.

Antioxidants protect cells from the damaging effects of reactive oxygen molecules such as free radicals (see Chapter 7). Vitamins C and E directly neutralize free radicals, whereas the antioxidant role of minerals such as selenium, zinc, copper (discussed above), and manganese is through their role as cofactors for antioxidant enzyme systems.

Selenium

Selenium is a trace mineral that is incorporated into the structure of certain proteins. One of these proteins is the antioxidant enzyme **glutathione peroxidase**. Glutathione peroxidase neutralizes peroxides before they can form free radicals, which cause oxidative damage (**Figure 8.28**). In addition to its antioxidant role in glutathione peroxidase, selenium is part of a protein needed for the synthesis of the **thyroid hormones**, which regulate metabolic rate.

---

glutathione peroxidase A selenium-containing enzyme that protects cells from oxidative damage by neutralizing peroxides.

FIGURE 8.28 Glutathione peroxidase
Glutathione peroxidase is a selenium-containing enzyme that neutralizes peroxides before they can form free radicals. Selenium therefore can reduce the body's need for vitamin E, which neutralizes free radicals.

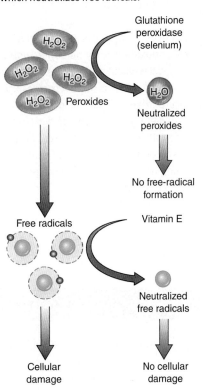

Meeting selenium needs The amount of selenium in food varies greatly, depending on the concentration of selenium in the soil where the food is produced. In regions of China with low soil selenium levels, a form of heart disease called **Keshan disease** occurs in children and young women. This disease can be prevented and cured with selenium supplements. In contrast, people living in regions of China with very high selenium in the soil may develop symptoms of selenium toxicity (**Figure 8.29**).

Selenium deficiencies and excesses are not a concern in the United States because the foods we consume come from many different locations around the country and around the world. The RDA for selenium for adults is 55 µg/day.[59] The average intake in the United States meets, or nearly meets, this recommendation for all age groups. Seafood, kidney, liver, and eggs are excellent sources of selenium. Grains, nuts, and seeds can be good sources, depending on the selenium content of the soil in which they were grown. Fruits, vegetables, and drinking water are generally poor sources. The UL for adults is 400 µg/day from food and supplements.[59]

Selenium and cancer An increased incidence of cancer has been observed in regions where selenium intake is low, suggesting that selenium plays a role in preventing cancer. In 1996, a study investigating the effect of selenium supplements on people with a history of skin cancer found

FIGURE 8.29 Soil selenium and health The amount of selenium in the soil varies widely from one region of China to another. When the diet consists primarily of locally grown food, these differences affect selenium intake and, therefore, health. When the diet includes foods from many different locations, the low selenium content of foods grown in one geographic region is offset by the high selenium content of foods from other regions.

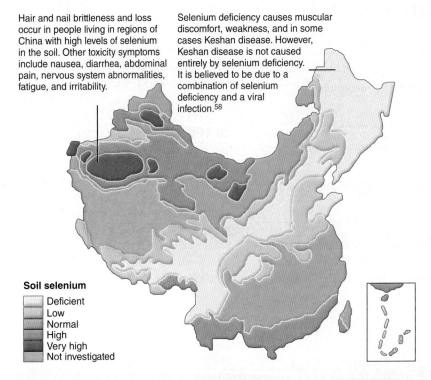

that the supplement had no effect on the recurrence of skin cancer but that the incidence of lung, prostate, and colon cancer decreased in the selenium-supplemented group.[60] This result caused speculation that selenium supplements could reduce the risk of cancer. Subsequent research, however, has not supported this result. Evidence now suggests that selenium supplements actually increase the incidence of certain types of cancer.[61] The reduction in the incidence of lung and prostate cancer seen in the 1996 study is now believed to have occurred primarily in people who began the study with low levels of selenium. Selenium deficiency can increase the risk of cancer, so selenium supplements would benefit people with low selenium status, but in those with adequate selenium status supplementation may actually increase the risk of cancer.[61]

Zinc

Zinc is the most abundant intracellular trace mineral. One way zinc reduces oxidation is by stimulating the synthesis of metal-binding proteins, which scavenge free radicals.[62] Zinc, along with copper, also protects cells from free-radical damage as a cofactor for a form of the antioxidant enzyme *superoxide dismutase*. Zinc is involved in the functioning of hundreds of other enzymes and proteins. These include enzymes that function in the synthesis of DNA and RNA, in carbohydrate metabolism, in acid–base balance, and in a reaction that is necessary for the absorption of folate from food. Zinc plays a role in the storage and release of insulin, the mobilization of vitamin A from the liver, and the stabilization of cell membranes. It influences hormonal regulation of cell division and is therefore needed for the growth and repair of tissues, the activity of the immune system, and the development of sex organs and bone. Together, these roles make zinc important for the regulation of protein synthesis and many other aspects of cellular metabolism. Some of the functions of zinc can be traced to its role in gene expression (**Figure 8.30**).[63]

Zinc transport from the mucosal cells of the intestine into the blood is regulated. When zinc intake is high, more zinc is held in the mucosal cells and lost in the feces when these cells die. When zinc intake is low, more dietary zinc passes into the blood for delivery to tissues.

Meeting zinc needs We consume zinc in red meat, liver, eggs, dairy products, vegetables, and seafood. Zinc from animal sources is better absorbed than is zinc from plant sources because zinc in plant foods is often bound by phytates (**Figure 8.31**).

Zinc in health and disease Severe zinc deficiency is relatively uncommon in North America, but in developing countries, it is more common and has important health and developmental consequences. Zinc deficiency interferes with growth and development, impairs immune function, and causes skin rashes and diarrhea. Mild zinc deficiency is a concern in the United States, particularly among older adults, who are at risk due to low dietary intake as well as decreases in absorption and utilization.[64] It is estimated that 40% of men and 45% of women over 50 consume less than the EAR for zinc.[52] Mild zinc deficiency results in diminished immune function and increased inflammation.[64] It has been hypothesized to play a role in the development of age-related diseases such as cardiovascular disease, type 2 diabetes, cancer, and autoimmune diseases.[65]

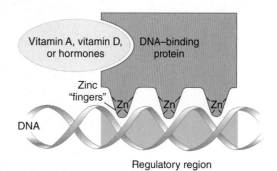

FIGURE 8.30 Zinc and gene expression One of zinc's most important roles is in gene expression. Zinc-containing DNA-binding proteins allow vitamin A, vitamin D, and a number of hormones to interact with DNA. The zinc forms "fingers" in the protein structure. When the vitamins or hormones bind to the protein, the zinc fingers bind to regulatory regions of DNA, increasing or decreasing the expression of specific genes and thus the synthesis of the proteins for which they code. Without zinc, these vitamins and hormones cannot function properly.

Ask Yourself

How could a zinc deficiency lead to a secondary vitamin A deficiency?

The risk of zinc deficiency is greater in areas where the diet is high in phytates, fiber, tannins, and oxalates, which limit zinc absorption. In the 1960s, a syndrome of growth depression and delayed sexual development was observed in Iranian and Egyptian men consuming a diet based on plant protein. The diet was not low in zinc, but it was high in grains containing phytates, which interfered with zinc absorption, thus causing the deficiency.

It is difficult to consume a toxic amount of zinc from food. However, high doses from supplements can cause toxicity symptoms. A single dose of 1 to 2 g can cause gastrointestinal irritation, vomiting, loss of appetite, diarrhea, abdominal cramps, and headaches. High intakes have been shown to decrease immune function, reduce concentrations of HDL cholesterol in the blood, and interfere with the absorption of copper. High doses of zinc can also interfere with iron absorption because iron and zinc are transported through the blood by the same protein. The converse is also true: Too much iron can limit the transport of zinc. Zinc and iron are often found together in

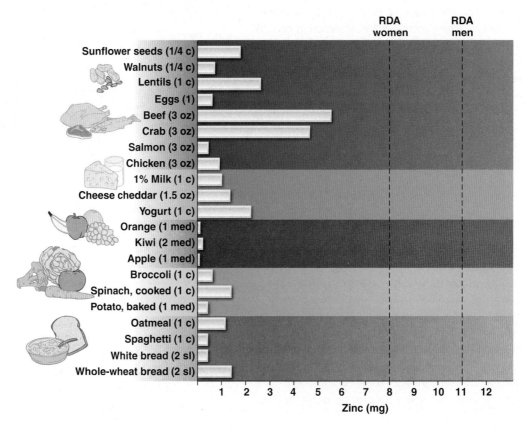

FIGURE 8.31 Food sources of zinc Meat, seafood, dairy products, legumes, and seeds are good sources of zinc. Refined grains are not good sources because zinc is lost in milling and not added back. Yeast-leavened grain products are better sources of zinc than are unleavened products because yeast leavening reduces phytate content. The dashed lines represent the RDAs for adult women and men, which are 8 and 11 mg/day, respectively.

foods, but food sources do not contain large enough amounts of either to cause imbalances.

Zinc supplements are marketed to improve immune function, enhance fertility and sexual performance, and cure the common cold. For individuals consuming adequate zinc, there is no evidence that extra zinc enhances immune function, fertility, or sexual performance. However, in individuals with a mild zinc deficiency, supplementation can have beneficial immune, antioxidant, and anti-inflammatory effects that can help in the treatment of a variety of diseases, including diarrhea, pneumonia, and tuberculosis. In older adults, improving zinc status with supplements can decrease the incidence of infections and a type of age-associated blindness; in children, it can reduce the incidence of respiratory infections and improve growth.[66,67] Zinc supplements are most often taken to treat colds. When administered within 24 hours of the onset of cold symptoms, zinc supplements have been found to reduce the duration and severity of the common cold in healthy people; the mechanism by which they work is unclear.[68] A UL has been set at 40 mg/day from all sources.[52]

Manganese

Manganese is a trace mineral that is a constituent of some enzymes and an activator of others. Enzymes that require manganese are involved in carbohydrate and cholesterol metabolism, bone formation, and the synthesis of urea. Manganese also helps prevent oxidative damage because it is a component of a form of superoxide dismutase found in the mitochondria. The recommended intake for manganese is 2.3 mg/day for adult men and 1.8 mg/day for adult women. The best dietary sources of manganese are whole grains, nuts, legumes, and leafy green vegetables.

Sulfur

Sulfur is a major mineral that is incorporated into the proteins in the body; the amino acids methionine and cysteine contain sulfur. Cysteine is also part of glutathione, a molecule essential for the action of the antioxidant enzyme glutathione peroxidase. Glutathione therefore plays an important role in protecting cells from oxidative damage and detoxifying drugs. The vitamins thiamin and biotin, which are essential for energy metabolism, also contain sulfur, and sulfur-containing ions are important in regulating acidity in the body.

We consume sulfur as a part of dietary proteins and the sulfur-containing vitamins. Sulfur is also found in some food preservatives, such as sulfur dioxide, sodium sulfite, and sodium and potassium bisulfite. There is no recommended intake for sulfur, and no deficiencies are known when protein needs are met.

Molybdenum

The trace mineral molybdenum is a cofactor that functions in the metabolism of the sulfur-containing amino acids methionine and cysteine and nitrogen-containing compounds that are present in DNA and RNA, in the production of a waste product called uric acid, and in the oxidation and detoxification of various other compounds. The recommended intake for molybdenum is 45 μg/day for adult men and women. The molybdenum content of food varies with the molybdenum content of the soil in the regions where the food is produced. The most reliable sources include milk and milk products, organ meats, breads, cereals, and legumes. Molybdenum is readily absorbed from foods; the amount in the body is regulated by varying the amount excreted in the urine and bile.

Concept Check

1. **Why** is selenium said to spare vitamin E?
2. **How** does zinc facilitate that action of vitamin A?
3. **What** do zinc, copper, and manganese have in common?
4. **Why** is zinc not absorbed as well from plant sources as from animal sources?

8.7 Minerals and Energy Metabolism

LEARNING OBJECTIVES

1. **Explain** the relationship between iodine and the thyroid hormones.
2. **Discuss** how iodine deficiency impacts the productivity of populations.
3. **List** dietary sources of iodine.
4. **Describe** the relationship between chromium and blood glucose regulation.

Many minerals are needed to produce ATP from carbohydrate, fat, and protein. For example, iron is needed to transport oxygen to cells for aerobic metabolism, and a number of iron-containing proteins help transport electrons in cellular respiration. Magnesium helps stabilize the ATP molecule and so is needed for all reactions that synthesize or require ATP. Zinc, copper, and manganese are cofactors in reactions necessary for energy metabolism, and selenium is a cofactor needed to synthesize the thyroid hormones. These have all been discussed above with other essential functions. Discussed below are the trace minerals iodine, a component of the thyroid hormones, which regulate metabolic rate, and chromium, which helps insulin get glucose into cells, where it can be used to provide energy.

Iodine

About three-fourths of the iodine in the body is found in a small gland in the neck called the **thyroid gland**. Iodine is concentrated in this gland because it is an essential component of the thyroid hormones, which are produced here. Thyroid hormones regulate metabolic rate, growth, and development, and they promote protein synthesis.

Iodine in health and disease Thyroid hormone levels are carefully regulated. When iodine is adequate, thyroid hormones are produced; if iodine is deficient, thyroid hormones cannot be synthesized, and the thyroid gland enlarges, forming a **goiter** (**Figure 8.32**). Without sufficient thyroid hormones, the metabolic rate slows, causing fatigue and weight gain. Iodine deficiency disorders have their most serious impact during growth and development. If iodine is

---

goiter An enlargement of the thyroid gland caused by a deficiency of iodine.

NUTRITION INSIGHT

FIGURE 8.32 Iodine deficiency disorders Iodine deficiency is a problem in developing countries around the world; this deficiency causes goiter and impairs cognitive development.

▶ When thyroid hormone levels drop too low, **thyroid-stimulating hormone (TSH)** is released and stimulates the thyroid gland to take up iodine and synthesize more hormones (blue arrows). If iodine is not available (purple arrows), thyroid hormones cannot be made, and the stimulation continues, causing the thyroid gland to enlarge.

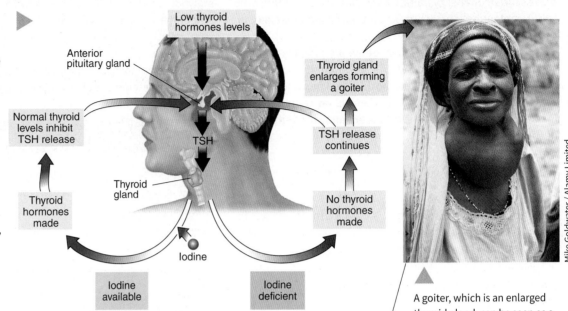

▶ A goiter, which is an enlarged thyroid gland, can be seen as a swelling in the neck. In milder cases of goiter, treatment with iodine causes the thyroid gland to return to its normal size, but it may remain enlarged in more severe cases.

▶ Iodine deficiency impairs cognitive and motor development, which impairs school performance. Iodine deficient individuals may forfeit 15 IQ points.[69]

▶ Although iodine deficiency remains a global health issue, its incidence has declined dramatically since universal salt iodization was adopted in 1993. An estimated 70% of households worldwide now have access to iodized salt.[70] Excessive iodine intake can occur if the level of iodine added to salt is too high.

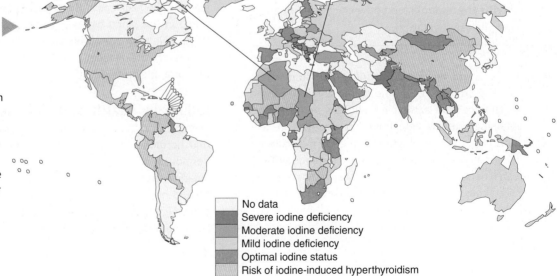

Ask Yourself

Why does the thyroid gland enlarge when iodine is deficient?

deficient during pregnancy, the risk of stillbirth and spontaneous abortion increases and the baby may be born with an underactive thyroid gland. This causes a deficiency of the thyroid hormones, which if left untreated can result in **cretinism**. Cretinism is characterized by symptoms that include impaired mental development, deaf mutism, and growth failure. In children and adolescents, iodine deficiency impairs mental function and reduces intellectual capacity. Though easily prevented, iodine deficiency is the world's most prevalent cause of brain damage.[70]

Iodine deficiency is most common in regions where the soil is low in iodine and there is little access to fish and seafood. The risk of iodine deficiency is also increased by the consumption of foods that contain **goitrogens**, substances that interfere with iodine utilization or with thyroid function. Goitrogens are found in turnips, rutabaga, cabbage, millet, and cassava. When these foods are boiled, the goitrogen content is reduced because some of these compounds leach into the cooking water. Goitrogens are primarily a problem in African countries where cassava is a dietary staple. Goitrogens are not a problem in the United States because the typical diet does not include large amounts of foods containing them.

Chronically high intakes or a sudden increase in iodine intake can also cause an enlargement of the thyroid gland. For example, in a person with a marginal intake, a large dose from supplements could cause thyroid enlargement, even at levels that would not be toxic in a healthy person. The UL for adults is 1100 µg/day of iodine from all sources.[52]

Meeting iodine needs The iodine content of food varies, depending on the soil in which plants are grown or where animals graze. When the Earth was formed, all soils were high in iodine, but today mountainous areas and river valleys have little iodine left in the soil because it has been washed out by glaciers, snow, rain, and floodwaters. The iodine washed from the soil has accumulated in the oceans. Therefore, foods from the sea, such as fish, shellfish, and seaweed, are the best sources of iodine.

The largest contributors of iodine to the North American diet include dairy products, grains, and **iodized salt** (**Figure 8.33**). We also obtain iodine from additives used in cattle feed and food dyes, from dough conditioners used in baking, and from sterilizing agents used in restaurants and on dairy farm equipment. Intakes from these sources of iodine have been declining due to reductions in the amount of iodine added to cattle feed and elimination of some iodine-containing dough conditioners.[71] Iodine intake from iodized salt used at home has also declined due to recommendations to reduce salt intake. Reducing the amount of salt consumed from processed foods does not impact iodine

FIGURE 8.33 Iodized salt Iodized salt was first introduced in Switzerland in the 1920s as a way to combat iodine deficiency. Salt was chosen because it is readily available, inexpensive, and consumed in regular amounts throughout the year. It takes only about half a teaspoon of iodized salt to provide the recommended amount of iodine. Iodized salt should not be confused with sea salt, which is a poor source of iodine because its iodine is lost in the drying process.

George Semple

intake because the salt used in processed foods is almost always noniodized.[72]

The RDA for iodine for adults is 150 µg/day. Overall, it appears that the general U.S. population has adequate iodine intake but that some pregnant women may be at risk for iodine deficiency, particularly those who do not use iodized salt or who do not consume dairy products.[72, 73] Adequate iodine is particularly important during pregnancy because it is essential for normal brain development.

Chromium

Chromium is required to maintain normal blood glucose levels. It is believed to act by enhancing the effects of insulin.[74] Insulin facilitates the entry of glucose into cells and stimulates the synthesis of proteins, lipids, and glycogen. When chromium is deficient, more insulin is required to produce the same effect. A deficiency of chromium therefore affects the body's ability to regulate blood glucose, causing diabetes-like symptoms such as elevated blood glucose levels and increased insulin levels.

---

cretinism A condition resulting from poor maternal iodine intake during pregnancy that impairs mental development and growth in the offspring.

iodized salt Table salt to which a small amount of sodium iodide or potassium iodide has been added in order to supplement the iodine content of the diet.

> **What Should I Eat?**
>
> **Trace Minerals**
>
> **Add more iron**
> - Have some meat; red meat, poultry, and fish are all good sources of heme iron.
> - Fortify your breakfast by choosing an iron-fortified cereal and adding some raisins.
> - Cook in an iron skillet to add iron to food.
> - Have some beans; they are a good source of iron.
>
> **Increase iron absorption**
> - Have orange juice with your iron-fortified cereal.
> - Don't take your calcium supplement with your iron sources.
>
> **Think zinc**
> - Scramble some eggs.
> - Beef up your zinc by having a few ounces of meat.
> - Eat whole grains but make sure they are yeast leavened.
>
> **Trace down your minerals**
> - Check to see if your salt is iodized.
> - Replace refined grains with whole grains to increase your chromium intake.
> - Have some seafood to add selenium and iodine to your diet.
>
> Use iProfile to find the iron and zinc content of 1 cup of pinto beans.

Dietary sources of chromium include liver, brewer's yeast, nuts, and whole grains. Milk, vegetables, and fruits are poor sources. Refined carbohydrates such as white breads, pasta, and white rice are also poor sources because chromium is lost in milling and is not added back in the enrichment process (see *What Should I Eat? Trace Minerals*). Chromium intake can be increased by cooking in stainless-steel cookware because chromium leaches from the steel into the food. The recommended intake for chromium is 35 µg/day for men ages 19 to 50 and 25 µg/day for women ages 19 to 50.[52]

Overt chromium deficiency is not a problem in the United States; nevertheless, chromium, in the form of chromium picolinate, is a common dietary supplement. Because chromium is needed for insulin action and insulin promotes protein synthesis, chromium picolinate is popular with athletes and dieters who take it to reduce body fat and increase muscle mass. However, studies of chromium picolinate and other chromium supplements in healthy human subjects have not found them to have beneficial effects on muscle strength, body composition, or weight loss.[75] Toxicity is always a concern with nutrient supplements, but in the case of chromium, there is little evidence of dietary toxicity in humans. The DRI committee concluded that there was insufficient data to establish a UL for chromium.

> **Concept Check**
>
> 1. **Why** does iodine deficiency cause the thyroid gland to enlarge?
> 2. **How** does iodine deficiency affect infants and growing children?
> 3. **Why** is salt fortified with iodine?
> 4. **What** effect can a chromium deficiency have on blood glucose levels?

Summary

1 Water 220

- Water, which is found intracellularly and extracellularly, accounts for about 60% of adult body weight. The amount of water in the blood is a balance between the forces of **blood pressure** and osmosis.
- Because water isn't stored in the body, intake from fluids and foods, as shown in the illustration, must replace losses in urine, feces, sweat, and evaporation. Water intake is stimulated by the sensation of **thirst**. The kidneys regulate urinary water losses.

Figure 8.2 Water balance

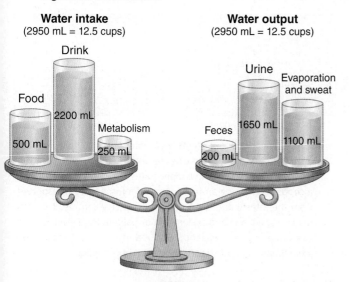

- In the body, water is a **solvent** where chemical reactions occur; it also transports nutrients and wastes, provides protection, helps regulate temperature, and participates in chemical reactions and acid–base balance.
- **Dehydration** occurs when there is too little water in the body. **Water intoxication** causes **hyponatremia**, which can result in abnormal fluid accumulation in body tissues.
- The recommended intake of water is 2.7 L/day for women and 3.7 L/day for men; needs vary depending on environmental conditions and activity level.

2 An Overview of Minerals 226

- **Major minerals** and **trace minerals** are distinguished by the amounts needed in the diet and found in the body. Both plant and animal foods are good sources of minerals.
- Mineral bioavailability is affected by the food source of the mineral, the body's need, and interactions with other minerals, vitamins, and dietary components, such as fiber, phytates, oxalates, and tannins, which are plentiful in the foods shown here.

Figure 8.9 Compounds that interfere with mineral absorption

- Minerals are needed to provide structure and to regulate biochemical reactions, often as **cofactors**. DRI recommendations have been established for the seven major minerals and for nine of the trace minerals.

3 Electrolytes: Sodium, Potassium, and Chloride 232

- The minerals sodium, potassium, and chloride are **electrolytes** important in the maintenance of fluid balance and the functioning of nerves and muscles. The kidneys are the primary regulator of electrolyte and fluid balance.
- Sodium, potassium, and chloride depletion can occur when losses are increased by heavy and persistent sweating, diarrhea, or vomiting. Diets high in sodium and low in potassium are associated with an increased risk of **hypertension**. The **DASH eating plan**—a dietary pattern moderate in sodium; high in potassium, magnesium, calcium, and fiber; and low in fat, saturated fat, and cholesterol—lowers blood pressure.
- As shown in the graph, processed foods add sodium to the diet. Potassium is highest in unprocessed foods such as fresh fruits and vegetables. Recommendations for health suggest that we increase our intake of potassium and consume less sodium.

Figure 8.15b Processing adds sodium

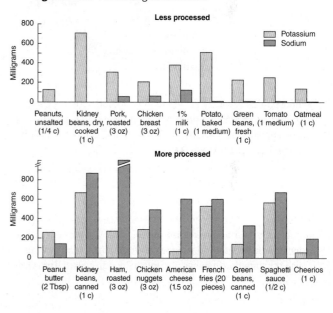

4 Minerals and Bone Health 240

- Bone is living tissue that is constantly remodeled. **Peak bone mass** occurs in young adulthood. **Age-related bone loss** occurs in adults when more bone is broken down than is made. Bone loss is accelerated in women after **menopause**. **Osteoporosis** occurs when bone mass is so low that it increases the risk of bone fractures.
- Most of the calcium in the body is found in bone, but calcium is also needed for nerve transmission, muscle contraction, blood clotting, and blood pressure regulation. Good sources of calcium in the U.S. diet include dairy products, fish consumed with bones, and leafy green vegetables. As shown in the illustration, low blood calcium causes the release of parathyroid hormone (PTH), which affects the amount of calcium excreted in the urine, absorbed from the diet, and released from bone. When blood calcium is high, **calcitonin** blocks calcium release from bone.

Figure 8.18 Raising blood calcium levels

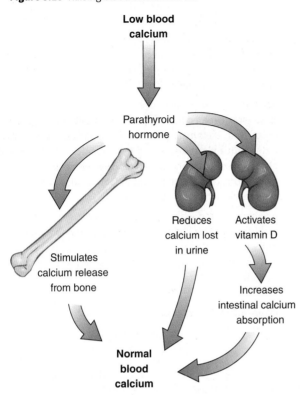

Figure 8.26 Iron deficiency

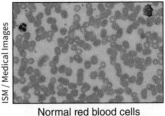
Normal red blood cells

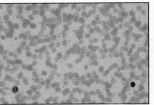
Iron deficiency anemia

- Copper functions in proteins that affect iron and lipid metabolism, synthesis of connective tissue, antioxidant capacity, and iron transport. Iron status and copper status are interrelated because a copper-containing protein is needed for iron to be transported from the intestinal cells. A copper deficiency can result in iron deficiency anemia and skeletal abnormalities. High levels of zinc can cause copper deficiency. Seafood, nuts, seeds, and whole-grain breads and cereals are good sources of copper.

6 Antioxidant Minerals 253

- Selenium is part of the antioxidant enzyme **glutathione peroxidase**. As shown here, glutathione peroxidase neutralizes reactive peroxides and can spare some of the need for vitamin E. Selenium is also needed for the synthesis of **thyroid hormones**. Dietary sources include seafood, eggs, organ meats, and plant foods grown in selenium-rich soils. Selenium deficiency causes muscle discomfort and weakness and is associated with **Keshan disease**. Low selenium intake has been linked to increased cancer risk.

Figure 8.28 Glutathione peroxidase

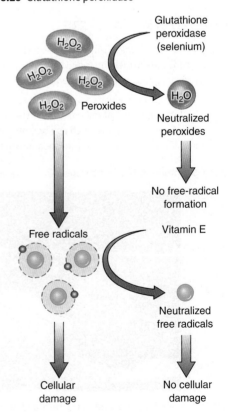

- Phosphorus is widely distributed in the food supply. It plays an important structural role in bones and teeth. Phosphorus helps prevent changes in acidity and is an essential component of phospholipids, ATP, DNA, and RNA.
- Magnesium is important for bone health and blood pressure regulation, and it is needed as a cofactor and to stabilize ATP. The best dietary sources are whole grains and green vegetables.
- Fluoride is necessary for the maintenance of bones and teeth and the prevention of **dental caries**. Dietary sources of fluoride include fluoridated drinking water, toothpaste, tea, and marine fish consumed with bones. Too much fluoride causes **fluorosis** in children.

5 Minerals and Healthy Blood 249

- Iron functions as part of **hemoglobin**, **myoglobin**, and proteins involved in energy metabolism. The amount of iron absorbed depends on the form of iron and other dietary components. **Heme iron**, found in meats, is more absorbable than nonheme iron, which is the only form found in plant foods. **Iron deficiency anemia** is characterized by small, pale red blood cells such as those in the right-hand photo. It causes fatigue and decreased work capacity. The amount of iron transported from the intestinal cells to body cells depends on the amount needed. If too much iron is absorbed, as in **hemochromatosis**, the heart and liver can be damaged, and diabetes and cancer are more likely. A single large dose of iron is toxic and can be fatal.

- Zinc, along with copper, is needed for the activity of a form of the antioxidant enzyme superoxide dismutase. Zinc's roles in gene expression and as a cofactor for many enzymes make it important for the regulation of protein synthesis and many other aspects of cellular metabolism. Good sources of zinc include red meats, eggs, dairy products, and whole grains. The amount of zinc in the body is regulated primarily by the amount absorbed and lost through the small intestine. Zinc deficiency depresses immunity. Too much zinc depresses immune function and contributes to copper and iron deficiency.
- Manganese is needed for the activity of a form of superoxide dismutase found in the mitochondria.
- Sulfur is a component of the sulfur-containing amino acids cysteine and methionine. Cysteine is part of the structure of glutathione, a molecule essential for the action of the antioxidant enzyme glutathione peroxidase.
- Molybdenum is needed for sulfur metabolism and is an enzyme cofactor.

7 Minerals and Energy Metabolism 257

- Iodine is an essential component of thyroid hormones, which regulate metabolic rate. The best sources of iodine are seafood, foods grown near the sea, and **iodized salt**. When iodine is deficient, the **thyroid gland** enlarges, forming a **goiter**, shown here. Iodine deficiency also affects growth and development. The use of iodized salt has reduced the incidence of iodine deficiency worldwide.

Figure 8.32 Iodine deficiency disorders

- Chromium affects metabolism because it is needed for normal insulin action and glucose utilization. It is found in liver, brewer's yeast, nuts, and whole grains.

Key Terms

- age-related bone loss 240
- aldosterone 234
- antidiuretic hormone (ADH) 222
- blood pressure 220
- bone remodeling 240
- bone resorption 242
- calcitonin 242
- cofactor 230
- cretinism 259
- DASH (Dietary Approaches to Stop Hypertension) eating plan 235
- dehydration 224
- dental caries 246
- diuretic 226
- electrolyte 232
- fluorosis 248
- glutathione peroxidase 253
- goiter 257
- goitrogen 259
- heme iron 249
- hemochromatosis 252
- hemoglobin 249
- hypertension 234
- hyponatremia 224
- iodized salt 259
- ion 230
- iron deficiency anemia 251
- iron overload 252
- Keshan disease 254
- major mineral 226
- menopause 240
- mineral 226
- myoglobin 249
- osteopenia 241
- osteoporosis 240
- peak bone mass 240
- postmenopausal bone loss 240
- prehypertension 235
- sodium chloride 232
- solute 220
- solvent 223
- thirst 222
- thyroid gland 257
- thyroid hormone 253
- thyroid-stimulating hormone (TSH) 258
- trace mineral 226
- water intoxication 224

What is happening in this picture?

This photo shows astronaut Dan Burbank exercising using the advanced Resistive Exercise Device (aRED) aboard the International Space Station. Resistance exercise and adequate nutrient intake are important for the maintenance of bone.

Think Critically

1. Why is scheduled exercise even more important for bone health in space than it is on Earth?
2. What nutrients other than calcium might be of particular concern for bone health in an astronaut? Why?
3. If calcium were lost from the bones of astronauts in space, what would happen to their blood calcium levels? Why is this a concern?

CHAPTER 9

Energy Balance and Weight Management

You overslept, and you're late for work. There's just enough time to shower, brush your teeth, and get dressed. Breakfast is out of the question.

You hop in your car, race to work, and realize you have five minutes to spare, just enough time to grab something to eat at that fast-food place nearby. You scan the menu, searching for something that matches your low-calorie, whole-grain, nondairy preferences. But then you order the steak, egg, and cheese croissant sandwich.

It's satisfying in the moment, savory and indulgent, and you're relieved you got something to carry you to lunchtime, when you can repent and eat the salad you left in the fridge the day before.

In America's obesogenic environment, where tempting food surrounds us and convenience is king, we as consumers make concessions in order to get the fuel we need in the time we have. At the same time, our jobs and lives have become less physically demanding.

Having easy access to food and less strenuous jobs may seem advantageous, but there are consequences. More than one in three adult Americans are obese, increasing their risks of developing diabetes, heart disease, and cancer. The economic effect is also astounding: The annual medical cost of obesity is estimated to be about $150 billion.[1]

CHAPTER OUTLINE

Body Weight and Health 266
- Why Are We Getting Fatter?
- What's Wrong with Having Too Much Body Fat?
- What Is a Healthy Weight?

Energy Balance 273
- Balancing Energy Intake and Expenditure
- Estimated Energy Requirements

What Determines Body Size and Shape? 278
- Genes Versus Environment
- Regulation of Food Intake and Body Weight
What a Scientist Sees: Leptin and Body Fat
- Why Do Some People Gain Weight More Easily?

Managing Body Weight 283
- Weight-Loss Goals and Guidelines
What Should I Eat? Weight Management
- Managing America's Weight
- Suggestions for Weight Gain
- Diets and Fad Diets
Thinking It Through: A Case Study on Food Choices and Body Weight

Medications and Surgery for Weight Loss 290
- Weight-Loss Medications and Supplements
- Weight-Loss Surgery
Debate: Is Surgery a Good Solution to Obesity?

Eating Disorders 293
- Types of Eating Disorders
- What Causes Eating Disorders?
- Anorexia Nervosa
- Bulimia Nervosa
- Binge-Eating Disorder
- Eating Disorders in Special Groups
- Preventing and Getting Treatment for Eating Disorders

9.1 Body Weight and Health

LEARNING OBJECTIVES

1. **Discuss** the growing obesity epidemic.
2. **Identify** environmental and lifestyle factors that have led to weight gain among Americans.
3. **Describe** the health consequences of excess body fat.
4. **Calculate** your BMI and determine whether it indicates you are at a healthy weight.
5. **Discuss** how the amount and location of body fat affect the health risks associated with being overweight.

In the United States today, more than 70% of adults are either **overweight** or **obese**.[2] Carrying excess body weight usually means excess body fat, which increases the risk of a host of chronic diseases. The number of people who carry excess fat has increased dramatically over the past five decades. In 1960, only 13.4% of American adults were obese. By 1990, about 23% were obese, and today that has increased to 37.9% (**Figure 9.1**).[2] Excess body weight and fat are a major public health concern that affects both men and women of all ages and all racial and ethnic groups. Obesity rates for minorities often exceed those in the general population: 48.4% of African Americans and 42.6% of Hispanic Americans are obese.[3]

FIGURE 9.1 Obesity across America This graph shows the dramatic rise in obesity that has occurred in the United States over the past few decades. It has led medical and public health officials to label the situation an epidemic. The map shows the percentage of the adult population currently classified as obese in each state.

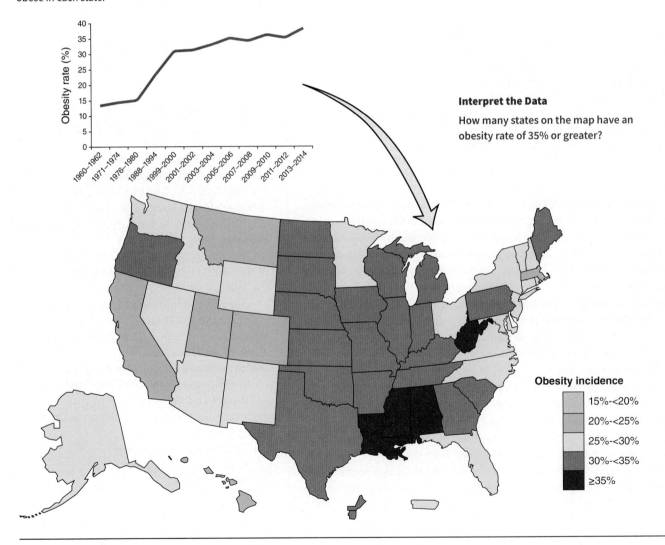

Interpret the Data
How many states on the map have an obesity rate of 35% or greater?

Obesity incidence
- 15%–<20%
- 20%–<25%
- 25%–<30%
- 30%–<35%
- ≥35%

overweight Being too heavy for one's height, usually due to an excess of body fat. Overweight is defined as having a body mass index (ratio of weight to height squared) of 25 to 29.9 kilograms/meter² (kg/m²).

obese Having excess body fat. Obesity is defined as having a body mass index (ratio of weight to height squared) of 30 kg/m² or greater.

Excess body weight is not just an American problem but also a growing concern worldwide. The worldwide incidence of obesity more than doubled between 1980 and 2014. It is such an important trend that the term *globesity* has been coined to reflect the escalation of global obesity and overweight. Around the world 39% of adults are overweight and 13% are obese.[4] Once considered a problem only in high-income countries, overweight and obesity are now on the rise in low- and middle-income countries, particularly in urban settings.

Why Are We Getting Fatter?

Americans today live in an **obesogenic environment**, one that promotes weight gain and is not conducive to weight loss. This environment has affected what we eat, how much we eat, and how much exercise we get. Simply put, more Americans are overweight than ever before because we are eating more and burning fewer calories than we did 40 or 50 years ago.[5] Food is plentiful and continuously available, and little activity is required in our daily lives.

Eating more Supermarkets, fast-food restaurants, and convenience marts make palatable, affordable food readily available to the majority of the population 24 hours a day. We are constantly bombarded with cues to eat: Advertisements entice us with tasty, inexpensive foods, and convenience stores, food courts, and vending machines tempt us with the sights and smells of fatty, sweet, high-calorie snacks. These external cues, such as the sight or smell of food, stimulate **appetite**. Because appetite is triggered by external cues, it is usually appetite, and not **hunger**, that makes us stop for an ice cream cone on a summer afternoon or give in to the smell of freshly baked chocolate chip cookies while strolling through the mall.

As a result, since 1970 the amount of energy available to us has increased by about 600 Calories per day, with the greatest increases in added fats, grains, dairy products, and sweeteners.[5] Living in an obesogenic environment promotes eating and hence weight gain and makes it difficult to lose weight. Studies examining the relationship between the food environment and BMI have found that people in communities with more fast-food or quick-service restaurants tend to have higher BMIs.[5]

In addition to having more enticing choices available to us, we consume more calories today because portion sizes have increased (**Figure 9.2**). The more food that is put in front of people, the more they eat.[5] Portion size is associated with body weight; being served and consuming larger portions is associated with weight gain, whereas small portions are associated with weight loss.[5]

Social changes over the past few decades have also contributed to the increase in the number of calories Americans

FIGURE 9.2 Portion distortion The burger and French fry portions served in fast-food restaurants today are two to five times larger than they were when fast food first appeared about 50 years ago. Soft-drink portion sizes have also escalated. A large fast-food soft drink today contains 32 ounces, providing about 300 Calories, and 20-oz bottles have replaced 12-oz cans in many vending machines.

50 years ago Today

Soft drinks
62%

French fries
57%

Cheeseburgers
24%

Percentage increase in portion size

consume. Busy schedules and an increase in the number of single-parent households and households with two working parents mean that families are often too rushed to cook meals at home. As a result, prepackaged, convenience, and fast-food meals have become mainstays. These foods are typically higher in fat and energy than foods prepared at home.

Moving less Along with America's rising energy intake, there has been a decline in the amount of energy Americans expend, both at work and at play. Fewer American adults today work in jobs that require physical labor. People drive to work rather than walk or bike, take elevators instead of stairs, use dryers rather than hang clothes outside, and cut the lawn with riding mowers rather than with push mowers. All these modern conveniences reduce the amount of energy expended

---

appetite A desire to consume specific foods that is independent of hunger.

hunger A desire to consume food that is triggered by internal physiological signals.

FIGURE 9.3 **Activity reduces the risk of obesity** Men and women who walk or ride a bicycle to work are more active and have healthier body weights and less body fat than their counterparts who drive to work.[7]

T photography / Shutterstock

daily (**Figure 9.3**). Americans are also less active during their leisure time because busy schedules and long days at work and commuting leave little time for active recreation. Instead, at the end of the day, people tend to sit in front of television sets, video games, tablets, and computers.

Inactivity is also contributing to excess body weight among children. In the 1960s, schools provided daily physical education classes, and children spent their after-school hours playing outdoors; today, they are more likely to spend their afternoons indoors watching television, texting with friends, and playing video games. As a result, they burn fewer calories, snack more, and consequently gain weight. In the United States, about 17% of children and adolescents ages 2 through 19 are obese.[6]

What's Wrong with Having Too Much Body Fat?

Having too much body fat increases the risk of developing chronic health problems, including high blood pressure, heart disease, high blood cholesterol, diabetes, gallbladder disease, liver disease, arthritis, sleep disorders, respiratory problems, menstrual irregularities, and cancers of the breast, uterus, prostate, and colon (**Figure 9.4**). Obesity

FIGURE 9.4 **Health consequences of excess body fat** Obesity-related health complications such as those highlighted here have reached epidemic proportions in the United States and around the world.

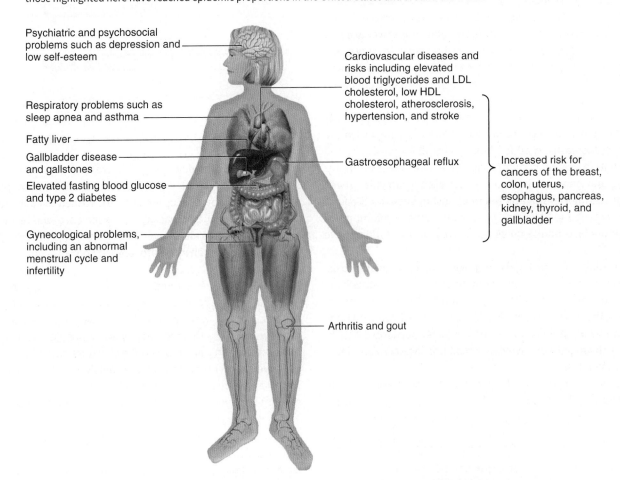

also increases the incidence and severity of infectious disease and has been linked to poor wound healing and surgical complications. The more excess body fat you have, the greater your health risks. The longer you carry excess fat, the greater the risks; individuals who gain excess weight at a young age and remain overweight throughout life face the greatest health risks.

Being overweight or obese also has psychological and social consequences. Depression is more common among obese individuals, and they experience discrimination in college admissions, in the workplace, and even on public transportation.[8] The physical health consequences of excess body fat may not manifest themselves as disease for years, but the psychological and social problems are experienced every day.

Because obesity increases health problems, it increases health-care costs. The greater the number of obese people, the higher the nation's health-care expenses and the higher the cost to society as a whole in terms of lost wages and productivity. It is estimated that obesity increases annual medical expenditures by about 7 to 10%.[9]

What Is a Healthy Weight?

A healthy weight is a weight that minimizes the health risks associated with too much or too little body fat. Your body weight is the sum of the weight of your fat and your **lean body mass**. Some body fat is essential for health; too much or too little increases health risks (see Figure 9.4). How much weight and fat is too much or too little depends on your age, gender, and lifestyle and where your fat is located.

Body mass index (BMI) The current standard for assessing the healthfulness of body weight is **body mass index (BMI)**, which is determined by dividing body weight (in kilograms) by height (in meters) squared. A healthy BMI for adults is between 18.5 and 24.9 kg/m². People with a BMI in this range have the lowest health risks. Although BMI is not actually a measure of body fat, it is recommended as a way to assess body fatness that is better than measuring weight alone.[10] You can use **Figure 9.5** to determine your BMI or calculate it according to either of these equations:

$$BMI = \text{Weight in kilograms}/(\text{Height in meters})^2$$

or

$$BMI = [\text{Weight in pounds}/(\text{Height in inches})^2] \times 703$$

BMI can be a useful tool, but other information is also needed to assess the health risks that are associated with one's BMI. For example, having a BMI in the underweight category is associated with increased risk of early death, but this does not mean that all thin people are at risk.[11] People who are naturally lean have a lower incidence of certain chronic diseases and do not face increased health risks due to their low body weight. However, low body fat due to starvation, eating disorders, or a disease process decreases energy reserves and the ability of the immune system to fight disease. Likewise, a BMI in the overweight category does not always increase risk. Someone who is overweight based on BMI but consumes a healthy diet and exercises regularly may be more fit and at lower risk for chronic diseases than someone with a BMI in the healthy range who is sedentary and eats a poor diet. Another reason that someone with a high BMI may not be at risk is if the high BMI is due to a large amount of muscle rather than too much body fat (**Figure 9.6**).

Body composition **Body composition**, which refers to the relative proportions of fat and lean tissue that make up the body, affects the risks associated with excess body weight. Having more than the recommended percentage of body fat increases health risks, whereas having more lean body mass does not (see Figure 9.6). In general, women store more body fat than men do, so the level that is healthy for women is somewhat higher than the level that is healthy for men. A healthy level of body fat for young adult females is between 21 and 32% of total weight; for young adult males, it is between 8 and 19%.[12] With aging, lean body mass decreases and body fat increases, even if body weight remains the same. Some of this change may be prevented through exercise. Body composition can be measured using a variety of techniques (**Figure 9.7**).

Location of body fat The location of body fat stores affects the risks associated with having too much fat (**Figure 9.8**). Excess **subcutaneous fat**, which is adipose tissue located under the skin, does not increase health risk as much as does excess **visceral fat**, which is adipose tissue located around the organs in the abdomen. Generally, fat in the hips and lower body is subcutaneous, whereas fat in the abdominal region is primarily visceral. An increase in visceral fat is associated with a higher incidence of heart disease, high blood cholesterol, high blood pressure, stroke, type 2 diabetes, and some types of cancer.[13]

Where extra fat is deposited is determined primarily by genetics. Age, gender, ethnicity, and lifestyle also influence where fat is stored.[13] Visceral fat storage increases with age. Excess visceral fat is more common in men than in women,

---

lean body mass Body mass attributed to nonfat body components such as bone, muscle, and internal organs; also called *fat-free mass*.

body mass index (BMI) A measure of body weight relative to height that is used to compare body size with a standard.

FIGURE 9.5 What's your BMI? To find your BMI, locate your height in the leftmost column and read across to your weight. Follow the column containing your weight up to the top line to find your BMI. For example, as highlighted, someone who is 5'6" and weighs 142 pounds would have a BMI of 23 kg/m². A BMI < 18.5 kg/m² is classified as **underweight**, a BMI ≥ 25 and < 30 kg/m² is classified as overweight, and a BMI of ≥ 30 kg/m² is classified as obese. A BMI ≥ 40 kg/m² is considered **extreme obesity** or **morbid obesity**.[10]

	UNDER-WEIGHT		NORMAL						OVERWEIGHT					OBESE										EXTREME OBESITY		
BMI	17	18	19	20	21	22	23	24	25	26	27	28	29	30	31	32	33	34	35	36	37	38	39	40	41	42
Height (feet/inches)	Body Weight (pounds)																									
4'10"	81	86	91	96	100	105	110	115	119	124	129	134	138	143	148	153	158	162	167	172	177	181	186	191	196	201
4'11"	84	89	94	99	104	109	114	119	124	128	133	138	143	148	153	158	163	168	173	178	183	188	193	198	203	208
5'0"	87	92	97	102	107	112	118	123	128	133	138	143	148	153	158	163	168	174	179	184	189	194	199	204	209	215
5'1"	90	95	100	106	111	116	122	127	132	137	143	148	153	158	164	169	174	180	185	190	195	201	206	211	217	222
5'2"	93	98	104	109	115	120	126	131	136	142	147	153	158	164	169	175	180	186	191	196	202	207	213	218	224	229
5'3"	96	102	107	113	118	124	130	135	141	146	152	158	163	169	175	180	186	191	197	203	208	214	220	225	231	237
5'4"	99	105	110	116	122	128	134	140	145	151	157	163	169	174	180	186	192	197	204	209	215	221	227	232	238	244
5'5"	102	108	114	120	126	132	138	144	150	156	162	168	174	180	186	192	198	204	210	216	222	228	234	240	246	252
5'6"	105	112	118	124	130	136	142	148	155	161	167	173	179	186	192	198	204	210	216	223	229	235	241	247	253	260
5'7"	108	115	121	127	134	140	146	153	159	166	172	178	185	191	198	204	211	217	223	230	236	242	249	255	261	268
5'8"	112	119	125	131	138	144	151	158	164	171	177	184	190	197	203	210	216	223	230	236	243	249	256	262	269	276
5'9"	115	122	128	135	142	149	155	162	169	176	182	189	196	203	209	216	223	230	236	243	250	257	263	270	277	284
5'10"	119	126	132	139	146	153	160	167	174	181	188	195	202	209	216	222	229	236	243	250	257	264	271	278	285	292
5'11"	122	129	136	143	150	157	165	172	179	186	193	200	208	215	222	229	236	243	250	257	265	272	279	286	293	301
6'0"	125	133	140	147	154	162	169	177	184	191	199	206	213	221	228	235	242	250	258	265	272	279	287	294	302	309
6'1"	129	137	144	151	159	166	174	182	189	197	204	212	219	227	235	242	250	257	265	272	280	288	295	302	310	318
6'2"	132	140	148	155	163	171	179	186	194	202	210	218	225	233	241	249	256	264	272	280	287	295	303	311	319	326
6'3"	136	144	152	160	168	176	184	192	200	208	216	224	232	240	248	256	264	272	279	287	295	303	311	319	327	335
6'4"	140	148	156	164	172	180	189	197	205	213	221	230	238	246	254	263	271	279	287	295	304	312	320	328	336	344

FIGURE 9.6 BMI and muscle mass In muscular athletes BMI does not provide an accurate estimate of health risk. Both of these individuals have a BMI of 33 kg/m², but only the man on the right has excess body fat. The high body weight of the man on the left is due to his large muscle mass. His body fat, and hence his risk of obesity-related health problems, is low.

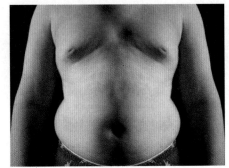

Jodi Cobb / National Geographic Creative

Joel Sartore / National Geographic Creative

FIGURE 9.7 Techniques for measuring body composition
The techniques available for assessing body composition differ in ease, availability, cost, and accuracy.

Skinfold thickness uses measurements of the thickness of the fat layer under the skin at several locations to calculate total body fat. This technique assumes that the amount of fat under the skin is representative of total body fat. It is fast, easy, and inexpensive but can be inaccurate if not performed by a trained professional.

Underwater weighing relies on the fact that lean tissue is denser than fat tissue. The difference between a person's weight on land and his or her weight underwater is used to calculate body density; the higher a person's body density, the less fat he or she has. Underwater weighing is accurate but can't be used for small children or for ill or frail adults.

Air displacement measures the amount of air displaced by the body in a closed chamber. This along with body weight is used to determine body density, which is related to body fat mass. This is accurate and easy for the subject but expensive and not readily available.

Bioelectric impedance analysis measures an electric current traveling through the body. It is based on the fact that current moves easily through lean tissue, which is high in water, but is slowed by fat, which resists current flow. Bioelectric impedance measurements are fast, easy, and painless but can be inaccurate if the amount of body water is higher or lower than typical. For example, in someone who has been sweating heavily, the estimate of percentage body fat obtained using bioelectric impedance will be artificially high.

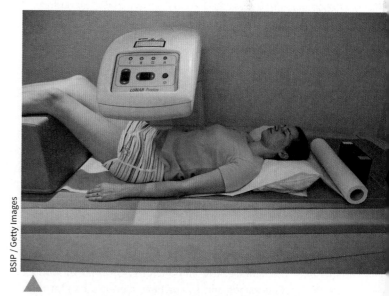

Dual-energy X-ray absorptiometry (DXA) distinguishes among various body tissues by measuring differences in levels of X-ray absorption. A single investigation can accurately determine total body mass, bone mineral mass, and body fat percentage, but the apparatus is expensive and not readily available.

NUTRITION INSIGHT | **FIGURE 9.8** **Body Fat: Apples and Pears** Excess fat in the visceral region increases health risks. Waist circumference measurements can help assess risk.

a. People who carry their excess fat around and above the waist have more visceral fat. Those who carry their extra fat below the waist, in the hips and thighs, have more subcutaneous fat. In the popular literature, these body types have been dubbed "apples" and "pears," respectively.

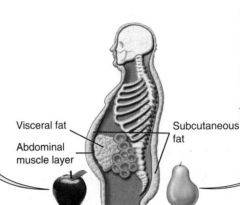

b. Waist circumference is indicative of the amount of visceral fat, the type of fat that is associated with increased health risk. Waist measurements along with BMI are used to estimate the health risk associated with excess body fat. These waist circumference "cutpoints" are not useful in patients with a BMI of 35 kg/m² or greater.

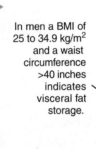

In men a BMI of 25 to 34.9 kg/m² and a waist circumference >40 inches indicates visceral fat storage.

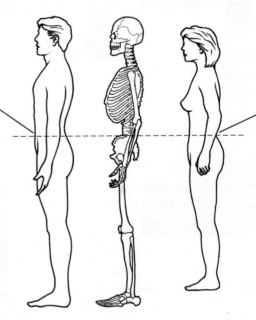

In women a BMI of 25 to 34.9 kg/m² and a waist circumference >35 inches indicates visceral fat storage.

but after menopause, the amount of visceral fat in women increases. Caucasians and Asians have more visceral fat than African Americans with similar amounts of body fat.[13] Stress, tobacco use, and alcohol consumption predispose people to visceral fat deposition, and weight loss and exercise reduce the amount of visceral fat.

Concept Check

1. **How** has the incidence of overweight and obesity changed in the United States over the past five decades?
2. **Why** does food environment affect body weight?
3. **What** are two chronic disorders that are more common in obese individuals than in lean individuals?
4. **When** is a high BMI not associated with an increased health risk?
5. **What** pattern of fat distribution increases health risks?

9.2 Energy Balance

LEARNING OBJECTIVES

1. **Explain** the principle of energy balance.
2. **Describe** the components of energy expenditure.
3. **Calculate** your EER at various levels of activity.

The high rate of overweight and obesity in virtually every population group in the United States demonstrates that many Americans have been in energy imbalance. According to the principle of **energy balance**, if you consume the same amount of energy—or calories—as you expend, your body weight will remain the same. If you consume more energy than you expend, you will gain weight, and if you expend more energy than you consume, you will lose weight. For many in the United States today, energy intake has been exceeding energy expenditure, leading to weight gain. Bringing it back into balance requires an understanding of how many calories we need and how we use energy.

Balancing Energy Intake and Expenditure

The energy needed to fuel your body comes from the food you eat and the energy stored in your body. You use this energy to stay alive, process your food, move, and grow.

Energy intake The amount of energy you consume depends on what and how much you eat and drink. The carbohydrate, fat, protein, and alcohol consumed in food and drink all contribute energy: 4, 9, 4, and 7 Calories/gram, respectively (**Figure 9.9a**). Vitamins, minerals, and water, though essential nutrients, do not provide energy. You can determine your calorie intake by looking up values in a nutrient composition table or database and using food labels (**Figure 9.9b**).

Energy expenditure The total amount of energy used by the body each day is called **total energy expenditure**. It includes the energy needed to maintain basic body functions as well as that needed to fuel physical activity and process food. In individuals who are growing or pregnant, total energy expenditure also includes the energy used to deposit new tissues. In women who are lactating, it includes the energy used to produce milk. A small amount of energy is also used to maintain body temperature in a cold environment.

For most people, about 60 to 75% of total energy expenditure is used for **basal metabolism**. Basal metabolism includes all the essential metabolic reactions and life-sustaining functions needed to keep you alive, such as breathing, circulating blood, regulating body temperature, synthesizing tissues, removing waste products, and sending nerve signals. The rate at which energy is used for these basic functions is the **basal metabolic rate (BMR)**. The energy expended for basal metabolism does *not* include the energy needed for physical activity or for the digestion of food and absorption of nutrients.

BMR increases with increasing body weight and is affected by a number of factors (**Figure 9.9c**). It is higher in people with more muscle because it takes more energy to maintain lean tissue than to maintain body fat. It is generally higher in men than in women because men have a greater amount of lean body mass. It decreases with age, partly because of the decrease in lean body mass that occurs as we get older. BMR is also lower when calorie intake is consistently below the body's needs.[14] This drop in BMR reduces the amount of energy needed to maintain body weight. It is a beneficial adaptation in someone who doesn't have access to sufficient food, but in someone who is trying to lose weight, it is frustrating because it makes weight loss more difficult.

Physical activity is the second major component of total energy expenditure. In most people, physical activity accounts for a smaller proportion of total energy expenditure than basal metabolism does—about 15 to 30% of energy requirements (**Figure 9.9d**). The energy we expend in physical activity includes both planned exercise and daily activities such as walking to work, typing, performing yard work, work-related activities, and even fidgeting. The energy expended for daily activities is called **nonexercise activity thermogenesis (NEAT)** and includes the energy expended for everything that is not sleeping, eating, or sports-like exercise. In most people NEAT accounts for the majority of the energy expended for activity, but it varies enormously, depending on an individual's occupation and daily movements.

The amount of energy used for activity depends on the size of the person, how strenuous the activity is, and the length of time it is performed. Because it takes more energy to move a heavier object, the amount of energy expended for many

---

energy balance The amount of energy consumed in the diet compared with the amount expended by the body over a given period.

basal metabolism The energy expended to maintain an awake, resting body that is not digesting food.

basal metabolic rate (BMR) The rate of energy expenditure under resting conditions. It is measured after 12 hours without food or exercise.

CHAPTER 9 Energy Balance and Weight Management

NUTRITION INSIGHT | **FIGURE 9.9** **Energy balance: Energy in and energy out** What you weigh is determined by the balance between how much energy you take in and how much energy you expend.

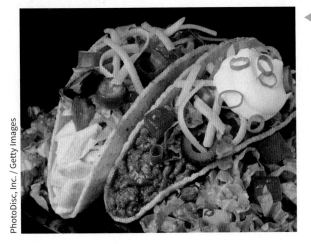

a. The energy we take in comes from the food and beverages we consume. The number of calories in a food depends on how much carbohydrate, fat, and protein it contains. Each of these tacos contains 9 g of protein, 16 g of carbohydrate, and 13 g of fat. The energy content of each is:

$$(9 \text{ g} \times 4 \text{ Cal/g protein})$$
$$+ (16 \text{ g} \times 4 \text{ Cal/g carbohydrate})$$
$$+ (13 \text{ g} \times 9 \text{ Cal/g fat})$$
$$= 217 \text{ Cal.}$$

b. Food labels provide information about the calories in foods and beverages. On the current food label (left) the serving size for a 20 oz beverage is listed as 8 fl oz, so if you drink the whole bottle, you will consume 2.5 servings, and thus 2.5 times the Calories per serving. The planned Nutrition Facts label (right) will prominently show the Calories in what is typically consumed—the whole bottle.

Ask Yourself

Which Nutrition Facts label would better help you recognize how many Calories you get from drinking a bottle of iced tea?

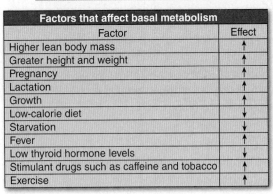

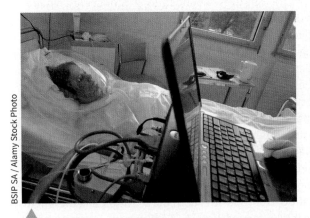

Factors that affect basal metabolism	
Factor	Effect
Higher lean body mass	↑
Greater height and weight	↑
Pregnancy	↑
Lactation	↑
Growth	↑
Low-calorie diet	↓
Starvation	↓
Fever	↑
Low thyroid hormone levels	↓
Stimulant drugs such as caffeine and tobacco	↑
Exercise	↑

c. To determine basal energy expenditure a person's BMR can be measured by collecting expired gases in a metabolic hood, as shown. Measurements of the oxygen consumed and carbon dioxide produced by aerobic metabolism can be used to estimate the amount of energy that is being expended. BMR is measured in the morning, in a warm room, before rising, and at least 12 hours after food intake or activity. **Resting metabolic rate (RMR)** is measured after 5 to 6 hours without food or exercise and yields values about 10 to 20% higher than BMR.[15] A variety of factors affect BMR and RMR (see table).

(continues)

FIGURE 9.9 *(continued)*

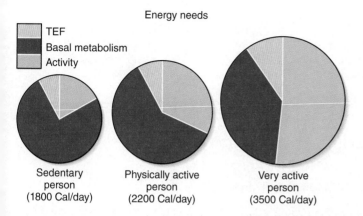

d. As shown in these pie charts, sedentary people expend little energy in activity; they must plan their intake carefully so it does not exceed energy expenditure. More active people burn more calories for activity, so they can eat more and still maintain their weight. Very active people, such as professional athletes, can actually burn more calories for activity than they do for basal metabolism.

e. The amount of energy used to process the food we eat varies with the size and composition of the meal. A bigger meal requires more energy to process so has a higher TEF. A high-fat meal yields a lower TEF than one high in carbohydrate or protein because dietary fat is used and stored more efficiently.[16]

activities increases as body weight increases. More strenuous activities, such as jogging, use more energy than do less strenuous activities, such as walking, but if you walk for an hour, you will probably burn as many calories as you would by jogging for 30 minutes (see Appendix D).

We also use energy to digest food and to absorb, metabolize, and store the nutrients from this food. The energy used for these processes is called either the **thermic effect of food (TEF)** or **diet-induced thermogenesis**. This energy expenditure causes body temperature to rise slightly for several hours after a person has eaten. The energy required for TEF is estimated to be about 10% of energy intake but can vary, depending on the amounts and types of nutrients consumed (**Figure 9.9e**).

---

thermic effect of food (TEF) or **diet-induced thermogenesis**
The energy required for the digestion of food and absorption, metabolism, and storage of nutrients.

FIGURE 9.10 Energy balance: Storing and retrieving energy When calories are consumed in excess of needs, they are stored, mostly as fat. If the excess calories are consumed as fat, they are easily stored as body fat. If the excess calories are consumed as carbohydrate, they are stored as glycogen or converted into fat. If excess calories are consumed as protein, they are converted into body fat. When calorie intake is less than needs, energy can be retrieved from stores. Glycogen and body proteins can be broken down to supply glucose, and triglycerides in adipose tissue can be broken down to supply fatty acids.

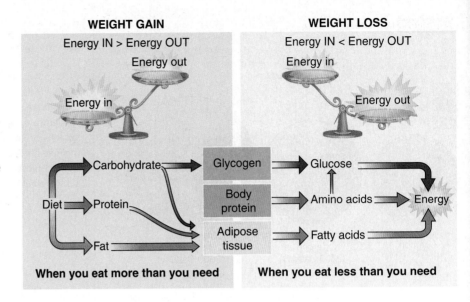

The basics of weight gain and weight loss

If you consume more energy than you expend, the excess energy is stored for later use (**Figure 9.10**). A small amount of energy is stored as glycogen in liver and muscle, but most is stored as triglycerides in **adipocytes**, which make up adipose tissue. Adipocytes contain large fat droplets (see Figure 5.12b in Chapter 5). The cells increase in size as they accumulate more fat, and they shrink as fat is removed. If intake exceeds needs over the long term, adipocytes enlarge, and the amount of body fat increases, causing weight gain. The larger the number of adipocytes, the greater the body's ability to store fat. Most adipocytes are formed during infancy and adolescence, but excessive weight gain can cause the formation of new adipocytes at any time of life.

Stored energy is used when energy intake is low, both in the short term, such as when you haven't eaten a meal for a few hours, and in the long term, such as when you are trying to lose weight. To maintain a steady supply of blood glucose, liver glycogen is broken down (see Figure 9.10). Although protein is not considered a form of stored energy (see Chapter 6), when energy needs are not met, body protein, primarily muscle protein, can be broken down to yield amino acids, which can then be used to make glucose or produce ATP. Energy for tissues that don't require glucose is provided by the breakdown of stored fat (triglycerides). Nutrients consumed in the next meal replenish these stores, but with prolonged energy restriction, fat and protein are lost, and body weight is reduced. It is estimated that an energy deficit of about 3500 Calories results in the loss of a pound of adipose tissue.

Estimated Energy Requirements

The current recommendations for energy intake in the United States are the *Estimated Energy Requirements* (*EER*; see Chapter 2), the number of calories needed for a healthy individual to maintain his or her weight.[15] They are calculated using equations that take into account gender, age, height, weight, activity level, and life stage, all of which affect calorie needs.

To calculate your EER, you must first determine your physical activity level.[15] You can do this by keeping a daily log of your activities and recording the amount of time spent at each. Use **Figure 9.11** to help translate the amount of time you spend engaged in moderate-intensity or vigorous activity into an activity level (sedentary, low active, active, or very active). Each activity level corresponds to a numerical physical activity (PA) value that can be used to calculate your EER (**Table 9.1**). For example, if you spend about an hour a day walking (a moderate-intensity activity) or about 30 minutes jogging (a vigorous activity), you are in the active category and should use the active PA value corresponding to your age and gender when calculating your EER.

Activity level has a significant effect on calorie needs. For example, a 22-year-old man who is 6 feet tall and weighs 185 pounds needs about 2770 Calories/day to maintain his weight if he is sedentary but almost 600 Calories/day more if he is at the active physical activity level.

Once you have determined your physical activity level, you can calculate your EER by entering your age, weight, height,

---

adipocyte A cell that stores fat.

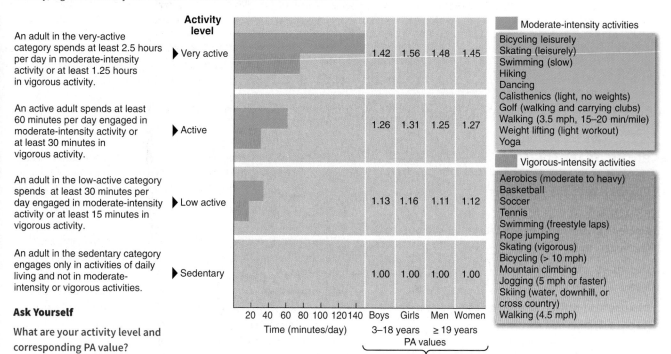

FIGURE 9.11 Physical activity level and PA value Physical activity level, which is used to calculate EER, is categorized as sedentary, low active, active, or very active. A sedentary person spends about 2.5 hours per day engaged in the activities of daily living, such as dressing, bathing, housework, and yard work. Adding activity moves the person into the low-active, active, or very-active category. Activity can be moderate or vigorous or a combination of the two; compared to moderate-intensity activity, vigorous activity will burn the same number of calories in less time.

An adult in the very-active category spends at least 2.5 hours per day in moderate-intensity activity or at least 1.25 hours in vigorous activity.

An active adult spends at least 60 minutes per day engaged in moderate-intensity activity or at least 30 minutes in vigorous activity.

An adult in the low-active category spends at least 30 minutes per day engaged in moderate-intensity activity or at least 15 minutes in vigorous activity.

An adult in the sedentary category engages only in activities of daily living and not in moderate-intensity or vigorous activities.

Ask Yourself

What are your activity level and corresponding PA value?

Each physical activity level is assigned a numerical physical activity (PA) value that can then be used in the EER calculation.

TABLE 9.1 Calculating Your EER

Find your weight in kilograms (kg) and your height in meters (m):

Weight in kilograms = weight in pounds ÷ 2.2 lb/kg
Height in meters = height in inches × 0.0254 m/in.
For example:
  160 pounds = 160 lb ÷ 2.2 lb/kg = 72.7 kg
  5'9" = 69 in. × 0.0254 m/in. = 1.75 m

Estimate the amount of physical activity you get per day and use Figure 9.11 to find the PA value for someone your age, gender, and activity level:

For example, if you are a 19-year-old male who performs 40 minutes of vigorous activity a day, you are in the active category and have a PA value of 1.25.

Choose the appropriate EER prediction equation below and calculate your EER:

For example, if you are an active 19-year-old male:
  EER = 662 − (9.53 × age in yr) + PA [(15.91 × weight in kg) + (539.6 × height in m)]
where age = 19 yr, weight = 72.7 kg, height = 1.75 m, active PA value = 1.25
  EER = 662 − (9.53 × 19) + 1.25 ([15.91 × 72.7] + [539.6 × 1.75]) = 3107 kcal/day

Life stage	EER prediction equation[a]
Boys 9–18 yr	EER = 88.5 − (61.9 × age in yr) + PA[(26.7 × weight in kg) + (903 × height in m)] + 25
Girls 9–18 yr	EER = 135.3 − (30.8 × age in yr) + PA[(10.0 × weight in kg) + (934 × height in m)] + 25
Men ≥ 19 yr	EER = 662 − (9.53 × age in yr) + PA[(15.91 × weight in kg) + (539.6 × height in m)]
Women ≥ 19 yr	EER = 354 − (6.91 × age in yr) + PA[(9.36 × weight in kg) + (726 × height in m)]

[a]These equations are appropriate for determining energy expenditure in normal-weight individuals. Equations that predict the amount of energy needed for weight maintenance in overweight and obese individuals are also available (see Appendix A).

CHAPTER 9 Energy Balance and Weight Management

and PA value (see Figure 9.11) into the appropriate EER prediction equation. Table 9.1 provides equations for normal-weight adults and children age 9 and older. Equations for other groups are in Appendix A.

Concept Check

1. **What** happens to energy stores when energy intake exceeds expenditure?
2. **Which** component of energy expenditure is easiest to modify?
3. **What** is your EER?

9.3 What Determines Body Size and Shape?

LEARNING OBJECTIVES

1. **Discuss** genetic and environmental factors that affect body weight.
2. **List** four physiological signals that determine whether you feel hungry or full.
3. **Describe** some ways in which hormones regulate body fat levels.
4. **Discuss** factors that cause some people to gain weight more easily than others.

You are probably shaped like your mother or your father. This is because much of the information that determines body size and shape is contained in the genes you inherit from your parents. Some of us inherit long, lean bodies, and others inherit huskier builds and the tendency to put on pounds (**Figure 9.12**). The genes involved in determining body shape and regulating body weight have been called **obesity genes**. Numerous obesity genes have been identified. They are responsible for the production of proteins that affect how much food you eat, how much energy you expend, and the way fat is stored in your body. Despite our growing knowledge, scientists still do not completely understand how these genes interact with each

FIGURE 9.12 Genes and body shape   The genes we inherit from our parents are important determinants of our body size and shape. The boy on the left inherited his father's long, lean body, whereas the boy on the right has his father's huskier build and will likely have a tendency to be overweight throughout his life.

other and with your environment and your lifestyle choices to determine body weight and size.[17]

Genes Versus Environment

The genes you inherit play a major role in determining your body weight. If one or both of your parents is obese, your risk of becoming obese is increased, and the risk increases with the magnitude of the obesity. By studying twins, researchers have been able to determine that about 40 to 70% of the variation in BMI between individuals can be attributed to genes.[18] The rest is determined by the environment in which you live and the lifestyle choices you make. So even if you inherit genes that predispose you to being overweight, your actual weight is determined by the balance between the genes you inherit and your lifestyle choices.

When individuals who are genetically susceptible to weight gain find themselves in an obesogenic environment where appealing food is plentiful and physical activity is easily avoided, obesity is a likely outcome but not the only possible one. If you inherit genes that predispose you to being overweight but carefully monitor your diet and exercise regularly, you can maintain a healthy weight. It is also possible for individuals with no genetic tendency toward obesity to end up overweight if they consume a high-calorie diet and get little exercise. The interplay between genetics and lifestyle is illustrated by the higher incidence of obesity in Pima Indians living in Arizona than in a genetically similar group of Pima Indians living in Mexico (**Figure 9.13**).[19]

Regulation of Food Intake and Body Weight

What we eat and how much we exercise vary from day to day, but body weight tends to stay relatively constant for long periods. The body compensates for variations in diet and exercise by adjusting energy intake and expenditure to keep weight at a particular level, or **set point**. This set point, which is believed to be determined in part by genes, explains why your weight remains fairly constant, despite the added activity of a weekend hiking trip, or why most people gain back the weight they lose when they follow a weight-loss diet.[22]

To regulate weight and fatness at a constant level, the body must be able to respond both to short-term changes in food intake and to long-term changes in the amount of stored body fat. Signals related to food intake affect hunger and **satiety** over a short period—from meal to meal—whereas signals from adipose tissue trigger the brain to adjust both food intake and energy expenditure for long-term weight regulation.

Regulating how much we eat at each meal How do you know how much to eat for breakfast or when it is time to eat lunch? The physical sensations of hunger and satiety

FIGURE 9.13 **Genes versus lifestyle** All Pima Indians carry genes that increase their risk of obesity, but differences in lifestyle between different Pima populations cause their obesity rates to diverge.

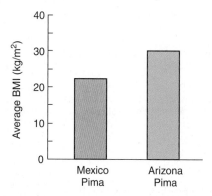

a. Genetic analysis of the Pima Indian population living in Arizona has identified a number of genes that may be responsible for this group's tendency to store excess body fat.[20] This, combined with an environment that fosters a sedentary lifestyle and consumption of high-calorie, high-fat processed foods, has resulted in a strikingly high incidence of obesity.

b. The Pima Indians of Mexico have the same genetic susceptibility to obesity as the Arizona Pimas but are farmers who work in the fields and consume the food they grow.[21] They still have higher rates of obesity than would be predicted from their diet and exercise patterns, suggesting that they possess genes that favor fat storage, but they are significantly less obese than the Arizona Pimas.

satiety The feeling of fullness and satisfaction caused by food consumption that eliminates the desire to eat.

that determine how much you eat at each meal are triggered by neural and hormonal signals from the gastrointestinal tract, levels of nutrients and hormones circulating in the blood, and messages from the brain.[23] Some signals are sent before you eat to tell you that you are hungry, some are sent while food is in the gastrointestinal tract, and some occur when nutrients are circulating in the bloodstream (**Figure 9.14**).

The gastrointestinal tract releases hormones that are involved in the control of food intake. The hormone **ghrelin**, which is secreted by the stomach, increases appetite and is believed to stimulate the desire to eat meals at usual times. Blood levels of ghrelin rise an hour or two before a meal and drop very low after a meal. Peptide YY is one of a number of hormones that causes a reduction in appetite. It is released from the gastrointestinal tract after a meal, and the amount released is proportional to the number of calories in the meal.[23]

Psychological factors can also affect hunger and satiety. Some people eat for comfort and to relieve stress. Others lose their appetite when they experience these emotions. Psychological distress can alter the mechanisms that regulate food intake.

Regulating body fat over the long term Sometimes we don't pay attention to how full we are after a meal, and we make room for dessert anyway. If this happens often enough, it can cause an increase in body weight and fatness. To return fatness to a set level, the body must be able to monitor how much fat is present. Some of this information comes from hormones.

Leptin is one hormone involved in regulating body fatness in the long term. Leptin is produced by the adipocytes, and the amount produced is proportional to the size of the adipocytes. The effect of leptin on energy intake and expenditure, and hence body weight, depends on the amount of leptin released (see *What a Scientist Sees: Leptin and Body Fat*). Unfortunately, leptin regulation, like other regulatory mechanisms, is much better at preventing weight loss than at preventing weight gain. Obese individuals generally have high levels of leptin, but these levels are not effective at reducing calorie intake and increasing energy expenditure.[24]

Despite regulatory mechanisms that act to keep our weight stable, changes in physiological, psychological, and environmental circumstances cause the level at which body

FIGURE 9.14 **Short-term regulation of food intake** Hunger is affected by sensations from the environment, signals from the gastrointestinal tract, and levels of nutrients and hormones circulating in the blood.

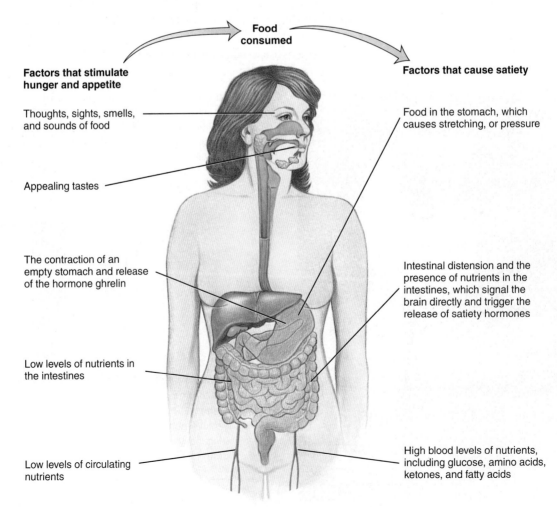

What a Scientist Sees

Leptin and Body Fat

The average person looking at the photo below sees a normal mouse and a very fat mouse. A scientist sees a clue to how body weight is regulated in mice and in humans. The hormone leptin acts in a part of the brain called the hypothalamus to help maintain body fat at a normal level. As shown in the diagram, the effect of leptin depends on how much of it is present. If weight is lost, fat is lost from adipocytes, and less leptin is released, causing an increase in food intake and a decrease in energy expenditure. If weight is gained, the adipocytes accumulate fat, and more leptin is released, triggering events that decrease food intake and increase energy expenditure.

The obese mouse inherited a defective leptin gene, so it produces no leptin. The lack of leptin continues to signal the mouse to eat more and expend less energy. The mouse on the right also inherited a defective leptin gene, but treatment with leptin injections returned its weight to normal.

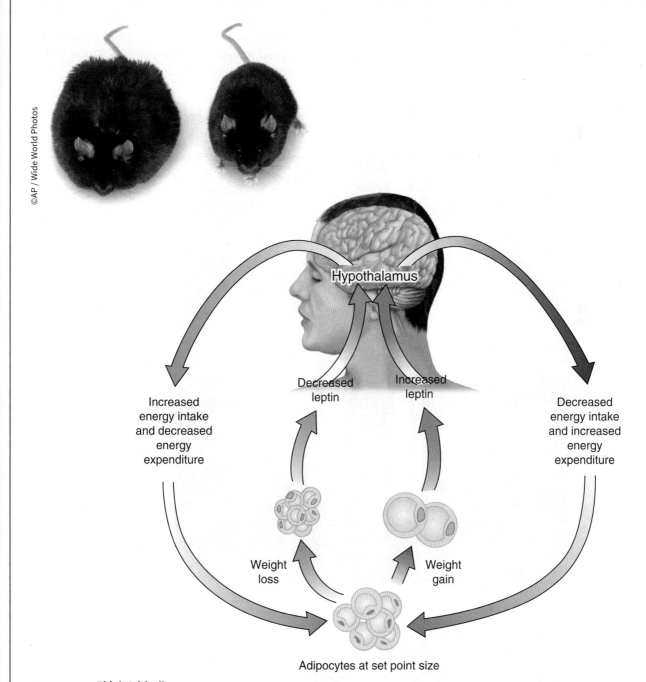

Think Critically

What might happen to someone who does not produce enough leptin? How about a person with a defect that causes overproduction of leptin?

weight is maintained to change, usually increasing it over time. This supports the hypothesis that the mechanisms that defend against weight loss are stronger than those that prevent weight gain.[25]

Why Do Some People Gain Weight More Easily?

A few cases of human obesity have been linked directly to defects in specific genes,[20] but mutations in single genes are not responsible for most human obesity.[17] Rather, variations in many genes interact with one another and affect metabolic rate, food intake, fat storage, and activity level. These in turn affect body weight, determining why some of us stay lean and others put on pounds.

Some people, such as the Pima Indians discussed earlier, may gain weight more easily because they inherited genes that make them more efficient at using energy and storing fat. Individuals with these "thrifty genes" would have had a survival advantage in times of food scarcity. However, in the United States today, where food is abundant, people who inherited these thrifty genes are more likely to be overweight or obese.

Some people may gain weight more easily because they inherit a tendency to expend less energy on activity. Even if they expend the same amount of energy engaged in planned exercise as a lean person, a heavier person's total energy expenditure may be lower because he or she expends less energy for NEAT activities, such as housework, walking between classes, fidgeting, and moving to maintain posture (**Figure 9.15**). One of the factors hypothesized to contribute to weight gain is the

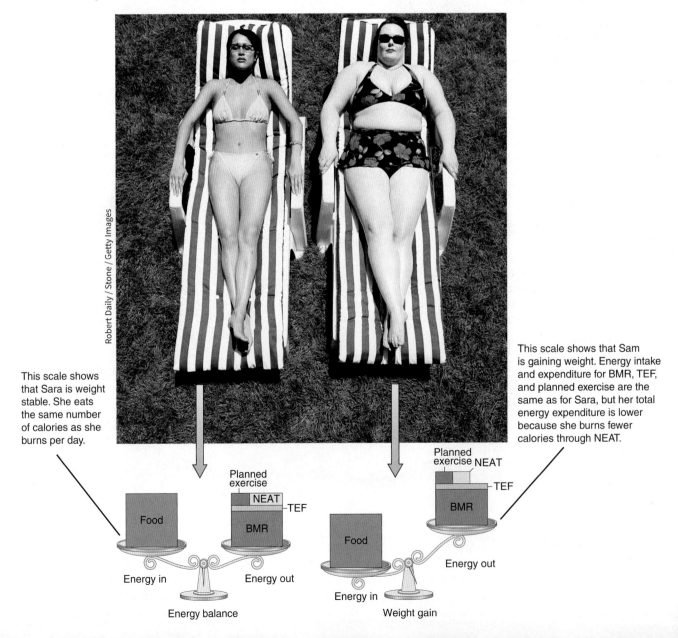

FIGURE 9.15 **Differences in NEAT can affect body weight** Sara and Sam both eat about 2000 Calories daily and do the same workouts at the gym. There are many possible reasons that Sam weighs more than Sara. One may be a difference in NEAT.

This scale shows that Sara is weight stable. She eats the same number of calories as she burns per day.

This scale shows that Sam is gaining weight. Energy intake and expenditure for BMR, TEF, and planned exercise are the same as for Sara, but her total energy expenditure is lower because she burns fewer calories through NEAT.

inability of susceptible individuals to compensate for increases in energy intake by increasing energy expenditure through NEAT. When obese and lean individuals who were overfed by 1000 Cal/day were compared, the lean people walked more and sat less than the obese study subjects, thus burning off more of the extra calories.[26]

Concept Check

1. **What** is the role of genes in regulating body weight?
2. **Why** do we feel full after eating?
3. **What** happens to leptin levels when you lose weight?
4. **How** would differences in NEAT affect body weight?

9.4 Managing Body Weight

LEARNING OBJECTIVES

1. **Evaluate** an individual's weight and risk factors to determine whether weight loss is recommended.
2. **Discuss** the recommendations for the rate and amount of weight loss.
3. **Distinguish** between a sound weight management program and a fad diet.

Managing your body weight to keep it in the healthy range requires maintaining a balance between energy intake and energy expenditure. For some people, weight management may mean avoiding weight gain as they age by making healthy food choices, controlling portion size, and maintaining an active lifestyle. For others, it may mean making major lifestyle changes in order to reduce their weight into the healthy range and keep it there.

Weight-Loss Goals and Guidelines

Just about everybody wants to lose a few pounds or more, but not everyone who is concerned about weight needs to lose weight in order to be healthy (**Figure 9.16**). The medical goal

PROCESS DIAGRAM | **FIGURE 9.16** **Weight-loss decision tree** This decision tree can be used to determine whether someone would benefit from weight loss.[10]

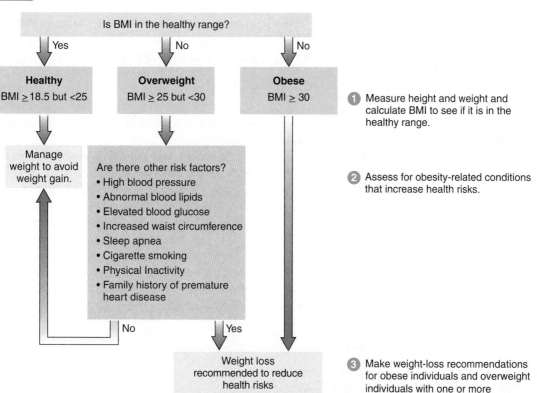

for weight loss is for overweight or obese people to reduce the health risks associated with their excess body fat. These risks and the benefits of losing weight are related to the degree of the excess, the location of the excess fat, and the presence of other diseases or risk factors that often accompany excess body fat. Weight loss is recommended for those with a BMI $\geq$ 30 kg/m² or a BMI $\geq$ 25 kg/m² who have obesity-related conditions such as elevated blood pressure, blood glucose, or blood lipids.[27]

Weight loss is achieved by reducing energy intake and increasing energy expenditure. The approach that is best for each individual depends on personal preferences and readiness and ability to make lifestyle changes. In general a weight loss of about 10% of body weight over 6 months is sufficient to reduce some of the health risks associated with excess body fat. This amount of weight loss can lower blood pressure, reduce the risk of developing type 2 diabetes, improve blood cholesterol profiles, and reduce the need for medications to control blood pressure, blood glucose, and blood lipid levels.[10, 27] After 6 months, risk factors can be reassessed to determine the need for additional weight loss.

Whatever the approach, losing weight slowly, at a rate of ½ to 2 pounds per week, helps ensure that most of what is lost is fat and not lean tissue. More severe energy restriction causes more rapid weight loss but leads to greater losses of water and protein and causes a more significant drop in BMR thus reducing energy needs. When weight loss is rapid, weight regain is more likely.[28] People who cut calories to lose weight generally achieve a maximal weight loss at about 6 months, after which they regain some or all of the weight. This weight gain often leads to repeated cycles of weight loss and regain (**Figure 9.17**).

Successful long-term weight management requires changing the eating and activity behavior patterns that led to weight gain in the first place. Compared to diet alone, combining diet, activity, and **behavior modification** results in greater short-term weight loss and better weight maintenance in the long term.[10, 29]

Decreasing energy intake

There are thousands of different weight-loss plans; whether a plan is based on solid nutrition principles or the latest fad, all weight-loss plans promote loss by restricting energy intake. Because a pound of adipose tissue provides about 3500 Calories, losing a pound of fat requires decreasing intake and/or increasing expenditure by this amount. For example, reducing energy intake by 500 Calories per day for a week would result in the loss of a pound of fat (500 Cal × 7 days = 3500 Cal). For healthy weight loss, intake must be low in energy but high in nutrients in order to provide for all the body's nutrient needs (see *What*

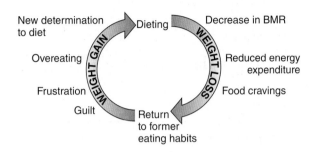

FIGURE 9.17 **Weight Cycling** Managing weight is an ongoing process that requires long-term changes in lifestyle. Often, weight loss leads to a pattern of repeated cycles of weight loss and gain, referred to as *weight cycling*, or *yo-yo dieting*. Although this cycle was once thought to increase visceral fat storage and make future weight loss more difficult, current research shows that the benefits of weight loss outweigh any risk associated with weight cycling.[30]

Should I Eat?). Even when choosing nutrient-dense foods, it is difficult to meet nutrient needs with an intake of fewer than 1200 Calories/day; therefore, dieters consuming less than this amount should take a multivitamin/multimineral supplement. Medical supervision is recommended if intake is below 800 Calories/day.

Increasing physical activity

Exercise increases energy expenditure and therefore makes weight loss easier. If food intake stays the same, adding enough exercise to expend 200 Calories five days a week will result in the loss of a pound in about three and a half weeks. Exercise also promotes muscle development, and because muscle is metabolically active tissue, increased muscle mass increases energy expenditure. In addition, physical activity improves overall fitness and relieves boredom and reduces the risk of stress-induced weight gain.[31] Weight loss is maintained better when physical activity is included. The benefits of exercise are discussed more fully in Chapter 10.

To achieve and maintain a healthy body weight, it is recommended that adults engage in the equivalent of 150 minutes of moderate-intensity aerobic activity per week.[32] Those who are overweight or obese may need to gradually increase their weekly minutes of aerobic physical activity over time and decrease calorie intake to achieve a negative energy balance and reach a healthy weight. The amount of activity needed to achieve and maintain a healthy body weight varies; some may need more than the equivalent of 300 minutes per week of moderate-intensity activity.

---

behavior modification A process that is used to gradually and permanently change habitual behaviors.

> ## What Should I Eat?
>
> ### Weight Management
>
> **Plan to balance your intake and output**
> - Know your calorie needs and monitor what you eat.
> - Weigh yourself frequently; if the number goes up, cut down your calories.
> - When you add dessert, add extra exercise.
>
> **Moderate your intake**
> - Replace your sugar-sweetened soft drink with a glass of water with lemon.
> - Put less dressing on your salad, less mayonnaise on your sandwich, and less butter on your toast.
> - Control your snacks: Rather than eat right from the bag or box, pour chips or crackers into a one-serving bowl.
> - Fill up on veggies. Add less cheese and more vegetables to your pizza or omelet to fill up with fewer calories.
>
> - When you eat out, share an entrée with a friend or take some home for lunch the next day.
> - Watch your alcohol consumption and count the calories in alcoholic beverages.
>
> **Expand your expenditure**
> - Go for a bike ride or walk after dinner.
> - Make your leisure time active. Try bowling or miniature golf instead of watching TV on Friday nights.
> - Use your lunch break to burn a few extra calories with a walk.
> - Take up a sport like tennis; you don't have to be good to get plenty of exercise.
> - Boost your NEAT calories; small changes such as getting off the bus one stop early or parking at the back of the lot all add up.
>
> Use iProfile to calculate the number of calories you consume as between-meal snacks.

Modifying behavior To successfully lose weight and keep it off, the food consumption and exercise patterns that led to weight gain need to be identified and replaced with new ones that promote and maintain weight loss (**Figure 9.18**). Successful behavior modification includes regular self-monitoring of food intake, physical activity, and weight. Behavior modification can help you to establish patterns of food intake and exercise that you can maintain throughout your life without gaining weight (see *Thinking it Through*).

Managing America's Weight

To become a thinner nation, we need strategies that can help all Americans improve their food choices, reduce portion sizes, and increase their physical activity.[33] Although successful weight management ultimately depends on an individual's choices, as suggested by the 2015–2020 Dietary Guidelines, all segments of society have a role in managing our weight.[5] Food manufacturers and restaurants can help us cut calories by offering healthier foods and by packaging or serving foods in smaller portions. Communities can help increase activity by providing parks, bike paths, and other recreational facilities for people of all ages. Businesses and schools can contribute by offering more opportunities for physical activity at the workplace and during the school day. These changes would make our environment less obesogenic and would make it easier for people to manage their weight. Small changes in eating and exercise habits, if they are consistent, can help slow or stop the rising incidence of obesity in the population. It has been estimated that a shift in energy balance by as little as 100 to 200 Calories/day, the equivalent of skipping one sugar-sweetened beverage, cutting out a scoop of ice cream, or walking an extra mile a day, would prevent further weight gain in 90% of the population.[5, 29]

Suggestions for Weight Gain

As difficult as weight loss is for some people, weight gain can be equally elusive for underweight individuals. The first step toward weight gain is to rule out medical reasons for low body weight. This is particularly important when weight loss occurs unexpectedly. If low body weight is due to low energy intake or high energy expenditure, gradually increasing consumption of energy-dense foods is suggested. Energy intake can be increased by eating meals more frequently; adding healthy high-calorie snacks, such as nuts, peanut butter, or milkshakes, between meals; and replacing low-calorie drinks such as water and diet beverages with 100% fruit juices and milk.

To encourage a gain in muscle rather than fat, muscle-strengthening exercise should be a component of any weight-gain program. This approach requires extra calories to fuel the activity needed to build muscles. These weight-gain recommendations apply to individuals who are naturally thin and have trouble gaining weight on the recommended energy intake. This dietary approach may not promote weight gain for those who limit intake because of an eating disorder.

Diets and Fad Diets

Want to lose 10 lb in just five days? What dieter wouldn't? People who are desperate to lose weight fall prey to all sorts of diets that promise quick fixes. They willingly eat a single food for days at a time, select foods on the basis of special fat-burning

PROCESS DIAGRAM **FIGURE 9.18** **ABCs of behavior modification** Behavior modification is based on the theory that behaviors involve three factors: antecedents or cues that lead to a behavior, the behavior itself, and the consequences of the behavior. These are referred to as the ABCs of behavior modification. The example shown here illustrates how behavior modification can help reduce food intake to manage weight.

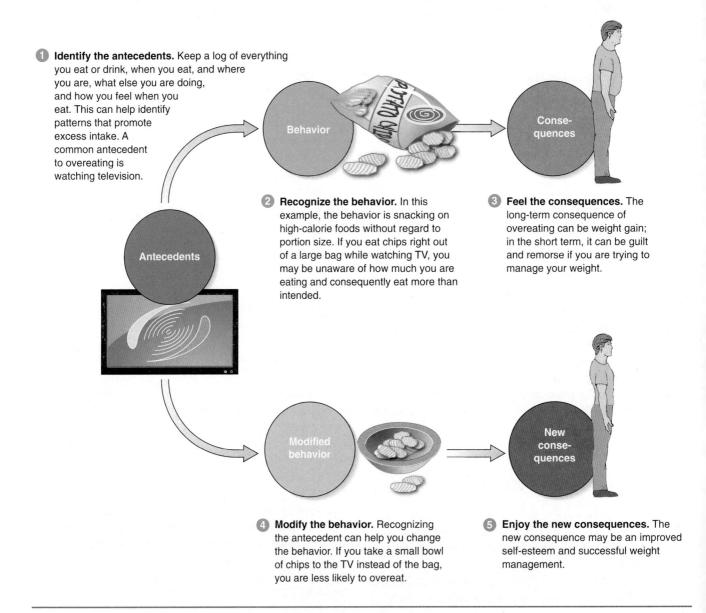

1. **Identify the antecedents.** Keep a log of everything you eat or drink, when you eat, and where you are, what else you are doing, and how you feel when you eat. This can help identify patterns that promote excess intake. A common antecedent to overeating is watching television.

2. **Recognize the behavior.** In this example, the behavior is snacking on high-calorie foods without regard to portion size. If you eat chips right out of a large bag while watching TV, you may be unaware of how much you are eating and consequently eat more than intended.

3. **Feel the consequences.** The long-term consequence of overeating can be weight gain; in the short term, it can be guilt and remorse if you are trying to manage your weight.

4. **Modify the behavior.** Recognizing the antecedent can help you change the behavior. If you take a small bowl of chips to the TV instead of the bag, you are less likely to overeat.

5. **Enjoy the new consequences.** The new consequence may be an improved self-esteem and successful weight management.

qualities, and consume odd combinations at specific times of the day. Most diets, no matter how outlandish, will promote weight loss because they reduce energy intake. Even diets that focus on modifying fat or carbohydrate intake or promise to allow unlimited amounts of certain foods work because intake is reduced. The true test of the effectiveness of a weight-loss plan is whether it promotes weight loss that can be maintained over the long term.

People often don't recognize that once you lose weight, you need to eat less to stay at the lower weight. For example, an inactive 30-year-old, 5′4″ woman who weighs 170 lb needs to consume about 2100 Calories/day to maintain her weight. If she loses 40 lb but does not change her activity level, she will need to consume only about 1880 Calories to maintain her healthier reduced weight. If, once the weight is lost, she resumes her pre-weight-loss dietary pattern, eating 2100 Calories/day, she will regain all the lost weight.

Effective weight-management programs promote healthy weight-loss diets and encourage changes in the lifestyle patterns that led to weight gain. When selecting a

Thinking It Through

A Case Study on Food Choices and Body Weight

At a visit to her doctor after she finished her freshman year in college, Leticia learned that she had gained 15 pounds. She can't figure out why. She didn't think she was eating any differently than she did at home, and she exercises for an hour three evenings a week. She is 5 feet 4 inches tall and now weighs 155 lb.

What is her BMI? Is it in the healthy range?
Your answer:

Leticia's blood pressure and all of the lab values are in the normal range. She doesn't smoke, and no one in her family has had heart disease.

Does she need to lose weight to improve her health? (Hint: Use the decision tree in Figure 9.16.)
Your answer:

Leticia decides to keep a diary of what and when she eats. The diary shows that she eats a lot at night while studying. She tends to keep an open bag of chips or a package of cookies and bottle of cola at her desk. She estimates that her evening snacking adds about 500 Calories to her daily intake.

Suggest some behavioral strategies to reduce the number of calories she consumes while studying.
Your answer:

Use iProfile to suggest three snack options with 200 Calories or less that Leticia could keep in her dorm room refrigerator to replace the high-calorie ones she has been eating.
Your answer:

Leticia's food diary shows that her highest-calorie meals are the ones she eats when she leaves campus to have lunch with friends at fast-food restaurants. Meal A, shown below, is her typical lunch, and Meal B is the lower-calorie lunch that she plans to replace it with.

How can Meal B have fewer calories even though it looks like more food?
Your answer:

Why might Meal B satisfy hunger just as well as or better than Meal A?
Your answer:

(Check your answers in Appendix L.)

program, look for one that is based on sound nutrition and exercise principles, suits your individual preferences in terms of food choices as well as time and costs, and promotes long-term lifestyle changes. Quick fixes are tempting, but if the program's approach is not one that can be followed for a lifetime, it is unlikely to promote successful weight management (**Table 9.2**).

Some of the most common methods for reducing calorie intake include using food guides such as MyPlate for diet planning, eating a moderate- to high-protein diet, choosing prepackaged meals or liquid meals, and reducing the fat or carbohydrate content of the diet (**Figure 9.19**). All these methods cause weight loss by limiting, in one way or another, the number of calories consumed.

TABLE 9.2 Distinguishing between healthy diets and fad diets

A healthy diet...	A fad diet...
Promotes a healthy dietary pattern that meets nutrient needs, includes a variety of foods, suits food preferences, and can be maintained throughout life.	Limits food selections to a few food groups or promotes rituals such as eating only specific food combinations. As a result, it may be limited in certain nutrients and in variety.
Promotes a reasonable weight loss of 1/2 to 2 lb per week.	Promotes rapid weight loss of much more than 2 lb/week
Promotes or includes physical activity.	Suggests weight loss without the need to exercise.
Is flexible enough to be followed when eating out and includes foods that are easily obtained.	May require a rigid menu or avoidance of certain foods or may include "magic" foods that promise to burn fat or speed up metabolism.
Is based on whole foods, not costly supplements or other products.	May require the purchase of special foods, weight-loss patches, expensive supplements, creams, or other products.
Promotes a change in behavior. Teaches new eating habits. Provides social support.	Recommends an eating pattern that is difficult to follow in the long term.
Is based on sound scientific principles and may include monitoring by qualified health professionals.	Makes outlandish and unscientific claims, does not support claims that it is clinically tested or scientifically proven, claims that it is new and improved or based on some new scientific discovery, or relies on testimonials from celebrities or connects the diet to trendy places such as Beverly Hills.

FIGURE 9.19 Common dieting methods Weight-loss diets take a variety of forms, but ultimately all promote weight loss by reducing energy intake.

a. Healthy food patterns
Some diet plans such as the MyPlate My Weight Manager shown here and the DASH eating plan limit energy intake by recommending a healthy food pattern along with portion control. These plans are varied and are likely to meet nutrient needs. They teach meal-planning skills that are easy to apply away from home and can be used over the long term.

b. Moderate- to high-protein diets
Many popular weight-loss diets promote a protein intake of 25% or more of calories. These high-protein diets rely on meat and eggs as the primary source of calories. The high protein content promotes satiety and preserves muscle mass, but this approach has not been shown to be more effective for weight loss than other calorie-restricted diets.[10,34] These diets are difficult to stick with in the long term, and low intakes of fruits, vegetables, and grains reduce nutritional adequacy.

Images of Africa / Getty Images

High protein menu

Breakfast
8 oz. nonfat yogurt
Ham slices
Vanilla oatbran porridge

Snack
4 oz. nonfat cottage cheese

Lunch
Hard-boiled egg with herb mayonnaise
5 oz. steak
8 oz. nonfat yogurt

Snack
4 oz. nonfat yogurt

Dinner
1 lb. shrimp sauteed in herbs
8 oz. tandoori chicken cutlets

(continues)

FIGURE 9.19 *(continued)*

c. Liquid and prepackaged meals
Liquid formulas and prepackaged meals make portion control easy, but they are not practical when traveling or eating out, and they do not teach the food-selection skills needed to make a long-term lifestyle change. Programs that rely exclusively on liquid formulas are not recommended without medical supervision.

d. Reduced fat diets
Low-fat diets typically reduce calorie intake because fat is high in calories. Low-fat diets can include large quantities of fresh fruits and vegetables, which are low in fat and calories and high in nutrients. But just because a food is low in fat does not mean it is low in calories. In the 1990s, low-fat cookies, crackers, and cakes flooded the market. These foods were low in fat but not in calories. When eaten in excess, they contributed to weight gain.

e. Carbohydrate-restricted diets
Carbohydrate-restricted diets limit total carbohydrate or allow only low glycemic index carbs, such as legumes and vegetables. Low-carb diets promote weight loss because people eat less. This may be due to metabolic changes that suppress appetite, but intake is also reduced because of the monotony of the food choices.[35] Low-carb diets that restrict choices from entire food groups are not nutritionally adequate and are rarely adhered to in the long term.

Think Critically

Which of these diet plans will promote weight loss? Which are best for long-term weight management? Why?

Concept Check

1. **Why** is weight loss not recommended for everyone with a BMI above the healthy range?
2. **What** is the recommended rate of weight loss?
3. **What** are some characteristics of a good weight-loss program?

9.5 Medications and Surgery for Weight Loss

LEARNING OBJECTIVES

1. **Explain** how medications and supplements can promote weight loss.
2. **Describe** the criteria for recommending weight loss medications and surgery.
3. **Describe** how bariatric surgery promotes weight loss.

For most people weight management means managing how much they eat and how much they move. But for some people, comprehensive lifestyle modifications are not enough to bring weight into the healthy range and keep it there. In these cases, medications can help with weight management; some of them require a prescription, and others are medications or dietary supplements available over-the-counter. Surgery is another option for reducing body weight.[29]

Weight-Loss Medications and Supplements

Prescription medications approved by the FDA for the treatment of obesity include those that reduce appetite and increase the sense of fullness (for example, phentermine and lorcaserin) and those that decrease the absorption of fat in the intestine (for example, orlistat). In general these are recommended for individuals with a BMI ≥ 30 kg/m² or a BMI ≥ 27 kg/m² with one or more obesity-related risk factors. They are prescribed as part of a weight management program that includes diet and exercise. As with any other treatment, the risks should be weighed against the benefits before starting a medication for weight loss. One of the major disadvantages of drug treatment is that even if the drug promotes weight loss, the weight is usually regained when the drug is discontinued.

Like prescription drugs, over-the-counter weight-loss medications are regulated by the FDA and must adhere to strict guidelines regarding the dose per pill and the effectiveness of the ingredients. The FDA has approved only a limited number of substances for sale as nonprescription weight-loss medications. One of these is a nonprescription version of orlistat (**Figure 9.20**). As with prescription drugs, any weight loss that occurs with over-the-counter weight-loss medications is usually regained when the product is no longer consumed.

In addition to weight-loss medications, hundreds of dietary supplements claim to promote weight loss. As with other dietary supplements, weight-loss supplements are not strictly regulated by the FDA, so their safety and effectiveness may not have been carefully tested. Some products claiming to be weight-loss supplements have been found to contain hidden prescription drugs or compounds that have not been adequately studied in humans.[36] It cannot be assumed that a product is safe simply because it is labeled a dietary supplement or as "herbal" or "all natural."

Weight-loss supplements that contain soluble fiber promise to reduce the amount you eat by filling up your stomach.

FIGURE 9.20 **Alli: Blocking fat absorption** Alli is an over-the-counter version of the prescription drug orlistat. It acts by disabling the enzyme lipase, which breaks triglycerides into fatty acids and monoglycerides. When Alli is present, the triglycerides are not broken down, so they cannot be absorbed. The undigested fat is eliminated in the feces. This cuts the number of calories from dietary fat that are absorbed, but it is not without side effects. Fat in the colon may cause gas, diarrhea, and more frequent and hard-to-control bowel movements.

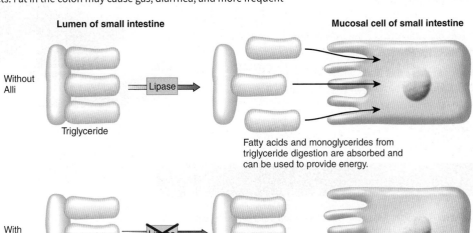

Think Critically

Will Alli be an effective weight-loss aid for someone who eats a low-fat diet? Why or why not?

Although they are safe, their use may contribute to only small amounts of weight loss.[37] Hydroxycitric acid, conjugated linoleic acid, and chromium picolinate are weight-loss supplements that promise to enhance fat loss by altering metabolism to prevent the synthesis and deposition of fat. None of these supplements has been shown to be effective for promoting weight loss in humans.[37] Supplements that boost energy expenditure, often called "fat burners," can be effective but have serious and potentially life-threatening side effects. One of the most popular and controversial herbal fat burners is ephedra, a stimulant that increases blood pressure and heart rate and constricts blood vessels. Due to safety concerns, the FDA banned it in 2004. After the ban was instituted, supplement manufacturers began substituting other herbal products, such as bitter orange, that contain similar stimulants and therefore may have similar side effects.[37] Fat burners also typically contain guarana, an herbal source of caffeine. Green tea extract is another popular supplement used to boost metabolism and aid weight loss. It is also a source of caffeine as well as phytochemicals called catechins. Studies have shown it to have only a small effect on weight loss in certain populations.

Consuming green tea is safe, but green tea extract has been associated with liver damage.[37]

Taking some dietary supplements results in weight loss through water loss—either because these supplements are diuretics or because they cause diarrhea. Water loss decreases body weight but does not cause a decrease in body fat. Herbal laxatives in weight-loss teas and supplements include senna, aloe, buckthorn, rhubarb root, cascara, and castor oil. Overuse of these substances can have serious side effects, including diarrhea, electrolyte imbalances, and liver and kidney toxicity.[38]

Weight-Loss Surgery

For individuals who have a BMI ≥ 40 kg/m² or a BMI ≥ 35 kg/m² with one or more obesity-related risk factors and who have failed to lose sufficient weight to improve their health using lifestyle interventions, there are a number of surgical weight loss options. **Adjustable gastric banding** and **gastric sleeve surgery** both restrict the amount of food that can be consumed (**Figure 9.21**). Other surgical procedures, such

FIGURE 9.21 **Weight-loss surgery** Gastric banding, gastric sleeve surgery, and gastric bypass are three common surgical approaches to treating obesity.

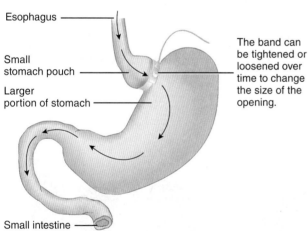

a. Gastric banding involves surgically placing an adjustable band around the upper part of the stomach, creating a small pouch. The narrow opening between the stomach pouch and the rest of the stomach slows the rate at which food leaves the pouch. This promotes weight loss by reducing the amount of food that can be consumed at one time and slowing digestion. Gastric banding entails less surgical risk and is more easily reversible than other types of weight-loss surgery.

b. Gastric sleeve surgery involves removing a large part of the stomach. The procedure is not reversible, and since it is relatively new, the long-term risks and effectiveness are not known.

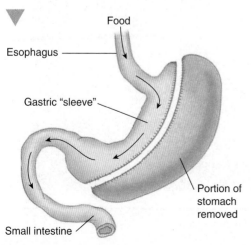

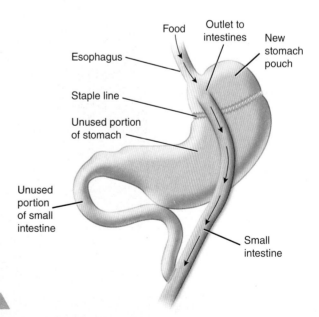

c. The type of gastric bypass surgery shown here involves bypassing part of the stomach and small intestine by connecting the intestine to the upper portion of the stomach. Food intake is reduced because the stomach is smaller, and absorption is reduced because the small intestine is shortened. Gastric bypass entails short-term surgical risks and a long-term risk of nutrient deficiencies, particularly of vitamin B_{12}, folate, calcium, and iron, because absorption of these nutrients is reduced.

Debate

Is Surgery a Good Solution to Obesity?

The Issue

The medical, social, and financial costs of obesity are well known. So is the fact that eating less and moving more will promote weight loss. However, these conventional methods do not always work. When they fail, is surgery a good weight-loss option?

Obesity surgery, known as bariatric surgery, is becoming increasingly common; over the past decade, there has been a 15-fold increase in the frequency of these procedures.[39] They are effective, and for some they can be lifesaving. When compared with conventional treatment, weight loss was greater at 2 years and at 10 years in those who had bariatric surgery. Weight loss surgery reduced the incidence of and improved the control of type 2 diabetes, hypertension, high blood cholesterol, gastrointestinal reflux disease, and sleep apnea. In addition, quality of life improved in those who had surgery, and medication costs and overall mortality decreased.[39-41]

About 1 in 20 Americans is severely obese and meets the criteria for treatment with bariatric surgery, but only 0.6% of eligible patients have the surgery each year. This is because bariatric surgery is not without costs and complications.[42] Bariatric surgery costs $20,000 to $30,000 and upwards of $65,000 if there are complications.[43] Complications are common; although fewer than 1% of patients die during or immediately after surgery, 17% experience adverse events.[44] The major complications, some of which require follow-up surgery, include ulcers, blockage of the opening from the stomach, leakage at the connection between the stomach and intestine, pneumonia, and blood clots. The incidence of gallstones and subsequent gallbladder removal is high because of the rapid weight loss that occurs following surgery. The complication rate is high not only because anesthesia and surgery are risky in obese individuals but also because only a limited number of surgeons and surgical centers have experience with these procedures.

In addition to surgical complications, long-term issues affect digestion and lifestyle. Procedures that reduce the length of the small intestine reduce nutrient absorption and can lead to malnutrition if dietary supplements are not consumed for life. Some people feel very sleepy after eating, and many experience chronic diarrhea, gas, foul-smelling stools, and other changes in bowel habits. Anyone who has this surgery must be willing to commit to changes in the types and amounts of food they consume. Eating too much or eating the wrong foods may induce vomiting or cause food to dump rapidly into the small intestine, leading to symptoms such as nausea, rapid pulse, and diarrhea.

So is bariatric surgery a good option? It is expensive and risky, but results are impressive in the first decade after the procedures. For those who have high blood pressure, diabetes, high blood cholesterol, and arthritis, all of which are helped by weight loss, the surgery can enhance quality of life and even be lifesaving. But we still do not fully understand what the consequences of rearranging the GI anatomy will be in 20 or 30 years.

Think Critically

In patients who have undergone bariatric surgery, some weight gain is common after 2 to 5 years. Why might this weight gain occur?

Significant weight loss is usually achieved 18 to 24 months after weight-loss surgery. NBC's *Today Show* weather anchor Al Roker lost 100 lb after undergoing gastric bypass surgery and has maintained his leaner physique.

as **gastric bypass**, limit both the amount of food that can be consumed and the amounts of nutrients that can be absorbed. These surgical approaches are recommended only in cases in which the health risks of obesity are greater than the health risks of the surgery (see *Debate: Is Surgery a Good Solution to Obesity?*).[10]

Another popular surgical procedure for reducing body fat is **liposuction**. This procedure involves inserting a large hollow needle under the skin into a fat deposit and literally vacuuming out the fat. Liposuction is considered a cosmetic procedure. It can reduce the amount of fat in a specific location, but it does not significantly reduce overall body weight.

Concept Check

1. **What** are two ways in which weight loss medications promote weight loss?
2. **When** is weight loss surgery recommended?
3. **How** does gastric bypass cause weight loss?

9.6 Eating Disorders

LEARNING OBJECTIVES

1. **Distinguish** between anorexia, bulimia, and binge-eating disorder.
2. **Describe** demographic and psychological factors associated with increased risk of developing an eating disorder.
3. **Discuss** how body ideal and the media affect the incidence of eating disorders.
4. **Explain** what is meant by the binge/purge cycle.

What and how much people eat vary, depending on social occasions, emotions, time limitations, hunger, and the availability of food, but generally people eat when they are hungry, choose foods that they enjoy, and stop eating when they are satisfied. Abnormal or disordered eating occurs when a person is overly concerned with food, eating, and body size and shape. When the emotional aspects of food and eating overpower the role of food as nourishment, an **eating disorder** may develop. Eating disorders affect the physical and nutritional health and psychosocial functioning of up to 11 million Americans.[45] If untreated, eating disorders can be fatal; they have the highest mortality rate of any mental illness.[46]

Types of Eating Disorders

Mental health guidelines define a number of eating disorders, including **anorexia nervosa**, **bulimia nervosa**, and **binge-eating disorder** (Table 9.3).[47]

What Causes Eating Disorders?

We do not completely understand what causes eating disorders, but we do know that genetic, psychological, and sociocultural factors contribute to their development (**Figure 9.22**). Eating disorders can be triggered by traumatic events such as sexual abuse or by day-to-day occurrences such as teasing or judgmental comments from a friend or a coach. Eating disorders occur in people of all ages, races, and socioeconomic backgrounds, but some groups are at greater risk than others. Women are more likely than men to develop eating disorders. Professional dancers, models, and others who are concerned about maintaining a low body weight are most likely to develop eating disorders. Eating disorders commonly begin in adolescence, when physical, psychological, and social development is occurring rapidly.

Psychological issues People with eating disorders often have low self-esteem. *Self-esteem* refers to the judgments people make and maintain about themselves—a general attitude of approval or disapproval about worth and capability. A poor **body image** contributes to low self-esteem. Eating disorders are characterized not only by dissatisfaction with one's body but also with distorted body image. Someone with a distorted body image is unable to judge the size of his or her own body. Thus, even if a young woman achieves a body weight comparable to that of a fashion model, she may continue to see herself as fat and strive to lose more weight.

People with eating disorders are often perfectionists who set very high standards for themselves and strive to be in control of their bodies and their lives. Despite their many

---

eating disorder A psychological illness characterized by specific abnormal eating behaviors, often intended to control weight.

anorexia nervosa An eating disorder characterized by self-starvation, a distorted body image, abnormally low body weight, and a pathological fear of becoming fat.

bulimia nervosa An eating disorder characterized by the consumption of a large amount of food at one time (binge eating)

followed by purging behaviors such as self-induced vomiting to prevent weight gain.

binge-eating disorder An eating disorder characterized by recurrent episodes of binge eating accompanied by a loss of control over eating in the absence of purging behavior.

body image The way a person perceives and imagines his or her body.

TABLE 9.3 Distinguishing among eating disorders

Characteristic	Anorexia nervosa	Bulimia nervosa	Binge-eating disorder
Body weight	Below normal	Normal or slightly overweight	Above normal
Binge eating	Possibly	Yes, at least once a week for three months	Yes, at least once a week for three or more months
Purging	Possibly	Yes, at least once a week for three months	No
Restricts food intake	Yes	Yes	Yes
Body image	Dissatisfaction with body and distorted image of body size	Dissatisfaction with body and distorted image of body size	Dissatisfaction with body
Fear of being fat	Yes	Yes	Not excessive
Self-esteem	Low	Low	Low
Typical age of onset	Adolescence/young adulthood	Adolescence/young adulthood	Adolescence and any age during adulthood

FIGURE 9.22 Factors contributing to eating disorders Eating disorders are caused by a combination of genetic, psychological, and sociocultural factors. Although these disorders are not necessarily passed from parent to child, the genes that a person inherits contribute to psychological and biological characteristics that can predispose him or her to developing an eating disorder. When placed in a conducive sociocultural environment, an individual who carries such genes will be more likely than others to develop an eating disorder.

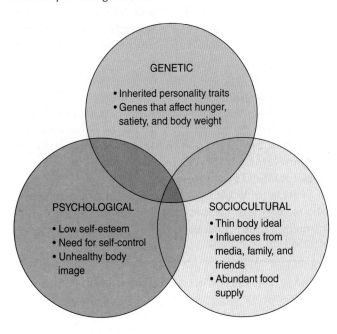

achievements, they feel inadequate, defective, and worthless. They may use their relationship with food to gain control over their lives and boost their self-esteem. Controlling their food intake and weight demonstrates their ability to control other aspects of their lives, and they can associate this control with success.

Sociocultural issues What is viewed as an "ideal" body differs across cultures and has changed throughout history (**Figure 9.23**). Cultural ideals about body size are linked to body image and the incidence of eating disorders.[52] Eating disorders occur in societies where food is abundant and the body ideal is thin. They are rare in societies where food is scarce and people must worry about where their next meal is coming from.

U.S. culture today is a culture of thinness. Messages about what society views as a perfect body—the ideal that we should strive for—are constantly delivered by the media. Tall, lean women are featured in movies and advertisements. Thinness is associated with beauty, success, intelligence, and vitality. A young woman facing a future in which she must be independent, have a prestigious job, maintain a successful love relationship, bear and nurture children, manage a household, and keep up with fashion trends can become overwhelmed. Unable to master all these roles, she may look for some aspect of her life that she can control. Food intake and body weight are natural choices because thinness is associated with success. These messages about how we should look are hard to ignore and can create pressure to achieve this ideal body. But it is a standard

Eating Disorders 295

NUTRITION INSIGHT **FIGURE 9.23** **Body ideal and body weight** What is viewed as a desirable body size and shape is influenced by society and culture.

a. A fuller figure is still desirable in many cultures. Young women in these cultures, such as the Zulu of South Africa, may struggle to gain weight in order to achieve what is viewed as the ideal female body. As television images of very thin Western women become more accessible, the Zulu cultural view of plumpness as desirable may be changing.[48]

Actress Lillian Russell **1900** is considered a beauty at about 200 pounds

The thinner flapper **1920s** look becomes popular

The curvy figure of **1950s** Marilyn Monroe becomes the beauty standard

Twiggy, who weighs **1960s** less than 100 pounds, is the leading model

Jane Fonda's workout **1980s** book is a best seller

The fashion ideal today is thin but well muscled **Today**

Lillian Russell

Marilyn Monroe

Twiggy

c. The way the media presents female beauty can affect body image.[49] The bodies the media portrays as "ideal" are frequently atypical of normal, healthy women; a typical fashion model is 5'9" and weighs only about 100 lb, whereas an average woman is about 5'4" and weighs 166 pounds.[50]

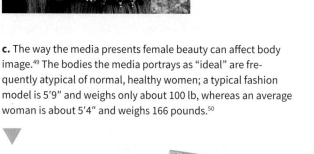

b. Thinness has not always been the beauty standard in the United States. This time line shows how the female body ideal has changed over the years. As female models, actresses, and other cultural icons have become thinner over the past several decades, the incidence of eating disorders has increased.

d. The toys that children play with set a cultural standard for body ideal. One study showed that little girls who played with thin dolls had greater body dissatisfaction than girls who played with full-figured dolls.[51] Playing with unrealistically thin dolls may encourage girls to aspire to a thinner shape; likewise, playing with muscular action figures may encourage boys to aspire to the muscular arms and legs and six-pack abs seen on male action figures. In response to such concerns, the original thin Barbie doll has been joined by petite, tall, and curvy dolls.

that is very difficult to meet—a standard that is contributing to disturbances in body image and eating behavior.

Although men currently represent a smaller percentage of people with eating disorders, the numbers are increasing, and it is now estimated that 25% of people with anorexia are male.[53] This is likely due to increasing pressure to achieve an ideal male body. Advertisements directed at men are showing more and more exposed skin, with a focus on well-defined abdominal and chest muscles.

Anorexia Nervosa

Anorexia means lack of appetite, but in the case of the eating disorder anorexia nervosa, it is a desire to be thin, rather than a lack of appetite, that causes individuals to decrease their food intake. Anorexia nervosa is characterized by a distorted body image, excessive dieting that leads to severe weight loss, and a pathological fear of being fat. It typically begins in adolescence and affects about 1% of women in their lifetime.[46] The death rate is about 5 deaths per 1000 persons per year, and 1 in 5 of those who die commit suicide.[54]

The psychological component of anorexia nervosa revolves around an overwhelming fear of gaining weight, even in individuals who are already underweight. It is not uncommon for individuals with anorexia to feel that they would rather be dead than fat. Anorexia is also characterized by disturbances in body image or perception of body size that prevent those with this disorder from seeing themselves as underweight even when they are dangerously thin. People with this disorder may use body weight and shape as a means of self-evaluation: "If I weren't so fat, everyone would like and respect me, and I wouldn't have other problems." However, no matter how much weight they lose, they do not gain self-respect, inner assurance, or the happiness they seek. Therefore, they continue to restrict their intake and use other behaviors to lose weight.

The most obvious behaviors associated with anorexia nervosa are those that contribute to the maintenance of a body weight that is less than minimally normal. These behaviors include restriction of food intake, binge-eating and purging episodes in some patients, strange eating rituals, and excessive activity (**Figure 9.24**). For some individuals with anorexia, the increase in activity is surreptitious, such as going up and down stairs repeatedly or getting off the bus a few stops too early. For others, the activity takes the form of strenuous physical exercise. They may become fanatical athletes and feel guilty if they cannot exercise. They link exercise and eating, so a certain amount of exercise earns them the right to eat, and if they eat too much, they must pay the price by adding extra exercise. They do not stop when they are tired; instead, they train compulsively beyond reasonable endurance.

The first obvious physical manifestation of anorexia nervosa is weight loss. As weight loss becomes severe, symptoms of starvation begin to appear. Starvation affects mental function, causing the person to become apathetic, dull, exhausted, and depressed. Fat stores are depleted. Other symptoms that appear include muscle wasting, inflammation and swelling of the lips, flaking and peeling of the skin, growth of fine hair (called lanugo) on the body, and dry, thin, brittle hair on the head. In females, estrogen levels drop, and in some women

FIGURE 9.24 A day in the life of a person with anorexia For individuals with anorexia nervosa, food and eating become an obsession. In addition to restricting the total amount of food they consume, people with anorexia develop personal diet rituals, limiting certain foods and eating them in specific ways. Although they do not consume very much food, they are preoccupied with food and spend an enormous amount of time thinking about it, talking about it, and preparing meals for others. Instead of eating, they move the food around the plate and cut it into tiny pieces.

menstruation becomes irregular or stops. In males, testosterone levels decrease. In the final stages of starvation, the person experiences abnormalities in electrolyte and fluid balance and cardiac irregularities. Suppression of immune function leads to infection, which further increases nutritional needs.

The goal of treatment for anorexia nervosa is to help resolve the underlying psychological and behavioral problems while providing for physical and nutritional rehabilitation. Treatment requires an interdisciplinary team of nutritional, psychological, and medical specialists and typically requires years of therapy. The goal of nutrition intervention is to promote weight gain by increasing energy intake and expanding dietary choices.[55] In more severe cases of anorexia, hospitalization is required so that food intake and exercise behaviors can be controlled. Intravenous nutrition may be necessary to keep a patient with anorexia alive. Some people with anorexia make full recoveries, but about half have poor long-term outcomes—remaining irrationally concerned about weight gain and never achieving normal body weight.

Bulimia Nervosa

The word *bulimia* comes from the Greek *bous*, meaning "ox," and *limos*, meaning "hunger," denoting hunger of such intensity that a person could eat an entire ox. The term *bulimia nervosa* was coined in 1979 by a British psychiatrist who suggested that bulimia consists of powerful urges to overeat in combination with a morbid fear of becoming fat and avoidance of the fattening effects of food by inducing vomiting and/or abusing purgatives.[56]

Bulimia is more common than anorexia nervosa, occurring in about 1.5% of women during their lifetime.[46] It is characterized by frequent episodes of binge eating followed by inappropriate behaviors such as self-induced vomiting to avoid weight gain. Like individuals with anorexia, those with bulimia have an intense fear of becoming fat and a negative body image, accompanied by a distorted perception of body size. Because self-esteem is highly tied to impressions of body shape and weight, people with bulimia may blame all their problems on their appearance. They are preoccupied with the fear that once they start eating, they will not be able to stop. They may engage in continuous dieting, which leads to a preoccupation with food. They are often socially isolated and may avoid situations that will expose them to food, such as going to parties or out to dinner; thus they become further isolated.

Bulimia typically begins with food restriction motivated by the desire to be thin. Overwhelming hunger may finally cause the dieting to be interrupted by a period of overeating. Eventually a pattern develops that consists of semistarvation interrupted by periods of gorging. During a binge-eating episode, a person with bulimia experiences a sense of lack of control. Binges usually last less than two hours and occur in secrecy. Eating stops when the food runs out or when pain, fatigue, or an interruption intervenes. The amount of food consumed in a binge may not always be enormous, but the individual perceives it as a binge episode (**Figure 9.25**).

After binge episodes, individuals with bulimia use various behaviors to eliminate the extra calories and prevent weight gain. Some use behaviors such as fasting or excessive exercise, but most use purging behaviors such as vomiting or taking laxatives, diuretics, or other medications. Self-induced vomiting

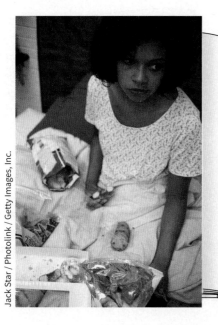

FIGURE 9.25 **A day in the life of a person with bulimia** The amount of food consumed during a binge varies but is typically on the order of 3400 Calories, while a normal young woman may consume only about 2000 Calories in an entire day. Self-induced vomiting is the most common purging behavior. At first, a physical maneuver such as sticking a finger down the throat is needed to induce vomiting, but patients eventually learn to vomit at will. Bingeing and purging are followed by intense feelings of guilt and shame.

Dear Diary,
Today started well. I stuck to my diet through breakfast, lunch, and dinner, but by 8 PM I was feeling depressed and bored. I thought food would make me feel better. Before I knew it I was at the convenience store buying two pints of ice cream, a large bag of chips, a one pound package of cookies, a half dozen candy bars, and a quart of milk. I told the clerk I was having a party. But it was a party of one. Alone in my dorm room I started by eating the chips, then polished off the cookies and the candy bars, washing them down with milk and finishing with the ice cream. Luckily no one was around so I was able to vomit without anyone hearing. I feel weak and guilty but also relieved that I got rid of all those calories. Tomorrow, I will start a new diet.

does eliminate some of the food before the nutrients have been absorbed, preventing weight gain, but laxatives and diuretics cause only water loss. Nutrient absorption is almost complete before food enters the colon, where laxatives have their effect. The weight loss associated with laxative abuse is due to dehydration. Diuretics also cause water loss, but via the kidney rather than the GI tract. They do not cause fat loss.

It is the purging portion of the binge/purge cycle that is most hazardous to health. Vomiting brings stomach acid into the mouth. Frequent vomiting can cause tooth decay, sores in the mouth and on the lips, swelling of the jaw and salivary glands, irritation of the throat and esophagus, and changes in stomach capacity and the rate of stomach emptying.[55] It also causes broken blood vessels in the face due to the force of vomiting, as well as electrolyte imbalance, dehydration, muscle weakness, and menstrual irregularities. Laxative and diuretic abuse can also lead to dehydration and electrolyte imbalance.

The overall goal of therapy for people with bulimia nervosa is to reduce or eliminate bingeing and purging behavior by separating the patients' eating behavior from their emotions and their perceptions of success and promoting eating in response to hunger and satiety. Psychological issues related to body image and a sense of lack of control over eating must be resolved. Nutritional therapy must address physiological imbalances caused by purging episodes as well as provide education on nutrient needs and how to meet them. Antidepressant medications may be beneficial in reducing the frequency of binge episodes. Treatment has been found to speed recovery, but long-term follow-up indicates that in many individuals, symptoms diminish even without treatment.[47]

Binge-Eating Disorder

Binge-eating disorder is the most common eating disorder. Unlike anorexia and bulimia, binge-eating disorder is not uncommon in men, who account for about 40% of cases. It is most common in overweight individuals (**Figure 9.26**). Individuals with binge-eating disorder engage in recurrent episodes of binge eating and experience a loss of control over eating but do not regularly engage in purging behaviors.

The major complications of binge-eating disorder are the health problems associated with obesity, which include diabetes, high blood pressure, high blood cholesterol levels, gallbladder disease, heart disease, and certain types of cancer. Treatment of binge-eating disorder involves counseling to improve body image and self-acceptance, a nutritious reduced-calorie diet and increased exercise to promote weight loss, and behavior therapy to reduce bingeing.

Eating Disorders in Special Groups

Although anorexia and bulimia are most common in women in their teens and 20s, eating disorders occur in all age groups, among both men and women. Both male and female athletes are at high risk for eating disorders, with an incidence that exceeds that seen in nonathletes.[57] Eating disorders occur during pregnancy and are becoming more frequent among younger children due to social values about food and body weight. They also occur in individuals with diabetes. A number of less common eating disorders appear in special groups in the general population (**Table 9.4**).

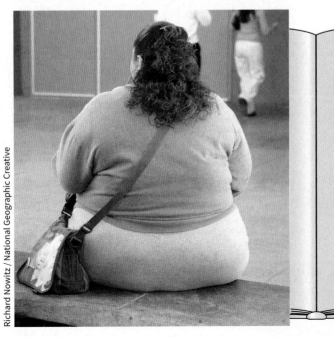

FIGURE 9.26 **A day in the life of a person with binge-eating disorder** People with binge-eating disorder often seek help for their weight rather than for their disordered eating pattern. It is estimated that about 2.8% of adults in the United States suffer from binge-eating disorder in their lifetime.[46]

TABLE 9.4 Other eating disorders and abnormal eating behaviors

Eating disorder	Who is affected	Characteristics and consequences
Anorexia athletica	Athletes	Engaging in compulsive exercise to lose weight or maintain a very low body weight. Can lead to more serious eating disorders and serious health problems, including kidney failure, heart attack, and death.
Avoidant/restrictive food intake disorder	Infants, children, and adults	Similar to anorexia nervosa in that the individual avoids eating and experiences weight loss and the other physical symptoms of anorexia. However, there is no distorted body image or fear of weight gain.
Diabulimia (insulin misuse)	People taking insulin to control diabetes	A disorder in which insulin doses are manipulated to control weight. Without insulin, glucose cannot enter cells to provide fuel, blood levels rise, and weight drops. However, over the long-term elevated blood sugar can lead to blindness, kidney disease, heart disease, nerve damage, and amputations.
Female athlete triad	Female athletes in weight-dependent sports	A syndrome involving energy restriction, along with high levels of exercise, that causes low estrogen levels. Low estrogen levels lead to amenorrhea and interfere with calcium balance, eventually causing reductions in bone mass and an increased risk of bone fractures. It is a component of relative energy deficiency in sport (discussed further in Chapter 10).
Muscle dysmorphia (megarexia, or reverse anorexia)	Bodybuilders and avid gym-goers; more common in men than in women	An obsession with being small and underdeveloped. Those affected believe that their muscles are inadequate, even when they have normal or above average muscle mass. They become avid weightlifters and may use anabolic steroids or other muscle-enhancing drugs.
Night-eating syndrome	Obese adults and those experiencing stress	A disorder that involves consuming most of the day's calories after the evening meal or at night after awakening from sleep. People with this disorder—which contributes to weight gain—are tense, anxious, upset, or guilty while eating. A similar disorder, in which a person may eat while asleep and have no memory of the events, is called nocturnal sleep-related eating disorder (NS-RED) and is considered a sleep disorder, not an eating disorder.
Pica	Pregnant women, children, people whose family or ethnic customs include eating certain nonfood substances	Persistent ingestion of nonfood items such as dirt, clay, paint chips, plaster, chalk, laundry starch, coffee grounds, and ashes. Depending on the items consumed, pica can cause perforated intestines and contribute to mineral deficiencies or intestinal infections (discussed further in Chapter 11).
Rumination disorder	Infants, children, adolescents, and adults	Eating, swallowing, and then regurgitating food back into the mouth, where it is chewed and ejected or reswallowed. It causes bad breath, indigestion, chapped lips, damage to the teeth and other tissues in the mouth, aspiration of food leading to pneumonia, weight loss and failure to grow (children), electrolyte imbalance, and dehydration.
Selective eating disorder	Children	Eating only a few foods, mostly those high in carbohydrate. If the disorder continues for long periods, it increases the risk of malnutrition.

Preventing and Getting Treatment for Eating Disorders

Because eating disorders are often triggered by weight-related criticism, elimination of this type of behavior can help prevent them. Another important target for reducing the incidence of eating disorders is the media. If the unrealistically thin body ideal presented by the media were altered, the incidence of eating disorders would likely decrease. Even with these interventions, however, eating disorders are unlikely to go away entirely. Education through schools and communities about the symptoms and complications of eating disorders can help people identify friends and family members who are at risk and persuade those with early symptoms to seek help.

The first step in preventing individuals from developing eating disorders is to recognize those who are at risk. Early intervention can help prevent at-risk individuals from developing serious eating disorders. Excessive concerns about body weight, having friends who are preoccupied with weight, being teased by peers about weight, and family problems all predispose a person to developing an eating disorder.

Once an eating disorder has developed, the person usually does not get better on his or her own. The actions of family members and friends can help people suffering from eating disorders get help before their health is impaired. But it is not always easy to persuade a friend or relative with an eating disorder to agree to seek help. People with eating disorders are good at hiding their behaviors and denying the problem, and often they do not want help. When confronted, one person might be relieved that you are concerned and willing to help, whereas another might be angry and defensive. When approaching someone about an eating disorder, it is important to make it clear that you are not forcing the person to do anything he or she doesn't want to do. Continued encouragement can help some people agree to seek professional help.

Concept Check

1. **Which** eating disorder is characterized by extreme weight loss?
2. **What** factors contribute to the higher incidence of eating disorders among women than among men?
3. **What** is meant by body image?
4. **How** does a food binge differ from "normal" overeating?

Summary

1 Body Weight and Health 266

- The number of Americans who are **overweight** and **obese** has reached epidemic proportions in the United States, as seen in this obesity map, as well as elsewhere around the world.

Figure 9.1 Obesity across America

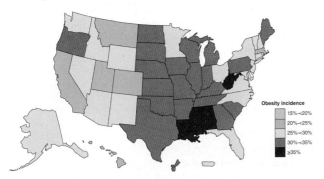

- America's **obesogenic environment** is contributing to the obesity epidemic. Americans have gotten fatter because they are consuming more calories due to poor food choices and larger portion sizes and moving less due to modern lifestyles in which computers, cars, and other conveniences reduce the amount of energy expended in work and play.
- Excess body fat increases the risk of chronic diseases such as diabetes, heart disease, high blood pressure, and certain types of cancer. Excess body fat also creates psychological and social problems.
- **Body mass index (BMI)** can be used to evaluate the health risks of a particular body weight and height. Measures of **body composition** can be used to determine the proportion of a person's weight that is due to fat. Excess **visceral fat** is a greater health risk than excess **subcutaneous fat**.

2 Energy Balance 273

- The principle of **energy balance** states that if energy intake equals energy expenditure, body weight will remain constant. Energy is provided to the body by the carbohydrate, fat, and protein in the food we eat. This energy is used to maintain **basal metabolic rate (BMR)**, to support activity, and to digest food and to absorb, metabolize, and store the nutrients (**thermic effect of food [TEF]**). When excess energy is consumed, it is stored, primarily as fat in **adipocytes**, causing weight gain. When energy in the diet does not meet needs, energy stores in the body are used, and weight is lost, as shown in the illustration.

Figure 9.10 Energy balance: Storing and retrieving energy

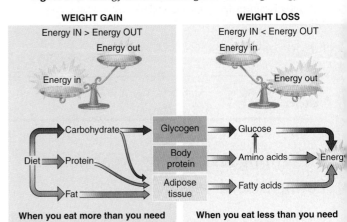

- The energy needs of healthy people can be predicted by calculating their Estimated Energy Requirements (EERs). A person's EER depends on gender, age, life stage, height, weight, and level of physical activity.

3 What Determines Body Size and Shape? 278

- The genes people inherit affect their body size and shape, as illustrated by the father and son shown here, but environmental factors and personal choices concerning the amount and type of food consumed and the amount and intensity of exercise performed also affect body weight.

Figure 9.12 Genes and body shape

- **Hunger** and **satiety** from meal to meal are regulated by signals from the gastrointestinal tract, hormones, and levels of circulating nutrients. Signals from fat cells, such as the release of **leptin**, regulate long-term energy intake and expenditure.
- Some people gain weight more easily because they inherit genes that make them more efficient at using energy and storing fat and/or because they expend less energy through **nonexercise activity thermogenesis (NEAT)**.

4 Managing Body Weight 283

- As seen in the decision tree, weight loss is recommended for those with a BMI above the healthy range who have excess body fat and risk factors associated with obesity, such as diabetes, hypertension, abnormal lipid levels, and an increased waist circumference. The goal is a loss of 10% of body weight within the first 6 months.

Figure 9.16 Weight-loss decision tree

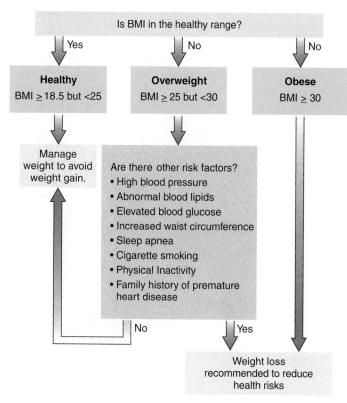

- Successful weight loss involves a reduction in energy intake, an increase in energy expenditure, and **behavior modification** to change behaviors that led to weight gain and help keep weight in a healthy range over the long term. To lose a pound of adipose tissue, energy expenditure must be increased or intake decreased by approximately 3500 Calories. Slow, steady weight loss of ½ to 2 lb per week is more likely to be maintained than more rapid weight loss.
- To become a thinner nation, we need strategies that can help all Americans improve their food choices, reduce portion sizes, and increase their physical activity. This requires action by all segments of society, including food manufacturers, restaurants, communities, businesses, and individuals.
- If being **underweight** is not due to a medical condition, weight gain can be accomplished by increasing energy intake and lifting weights to increase muscle mass.
- A good weight-loss program is one that promotes physical activity and a wide variety of nutrient-dense food choices, does not require the purchase and consumption of special foods or combinations of foods, and can be followed for life.

5 Medications and Surgery for Weight Loss 290

- Prescription weight-loss medications are recommended only for individuals who are significantly overweight or have accompanying health risks. Nonprescription weight loss medications and dietary supplements are also available. Weight is usually regained when medications are discontinued.

- Weight-loss surgery is recommended for those whose health is seriously compromised by their body weight and for whom conventional weight-loss methods have failed. Some common surgical approaches to treating obesity are **gastric bypass**, shown here, **gastric sleeve surgery**, and **adjustable gastric banding**.

Figure 9.21c Weight-loss surgery

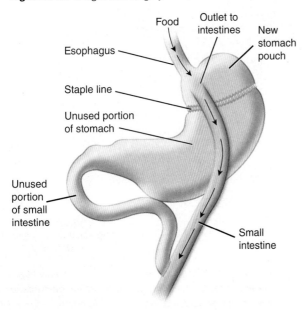

here. **Bulimia nervosa** is characterized by repeated cycles of binge eating followed by purging and other behaviors to prevent weight gain. **Binge-eating disorder** is characterized by bingeing without purging. People with this disorder are typically overweight.

Figure 9.24 A day in the life of a person with anorexia

- Eating disorders are caused by a combination of genetic, psychological, and sociocultural factors. The lean body ideal in the United States is believed to contribute to disturbances in **body image** that lead to eating disorders. Treatment involves medical, psychological, and nutritional intervention to stabilize health, change attitudes about body size, and improve eating habits while supplying an adequate diet.

6 Eating Disorders   293

- **Eating disorders** are psychological disorders that involve dissatisfaction with body weight. **Anorexia nervosa** involves self-starvation, resulting in an abnormally low body weight, as shown

---

Key Terms

- adipocyte   276
- adjustable gastric banding   291
- anorexia nervosa   293
- appetite   267
- basal metabolic rate (BMR)   273
- basal metabolism   273
- behavior modification   284
- binge-eating disorder   293
- body composition   269
- body image   293
- body mass index (BMI)   269
- bulimia nervosa   293
- eating disorder   293

- energy balance   273
- extreme obesity or morbid obesity   270
- gastric bypass   293
- gastric sleeve surgery   291
- ghrelin   280
- hunger   267
- lean body mass   269
- leptin   280
- liposuction   293
- nonexercise activity thermogenesis (NEAT)   273
- obese   266
- obesity genes   278

- obesogenic environment   267
- overweight   266
- resting metabolic rate (RMR)   274
- satiety   279
- set point   279
- subcutaneous fat   269
- thermic effect of food (TEF) or diet-induced thermogenesis   275
- total energy expenditure   273
- underweight   270
- visceral fat   269

What is happening in this picture?

Sumo wrestlers train for many hours each day and eat huge amounts of food. The result is a high BMI and a large waist but surprisingly little visceral fat.

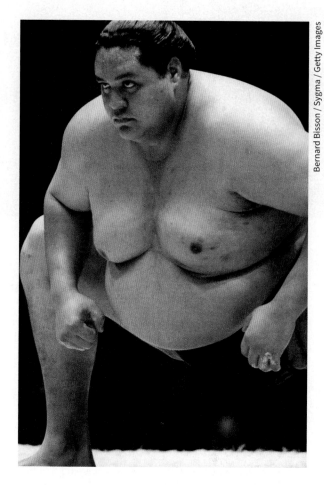

Think Critically

1. Why do these individuals have a low level of visceral fat?
2. Do you think they are at risk for diabetes and heart disease?
3. What type of fat is hanging over the belt of this wrestler?
4. What may happen if this wrestler retires and stops exercising but keeps eating large amounts of food?

CHAPTER 10

Nutrition, Fitness, and Physical Activity

Professional athletes, like these cyclists, have no trouble getting enough exercise. During training rides and races, long-distance cyclists may be on their bikes for six hours a day and require over 6000 Calories to maintain their weight. If cyclists want to stay healthy and perform at their best, these calories need to come at the right times and from the right kinds of foods. Most of us don't face this type of exercise and nutritional challenge; many of us struggle with the challenge of getting enough physical activity in a world that demands less and less physical exertion.

As advances in technology have made our jobs and lives less labor intensive, to stay in shape, people have had to allot time for physical exertion and seek out new ways of staying fit. Fitness crazes and fad diets litter our recent history—from Pilates and kickboxing to spin classes and Zumba; from eating your Wheaties to a high-carb diet to an all-meat diet; and now fitness trackers follow every move you make and morsel you consume.

As innovation continues to ease the daily physical challenges and exertions we face, to remain healthy, we have to pay a lot more attention to what we eat and how much activity we get.

CHAPTER OUTLINE

Food, Physical Activity, and Health 305
- Physical Activity Reduces the Risk of Chronic Disease • Physical Activity Makes Weight Management Easier

The Four Components of Fitness 307
- Cardiorespiratory Endurance
- Muscle Strength and Endurance
- Flexibility • Body Composition

Physical Activity Recommendations 309
- What to Look for in a Fitness Program
- Creating an Active Lifestyle

Fueling Activity 313
- Exercise Duration and Fuel Use
- Exercise Intensity and Fuel Use
  What a Scientist Sees: The Fat-Burning Zone
- Fitness Training and Fuel Use

Energy and Nutrient Needs for Physical Activity 319
- Energy Needs
- Carbohydrate, Fat, and Protein Needs
- Vitamin and Mineral Needs
- Water and Electrolyte Needs

Food and Drink to Optimize Performance 325
- What to Eat and Drink Before Exercise
- What to Eat and Drink During Exercise
- What to Eat and Drink After Exercise
  Thinking It Through: A Case Study on Snacks for Exercise

Ergogenic Aids 329
- Vitamin and Mineral Supplements
- Supplements to Build Muscle
- Supplements to Enhance Performance in Short, Intense Activities
- Supplements to Enhance Endurance
- Diet, Supplements, and Performance
  Debate: Energy Drinks for Athletic Performance?

10.1 Food, Physical Activity, and Health

LEARNING OBJECTIVES

1. **Discuss** how food and physical activity interact to promote health.
2. **Explain** the impact of physical activity on chronic disease.
3. **Discuss** the role of physical activity in weight management.

Food and physical activity are both necessary to achieve optimal health. The right foods provide the energy, nutrients, and other substances needed to promote health and reduce disease risk. Adequate physical activity improves your **fitness** and overall health. However, the link between food and physical activity goes beyond the fact that both promote health and reduce disease risk. Physical activity burns calories and utilizes nutrients, which must be supplied by the diet. Therefore, the foods you eat are necessary to fuel your activity and optimize athletic performance (**Figure 10.1**). The connection between diet and physical activity holds true whether your fitness goal is to keep your weight in the healthy range, to reduce your risk of chronic disease, to be able to complete the activities of your daily life, or to perform optimally in athletic competition.

Physical Activity Reduces the Risk of Chronic Disease

Physical activity includes both planned exercise and daily activities such as cleaning, cooking, yard work, and recreation. Fitness achieved through regular physical activity not only makes everyday tasks easier but also can prevent or delay the onset of chronic conditions such as cardiovascular disease, hypertension, type 2 diabetes, breast and colon cancer, and bone and joint disorders (**Figure 10.2**).[1] Adequate physical activity reduces overall mortality, regardless of whether the person is lean, normal weight, or obese.[1] The health benefits of physical activity are so great that they can even overcome some of the health risks of carrying excess body fat. So even if you can't take off extra pounds, you'll still benefit from being active. If you are in a profession that requires high levels of activity, such as a construction worker or ski instructor, your everyday activities may be enough to optimize your health and prevent weight gain, but for most of us to achieve these benefits, we must add exercise to our days.

In addition to decreasing the risk of chronic disease, adequate physical activity improves mood and self-esteem and increases vigor and overall well-being. Physical activity has also been shown to reduce depression and anxiety, as well as to improve the quality of life.[2] The mechanisms involved are not clear, but one hypothesis focuses on the production of **endorphins**. Certain types of activity stimulate the release of these chemicals, which are thought to be natural mood enhancers that play a role in triggering what athletes describe as an "exercise high." In addition to causing this state of exercise euphoria, endorphins are thought to aid in relaxation, pain tolerance, and appetite control.

Physical Activity Makes Weight Management Easier

Physical activity makes weight management easier because it increases both energy needs and lean body mass. During

FIGURE 10.1 Food and physical activity benefit health The energy and nutrients in food fuel our activity; physical activity in turn affects our energy and nutrient needs. Both food and physical activity are necessary for optimal health.

George Shelly Productions / Getty Images

---

fitness A set of attributes related to the ability to perform routine physical activities without undue fatigue.

endorphins Compounds that cause a natural euphoria and reduce the perception of pain under certain stressful conditions.

FIGURE 10.2 **Health benefits of physical activity** Engaging in enough of the right types of activity improves strength and endurance, reduces the risk of chronic disease, aids weight management, reduces sleeplessness, improves self-image, and helps relieve stress, anxiety, and depression.

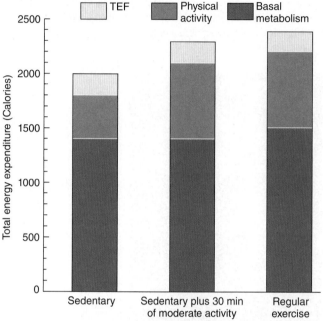

FIGURE 10.3 **Physical activity increases energy expenditure**
The total amount of energy we expend each day is the sum of the energy used for basal metabolism, physical activity, and the thermic effect of food (TEF). Adding 30 minutes of moderate activity to a sedentary lifestyle can increase energy expenditure by as much as 300 Calories. A program of regular exercise increases muscle mass, which increases basal metabolism, further increasing total energy expenditure.

Ask Yourself
Why does regular exercise cause an increase in basal metabolism?

moderate to strenuous physical activity, energy expenditure can rise well above the resting rate, and some of this increase persists for many hours after activity slows.[3] Over time, regular exercise increases lean body mass. Even at rest, lean tissue uses more energy than fat tissue; therefore, the increase in lean body mass increases basal metabolism. The combination of increased energy output during physical activity, the rise in energy expenditure that persists for a period after activity, and the increase in basal needs over the long term can have a major impact on total energy expenditure (**Figure 10.3**). The more energy you expend, the more food you can consume while maintaining a healthy weight. As discussed in Chapter 9, exercise is an essential component of weight management: It increases energy needs, promotes loss of body fat, and slows the loss of lean tissue that occurs with energy restriction.[4] Regular physical activity has also been reported to reduce the risk of stress-induced weight gain.[5]

Concept Check

1. **What** is the relationship between food and physical activity?
2. **How** does physical activity affect heart health?
3. **Why** does physical activity help in managing weight?

10.2 The Four Components of Fitness

LEARNING OBJECTIVES

1. **List** the characteristics of a fit individual
2. **Describe** the overload principle.
3. **Explain** how aerobic exercise affects heart rate and aerobic capacity.
4. **Compare** the amount of muscle in fit and unfit individuals.

Fitness is defined by endurance, strength, flexibility, and body composition (**Figure 10.4**). A fit person can continue an activity for a longer period than an unfit person before fatigue forces them to stop. Fitness is achieved by regular exercise. When you exercise, changes occur in your body: You breathe harder, your heart beats faster, and your muscles stretch and strain. Over time you adapt to the exercise you perform and can continue for a few minutes longer, lift a heavier weight, or stretch a millimeter farther. This is known as the **overload principle**. The more you do, the more you are capable of doing and the more fit you become.

Cardiorespiratory Endurance

How long you can jog or ride your bike depends on the ability of your cardiovascular and respiratory systems, referred to jointly as the cardiorespiratory system, to deliver oxygen and nutrients to your tissues and remove wastes. **Cardiorespiratory endurance** is enhanced by regular **aerobic activity** (**Figure 10.4a**).

Regular aerobic activity strengthens the heart muscle and increases the amount of blood pumped with each heartbeat.

overload principle The concept that the body adapts to the stresses placed on it.

cardiorespiratory endurance The efficiency with which the body delivers to cells the oxygen and nutrients needed for muscular activity and transports waste products from cells.

aerobic activity Endurance activity that increases heart rate and uses oxygen to provide energy as ATP.

NUTRITION INSIGHT | **FIGURE 10.4** **The components of fitness** Your fitness is determined by your cardiorespiratory endurance, muscle strength and endurance, flexibility, and the proportion of your body that is lean tissue versus fat.

a. Aerobic activity such as jogging, bicycling, and swimming strengthens the cardiovascular and respiratory systems. A quick way to keep your activity in the aerobic range is to proceed at a pace that is slow enough to allow you to carry on a conversation but fast enough that you cannot sing while exercising.

b. Weightlifting stresses the muscles, causing them to adapt by increasing in size and strength—a process called **hypertrophy**. The larger, stronger muscles can lift the same weight more easily. Muscles that are not used due to a lapse in weight training, injury, or illness become smaller and weaker. This process is called **atrophy**. Thus, there is truth to the saying "Use it or lose it."

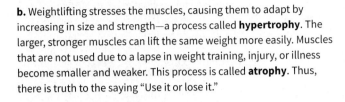

c. Flexibility exercises can be static or dynamic. In a static stretch, such as that shown on the left, a position that stretches a muscle or group of muscles to its farthest point is held for about 30 seconds. Dynamic stretching, shown on the right, involves motion. It uses controlled leg and arm swings, torso twists, and side lunges to extend muscles gently to the limits of their range of motion. The combination of static and dynamic stretching that is best depends on the person and his or her sport.[6]

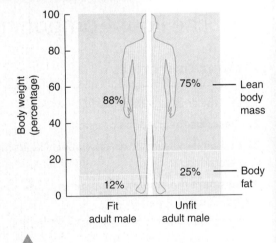

d. A fit person has more muscle mass than an unfit person of the same height and weight. Becoming fit by engaging in aerobic activity and muscle-strengthening exercise has a positive impact on body composition, reducing body fat and increasing the proportion of lean tissue.

This in turn decreases **resting heart rate**, the rate at which the heart beats when the body is at rest to supply blood to the tissues. The more fit you are, the lower your resting heart rate and the more blood your heart can pump to your muscles during exercise. In addition to increasing the amount of oxygen-rich blood that can be pumped to your muscles, regular aerobic activity increases the ability of your muscles to use oxygen to produce ATP. Your body's maximum ability to generate ATP using aerobic metabolism is called your **aerobic capacity**, or VO_2 max. Aerobic capacity is a function of the ability of the cardiorespiratory system to deliver oxygen to the cells and the ability of the cells to use oxygen to produce ATP. The greater your aerobic capacity, the more intense activity you can perform before lack of oxygen affects your performance.

Muscle Strength and Endurance

Greater **muscle strength** enhances the ability to perform tasks such as pushing or lifting. In daily life, this could mean lifting a gallon of milk off the top shelf of the refrigerator with one hand, carrying a full trash can out to the curb, or moving a couch into your new apartment. Greater **muscle endurance** enhances your ability to continue repetitive muscle activity, such as shoveling snow or raking leaves. Muscle strength and endurance are increased by repeatedly using muscles in activities that require moving against a resisting force. This type of exercise is called **muscle-strengthening exercise**, strength-training exercise, or resistance-training exercise and includes activities such as weightlifting and calisthenics (**Figure 10.4b**).

Flexibility

When you think of fitness, you may picture someone with bulging muscles, but fitness also involves flexibility. Flexibility determines your range of motion—how far you can bend and stretch muscles and ligaments. Regularly moving limbs, the neck, and the torso through their full range of motion helps increase and maintain flexibility. If your flexibility is poor, you cannot easily bend to tie your shoes or stretch to remove packages from the car. Being flexible can enhance postural stability and balance.[7] Stretching can improve flexibility but has not been found to improve athletic performance. An exercise warm-up that includes stretching followed by exercises that involve more movement, such as easy jogging, improves range of motion and reduces the risk of injury (**Figure 10.4c**).[8]

Body Composition

Individuals who are physically fit have a greater proportion of muscle and a smaller proportion of fat than do unfit individuals of the same height and weight (**Figure 10.4d**). The amount of body fat a person has is also affected by gender and age. In general, women have more stored body fat than men. For young adult women, a healthy amount of body fat is 21 to 32% of total weight; in adult men, a healthy amount is 8 to 19%.[9]

Concept Check

1. **What** distinguishes a fit person from an unfit one?
2. **Why** do your muscles get bigger when you lift weights?
3. **How** does aerobic activity affect heart rate?
4. **Why** do fit individuals have a greater percentage of lean tissue than unfit individuals?

10.3 Physical Activity Recommendations

LEARNING OBJECTIVES

1. **Describe** the amounts and types of activity recommended to optimize health.
2. **Classify** activities as aerobic or anaerobic.
3. **Plan** a fitness program that can be integrated into your daily routine.
4. **Explain** overtraining syndrome.

To reduce the risk of chronic disease, public health guidelines advise at least 150 minutes of moderate-intensity or 75 minutes of vigorous-intensity aerobic physical activity each week or an equivalent combination of both.[7, 10, 11] Even a small amount of exercise is better than none, and greater health benefits can be obtained by exercising more vigorously or for a longer duration. Moderate-intensity exercise is the equivalent of walking 3 miles in about an hour or bicycling 8 miles in about an hour. Vigorous-intensity exercise is equivalent to jogging at a rate of 5 miles per hour or faster or bicycling at 10 miles per hour or faster.[11] Adults should also include muscle-strengthening activities on two or more days

---

aerobic capacity The maximum amount of oxygen that can be consumed by the tissues during exercise. Also called maximal oxygen consumption, or VO_2 max.

muscle-strengthening exercise Activities that are specifically designed to increase muscle strength, endurance, and size; also called strength-training exercise or resistance-training exercise.

per week, but time spent in muscle-strengthening activities does not count toward meeting the aerobic activity guidelines. Fewer than half of U.S. adults 18 years and older currently meet the physical activity guidelines for aerobic physical activity, and only 21% meet muscle-strengthening activity guidelines.[12]

What to Look for in a Fitness Program

A complete fitness program includes aerobic activity for cardiovascular conditioning, stretching exercises for flexibility, and muscle-strengthening exercises to increase muscle strength and endurance and maintain or increase muscle mass.[11, 13] The program should be integrated into an active lifestyle that includes a variety of everyday activities, enjoyable recreational activities, and a minimum amount of time spent in sedentary activities (**Figure 10.5**).

Moderate or vigorous aerobic activity should be performed most days of the week. An activity is aerobic if it raises your heart rate to 60 to 85% of your **maximum heart rate**; when you are exercising at an intensity in this range, you are said to be in your **aerobic zone** (**Figure 10.6**). For a sedentary individual

FIGURE 10.5 **Physical activity recommendations** A healthy lifestyle minimizes sedentary activities and includes some planned exercise and a variety of everyday activities. At least 150 minutes of aerobic activity is recommended, along with activities that improve muscle strength and flexibility. Achieving the recommended amounts and types of activity will help improve and maintain fitness and health.[7, 8] In general, more exercise is better than less.

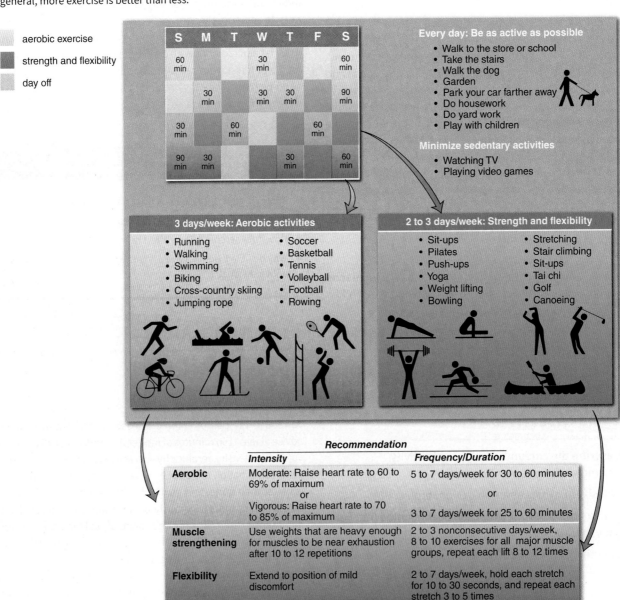

maximum heart rate The maximum number of beats per minute that the heart can attain.

FIGURE 10.6 Finding your aerobic zone Checking your heart rate during your exercise session can help you determine if you are exercising in your aerobic zone.

a. You can check your heart rate by feeling the pulse at the side of your neck, just below the jawbone. A pulse is caused by the heart beating and forcing blood through the arteries. The number of pulses per minute equals heart rate.

b. You can determine your aerobic zone by calculating 60% and 85% of your maximum heart rate. Maximum heart rate is dependent on age and can be estimated in men by subtracting age from 220 and in women by subtracting 88% of age from 206, according to the following equations.[14]
Men: Maximum heart rate = 220 − age
Women: Maximum heart rate = 206 − (0.88 × age)
For example, a 20-year-old man would have a maximum heart rate of 200 (220 − 20) beats per minute. If he exercises at a pace that keeps his heart rate between 120 (0.6 × 200) and 170 (0.85 × 200) beats per minute, he is in his aerobic zone.

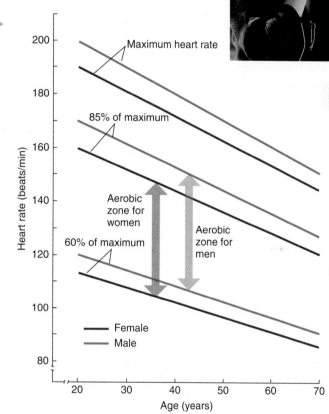

Interpret the Data

What is the aerobic zone for a 30-year-old woman? What happens to this range when she turns 40?

who is beginning a fitness program, mild exercise such as walking can raise the heart rate into the aerobic zone. As fitness improves, an exerciser must perform more intense activity to raise the heart rate to this level.

Aerobic activities of different intensities can be combined to meet recommendations and achieve health benefits. The total amount of energy expended in physical activity depends on the intensity, duration, and frequency of the activity. Vigorous physical activity, such as jogging, that raises heart rate to the high end of the aerobic zone (70 to 85%) improves fitness more and burns more calories per unit of time than does moderate-intensity activity, such as walking, which raises heart rate only to the low end of the zone (60 to 69%).

Individuals should structure their fitness program based on their needs, goals, and abilities. For example, some people might prefer a short, intense workout such as a 30-minute run, while others would rather work out for a longer time, at a lower intensity, such as a one-hour walk. Some may choose to complete all their exercise during the same session, while others may spread their exercise throughout the day, in shorter bouts. Three short bouts of 10-minute duration can be as effective as a continuous bout of 30 minutes for reducing the risk of chronic disease.[13] It is preferable to spread your aerobic activity throughout the week rather than cram it all into the weekend. Exercising at least three days per week produces health benefits and reduces the risk of injury and fatigue. A combination of intensities, such as a brisk 30-minute walk twice during the week in addition to a 20-minute jog on two other days, can meet recommendations.

Muscle strengthening and stretching can be performed less often than aerobic activities. Muscle strengthening is needed only two to three days a week at the start of a fitness program and two days a week after the desired strength has been achieved. Muscle strengthening should not be done on consecutive days. The rest between sessions gives the muscles time to respond to the stress by getting stronger. Increasing the amount of weight lifted increases muscle strength, whereas increasing the number of repetitions improves muscle endurance. Flexibility exercises can be performed two to seven days per week. Time spent stretching does not count toward meeting aerobic or strength-training guidelines.

In addition to following these physical activity recommendations, it is important to minimize sedentary time. A few hours of planned exercise cannot compensate for extended periods of time spent in sedentary pastimes. Therefore, even people who exercise enough to meet physical activity guidelines may be at increased risk of cardiovascular disease, depression, increased waist circumference, and other adverse effects if they spend long periods sitting in a car, at a desk, or in front of the television.[7, 15] Reducing total time spent in sedentary pursuits and breaking up periods of sedentary activity with short bouts of physical activity and standing, which can attenuate the adverse effects of sedentary behavior, should be a goal for all adults, regardless of their exercise habits.

Creating an Active Lifestyle

Incorporating activity into your day-to-day life may require a change in lifestyle, which is not always easy. Many people avoid exercise because they do not enjoy it, think it requires them to join an expensive health club, have little motivation to do it alone, or find it inconvenient or uncomfortable. Finding an activity you enjoy, setting aside a time that is realistic and convenient, and finding a place that is appropriate and safe are important steps in starting and maintaining a fitness program (**Table 10.1**). Riding your bike to class or work rather than driving, taking a walk during your lunch break, and enjoying a game of catch or tag with your friends or family are all effective ways to increase your everyday activity level. The goal is to gradually make lifestyle changes that increase physical activity.

Before beginning a fitness program, check with your physician to be sure that the activities are appropriate for you, considering your medical history (**Figure 10.7**). If you choose to exercise outdoors rather than in a gym, reduce or curtail exercise in hot, humid weather in order to avoid heat-related illness. In cold weather, wear clothing that allows for evaporation of sweat while providing protection from the cold. Start each exercise session with a warm-up, such as mild stretching or easy jogging, to increase blood flow to the muscles. End with a cool-down period, such as walking or stretching, to prevent muscle cramps and slowly reduce heart rate.

Don't overdo it. If you don't rest enough between exercise sessions, fitness and performance will not improve. During rest, the body replenishes energy stores, repairs damaged tissues, and builds and strengthens muscles. In athletes, excessive training without sufficient rest to allow for recovery can lead to **overtraining syndrome**. The most common symptoms of this condition are fatigue,

TABLE 10.1 Suggestions for starting and maintaining an exercise program

Start slowly. Set specific, attainable goals. Once you have met them, add more
- Walk around the block after dinner twice a week.
- Get off the bus or subway one stop early every day.
- Use half of your lunch break to exercise on Tuesdays and Thursdays.
- Do a few biceps curls each time you take the milk out of the refrigerator.

Make your exercise fun and convenient
- Opt for activities you enjoy: Bowling and dancing may be more fun for you than using a treadmill at the gym.
- Find a partner to exercise with you.
- Choose times that fit your schedule.

Stay motivated
- Vary your routine: Swim one day and mountain bike the next.
- Challenge your strength or endurance twice a week and do moderate workouts on other days.
- Track your progress by recording all your activity.
- Reward your success with a book, movie, or new workout clothes each time you reach a goal.

Keep your exercise safe
- Warm up before you start and cool down when you are done.
- Wear light-colored or reflective clothing that is appropriate for the environmental conditions.
- Don't overdo it: Alternate hard days with easy days and take a day off when you need it.
- Listen to your body and stop before an injury occurs.

---

overtraining syndrome A collection of emotional, behavioral, and physical symptoms that occurs when the amount and intensity of exercise exceeds an athlete's capacity to recover.

FIGURE 10.7 **Physical activity is for everyone** Almost anyone of any age can be active, no matter where they live, how old they are, or what physical limitations they have.

performance decline, and mood disturbances.[16] Athletes may become moody, easily irritated, or depressed; experience altered sleep patterns; or lose their competitive desire and enthusiasm. It may take weeks or months to recover. In addition to the physical demands of training, psychological factors such as excessive expectations from a coach or family members, competitive stress, personal or emotional problems, and school- or work-related demands contribute to the development of overtraining syndrome. Overtraining syndrome occurs only in serious athletes who are training extensively, but rest is essential for anyone who is working to increase fitness.

Concept Check

1. **How** much aerobic activity is recommended to reduce the risk of chronic disease?
2. **What** is your aerobic zone?
3. **What** types of activities should be part of a fitness program?
4. **Who** is at risk for overtraining syndrome?

10.4 Fueling Activity

LEARNING OBJECTIVES

1. **Compare** the fuels used to generate ATP by anaerobic and aerobic metabolism.
2. **Discuss** the effect of exercise duration and intensity on whether aerobic or anaerobic metabolism predominates.
3. **Describe** the physiological changes that occur in response to exercise training.

The body runs on energy from the carbohydrate, fat, and protein in food and body stores. These fuels are needed whether you are writing a term paper, riding your bike to class, or running a marathon. Before they can be used to fuel activity, their energy must be transferred to the high-energy compound ATP, the immediate source of energy for body functions. ATP can be generated both in the absence of oxygen, by **anaerobic metabolism**, and in the presence of oxygen, by **aerobic metabolism** (**Figure 10.8**). The type of metabolism that predominates during an activity determines how much carbohydrate, fat, and protein are used to fuel the activity.

The availability of oxygen determines whether ATP is produced predominantly by anaerobic metabolism or aerobic metabolism. Oxygen is taken in by the respiratory system and delivered to the muscles by the blood (**Figure 10.9**). When you are at rest, your muscles do not need much energy, and your heart and lungs are able to deliver enough oxygen to meet your energy needs using aerobic metabolism. When you exercise, your muscles need more energy. To increase the amount of energy provided by aerobic metabolism, you must increase the amount of oxygen delivered to the muscles. Your body accomplishes this by increasing both heart rate and breathing rate. The amount of oxygen the circulatory and respiratory systems deliver to tissues during exercise is affected by how long an activity is performed, the intensity of the activity, and the physical conditioning of the exerciser.

Exercise Duration and Fuel Use

When you take the first steps of your morning jog, your muscles increase their activity, but your heart and lungs have not had time to step up their delivery of oxygen to them. To get the energy they need, the muscles rely on the small amount of ATP that is stored in resting muscle. This is enough to sustain activity

FIGURE 10.8 **Anaerobic versus aerobic metabolism** ATP is produced in the cytosol by anaerobic metabolism when no oxygen is available. Anaerobic metabolism produces ATP very rapidly but uses only glucose as a fuel. The **lactic acid** that is produced can be used as a fuel for aerobic metabolism. Aerobic metabolism requires oxygen, takes place in the mitochondria, and can use carbohydrate, fat, or protein to produce ATP. Aerobic metabolism produces the majority of ATP; it is slower but more efficient at generating ATP than anaerobic metabolism.

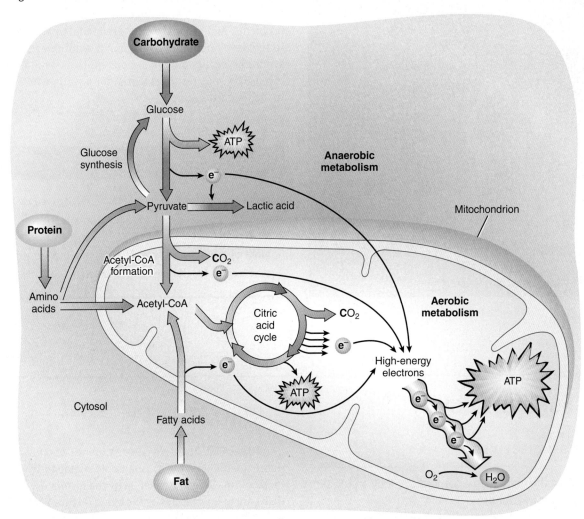

for a few seconds. As the stored ATP is used up, enzymes break down another high-energy compound, **creatine phosphate**, to convert ADP (adenosine diphosphate) to ATP, allowing your activity to continue. But, like the amount of ATP, the amount of creatine phosphate stored in the muscle at any time is small and soon runs out (**Figure 10.10**).

Short-term energy: Anaerobic metabolism

After about 15 seconds of exercise, the ATP and creatine phosphate in your muscles are used up, but your heart rate and breathing have not increased enough to deliver more oxygen to the muscles. To get more energy at this point, your muscles must produce the additional ATP without oxygen. This anaerobic metabolism can produce ATP very rapidly but can use only glucose as a fuel (see Figure 10.8). The amount of glucose is limited, so anaerobic metabolism cannot continue indefinitely.

Long-term energy: Aerobic metabolism

After you have been exercising for two to three minutes, your breathing and heart rate have increased to supply more oxygen to your muscles. This allows aerobic metabolism to predominate. Aerobic metabolism produces ATP at a slower rate than does anaerobic metabolism, but it is much more efficient, producing about 18 times more ATP for each molecule of glucose. As a result, glucose is used more slowly

---

creatine phosphate A compound stored in muscle that can be broken down quickly to make ATP.

Fueling Activity 315

PROCESS DIAGRAM   FIGURE 10.9 **Getting oxygen to muscle cells** When you exercise, your muscles need more oxygen. Your body responds to this need by breathing faster and deeper in order to take in more oxygen through the lungs and by increasing heart rate in order to deliver the additional oxygen to your muscles.

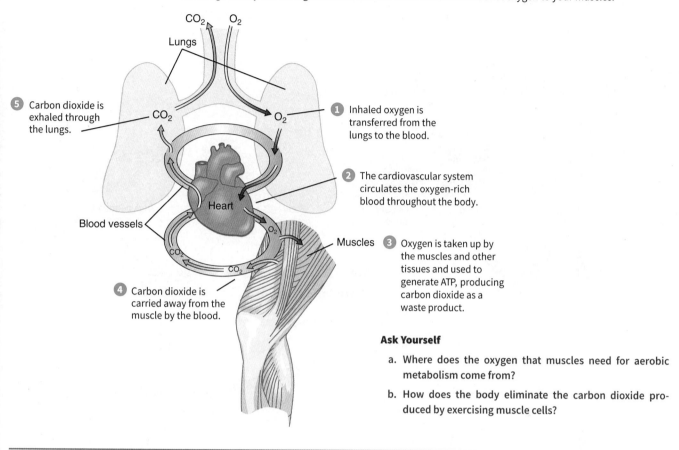

1. Inhaled oxygen is transferred from the lungs to the blood.
2. The cardiovascular system circulates the oxygen-rich blood throughout the body.
3. Oxygen is taken up by the muscles and other tissues and used to generate ATP, producing carbon dioxide as a waste product.
4. Carbon dioxide is carried away from the muscle by the blood.
5. Carbon dioxide is exhaled through the lungs.

Ask Yourself

a. Where does the oxygen that muscles need for aerobic metabolism come from?

b. How does the body eliminate the carbon dioxide produced by exercising muscle cells?

FIGURE 10.10 **Changes in the source of ATP over time** The source of the ATP that fuels muscle contraction changes over the first few minutes of exercise. If the intensity of activity remains moderate, aerobic metabolism will predominate after about 5 minutes.

Instant energy
During the first few seconds of exercise, the muscles get energy from stored ATP. Then, for the next 10 seconds or so, creatine phosphate stored in the muscles is broken down to form more ATP.

Short-term energy
Anaerobic metabolism of glucose, obtained either from the blood or from muscle glycogen, becomes the predominant source of ATP when creatine phosphate stores have been depleted. Thirty seconds into the activity, anaerobic pathways are operating at full capacity.

Long-term energy
After about 2 to 3 minutes, oxygen delivery to the muscles has increased enough to support aerobic metabolism, which uses fatty acids and glucose to produce ATP.

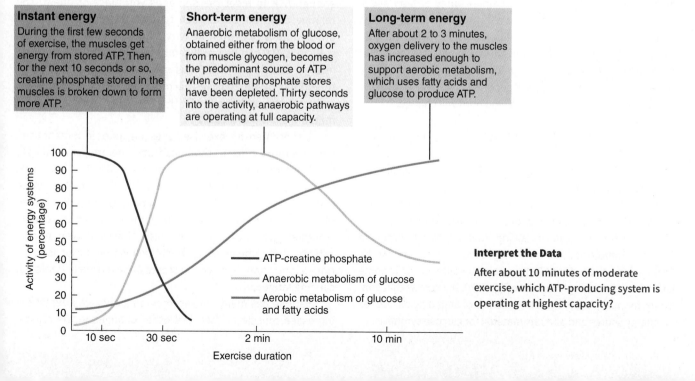

Interpret the Data

After about 10 minutes of moderate exercise, which ATP-producing system is operating at highest capacity?

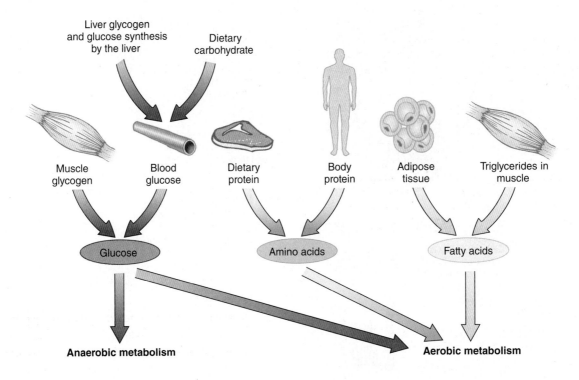

FIGURE 10.11 Fuels for anaerobic and aerobic metabolism The glucose used to fuel muscle contraction comes from muscle glycogen breakdown or blood glucose. Blood glucose is supplied by the breakdown of liver glycogen, glucose synthesis by the liver, and carbohydrate consumed during exercise. Some of the fatty acids used as fuel come from triglycerides stored in the muscle, but most come from adipose tissue. The amino acids available to the body come from the digestion of dietary proteins and from the breakdown of body proteins.

than in anaerobic metabolism. In addition, aerobic metabolism can use fatty acids and amino acids to generate ATP (**Figure 10.11**).

In a typical adult, about 90% of stored energy is found in adipose tissue; this provides an ample supply of fatty acids. When you continue to exercise at a low to moderate intensity, aerobic metabolism predominates, and fatty acids become the primary fuel source for your exercising muscles (see *What a Scientist Sees: The Fat-Burning Zone*). When you pick up the pace, the relative amount of ATP generated by anaerobic versus aerobic metabolism and the fuels you burn will change.

Protein as a fuel for exercise Although protein is not considered a major energy source for the body, even at rest, small amounts of amino acids are used for energy. The amount increases if your diet does not provide enough total energy to meet needs, if you consume more protein than you need, or if you are involved in endurance exercise (see Chapter 6).

When the nitrogen-containing amino group is removed from an amino acid, the remaining carbon compound can be broken down to produce ATP by aerobic metabolism or, in some cases, used to make glucose (see Figure 10.8). Exercise that continues for many hours increases the use of amino acids both as an energy source and as a raw material for glucose synthesis.

Exercise Intensity and Fuel Use

The energy contributions made by anaerobic and aerobic metabolism combine to ensure that your muscles get enough ATP to meet the demands you place on them. The relative contribution of each type of metabolism depends on the intensity of your activity. With low-intensity activity, sufficient ATP can be produced by aerobic metabolism. With intense exercise, more ATP is needed, but oxygen delivery to and use by the muscles becomes limited, so the muscles must get the additional ATP they need by using anaerobic metabolism (**Figure 10.12**).

Lower-intensity exercise relies on aerobic metabolism, which is more efficient than anaerobic metabolism and uses both glucose and fatty acids to produce ATP. The body's fat reserves are almost unlimited, so if fat is the fuel, exercise can theoretically continue for a very long time. For example, it is estimated that a 130-lb woman has enough energy stored as body fat to run 1000 miles. However, even aerobic activity uses some glucose, which means that if exercise continues long enough, glycogen stores are eventually depleted, causing fatigue.

Fatigue has many causes, including glycogen depletion, increased muscle acidity, and other changes in the muscle

What a Scientist Sees

The Fat-Burning Zone

Have you ever jumped onto a treadmill and chosen the workout that puts you in the "fat-burning zone" rather than the one that puts you in the "cardio zone" because your goal was to lose weight? The fat-burning zone is a lower-intensity aerobic workout that keeps your heart rate between about 60 and 69% of maximum. The cardio zone is a higher-intensity aerobic workout that keeps your heart rate between about 70 and 85% of maximum.

However, do you really burn more fat during a slow 30-minute jog in the fat-burning zone than during a vigorous 30-minute run in the cardio zone? A scientist sees that you do burn a higher percentage of calories from fat during a lower-intensity aerobic workout, but that's not the whole story. When you pick up the pace and exercise in what the treadmill calls the cardio zone, you continue to burn fat. The graph shows that 50% of the calories burned come from fat during the lower-intensity workout (that is, in the fat-burning zone) and only 40% come from fat during the higher-intensity workout. Looking at the actual numbers of calories burned, however, the scientist sees that at the higher intensity, you burn just as much fat (about 150 Calories/hour) but a much greater number of calories overall.

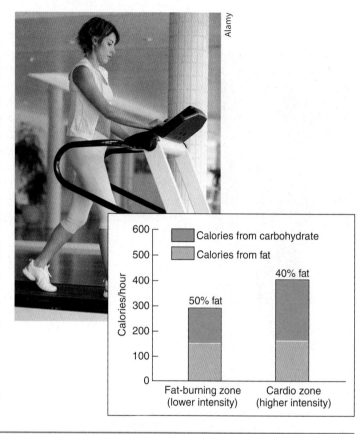

Think Critically

Which workout will help you lose the most weight: 30 minutes in the cardio zone or 30 minutes in the fat-burning zone? Why?

FIGURE 10.12 The effect of exercise intensity on fuel use Exercise intensity determines the contributions of carbohydrate, fat, and protein as fuels for ATP production. At rest and during low- to moderate-intensity exercise, aerobic metabolism predominates, so fatty acids are an important fuel source. As exercise intensity increases, the proportion of energy supplied by anaerobic metabolism increases, and glucose becomes the predominant fuel. Keep in mind, however, that during exercise, the total amount of energy expended is greater than the amount expended at rest.

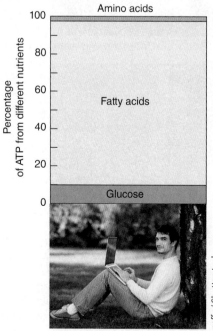
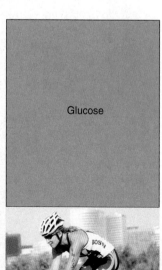

Rest | Moderate-intensity activity | High-intensity activity

FIGURE 10.13 Fatigue: "Hitting the wall" Glycogen depletion is a concern for athletes because the amount of stored glycogen available to produce glucose during exercise is limited. When athletes run out of glycogen, they experience a feeling of overwhelming fatigue that is sometimes referred to as "hitting the wall" or "bonking."

Between 60 and 120 grams of glycogen are stored in the liver; glycogen stores are highest just after a meal. Liver glycogen is used to maintain blood glucose between meals and during the night. Eating a high-carbohydrate breakfast will replenish the liver glycogen you used while you slept.

There are about 200 to 500 g of glycogen in the muscles of a 70-kg (154-lb) person. The glycogen in a muscle is used to fuel that muscle's activity.

cells and the concentrations of molecules involved in muscle metabolism.[17] Fatigue occurs much more quickly with high-intensity exercise than with lower-intensity exercise because more intense exercise relies more on anaerobic metabolism, which can use only glucose for fuel. Glycogen stores thus are rapidly depleted (**Figure 10.13**). Anaerobic metabolism also produces lactic acid. With low-intensity exercise, the small amounts of lactic acid produced are carried away from the muscles and used by other tissues as an energy source or converted back into glucose by the liver. During high-intensity exercise, the amount of lactic acid produced exceeds the amount that can be used, and the lactic acid builds up in the muscle and subsequently in the blood. Lactic acid buildup accompanies exercise-associated muscle fatigue, but it is only one of many metabolic changes associated with fatigue.[17, 18]

Fitness Training and Fuel Use

When you exercise regularly to improve your fitness, the training causes physiological changes in your body. The changes caused by repeated bouts of aerobic exercise increase the amount of oxygen that can be delivered to the muscles and the ability of the muscles to use oxygen to generate ATP by aerobic metabolism (**Figure 10.14**).[19] This increased aerobic capacity

NUTRITION INSIGHT

FIGURE 10.14 Physiological changes caused by aerobic training Aerobic training causes physiological changes in the cardiovascular system that increase the delivery of oxygen to cells. It also causes changes in the muscle cells that increase glycogen storage and the ability to use oxygen to generate ATP.[19]

Training causes the heart to become larger and stronger so that the amount of blood pumped with each beat is increased. As shown in the graph, the heart of a trained athlete can pump more blood per minute than can the heart of an untrained individual.

Training causes blood volume and the number of red blood cells to expand, increasing the amount of hemoglobin so that more oxygen can be transported. It also causes the number of capillary blood vessels in the muscles to increase so that blood is delivered to muscles more efficiently.

Training enhances the ability to store muscle glycogen and increases the number and size of muscle-cell mitochondria. Because aerobic metabolism occurs in the mitochondria, the greater size and number of mitochondria increases the capacity of muscle cells to burn fatty acids to produce ATP.

allows fatty acids to be used for fuel during higher-intensity activity so that glycogen is spared and the onset of fatigue is delayed. Training aerobically also increases the amount of glycogen stored in the muscles. Because trained athletes store more glycogen and use it more slowly, they can sustain aerobic exercise for longer periods at higher intensities than can untrained individuals.

> **Concept Check**
>
> 1. **What** fuels are used in anaerobic metabolism?
> 2. **What** type of metabolism does a marathon runner rely on?
> 3. **Why** is a trained athlete able to perform at a higher intensity for a longer time than an untrained person?

10.5 Energy and Nutrient Needs for Physical Activity

LEARNING OBJECTIVES

1. **List** the factors that affect the energy needs of athletes.
2. **Compare** the macronutrient needs of athletes and nonathletes.
3. **Discuss** micronutrients that may be at risk in athletes.
4. **Explain** why athletes are at risk for dehydration and hyponatremia.

Good nutrition is essential to performance, whether you are a marathon runner or a mall walker. Your diet must provide enough energy to fuel activity, enough protein to support protein synthesis, sufficient micronutrients to metabolize the energy-yielding nutrients, and enough water to transport nutrients and cool your body. The major differences between the nutritional needs of a serious athlete and those of a casual exerciser are the amount of energy and fluid they need and the quantity and timing of their protein intake.

Energy Needs

Energy is needed to support body functions, provide enough of each of the macronutrients, and assist in maintaining the desired body composition.[20] Everyone's energy needs are affected by his or her age, weight, and body composition, but for most of us, these needs are fairly stable and determined primarily by basal needs. In athletes, however, energy needs can be increased dramatically depending of their training and competition schedule. The energy expended for physical activity is determined by the intensity, duration, and frequency of the activity and the environmental conditions under which the exercise is performed (**Figure 10.15**). While a casual

FIGURE 10.15 **Factors affecting energy expenditure** The number of calories expended per hour during exercise depends on the characteristics of the exerciser, the exercise, and the environment. For example, body weight affects energy needs because moving a heavier body requires more energy than moving a lighter one. Therefore, if the pace is the same, a 170-lb woman requires more energy to run for an hour than does a 125-lb woman.

Environmental factors: Energy needs are greater when exercising in extreme cold or heat and at high altitudes.

Age: Energy needs decrease as adults age.

Body weight: Energy needs increase with increasing body weight.

Emotions: Fear and stress can increase energy needs.

Body composition Energy needs increase with increasing lean body mass.

Exercise intensity: Energy needs per hour increase as exercise intensity increases.

Exercise duration: Energy needs are greater the longer the activity continues.

David Grossman / Alamy Stock Photo

exerciser may burn only 100 additional Calories/day, the training required for an endurance athlete, such as a marathon runner, may increase energy expenditure by 2000 to 3000 Calories/day. Some athletes require 6000 Calories/day or more to maintain their body weight. In general, the more intense the activity, the more energy it requires, and the more time spent exercising, the more energy is expended (see Appendix D). Running for 60 minutes, for instance, involves more work than walking for the same amount of time and therefore requires more energy.

Energy intake to optimize body weight and composition
Body weight and composition can affect exercise performance. In sports such as football and weightlifting, having a large amount of muscle is advantageous, and athletes may try to build muscle and increase body weight. An increase in muscle mass can be achieved through a combination of increased energy intake, adequate protein intake, and muscle-strengthening exercise to promote an increase in lean tissue rather than fat.

In sports such as ballet, gymnastics, and certain running events, small, light bodies offer an advantage, so athletes may restrict energy intake in order to reduce body fat and maintain a low body weight. While a slightly leaner physique may be beneficial in these sports, dieting to maintain an unrealistically low weight may decrease the amount of energy available to maintain health and performance. An athlete who needs to lose weight should do so in advance of the competitive season to prevent the calorie restriction from affecting performance. The general guidelines for healthy weight loss should be followed: Reduce energy intake by 200 to 500 Calories/day, increase activity, and change the behaviors that led to weight gain (see Chapter 9).

Relative energy deficiency in sport
When looking at what most athletes eat, it may seem unlikely that they are not getting enough calories. However, a syndrome referred to as **relative energy deficiency in sport (RED-S)** is not uncommon (**Figure 10.16**). It occurs when athletes do not consume

FIGURE 10.16 **Low energy availability** Relative energy deficiency in sport (RED-S) can affect the health and performance of all athletes. In females, it may manifest as the female athlete triad.

a. When energy intake does not meet the high demands of exercise, low energy availability can lead to RED-S, which has negative effects on both performance and health.

b. The set of RED-S symptoms known as the female athlete triad affect reproductive and bone health. Low calcium intake and low estrogen levels lead to low peak bone mass, premature bone loss, and increased risk of stress fractures and osteoporosis. Neither adequate dietary calcium nor the increase in bone mass caused by weight-bearing exercise can compensate for the bone loss caused by low estrogen levels.

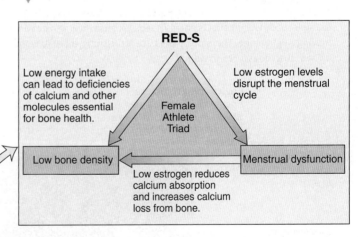

Think Critically

Why is bone loss accelerated in young girls who are not menstruating?

enough to support their energy expenditure, so they do not have enough energy available to maintain their heath and support body functions. For example, if a female athlete has low energy availability, it may progress to RED-S, causing changes in hormone levels that affect the menstrual cycle and lead to low bone mineral density. This triad of low energy availability, menstrual irregularities, and poor bone health was previously known as the **female athlete triad**. It is now recognized as part of RED-S (see Figure 10.16). RED-S can affect any athlete, male or female, and can alter many other aspects of physiology that impair health and exercise performance.[21] RED-S is more common in athletes who intentionally restrict their energy intake and weight in order to optimize performance and in those involved in sports that have weight classes, such as wrestling and boxing. These athletes are also vulnerable to eating disorders such as anorexia and bulimia.[22]

Athletes in sports with weight classes may put their health at risk by trying to lose weight rapidly. Competing at the high end of a weight class is thought to offer an advantage over smaller opponents, and so athletes may try to lose weight by using sporadic diets that severely restrict energy intake or dehydrating themselves through such practices as vigorous exercise, fluid restriction, wearing of vapor-impermeable suits, or use of hot environments, such as saunas and steam rooms, to increase sweat loss. They may also resort to even more extreme measures, such as vomiting and the use of diuretics and laxatives. These practices can be dangerous and even fatal (**Figure 10.17**). They may impair performance and can adversely affect heart and kidney function, temperature regulation, and electrolyte balance.

Carbohydrate, Fat, and Protein Needs

The composition of the diet is also a factor in maximizing athletic performance. Carbohydrate is an important fuel source for brain and muscle during exercise, and body stores of carbohydrate are relatively limited. Therefore, getting enough dietary carbohydrate is important for athletic performance. Carbohydrate recommendations range widely from 3 to 12 g/kg/day, depending on the type of training or competition.[20] The timing of carbohydrate intake can be manipulated so carbohydrate is available to promote optimal performance during competition or key training sessions (discussed below).

Dietary fat is needed to provide energy and essential fatty acids, as well as to facilitate the absorption of fat-soluble vitamins. The amount of fat recommended in an athlete's diet is the same as that for the general population—between 20 and 35% of energy (**Figure 10.18**). To allow for enough carbohydrate, fat intakes at the lower end of this range may be needed for some athletes, but diets that are very low in fat (less than 20% of calories) do not benefit performance and may reduce intake of fat-soluble vitamins and essential fatty acids.[20]

Although protein is not a significant energy source, accounting for only about 5% of energy expended, dietary protein is important as both a trigger and a substrate for the synthesis of proteins needed for exercise. Protein is needed for the maintenance, growth, and repair of muscles and non-muscle tissues such as tendons and bones.[20] Protein is also needed to support the metabolic reactions that generate ATP.

FIGURE 10.17 Making weight After three young wrestlers died while exercising in plastic suits in order to sweat off water, wrestling guidelines were changed to improve safety.[23] Weight classes were altered to eliminate the lightest class, plastic sweat suits were banned, weigh-ins were moved to one hour before competition, and mandatory weight-loss rules were instituted. The percentage of body fat can be no less than 5% for college wrestlers and 7% for high school wrestlers.

Carolyn Kaster / ©AP / Wide World Photos

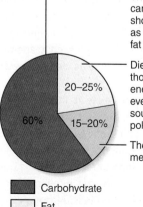

FIGURE 10.18 Proportions of energy-yielding nutrients in an athlete's diet The proportions of carbohydrate, fat, and protein recommended in the diets of athletes, shown in the pie chart are within the ranges recommended for the general public: 45 to 65% of total energy from carbohydrate, 20 to 35% of energy from fat, and 10 to 35% of energy from protein.

- The majority of calories should come from carbohydrate. Most of the carbohydrate should be from nutrient-dense choices such as whole grains, fruits, vegetables, and low-fat dairy products.
- Dietary fat is essential for health even though body fat stores contain enough energy to fuel even the longest endurance events. Most dietary fat should be from sources high in heart-healthy mono- and polyunsaturated fats.
- The protein needs of athletes can be met with either plant or animal sources.

■ Carbohydrate
□ Fat
▨ Protein

These reactions, which increase during exercise, depend on proteins such as enzymes and transport proteins. The amount of dietary protein needed to support changes in metabolism and maintain and repair lean tissues ranges from 1.2 to 2.0 g/kg/day.[20] Even higher intakes may be needed for short periods of intense training or when energy intake is low. While this amount is greater than the RDA (0.8 g/kg per day), it is not greater than the amount of protein habitually consumed by most athletes.[24] For example, an 85-kg man consuming 3000 Calories, of which 15 to 20% is from protein, would be consuming 1.6 g of protein/kg of body weight. The timing of protein intake is important for optimizing the muscle protein synthesis that occurs in response to exercise; protein should be consumed in the immediate postexercise period (discussed below).

Vitamin and Mineral Needs

An adequate intake of vitamins and minerals is essential for optimal performance. These micronutrients are needed for energy metabolism, oxygen delivery, protection against oxidative damage, and repair and maintenance of body structures.

Exercise increases the amounts of many vitamins and minerals used both in metabolism during exercise and in repairing tissues after exercise. In addition, exercise may increase losses of some micronutrients. Nevertheless, special micronutrient recommendations have not been made for athletes, and most athletes can meet their needs by consuming a balanced diet that meets their energy needs. Because athletes must eat more food to satisfy their higher energy needs, they consume extra vitamins and minerals with these foods, particularly if they choose nutrient-dense foods. Athletes who restrict their intake in order to maintain a low body weight, eliminate one or more food groups from their diets, or make poor food choices may benefit from vitamin and/or mineral supplements. Nutrients of key concern for athletes include iron, calcium, vitamin D, and some antioxidants.[20]

Iron Iron deficiency is common in athletes. They are at risk because exercise increases both the demand for iron and iron losses; reduced iron intake and absorption can also affect iron status.[25] Even iron deficiency that has not progressed to anemia can impair athletic and mental performance as well as overall health.[20]

Exercise increases iron needs because it increases the need for a number of iron-containing proteins such as hemoglobin and myoglobin. Intense training increases iron losses in feces, urine, and sweat.[20] A phenomenon called foot-strike hemolysis, which refers to the breaking of red blood cells due to the contraction of large muscles or impact in events such as running, has also been blamed for iron loss. Although this hemolysis does occur, the iron from these cells is recycled, and the breaking of red blood cells stimulates the production of new ones, so it does not contribute to poor iron status.[26] Iron losses are greater in female athletes due to the iron lost in menstrual blood. This increases the risk of iron deficiency; more than 50% of female adolescent athletes may experience iron deficiency.[25, 26]

Inadequate iron intake may also contribute to low iron stores in athletes of both genders who are attempting to keep their body weight down and/or do not eat meat, which is an excellent source of readily absorbable heme iron. The ability to obtain adequate iron is further impaired by a reduction in iron absorption from the GI tract that occurs for several hours after exercise.[25, 27]

Iron deficiency anemia (see Chapter 8) should not be confused with **sports anemia**, which is an adaptation to training that does not seem to impair the delivery of oxygen to tissues (**Figure 10.19**).[19] Although a specific iron RDA has not been set for athletes, the DRIs acknowledge that the requirement may be 30 to 70% higher for athletes than for the general population.[28]

Calcium and vitamin D Adequate calcium and vitamin D are needed for bone health. Low calcium intake can occur in athletes with disordered eating and in those who restrict energy intake or avoid dairy products. The risk of low bone mineral density and stress fractures is increased by RED-S, particularly in female athletes who experience menstrual dysfunction. Regulation of calcium absorption and metabolism depends on adequate vitamin D. There is also growing evidence that vitamin D is important for other aspects of athletic performance.[20] Athletes who live at latitudes above the 35th parallel or who train and compete indoors are at greater risk for low blood vitamin D levels. Other factors such as dark

FIGURE 10.19 Sports anemia Training causes blood volume to expand in order to increase oxygen delivery, but the synthesis of red blood cells lags behind the increase in blood volume.[19] The result is a decrease in the percentage of blood volume that is red blood cells. However, the total number of red blood cells stays the same or increases slightly, so the transport of oxygen is not impaired. As training progresses, the number of red blood cells increases to catch up with the increase in total blood volume.

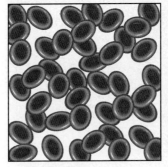
Normal

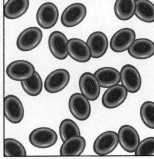
Sports anemia

skin color, training in the early morning or evening hours, and covering the skin with clothing, equipment, or sunblock also increase risk (see Chapter 7). Unless the diet is carefully planned to include foods high in vitamin D, such as milk, salmon, and swordfish, supplements may be needed to maintain sufficient vitamin D status.[20]

Antioxidant nutrients Exercise increases the amount of oxygen used by the muscles and the rate of ATP-producing metabolic reactions. This increased oxygen use increases the production of free radicals, but rather than causing damage, it causes an increase in antioxidant systems in the body and benefits health.[29] Some of these antioxidant defenses rely on dietary antioxidants such as vitamin C, vitamin E, β-carotene, and selenium. Despite the importance of antioxidants for health and performance, there is little evidence that supplementation with antioxidants improves human performance.[20] The best way to ensure adequate intake of dietary antioxidants is to consume a varied nutrient-dense diet.

Water and Electrolyte Needs

During exercise, water is needed to cool the body and to transport both oxygen and nutrients to the muscles and remove waste products from them. Extra fluids are needed during exercise because water losses from evaporation and sweat are increased.

If fluid intake does not keep up with losses, dehydration can occur. Dehydration can be hazardous to the performance and health of even the most casual exerciser (**Figure 10.20**). Most people drink only enough to assuage their thirst while exercising, but because thirst is not a reliable short-term indicator of the body's water needs, this amount typically is not enough to replace water losses. It is therefore important to schedule regular fluid breaks. However, even when fluids are consumed at regular intervals throughout exercise, it may not be possible to drink enough to compensate for losses. This is a particular problem when exercising in the heat. The risk of dehydration is greater in hot environments. However, dehydration may also occur when exercising in the cold because cold air tends to be dry, so evaporative losses from the lungs are greater. In addition, insulated clothing worn in cold weather may increase sweat loss, and fluid intake may be reduced because a chilled athlete may be reluctant to drink a cold beverage. Female athletes training in cold weather may also limit fluid intake in order to avoid the inconvenience of removing clothing in order to urinate.

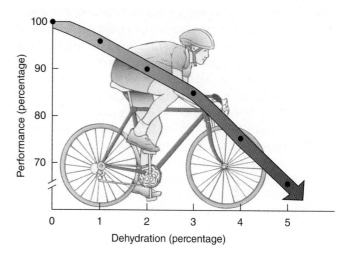

FIGURE 10.20 **Dehydration and performance** As the severity of dehydration increases, exercise performance declines. Even mild dehydration—a water loss of 1 to 2% of body weight—can impair exercise performance. A 3% reduction in body weight can significantly reduce the amount of blood pumped with each heartbeat because the blood volume is decreased. This, in turn, reduces the circulatory system's ability to deliver oxygen and nutrients to cells and remove waste products.

Interpret the Data

If a person loses 4% of his body weight as water during a competition, by what percentage will his performance be decreased by the end of the event?

a. 10%   b. 20%
c. 25%   d. 30%

Anyone exercising should consume enough fluid before exercise to start out well hydrated, enough during exercise to minimize weight loss, and enough after exercise to restore fluid balance during the remainder of the day. Recommendations for the specific amounts and types of beverages are given below.

Dehydration and heat-related illnesses Dehydration occurs when water loss is great enough for blood volume to decrease, thereby reducing the ability of the circulatory system to deliver oxygen and nutrients to exercising muscles (see Chapter 8). A decrease in blood volume also reduces blood flow to the skin and the amount of sweat produced, thus limiting the body's ability to cool itself. As a result, core body temperature can increase, and with it the risk of various **heat-related illnesses**.

---

heat-related illnesses Conditions, including heat cramps, heat exhaustion, and heat stroke, that can occur due to an unfavorable combination of exercise, hydration status, and climatic conditions.

Heat cramps are involuntary muscle spasms that occur during or after intense exercise, usually in the muscles involved in the exercise. They are a form of heat-related illness caused by an imbalance of electrolytes at the muscle cell membranes. They can occur when water and salt are lost during extended exercise.

Heat exhaustion occurs when water loss causes blood volume to decrease so much that it is not possible both to cool the body and to deliver oxygen to active muscles. It is a form of heat-related illness characterized by a rapid but weak pulse, low blood pressure, disorientation, profuse sweating, and fainting. A person who is experiencing symptoms of heat exhaustion should stop exercising and move to a cooler environment.

Heat exhaustion can progress to **heat stroke**, the most serious form of heat-related illness. It occurs when core body temperature rises above 105°F, causing the brain's temperature-regulatory center to fail. When this occurs, the individual does not sweat even though body temperature is rising. Heat stroke is characterized by elevated body temperature; hot, dry skin; extreme confusion; and unconsciousness. It requires immediate medical attention.

Exercising in hot, humid weather increases the risk of heat-related illnesses. As environmental temperature rises, the body has more difficulty dissipating heat, and as humidity rises, the body's ability to cool itself through evaporation decreases (**Figure 10.21**).

Low blood sodium Sweating helps us stay cool. Sweat is mostly water, but excessive sweating also causes losses of sodium and small amounts of potassium, calcium, and magnesium.[20] If the water and sodium lost in sweat are not replaced in the right proportions, low blood sodium, or hyponatremia, may result (see Chapter 8). For most activities, sodium losses are small, so sweat losses can be replaced with plain water, and lost electrolytes can be replaced during the meals following exercise. However, during endurance events such as marathons and triathlons, when sweating continues for many hours, both water and sodium need to be replenished. If an athlete replaces the lost fluid with plain water, the sodium that remains in the blood is diluted, causing hyponatremia (**Figure 10.22**). As sodium concentrations in the blood decrease, water moves into body tissues by osmosis, causing swelling. Fluid accumulation in the lungs interferes with gas exchange, and fluid accumulation in the brain causes disorientation, seizure, coma, and death.

The risk of hyponatremia can be reduced by consuming sodium-containing sports drinks during long-distance events, increasing sodium intake several days prior to a competition, and avoiding acetaminophen, aspirin, ibuprofen, and other nonsteroidal anti-inflammatory drugs, which may contribute to the development of hyponatremia by interfering with kidney function. The early symptoms of hyponatremia may be similar to those of dehydration: nausea, muscle cramps,

FIGURE 10.21 Heat index[30] Exercise in extreme conditions increases the risk of heat-related illness. *Heat index*, or *apparent temperature*, is a measure of how hot it feels when the relative humidity is added to the air temperature. To find the heat index, find the intersection of the temperature on the left side of the table and the relative humidity across the top. The colored zones correspond to heat index levels that contribute to increasingly severe heat illnesses with continued exposure and/or physical activity.

°F / RH	40	45	50	55	60	65	70	75	80	85	90	95	100
110	136												
108	130	137											
106	124	130	137										
104	119	124	131	137									
102	114	119	124	130	137								
100	109	114	118	124	129	136							
98	105	109	113	117	123	128	134						
96	101	104	108	112	116	121	126	132					
94	97	100	102	106	110	114	119	124	129	135			
92	94	96	99	101	105	108	112	116	121	126	131		
90	91	93	95	97	100	103	106	109	113	117	122	127	132
88	88	89	91	93	95	98	100	103	106	110	113	117	121
86	85	87	88	89	91	93	95	97	100	102	105	108	112
84	83	84	85	86	88	89	90	92	94	96	98	100	103
82	81	82	83	84	84	85	86	88	89	90	91	93	95
80	80	80	81	81	82	82	83	84	84	85	86	86	87

With prolonged exposure and/or physical activity:

- **Extreme danger** — Heat stroke highly likely
- **Danger** — Heat stroke, heat cramps, and/or heat exhaustion likely
- **Extreme caution** — Heat stroke, heat cramps, and/or heat exhaustion possible
- **Caution** — Fatigue possible

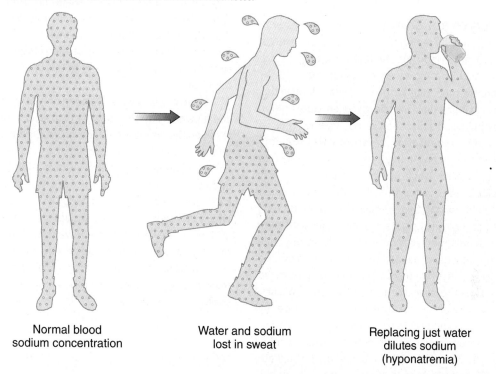

FIGURE 10.22 **Diluting blood sodium** Water and sodium are lost in sweat. Drinking plain water during extended periods of excessive sweating can dilute the sodium remaining in the blood. Hyponatremia occurs in about 6% of male ultra-endurance athletes.[31]

Normal blood sodium concentration

Water and sodium lost in sweat

Replacing just water dilutes sodium (hyponatremia)

disorientation, slurred speech, and confusion. A proper diagnosis is important because drinking water alone will make the problem worse. Mild symptoms of hyponatremia can be treated by eating salty foods or drinking a sodium-containing beverage, such as a sports drink. More severe symptoms require medical attention.

Concept Check

1. **How** do exercise intensity and duration affect the energy needs of athletes?
2. **Why** do athletes need more carbohydrate and protein than nonathletes?
3. **What** factors increase an athlete's risk of iron deficiency?
4. **What** environmental conditions increase the risk of dehydration?

10.6 Food and Drink to Optimize Performance

LEARNING OBJECTIVES

1. **Discuss** the amount of fluid recommended for athletes.
2. **Plan** a precompetition meal for a marathon runner.
3. **Describe** the recommendations for food and drink during extended exercise.
4. **Explain** the role of carbohydrate and protein in postcompetition meals.

For most of us, a trip to the gym requires no special nutritional planning, but for competitive athletes, when and what they eat and drink before, during, and after competition are as important as a balanced overall diet. The type and amount of fluids and foods consumed at these times may give or take away the extra seconds that can mean victory or defeat.

What to Eat and Drink Before Exercise

Preparing for an athletic event involves not just physical training but also careful consideration of what to eat and drink in the days and hours before the event. Because the availability of carbohydrate is essential for exercise performance, strategies to increase glycogen stores may begin several days before an event; the food and fluids consumed just before an event

Maximizing glycogen stores

Glycogen stores are a source of glucose, and larger glycogen stores allow exercise to continue for longer periods (**Figure 10.23**). Glycogen stores and hence endurance are increased by increasing carbohydrate intake. Athletes preparing for events lasting longer than 90 minutes can maximize their muscle glycogen stores before a competition by following a diet and exercise regimen referred to as **glycogen supercompensation**, or **carbohydrate loading**. Such a regimen involves consuming a very high-carbohydrate diet while resting for 36 to 48 hours before competition.[20, 33] The diet should provide 10 to 12 g of carbohydrate/kg of body weight per day. For a 150-lb person, this is equivalent to about 700 g of carbohydrate per day. Having a stack of pancakes with syrup and a glass of milk only provides about 150 g of carbohydrate, less than a quarter of what is needed for the day. A number of commercial high-carbohydrate beverages (50 to 60 g of carbohydrate in 8 fluid oz) are available to help athletes consume the amount of carbohydrate recommended to maximize glycogen stores. These should not be confused with sports drinks designed to be consumed during competition, which contain only about 10 to 20 g of carbohydrate in 8 fluid oz. Trained athletes who follow a carbohydrate-loading regimen can double their muscle glycogen content.[34]

Although glycogen supercompensation is beneficial to endurance athletes, it provides no benefit, and even has some disadvantages, for those exercising for less than 90 minutes. For every gram of glycogen in the muscle, about 3 g of water is also deposited. This water will cause weight gain and may cause some muscle stiffness. As glycogen is used, the water is released. This can be an advantage when exercising in hot weather, but the extra weight is a disadvantage for individuals competing in short-duration events.

The precompetition meal

The beverages and foods consumed before competition should provide plenty of fluid and carbohydrate and not cause gastrointestinal upset. To ensure hydration, athletes should drink about 5 to 10 mL/kg of body weight (about 1.5 to 3 cups for a 70 kg athlete) two to four hours before an exercise session (**Table 10.2**). Sodium consumed in this fluid or along with this in food can help with fluid retention.

Muscle glycogen is depleted only by activity, but liver glycogen, which supplies blood glucose, is depleted even during rest if no food is ingested. Therefore, high-carbohydrate meals or snacks should be consumed one to four hours before exercise (see Table 10.2). This will help maintain blood glucose and continue to increase glycogen stores. Spicy foods, which can cause heartburn, and large amounts of simple sugars, which can cause diarrhea, should also be avoided unless the athlete is accustomed to eating these foods. The effects of different foods should be tested during training, not during competition. In addition to providing nutritional clout, a meal that includes "lucky" foods may provide an added psychological advantage.

What to Eat and Drink During Exercise

During exercise, whether casual or competitive, exercisers should try to drink enough fluid to limit weight loss to less than 2% of body weight.[20] For most athletes, adequate hydration can be achieved by drinking about 1.5 to 3.5 cups (400 to 800 mL) of fluid per hour for the duration of the exercise.

Most people don't need to eat while they exercise. For exercise lasting less than 45 minutes, no food is needed, and water is the only fluid needed (see Table 10.2). If the activity lasts 45 minutes or longer, however, it is important to consume carbohydrate during exercise in order to maintain glucose supplies

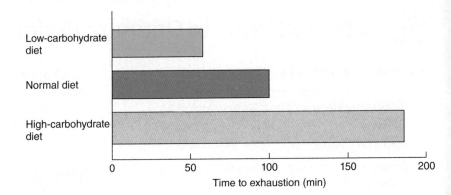

FIGURE 10.23 **Dietary carbohydrate and endurance** The amount of carbohydrate consumed in the diet affects the level of muscle glycogen and hence an athlete's endurance. This graph shows endurance capacity during cycling exercise after three days of a low-carbohydrate diet (less than 5% of calories from carbohydrate), a normal diet (about 55% of calories from carbohydrate), and a high-carbohydrate diet (82% of calories from carbohydrate).[32]

---

glycogen supercompensation or **carbohydrate loading** A regimen designed to increase muscle glycogen stores beyond their usual level.

TABLE 10.2 A summary of food and drink for exercise

When to eat and drink	What to eat	What to drink
1 to 4 hrs before exercise	• Calories: about 300 to 500 • Carbohydrate: 1 to 4 g/kg of body weight • Protein: 8 to 16 g (10 to 20% of calories) • Fat: 3 to 8 g (10 to 25% of calories) • Fiber: low	2 to 4 hrs before exercise: Drink 5 to 10 mL/kg of body weight
During exercise	For exercise lasting: • < 45 min: No food needed • 45–75 min: small amounts of carbohydrate from food or beverages • 60–150 min: 30 to 60 g of carbohydrate per hour • > 150 min: 90 g of carbohydrate per hour	Drink 400 to 800 mL/hr For exercise lasting: • < 45 min: drink plain water • ≥ 45 min: sports drink containing carbohydrate and sodium
After exercise	• Carbohydrate: 1.0 to 1.2 g/kg/hour for 4 to 6 hours • Protein: 0.3 g/kg of body weight within 2 hrs after exercise; consume protein in meals for 24 hrs after exercise	Replace each kg of body weight lost with 1.25 to 1.5 L of fluid (about 2.5 to 3 cups/lb lost)

and thus delay fatigue. Liver glycogen stores are depleted by the overnight fast, so carbohydrate consumption during exercise is particularly important for athletes who exercise in the morning, before eating breakfast.

For high-intensity exercise that lasts 45 to 75 minutes, only small amounts of carbohydrate should be consumed during the activity. In addition to providing fuel, frequent contact of carbohydrate with the mouth enhances perceptions of well-being and increases work output. During endurance exercise lasting 1 to 2.5 hours, 30 to 60 g of carbohydrate (the amount in a large banana or an energy bar) should be consumed per hour. For ultra endurance exercise (greater than 2.5 to 3 hours), carbohydrate intake should increase to 90 g/hour. This carbohydrate can be obtained from a sports drink, but consuming a solid-food snack or a carbohydrate gel with water is also appropriate.[20] Athletes consuming solid food should choose high-carbohydrate snacks that are low in fat and fiber and moderate in protein to speed absorption and minimize GI distress (see *Thinking It Through*).

One way to obtain both fluid and carbohydrate during exercise is with a sports drink (**Figure 10.24**). Commercial sports drinks contain small amounts (about 10 to 20 g of carbohydrate/cup) of rapidly absorbed sources of carbohydrate, such as glucose, sucrose, or glucose polymers (chains of glucose molecules). The right proportion of carbohydrate to water is important. If the concentration of carbohydrate is too low, it will not help performance; if it is too high, it will delay stomach emptying. Water and carbohydrate trapped in the stomach do not benefit the athlete and may cause stomach cramps. Because fruit juices and soft drinks contain twice as much sugar as sports drinks, they are not recommended unless they are diluted with an equal volume of water. Flavored beverages also tempt athletes to drink more, helping to ensure adequate hydration.

Snacks and sports drinks also provide sodium. When sodium losses are high, such as in athletes with high sweat rates, salty sweat, or those exercising longer than two hours, sodium-containing fluids are recommended. Although the amount of sodium lost in sweat during exercise lasting less than two hours is usually minimal, having a sodium-containing snack or beverage during exercise lasting more than about an hour will reduce the risk of hyponatremia, improve glucose and water absorption, and stimulate thirst.

What to Eat and Drink After Exercise

When you stop exercising, your body must shift from the task of breaking down glycogen, triglycerides, and muscle proteins for fuel to the job of restoring muscle and liver glycogen, depositing lipids, and synthesizing muscle proteins. Meals eaten after exercise should replenish lost fluid, electrolytes, and glycogen and provide protein for building and repairing muscle tissue.

After exercise, the first priority for all exercisers is to replace fluid losses. To restore hydration, water and sodium should be consumed at a modest rate. Because losses in sweat and urine continue after exercise has stopped, rehydration requires ingestion of a greater volume of fluid than the deficit. Each kilogram of weight lost should be replaced with 1.25 to 1.5 L of fluid (about 2.5 to 3 cups/lb lost).[20]

For serious athletes competing on consecutive days, glycogen replacement is also a priority. To maximize glycogen replacement, a snack or beverage providing about 1.0 to 1.2 g/kg body weight of easily absorbed carbohydrate should be consumed per hour for about four to six hours.[20] This is about 70 to 85 g of carbohydrate for a 70-kg (154-lb) person. This glycogen-restoring regimen can replenish muscle and liver glycogen within 24 hours

Thinking It Through

A Case Study on Snacks for Exercise

Mark enjoys long-distance cycling. On weekends, he often goes on a 40- or 50-mile ride, which takes him three to four hours. Even though he consumes a sports drink from his bike bottle, after about two hours, he gets hungry and fatigued, and so he is looking for a snack that's easy to carry.

Should Mark choose a high-carbohydrate, high-protein, or high-fat snack to provide the energy he needs to continue his ride? Why?

Answer: Mark should choose a snack that is high in carbohydrate and low in fat because carbohydrate is the fuel that is depleted during prolonged exercise, and too much fat can delay stomach emptying and cause stomach upset.

One option for Mark is half of a turkey sandwich and a cup of grapes. This provides 270 Calories with 45 g of carbohydrate and 3 g of fat.

Suggest another snack that Mark could put together using items he might have in his kitchen cupboard that would provide 250 to 300 Calories, about 50 grams of carbohydrate, and less than 8 grams of fat.

Your answer:

The bike shop sells a variety of energy bars, or *endurance bars*, that claim to prevent hunger and maintain blood glucose during extended activity. These provide about 45 g of carbohydrate and no more than 10% fat. Mark wonders if he should just bring a candy bar instead.

Compare the amounts of carbohydrate and fat in the sandwich and grapes with the amounts in the endurance bars and the candy bars shown here. Which is a better option for Mark during his ride? Why?

Your answer:

Compare the amounts of added sugars, saturated fat, and fiber in the endurance bar with the amounts in the candy bar. What would you tell Mark about these?

Your answer:

Portability and cost are important in choosing a snack for exercise. Which would you choose—a sandwich and grapes, the endurance bar, or the candy bar? Why?

Your answer:

(Check your answers in Appendix L.)

Sports bar
1 bar (65 g)

Nutritional information

Amount/ sports bar		Amount/ candy bar
230	Energy (Calories)	271
2	Total fat (g)	14
0.5	Saturated fat (g)	5
45	Total carbohydrate (g)	35
3	Fiber (g)	1
14	Added sugars (g)	30
10	Protein	4

Candy bar
1 bar (57 g)

---

of an athletic event and is critical for optimizing performance on the following day. However, most of us are not competitive athletes, so we don't need a special glycogen replacement strategy to ensure that our glycogen stores are replenished before our next visit to the gym. If your routine includes 30 to 60 minutes at the gym, a typical diet that provides about 55% of calories from carbohydrate will replace the glycogen used so that you will be ready for a workout again the next day.

Protein is also important during the recovery period. Including high-quality protein in postexercise foods or beverages stimulates muscle protein synthesis and provides the amino acids needed for protein synthesis and repair.[35]

FIGURE 10.24 **Consider the calories in sports drinks** Sports drinks are unnecessary for short-term exercise, so if you are exercising to lose weight, consider the calories. A typical 20-oz sports drink provides about 150 Calories, so it will replace more than half of the calories expended during a 40-minute ride on a stationary bicycle.

Enhancement of muscle protein synthesis can be achieved by consuming 0.3 g of protein/kg body weight within 2 hours after exercise; balanced meals containing protein should be consumed during the 24-hour period after exercise.[20]

Concept Check

1. **How** much of what fluid should you drink during a two-hour bike ride?
2. **What** should an athlete eat as a precompetition meal, and why?
3. **Why** might a low-carbohydrate diet be a poor choice for an endurance athlete?
4. **When** should athletes consume protein?

10.7 Ergogenic Aids

LEARNING OBJECTIVES

1. **Explain** why certain vitamins and minerals are used as ergogenic aids.
2. **Discuss** the risks associated with using anabolic steroids.
3. **Explain** how supplements might enhance performance in short, intense activities.
4. **Describe** how a supplement might improve endurance.

Citius, altius, fortius—"faster, higher, stronger"—is the motto of the Olympic Games. For as long as there have been competitions, athletes have been willing to go to great lengths to improve their performance. Athletes yearning for a competitive edge are willing to try all kinds of supplements and other **ergogenic aids**. Many of the vitamins, minerals, and other substances in ergogenic supplements are involved in providing energy for exercise or promoting recovery from exercise. Most of these are not effective; only a few have a small benefit for some aspect of athletic performance. When considering whether to use an ergogenic supplement or any other type of supplement, a risk–benefit analysis should be used to determine whether the supplement is appropriate for you.

Vitamin and Mineral Supplements

Many of the promises made to athletes about the benefits of vitamin and mineral supplements are extrapolated from the biochemical functions of these micronutrients. For example, B vitamins are promoted to enhance ATP production because of their roles in muscle energy metabolism. Vitamin B_6, vitamin B_{12}, folic acid, and iron are promoted for aerobic exercise because they are involved in the synthesis of red blood cells, which transport oxygen to exercising muscles. Supplements of antioxidant nutrients such as vitamin E, vitamin C, and selenium are suggested to reduce oxidative stress because exercise increases the production of free radicals. Supplements of chromium (chromium picolinate) and vanadium (vanadyl sulfate) are marketed to increase lean body mass and decrease body fat because they are needed for insulin action, and insulin promotes protein synthesis. These micronutrients are important for optimal athletic performance, but unless the athlete is deficient in one or more of them, there is no evidence that supplements enhance athletic performance (**Table 10.3**).

ergogenic aid A substance, appliance, or procedure that improves athletic performance.

TABLE 10.3 Claims, benefits, and risks of ergogenic supplements

Supplement	Claims	Effectiveness	Risks
Amino acids (ornithine, arginine, and lysine)	Stimulate growth hormone release; enhance muscle growth.	Large doses stimulate the release of growth hormone, possibly to higher levels than from exercise alone, but the effect on exercise performance is unclear.[36, 37]	Large doses of one amino acid can interfere with absorption of other amino acids.
Antioxidant nutrients: vitamin E, vitamin C, and selenium	Reduce oxidative stress; improve performance.	Exercise increases oxidative processes but also boosts endogenous antioxidant systems. Antioxidant supplements have not been found to improve performance.[38]	None if less than UL
B vitamins: B_6, B_{12}, folate	Increase ATP production, improve performance.	Needed for energy metabolism and red blood cell synthesis. A deficiency interferes with ATP production and oxygen availability and impairs athletic performance. Supplements are effective only if the vitamin is deficient in the diet.	None if less than UL.
Beetroot juice (nitrate)	Improves endurance by increasing production of nitric oxide, which dilates blood vessels and makes mitochondria more efficient.	Improves cardiorespiratory endurance in athletes.[39]	Reddish urine and stools, which is harmless, and increased risk of kidney stones in susceptible individuals. Long-term safety is unknown.[40]
β-alanine	Neutralizes acid. Increases strength, power, and endurance during high-intensity exercise.	Supplements can delay fatigue and increase the ability to sustain muscle power output during high-intensity exercise lasting between 1 and 4 minutes.[41]	Safety has not been adequately studied, so caution is recommended in using this supplement.[42]
β-hydroxy-β-methylbutyrate (HMB)	Increases strength and muscle growth and improves muscle recovery.	Some studies have found that supplementation in trained athletes has a beneficial effect on body composition and aerobic capacity.[43]	No demonstrated toxicity in humans.[44]
Bicarbonate	Neutralizes acid; preserves muscle function; delays fatigue; improves performance.	Moderately positive effect in intense exercise lasting only 1 to 7 minutes; may be beneficial in sports involving intermittent or sustained periods of high-intensity exercise.[45]	Abdominal cramps and diarrhea. Other possible side effects have not been carefully researched.
Caffeine	Increases release of fatty acids from adipose tissue; increases alertness; spares glycogen; enhances endurance.	Improves performance for sustained endurance exercise as well as intermittent activity of prolonged duration, as occurs in a team sport such as soccer.[46]	Nervousness, anxiety, insomnia, digestive discomfort, abnormal heartbeat. May impair performance by causing gastrointestinal upset or caffeine intoxication.
Carnitine	Increases endurance by enhancing fat utilization during exercise.	Enough carnitine is made in the body to ensure efficient use of fatty acids. Carnitine supplements have not been shown to increase endurance.[47]	D,L-carnitine and D-carnitine can be toxic.
Chromium (chromium picolinate)	Increases lean body mass; decreases body fat; delays fatigue.	Needed for insulin action, and insulin promotes protein synthesis, but supplements have not consistently been found to affect muscle strength or body composition.[48]	No adverse effects have been identified.

(continues)

TABLE 10.3	Claims, benefits, and risks of ergogenic supplements *(continued)*		
Supplement	**Claims**	**Effectiveness**	**Risks**
Creatine	Increases strength, power, and muscle mass when combined with strength training.	Supplementation increases strength, power, and muscle mass more than strength training alone. The greatest benefit is for short-duration high-intensity intermittent exercise such as sprinting and weight lifting.[49]	Weight gain and GI discomfort. Should not be used by individuals with or at risk for renal disease.[50, 51] Consult a physician before using
Iron	Increases availability of oxygen.	Needed for red blood cell synthesis and delivery of oxygen to cells. A deficiency will interfere with oxygen availability and impair athletic performance. Supplements are effective only if iron is deficient.	None if less than UL.
Medium-chain triglycerides (MCTs)	Increase the availability of fat as a fuel for exercise, sparing glycogen and delaying fatigue.	Supplementation has not been found to increase endurance, spare glycogen, or enhance performance.[52]	None known.
Protein (whey, casein, soy, pea)	Optimizes the exercise-induced enhancement of muscle protein synthesis if consumed after exercise.	Supplements are effective but no more effective than food sources of protein.	None.
Vanadium (vanadyl sulfate)	Assists insulin action; increases muscle growth and lean body mass.	There is no evidence that supplemental vanadium increases lean body mass.[53]	Safe at doses of less than 1.8 mg per day. Higher doses cause abdominal discomfort, diarrhea, nausea, gas, and potentially kidney damage over the long term.

Supplements to Build Muscle

Bigger stronger muscles often lead to greater athletic success. A number of ergogenic aids are used to increase muscle size and strength. Protein supplements and hormones that promote muscle growth are discussed below, and supplements containing amino acids or β-hydroxy-β-methylbutyrate are summarized in Table 10.3.

Protein supplements are marketed to athletes with the promise of enhancing muscle growth and strength. Consuming high-quality protein after exercise optimizes the muscle protein synthesis stimulated by exercise, but this protein does not need to come from supplements. Foods such as lean meat, eggs, milk, or soymilk consumed after exercise can provide the protein needed for muscle growth. When these whole food sources of protein are not available, protein supplements can help athletes optimize protein synthesis after exercise.[20] Protein supplements should be used to optimize exercise recovery, not to replace nutrient-dense whole foods.

Some athletes also use hormones to increase muscle mass. **Anabolic steroids** are synthetic versions of the human steroid hormone testosterone. Natural testosterone stimulates and maintains the male sexual organs and promotes the development of bones and muscles and the growth of skin and hair. The synthetic testosterone in anabolic steroids has a greater effect on muscle, bone, skin, and hair than it does on sexual organs. When taken in conjunction with exercise and an adequate diet, anabolic steroids accelerate protein synthesis, causing an increase in muscle size and strength. However, they have extremely dangerous side effects. The use of synthetic testosterone shuts down the production of natural testosterone (**Figure 10.25**). Without natural testosterone, the sexual organs are not maintained; this leads to shrinkage of the testicles and a decrease in sperm production.[54] In adolescents, the use of anabolic steroids causes cessation of bone growth and stunted height. Anabolic steroid use may also cause oily skin and acne, water retention in the tissues, yellowing of the eyes and skin, coronary artery

anabolic steroids Synthetic fat-soluble hormones that mimic testosterone and are used to increase muscle strength and mass.

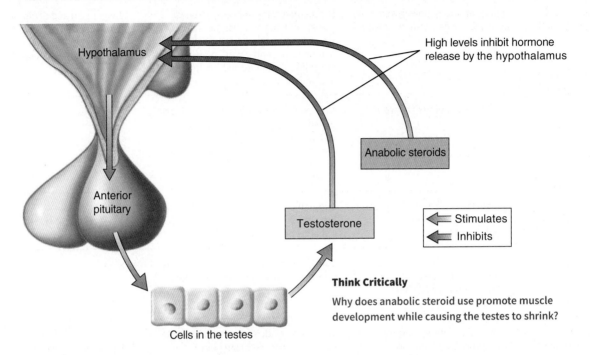

FIGURE 10.25 **Anabolic steroids** When testosterone levels are low, the hypothalamus releases a hormone that stimulates the anterior pituitary to secrete a hormone that increases the production of testosterone by the testes. High levels of either natural or synthetic testosterone inhibit the release of the stimulatory hormone from the hypothalamus, shutting down the synthesis of natural testosterone.

Think Critically
Why does anabolic steroid use promote muscle development while causing the testes to shrink?

disease, liver disease, and sometimes death. Users may experience psychological and behavioral side effects such as violent outbursts, insomnia, and depression, possibly leading to suicide.[54] Anabolic steroids are controlled substances; possession without a prescription is illegal. They are prohibited by the International Olympic Committee, the National Collegiate Athletic Association (NCAA), and most other sporting organizations. Despite the physical and legal risks, between 1 million and 3 million athletes in the United States have used anabolic steroids.[55]

Steroid precursors are compounds that can be converted into steroid hormones in the body. Like anabolic steroids, they are controlled substances. The best known of these is androstenedione, often referred to as "andro." Contrary to marketing claims, the use of andro or other steroid precursors has not been found to increase testosterone levels or produce any ergogenic effects, and these substances may cause some of the same side effects as anabolic steroids.[56]

Growth hormone is another hormone used to increase muscle protein synthesis. Despite this physiological effect, however, it has not been shown to enhance muscle strength, power, or aerobic exercise capacity, but there is evidence that it improves anaerobic exercise capacity.[57] Prolonged use of growth hormone can cause heart dysfunction, high blood pressure, and excessive growth of some body parts, such as hands, feet, and facial features.[57] Growth hormone is on the World Anti-Doping Agency's list of prohibited substances.

Supplements to Enhance Performance in Short, Intense Activities

A number of supplements are marketed to athletes who seek to improve performance in high-intensity sports that rely on anaerobic metabolism. Some, such as bicarbonate and β-alanine act by neutralizing acid (see Table 10.3). Acid produced during exercise is believed to interfere with muscle function and contribute to fatigue. Neutralizing the acid is believed to preserve muscle function, delay fatigue, and improve performance. Bicarbonate directly buffers acid, whereas β-alanine is used in the muscle to synthesize a compound called carnosine. Carnosine is a buffer that prevents changes in acidity within the muscle.

One of the most popular ergogenic supplements is **creatine**. This nitrogen-containing compound is found primarily in muscle, where it is used to make creatine phosphate (**Figure 10.26**). Higher levels of creatine and creatine phosphate provide more quick energy for short-term muscular activity. Combining creatine supplementation with strength training has been shown to increase strength, power, and muscle mass more than strength training alone. Intakes of 3 g/day are considered safe in healthy adults, but creatine supplements should not be taken by individuals with or at risk for renal disease.[50, 51] The FDA has advised consumers to consult a physician before using creatine.

FIGURE 10.26 **Creatine boosts creatine phosphate** Creatine can be synthesized in the liver and kidneys and is consumed in the diet in meat and milk. The more creatine consumed, the greater the amount of creatine stored in the muscles. Increasing creatine intake with supplements has been shown to increase levels of muscle creatine and creatine phosphate, which is made from it.[49] During short bursts of intense activity, creatine phosphate can transfer a phosphate group to ADP, forming creatine, and ATP that can be used for muscle contraction.

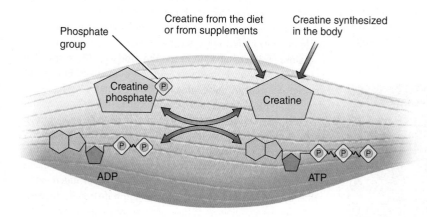

Ask Yourself

Why might creatine supplements have a greater ergogenic effect in someone who consumes a vegetarian diet than in someone who consumes a diet high in meat and dairy products?

Supplements to Enhance Endurance

Endurance athletes are concerned about running out of glycogen. Glycogen is spared when fat is used as an energy source, allowing exercise to continue for a longer time before glycogen is depleted and fatigue sets in. A number of substances, including carnitine and medium-chain triglycerides (MCTs), are taken to increase endurance because they are hypothesized to increase the amount of fat available to the muscle cells. Carnitine is found in the diet in red meats and dairy products, and it is synthesized in the body. It is needed to transport fatty acids into the mitochondria, where they are used to produce ATP by aerobic metabolism. Although supplements are promoted to enhance fatty acid availability in the mitochondria, there is no evidence that supplemental carnitine increases fatty acid transport. MCTs are composed of fatty acids with medium-length chains of 8 to 10 carbons. These fatty acids can be absorbed directly into the blood without first being incorporated into chylomicrons. They are therefore absorbed quickly, causing blood fatty acids levels to rise, increasing the availability of fat as a fuel for exercise. Neither carnitine nor MCTs have been found to increase endurance (see Table 10.3).

Athletes also use caffeine to increase aerobic capacity. Caffeine is a stimulant found in coffee, tea, soft drinks, and energy drinks (see *Debate: Energy Drinks for Athletic Performance?*). Consuming 3 to 6 mg of caffeine per kilogram of body weight, an amount equivalent to about 2.5 cups of percolated coffee, up to an hour before exercising has been shown to improve performance in endurance exercise and intermittent activity of prolonged duration, such as soccer.[58] Caffeine enhances performance both through its stimulatory effect on the central nervous system, which increases vigilance and alertness, and by increasing the release of fatty acids.[58, 59] When fatty acids are used as a fuel source, less glycogen is used, and the onset of fatigue is delayed. Caffeine is a diuretic but does not cause significant dehydration during exercise. Athletes who are unaccustomed to caffeine respond better to it than do those who consume caffeine routinely. In the past caffeine intake was regulated by the International Olympic Committee, but intake is no longer restricted.

Athletes also use the hormone erythropoietin, known as EPO, to enhance endurance. Natural erythropoietin is produced by the kidneys and stimulates cells in the bone marrow to differentiate into red blood cells. EPO can enhance endurance by increasing the number of red blood cells and hence the ability to transport oxygen to the muscles. It therefore increases aerobic capacity and spares glycogen. However, too much EPO can cause production of too many red blood cells, which can lead to excessive blood clotting, heart attacks, and strokes. EPO was banned in 1990, after it was linked to the deaths of more than a dozen cyclists.[66]

Diet, Supplements, and Performance

The bottom line with ergogenic aids is to do no harm. If they are used, careful consideration should be given to the risks compared with the benefits. Supplements may garner most of the press when it comes to exercise performance, but their impact is small compared to the benefits of a well-planned healthy diet (**Figure 10.27**).

Debate

Energy Drinks for Athletic Performance?

The Issue

Energy drinks are sold alongside sports drinks, and manufacturers of these beverages often sponsor athletes and athletic events. Should they be used as ergogenic aids? Is drinking them a safe way to improve your game?

Tony Cenicola / Redux Pictures

The popularity of energy drinks with names like Red Bull, Monster, and Full Throttle has soared over the past decade. They promise to keep you alert to study, work, drive, party all night, and perhaps excel at your next athletic competition. The main ingredients in these drinks are sugar and caffeine. Glucose is an important fuel for exercise, and caffeine is known to enhance endurance, so these drinks may seem like an ideal ergogenic aid.

A traditional sports drink, like Gatorade, contains about 28 g of sugar in 16 oz; a typical energy drink provides twice this much—55 to 60 g (or about 14 teaspoons). Since carbohydrate fuels activity, it may seem that the additional sugar would provide energy for prolonged exercise. But more is not always better during activity. Providing more sugar does not increase the rate at which it is absorbed, so energy drinks don't get any more glucose to the muscles than sports drinks do. The unabsorbed sugar in the stomach can cause GI distress and also slow fluid absorption.

The caffeine content of energy drinks is similar to coffee and ranges from 50 to about 500 mg per can or bottle (see Table). Caffeine is an effective ergogenic aid that enhances endurance when consumed in moderate doses (3 to 6 mg/kg body weight, or about 200 to 400 mg for a 150-lb person).[58] So, the caffeine in energy drinks is likely responsible for the ergogenic effects of these beverages.[60] But too much caffeine, referred to as *caffeine intoxication*, causes nervousness, anxiety, restlessness, insomnia, gastrointestinal upset, tremors, increased blood pressure, and rapid heartbeat. This is a particular concern in young athletes because their lower body weight means that the energy drink supplies more caffeine per kilogram than in an adult. A number of cases of caffeine-associated death, seizure, and cardiac arrest have occurred after consumption of energy drinks.[61-63] The timing of energy drink consumption may be important; caffeine ingested in the evening has less of an ergogenic effect and more side effects.[64] If energy drinks are used as a fluid replacement beverage during exercise, particularly in a hot environment, enough caffeine may be consumed to alter fluid and mineral balance and increase core body temperature.[60] The FDA limits the amount of caffeine in soft drinks to 0.02% (about 71 mg in 12 oz), but energy drinks are considered dietary supplements, and therefore the caffeine content is not regulated.

Energy drinks often also contain other ingredients that promise to improve performance, such as B vitamins, taurine, guarana, and ginseng. B vitamins are needed to produce ATP, so they are marketed to enhance energy production from sugar. But unless you are deficient in these vitamins, drinking them in an energy drink will not enhance your ATP production. Taurine is an amino acid that may reduce the amount of muscle damage and improve exercise performance and capacity, but not all research supports these claims.[63] Guarana is an herbal ingredient that contains caffeine as well as small amounts of the stimulants theobromine and theophylline. The extra caffeine from guarana (not included in the caffeine listed for these beverages) may contribute to caffeine toxicity. Ginseng is also claimed to have performance-enhancing effects, but these effects have not been demonstrated scientifically.[62, 65] In general, the amounts of these ingredients are too small to have much effect, and the safety of consuming them in combination with caffeine prior to or during exercise has yet to be established.[62]

So should you down an energy drink before your next competition? These drinks do provide a caffeine boost, but is it so much caffeine that you risk dehydration, high blood pressure, and heart problems? Energy drinks provide sugar to fuel activity, but will they upset your stomach? What about the herbal ingredients—do they offer a benefit you are looking for?

Caffeine in beverages (per 16 oz serving)

Beverage	Caffeine (mg)	Sugar (g)
Coffee	180–300	0
Coca-Cola Classic	46	52
Mountain Dew	72	62
Monster or Jolt	160	54
Arizona Caution Extreme Energy Shot	200	66
Red Bull	160	56

Think Critically

Describe the risks and benefits of drinking the beverages in the table during exercise compared with drinking a noncaffeinated sports drink.

FIGURE 10.27 **The impact of diet and supplements on performance** This figure illustrates the relative importance of various nutrition strategies for exercise performance. Along with talent and training, eating an overall healthy diet provides the most significant benefit. Sports foods and beverages can supply energy and ensure hydration during an athletic event; most ergogenic supplements provide little or no performance boost.

Concept Check

1. **When** would an iron supplement improve performance?
2. **How** do anabolic steroids affect the production of testosterone?
3. **Why** are creatine supplements beneficial for sprint and strength athletes?
4. **How** does caffeine increase endurance?

Summary

1 Food, Physical Activity, and Health 305

- The right type and amount of food and enough physical activity are needed to optimize health as seen in the illustration. Physical activity uses energy and nutrients, so the foods you eat are necessary to fuel your activity and optimize athletic performance.

Figure 10.1 Food and physical activity benefit health

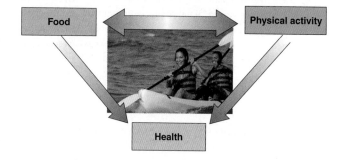

- Regular physical activity can reduce the risk of chronic diseases such as obesity, heart disease, diabetes, and osteoporosis. It can also improve mood and self-esteem and increase vigor and overall well-being.

- Physical activity helps manage body weight by increasing energy expenditure and by increasing the proportion of body weight that is lean tissue.

2 The Four Components of Fitness 307

- **Fitness** is defined by **cardiorespiratory endurance**, muscle strength, flexibility, and body composition. Regular **aerobic activity**, such as the swimming shown here, improves **aerobic capacity**.

Figure 10.4a The components of fitness

- **Muscle-strengthening exercise** increases muscle strength and endurance.
- Stretching improves flexibility, which can enhance postural stability and balance and reduce the risk of injury.
- Fit individuals have a higher percentage of lean tissue than unfit individuals of the same height and weight.

3 Physical Activity Recommendations 309

- To reduce the risk of chronic disease, public health guidelines advise at least 150 minutes of moderate-intensity or 75 minutes of vigorous-intensity aerobic physical activity each week or an equivalent combination of both, as shown by the calendar.

Figure 10.5 Physical activity recommendations

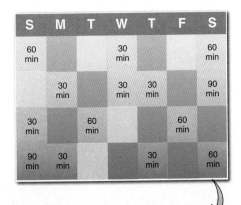

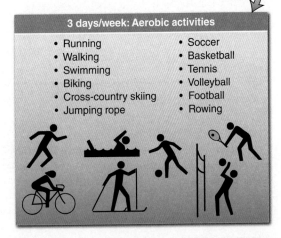

- A complete fitness program should be part of an active lifestyle and should include aerobic activity for cardiovascular conditioning, stretching exercises for flexibility, and muscle-strengthening exercises to increase muscle strength and endurance and maintain or increase muscle mass. In serious athletes, adequate rest is needed to prevent **overtraining syndrome**.

4 Fueling Activity 313

- The graph illustrates that during the first 10 to 15 seconds of exercise, ATP and **creatine phosphate** stored in the muscle provide energy to fuel activity. During the next 2 to 3 minutes, the amount of oxygen at the muscle remains limited, so ATP is generated by the **anaerobic metabolism** of glucose. After a few minutes, the delivery of oxygen at the muscle increases, and ATP can be generated by **aerobic metabolism**. Aerobic metabolism is more efficient than anaerobic metabolism and can utilize glucose, fatty acids, and amino acids as energy sources. The use of protein as an energy source increases when exercise continues for many hours.

Figure 10.10 Changes in the source of ATP over time

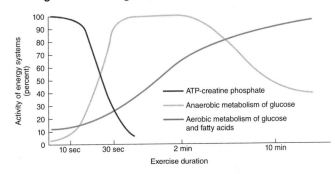

- For short-term, high-intensity activity, ATP is generated primarily from the anaerobic metabolism of glucose from glycogen stores. Anaerobic metabolism uses glucose rapidly and produces **lactic acid**. Both of these factors are associated with the onset of fatigue. For lower-intensity exercise of longer duration, aerobic metabolism predominates, and both glucose and fatty acids are important fuel sources.
- Fitness training causes changes in the cardiovascular system and muscles that improve oxygen delivery and utilization, allowing aerobic activity to be sustained for longer periods at higher intensity.

5 Energy and Nutrient Needs for Physical Activity 319

- Your intake must provide enough energy to fuel activity, enough protein to develop and maintain muscle mass, sufficient micronutrients to metabolize the energy-yielding nutrients, and enough water to transport nutrients and cool your body. Serious athletes need more energy, fluid, and protein than casual exercisers, and the timing of protein intake is also important.

- Enough energy must be available to fuel activity and maintain body functions. The pressure to compete and maintain a body weight that is optimal may cause some athletes to restrict energy intake. A combination of excessive exercise and energy restriction puts athletes at risk for the **relative energy deficiency in sport (RED-S)**, shown here.

Figure 10.16 Low energy availability

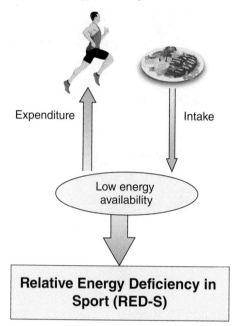

- Carbohydrate is an important fuel for the brain and muscles during exercise. The diet should provide enough carbohydrate to maintain blood glucose and glycogen stores. Dietary fat is needed to provide energy and essential fatty acids and to facilitate the absorption of fat-soluble vitamins. The amount recommended is the same as for the general population. The protein needs of athletes are higher because dietary protein is needed to support changes in metabolism and maintain and repair lean tissues, including muscle. Protein consumed after exercise optimizes the muscle protein synthesis that occurs in response to exercise.
- Vitamin and mineral intake must be adequate to allow metabolism to proceed and to maintain muscle and bone. Nutrients of key concern for athletes include iron, calcium, vitamin D, and some antioxidants. Most athletes who consume a varied diet that meets their energy needs also meet their vitamin and mineral needs from their diet alone. Those who restrict their food intake may be at risk for deficiencies. Increased iron needs and greater iron losses due to fitness training put athletes, particularly female athletes, at risk for iron deficiency.
- Water is needed to ensure that the body can be cooled and that nutrients and oxygen can be delivered to body tissues. Fluid requirements of athletes are increased due to increased losses from evaporation and sweat. Exercising in a hot environment increases the risk of dehydration and **heat-related illnesses**. The risk of hyponatremia is increased by drinking plain water to replace large sweat losses.

6 Food and Drink to Optimize Performance 325

- Preparing for an athletic event involves considering what you will eat and drink in the days as well as hours before the event. Because the availability of carbohydrate is essential for exercise performance, competitive endurance athletes may benefit from **glycogen supercompensation** (**carbohydrate loading**), which maximizes glycogen stores before an event. Larger glycogen stores allow exercise to continue for longer periods. Glycogen stores and hence endurance are increased by increasing carbohydrate intake, as seen in the graph.

Figure 10.23 Dietary carbohydrate and endurance

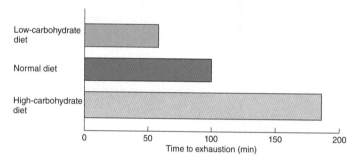

- The meals and snacks consumed before competition should provide plenty of fluid so the athletes begin well hydrated. They should be high in carbohydrate to fill glycogen stores, moderate in protein, and low in fat to increase stomach emptying, and limited in foods and nutrients that cause gastrointestinal upset.
- During exercise, athletes should drink enough fluid to limit weight loss to less than 2% of body weight. For exercise lasting less than 45 minutes, no food is needed, and water is the only fluid needed. If the activity lasts 45 minutes or longer, however, it is important to consume carbohydrate during exercise in order to maintain glucose supplies and thus delay fatigue. A sodium-containing snack or beverage during exercise lasting more than about an hour will reduce the risk of hyponatremia, improve glucose and water absorption, and stimulate thirst.
- Postcompetition snacks and meals should replace lost fluids and electrolytes, provide carbohydrate to restore muscle and liver glycogen, and provide protein to both stimulate muscle protein synthesis and provide the amino acids needed for protein synthesis and repair.

7 Ergogenic Aids 329

- Many types of **ergogenic aids** are marketed to improve athletic performance. An individual risk–benefit analysis should be used to determine whether a supplement is appropriate for you.
- Athletes may take protein supplements or use hormones to build muscle. Protein supplements are no more effective than food sources of protein. **Anabolic steroids** are hormones that, when combined with muscle-strengthening exercise, increase muscle size and strength, but these supplements are illegal and have dangerous side effects.
- **Creatine** supplementation increases muscle creatine phosphate levels and has been shown to increase muscle mass and improve performance in short-duration, high-intensity exercise.
- Caffeine use can improve performance in endurance activities, but high doses can cause caffeine toxicity and contribute to dehydration in some athletes. EPO can enhance endurance by increasing the number of red blood cells, but it is dangerous and is a prohibited substance.
- A healthy diet is the basis for successful athletic performance. Beverages and foods that supply fluids and energy can enhance performance. Some ergogenic supplements are beneficial for certain types of activity, but many offer little or no benefit, and as shown in the illustration, their overall impact is small.

Figure 10.27 The impact of diet and supplements on performance

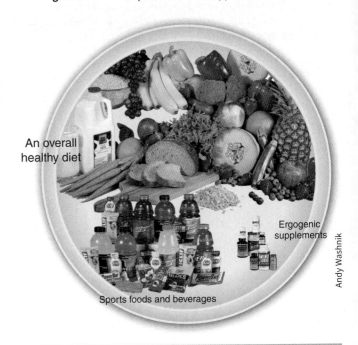

Key Terms

- aerobic activity 307
- aerobic capacity 309
- aerobic metabolism 313
- aerobic zone 310
- anabolic steroids 331
- anaerobic metabolism 313
- atrophy 308
- cardiorespiratory endurance 307
- creatine 332
- creatine phosphate 314
- endorphins 305
- ergogenic aids 329
- female athlete triad 321
- fitness 305
- glycogen supercompensation or carbohydrate loading 326
- heat cramps 324
- heat exhaustion 324
- heat stroke 324
- heat-related illnesses 323
- hypertrophy 308
- lactic acid 314
- maximum heart rate 310
- muscle endurance 309
- muscle strength 309
- muscle-strengthening exercise 309
- overload principle 307
- overtraining syndrome 312
- relative energy deficiency in sport (RED-S) 320
- resting heart rate 309
- sports anemia 322
- steroid precursors 332

What is happening in this picture?

During competitive events, cyclists often spend six or more hours a day riding their bikes. This rider is picking up a musette bag, which contains water bottles and snacks. He will carry the bottles on his bike and the snacks in his jersey pockets to have water and fuel available during the ride.

Think Critically

1. How much might someone need to drink during six hours of cycling?
2. What type of fluid do you think is in the water bottles? Why?
3. What type of food might the riders want to have in the musette bags?

CHAPTER 11

Paul Damien / National Geographic Creative

Nutrition During Pregnancy and Infancy

"You're eating for two now!" Optimal maternal nutrition during pregnancy can help ensure the birth of a healthy baby; too much or too little can compromise fetal health.

Gaining too little weight during pregnancy can produce an undernourished infant who may never catch up with healthy counterparts. Overnutrition poses risks to both mother and child during birth. A fetus may also be subject to dangerous effects from deficiencies or excesses of vitamins and minerals. For example, lack of sufficient iodine during pregnancy has been linked to brain damage in the baby, insufficient iron creates an increased risk of anemia for an infant, and too much vitamin A causes an increased risk of birth defects of the head, face, heart, and brain.

The responsibility of the mother does not stop when the infant is born. The 40 weeks of a pregnancy produce a helpless newborn who requires a lengthy period of care. Breast milk provides optimal nutrition for a growing infant, and a breast-feeding mother's diet continues to affect the nutrition of her child. Maternal and infant nutrition can affect both growth in the early years and the potential for developing chronic diseases later in life. Healthier pregnancies yield healthier babies who become healthier children and ultimately healthier adults.

CHAPTER OUTLINE

Changes in the Body During Pregnancy 340
- Nourishing the Embryo and Fetus
- Maternal Weight Gain During Pregnancy
- Physical Activity During Pregnancy
- Discomforts of Pregnancy
- Complications of Pregnancy

Nutritional Needs During Pregnancy 345
- Energy and Macronutrient Needs
- Fluid and Electrolyte Needs
- Vitamin and Mineral Needs

What a Scientist Sees: Folic Acid Fortification and Neural Tube Defects
- Meeting Nutrient Needs with Food and Supplements
- Food Cravings and Aversions

Thinking It Through: A Case Study on Nutrient Needs for a Successful Pregnancy
What Should I Eat? During Pregnancy

Factors That Increase the Risks Associated with Pregnancy 351
- Maternal Nutritional Status
- Maternal Age and Health
- Poverty
- Exposure to Toxic Substances

Lactation 356
- Milk Production and Let-Down
- Energy and Nutrient Needs During Lactation

Nutrition for Infants 359
- Infant Growth and Development
- Energy and Nutrient Needs of Infants

Debate: DHA/ARA-Fortified Infant Formulas
- Meeting Needs with Breast Milk or Formula
- Safe Feeding for the First Year

11.1 Changes in the Body During Pregnancy

LEARNING OBJECTIVES

1. **Describe** how the embryo and fetus are nourished.
2. **Discuss** why appropriate weight gain is important during pregnancy.
3. **Explain** why morning sickness, heartburn, and constipation are common during pregnancy.
4. **Review** the risks associated with the hypertensive disorders of pregnancy and gestational diabetes.

Whether you end up 6 feet 4 inches or 5 feet 3 inches tall, you begin as a single cell that arises from the union of a sperm and an egg. Over the course of 40 weeks, this cell grows and develops into a fully formed human baby. Prenatal growth and development are carefully orchestrated processes that require adequate supplies of all the essential nutrients in order to progress normally.

In the days after **fertilization**, the single cell divides rapidly to form a ball of cells (**Figure 11.1**). The cells then begin to differentiate and move to form body structures. During these early steps in development, this ball of cells obtains the nutrients it

PROCESS DIAGRAM

FIGURE 11.1 Prenatal development This cross section shows the path of the egg and developing embryo from the ovary, where the egg is produced, through the oviduct to the uterus, where most prenatal development occurs.

② Fertilization occurs in the oviduct 12 to 24 hours after ovulation.

③ About 30 hours after fertilization, the fertilized egg has completed its first cell division.

④ About 3 or 4 days after fertilization, the developing embryo is a ball of about 100 cells.

① Ovulation releases an egg from the woman's ovary.

⑤ About 6 days after fertilization, the developing embryo begins to implant itself in the uterine lining. Implantation is complete by 14 days after fertilization.

⑥ During the embryonic stage of development (2 to 8 weeks) cells differentiate and arrange themselves in the proper locations to form the major organ systems. The embryo shown here is about 5 to 6 weeks old and less than 3 cm long. The organ systems and external body structures are not fully developed.

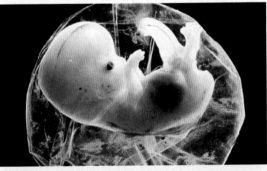

⑦ During the fetal stage of development (9 weeks until birth) the fetus grows, and internal and external body structures continue to develop. This fetus is about 16 weeks old and about 16 cm long.

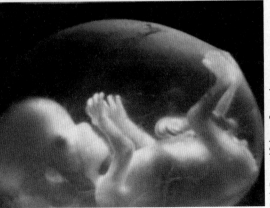

fertilization The union of a sperm and an egg.

needs from the fluids around it. About a week after fertilization, the developing embryo begins burrowing into the lining of the uterus; and by two weeks, **implantation** is complete, and the cluster of cells has become an **embryo**.

Nourishing the Embryo and Fetus

The embryonic stage of development lasts until the eighth week after fertilization. During this time, the cells differentiate to form the multitude of specialized cell types that make up the human body. At the end of this stage, the embryo is about 3 cm long and has a beating heart. The rudiments of all major external and internal body structures have been formed.

The early embryo gets its nourishment by breaking down the lining of the uterus, but soon this source is inadequate to meet its growing needs. After about five weeks, the **placenta** takes over the role of nourishing the embryo (**Figure 11.2**). The placenta also secretes hormones that are necessary to maintain pregnancy.

From the ninth week on, the developing offspring is a **fetus**. During the fetal period, structures formed during the embryonic period grow and mature. The placenta continues to nourish the fetus until birth. During this time, the length of the fetus increases from about 3 cm to around 50 cm. The fetal period usually ends after 40 weeks, with the birth of an infant weighing 3 to 4 kg (6.5 to 9 lb).[1]

Infants who are born on time but have failed to grow adequately in the uterus are said to be **small for gestational age**. Those born before 37 weeks of gestation are said to be **preterm**, or **premature**. Whether born too soon or too small, **low-birth-weight** infants and **very-low-birth-weight** infants are at increased risk for illness and early death (**Figure 11.3**).

Maternal Weight Gain During Pregnancy

During pregnancy, a woman's body undergoes many changes to support the growth and development of her child. Her blood volume increases by 50%. The placenta develops in order to allow nutrients to be delivered to the fetus and produce hormones that orchestrate other changes in the mother's body. The amount of body fat increases to provide the energy needed late in pregnancy. The uterus enlarges, muscles and ligaments relax to accommodate the growing fetus and allow for childbirth, and the breasts develop in preparation for **lactation**. All these changes naturally result in weight gain.

FIGURE 11.2 The placenta The placenta is made up of branchlike projections that extend from the embryo into the uterine lining, placing maternal and fetal blood in close proximity. The placenta allows nutrients and oxygen to pass from maternal blood to fetal blood and waste products to be transferred from fetal blood to maternal blood. Fetal blood travels to and from the placenta via the umbilical cord.

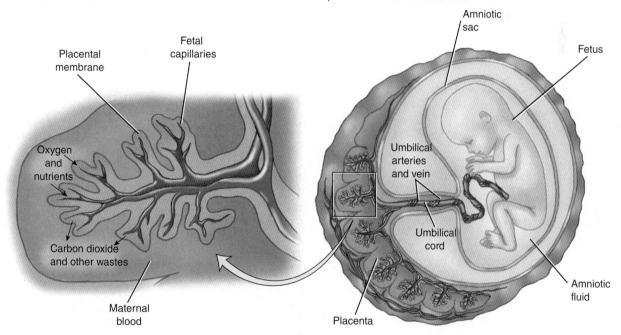

implantation The process through which a developing embryo embeds itself in the uterine lining.

embryo A developing human from two through eight weeks after fertilization.

placenta An organ produced from maternal and embryonic tissues. It secretes hormones, transfers nutrients and oxygen from the mother's blood to the fetus, and removes metabolic wastes.

fetus A developing human from the ninth week after fertilization to birth.

low birth weight A birth weight less than 2.5 kg (5.5 lb).

very low birth weight A birth weight less than 1.5 kg (3.3 lb).

lactation Production and secretion of milk.

FIGURE 11.3 **Low-birth-weight infants** Low-birth-weight and very-low-birth-weight infants require special medical and nutritional care in order to continue to grow and develop. Today, with advances in medical and nutritional care, infants born as early as 25 weeks of gestation and those weighing as little as 1 kg (2.2 lb) can survive.[2]

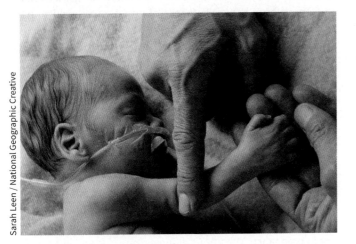

A healthy, normal-weight woman should gain 11 to 16 kg (25 to 35 lb) during pregnancy (**Figure 11.4a**).[1] The rate of weight gain is as important as the amount gained. Little gain is expected in the first 3 months, or **trimester**—usually about 1 to 2 kg (2 to 4 lb). In the second and third trimesters, the recommended maternal weight gain is about 0.5 kg (1 lb)/week. Women who are underweight and women who are overweight or obese at conception should also gain weight at a slow, steady rate, but the total amount of weight gain recommended is higher and lower than for normal-weight women, respectively (**Figure 11.4b**).[1]

Being underweight (BMI < 18.5 kg/m²) at the onset of pregnancy or gaining too little weight during pregnancy increases the risk of preterm birth and of delivering a low-birth-weight baby.[3,4] Excess weight, whether present before conception or gained during pregnancy, can also compromise pregnancy outcome. The mother's risks for high blood pressure, diabetes, a difficult delivery, and need for a **cesarean section** are increased by excess weight, as are the risks of preterm delivery, fetal and infant death, birth defects, and delivering a

FIGURE 11.4 **Rate and composition of weight gain during pregnancy** Both maternal and fetal tissues contribute to the weight gained during pregnancy. Therefore, all women should gain weight during pregnancy; the recommended amount depends on a woman's prepregnancy weight.

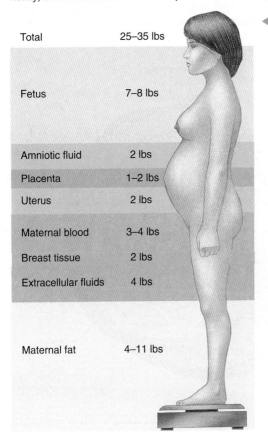

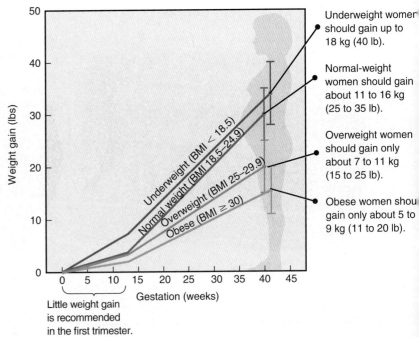

a. The weight of an infant at birth accounts for about 25% of the mother's weight gain during pregnancy. The placenta, amniotic fluid, and changes in maternal tissues, including enlargement of the uterus and breasts, expansion of the volume of blood and other extracellular fluids, and increased fat stores, account for the rest.

b. A similar pattern of weight gain is recommended for women who are normal weight, underweight, overweight, or obese at the start of pregnancy, but the recommendations for total weight gain are different.[1]

Interpret the Data

How much weight should a woman with a prepregnancy BMI of 27 kg/m² gain during her pregnancy?

large-for-gestational-age baby.[5, 6] Excessive prenatal weight gain also increases the mother's long-term risk for obesity, and there is evidence that it may increase the risk that the baby will be overweight in childhood.[7] However, dieting during pregnancy is not advised, even for obese women. If possible, excess weight should be lost before the pregnancy begins or, alternatively, after the child has been born and weaned.

Physical Activity During Pregnancy

During pregnancy, women gain weight and carry the extra weight in the front of the body, where it can interfere with balance and put stress on bones, joints, and muscles, increasing the risk of exercise-related injury. However, this doesn't mean that a pregnant woman should give up her regular exercise routines. Physical activity during pregnancy can improve digestion; prevent excess weight gain, low back pain, and constipation; reduce the risk of diabetes and high blood pressure; and speed recovery from childbirth. Guidelines for exercise during pregnancy have been developed to maximize the benefits and minimize the risks of injury to mother and fetus. In general, women who were physically active before becoming pregnant can continue a program of about 30 minutes of carefully chosen moderate exercise per day.[8, 9] Women who weren't active before pregnancy should slowly add low-intensity, low-impact activities.[9] Because intense exercise can limit the delivery of oxygen and nutrients to the fetus, intense exercise should be limited (**Table 11.1**).

Discomforts of Pregnancy

The physiological changes that occur during pregnancy can cause uncomfortable side effects. For example, the expansion in blood volume necessary to nourish the fetus often causes an accumulation of extracellular fluid in the tissues, a condition known as **edema**. Edema can be uncomfortable but does not increase medical risks unless it is accompanied by a rise in blood pressure.

Nausea and vomiting affect 50 to 80% of women during the first trimester of pregnancy.[10] This is referred to as **morning sickness**, but symptoms can occur at any time during the day or night. Morning sickness is thought to be related to hormones that are released early in pregnancy. The symptoms may be alleviated by eating small, frequent snacks of dry, starchy foods, such as plain crackers or bread. In most women, the symptoms of morning sickness decrease significantly after the first trimester, but in some they last for the entire pregnancy and, in severe cases, may require intravenous nutrition.[11]

The hormones produced during pregnancy to relax uterine muscles also relax the muscles of the gastrointestinal tract. This relaxation, along with crowding of the organs by the growing baby, can cause heartburn and constipation (**Figure 11.5**). Heartburn can be reduced by limiting high-fat foods, which leave the stomach slowly; avoiding substances, such as caffeine and chocolate, that are known to cause heartburn; eating small, frequent meals; and remaining upright after eating. Constipation can be prevented by maintaining a moderate level of physical activity and consuming plenty of fluids and high-fiber foods. Hemorrhoids are common during pregnancy as a result of both constipation and changes in blood flow.

Complications of Pregnancy

Most of the 3.95 million women who give birth every year in the United States have healthy pregnancies. However, about

TABLE 11.1 Guidelines for physical activity during pregnancy

Do...	Don't...
Obtain permission from your health-care provider before beginning an exercise program.	Exercise strenuously during the first trimester.
Increase activity gradually if you were inactive before pregnancy.	Exercise strenuously for more than 15 minutes at a time during the second and third trimesters.
Exercise regularly rather than intermittently.	Exercise to the point of exhaustion.
Stop exercising when fatigued.	Exercise lying on your back after the first trimester.
Choose non-weight-bearing activities, such as swimming, that entail minimal risk of falls or abdominal injury.	Scuba dive or engage in activities that entail risk of abdominal trauma, falls, or joint stress.
Drink plenty of fluids before, during, and after exercise.	Exercise in hot or humid environments.

Heather Perry / National Geographic Creative

large for gestational age   Weighing more than 4 kg (8.8 lb) at birth.

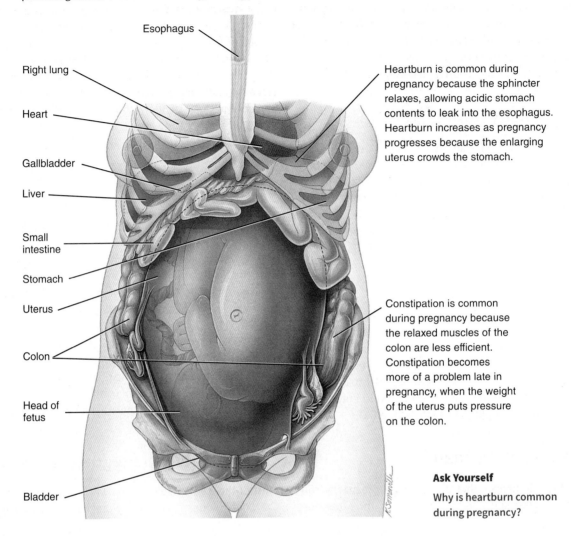

FIGURE 11.5 Crowding of the gastrointestinal tract During pregnancy, the uterus enlarges and pushes higher into the abdominal cavity, exerting pressure on the stomach and intestines.

Heartburn is common during pregnancy because the sphincter relaxes, allowing acidic stomach contents to leak into the esophagus. Heartburn increases as pregnancy progresses because the enlarging uterus crowds the stomach.

Constipation is common during pregnancy because the relaxed muscles of the colon are less efficient. Constipation becomes more of a problem late in pregnancy, when the weight of the uterus puts pressure on the colon.

Ask Yourself
Why is heartburn common during pregnancy?

10% of babies are born too soon, 8% have low or very low birth weights, and about 6 out of 1000 of those born alive die in their first year of life.[12] The complications of pregnancy and childbirth also take the lives of about 21 women out of every 100,000 live births.[13] If complications that occur during pregnancy are caught early, they can usually be managed, resulting in a safe delivery and a healthy baby.

High blood pressure
A spectrum of conditions involving elevated blood pressure affect up to 10% of pregnant women in the United States.[14] These **hypertensive disorders of pregnancy** account for more than 12% of pregnancy-related maternal deaths.[15]

High blood pressure that is present before the pregnancy or diagnosed before 20 weeks of gestation is referred to as chronic hypertension; it complicates as many as 5% of pregnancies.[16, 17] **Gestational hypertension** is an abnormal rise in blood pressure that occurs after the 20th week of pregnancy and resolves within 12 weeks of birth; it complicates about 6% of pregnancies and may signal the potential for a more serious condition called **preeclampsia**.[16, 17] Preeclampsia is characterized by high blood pressure, protein in the urine, severe headaches, changes in vision, rapid weight gain, abdominal pain, and swelling of the hands and feet. It is dangerous to the baby because blood flow to the placenta is reduced, and it is dangerous to the mother because it can progress to a more severe condition called **eclampsia**, in which life-threatening seizures occur. Preeclampsia can often be managed with bed rest and careful medical monitoring, but in some cases may necessitate early delivery of the baby. The condition usually resolves after delivery. Preeclampsia and eclampsia are most common in mothers under 18 and over 35 years of age, low-income mothers, obese mothers, and mothers with chronic hypertension or kidney disease.

---

preeclampsia A condition characterized by elevated blood pressure, a rapid increase in body weight, protein in the urine, and edema. Also called *toxemia*.

eclampsia Convulsions or seizures that occur in association with preeclampsia. Untreated, it can lead to coma or death.

The causes of hypertension during pregnancy are not fully understood. At one time, low-sodium diets were prescribed to prevent preeclampsia, but studies have not found such diets to be beneficial in lowering blood pressure or preventing this condition.[18] Calcium may play a role in preventing hypertensive disorders of pregnancy. Women with a high intake of calcium have a low incidence of these disorders, and calcium supplements have been found to reduce the risk of preeclampsia in high-risk women with low calcium intakes.[14] Calcium supplements are not routinely recommended for healthy pregnant women.

Gestational diabetes Diabetes that develops in the second or third trimester of pregnancy is known as **gestational diabetes**. Glucose passes freely across the placenta, so when the mother's blood glucose levels are high, the growing fetus receives extra glucose and hence extra calories. The baby grows rapidly and is at risk for being large for gestational age. Gestational diabetes also increases the risk of difficult delivery, preterm delivery, and birth defects.[5] Controlling the mother's blood glucose through changes in diet and activity, and in some cases medication, reduces the risks to the mother and baby. Gestational diabetes is common in obese women and those with a family history of type 2 diabetes and occurs more frequently among Asian, African American, Hispanic/Latino, and Native American women than among Caucasian women.[5] It usually resolves after the birth, but 35 to 60% of women who have had gestational diabetes develop type 2 diabetes within the next 5 to 10 years.[19] Babies born to mothers with gestational diabetes are at increased risk of developing diabetes as adults.[19]

Concept Check

1. **What** is the role of the placenta?
2. **How** does low maternal weight gain during pregnancy affect the health of the child?
3. **Why** do heartburn and constipation tend to increase later in pregnancy?
4. **How** does gestational diabetes in a mother affect the baby?

11.2 Nutritional Needs During Pregnancy

LEARNING OBJECTIVES

1. **Compare** the energy and protein needs of pregnant and nonpregnant women.
2. **Explain** why pregnancy increases the need for many vitamins and minerals.
3. **Discuss** the need for dietary supplements during pregnancy.

During pregnancy, the mother's diet must provide all of her nutrients as well as those needed for the baby's growth and development. Because the increase in nutrient needs is greater than the increase in energy needs, a nutrient-dense diet is essential.

Energy and Macronutrient Needs

Although pregnant women are eating for two, they don't need to eat twice as much as they normally do. During the first trimester, energy needs are not increased above levels for nonpregnant women. During the second and third trimesters, a pregnant woman should consume an additional 340 and 452 Calories/day, respectively (**Figure 11.6**).[20]

Protein needs are increased during pregnancy because protein is needed for the synthesis of new blood cells, formation of the placenta, enlargement of the uterus and breasts, and growth of the baby. An additional 25 g of protein above the RDA for nonpregnant women, or 1.1 g/kg/day, is recommended during the second and third trimesters of pregnancy (see Figure 11.6).

To ensure sufficient glucose to fuel the fetal and maternal brains during pregnancy, the RDA for carbohydrate is increased by 45 g/day to 175 g/day. If this carbohydrate comes from whole grains, fruits, and vegetables, it will also provide the additional 3 g/day of fiber recommended during pregnancy.

Although it is not necessary to increase total fat intake during pregnancy, additional amounts of the essential fatty acids linoleic and α-linolenic acid are recommended because they are incorporated into the placenta and fetal tissues. The long-chain polyunsaturated fatty acids docosahexaenoic acid (DHA) and arachidonic acid (ARA) are important because they not only support maternal health but are essential for development of the eyes and nervous system in the fetus.[21] These fatty acids can be obtained from a varied diet that includes vegetable oils and fish or fish oil supplements.

---

gestational diabetes A condition characterized by high blood glucose levels that develop during pregnancy.

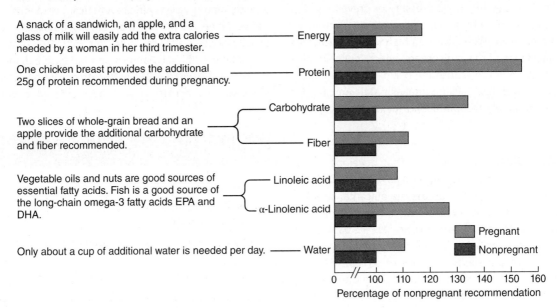

FIGURE 11.6 Energy and macronutrient recommendations This graph illustrates the difference between the recommended daily intakes of energy, protein, carbohydrate, fiber, essential fatty acids, and water for 25-year-old nonpregnant women and 25-year-old pregnant women during the third trimester.

Despite increases in the recommended intakes of protein, carbohydrate, and specific fatty acids during pregnancy, the distribution of calories from protein, carbohydrate, and fat should be about the same as that recommended for the general population.

Fluid and Electrolyte Needs

During pregnancy, a woman will accumulate 6 to 9 L of water. Some of this water is intracellular, resulting from the growth of tissues, but most is due to increases in the volume of blood and the fluid between cells. The need for water increases from 2.7 L/day in nonpregnant women to 3 L/day during pregnancy (see Figure 11.6).[18] Despite changes in the amount and distribution of body water during pregnancy, there is no evidence that the requirements for potassium, sodium, and chloride are different for pregnant women than for nonpregnant women.

Vitamin and Mineral Needs

Many vitamins and minerals are needed for the growth of new tissues in both mother and child (**Figure 11.7**). For many of these nutrients, the increased need is easily met with the extra food the mother consumes. For others, the increased need is met because their absorption is increased during pregnancy. For a few, including calcium, vitamin D, folate, vitamin B_{12}, iron, zinc, and iodine, there is a risk that some women will not consume adequate amounts without supplementation.[22]

Calcium and vitamin D During gestation, the fetus accumulates about 30 g of calcium, mostly during the third trimester, when the bones are growing rapidly and the teeth are forming.

However, the RDA is not increased during pregnancy because calcium absorption doubles.[23] The RDA for calcium can be met by consuming three to four servings of dairy products daily. Women who are lactose intolerant can meet their calcium needs with yogurt, cheese, reduced-lactose milk, calcium-rich vegetables, calcium-fortified foods and beverages, and calcium supplements. Many pregnant women fail to consume adequate amounts of calcium. Women with low intakes may need supplements, and calcium supplements may help prevent preeclampsia in pregnant teens, individuals with inadequate calcium intake, and women who are known to be at risk for developing preeclampsia.[16, 23, 24]

Adequate vitamin D is essential to ensure efficient calcium absorption. The RDA for vitamin D during pregnancy is 600 IU (15 μg)/day, the same as it is for nonpregnant women.[23] This is based on the assumption of minimal sun exposure. There are few food sources of vitamin D and, depending on latitude and skin color, 30 to 96% of pregnant women have low blood levels. Therefore, supplements may be needed to meet needs.[24]

Folate (folic acid) and vitamin B_{12} Folate is needed for the synthesis of DNA and hence for cell division. Adequate folate intake before conception and during early pregnancy is crucial because rapid cell division occurs in the first days and weeks of pregnancy.

Low folate levels increase the risk of abnormalities in the formation of the *neural tube*, which forms the baby's brain and spinal cord (see Figure 7.17). Neural tube closure, a critical step in neural tube development, occurs between 21 and 28 days after conception, often before a woman knows she is pregnant. Therefore, it is recommended that all women capable of becoming pregnant consume 400 μg daily of synthetic folic

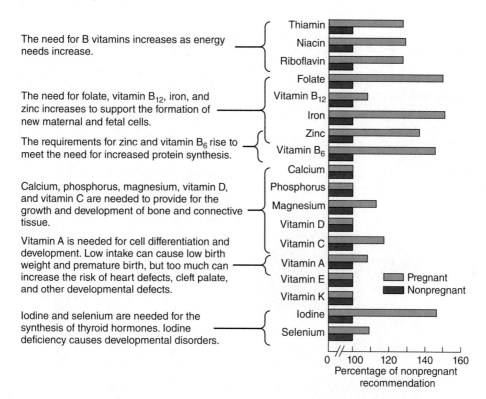

FIGURE 11.7 Micronutrient needs during pregnancy The graph compares the recommended micronutrient intakes for 25-year-old nonpregnant women and 25-year-old women during the third trimester of pregnancy.

acid from fortified foods, supplements, or a combination of the two, in addition to a varied diet that is rich in natural sources of folate, such as leafy greens, legumes, and orange juice (see *What a Scientist Sees: Folic Acid Fortification and Neural Tube Defects*).

Folate continues to be important even after the neural tube closes. Folate deficiency can cause macrocytic anemia in the mother, and inadequate folate intake is associated with prematurity, low birth weight, and birth defects.[26] During pregnancy, the RDA is 600 μg/day.[27]

Vitamin B_{12} is essential for the regeneration of active forms of folate. A deficiency of vitamin B_{12} can therefore result in macrocytic anemia in the mother and impaired growth and cognitive development in the fetus.[28] Based on the amount of vitamin B_{12} transferred from mother to fetus during pregnancy and on the increased efficiency of vitamin B_{12} absorption during pregnancy, the RDA for pregnancy is set at 2.6 μg/day.[27] This recommendation is easily met by consuming a diet containing even small amounts of animal products. Pregnant women who consume vegan diets must include vitamin B_{12} supplements or foods or beverages fortified with vitamin B_{12} to meet their needs as well as those of the fetus.

Iron and zinc
Iron needs are high during pregnancy because iron is required for the synthesis of hemoglobin and other iron-containing proteins in both maternal and fetal tissues. The RDA for pregnant women is 27 mg/day, 50% higher than the recommended amount for nonpregnant women.[29] Many women start pregnancy with diminished iron stores and quickly become iron deficient. This occurs even though iron absorption is increased during pregnancy and iron losses decrease because menstruation ceases. Iron deficiency anemia during pregnancy is associated with low birth weight, preterm delivery, and impaired cognitive development.[30]

Meeting iron needs during pregnancy requires a well-planned diet. Red meat is a good source of the more absorbable heme iron, and leafy green vegetables and fortified cereals are good sources of nonheme iron. Consuming citrus fruit, which is high in vitamin C, or meat, which contains heme iron, along with foods that are good sources of nonheme iron, can enhance nonheme iron absorption. Iron supplements are typically recommended, and iron is included in prenatal supplements.

Zinc is involved in the synthesis and function of DNA and RNA and the synthesis of proteins. It is therefore extremely important for growth and development. Because zinc absorption is inhibited by high iron intake, iron supplements may compromise zinc status if the mother's diet is low in zinc. The RDA for zinc is 13 mg/day for pregnant women age 18 and younger and 11 mg/day for pregnant women age 19 and older.[29] As is the case with iron, the zinc in red meat is more absorbable than the zinc from other sources. Pregnant vegetarian women consume less zinc than nonvegetarians, and both groups consume less than the RDA.[31]

What a Scientist Sees

Folic Acid Fortification and Neural Tube Defects

Consumers reading the ingredient list on this box of pasta see that it is made from a coarse flour called durum semolina, with added niacin, thiamin, riboflavin, iron, and folic acid. In 1998, the United States and Canada began requiring the addition of folic acid to pasta and other enriched grain products; their goal was to increase folic acid intake in women of childbearing age to reduce the incidence of neural tube defects. Enriched grains were chosen for fortification because they are commonly consumed in regular amounts by this target population. Because high folic acid intake can mask the symptoms of vitamin B_{12} deficiency, the amount added to enriched grains was kept low enough to avoid this problem in any segment of the population but high enough to reduce the risk of neural tube defects. A scientist can see that this public health measure has succeeded. Since the initiation of folic acid fortification, the incidence of neural tube defects has been reduced by 36% in the United States, and reductions ranging from 31 to 50% have been observed in other countries where folic acid fortification is mandatory (see graph).[25]

Think Critically

Suggest some possible reasons why fortifying grains with folic acid has not completely eliminated neural tube defects.

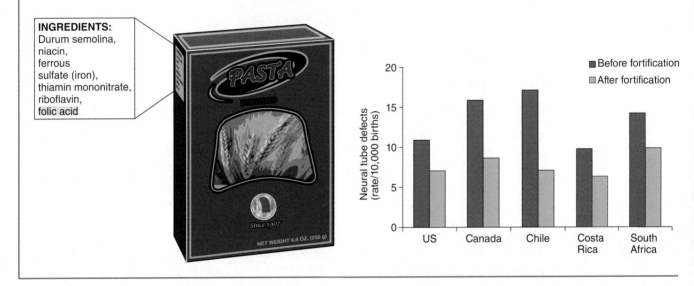

Iodine

Iodine requirements are increased during pregnancy because there is a 50% increase in maternal thyroid hormone production as well as an increase in iodine loss in the urine.[24] Iodine deficiency during pregnancy causes brain damage in the fetus as well as fetal goiter, hypothyroidism, and cretinism.[24] For many decades iodine intake was not a concern in the United States, but intakes have been declining (see Chapter 8); between 1970 and 1990, intake dropped by 50%.[32] Recent surveys indicate that some pregnant women in the United States, particularly African American women, have inadequate intakes of iodine.[33] The RDA for pregnant women is 220 µg/day from food or supplemental sources.[29]

Meeting Nutrient Needs with Food and Supplements

The energy and nutrient needs of pregnancy can be met by following healthy dietary patterns such as the Mediterranean-Style Eating Pattern, the DASH Eating Plan, or MyPlate (**Figure 11.8**).

Even when a healthy diet is consumed, it is difficult to meet all the vitamin and mineral needs of pregnancy. Therefore, prenatal supplements are generally prescribed for pregnant women. These supplements include additional folic acid, iron, iodine, calcium, and a host of other micronutrients. Prenatal supplements, however, must be taken along with, not in place of, a carefully planned diet (**Figure 11.9**) (see *Thinking It Through*).

Food Cravings and Aversions

Most women experience some food cravings and aversions during pregnancy. The foods most commonly craved are ice cream, sweets, candy, fruit, and fish. Common aversions include coffee, highly seasoned foods, and fried foods. It is not known why women have these cravings and aversions. It has been suggested that hormonal or physiological changes during pregnancy—in particular, changes in taste and smell—may be the cause, but psychological and behavioral factors are also involved.

Nutritional Needs During Pregnancy **349**

FIGURE 11.8 **MyPlate Recommendations for Moms** Shown here are the food group amounts based on the MyPlate recommendations (www.choosemyplate.gov) for a 26-year-old expectant mother who is 5 feet 4 inches tall, gets 30 to 60 minutes of exercise a day, and weighed 125 lb before she became pregnant. Energy needs are not increased during the first trimester, so the recommended amounts from each group for the first trimester are the same as for a nonpregnant woman.

Ask Yourself

Compared to a nonpregnant woman, how many more cups of milk should a woman in her third trimester be consuming? Why?

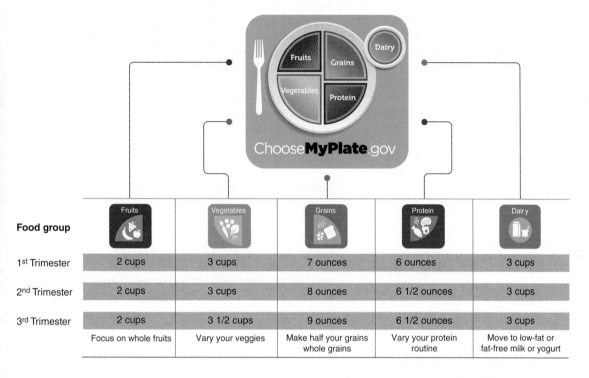

Food group	Fruits	Vegetables	Grains	Protein	Dairy
1st Trimester	2 cups	3 cups	7 ounces	6 ounces	3 cups
2nd Trimester	2 cups	3 cups	8 ounces	6 1/2 ounces	3 cups
3rd Trimester	2 cups	3 1/2 cups	9 ounces	6 1/2 ounces	3 cups
	Focus on whole fruits	Vary your veggies	Make half your grains whole grains	Vary your protein routine	Move to low-fat or fat-free milk or yogurt

FIGURE 11.9 **Prenatal supplements** Prenatal supplements cannot take the place of a well-planned diet. They typically do not provide enough calcium to meet the needs of pregnant women; to do so, the tablet would have to be very large. They also lack protein needed for tissue synthesis, complex carbohydrates needed for energy, essential fatty acids for brain and nerve tissue development, fiber and fluid to help prevent constipation, and the phytochemicals found in a healthy diet.

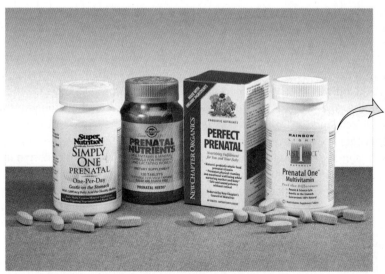

© 2012 John Wiley & Sons, Inc. All rights reserved. Photo: Andy Washnik

Supplement Facts	
Serving Size 1 Tablet	
Servings Per Container 60	
Amount Per 1 Tablet	**% Daily Value**
Vitamin A (as beta carotene) 5000 IU	63%
Vitamin C (as ascorbic acid) 85 mg	100%
Vitamin D (as cholecalciferol) 400 IU	200%
Vitamin E (as d-alpha tocopheryl acetate) 22 IU	67%
Vitamin K 90 mcg	100%
Thiamin 1.4 mg	100%
Riboflavin 1.6 mg	100%
Niacin (as niacinamide) 17 mg	100%
Vitamin B_6 (as pyridoxine HCl) 2.6 mg	137%
Folic acid 1000 mcg	167%
Vitamin B_{12} (as cyanocobalamin) 2.6 mg	100%
Pantothenic Acid (as as d-calcium pantothenate) 6 mg	100%
Iron (as iron fumarate) 27 mg	100%
Iodine (kelp) 220 mcg	100%
Zinc (as monomethionine & gluconate) 11mg	100%
Selenium (as sodium selenate) 60 mcg	100%
Copper (as copper sulfate) 1000 mcg	100%
Calcium (as calcium carbonate) 200 mg	20%
* **Daily Values based on RDAs for pregnant women ages 19-50**	
Other ingredients: stearic acid, vegetable stearate, silicon dioxide, croscarmellose sodium, microcrystalline cellulose, natural coating (contains hydroxypropyl methylcellulose, titanium dioxide, riboflavin, polyethylene glycol and polysorbate)	

Thinking It Through

A Case Study on Nutrient Needs for a Successful Pregnancy

Tina is a moderately active 26-year-old at the end of her fourth month of pregnancy. She is 5 feet 4 inches tall and weighed 125 lb at the start of her pregnancy. She is concerned about gaining too much weight because her sister gained 50 lb during her pregnancy and has had difficulty losing the excess weight since her baby was born. Tina's obstetrician recommends that she gain 25 to 35 lb during her pregnancy, but she would like to keep her weight gain to only 15 lb.

Based on the graph below, why would gaining only 15 lb put the baby at risk? Why would gaining 50 lb put the baby at risk?

Your answer:

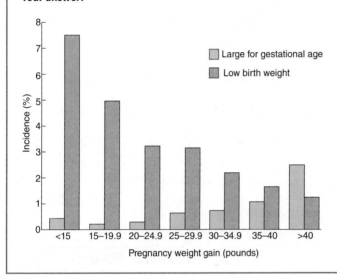

Tina's typical diet provides enough calories for a nonpregnant woman, but during her second and third trimesters, she needs additional calories and nutrients. To decide what to add to her diet, she compares her typical intake to the MyPlate recommendations for pregnant women. Her current diet includes 1 cup of fruit, 1.5 cups of vegetables, 8 ounces (oz) of grains, 6 oz of protein foods, and 2 cups of dairy.

Use Figure 11.8 to compare Tina's current intake to the MyPlate recommendations for the second trimester of pregnancy. What changes does she need to make to meet recommendations?

Your answer:

Tina is taking a prenatal supplement but is curious about whether her diet alone can meet her nutrient needs. She analyzes her diet and finds that she is meeting her needs for all nutrients except iron. Her current diet provides only 13.5 mg of iron—significantly less than the RDA of 27 mg for pregnant women.

Use iProfile to find foods that Tina can add to her diet to increase her iron intake by 13.5 mg/day. How many calories would your suggestions add?

Your answer:

Considering the foods you suggested to increase her iron intake, do you think it is reasonable for Tina to consume 27 mg of iron each day from her diet alone? Why or why not?

Your answer:

(Check your answers in Appendix L.)

Usually, the foods that pregnant women crave are not harmful and can be safely included in the diet to meet not only nutritional needs but also emotional needs and individual preferences (see *What Should I Eat?*). However, the ingestion of nonfood substances such as clay, laundry starch, and ashes, known as **pica**, can have serious health consequences during pregnancy. Hypotheses as to why pica occurs include cultural expectations; protection against harmful pathogens and toxins; suppression of nausea, vomiting, and diarrhea; and contribution of micronutrients.[34, 35] However, the overall risks overshadow any of these potential benefits. Consuming large amounts of these substances can cause intestinal obstructions, reduce intake of nutrient-dense foods, inhibit absorption of minerals such as iron, and increase the risk of consuming parasites and other harmful microoganisms and toxins such as lead. Complications of pica include iron deficiency anemia, lead poisoning, and parasitic infestations. Anemia and high blood pressure are more common in those with pica than among other pregnant women, but it is not clear whether pica is a result of these conditions or a cause. In newborns, anemia and low birth weight are often related to pica in the mother (**Figure 11.10**).

---

pica An abnormal craving for and ingestion of nonfood substances that have little or no nutritional value.

What Should I Eat?

During Pregnancy

Make nutrient-dense choices
- Have yogurt for a midmorning snack.
- Put some peanut butter on your banana to add some protein to your snack.
- Have a cup of pasta Florentine (with spinach): It provides both a natural and a fortified source of folate.

Drink plenty of fluids
- Have a glass of low-fat milk to boost fluid, calcium, and vitamin D intake.
- Keep a bottle of water at your desk or in your car.
- Relax with a cup of tea.

Indulge your cravings, within reason
- Enjoy an ice cream cone; it provides calcium and protein.
- Enjoy your cookies with a glass of low-fat milk.

Use iProfile to plan a nutritious 300-Calorie snack for a pregnant woman.

FIGURE 11.10 Pica This African American woman in Georgia is eating a white clay called kaolin, which some women crave during pregnancy. Eating kaolin is also a traditional remedy for morning sickness.

Concept Check

1. **What** snack could a pregnant woman in her 3rd trimester add to her day to meet her increased energy and protein needs?
2. **Why** isn't the recommendation for dietary calcium increased during pregnancy?
3. **Why** are iron supplements recommended during pregnancy?

11.3 Factors That Increase the Risks Associated with Pregnancy

LEARNING OBJECTIVES

1. **Explain** what is meant by critical periods of prenatal development.
2. **Discuss** how nutritional status can influence the outcome of pregnancy.
3. **Explain** how a pregnant woman's age and health status affect the risks associated with pregnancy.
4. **Describe** the effects of alcohol, mercury, and cocaine on the outcome of pregnancy.

Anything that interferes with embryonic or fetal development can cause a baby to be born too soon or too small or can result in birth defects. The embryo and fetus are particularly vulnerable to damage because their cells are dividing rapidly, differentiating, and moving to form organs and other structures. Developmental errors can be caused by deficiencies or excesses in the maternal diet and by harmful substances that are present in the environment, consumed in the diet, or taken as medications or recreational drugs. Any chemical, biological, or physical agent that causes a birth defect is called a **teratogen**. And because each organ system develops at a different time and rate, each has a **critical period** during which exposure to a teratogen is most likely to disrupt development and cause irreversible damage (**Figure 11.11**). Severe damage can result in miscarriage. Some women are at increased risk for complications during pregnancy because of their nutritional status, age, or preexisting health problems. Others are at risk because of exposure to harmful substances.

Maternal Nutritional Status

Before pregnancy, proper nutrition is important to allow conception and maximize the likelihood of a healthy pregnancy. Women with reduced body fat—due to starvation diets, anorexia nervosa, or excessive exercise—may have abnormal hormone levels because adipose tissue plays a role in the metabolism of the sex hormones. When hormone levels are too low, ovulation does not occur, and conception is not possible. Too much body fat has also been associated with reduced fertility.[5] Deficiencies or excesses of specific nutrients can affect both fertility and the outcome of pregnancy.

During pregnancy, the effects of malnutrition vary, depending on when during the pregnancy it occurs.[37] Poor nutrition early in pregnancy affects embryonic development and the potential of the embryo to survive, and poor nutrition in the latter part of pregnancy affects fetal growth. Insufficient energy intake during early pregnancy is not likely to interfere with fetal growth because the energy demands of the embryo are small. However, if the embryo does not receive adequate amounts of the nutrients needed for cell division and differentiation, such as folate and vitamin A, malformations or death can result. After the first trimester, most organs and structures have already formed, so malformations are less likely, but undernutrition in the mother can interfere with fetal growth and reduce birth weight.

Maternal malnutrition not only interferes with fetal growth and development but also causes changes that can affect the child's risk of developing obesity and other chronic diseases later in life. The mechanism that allows nutrition early in development to trigger permanent responses later in life is called **nutritional programming**. There is strong evidence that maternal under- and overnutrition during pregnancy can lead to alterations in metabolism and body composition in the offspring later in life. Nutritional programming that occurs before birth is now thought to be important in the development of obesity, type 2 diabetes, and cardiovascular disease (**Figure 11.12**).[38]

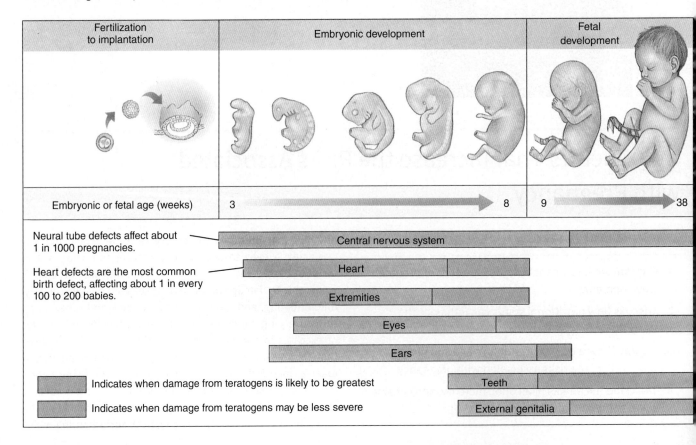

FIGURE 11.11 Critical periods of development The critical periods of development are different for different body systems. Because the majority of cell differentiation occurs during the embryonic period, this is the time when exposure to teratogens can do the most damage, but vital body organs can still be affected during the fetal period. One in every 33 babies born in the United States has a birth defect.[36]

FIGURE 11.12 **Effects of the Dutch famine** During World War II, an embargo on food transport to the Netherlands in the winter of 1944–1945 caused the average intake per person per day to drop below 1000 Calories for about 6 months. Babies born to women exposed to the famine were more likely than others to develop diabetes, heart disease, obesity, and other chronic diseases in adulthood.[38]

Maternal Age and Health

Pregnancy places stresses on the body at any age, but the stresses have a greater impact on women in their teens and those age 35 or older than on women between these ages. Pregnant teens are still growing, so their nutrient intake must meet their needs for growth as well as for pregnancy.[39] Pregnant teenagers are at increased risk of hypertensive disorders and are more likely than pregnant adults to deliver preterm and low-birth-weight babies. To produce a healthy baby, a pregnant teenager needs early medical intervention and nutrition counseling (**Figure 11.13**). Although the rate of teenage pregnancy has decreased over the past two decades—from 62 babies/1000 teens in 1991 to about 24/1000 in 2014—it remains a major public health problem.[40]

The nutritional requirements for older women during pregnancy are no different from those for women in their 20s, but pregnancy after age 35 carries additional risks. Older women are more likely to already have one or more medical conditions, such as cardiovascular disease, kidney disorders, obesity, or diabetes. These preexisting conditions increase the risks associated with pregnancy. For example, a woman who has preexisting high blood pressure is at increased risk for having a low-birth-weight baby, and a woman who has diabetes is more likely to have a baby who is large for gestational age. Older women are also more likely to develop gestational diabetes, hypertensive disorders of pregnancy, and other complications. Older women have a higher incidence of low-birth-weight babies, preterm birth, stillbirth, and perinatal death; however, the actual rate of stillbirth/neonatal death is still very low.[41]

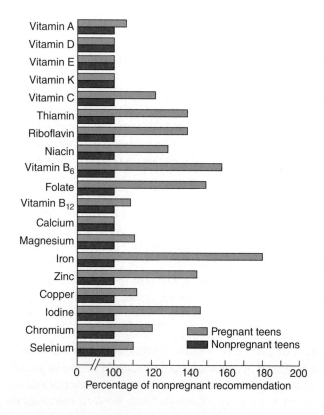

FIGURE 11.13 **Nutrient needs of pregnant teens** Because the nutrient needs of pregnant teens are different from those of pregnant adults, the DRIs include a special set of nutrient recommendations for this group. This graph illustrates the recommended daily micronutrient intake for 14- to 18-year-old girls during the second and third trimesters of pregnancy compared to the recommendations for 14- to 18-year-old nonpregnant girls.

Greater maternal age also increases the risk of having a baby with chromosomal abnormalities, especially **Down syndrome** (**Figure 11.14**). The frequency of twins and triplets is higher among older mothers in part because of their greater use of fertility treatments. Multiple pregnancies increase nutrient needs and the risk of preterm delivery.

Women with a history of miscarriage or birth defects are also at increased risk. For example, a woman who has had a number of miscarriages is more likely to have another, and a woman who has had one child with a birth defect has an increased risk for defects in subsequent pregnancies. Another factor that increases risks is frequent pregnancies. An interval of less than 18 months between pregnancies increases the risk of delivering a preterm or small-for-gestational-age baby as well the risk of neonatal or infant mortality.[42] One reason for these increased risks is that the mother may not have replenished the nutrient stores depleted in the first pregnancy before becoming pregnant again.

Poverty

One of the greatest risk factors for poor pregnancy outcome is poverty, which limits access to food, education, and health care.[44] Low-income women are unlikely to receive medical care until late in pregnancy and consequently have a high incidence of low-birth-weight and preterm infants. A federally funded program that addresses the nutritional needs of pregnant women is the **Special Supplemental Nutrition Program for Women, Infants, and Children (WIC)**. This program provides nutrition counseling, funds to purchase nutritious foods, and referrals to health and other services for low-income women who are pregnant, postpartum, or breastfeeding, as well as for infants and children until age 5 years.

Exposure to Toxic Substances

A pregnant woman's exposure to toxins in food, water, and the environment can affect the developing fetus. Even substances that seem innocuous can be dangerous. For example, herbal remedies, including herbal teas, should be avoided unless their consumption during pregnancy has been determined to be safe.

Caffeine When consumed in excess, coffee and other caffeine-containing beverages have been associated with increased risks of miscarriage or low birth weight. However, moderate caffeine consumption, less than 200 mg/day, the amount in one to two cups of coffee or about two or three 20-oz caffeinated soft drinks, does not appear to increase the risk of miscarriage or low birth weight.[16, 45]

Mercury in fish Fish has both benefits and risks for pregnant women. It is a source of lean protein for tissue growth and of omega-3 fatty acids and iodine needed for brain development, but if it is contaminated with mercury, consumption

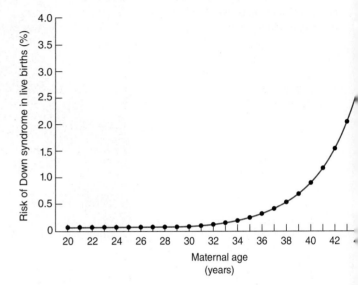

FIGURE 11.14 Incidence of Down syndrome Down syndrome is caused by a defect during egg or sperm formation that results in extra genetic material in the baby. It causes distinctive facial characteristics, developmental delays, and other health problems. The incidence of Down syndrome rises significantly with maternal age of 35 years or more. However, while older mothers are more likely to conceive a Down syndrome baby, 80% of Down syndrome births are to mothers under age 35, as younger women have more babies overall.[43]

during pregnancy can cause developmental delays and brain damage. Rather than avoid fish, pregnant women should be informed consumers. The Dietary Guidelines recommend that pregnant women consume 8 to 12 oz/week of seafood from a variety of low-mercury fish and shellfish.[46] Exposure to mercury from fish can be controlled by avoiding varieties that are high in mercury and limiting intake of fish that contain lower amounts of mercury (**Table 11.2**).

Food-borne illness The immune system is weakened during pregnancy, increasing susceptibility to and the severity of certain food-borne illnesses. *Listeria* infection is 18 times more common in pregnant women than in the general population. Infection during pregnancy is associated with an increased risk of miscarriage and stillbirth; the infection can be transmitted to the fetus, causing meningitis and serious blood infections.[49] About one-third of fetuses with *Listeria* infections do not survive.[50] The bacteria are commonly found in unpasteurized milk, soft cheeses, and uncooked hot dogs and lunch meats (see Table 11.2).

Toxoplasmosis is an infection caused by a parasite. If a pregnant woman becomes infected, the infection can be passed to her unborn baby. Infected babies may develop vision and hearing loss, intellectual disability, and/or seizures, and some die within a few days of birth. The toxoplasmosis parasite is found in cat feces, soil, and undercooked infected meat (see Table 11.2). Pregnant women should follow the safe food-handling recommendations discussed in Chapter 13.

TABLE 11.2 Food safety during pregnancy[47, 48]

Avoid swordfish, shark, king mackerel, tilefish, marlin, orange roughy, and bigeye tuna, which can be high in mercury.
Eat 2 to 3 servings (8 to 12 oz) per week of low mercury fish such as salmon, shrimp, canned light tuna, pollock, catfish, and cod or one serving/week (4 oz) of fish higher in mercury, such as halibut, mahi mahi, or canned albacore or chunk white tuna. (See Appendix C for complete fish lists)
Eat a variety of fish.
Check local advisories about the safety of fish caught in local waters. If no advice is available, eat up to 4 oz per week but don't consume any other fish during that week.
Don't drink raw (unpasteurized) milk or consume products made with unpasteurized milk, such as certain Mexican-style soft cheeses.
Don't eat refrigerated smoked fish, cold deli salads, or refrigerated pâtés or meat spreads.
Don't eat hot dogs and lunch meats unless they have been reheated to steaming hot.
Don't eat raw or undercooked meat, poultry, fish, shellfish, or eggs.
Don't eat unwashed fruits and vegetables, raw sprouts, or unpasteurized juice.

Alcohol Alcohol consumption during pregnancy is one of the leading causes of preventable birth defects. Alcohol is a teratogen that is particularly damaging to the developing nervous system. It also indirectly affects fetal growth and development because it is a toxin that reduces blood flow to the placenta, thereby decreasing the delivery of oxygen and nutrients to the fetus. Use of alcohol can also impair maternal nutritional status, further increasing the risks to the fetus.

Prenatal exposure to alcohol can cause a spectrum of disorders, depending on the amount, timing, and duration of the exposure. One of the most severe outcomes of alcohol use during pregnancy is delivery of a baby with **fetal alcohol syndrome (FAS)** (Figure 11.15). Not all babies who are exposed to alcohol while in the uterus have FAS. The term **fetal alcohol spectrum disorders (FASDs)** is used to refer to all the physical and behavioral disorders or conditions and functional or mental impairments linked to prenatal alcohol exposure.[51] As many as 2 to 5% of young schoolchildren in the United States are affected by FASDs.[52] Because alcohol consumption in each trimester has been associated with fetal abnormalities, and there is no level of alcohol consumption that is known to be safe, complete abstinence from alcohol is recommended during pregnancy.

Tobacco use If a woman uses tobacco products during pregnancy, her baby will be affected before birth and throughout life. The carbon monoxide in tobacco smoke binds to hemoglobin in maternal and fetal blood, reducing the amount

FIGURE 11.15 Dangers of alcohol use during pregnancy The facial characteristics shared by children with fetal alcohol syndrome (FAS) include a low nasal bridge, short nose, distinct eyelids, and a thin upper lip. Newborns with FAS may be shaky and irritable, with poor muscle tone. Other problems associated with FAS include heart and urinary tract defects, impaired vision and hearing, and delayed language development. Below-average intellectual function is the most common and most serious effect.

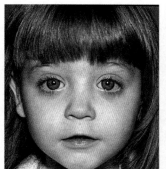
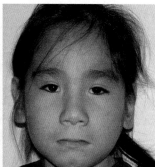
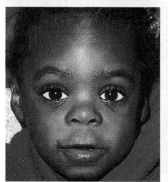
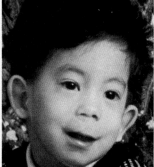

© 2012 Susan Astley PhD, University of Washington

---

fetal alcohol syndrome (FAS) A characteristic group of physical and mental abnormalities in an infant resulting from maternal alcohol consumption during pregnancy.

of oxygen delivered to fetal tissues. The nicotine absorbed from any form of tobacco product, including e-cigarettes, is a teratogen that can affect brain development.[53, 54] Nicotine also constricts arteries and limits blood flow, reducing the amounts of oxygen and nutrients delivered to the fetus.[55] Tobacco use during pregnancy reduces birth weight and increases the risks of stillbirth, preterm delivery, birth defects, and early infant death.[53] In women who don't smoke, environmental exposure to cigarette smoke has been found to increase the risk of having a low-birth-weight baby. The risks of **sudden infant death syndrome (SIDS)**, or **crib death**, and respiratory problems are also increased in children exposed to cigarette smoke both in the uterus and after birth.[56] The effects of maternal smoking follow children throughout life; these children are more likely to develop lung problems later in life.[57]

Legal and illicit drug use Certain drugs—whether over-the-counter, prescribed, or illegal—can affect fetal development. For example, the acne medications Accutane and Retin-A are derivatives of vitamin A that can cause birth defects if used during pregnancy. A woman who is considering becoming pregnant should discuss her plans with her physician in order to determine the risks associated with any medication she is taking.

Abuse of illegal substances during pregnancy is a national health issue. It is estimated that about 4.4% of pregnant women use illicit drugs.[58] The risk to the infant depends on the drug. Marijuana and cocaine can cross the placenta and enter the fetal blood. Prenatal marijuana exposure may cause subtle abnormalities in infant neurobehavior, such as increased irritability and muscle tremors, but it has not been shown to affect fetal growth or cause birth defects.[58] Cocaine use creates problems for both the mother and the infant before, during, and after delivery. Cocaine is a teratogen that interferes with nervous system development and may cause permanent changes in brain structure and function.[58] Cocaine also constricts blood vessels, thereby reducing the delivery of oxygen and nutrients to the fetus. Cocaine use during pregnancy is associated with an increased risk of miscarriage, fetal growth retardation, premature labor and delivery, low birth weight, and birth defects.[59] Exposure to cocaine and other illegal drugs before birth has also been shown to affect infant behavior and influence learning and attention span during childhood.[58] Newborns exposed to drugs in the womb may be born addicted to these drugs and experience withdrawal symptoms after birth.[60] This is particularly a problem with opioids such as heroin, codeine, and oxycodone.

> ### Concept Check
>
> 1. **Why** does the effect of a given teratogen vary, depending on when a pregnant woman is exposed to it?
> 2. **What** is meant by nutritional programming?
> 3. **Why** are the requirements for some nutrients different in pregnant teenage girls than in pregnant adult women?
> 4. **How** much alcohol can be safely consumed during pregnancy?

11.4 Lactation

LEARNING OBJECTIVES

1. **Describe** the events that trigger milk production and let-down.
2. **Discuss** the energy and water needs of lactating women.
3. **Compare** the micronutrient needs of lactating women with those of nonpregnant, nonlactating women.

The need for many nutrients is even greater during lactation than during pregnancy. The milk produced by a breast-feeding mother must meet all the nutrient needs of her baby, who is bigger and more active than he or she was in the womb. To meet these needs, a lactating woman must choose a varied, nutrient-dense diet.

Milk Production and Let-Down

Lactation involves both the synthesis of milk components—proteins, lactose, and lipids—and the movement of these components through the milk ducts to the nipple (**Figure 11.16**). Milk production and **let-down** are triggered by hormones that are released in response to an infant's

---

sudden infant death syndrome (SIDS) or **crib death** The unexplained death of an infant, usually during sleep.

let-down The release of milk from the milk-producing glands and its movement through the ducts and storage sinuses.

FIGURE 11.16 **Anatomy of milk production** Throughout pregnancy, hormones prepare the breasts for lactation by stimulating the enlargement and development of the milk ducts and milk-producing glands, called *alveoli* (singular *alveolus*). During lactation, milk travels from the alveoli through the ducts to the milk storage sinuses and then to the nipple.

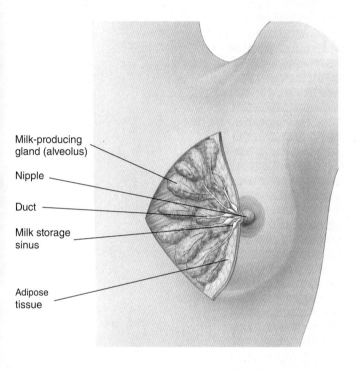

suckling. The pituitary hormone **prolactin** stimulates milk production; the more the infant suckles, the more milk is produced. Let-down is caused by **oxytocin**, another pituitary hormone. Oxytocin release is also stimulated by suckling, but as nursing becomes routine, oxytocin release and let-down may occur in response to just the sight or sound of an infant. Let-down can be inhibited by nervous tension, fatigue, or embarrassment. Because let-down is essential for breast-feeding and makes suckling easier for the child, slow let-down can make feeding difficult.

Energy and Nutrient Needs During Lactation

Human milk contains about 70 Calories/100mL (160 Calories/cup). During the first 6 months of lactation, an average infant consumes 600 to 900 mL (about 2.5 to 4 cups)/day, so approximately 500 Calories are required from the mother each day. Much of this energy must come from the diet, but some can come from maternal fat stores. Because some of the energy for milk production comes from fat stores, the increase in recommended energy intake is lower during lactation than during pregnancy, even though total energy demands are greater (**Figure 11.17**). An additional 330 Calories/day above nonpregnant, nonlactating needs are recommended during the first 6 months of lactation, and an additional 400 Calories are recommended for the second 6 months. Beginning 1 month after birth, most lactating women lose 0.5 to 1 kg (1 to 2 lb)/month for 6 months. Rapid weight loss is not recommended during lactation because it can decrease milk production.

To ensure adequate protein for milk production, the RDA for lactation is increased by 25 g/day. The recommended intakes of total carbohydrate, fiber, and the essential fatty acids linoleic and α-linolenic acid are also higher during lactation (see Figure 11.17).[20] Since the fatty acid composition of the mother's diet determines the fatty acid composition of her breast milk, adequate intake of omega-3 and omega-6 fatty acids ensures a healthy balance of these essential fatty acids in her breast milk.[61]

FIGURE 11.17 **Energy and macronutrient needs during lactation** This graph compares the energy and macronutrient recommendations for 25-year-old nonpregnant women and 25-year-old women during the third trimester of pregnancy and the first 6 months of lactation.

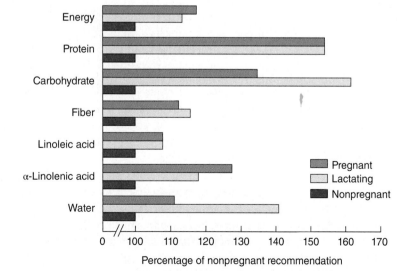

Ask Yourself

Why is the need for water greater during lactation than during pregnancy?

358 CHAPTER 11 Nutrition During Pregnancy and Infancy

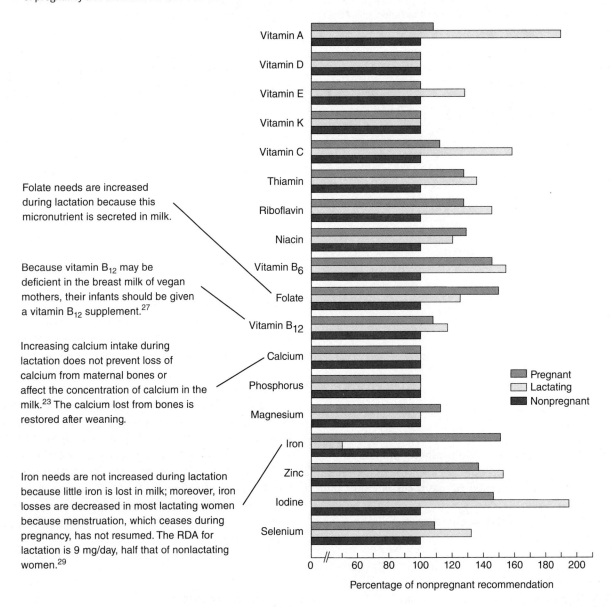

FIGURE 11.18 Micronutrient needs during lactation This graph compares the vitamin and mineral recommendations for 25-year-old nonpregnant women and 25-year-old women during the third trimester of pregnancy and the first 6 months of lactation.

Folate needs are increased during lactation because this micronutrient is secreted in milk.

Because vitamin B_{12} may be deficient in the breast milk of vegan mothers, their infants should be given a vitamin B_{12} supplement.[27]

Increasing calcium intake during lactation does not prevent loss of calcium from maternal bones or affect the concentration of calcium in the milk.[23] The calcium lost from bones is restored after weaning.

Iron needs are not increased during lactation because little iron is lost in milk; moreover, iron losses are decreased in most lactating women because menstruation, which ceases during pregnancy, has not resumed. The RDA for lactation is 9 mg/day, half that of nonlactating women.[29]

To avoid dehydration and ensure adequate milk production, lactating women need to consume about 1 L/day of additional water (see Figure 11.17).[18] Consuming an extra glass of milk, juice, or water at each meal and whenever the infant nurses can help ensure adequate water intake.

The recommended intakes for several vitamins and minerals are increased during lactation to meet the needs for synthesizing milk and to replace the nutrients secreted in the milk (**Figure 11.18**). For some nutrients, including thiamin, riboflavin, vitamin B_6, vitamin B_{12}, vitamin A, vitamin D, selenium, and iodine, low maternal intake reduces the amount secreted into the milk.[62, 63] For example, low levels of vitamin B_{12} in a mother's diet can result in low levels of vitamin B_{12} in her milk. For other nutrients, including folate, calcium, iron, copper, and zinc, levels in the milk are maintained at the expense of maternal stores.[62]

Concept Check

1. **What** causes milk let-down?
2. **Where** does the energy for milk production come from?
3. **Why** is the recommended calcium intake for a new mother not increased while she is lactating?

11.5 Nutrition for Infants

LEARNING OBJECTIVES

1. **Explain** how growth charts are used to monitor the nutritional well-being of infants.
2. **Contrast** the energy and macronutrient needs of infants and adults.
3. **Compare** the benefits of breast-feeding and formula-feeding.
4. **Discuss** the importance of choosing foods that are appropriate for a child's developmental stage.

After the umbilical cord is cut, a newborn must actively obtain nutrients rather than being passively fed through the placenta. The energy and nutrients an infant consumes must support his or her continuing growth and development, as well as his or her increasing activity level.

Infant Growth and Development

During **infancy**—the first year of life—growth and development are extremely rapid. Infants get bigger and develop physically, intellectually, and socially. Adequate nutrition is essential for achieving growth and developmental milestones.

Healthy infants follow standard patterns of growth—that is, whether a newborn weighs 6 lb or 8 lb at birth, the rate of growth should be approximately the same: rapid initially and slowing slightly as the infant approaches his or her first birthday. A rule of thumb is that an infant's birth weight should double by 4 months of age and triple by 1 year of age. In the first year of life, most infants increase their length by 50%. Growth is the best indicator of adequate nutrition in an infant.

Growth can be assessed in infants and children using growth charts. The growth charts used for infants were developed by the World Health Organization based on growth patterns of breast-fed infants living in optimal conditions. To use a growth chart, the infant's weight and/or height is plotted on the graph, and the resulting percentile indicates where the infant's growth falls in relation to optimal growth patterns for infants of the same age (**Figure 11.19**).[64] Children usually remain at the same percentile as they grow. For instance, a child who is at the 50th percentile for height and the 25th percentile for weight generally remains close to these percentiles throughout childhood. Small and premature infants often follow a pattern that is parallel to, but below, the growth curve for a period of time and then experience catch-up growth that brings them into the same range as children of the same age.

Slight variations in growth rate are normal, but a consistent pattern of not following the established growth curve or a sudden change in growth pattern is cause for concern and could indicate overnutrition or undernutrition. For example, a rapid increase in weight without an increase in height may indicate that the infant is being overfed. Just as there are critical periods in prenatal life, there are critical periods during infancy when nutrition can have permanent effects on growth and development and long-term health. Evidence suggests that accelerated weight gain in the first 2 years of life increases the risk of obesity, diabetes, high blood pressure, harmful blood lipid levels, and cardiovascular disease in adulthood.[65] A pattern of accelerated weight increase should be addressed early in life to reduce the likelihood of these chronic conditions later.

Growth that is slower than the predicted pattern indicates **failure to thrive**, a catchall term for any type of growth failure in a young child. The cause may be a congenital condition, disease, poor nutrition, neglect, abuse, or psychosocial problems. The treatment is usually an individualized plan that includes adequate nutrition and careful monitoring by physicians, dietitians, and other health-care professionals. Undernutrition during infancy can permanently affect growth and development as well as health later in life. Because the first year of life is a critical period for brain development, undernutrition interferes with learning and affects behavior. Undernutrition in childhood is a risk factor for diabetes, high blood pressure, and cardiovascular disease in adulthood.[66]

Energy and Nutrient Needs of Infants

The rapid growth rate of infants increases their need for energy, protein, and vitamins and minerals that are important for growth. Human milk and commercially produced formula are designed to meet infants' nutrient needs. Nevertheless, infants may still be at risk for deficiencies of iron, vitamin D, and vitamin K and for suboptimal levels of fluoride.

Energy and macronutrients Infants require more calories and protein per kilogram of body weight than do individuals at any other time of life (**Figure 11.20a**).[20] As infants grow older, their rate of growth slows, but they become more mobile, so the amount of energy they need for activity increases. Because infants change so much during the first year, energy recommendations are made for three age groups—0 to 3 months,

---

failure to thrive Inability of a child's growth to keep up with normal growth curves.

FIGURE 11.19 Growth charts
This growth chart can be used to compare a boy's length and weight at a particular age to standards for optimal growth. A similar chart is available for girls (see Appendix E). Growth charts for boys and girls from birth to 24 months of age are also available to monitor weight for length and head circumference.

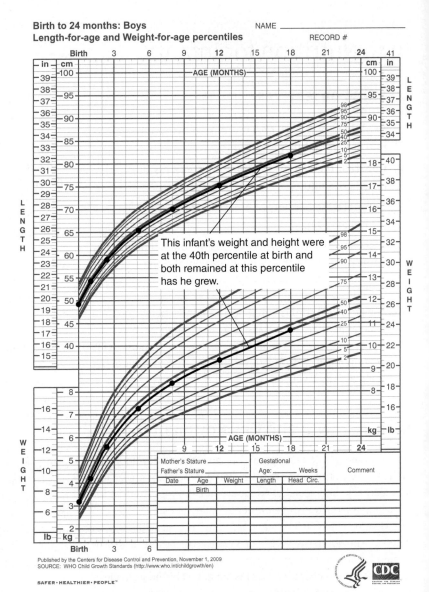

Interpret the Data

A 23-month-old boy who weighs 13 kg is at the _____ percentile.

a. 25th
b. 50th
c. 75th
d. 90th

4 to 6 months, and 7 to 12 months—and nutrient recommendations are made for two age groups—0 to 6 months and 7 to 12 months (see Appendix A).

The combination of high energy demands and a small stomach means that infants require an energy-dense diet. Healthy infants consume about 55% of their energy as fat during the first 6 months of life and 40% during the second 6 months. These percentages are far higher than the 20 to 35% of energy from fat recommended in the adult diet (**Figure 11.20b**). This high-fat diet allows the small volume of food that fits in an infant's stomach to provide enough energy to meet the infant's needs and provides essential fatty acids needed for growth and development (see Debate: DHA/ARA-Fortified Infant Formulas).

Fluid needs Infants have a higher proportion of body water than do adults, and they lose proportionately more water in urine and through evaporation. Urine losses are high because the kidneys are not fully developed and hence are unable to reabsorb as much water as adult kidneys. Water losses through evaporation are proportionately higher in infants than in adults because infants have a larger surface area relative to body weight. As a result, they need to consume more water per unit of body weight than do adults. Nevertheless, healthy infants who are exclusively breast-fed do not require additional water.[18] In older infants, some water is obtained from food and from beverages other than milk. When water losses are increased by diarrhea or vomiting, additional fluids may be needed.

Micronutrients at risk Iron is the nutrient that is most commonly deficient in infants who are consuming adequate energy and protein. In a full-term infant, iron deficiency is usually not a problem during the first 6 months of life because infants have iron stores at birth, and the iron in human milk, though not abundant, is very well absorbed. Because most of the transfer of iron from mother to fetus occurs during the third trimester, babies who are born prematurely may not have time to accumulate sufficient iron and are at greater risk of deficiency. The AI for iron from birth to 6 months is 0.27 mg/day.

FIGURE 11.20 Energy and macronutrient needs The amount of energy and the distribution of energy-yielding nutrients required by infants are strikingly different from what is recommended in the adult diet.

a. The total amount of energy required per day by a newborn is less than the amount needed by an adult. When this amount is expressed as Calories/kg of body weight, however, we see that infants require about three times more energy than an adult male.

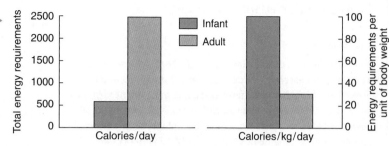

Carbohydrate is a major contributor to an infant's energy intake; most of this comes from lactose.

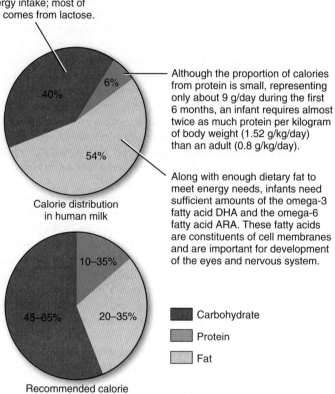

Calorie distribution in human milk

Although the proportion of calories from protein is small, representing only about 9 g/day during the first 6 months, an infant requires almost twice as much protein per kilogram of body weight (1.52 g/kg/day) than an adult (0.8 g/kg/day).

Along with enough dietary fat to meet energy needs, infants need sufficient amounts of the omega-3 fatty acid DHA and the omega-6 fatty acid ARA. These fatty acids are constituents of cell membranes and are important for development of the eyes and nervous system.

Recommended calorie distribution for the adult diet

- Carbohydrate
- Protein
- Fat

b. Comparing the distribution of calories from carbohydrate, fat, and protein in human milk with that recommended for an adult illustrates proportionally how much more fat infants need. As an infant grows and solid foods are introduced into the diet, the percentage of calories from carbohydrate in the diet increases, and the percentage from fat decreases.

Think Critically

If breast milk is the best food for infants, why is it so high in fat?

After 4 to 6 months, iron stores are depleted, but iron needs remain high. The RDA for infants 7 to 12 months old increases to 11 mg/day.[29] By this age, the diet of breast-fed infants should contain other sources of iron. Formula-fed infants should be fed iron-fortified formula.

Breast milk is low in vitamin D. Therefore, it is recommended that breast-fed and partially breast-fed infants be supplemented with 400 IU (10 μg)/day of vitamin D beginning in the first few days of life and continuing until they are consuming about 1 L (4 cups) of vitamin D–fortified formula or milk daily.[23, 67] Infant formulas contain at least 400 IU (10 μg)/L of vitamin D. Formula-fed infants who consume at least 1 L/day of formula can meet their vitamin D needs from the formula alone; those consuming less than 1 L/day should receive a vitamin D supplement of 400 IU (10 μg)/day. The amount of sun exposure necessary to maintain an adequate level of vitamin D in any given infant at any point in time is not easy to determine. Light-skinned infants are more likely than darker-skinned babies to meet their vitamin D needs from sunlight.[67]

Infants are at risk for vitamin K deficiency because little vitamin K crosses the placenta, and the infant's gut is sterile, so there are no bacteria to synthesize this vitamin. Because lack of vitamin K can cause bleeding, it is recommended that all newborns receive an intramuscular injection containing 0.5 to 1.0 mg of vitamin K after the first feeding is completed and within the first six hours of life.[68] This provides them with enough vitamin K to last until the intestines have been colonized with bacteria that synthesize it.

Fluoride is important for tooth development, even before the teeth erupt. Breast milk is low in fluoride, and formula manufacturers use unfluoridated water in preparing liquid formula. Therefore, breast-fed infants, infants who are fed premixed formula, and those who are fed formula mixed with low-fluoride water at home are often given fluoride supplements beginning at 6 months of age. In areas where drinking water is fluoridated, infants who are fed formula reconstituted with tap water should not be given fluoride supplements.

Debate

DHA/ARA-Fortified Infant Formulas

The Issue

The fatty acids docosahexaenoic acid (DHA) and arachidonic acid (ARA) are essential for development of the retina and brain. Breast milk provides these fatty acids, and most infant formulas in the United States are fortified with them. Advertisements suggest that these formulas provide an advantage for infant development. Will feeding babies formula fortified with DHA and ARA make them smarter and improve their vision?

DHA and ARA are polyunsaturated fatty acids that can be made in the body from the essential omega-3 fatty acid α-linolenic acid and the essential omega-6 fatty acid linoleic acid, respectively (see Figure 5.13). DHA and ARA are an integral structural part of cell membranes in the brain and retina and are essential for normal brain development and vision.[69] Accumulation of these fatty acids in the brain and retina occurs most rapidly between the third trimester of pregnancy and the first 24 months after birth, so adequate amounts are crucial during this developmental period.

DHA and ARA are transferred from mother to fetus across the placenta during pregnancy. They are also found in breast milk, so it seems logical that they should be added to infant formula. But, unlike most other nutrients, the amounts of DHA and ARA in breast milk are variable, depending on maternal diet, so it is unclear what constitutes optimal levels. The amounts of these fatty acids transferred to the fetus by the placenta during the third trimester may be enough to ensure adequate amounts for brain development even if it they are not provided in the diet.[69] Infants born at term are also capable of synthesizing DHA and ARA, so those fed unfortified formula may be able to meet their needs if they have enough α-linolenic acid and linoleic acid in their diet. But there is wide individual variation in the ability to convert α-linolenic acid to DHA, and in some infants, conversion may be too low for optimal brain and visual development.[69, 70] Infants born before term cannot synthesize enough of these fatty acids, so it is recommended that DHA and ARA be included in formula for premature infants.[71]

So, are higher intakes of these fatty acids beneficial for the brains and eyes of babies born at term? Numerous studies have explored the impact that postnatal intake of these fatty acids has on intelligence and vision. Some have found that higher intakes of DHA provided by supplemented formulas are associated with improvements in cognitive development compared to infants with lower intakes, but the majority have found that supplementing infant formula with DHA and ARA has little effect on the cognitive function of infants or children.[72] The results of studies on visual acuity have found that infants supplemented with ARA and DHA have an increase in the rate of visual acuity development compared to infants fed unsupplemented formula, but it is unclear whether this has a persistent effect on vision.[69]

A number of reasons may explain why studies on the effects of fortified formulas are inconsistent. Differences may be due to genetic differences in the infant's capacity to synthesize ARA and DHA, variations in the ARA and DHA content of the formulas, the duration of formula-feeding, and the methods used to assess visual acuity and cognitive development. Higher intelligence in breast-fed infants may be due more to factors such as maternal IQ, income level, and the amount of time mothers spend with their infants than to whether they are breast- or formula-fed.[72, 73]

When it comes to brain and eye development, no one knows exactly how much DHA or ARA an infant needs. A direct link between use of fortified formula and better vision or higher IQ compared to use of unfortified formula has yet to be established, but fortified formulas are generally recommended. Standards for these have been established by several international organizations, based on the average concentrations in human milk, and most infant formulas sold in the United States are now fortified with DHA and ARA.[70] They may not make your baby smarter, but published literature has not demonstrated these formulas to be harmful for infants.[74] Breast milk is always best, so the biggest downside to fortified formulas is that advertising may make new mothers believe that they are as good as or better for their babies than breast milk.

Think Critically

Why is breast milk still better than these fortified formulas?

Meeting Needs with Breast Milk or Formula

Because of its health and nutritional benefits, breast-feeding is the recommended choice for the newborn of a healthy, well-nourished mother. Health professionals in the United States recommend exclusive breast-feeding for the first 6 months of life and breast-feeding with complementary foods for at least the first year and as long thereafter as mutually desired.[75] Breast-feeding after the first year continues to provide nutrition, comfort, and an emotional bond between mother and child. As infants begin consuming other foods, their demand for milk is reduced, and milk production decreases, but lactation can continue as long as suckling is maintained.

Whether they are breast-fed or formula-fed, young infants should be fed frequently, on demand. For breast-fed infants, a feeding should last approximately 10 to 15 minutes at each breast. Bottle-fed newborns may consume only a few ounces at each feeding; as the infant grows, the amount consumed will increase to 4 to 8 oz (**Figure 11.21**). A well-fed newborn, whether breast-fed or bottle-fed, should urinate enough to soak six to eight diapers a day and gain about 0.15 to 0.23 kg (0.33 to 0.5 lb)/week.

Nutrients in breast milk and formula Human milk is tailored to meet the needs of human infants. The composition of milk changes continuously to suit the needs of a growing infant. The first milk, called **colostrum**, which is produced by the breast for up to a week after delivery, has beneficial effects on the gastrointestinal tract, acting as a laxative that helps the baby excrete the thick, mucusy stool produced during life in the womb. Colostrum looks watery, but the nutrients it supplies meet the infant's needs until mature milk production begins. Mature breast milk contains an appropriate balance of nutrients in forms that are easily digested and absorbed. Infant formulas try to replicate human milk as closely as possible in order to match the growth, nutrient absorption, and other benefits associated with breast-feeding (**Table 11.3**).

Health benefits of breast-feeding Despite a nutritional profile that is similar, infant formulas can never exactly duplicate the composition of human milk.[63] Antibody proteins and immune-system cells pass from the mother to her child in the breast milk, providing immune protection for the infant. A number of enzymes and other proteins in breast milk prevent the growth of harmful microorganisms, and several of the carbohydrates in breast milk protect against disease-causing microorganisms, including viruses that cause diarrhea. One carbohydrate favors the development of a healthy microbiota in the infant's colon, which inhibits the growth of harmful bacteria. Growth factors and hormones present in human milk promote maturation of the infant's digestive, cardiovascular,

FIGURE 11.21 Dos and don'ts of bottle-feeding Proper bottle-feeding technique supports the infant's nutritional and overall health.

a. During bottle feeding, the infant's head should be higher than his or her stomach, and the bottle should be tilted so that there is no air in the nipple. Just as breast-fed infants alternate breasts, bottle-fed infants should be held alternately on the left and right sides to promote equal development of the head and neck muscles.

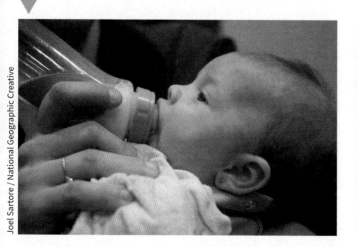

b. Infants should never be put to bed with a bottle because saliva flow decreases during sleep, and the formula remains in contact with the teeth for many hours. This causes **nursing bottle syndrome**, which is characterized by the rapid and serious decay of the upper teeth. Usually the lower teeth are protected by the tongue.

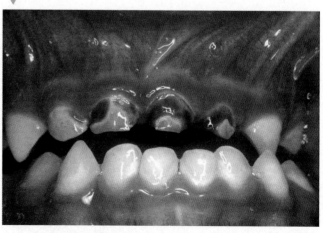

colostrum A substance, known as first milk, produced by the breast late in pregnancy and for up to a week after delivery. It contains more water, protein, immune factors, minerals, and vitamins and less fat than mature breast milk.

TABLE 11.3 Nutritional composition of breast milk and infant formula

Nutrient	Amount in breast milk	Amount in formula	Comparisons
Protein	1.8 g/100 mL	1.4 g/100 mL	The relatively low protein content of human milk and formula protects the immature infant kidneys from a too-high load of nitrogen wastes. Alpha-lactalbumin, the predominant protein in human milk, forms a soft, easily digested curd in the infant's stomach. Most formula is made from cow's milk that is modified to mimic the protein concentration and amino acid composition of human milk.
Fat	4 g/100 mL	4.8 g/100 mL	The fat in human milk is easily digested. Human milk is high in cholesterol and the essential fatty acids linoleic acid and α-linolenic acid, as well as their long-chain derivatives ARA and DHA, which are essential for normal brain development, eyesight, and growth. The fat in formula is derived from vegetable oils and provides linoleic and α-linolenic acid. Most formulas are also fortified with ARA and DHA.
Carbohydrate	7 g/100 mL	7.3 g/100 mL	Lactose, the primary carbohydrate in human milk and most formula, enhances calcium absorption. Because it is digested slowly, it stimulates the growth of beneficial acid-producing bacteria. Oligosaccharides in breast milk protect against respiratory and gastrointestinal disease.
Sodium	1.3 mg/100 mL	0.7 mg/100 mL	Because breast milk and formula are both low in sodium, the fluid needs of breast-fed and formula-fed infants can be met without an excessive load on the kidneys.
Calcium	22 mg/100 mL	53 mg/100 mL	The ratio of calcium to phosphorus in breast milk and formula enhances calcium absorption.
Phosphorus	14 mg/100 mL	38 mg/100 mL	
Iron	0.03 mg/100 mL	0.1 mg/100 mL	Iron and zinc are present in limited amounts in breast milk but are readily absorbed. Most infant formulas are fortified with iron and zinc because the forms present are less absorbable than those in breast milk.
Zinc	3.2 mg/100 mL	5.1 mg/100 mL	
Vitamin D	4 IU (0.1 µg)/100 mL	41 IU (1.0 µg)/100 mL	Formulas are fortified with vitamin D, which is present at low levels in breast milk.

nervous, and endocrine systems. In addition to the numerous health benefits of breast-feeding for infants, it has physical, emotional, and financial advantages for the mother (**Figure 11.22**).

When is formula-feeding better?

Despite the benefits of breast milk, formula-feeding may be the best option in some cases. Common illnesses such as colds and skin infections are not passed to the infant in breast milk, but a few, such as tuberculosis and HIV infection, are.[75] Some drugs can also pass from the mother to the baby in breast milk, which means that women who are taking medications should check with their physician about whether they can safely breast-feed while taking their medication. Because alcohol and drugs such as cocaine and marijuana can be passed to a baby in breast milk, alcoholic and drug-addicted mothers are counseled not to breast-feed. Nicotine from tobacco products is also rapidly transferred from maternal blood to milk, and heavy tobacco use may decrease the supply of milk.

Although alcoholic mothers are counseled not to breast-feed, occasional limited alcohol consumption while breast-feeding is probably not harmful if alcohol intake is timed so as to minimize the amount present in milk when the infant is fed. Consuming a single alcoholic drink is safe if the mother then waits at least 4 hours before breast-feeding. Alternatively, milk can be expressed before consuming the drink and fed to the infant later.

Feeding an infant with formula requires more preparation and washing than breast-feeding, but it can give the mother a break because other family members can share the responsibility. For preterm infants and those with genetic abnormalities, formula may be the best option because there are special formulas to meet these infants' unique needs. If an infant is too small or weak to take a bottle, pumped breast milk or formula can be fed through a tube.

FIGURE 11.22 **The benefits of breast-feeding**[75] Breast-feeding benefits both infants and their mothers.

Benefits for infants
- Provides optimum nutrition
- Enables strong bonding with mother
- Enhances immune protection
- Reduces allergies
- Decreases ear infections, respiratory illnesses, and asthma
- Reduces the likelihood of constipation, diarrhea, or chronic digestive diseases
- Reduces the risk for SIDS
- Lowers the risk for obesity, type 1 and type 2 diabetes, heart disease, hypertension, and childhood leukemia
- Aids in the development of facial muscles, speech development, and correct formation of the teeth
- Lessens the risk of overfeeding because the amount of milk consumed cannot be monitored visually

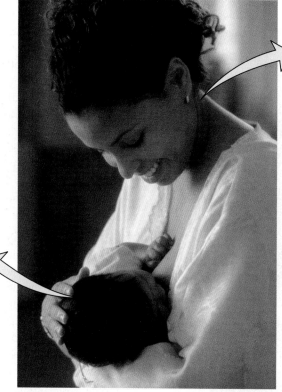

Cusp and Flirt / Masterfile

Benefits for mothers
- Provides relaxing, emotionally enjoyable interaction; strengthens bonding with infant
- Reduces financial costs
- Requires less preparation and clean-up time; always available
- Causes uterine contractions that help the uterus return to its normal size more quickly after delivery
- Increases energy expenditure, which may speed return to prepregnancy weight
- Lowers risk of developing type 2 diabetes and breast and ovarian cancers
- Improves bone density and decreases risk of fractures
- Inhibits ovulation, lengthening the time between pregnancies; however, breast-feeding cannot be relied on for birth control
- Decreases the risk of postpartum depression
- Enhances self-esteem in the maternal role

Safe Feeding for the First Year

Whether infants are breast-fed or formula-fed, care must be taken to ensure that their needs are met and their food is safe. If proper measurements are not used in preparing formula, the infant can receive an excess or a deficiency of nutrients and an improper ratio of nutrients to fluid (**Figure 11.23**). If the water and equipment used in preparing formula are not clean or if the prepared formula is left unrefrigerated, food-borne illness may result. Because sanitation is often lacking in developing nations, infections that lead to diarrhea and dehydration occur more frequently in formula-fed infants than in breast-fed infants.

Bacterial contamination is not a concern when a baby is breast-fed, but care must be taken to avoid contamination when milk is pumped from the breast and stored for later feedings. Hands, breast pumps, bottles, and nipples must be washed. Breast milk that is not immediately fed to the baby can be kept refrigerated for 24 to 48 hours. Warming breast milk in a microwave is not recommended because microwaving destroys some of its immune properties and may result in dangerously hot milk. The best way to warm milk is by running warm water over the bottle.

A concern that is unique to feeding breast milk is that substances in the maternal diet may pass into the milk and cause adverse reactions in the infant. For example, caffeine in the mother's diet can make the infant jittery and excitable, so a mother should avoid consuming large amounts of caffeine while breast-feeding. Most of these reactions seem to be unique to a particular mother and child, so as long as a food does not affect the infant's health or response to feeding, it can be included in the mother's diet.

Food allergies Food allergies are common in infants. Because infants' digestive tracts are immature, they allow the absorption of incompletely digested proteins, which in susceptible individuals can trigger a response from the immune system (see Chapters 3 and 6). After about 3 months of age, the risk of developing food allergies is reduced because incompletely digested proteins are less likely to be absorbed. Many children who develop food allergies before age 3 years eventually outgrow them. Allergies that appear after 3 years of age are more likely to be problematic throughout the individual's life.

Exclusive breast-feeding for the first 4 to 6 months reduces an infant's risk of developing a food allergy and is recommended for infants from families with a history of allergies.[76] Once a food allergy has developed, the only way to manage it is to avoid consuming the offending food, so if a formula-fed baby becomes allergic to the milk proteins used in the formula, soy formulas should be used. For infants who cannot tolerate milk or soy protein, formulas made from predigested proteins are an option.

FIGURE 11.23 **Safe infant feeding** To avoid bacterial contamination, wash hands before preparing formula. Clean bottles and nipples by washing them in a dishwasher or placing them in boiling water for five minutes. Boil water for one to two minutes and cool it before using it to mix powdered formula or dilute concentrates. Cover and refrigerate opened cans of ready-to-feed and liquid concentrate formula and use the formula within the period indicated on the can. Prepare formula immediately before a feeding and discard any excess.

Appropriate introduction of solid and semisolid foods can reduce the risk of an infant developing food allergies and help monitor for allergies that do develop. The most commonly recommended first food is iron-fortified infant rice cereal mixed with formula or breast milk, because this cereal is easily digested and rarely causes allergic reactions. After rice has been successfully included in the diet, other grains can be introduced, with wheat cereal given last because it is more likely than other cereals to cause an allergic reaction. Each new food should be offered for a few days without the addition of any other new foods. If an allergic reaction occurs, it is then easier to determine that the newly introduced food caused it. Foods that cause symptoms such as rashes, digestive upsets, or respiratory problems should be discontinued, and the symptoms should no longer be present before any other new foods are added.

Developmentally appropriate foods Although most of an infant's nutritional needs are met by breast milk or infant formula, solid and semisolid foods can be gradually introduced into the diet starting at 4 to 6 months of age. By this time, the infant's feeding abilities and gastrointestinal tract are mature enough to handle solid foods (**Figure 11.24**). Cow's milk should not be fed to infants less than 1 year of age because it is too high in protein and too low in iron.[77] At 1 year of age, whole cow's milk can be offered; it can be used until 2 years of age, after which reduced-fat or low-fat milk can be used. As a child becomes familiar with more variety, food choices should

NUTRITION INSIGHT

FIGURE 11.24 **Nourishing a developing infant** Foods that are offered to infants should be appropriate for their developmental and digestive abilities.

Age	Birth to 4 months	4 to 6 months	6 to 9 months	9 to 12 months
Developmental milestones	The infant takes milk by means of a licking motion of the tongue called *suckling*, which strokes or milks the liquid from the nipple. Solid food placed in the mouth at an early age is usually pushed out as the tongue thrusts forward.	The tongue is held farther back in the mouth, allowing solid food to be accepted without being expelled. The infant can hold his or her head up and is able to sit, with or without support.	The infant can sit without support, chew, hold food, and easily move hand to mouth.	The infant can drink from a cup and feed him/herself.
Foods	Breast milk or iron-fortified infant formula.	Breast milk or formula, iron-fortified infant cereal; rice cereal is usually the first solid food introduced because it is easily digested and less likely than other grains to cause allergies. After cereals, puréed vegetables and fruits can be introduced.	Breast milk or formula, iron-fortified infant cereal, puréed or strained vegetables, fruits, meats and beans, limited finger foods.	Breast milk or formula, iron-fortified infant cereal, chopped vegetables, soft fruits, meats and beans, nonchoking finger foods such as dry cereal, cooked pasta, and well-cooked vegetables.

be made from each of the food groups. To avoid choking, foods that can easily lodge in the throat, such as raw carrots, grapes, and hot dogs, should not be offered to infants or toddlers.

The American Academy of Pediatrics and the Dietary Guidelines recommend that fruit juice not be introduced until after 6 months of age and then be limited to 4 to 6 oz per day. Only 100% fruit juice should be offered. Fruit juice should be offered only from a cup; juice consumed from a covered cup or bottle encourages consumption over a long period of time and may allow prolonged contact with the teeth.[78] Honey should not be fed to children less than 1 year old because it may contain spores of *Clostridium botulinum*, the bacterium that causes botulism poisoning (discussed in Chapter 13). Older children and adults are not at risk from botulism spores because the environment in a mature gastrointestinal tract prevents the bacteria from growing.

Concept Check

1. **What** may be happening if a child whose birth weight was in the 50th percentile is now in the 30th percentile for growth?
2. **Why** do infants need proportionally more dietary fat than adults?
3. **Why** is breast milk the best choice for healthy mothers and babies?
4. **When** can solid food be introduced into an infant's diet?

Summary

1 Changes in the Body During Pregnancy 340

- During the first 8 weeks of development, all the organ systems necessary for life are formed in the **embryo**. Over the remaining weeks of pregnancy, the **fetus** grows, and organs continue to develop and mature. The **placenta** transfers nutrients and oxygen from maternal blood to fetal blood and removes wastes from the fetus. At birth, a healthy baby weighs 3 to 4 kg (6.5 to 9 lb). **Low-birth-weight** infants and infants who are **large for gestational age** are at increased risk of health problems.

- During pregnancy, the mother's body undergoes many changes to support the pregnancy and prepare for **lactation**. Recommended weight gain during pregnancy is 11 to 16 kg (25 to 35 lb) for normal-weight women. Too little or too much weight gain can place both mother and baby at risk, but weight loss should never be attempted during pregnancy. Normal-weight, underweight, overweight, and obese mothers should gain weight at a steady rate during pregnancy, but as shown here, the total amount of weight gain recommended depends on prepregnancy weight.

Figure 11.4b Rate and composition of weight gain during pregnancy

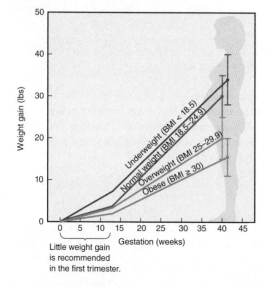

- During healthy pregnancies, moderate-intensity exercise that does not increase the risk of abdominal trauma, falls, or joint stress can be beneficial and safe.

- The hormones that direct changes in maternal physiology and fetal growth and development sometimes cause unwanted side effects, including **edema**, **morning sickness**, heartburn, constipation, and hemorrhoids.

- **Hypertensive disorders of pregnancy** increase risk for both mother and baby. **Gestational hypertension** involves an increase in blood pressure during pregnancy. **Preeclampsia** is a more severe condition, characterized by edema, weight gain, and protein in the urine, and can progress to life-threatening **eclampsia**. High blood glucose in the mother, called **gestational diabetes**, results in babies who are large for gestational age because extra glucose crosses the placenta from mother to fetus.

2 Nutritional Needs During Pregnancy 345

- During pregnancy, the requirements for energy, protein, carbohydrate, essential fatty acids, and water, shown in the graph, as well as for many vitamins and minerals increase above levels for nonpregnant women.

Figure 11.6 Energy and macronutrient recommendations

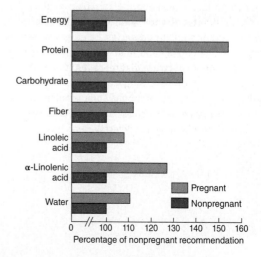

368 CHAPTER 11 Nutrition During Pregnancy and Infancy

- The RDA for calcium is not increased during pregnancy because the greater need is met by an increase in absorption. Vitamin D deficiency is a concern, particularly for darker-skinned women. Adequate folic acid early in pregnancy reduces the risk of neural tube defects. Iron needs are high, and deficiency is common during pregnancy. Iodine needs increase during pregnancy, and low intakes put infants at risk for brain damage.
- The recommendations of MyPlate can help pregnant women select foods to meet their nutritional needs. A prenatal supplement containing iron and folic acid and other vitamins and minerals is generally prescribed for all pregnant women.
- Food cravings are common during pregnancy. Most are harmless, but **pica** can reduce the intake of nutrient-dense foods, inhibit nutrient absorption, increase the risk of consuming toxins and infectious organisms, and cause intestinal obstructions.

3 Factors That Increase the Risks Associated with Pregnancy 351

- The rapidly developing embryo and fetus are susceptible to damage from poor nutrition and physical, chemical, or other environmental **teratogens**. Malnutrition during pregnancy can cause fetal growth retardation, low infant birth weight, birth defects, premature birth, miscarriage, and stillbirth. Maternal malnutrition also causes changes that can increase the child's risk of developing chronic diseases later in life.
- Pregnant teenagers are at increased risk because they are still growing themselves. Women over age 35 are at increased risk because they are more likely to have preexisting health conditions and to develop hypertensive disorders of pregnancy, gestational diabetes, or other complications during pregnancy.
- Poverty is one of the greatest risk factors for pregnancy. Low-income women typically do not receive adequate prenatal care. These women and their babies can benefit from the **Special Supplemental Nutrition Program for Women, Infants, and Children (WIC)**.
- Although moderate caffeine intake has little effect, excessive consumption can increase the risk of low birth weight and miscarriage. Following guidelines on fish consumption can minimize exposure to mercury, which can cause brain damage in a fetus. Pregnant women are particularly susceptible to foodborne illness and should follow safe food-handling recommendations. Alcohol consumption during pregnancy can cause a spectrum of physical and mental abnormalities. The more severe form, **fetal alcohol syndrome**, is characterized by distinct facial abnormalities depicted here; less severe physical, behavioral, functional, and mental impairment is referred to as **fetal alcohol spectrum disorders (FASDs)**. Use of tobacco products causes low birth weight and increases the risk of stillbirth, preterm delivery, and behavioral problems in the baby. The use of illegal drugs, such as cocaine, also increases the risk of low birth weight and birth defects.

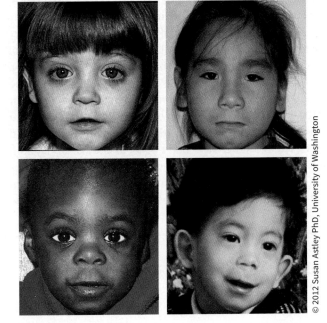

Figure 11.15 Dangers of alcohol use during pregnancy

© 2012 Susan Astley PhD, University of Washington

4 Lactation 356

- Milk production and **let-down** are triggered by hormones released in response to the suckling of an infant. During let-down, milk travels from the milk-producing glands shown here to the nipple.

Figure 11.16 Anatomy of milk production

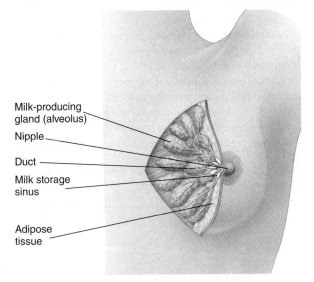

- Lactation requires energy and nutrients from the mother to produce adequate milk. The energy for milk production comes from the diet and maternal fat stores. Maternal needs for protein, water, and many vitamins and minerals are even greater during lactation than during pregnancy.

5 Nutrition for Infants 359

- Growth is the best indicator of adequate nutrition in an infant. Healthy infants follow standard patterns of growth. Growth charts can be used to compare an infant's growth with optimal growth patterns. Slow growth indicates **failure to thrive**, whereas excess weight gain may predispose a child to obesity. Undernutrition and overnutrition during infancy can have permanent effects on growth and development and the risk of chronic disease later in life.

- Infants require more calories and protein per kilogram of body weight than do individuals at any other time of life. Fat and fluid needs are also proportionately higher than in adults. Infants are at risk for deficiencies of iron, vitamin D, and vitamin K, as well as low fluoride intake.

- Breast milk is the ideal food for infants. It meets nutrient needs; it is always available; it requires no special equipment, mixing, or sterilization; and it provides immune protection. Infant formulas are patterned after human milk and provide adequate nutrition to a baby. Formula-feeding is the best option when the mother has certain infections or is taking prescription or illicit drugs, or when the infant has special nutritional needs.

- Introducing solid foods between 4 and 6 months of age, as shown here, adds iron and other nutrients to the diet. Newly introduced foods should be appropriate to the child's stage of development and offered one at a time in order to monitor for food allergies.

Figure 11.24 Nourishing a developing infant: 4 to 6 months

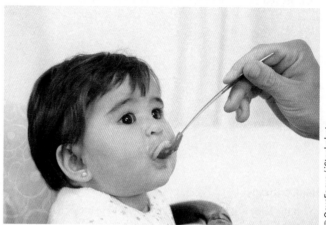

Key Terms

- cesarean section   342
- colostrum   363
- critical period   351
- Down syndrome   354
- eclampsia   344
- edema   343
- embryo   341
- failure to thrive   359
- fertilization   340
- fetal alcohol spectrum disorders (FASDs)   355
- fetal alcohol syndrome (FAS)   355
- fetus   341
- gestational diabetes   345
- gestational hypertension   344
- hypertensive disorders of pregnancy   344
- implantation   341
- infancy   359
- lactation   341
- large for gestational age   343
- let-down   356
- low birth weight   341
- morning sickness   343
- nursing bottle syndrome   363
- nutritional programming   352
- oxytocin   357
- pica   350
- placenta   341
- preeclampsia   344
- preterm or premature   341
- prolactin   357
- small for gestational age   341
- Special Supplemental Nutrition Program for Women, Infants, and Children (WIC)   354
- sudden infant death syndrome (SIDS), or crib death   356
- teratogen   351
- trimester   342
- very low birth weight   341

What is happening in this picture?

Ultrasound imaging, shown here, uses sound waves to visualize the embryo or fetus in the uterus. It can be used to assess the growth of the fetus and identify fetal and maternal health problems.

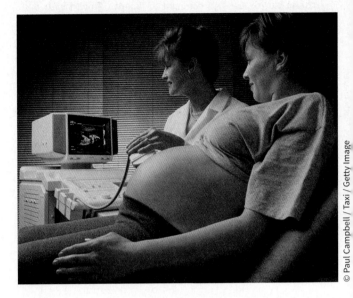

Think Critically

1. If the baby is larger than expected, what might be the problem?
2. If the baby is smaller than expected, what might be the cause?

CHAPTER 12

Nutrition from 1 to 100

In *As You Like It*, Shakespeare allotted a human lifetime seven acts to be played upon the world's stage, beginning with "the infant, mewling and puking in the nurse's arms" and ending in "second childishness . . . sans teeth, sans eyes, sans taste." While our senses and abilities do diminish as we age, with proper nutrition and health maintenance, we may not quite reach the state of helplessness described by The Bard.

For "the whining schoolboy"—and girl—"with . . . shining morning face," nutrition and healthy development go hand in hand. As the adolescent becomes "the lover," then "the soldier" of young adulthood, good nutrition is still necessary to maintain health and build a foundation of well-being for the mature "justice, in fair round belly with good capon lin'd"—Shakespeare's archetype of middle age. In his characterization of the closing years of life, "the lean and slipper'd pantaloon . . . a world too wide for his shrunk shank," Shakespeare highlights the muscle withering of old age that, along with other physical, social, and mental changes, presents a new set of challenges to maintaining nutritional health.

Good nutrition habits adopted in our early years will serve us well through the seven acts of our own "strange eventful history."

CHAPTER OUTLINE

The Nutritional Health of America's Youth 372
- What Are the Youth of America Eating?
- Poor Food Choices Are Impacting Children's Health

Nutrition for Children 376
- Monitoring Growth
- Energy and Nutrient Needs of Children
- Developing Healthy Eating Habits

What a Scientist Sees: Breakfast and School Performance

What Should I Eat? Childhood
- Special Concerns in Children

Thinking It Through: A Case Study on Under- and Overnutrition

Nutrition for Adolescents 383
- Energy and Nutrient Needs of Adolescents
- Meeting Teens' Nutritional Needs

What Should I Eat? Adolescence
- Special Concerns for Teens

Nutrition for the Adult Years 388
- What Is Aging? • Nutrition and Health Concerns Throughout Adulthood

Debate: Can Eating Less Help You Live Longer?
- Factors That Increase the Risk of Malnutrition in Older Adults • Keeping Healthy Throughout the Adult Years

What Should I Eat? Advancing Age

The Impact of Alcohol Throughout Life 397
- Alcohol Absorption, Transport, and Excretion • Alcohol Metabolism
- Adverse Effects of Alcohol
- Benefits of Alcohol Consumption

12.1 The Nutritional Health of America's Youth

LEARNING OBJECTIVES

1. **Discuss** the healthfulness of dietary patterns among children and teens in the United States.
2. **Explain** the impact of diet and lifestyle during childhood and adolescence on the risk of chronic disease throughout life.

The diets of America's youth are not as healthy as they could be, and as a result, our children and adolescents are not as healthy as they could be. About 17% of children and adolescents are obese, and only 22% of 6- to 19-year-olds meet the recommendation of 60 minutes of moderate to vigorous physical activity per day.[1,2] These patterns are contributing to a growing incidence of overweight and obesity as well as type 2 diabetes, hypertension, and early heart disease in children and young adults.

FIGURE 12.1 How good are children's diets? The Healthy Eating Index scores diets based on how well they meet the recommendations of the Dietary Guidelines and MyPlate. A score of 100% indicates that on average the recommendation was met or exceeded. Some recommendations target getting adequate amounts of foods and nutrients, and higher scores mean higher intakes; others focus on moderation, and higher scores indicate lower and hence more desirable intakes. For all, a higher percentage indicates a higher-quality diet. The results shown here indicate that the diets of children ages 2 to 17 need improvement.

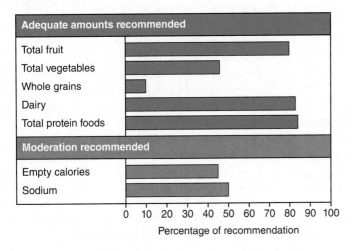

Interpret the Data

According to the graph, which of the following is consumed in an amount farthest from what is recommended?

a. sodium
b. total vegetables
c. total fruit
d. whole grains

What Are the Youth of America Eating?

Most children and adolescents in the United States consume more saturated fat, sodium, and added sugars and less calcium and fiber than is recommended. When analyzed using the Healthy Eating Index, which compares the quality of diets with recommendations, children ages 2 to 17 were found to consume too few vegetables and whole grains, too many empty calories, and too much sodium. Their diets came closest to meeting recommendations for fruits, protein foods, and dairy products (**Figure 12.1**).[3]

Many of the factors that contribute to poor dietary patterns among children can be traced to their food environment (**Figure 12.2a**). Children's food environment includes the foods they are exposed to in their homes, day-care centers, and schools, as well as throughout their community in grocery stores, restaurants, and other retail food establishments. Over the past few decades, changes in the way families eat have, in turn, changed the types of food children are exposed to and, therefore, what they eat. The family dinner, cooked by a stay-at-home parent and eaten with the entire family around the table, has been replaced with prepackaged or take-out food eaten in front of the television. Fewer meals are being prepared at home; they are brought home ready to eat from grocery stores, delis, or restaurants or are eaten away from home at restaurants and fast-food establishments. Compared to meals prepared at home, these meals are usually higher in calories, saturated fat, and sodium;[4,5] in addition, the portions tend to be larger, and these meals contain fewer fruits and vegetables.[6]

The food marketing and advertising to which children are exposed are also part of their unhealthy food environment. Watching television has a major influence on children's food choices and activity level (**Figure 12.2b**). Sitting in front of the television encourages snacking and reduces activity. Television also affects the quality of the diet because commercials introduce children to foods to which they might otherwise not be exposed (**Figure 12.2c**). When children are introduced to foods advertised on television, those foods are more likely to be purchased.[7] The American Academy of Pediatrics recommends that screen time, which includes watching television and using other digital media for entertainment, be banned for infants 18 months and younger. For children 18 to 24 months, parents who want to introduce media should choose high-quality programming and watch it with their children. For children 2 to 5 years of age, screen time should be limited to one hour per day of high-quality programs.[8] For children ages 6 and older, parents should develop a media plan that places limits on the time spent using media and the types of media, and they should ensure that media use does not interfere with adequate sleep, physical activity, family time, face-to-face social interactions, and other behaviors essential to health.[9,10]

FIGURE 12.2 Children's food environment Whether it is grocery store shelves strategically arranged to have sweets at eye level or a television advertisement for sweet, salty, or fatty foods, children's food environment contributes to unhealthy choices.

a. Grocery stores entice kids to make poor food choices by placing sugary cereals, candy, and other enticing snack foods lower on the shelves, where they are at eye level for young children.

b. Hours spent watching television are hours when physical activity is at a minimum. More time spent watching TV is associated with higher BMI among children and adolescents.[11]

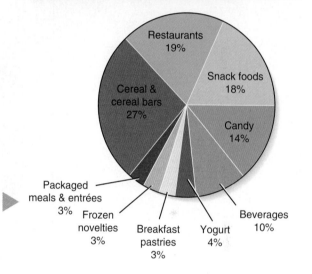

c. During children's television programming, food is the most frequently advertised product category. This chart, which illustrates the types of food advertised on Saturday morning children's television programming, shows that almost half of the commercials advertise candy, snack foods, beverages, and pastries.[12]

Poor Food Choices Are Impacting Children's Health

The unhealthy eating patterns and low-activity lifestyle that contributes to obesity and chronic disease in adults is having the same effect in children. As with adult obesity, childhood obesity increases the risks of a number of other chronic conditions, including high blood glucose and type 2 diabetes, high blood pressure, and elevated blood cholesterol, as well as social and psychological challenges.

Childhood obesity Today, about one in five school-aged children ages 6 to 19 years is obese.[1] Whether or not a child's weight is in the healthy range can be determined by measuring his or her body mass index (BMI) and plotting it on a gender-specific BMI-for-age growth chart. The percentile on the growth chart indicates whether the child is at healthy weight, overweight, obese, or underweight (**Figure 12.3**).[15]

Reducing the incidence of childhood overweight/obesity requires changes in the eating and activity patterns of individuals and the population as a whole. Best results are achieved by involving all levels of society – parents, caregivers and families, day cares and schools, the community, and healthcare providers – in creating healthy food environments and encouraging physical activity.[16]

It can be difficult to modify children's food consumption patterns. Denying food may promote further overeating by making the child feel that he or she will not obtain enough to satisfy hunger. Thus, restrictions on food intake should be relatively mild, and the focus instead should be on offering nutrient-dense foods. The goal is to promote healthy body weight in all children and help overweight and obese children stop gaining weight or lose weight. The intensity of the weight-loss programs recommended for children depends on their weight category and their health risks.[17] For overweight children at low risk, weight loss is not recommended; rather, weight gain should be slowed while they continue to grow taller. This allows them to "grow into" their weight. A child who is at the 85th percentile for BMI at age 7 and gains only a few pounds a year can be at the 75th percentile by age 9. For children in the obese BMI range, programs should be designed to promote slow weight loss of no more than 1 pound per month for children ages 2 to 5 and no more than 2 pounds per week for older children and adolescents. For severely obese youths (BMI > 99th percentile) for whom

FIGURE 12.3 BMI-for-age percentiles

This growth chart shows the BMI-for-age percentiles for boys 2 to 20 years of age. The colored areas represent BMI values associated with underweight, healthy weight, overweight, and obesity. A similar growth chart can be used to assess body weight in girls (see Appendix E).

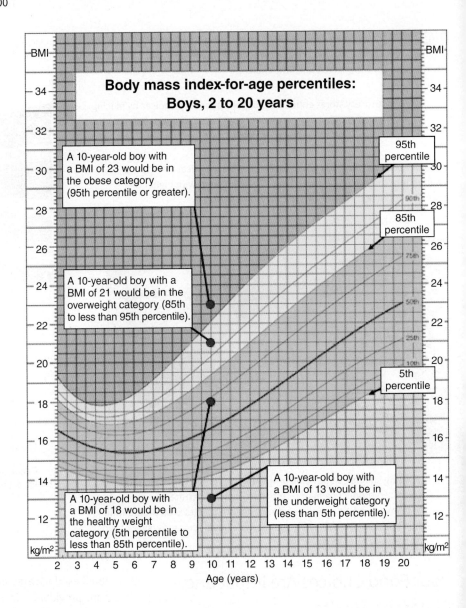

reduced intake, increased activity, and behavior change have not been successful, weight-loss medications and/or surgery may be recommended.[17]

Changing activity patterns is just as important for weight management as changing eating patterns. Public health recommendations, including the Dietary Guidelines, suggest that children be physically active for at least an hour each day.[18] But surveys show that fewer than a quarter of children and teens ages 6 to 19 meet these guidelines.[2] Encouraging activity may be particularly difficult in overweight children, who are often embarrassed by their bodies and shy away from group activities. Increases in physical activity need to be gradual in order to make exercise a positive experience. A good way to start is to limit time spent watching television and playing video and computer games. Less screen time allows more time for active games, walks after dinner, bike rides, hikes, swimming, volleyball, and other activities that the whole family can enjoy together.

Other chronic disease Obesity, along with the poor diet and lifestyle of America's youth, has led to a rising incidence of diabetes, high blood pressure, and high blood cholesterol (**Figure 12.4**). Until recently, type 2 diabetes was considered a disease that affected primarily adults over age 40, but more than 5000 children and adolescents are diagnosed with type 2 diabetes in the United States each year; the incidence is greatest in the 10- to 19-year-old group and is highest among minorities.[19] We are still learning about the impact of this disease in children, but we do know that it is a progressive disease that increases in severity with time from diagnosis (see Chapter 4).

The diets of children also increase their risk of high blood cholesterol and hypertension. Children in the United States currently consume more than the recommended maximum of 10% of their energy from saturated fat.[20] This can contribute to elevated blood cholesterol levels (see Figure 12.4). Elevated blood cholesterol levels during childhood and adolescence are associated with higher blood cholesterol and higher mortality

The Nutritional Health of America's Youth 375

NUTRITION INSIGHT

FIGURE 12.4 **Obesity and chronic disease among America's children** Obesity and related chronic diseases are a concern for America's children. Obese children are at risk for depression, unhealthy blood cholesterol levels, high blood pressure, and type 2 diabetes.

Results from the National Health and Nutrition Examination Survey indicate that the percentage of children who are obese increases with age.[1]

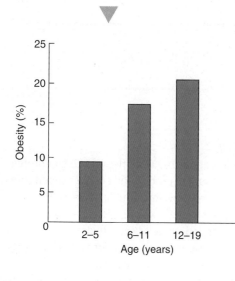

Psychosocial problems: Obese children are more likely than others to be socially isolated and to have depression, poor self-image, and low self-esteem. Social isolation, in turn, results in boredom, depression, inactivity, and withdrawal—all of which can increase eating and decrease activity, worsening the problem.

Elevated blood cholesterol: Blood cholesterol screening is recommended between the ages of 9 and 11 and again between 17 and 21 years of age.[14] Elevated blood cholesterol in childhood could lead to heart disease.

Cholesterol levels in children and adolescents 2 to 19 years old[13]		
	Total cholesterol (mg/100 mL)	LDL cholesterol (mg/100 mL)
Acceptable	<170	<110
Borderline	170–199	110–129
High	≥200	≥130

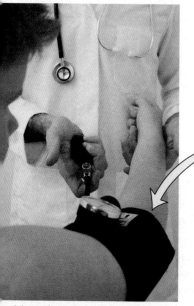

Science Photo Library / Science Source

Elevated blood pressure: The prevalence of hypertension and prehypertension has been increasing in children over the past 20 years. As with adults, blood pressure increases with obesity and can be reduced with a DASH-style eating plan, reductions in sodium intake, and increases in activity level.[14]

Type 2 diabetes: The longer an individual has diabetes, the greater the risk of complications that can lead to blindness, kidney failure, heart disease, or amputations (see Chapter 4).

Bambu Productions / Getty Images

rates from cardiovascular disease in adulthood.[21] Likewise, children who have blood pressure at the high end of normal as youngsters are more likely to develop high blood pressure as adults (see Figure 12.4).[22] Blood pressure can be affected by the amount of body fat, activity level, and sodium intake, as well as by the total pattern of dietary intake, so attention should be paid to these nutritional and lifestyle factors in children. This is particularly important if there is a family history of hypertension.

Concept Check

1. **What** food groups are most deficient in the diets of American children?
2. **How** does body weight during childhood affect chronic disease risk later in life?

12.2 Nutrition for Children

LEARNING OBJECTIVES

1. **Describe** how children's nutrient needs change as they grow.
2. **Discuss** how children's eating habits develop.
3. **Explain** how nutrition can affect the incidence of dental caries, attention-deficit/hyperactivity disorder, and lead toxicity.

Nutrient intake during childhood affects health throughout life. The food a child consumes must supply the energy and nutrients needed for growth and development as well as for maintenance and activity. Foods offered to children must also be appropriate for their stage of physical development and suit their developing tastes. A nutritious, well-balanced eating pattern and an active lifestyle allow children to grow to their potential and can prevent or delay the onset of the chronic diseases that plague adults. Therefore, learning healthy eating and exercise habits will benefit not only today's children but also tomorrow's adults.

Monitoring Growth

The best indicator that a child is receiving adequate nourishment—neither too little nor too much—is a normal growth pattern. Growth is most rapid in the first year of life, when an infant's length increases by 50%, or about 10 in. In the second year of life, children generally grow about 5 in; in the third year, 4 in; and thereafter, about 2 to 3 in/year. Although growth often occurs in spurts, growth patterns are predictable and can be monitored by comparing a child's growth pattern with standard patterns shown on growth charts (see Figure 12.3 and Appendix E).[23] When a child does not get enough calories, weight gain slows, and if the deficiency continues, growth in height slows.

Energy and Nutrient Needs of Children

As children grow and become more active, their requirements for energy and most nutrients increase. The average 2-year-old needs about 1000 Calories and 13 g of protein per day. By age 6, that child needs about 1600 Calories and 19 g of protein per day.[24] The total amount of protein and energy needed continue to increase as children grow into adults; however, the amounts needed per kilogram of body weight decrease (**Figure 12.5**).

The recommended range of carbohydrate intake for children over 1 year of age is the same as for adults: 45 to 65% of total energy intake. To provide enough energy to support rapid growth and development, the recommended range of fat intake is higher for children than for adults: 30 to 40% of total energy intake for 1- to 3-year-olds and 25 to 35% for 4- to 18-year-olds. As children grow, the recommended proportion of calories from fat decreases because a higher-fat diet is no longer needed to meet energy and developmental needs.

Infants have high water needs, but by 1 year of age, their evaporative losses have decreased, and their kidneys have matured, decreasing the loss of water in urine. Therefore, as with adults, in most situations, children can meet their water needs by drinking enough to satisfy thirst.[25] Water needs increase with illness, when the environmental temperature is high, and when activity increases losses from sweat.

Because children are smaller than adolescents and adults, the recommended amounts of most micronutrients are also

FIGURE 12.5 **Energy needs** As children grow, their larger body size causes an increase in total energy needs, but as growth slows, energy needs per kilogram of body weight decline.

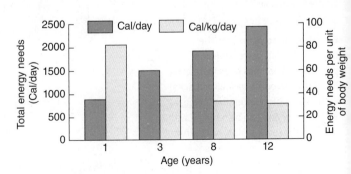

Interpret the Data

How many more Calories per day does a 12-year-old child need than an 8-year-old child?

smaller (see Appendix A). Recommended intakes do not differ for boys and girls until about age 9, at which time sexual maturation causes differences in nutrient requirements. Like adults, children who consume a varied, nutrient-dense diet can meet all their vitamin and mineral requirements with food.

The majority of toddlers and preschoolers in the United States meet recommendations for micronutrient intakes, particularly when fortified foods are included in the diet.[26–28] However, intakes of calcium and vitamin D are still at risk, as is the amount of iron, because of its importance for growth.

Calcium, vitamin D, and bone health Calcium intake in school-age children has been declining, primarily due to a decrease in the consumption of dairy products. Adequate calcium intake during childhood is essential for achieving maximum peak bone mass, which is important for preventing osteoporosis later in life (see Chapter 8). The RDA for calcium is 700 mg/day for toddlers (ages 1 to 3) and 1000 mg/day for young children (ages 4 to 8).[29]

Vitamin D, which is needed for calcium absorption, is also essential for bone health. Low intakes of milk combined with limited sun exposure contribute to low blood levels of vitamin D in many children in the United States.[30] The RDA is 600 IU (15 µg)/day for children and adolescents.[29]

Iron and anemia Children's iron needs are high because iron is required for growth. The RDA is 7 mg/day for toddlers and 10 mg/day for young children; the latter recommendation is higher than the RDA for adult men. The high needs and finicky eating habits of young children often lead to iron deficiency anemia, a condition that can impair learning ability and intellectual performance.[31] If anemia is diagnosed, iron supplements are usually prescribed until the child's iron stores have been replenished. These supplements should be kept out of the reach of children to prevent iron toxicity from a supplement overdose (see Chapter 8).

Developing Healthy Eating Habits

Much of what we choose to eat as adults depends on what we learned to eat as children. Children learn by example, and therefore the eating patterns, attitudes, and feeding styles of their caregivers influence what they learn to eat. Caregivers are responsible for deciding what foods should be offered to a child and when and where they should be eaten. The child must then decide whether to eat, what foods to eat, and how much to consume. As children grow older, their choices are increasingly affected by social activities, what they see at school, and what their friends are eating. The habits developed during childhood establish patterns that may last a lifetime.

What to offer? Children should be offered a balanced and varied diet that is adequate in energy and essential nutrients and is appropriate to their developmental needs. A healthy diet is based on whole grains, vegetables, and fruits; includes adequate dairy and other high-protein foods; and is low in saturated fat, sodium, and added sugars. MyPlate can be used as a guide for meeting the dietary goals of preschoolers and children (**Figure 12.6**).

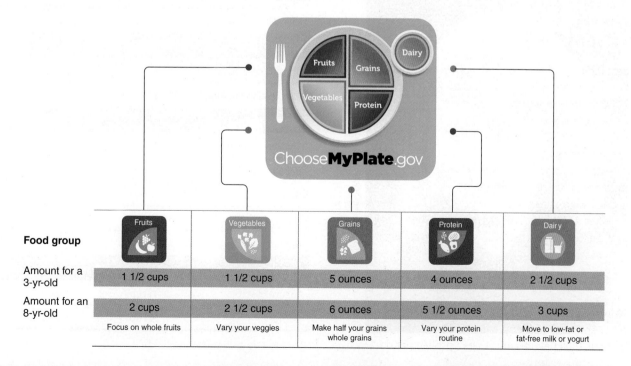

FIGURE 12.6 **MyPlate recommendations for children** MyPlate recommends amounts of food from each group that can be used to build healthy eating patterns for young children. Food choices to meet these amounts should be spread throughout the meals and snacks served each day.

Food group	Fruits	Vegetables	Grains	Protein	Dairy
Amount for a 3-yr-old	1 1/2 cups	1 1/2 cups	5 ounces	4 ounces	2 1/2 cups
Amount for an 8-yr-old	2 cups	2 1/2 cups	6 ounces	5 1/2 ounces	3 cups
	Focus on whole fruits	Vary your veggies	Make half your grains whole grains	Vary your protein routine	Move to low-fat or fat-free milk or yogurt

It isn't always easy to persuade a child to eat a variety of foods from all the food groups, as recommended by MyPlate. To increase variety, new foods should be introduced into a child's diet regularly. Children's food preferences are learned through repeated exposure to foods; a new food may need to be offered 8 or 10 times before the child will accept it. Children are also more likely to eat a new food if it is introduced at the beginning of a meal, when the child is hungry, and if the child sees his or her parents or peers eating it. Incorporating healthy foods into familiar dishes can also increase the variety of the diet. No matter how erratic children's food intake may be, caregivers should continue to offer a variety of healthy foods at each meal and let children select what and how much they will eat.

Getting children to consume the recommended amount of fruit is usually not difficult, but most servings should come from whole fruit, with limited amounts from 100% juice, not fruit drinks or juice cocktails. The American Academy of Pediatrics recommends limiting juice to 4 to 6 ounces/day for children ages 1 to 6 and 8 to 12 oz/day for children ages 7 and older (**Figure 12.7**).[32]

When and where to offer meals and snacks

Because children have small stomachs and high nutrient needs, they should consume small, nutrient-dense meals and snacks, ideally every 2 to 3 hours throughout the day (**Table 12.1**). Establishing a consistent meal pattern is important because children thrive on routine and feel secure when they know what to expect. Starting the day with a good breakfast is particularly important; children who eat breakfast are more likely than those who do not to meet their daily nutrient needs and do better in school (see *What a Scientist Sees: Breakfast and School Performance*).[37] Snacks should be as nutritious as meals to ensure that nutrient needs are met.

The setting in which a meal is consumed is also important. Children need companionship, conversation, and a pleasant location at mealtimes. Eating meals together helps children connect with family and culture and is associated with better

FIGURE 12.7 Limiting juice consumption More than half the fruit children consume is as juice rather than whole fruit. Although 100% fruit juice can be part of a healthy diet, too much can cause diarrhea, over- or undernutrition, and dental caries. It is recommended that juice not be offered to children in containers that can be carried around, encouraging continuous sipping.[32,33]

Tim Laman / National Geographic Creative

TABLE 12.1 Typical meal and snack patterns for 3- and 8-year-old children

Food	Amount	
	3-year-old	8-year-old
Breakfast		
Whole-grain cereal	½ cup	1 cup
Milk, 2%	½ cup	½ cup
Banana	½ medium	1 medium
Snack		
Peanut butter	2 Tbsp	2 Tbsp
Whole-wheat crackers	5	5
Milk, 2%	½ cup	1 cup
Lunch		
Vegetable soup	½ cup	1 cup
Grilled tuna sandwich	½	1
Tomato	¼ medium	½ medium
Orange	½ medium	1 medium
Milk, 2%	½ cup	1 cup
Snack		
Broccoli crowns	4	6
Ranch salad dressing	1 tsp	2 tsp
Dinner		
Chicken drumsticks	1	2
Baked sweet potato	½ cup	1 cup
Green beans	¼ cup	½ cup
Milk, 2%	½ cup	1 cup
Graham crackers	1	2
Snack		
Yogurt	½ cup	1 cup
Berries	½ cup	¾ cup

What a Scientist Sees

Breakfast and School Performance

This breakfast looks appealing, but regardless of what's on the plate, many children and teens do not make time for breakfast; they are more likely to skip breakfast than to skip any other meal.[34] What a scientist sees is the impact breakfast has on nutritional status and school performance. Skipping breakfast may result in a span of 15 or more hours without food. Because breakfast provides energy and nutrients to the brain, children who skip it are more likely to have academic, emotional, and behavioral problems than those who eat breakfast.[35] Studies have found that compared with non-breakfast eaters, children who routinely eat breakfast have better nutrient intakes, and these improvements in nutrient intakes are associated with improvements in academic performance, reductions in hyperactivity, better psychosocial behaviors, and less absence and tardiness (see graph).[36,37]

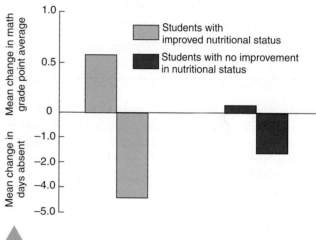

Implementation of a free school breakfast program in the Boston Public Schools improved the nutritional status of many students. Those whose nutritional status improved showed greater improvements in math grades and fewer school absences than students whose nutritional status was not improved.[36]

Think Critically

In addition to enhancing learning by fueling the brain after the overnight fast, in what other ways might breakfast contribute to better school performance?

---

school performance and decreased risk of unhealthy weight-loss practices and substance abuse.[38] Children learn by example; therefore, the eating patterns, attitudes, and feeding styles of their caregivers influence what they learn to eat. Children whose mothers' eat a healthy diet are more likely to have a healthy diet.[39] Children should be given plenty of time to finish eating. Slow eaters are unlikely to finish eating if they are abandoned by siblings who run off to play and adults who leave to wash dishes. Moreover, if mealtime is to be a nutritious, educational, and enjoyable experience, it should not be a battle zone. Food is not a reward or a punishment: It is simply nutrition. (see *What Should I Eat? Childhood*)

Meals at day care or school It is not easy to ensure that meals eaten away from home are nutritious because there is no guarantee that what is served at school or brought from home will be eaten. A packed lunch should contain foods that the child likes and that do not require refrigeration. For children who do not bring meals from home, federal school breakfast and lunch programs provide free or low-cost meals containing age-appropriate foods.

The **National School Lunch Program** provides free or reduced-cost lunches to eligible children. In 2014, school cafeterias in the United States served almost 5 billion lunches; more than two-thirds of them were free or at a reduced price.[40] While the program has provided meals for over 50 years, the specific requirements for these meals have changed over the years. The Healthy Hunger-Free Kids Act of 2010 provided a new set of school lunch standards based on the Dietary Guidelines and designed to promote nutrition education by teaching children

What Should I Eat?

Childhood

Serve children frequent nutritious meals and snacks
- Add peanut butter to a banana or an apple.
- Make a rainbow by including at least four colors in every meal.
- Cut and arrange foods in interesting shapes.

Sneak in more fruits and vegetables
- Bake bananas and berries into breads and muffins.
- Add vegetables to soups, tacos, and casseroles.
- Blend fruit into shakes and smoothies.

Include calcium and vitamin D where you can
- Make cream soup by adding milk.
- Use milk, rather than water, in your oatmeal.
- Serve pudding or custard for a calcium-rich dessert.

Add iron
- Make your spaghetti sauce with meat and cook it in an iron pot.
- Beef up your tacos and burritos.
- Serve iron-fortified breakfast cereal.

 Use iProfile to find snacks that are high in iron.

to make appropriate food choices (**Figure 12.8**).[41] In addition to lunches, federal guidelines regulate meals provided as part of the School Breakfast Program and foods sold in school vending machines. Vending machine snacks must have fewer than 200 Calories, less than 230 mg of sodium, less than 35% of calories from fat, and less than 35% of their weight from sugar. The only beverages that can be sold in vending machines are water, low-fat and fat-free milk, fruit and vegetable juices, and fruit and vegetable juices diluted with water but containing no sweeteners.[42]

Offering healthy meals at school can't compensate for poor food and activity choices at home (see *Thinking it Through*).

Special Concerns in Children

Some nutrition-related concerns that begin during childhood can impact nutritional health for a lifetime. These include dental caries, behavioral problems, and exposure to high levels of lead.

FIGURE 12.8 Building a healthy tray
School lunches are designed to meet needs as well as teach children healthy eating patterns. The Healthy Hunger Free Kids Act requires that students be offered five food components: milk, meat or meat alternatives, fruits, vegetables, and grains; they must take foods from at least three food groups, one of which must be a fruit or vegetable.[41]

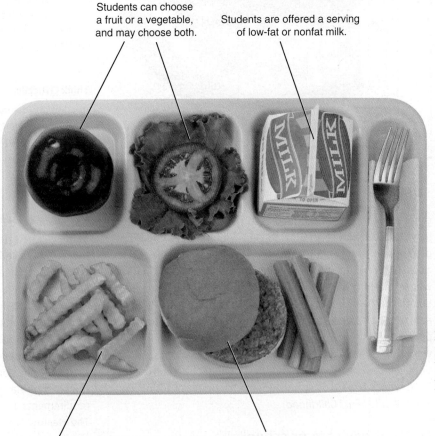

Students can choose a fruit or a vegetable, and may choose both.

Students are offered a serving of low-fat or nonfat milk.

Meals must be low in saturated fat and include only ingredients with 0 *trans* fat.

Grains should be rich in whole grains.

Thinking It Through

A Case Study on Under- and Overnutrition

Sam is 8 years old and has gained 5 pounds in the past three months. His parents are worried because all Sam wants to do is watch TV. Because Sam's parents are both overweight, they are concerned that he will also have a weight problem, so they take him to see his pediatrician.

Based on this growth chart, how has Sam's BMI percentile changed over the past year?

Your answer:

While reviewing Sam's diet and exercise patterns, the doctor learns that Sam has been watching TV or playing video games for about six hours a day. A recall of his intake shows that he eats donuts and milk for breakfast, gets lunch from the school lunch program, and then snacks so much on chips and candy when he gets home from school that he doesn't really eat dinner. He likes fruit, refuses to eat vegetables, and drinks 5 to 6 cups of whole milk daily.

What nutrients are likely to be excessive in this dietary pattern? Which are likely to be deficient?

Your answer:

A blood test reveals that Sam has iron deficiency anemia. The pediatrician prescribes an iron supplement and refers Sam and his parents to a dietitian. She recommends that Sam have a fortified cereal for breakfast to increase his intake of vitamins and minerals, particularly iron, and that he switch from whole to reduced-fat milk to lower his intake of calories and saturated fat. The dietitian also recommends changing Sam's after-school snacks to fruit so he will be hungry enough to eat a better evening meal.

Why might excessive consumption of dairy products contribute to Sam's anemia?

Your answer:

How might Sam's iron deficiency have contributed to his weight gain?

Your answer:

Sam needs to increase his activity to avoid further weight gain. What activity recommendations would you make for Sam?

Your answer:

(Check your answers in Appendix L.)

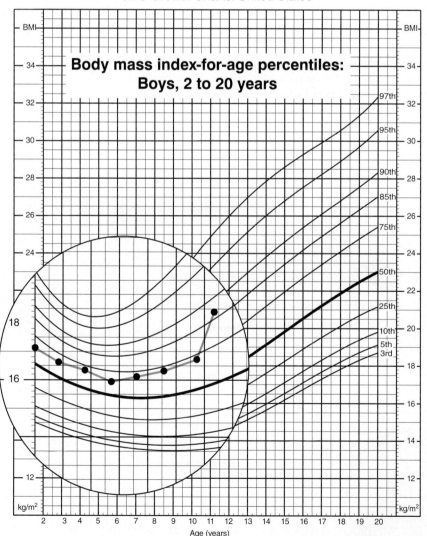

CDC Growth Charts: United States

Body mass index-for-age percentiles: Boys, 2 to 20 years

Dental caries

Dental caries are preventable but remain the most common chronic disease among children and adolescents. It is estimated that 42% of children ages 2 to 11 have had cavities in their primary teeth.[43] Because the primary teeth guide the growth of the permanent teeth, maintaining healthy primary teeth is just as important as preserving the permanent ones. Tooth decay is caused by prolonged contact between sugar and bacteria on the surface of the teeth (see Chapter 4). High-carbohydrate foods that stick to the teeth and sugar-sweetened beverages sipped throughout the day increase the risk of caries. In addition to eating a nutritious diet, brushing teeth regularly, visiting the dentist, and consuming fluoridated water help reduce the incidence of dental caries (see Chapter 8).

Attention-deficit/hyperactivity disorder (ADHD)

It is estimated that 11% of children 4 to 17 years of age in the United States have been diagnosed with attention-deficit/hyperactivity disorder (ADHD).[44] ADHD involves extreme physical activity, excitability, impulsiveness, distractibility, short attention span, and low tolerance for frustration. Children with ADHD have more difficulty learning but usually are of normal or above-average intelligence.

Although parents of children with ADHD report a worsening of symptoms after excessive ingestion of candy or soda, controlled studies have not found sugar intake to affect children's behavior or cognitive performance.[45] It has been suggested, however, that reactive hypoglycemia (see Chapter 4) following sugar ingestion may contribute to inattention and cognitive impairment in children with ADHD. The hyperactive behavior observed after sugar consumption may also be the result of situational factors such as the excitement of a birthday party rather than the sugar consumed.

Caffeine can also be a cause of hyperactive behavior in children. Caffeine is a stimulant that causes sleeplessness, restlessness, and irregular heartbeats. Beverages, foods, and medicines containing caffeine are often part of children's diets.

Other possible causes of hyperactivity include lack of sleep, overstimulation, desire for more attention, or lack of physical activity. Specific foods and food additives have also been implicated as causes of hyperactivity. Numerous studies have failed to provide sufficient evidence for the efficacy of any dietary treatment for ADHD. However, some children are sensitive to specific additives and may benefit from a diet that eliminates them.[45]

Lead toxicity

Lead is an environmental contaminant that can be toxic, especially in children under age 6 (**Figure 12.9**). Malnourished children are at particular risk because malnutrition increases lead absorption due to the fact that lead is better

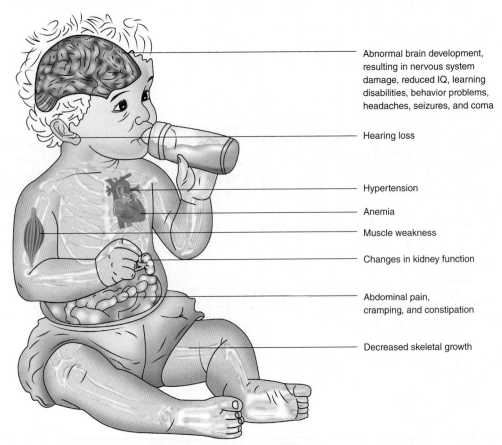

FIGURE 12.9 **How lead affects children** Lead toxicity can affect every organ system. It is particularly harmful to the developing bodies of young children. Lead disrupts the functioning of neurotransmitters and thus interferes with the functioning of the nervous system. Permanent neurological damage and behavioral disorders are associated with blood lead levels as low as or 5 μg/dL or less.

- Abnormal brain development, resulting in nervous system damage, reduced IQ, learning disabilities, behavior problems, headaches, seizures, and coma
- Hearing loss
- Hypertension
- Anemia
- Muscle weakness
- Changes in kidney function
- Abdominal pain, cramping, and constipation
- Decreased skeletal growth

absorbed from an empty stomach and when other minerals such as calcium, zinc, and iron are deficient. Air pollution, lead paints and plumbing, ceramic glazes, and contaminated water expose children to lead. Despite the recent press regarding water contaminated with lead in Flint, Michigan, the leading cause of lead poisoning in children is lead-based paint. When paint containing lead deteriorates into flakes, chips, or fine dust, it is easily inhaled or ingested by small children. Because children absorb lead much more efficiently than adults, they are more likely to develop toxic levels in their blood.

Once in the blood, lead accumulates in the bones and, to a lesser extent, the brain, teeth, and kidneys and affects systems throughout the body (see Figure 12.9). No safe blood lead level has been identified.[46] The effects of lead poisoning are permanent, but if high levels are detected early, the lead can be removed with medical treatment.

Because of the risks of lead toxicity from environmental contamination, lead is no longer used in house paint, gasoline, or solder. This has resulted in a decrease in the number of children with elevated blood lead levels, but low-income children who live in older buildings that still have lead paint and lead plumbing remain at particular risk.

Concept Check

1. **How** do children's energy requirements change as they age?
2. **What** factors affect children's food choices?
3. **Why** is lead exposure a particular concern for poorly nourished children?

12.3 Nutrition for Adolescents

LEARNING OBJECTIVES

1. **Describe** how puberty affects growth and body composition.
2. **Compare** the energy needs of adolescents with those of children and adults.
3. **Explain** why iron and calcium are of particular concern during the teen years.
4. **Use** MyPlate to plan a day's diet that would appeal to a teenager.

Adolescents are a unique population in many ways, and they have unique nutritional needs. The physical, emotional, mental, and social changes of adolescence transform a child into an adult. Organ systems develop and grow, **puberty** occurs, body composition changes, and the growth rates and nutritional requirements of boys and girls diverge (**Figure 12.10**). The physiological changes associated with sexual maturation affect nutrient requirements, and social and psychological changes that occur during adolescence influence nutrient intakes.

Energy and Nutrient Needs of Adolescents

The DRIs provide recommendations for adolescents in two age groups: 9 through 13 and 14 through 18. Separate recommendations are made for boys and girls because their needs begin to differ during the adolescent years.

Energy and energy-yielding nutrients The proportions of calories from carbohydrate, fat, and protein recommended for adolescents are similar to those for adults, but the total amount of energy needed by teenagers usually exceeds adult needs. Boys require more energy than girls because they have more muscle, and their bodies are larger. Protein requirements per kilogram of body weight are the same for boys and girls, but because boys are generally heavier, they require more total protein than do girls.

Vitamins The need for most of the vitamins rises to adult levels during adolescence. The requirements for B vitamins, which are involved in energy metabolism, are much higher in adolescence than in childhood because of higher energy needs. The rapid growth of adolescence further increases the need for vitamin B_6, which is important for protein synthesis, and for folate and vitamin B_{12}, which are essential for cell division. The high calorie intakes of teens help them meet most of their vitamin needs, but inadequate intakes of vitamin D put some teens at risk for deficiency. Vitamin D is important to support the rapid skeletal growth that occurs during adolescence. Low blood levels of vitamin D occur due to low intake as well as limited synthesis from sunlight in those with dark skin pigmentation or inadequate exposure to sunlight during the winter months (see Chapter 7).[47] Vitamin D deficiency is more prevalent in overweight and obese teens.[48]

---

puberty A period of rapid growth and physical changes that ends in the attainment of sexual maturity.

NUTRITION INSIGHT

FIGURE 12.10 Adolescent growth and development Growth and development occur rapidly during the adolescent years.

The **adolescent growth spurt** is an 18- to 24-month period of peak growth velocity that begins at about ages 10 to 13 in girls and 12 to 15 in boys. During a 1-year growth spurt, girls can gain 3.5 inches in height and boys can gain 4 inches.

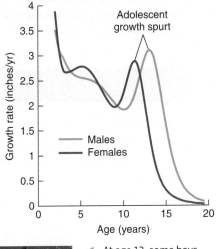

At age 13, some boys are physically still children, while others have matured sexually. Because there are large individual variations in the age at which these growth and developmental changes occur, the stage of maturation is often a better indicator of nutritional requirements than is chronological age.

During the adolescent growth spurt, boys gain some fat but add so much lean mass as muscle and bone that their percentage of body fat actually decreases. Girls gain proportionately more body fat and less lean tissue. By age 20, females have about twice as much adipose tissue as males and about 10% less lean tissue.

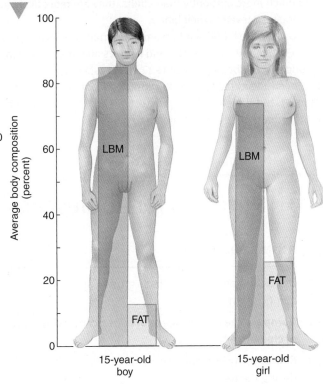

Iron The need for iron rises between childhood and adolescence. Iron is needed to synthesize hemoglobin for the expansion of blood volume and myoglobin for the increase in muscle mass. Because growth is more rapid in adolescent boys than in girls, they require more iron for the expansion of blood volume and for tissue synthesis. However, the onset of menstruation in girls increases their iron losses, making total iron needs greater in young women. The RDA is set at 11 mg/day for boys and 15 mg/day for girls ages 14 to 18. Girls are more likely than boys to consume less than the recommended amount because they require more iron, tend to eat fewer iron-rich foods, and consume fewer overall calories. As a result, iron deficiency is common in adolescent females, affecting about 9% of girls ages 12 to 15 and 16% of young women ages 16 to 19.[49]

Calcium The adolescent growth spurt increases both the length and mass of bones. Adequate calcium is essential for forming healthy bone. The RDA for calcium is 1300 mg/day for everyone between the ages of 9 and 18, but intake is typically below this amount in both sexes.[29] More than 50% of teenage girls do not consume the recommended amount of calcium.[50] Low calcium intake may reduce the level of peak bone mass achieved, increasing the risk of developing osteoporosis later in life.

Because milk and cheese, which are major sources of calcium in teen diets, can be high in saturated fat, adolescents should be encouraged to consume low-fat dairy products, calcium-fortified cereals, and vegetable sources of calcium. One of the contributors to low calcium intake is drinking soda and fruit juices, rather than milk (**Figure 12.11**).

Meeting Teens' Nutritional Needs

During adolescence, physiological changes dictate nutritional needs, but peer pressure may dictate food choices. Parents often have little control over what adolescents eat, and skipped meals and meals away from home are common. A food is more likely to be selected because it tastes good, it is easy to grab, or friends are eating it than because it is a healthy choice. No matter when foods are consumed throughout the day,

FIGURE 12.11 Beverage choices in children and teens[51]

Consumption of soft drinks by children and adolescents has increased since the 1970s, and milk consumption has decreased. Today, 46% of their beverage intake by weight is from soft drinks and other sweetened beverages, while only 29% is from milk. This pattern has increased the intake of added sweeteners and decreased the amounts of calcium, magnesium, potassium, protein, riboflavin, vitamin A, and zinc in the diet.

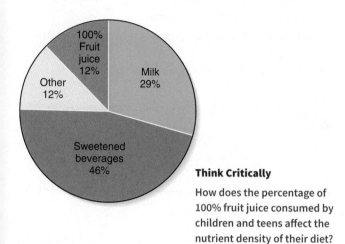

Think Critically
How does the percentage of 100% fruit juice consumed by children and teens affect the nutrient density of their diet?

an adolescent's diet should follow the recommendations of MyPlate for the appropriate age, gender, and activity level (see Figure 12.12 and *What Should I Eat? Adolescence*). The best indicators of adequate intake are satiety and a growth pattern that follows the curve of the growth charts.

Making fast food fit
A healthy diet can include an occasional fast-food meal, but a steady diet of burgers, fries, and pizza will likely contribute to an overall unhealthy diet. Fast food is typically low in fruits and vegetables and high in solid fats and sodium. About 35% of the empty calories consumed by U.S. children are from foods such as sugar-sweetened beverages, dairy desserts, French fries, and pizza served at fast-food restaurants.[52] Most teens in the United States consume more than the recommended amounts of added sugars, solid fat, and sodium and fewer fruits and vegetables than recommended. The lettuce and tomatoes that garnish a burger or taco are not enough to meet the MyPlate recommendations for vegetables. French fries, which are often high in *trans* fat and salt, are the most frequently consumed vegetable. To fit fast food into a healthy diet, make more nutrient-dense fast-food choices and make sure other meals and snacks eaten throughout the day supply the nutrients that

FIGURE 12.12 MyPlate recommendations for teens
The food group amounts recommended here are for 11- and 18-year-old boys and girls who engage in more than 60 minutes of activity daily. The 10 oz of grains recommended for an active 18-year-old boy may seem like a huge amount. But when spread over the course of a day (in the form of, say, a large bowl of cereal and toast for breakfast, two tacos for lunch, and a dinner that includes spaghetti and garlic bread), it is an amount that is easily consumed by a teenage boy.

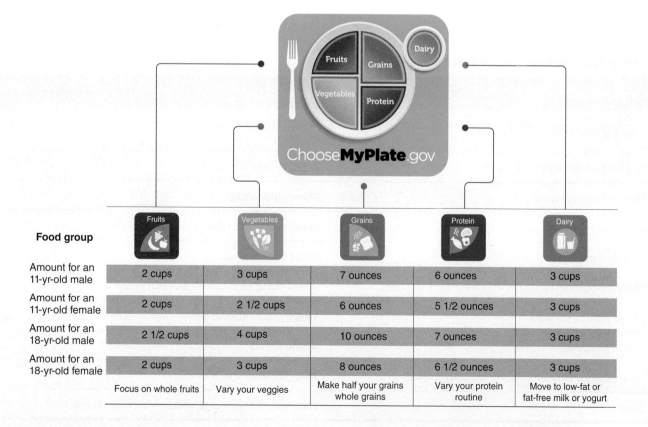

Food group	Fruits	Vegetables	Grains	Protein	Dairy
Amount for an 11-yr-old male	2 cups	3 cups	7 ounces	6 ounces	3 cups
Amount for an 11-yr-old female	2 cups	2 1/2 cups	6 ounces	5 1/2 ounces	3 cups
Amount for an 18-yr-old male	2 1/2 cups	4 cups	10 ounces	7 ounces	3 cups
Amount for an 18-yr-old female	2 cups	3 cups	8 ounces	6 1/2 ounces	3 cups
	Focus on whole fruits	Vary your veggies	Make half your grains whole grains	Vary your protein routine	Move to low-fat or fat-free milk or yogurt

What Should I Eat?

Adolescence

Balance unhealthy choices with healthy ones
- Have low-fat milk with your burger and fries.
- Put a variety of vegetables on your pizza.
- Dip your chips in salsa, guacamole, or hummus.

Eat breakfast
- Grab some whole-grain toast with peanut butter.
- Stick a cereal bar or muffin in your backpack.
- Have a yogurt on the go.

Snack well
- Reach for an apple, a pear, or an orange before the cookies and chips.
- Nibble on nuts and seeds.
- Crunch some baby carrots.

Count up your calcium and vitamin D
- Drink milk; low-fat milk has fewer calories than soda.
- Put extra low-fat milk on your cereal.
- Make a smoothie by mixing plain nonfat yogurt and fruit in a blender.
- Have cheese with crackers, on pizza, or in tacos.

 Use iProfile to calculate the calcium content of your favorite fast-food meal.

are not obtained from fast food. Many fast-food franchises now offer fruit, salads, yogurt, and low-fat milk. And some of the old standbys are not bad choices. A plain, single-patty hamburger provides a lot less fat and energy than one with two patties and a high-fat sauce. A chicken sandwich can be a healthy choice if it is grilled or barbecued, not breaded and fried (**Table 12.2**).

Keeping vegetarian choices healthy It is not uncommon for a teen to decide to consume a vegetarian diet even if the rest of the family does not. Some give up meat for health reasons or to lose weight, while others give up meat because they are concerned about animals and the environment. A vegetarian diet can be a healthy choice when it is carefully planned to meet nutrient needs, not just to eliminate meat. A poorly planned vegetarian diet will be no healthier than any other poorly planned diet. Adequate protein is generally not a problem, but meatless diets can be low in iron and zinc. Teenagers eating a vegan diet may also be at risk for vitamin B_{12} deficiency and inadequate calcium and vitamin D intake (see Chapter 6). We generally think of vegetarian diets as being low in fat and calories, but one that

TABLE 12.2 Make healthier fast-food choices

Instead of . . .	Choose . . .
Double-patty hamburger with cheese, mayonnaise, special sauce, and bacon	Regular single-patty hamburger without mayonnaise, special sauce, or bacon
Breaded and fried chicken sandwich	Grilled chicken sandwich
Chicken nuggets or tenders	Grilled chicken strips
Large French fries	Baked potato, side salad, or small order of fries
Fried chicken wings	Baked skinless wings
Crispy-shell chicken taco with extra cheese and sour cream	Grilled-chicken soft taco without sour cream
Nachos with cheese sauce	Tortilla chips with bean dip
12-in. meatball sub	6-in. turkey breast sub with lots of vegetables
Thick-crust pizza with extra cheese and meat toppings	Thin-crust pizza with extra veggies
Donut	Cinnamon and raisin bagel with low-fat cream cheese

Nutrition for Adolescents **387**

FIGURE 12.13 **Healthy vegetarian choices** The vegetarian meal on the left is high in saturated fat, cholesterol, and added sugar. In contrast, the whole-grain pita bread stuffed with chickpeas, corn, spinach, and tomatoes served with reduced-fat milk, shown on the right, is low in saturated fat, cholesterol, and added sugar and is high in fiber and a good source of calcium and iron.

is based on high-fat dairy products can be high in saturated fat, cholesterol, and calories (**Figure 12.13**).

Special Concerns for Teens

Peer pressure to fit in and concern about physical appearance probably have a greater impact on behavior during adolescence than at any other time in life. Many girls want to lose weight even if they are not overweight, and boys want to gain weight in order to achieve a strong, muscular appearance.

Eating disorders As discussed in Chapter 9, the excessive concern about weight, low self-esteem, and poor body image that are common during the teenage years contribute to the development of eating disorders. These disorders can be fatal, but even in less severe cases, the nutritional consequences of an eating disorder can affect growth and development during adolescence and can have a lifelong impact on bone health.

The impact of athletics Participation in competitive sports may affect adolescent nutrient needs and eating patterns. Like adult athletes, teen athletes require more water, protein, and energy than do their less active peers. Individuals involved in sports, such as football, that require the athlete to be large and heavy, usually do not have trouble eating enough to meet the additional energy and protein needs, but they may be tempted to experiment with anabolic steroids or other ergogenic supplements in an effort to "bulk up," and they may compromise their health as a result. As discussed in Chapter 10, anabolic steroids can stunt growth during adolescence as well as lead to sexual and reproductive disorders, heart disease, liver damage, acne, and aggressive, violent behavior. Teens participating in sports such as gymnastics that benefit from small, light bodies and sports with weight classes such as wrestling may restrict their food intake in order to stay lean. Weight restriction at this stage of life may affect nutritional status and maturation and increase the risk of developing an eating disorder and RED-S (see Chapter 10).

Tobacco use Tobacco use and addiction most commonly begin during youth and young adulthood. A recent survey found that one in four high school students currently use tobacco products, including e-cigarettes, which are the tobacco product most commonly used among high school students.[53] Tobacco use is linked to many disease conditions, including cardiovascular and respiratory diseases and several of types of cancer.[54] Smoking can limit appetite, and many teens start smoking in order to control their weight and are afraid to quit because they fear that they will gain weight if they do.[55] Smoking may also have an impact on nutrient intake; a study of smokers found that they eat more saturated fat and fewer fruits and vegetables than do nonsmokers.[56] This dietary pattern increases the risk of developing heart disease and cancer.

Alcohol use Alcohol is a drug that has short-term effects that occur soon after ingestion and long-term health consequences associated with overuse. These effects are discussed in greater depth in the last section of the chapter.

Although it is illegal to sell alcohol to adolescents, alcoholic beverages are commonly available at teen social gatherings, and peer pressure to consume them is strong. Surveys of American youth suggest that approximately 33% of the nation's high school students drank alcohol during the last month and 18% binge drank.[57] **Binge drinking** involves consumption of excessive amounts of alcohol in a short period of time. It often

---

binge drinking A pattern of drinking that brings a person's blood alcohol concentration to 80 mg/100 mL or above.

leads to dangerous risk-taking behaviors and blood alcohol levels that are high enough to cause loss of consciousness, coma, and even death.

Consumption of alcohol with caffeinated beverages is common among teens and is particularly dangerous.[58] The stimulant effects of the caffeine counter the depressant effects of the alcohol, but judgment and motor function are still impaired. The person is drunk but does not feel drunk so may drink to the point of alcohol toxicity or think it is safe to get behind the wheel of a car.

Concept Check

1. **How** does puberty affect body composition in males and females?
2. **How** do the energy needs of teens compare with those of adults?
3. **What** factors contribute to low calcium intake in teens?
4. **What** could a teen choose at a fast-food restaurant that would add to their vegetable intake?

12.4 Nutrition for the Adult Years

LEARNING OBJECTIVES

1. **Distinguish** life expectancy from active life expectancy.
2. **Compare** the energy and nutrient requirements of older and younger adults.
3. **Discuss** how the physical, mental, and social changes of aging increase nutritional risks.
4. **Track** the changes that occur in the MyPlate recommendations as adults age.

The benefits of a healthy diet do not stop when you stop growing. Good nutrition throughout your adult years can keep you healthy and active into your 80s and beyond. In the United States, **life expectancy** is 76.4 years for men and 81.2 years for women.[59] However, not all of these years are necessarily active and healthy. The average healthy or **active life expectancy** is 65.1 years for men and 68 years for women.[60] This means that, on average, men spend the last 11 years of life and women the last 13 years physically or socially restricted by disease and disability.

The goal of successful aging is to increase not only life expectancy but active life expectancy. Achieving this goal is particularly important because we live in a nation with an aging population (**Figure 12.14a**). Keeping older adults healthy will benefit not only the aging individuals themselves but also the family members who must find the time and resources to care

FIGURE 12.14 **The number of older adults is rising** The number of older adults is increasing, and there is great disparity in their health.

a. About 15% of the U.S. population is 65 years of age or over; by 2060 this percentage is expected to approach 30%.[61] In 2015, 35% of these individuals reported one or more disabilities. Only 3.1% of those over 65 live in institutional settings. However, the percentage increases dramatically with age; of those 85 years of age and older, 9% live in nursing homes.

b. Chronological age is not always the best indicator of a person's health. A person who is 75 may have the vigor and health of someone who is 55, or vice versa. Some older adults are healthy, independent, and active, while others are chronically ill, dependent, and at high risk for malnutrition.

life expectancy The average length of life for a particular population of individuals.

active life expectancy The number of years a person is able to live in an independent state, without being limited by chronic conditions.

for them and the public health programs that attempt to meet their needs. Although nutrition is not the key to immortality, a healthy diet can prevent malnutrition and delay the onset of chronic diseases that typically begin in middle age and reduce the quality of life in older adults (**Figure 12.14b**).

What Is Aging?

Aging is universal to all living things, but it is a process that we still don't fully understand. We know that as organisms grow older, the number of cells in their bodies decreases, and the functioning of the remaining cells declines. This loss of cells and cell function occurs throughout life, but the effects are not felt for many years because organisms start out life with more cells and cell function than they need. This reserve capacity allows an organism to continue functioning normally despite a decrease in the number and function of cells. In young adults, the reserve capacity of organs is 4 to 10 times that required to sustain life. As a person ages and reserve capacity decreases, the effects of aging become evident in all body systems. With this loss of function comes a reduction in the ability to repair damage and resist infection, so older people may die from diseases from which they could easily have recovered when they were younger.

The human **life span** is about 120 years, but how long individuals live and the rate at which they age are determined by the genes they inherit, their lifestyle, and the extent to which they are able to avoid accidents, disease, and environmental toxins (**Figure 12.15**). A person with a family history of heart disease who eats a healthy diet and exercises regularly may never be limited by heart disease. In contrast, someone with no family history of heart disease who is inactive, smokes, and eats a poor diet may develop heart problems that lead to disability and death (see *Debate: Can Eating Less Help You Live Longer?*).

Nutrition and Health Concerns Throughout Adulthood

The physiological and health changes that accompany aging affect energy and nutrient requirements, how some nutrient requirements must be met, and the risk of malnutrition. In order to best recommend nutrient intakes for adults of all ages, the DRIs include four adult age categories: young adulthood (ages 19–30), middle age (ages 31–50), adulthood (ages 51–70), and older adulthood (over age 70). These recommendations are designed to meet the needs of the majority of healthy individuals in each age group. Although the incidence of chronic diseases and disabilities increases with advancing age, these increases are not considered when making general nutrient intake recommendations.

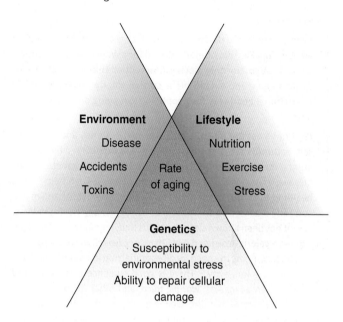

FIGURE 12.15 Factors that affect how fast we age Genes determine the efficiency with which cells are maintained and repaired and also determine our susceptibility to age-related diseases, such as cardiovascular disease and cancer, but lifestyle and environment also affect how fast we age.

Energy and energy-yielding nutrient recommendations Adult energy needs typically decline with age, due primarily to decreases in basal metabolic rate (BMR) and activity level (**Figure 12.16**). The need for most nutrients does not change, however, so in order to meet nutrient needs without exceeding energy needs, older adults must consume a nutrient-dense diet. For example, adult protein requirements do not change with age; therefore, older adults must consume a diet that is higher in protein relative to calories than that of younger adults.

The proportions of carbohydrate and fat recommended in the diet also remain the same in older adults as in younger adults. Most dietary carbohydrate should come from unrefined sources, which provide fiber and a variety of vitamins and minerals. Adequate fiber, when consumed with adequate fluid, helps prevent constipation, hemorrhoids, and diverticulosis—conditions that are common in older adults. High-fiber diets may also be beneficial in the prevention and management of diabetes, cardiovascular disease, and obesity.

Sources of dietary fat should also be chosen with nutrient density in mind. A diet with 20 to 35% of energy from fat that contains adequate amounts of the essential fatty acids and limits saturated fat and *trans* fat is recommended.

Meeting the water needs of older adults The recommended water intake for older adults is the same as for younger adults, but meeting these needs may be more

---

aging The inevitable accumulation of changes associated with and responsible for an ever-increasing susceptibility to disease and death.

life span The maximum age to which members of a species can live.

Debate

Can Eating Less Help You Live Longer?

The Issue

Even before Ponce de León searched for the fountain of youth in what is now Florida, humans longed for a way to preserve youth. So far, though, the only dietary intervention that has been shown to slow aging and extend life is cutting calorie intake. Is calorie restriction a good way for people to slow aging and extend their lives?

Most of us eat too much. Eating too many calories and too much salt, saturated fat, and sugar shortens life expectancy by increasing the incidence of obesity, high blood pressure, diabetes, heart disease, and cancer. Eliminating dietary excesses can help us live healthier lives, but many argue that eating even less than what is currently recommended will help us live longer and healthier lives.

It has been known since the 1930s that, for many animals, consuming a calorie-restricted diet (20 to 50% less than recommended intake) that meets the need for all essential nutrients will slow aging and increase longevity.[62] There is good evidence that short-lived species such as insects, worms, mice, and other rodents that are fed a calorie-restricted diet live longer—as much as 150% longer—than animals that eat more calories.[63] These calorie-restricted animals have a lower incidence of age-related chronic diseases, better immune function, lower blood glucose, and better overall organ function than do animals whose diets are not calorie restricted.

There is less evidence for the effectiveness of calorie restriction in humans and other primates. Two studies begun in the 1980s explored the effect of calorie restriction in rhesus monkeys. Both reported beneficial health effects in the calorie-restricted animals; however, only one found that the calorie-restricted group actually lived longer than controls.[64,65] Epidemiological evidence of the benefits of calorie restriction in humans comes from studying the people of Okinawa, who were the longest-lived people on Earth.[66] Until recently, Okinawans practiced calorie restriction by eating until they were only 80% full. In addition, they led active lives, worked as farmers, and consumed a nutrient-dense diet high in green and yellow vegetables, whole grains, soy, fish, and lean meat and low in refined grains, saturated fat, salt, and sugar.[67] Older Okinawans maintain a low BMI and have a lower incidence of chronic diseases than people living in the United States or on mainland Japan (see graph).[66] But, living an Okinawan lifestyle may not be enough to extend life; there is evidence that genetics and climate may contribute to greater life expectancy in Okinawans.[68–70]

So, is calorie restriction something that we should all practice in order to live longer? In a world in which obesity, diabetes, and other diet-related conditions are limiting our active life expectancy, the benefits of calorie restriction are intriguing. A study of calorie reduction in normal-weight individuals over a two-year period showed decreases in body weight, body fat, blood pressure, blood lipid levels, and insulin resistance. However, several of the other hormonal and physiological changes believed to contribute to longevity in calorie-restricted rodents did not occur.[62] One possible reason is that subjects were not able to maintain the 25% reduction in energy intake targeted by the study. During the first 6 months the reduction was 19.5%, and over the next 18 months it was only 9.1%. This shows that calorie restriction is far more difficult than drinking from the mythical

Many Okinawans who have followed the traditional dietary and lifestyle pattern have lived to be over 100 years old.

fountain of youth. A person whose EER is about 2000 Calories/day could eat only 1200 to 1500 Calories/day when practicing caloric restriction. Meeting nutrient needs would require carefully planned meals and snacks; a poorly planned diet would lead to malnutrition. In general, the two-year human trial was safe and well tolerated, but subjects did not sustain a 25% reduction.[62] Side effects included food cravings, a lack of energy, and in some cases psychological issues. Based on the Okinawan experience, this ascetic lifestyle would add only a few years of life. Would it be worth the sacrifices?

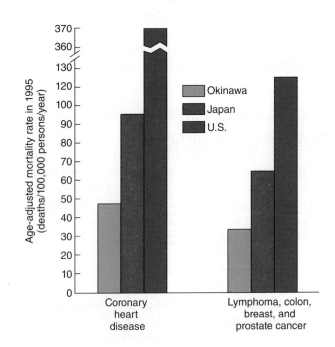

Think Critically

Use the graph to compare the incidence of heart disease in Okinawa with that in Japan and the United States. Suggest some possible explanations for these differences.

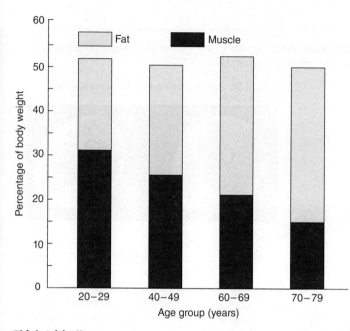

FIGURE 12.16 **Declining muscle mass** With age, the percentage of body weight that is muscle declines and the percentage that is fat increases.[71] Some of the decline in energy needs in older adults is due to this decrease in muscle mass; the less lean muscle a person has, the lower his or her BMR. The EER for an 80-year-old man is almost 600 Calories/day less than for a 20-year-old man of the same size and activity level. For women, the difference is about 400 Calories/day.

Think Critically

How much daily exercise would an 80-year-old woman need in order to expend about the same number of calories per day as a sedentary 20-year-old woman of the same height and weight?

challenging for older adults. Advanced age brings a reduction in the sense of thirst, which can decrease fluid intake. In addition, some older adults have mobility limitations that reduce their access to beverages. The risk of dehydration is further increased in older adults by greater water losses. The kidneys are no longer as efficient at conserving water as they once were, and many older adults also take medications that increase water loss.

The risk of vitamin and mineral deficiencies in older adults

The physiological changes of aging and the decrease in calorie needs put older adults at risk for deficiency of several vitamins and minerals. Recommended intakes are rarely higher than for younger adults, but for a few nutrients, special recommendations are made about how needs should be met.

Intakes of certain B vitamins are a concern for older adults. The RDA for vitamin B_6 is greater in adults ages 51 and older than for younger adults because higher dietary intakes are needed to maintain the same functional levels in the body. Folate intake is a concern because deficiency of folate alone or in combination with vitamin B_{12} and vitamin B_6 deficiencies may contribute to the development of cancer, cardiovascular disease, and cognitive dysfunction.[72]

The RDA for vitamin B_{12} is not increased in older adults, but it is recommended that people over 50 meet their RDA for vitamin B_{12} by consuming foods that are fortified with this vitamin or by taking a supplement containing vitamin B_{12}. This is because food-bound vitamin B_{12} is not absorbed efficiently in many older adults due to atrophic gastritis, an inflammation of the stomach lining that causes a reduction in stomach acid (see Chapter 7).[73,74] Reduced secretion of stomach acid also allows microbial overgrowth in the stomach and small intestine, and the greater number of microbes compete for available vitamin B_{12}, further reducing the amount of vitamin B_{12} that is absorbed. It is estimated that 10 to 30% of U.S. adults over age 50 and 40% of those in their 80s have atrophic gastritis. The vitamin B_{12} in fortified foods and supplements is not bound to proteins, so it is absorbed even when stomach acid levels are low. Atrophic gastritis may also reduce the absorption of iron, folate, calcium, and vitamin K.

Women over age 50 need less iron than younger women because they no longer lose iron through menstruation. The RDA for women 51 and older is 8 mg, the same as for adult men of all ages. Despite low iron needs, iron deficiency anemia is a concern among women and men in this age group. Common causes are chronic blood loss due to disease and medication and poor iron absorption due to antacid use and low stomach acid.

Calcium status is a concern in older adults because calcium intake is often low, and intestinal absorption decreases with age. Without sufficient calcium, bone mass decreases, and the risk of bone fractures due to osteoporosis increases. The reduction in estrogen that occurs with menopause further increases bone loss in women by increasing the rate of bone breakdown and decreasing the absorption of calcium from the intestine (see Chapter 8). The RDA for adult men ages 51 to 70 years is 1000 mg/day. Because of the accelerated bone loss in women during the postmenopausal period, the RDA for women 51 to 70 years is increased to 1200 mg/day. To reduce age-related bone loss, the RDA for men and women over 70 years of age is 1200 mg/day.[29]

Vitamin D, which is necessary for adequate calcium absorption, is also a concern in older adults. Intake is often low, and synthesis in the skin is reduced due to limited exposure to sunlight and because the capacity to synthesize vitamin D in the skin decreases with age. The RDA for people ages 51 to 70 years is 600 IU (15 µg)/day, the same as for younger adults. For individuals over age 70 years, the RDA is increased to 800 IU (20 µg)/day.[29]

Factors That Increase the Risk of Malnutrition in Older Adults

The aging process usually does not cause malnutrition in healthy, active adults, but nutritional health can be compromised by the physical changes that occur with age, the presence of disease, and economic, psychological, and social circumstances.[75] These factors can increase the risk of malnutrition by altering nutrient needs and decreasing the motivation to eat and the ability to acquire and enjoy food. Malnutrition then exacerbates some of these factors, contributing to a downward health spiral from which it is difficult to recover (**Figure 12.17a**).

Physiological changes With age comes a decrease in muscle size and strength (**Figure 12.17b**). This decrease affects both the skeletal muscles needed to move the body and the heart and respiratory muscles needed to deliver oxygen to the tissues. Some of this decline is due to changes in hormone levels and muscle protein synthesis, but lack of exercise is also an

NUTRITION INSIGHT

FIGURE 12.17 Causes and consequences of malnutrition Many of the physiological changes and health problems associated with age can affect nutritional status.

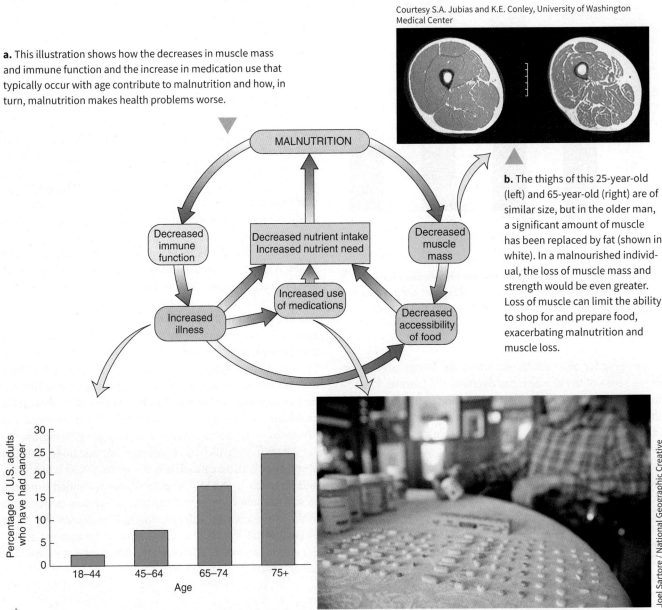

a. This illustration shows how the decreases in muscle mass and immune function and the increase in medication use that typically occur with age contribute to malnutrition and how, in turn, malnutrition makes health problems worse.

b. The thighs of this 25-year-old (left) and 65-year-old (right) are of similar size, but in the older man, a significant amount of muscle has been replaced by fat (shown in white). In a malnourished individual, the loss of muscle mass and strength would be even greater. Loss of muscle can limit the ability to shop for and prepare food, exacerbating malnutrition and muscle loss.

c. The incidence of cancer increases with age.[76] One reason for the higher incidence is that the immune system's ability to destroy cancer cells declines. Reduced immune function also increases the frequency of infectious diseases and reduces the ability to recover from these diseases.

d. It is common for older adults to take multiple medications. Shown here are the pills this 73-year-old man takes each week. Medications can affect nutritional status by interfering with taste, chewing, and swallowing; by causing loss of appetite, gastrointestinal upset, constipation, or nausea; and by increasing nutrient losses or decreasing nutrient absorption.

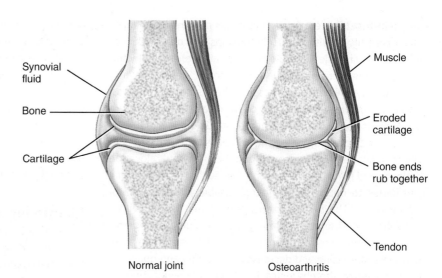

FIGURE 12.18 **Osteoarthritis** Osteoarthritis, the most common form of arthritis, occurs when the cartilage that cushions the joints degenerates, allowing the bones to rub together and cause pain. Anti-inflammatory medications help reduce the pain. Supplements of glucosamine and chondroitin are marketed to improve symptoms and slow the progression of osteoarthritis, but clinical studies on the effect of these supplements on pain and function have been equivocal.[80]

important contributor.[77] The decrease in muscle strength contributes not only to **physical frailty**, which is characterized by general weakness, impaired mobility and balance, and poor endurance, but also to the risk of falls and fractures. In those ages 85 years and older, loss of muscle strength is the limiting factor that determines whether they can continue to live independently.

The immune system's ability to fight disease declines with age. With this decline, the incidence of infections, cancers, and autoimmune diseases increases, and the effectiveness of immunizations decreases (**Figure 12.17c**). In turn, increases in infections and chronic disease can lead to increased use of medications that affect nutritional status (**Figure 12.17d**). Malnutrition exacerbates the decline in immune function.

Chronic illness About 66% of the older population suffers from multiple chronic conditions.[78] These conditions affect the ability to maintain good nutritional health because they can change nutrient requirements, decrease the appeal of food, and impair the ability to obtain and prepare an adequate diet.

Some illnesses change the type of diet that is recommended. For instance, kidney failure reduces the ability to excrete protein waste products, so the diet must be lower in protein. Blood pressure is affected by sodium intake, so a low-sodium diet is recommended for individuals with high blood pressure. Dietary restrictions such as these limit food choices and can affect the palatability of the diet and thereby contribute to malnutrition in older adults.

Physical disabilities can limit a person's ability to obtain and prepare food. The most common reason for physical disability among older adults is **arthritis**, a condition that causes pain in joints when they are moved (**Figure 12.18**). Arthritis, the nation's most common cause of disability, affects more than 53 million Americans.[79]

Visual disorders become more common as a person ages. **Macular degeneration** is the most common cause of blindness in older Americans. The **macula** is a small area of the retina of the eye that distinguishes fine detail. If the number of viable cells in the macula is reduced, visual acuity declines, ultimately resulting in blindness. **Cataracts** are another common cause of declining vision (**Figure 12.19**). Oxidative damage is believed to cause both macular degeneration and cataracts. Therefore, a diet that is high in foods containing antioxidant nutrients and phytochemicals might slow or prevent these eye disorders.

Changes in mental status can affect nutrition by interfering with the response to hunger and the ability to eat and to obtain and prepare food. The incidence of **dementia** increases with age. Dementia involves impairment in memory, thinking, or

FIGURE 12.19 **Cataracts** A cataract is a clouding of the lens of the eye that impairs vision. When cataracts obscure vision, the affected lens can be removed and replaced with an artificial plastic lens.

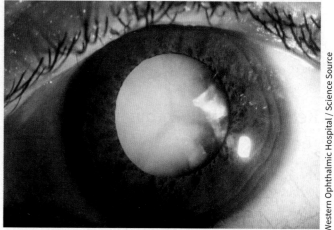

arthritis A disease characterized by inflammation of the joints, pain, and sometimes changes in structure.

macular degeneration Degeneration of a portion of the retina that results in loss of visual detail and eventually blindness.

dementia A deterioration of mental state that results in impaired memory, thinking, and/or judgment.

judgment that is severe enough to cause personality changes and affect daily activities and relationships with others. Causes of dementia include multiple strokes, alcoholism, dehydration, side effects of medication, vitamin B_{12} deficiency, and **Alzheimer's disease**. Regardless of the cause, neurological problems can affect the ability to consume a healthy diet.

Another cause of altered mental status in older adults is depression. Social, psychological, and physical factors all contribute to depression. Retirement and the death and relocation of friends and family members can cause social isolation, which contributes to depression. Physical disability causes loss of independence. The inability to engage in normal daily activities, easily visit with friends and family members, and provide for personal needs contributes to depression. Depression can make meals less appetizing and decrease the quantity and quality of foods consumed, thereby increasing the risk of malnutrition.

Use of medications Because older adults have an increased frequency of acute and chronic illnesses, they are likely to take multiple medications (see Figure 12.17d).[81] Medications can affect nutritional status, and nutritional status can alter the effectiveness of medications. The more medications taken, the greater the chance of side effects that affect nutritional status, such as decreased appetite, changes in taste, and nausea. Diet can also change the effectiveness of medications. For example, vitamin K hinders the action of warfarin, an anticoagulant drug taken to reduce the risk of blood clots. On the other hand, omega-3 fatty acids, such as those in fish oils, inhibit blood clotting and may intensify the effect of warfarin and cause bleeding.

Economic and social issues More than 4.2 million older Americans live below the poverty level, and many live on a fixed income, making it difficult to afford health care, especially medications, and a healthy diet.[61] Food costs and limited food preparation facilities can reduce the types of foods available. In 2015, 8.3% of households in the United States that included people 65 years of age and older experienced **food insecurity**,[82] which occurs when the availability of nutritionally adequate, safe food or the ability to acquire food in socially acceptable ways is limited. This in turn increases the risk of malnutrition.

Keeping Healthy Throughout the Adult Years

There is no secret dietary factor that will bestow immortality, but good nutrition and an active lifestyle are major determinants of successful aging. A well-planned, nutritionally adequate diet can extend an individual's years of healthy life by preventing malnutrition and delaying the onset of chronic diseases. Regular exercise can help maintain muscle mass, bone strength, and cardiorespiratory function, helping to prolong independent living. The foods consumed by older adults are determined not only by preferences and physiological changes but also by factors such as living arrangements, finances, transportation, and disability. For those with economic, social, or physical limitations, food assistance programs or assisted living can help prevent food insecurity.

Identifying older adults at risk To address concerns about the nutritional health of older adults, the U.S. Nutrition Screening Initiative promotes screening for nutrition-related problems in this group. This program is working to increase awareness of nutritional problems by involving practitioners and community organizations as well as relatives, friends, and others caring for older adults in evaluating the nutritional status of this group. This program developed the DETERMINE checklist (**Table 12.3**), based on an acronym for the physiological, medical, and socioeconomic situations that increase the risk of malnutrition among older adults. Older people themselves, as well as family members and caregivers, can use this tool to determine when malnutrition is a potential problem.

Meeting nutrient needs Meeting the nutrient needs of older adults can be challenging (see What Should I Eat? Advancing Age). Because energy needs are reduced while protein and most micronutrient needs remain the same or increase, food choices must be nutrient dense. Diets need to provide adequate fiber and fluid to prevent constipation; fiber intake in older adults is often below recommendations, and dehydration is a problem, especially in persons 85 and older.[75] The medical, social, and economic challenges that often accompany aging make it more difficult to meet nutrient needs. Many older adults need supplements of vitamin D, vitamin B_{12}, and calcium to meet their nutrient needs. However, supplements should not take the place of a balanced, nutrient-dense diet that is high in whole grains, fruits, and vegetables (**Figure 12.20**). In addition to essential nutrients, these foods contain fiber, phytochemicals, and other substances that may protect against disease.

Older adults who have physical limitations need to choose foods that they can easily prepare and consume. For those who have difficulty preparing foods, precooked foods, frozen dinners, and canned soup can provide a meal with almost no preparation. Medical nutritional products, such as Ensure or Boost, can also be used to supplement intake.

Alzheimer's disease A disease that results in a relentless and irreversible loss of mental function.

TABLE 12.3 DETERMINE: A checklist of the warning signs of malnutrition

Disease	Any disease, illness, or condition that causes changes in eating can predispose a person to malnutrition. Memory loss and depression can also interfere with nutrition if they affect food intake.
Eating poorly	Eating either too little or too much can lead to poor health.
Tooth loss/mouth pain	Poor health of the mouth, teeth, and gums interferes with the ability to eat.
Economic hardship	Having to, or choosing to, spend less than $25 to $30 per person per week on food interferes with nutrient intake.
Reduced social support	Not having contact with people on a daily basis has a negative effect on morale, well-being, and eating.
Multiple medicines	The more medicines a person takes, the greater the chances of side effects such as weakness, drowsiness, diarrhea, changes in taste and appetite, nausea, and constipation.
Involuntary weight loss or gain	Unintentionally losing or gaining weight is a warning sign that should not be ignored. Being overweight or underweight also increases the risk of malnutrition.
Needs assistance in self-care	Difficulty walking, shopping, and cooking increases the risk of malnutrition.
Elder above age 80	The risks of frailty and health problems increase with increasing age.

What Should I Eat?

Advancing Age

Consume plenty of fluids and fiber
- Drink a beverage with every meal.
- Keep a bottle of water handy and sip on it.
- Choose whole-grain breads and cereals.
- Add extra veggies to your soup.

Pay attention to vitamin B_{12}, calcium, and vitamin D
- Make sure your cereal is fortified with vitamin B_{12}.
- Drink milk; it provides both calcium and vitamin D.
- Spend a few minutes in the sun to get some vitamin D with no calories at all.
- Add some canned salmon to a salad for lunch.

Antioxidize
- Have a bowl of strawberries: They provide vitamin C and antioxidant phytochemicals.
- Choose colorful vegetables to boost carotenoids.
- Use a healthy vegetable oil and add nuts and seeds to your salads to get your vitamin E.
- Select seafood to boost selenium intake.

Work on meals for one
- Ask the grocer to break up larger packages of eggs and meats.
- Buy in bulk and share with a friend.
- Make a whole pot but freeze it in meal-size portions.
- Top a baked potato with leftover vegetables or sauces.

 Use iProfile to find foods that are fortified with vitamin B_{12}.

Physical activity for older adults Regular physical activity can extend years of active, independent life, reduce the risk of disability, and improve the quality of life for older adults. Exercise helps maintain bones and muscles. It lowers the risk of cardiovascular disease, diabetes, and cancer; reduces depression; and improves cognitive function.[75] Exercise also allows an increase in food intake without weight gain, so micronutrient needs are met more easily. A physical activity program for older adults should improve endurance, strength, flexibility, and balance and should be geared to the physical ability of the individual.[18,83]

Endurance activities such as walking, biking, and swimming provide protection against chronic disease.

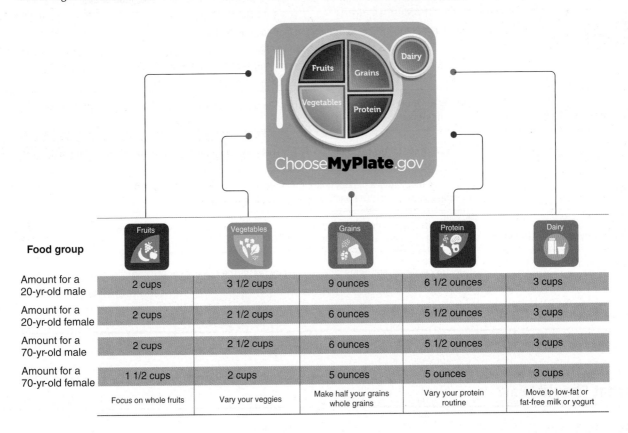

FIGURE 12.20 MyPlate for older adults The MyPlate recommendations shown here are for 20- and 70-year-old men and women who get less than 30 minutes of activity per day. The older adults need to consume fewer servings from most food groups and have less room for empty calories from foods high in added sugars or saturated fat.

Food group	Fruits	Vegetables	Grains	Protein	Dairy
Amount for a 20-yr-old male	2 cups	3 1/2 cups	9 ounces	6 1/2 ounces	3 cups
Amount for a 20-yr-old female	2 cups	2 1/2 cups	6 ounces	5 1/2 ounces	3 cups
Amount for a 70-yr-old male	2 cups	2 1/2 cups	6 ounces	5 1/2 ounces	3 cups
Amount for a 70-yr-old female	1 1/2 cups	2 cups	5 ounces	5 ounces	3 cups
	Focus on whole fruits	Vary your veggies	Make half your grains whole grains	Vary your protein routine	Move to low-fat or fat-free milk or yogurt

Recommendations are the same as for younger adults: a minimum of 150 minutes per week of moderate-intensity aerobic activity (**Figure 12.21**). Muscle-strengthening exercise two or more days per week is recommended to increase strength and lean body mass. Lifting small weights or stretching elastic bands at an intensity that requires some physical effort can provide strength training. Better flexibility enhances postural stability and balance, makes the tasks of everyday life easier, and may reduce the risk of injuries. Flexibility exercises should include those that move the muscles through a full range of motion, such as arm circles, as well as those that stretch muscles their full length. Improvements in strength, endurance, and flexibility all enhance balance, which reduces the risk of falls. Exercises that incorporate balance, coordination, gait, and agility such as backward walking, heel walking, tai chi, and yoga should be included in an exercise program for older adults.[83]

Overcoming economic and social issues Overcoming economic limitations may involve providing education about economics and food preparation or providing assistance with shopping and food preparation. Options for people with limited incomes include reduced-cost meals at senior centers, federal food assistance programs (discussed in Chapter 14), food banks, and soup kitchens.

Another problem that contributes to poor nutrient intake in older adults is loneliness. Living, cooking, and eating alone can decrease interest in food. Programs that provide nutritious meals in communal settings promote social interaction and can improve nutrient intake. For those who are unable to attend communal meals, home-delivered meals may be available.

Overcoming physical limitations: Assisted living The physical and psychological declines associated with aging eventually cause many people to require assistance in everyday living. Assisted living facilities allow individuals to live in their own apartments but provide help, as needed, with activities such as eating, bathing, dressing, housekeeping, and taking medications. These facilities provide an interim level of care for those who cannot live safely on their own but do not require the total care provided in a nursing home. Eventually, many older adults will need to live in nursing homes. Even though nursing homes provide access to food and medical care,

FIGURE 12.21 **Physical activity for older adults** Participating in exercise classes and other group-based activities can be a good way for older adults to start an exercise program. Water activities, such as water aerobics and swimming, do not stress the joints and hence can be used to improve endurance in individuals with arthritis or other bone and joint disorders. Some weight-bearing activity, such as walking, is encouraged to promote bone health.

Ira Block / National Geographic Creative

their residents are at increased risk for malnutrition because they are more likely to have medical conditions that increase nutrient needs or interfere with food intake or nutrient absorption; in addition, residents are at risk because they are dependent on others to provide for their care. Even when adequate meals are provided, many nursing home residents require assistance in eating and frequently do not consume all the food served, thus increasing their likelihood of developing deficits of energy, water, and other nutrients.[84]

Concept Check

1. **What** is the goal of successful aging?
2. **How** do the energy needs of older adults compare to those of younger adults?
3. **What** are three factors that increase the risk of malnutrition in older adults?
4. **How** would the MyPlate recommendations for an 80-year-old woman differ from those of a 30-year-old woman of the same height, weight, and activity level?

12.5 The Impact of Alcohol Throughout Life

LEARNING OBJECTIVES

1. **Define** moderate alcohol consumption.
2. **Explain** how alcohol is absorbed and metabolized.
3. **Describe** the short- and long-term problems of excess alcohol consumption.
4. **Discuss** the potential benefits of moderate alcohol consumption.

Since the dawn of civilization, almost every human culture has produced and consumed some type of alcoholic beverage. Depending on the times and the culture, alcohol use has been touted, casually accepted, denounced, and even outlawed.

Whether alcohol consumption represents a risk to health or provides some benefits depends on who is drinking and how much is consumed. When consumed by a pregnant woman, alcohol can cause birth defects in the developing child. When consumed during childhood and adolescence, when the brain is still developing and changing, alcohol can cause permanent reductions in learning and memory.[85] When consumed in excess by anyone, alcohol has medical and social consequences that negatively affect drinkers and those around them. It can reduce nutrient intake and affect the storage, mobilization, activation, and metabolism of nutrients. Its breakdown produces toxic compounds that damage tissues, particularly the liver. Moderate alcohol consumption by healthy adults, however, provides some health advantages. The Dietary Guidelines define moderate alcohol consumption as no more than one drink per day for women and two drinks per day for men (**Figure 12.22**).[33]

FIGURE 12.22 **Alcoholic beverages** Alcoholic beverages consist primarily of water, ethanol, and carbohydrate, with few other nutrients. A drink, defined as about 5 oz of wine, 12 oz of beer, or 1.5 oz of distilled spirits, contains 12 to 14 g of alcohol. The alcohol provides about 90 Calories (7 Calories/g), and carbohydrates provide the remaining calories in alcoholic beverages.

Alcohol Absorption, Transport, and Excretion

Chemically, any molecule that contains an –OH group is an **alcohol**, but we usually use the term to refer to **ethanol** and often to any beverage that contains ethanol (see Figure 12.22). Ethanol is a small molecule that is rapidly and almost completely absorbed in the stomach and small intestine. Because some alcohol is absorbed directly from the stomach, its effects are almost immediate, especially when consumed on an empty stomach. If there is food in the stomach, absorption is slowed because food dilutes the alcohol and slows the rate of stomach emptying.

Once it has been absorbed, alcohol enters the bloodstream and is rapidly distributed throughout all body water. Peak blood alcohol concentrations are attained approximately one hour after ingestion. Many variables affect blood alcohol level, including the kind and quantity of alcoholic beverage consumed, the speed at which the beverage is consumed, the food consumed with it, the weight of the consumer, whether the person is a male or a female, and the activity of alcohol-metabolizing enzymes in the body (**Figure 12.23a**).

Because alcohol is a toxin and cannot be stored in the body, it must be eliminated quickly. Absorbed alcohol travels to the liver, where it is given metabolic priority and is therefore broken down before other molecules. About 90% of alcohol consumed is metabolized by the liver. The remainder is excreted through the urine or eliminated via the lungs during exhalation (**Figure 12.23b**). The alcohol that reaches the kidney acts as a diuretic, increasing fluid excretion. Therefore, excessive alcohol intake can contribute to dehydration.

Alcohol Metabolism

In people who occasionally consume moderate amounts of alcohol, most of the alcohol is broken down in the liver by the enzyme alcohol dehydrogenase (ADH) (**Figure 12.24**). This enzyme has also been found in all parts of the gastrointestinal tract.[86] When greater amounts of alcohol are consumed, a second pathway in the liver, called the microsomal ethanol-oxidizing system (MEOS), also metabolizes alcohol. The rate at which ADH breaks down alcohol is fairly constant, but MEOS activity increases when more alcohol is consumed.[86] MEOS also metabolizes other drugs, so as activity increases in response to high alcohol intake, it can alter the metabolism of other drugs.

Adverse Effects of Alcohol

The consumption of alcohol has short-term effects that interfere with organ function for several hours after ingestion.

---

ethanol The alcohol in alcoholic beverages; it is produced by yeast fermentation of sugar.

FIGURE 12.23 Blood alcohol levels
Blood alcohol levels rise rapidly after consumption. The amount lost in expired air is proportional to blood levels.

a. Blood alcohol levels are higher in women than in men after consuming the same amount of alcohol. This may be due to lower levels of the enzymes that break down alcohol or to the fact that women have less body water than men, so the alcohol they consume is distributed in a smaller amount of body water.

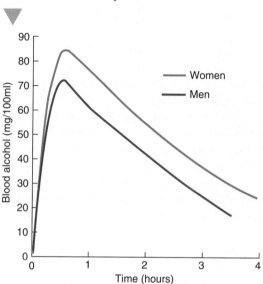

b. In the lungs, some alcohol diffuses out of the blood, into the air, and is exhaled. The amount of alcohol lost through the lungs is reliable enough to estimate blood alcohol level by using a Breathalyzer test.

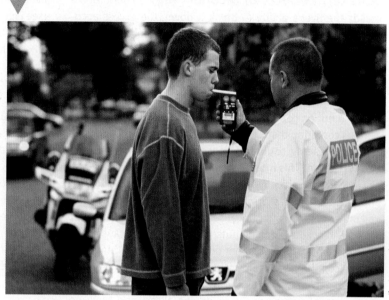

PROCESS DIAGRAM

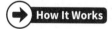

FIGURE 12.24 Metabolizing alcohol
The ADH pathway predominates when small amounts of alcohol are consumed. The MEOS becomes important when larger amounts of alcohol are consumed. The MEOS reaction requires oxygen and the input of energy (ATP) to break down alcohol. It also generates reactive oxygen molecules that can contribute to liver disease.

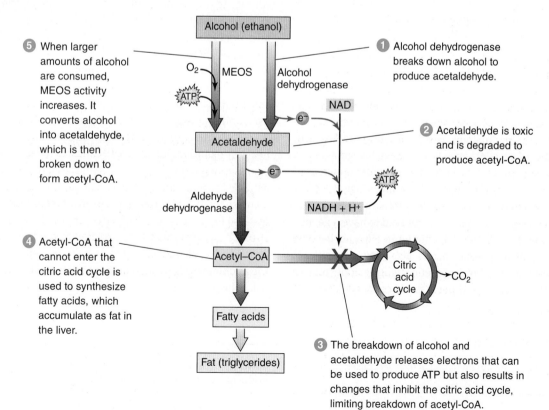

① Alcohol dehydrogenase breaks down alcohol to produce acetaldehyde.

② Acetaldehyde is toxic and is degraded to produce acetyl-CoA.

③ The breakdown of alcohol and acetaldehyde releases electrons that can be used to produce ATP but also results in changes that inhibit the citric acid cycle, limiting breakdown of acetyl-CoA.

④ Acetyl-CoA that cannot enter the citric acid cycle is used to synthesize fatty acids, which accumulate as fat in the liver.

⑤ When larger amounts of alcohol are consumed, MEOS activity increases. It converts alcohol into acetaldehyde, which is then broken down to form acetyl-CoA.

Chronic alcohol consumption has long-term effects that cause disease both because the alcohol interferes with nutritional status and because alcohol metabolism produces toxic compounds.

Short-term effects of alcohol

When alcohol intake exceeds the liver's ability to break it down, the excess accumulates in the bloodstream. The circulating alcohol acts as a depressant, impairing mental and physical abilities. First, alcohol affects reasoning; if drinking continues, the brain's vision and speech centers are affected. Next, large-muscle control becomes impaired, causing lack of coordination. Finally, if alcohol consumption continues, it can result in **alcohol poisoning**, a serious condition that can slow breathing, heart rate, and the gag reflex, leading to loss of consciousness, choking, coma, and even death. This most frequently occurs in cases of *binge drinking*. Even if an individual does not experience a loss of consciousness, excess drinking may still cause memory loss. Drinking enough alcohol to cause amnesia is called **blackout drinking**. Blackout drinking puts people at risk because they have no memory of events that occurred during the blackout. During alcohol-related memory blackouts, people may engage in risky behaviors such as having unprotected sexual intercourse, vandalizing property, or driving a car—and have no memory of it afterward.

The effects of alcohol on the central nervous system make driving while under the influence of alcohol very dangerous. Alcohol affects reaction time, eye–hand coordination, and balance. Not only does alcohol impair one's ability to operate a motor vehicle, it also impairs one's judgment in making the decision to drive. Abuse of alcohol also contributes to domestic violence and is a factor in more than 100,000 deaths per year, including almost 40% of all traffic fatalities.[87]

Alcoholism

A pattern of uncontrolled problematic alcohol consumption or the development of withdrawal symptoms when drinking is stopped is diagnostic of **alcohol use disorder**, commonly referred to as **alcoholism**.[88] Approximately 17 million adults in the United States ages 18 and older and an estimated 855,000 adolescents ages 12 to17 have been diagnosed with alcohol use disorder.[89] The risk of this disorder is increased in individuals who begin drinking at a younger age. Like any other drug addiction, alcohol use disorder is a physiological condition that needs treatment. It is believed to have a genetic component that makes some people more likely to become addicted, but environmental factors also play a significant role.[90] Thus, someone with a genetic predisposition toward alcoholism whose family and peers rarely consume alcohol is much less likely to become addicted than someone with the same genes who drinks regularly with friends.

Alcoholic liver disease

The most significant physiological effects of chronic alcohol consumption occur in the liver. The metabolism of alcohol by ADH promotes fat synthesis (see Figure 12.24), which leads to the accumulation of fat in the liver. Metabolism by the MEOS generates reactive oxygen molecules, which cause oxidation of lipids, membrane damage, and altered enzyme activities. Whether alcohol is broken down by ADH or the MEOS, toxic acetaldehyde is formed. Acetaldehyde binds to proteins and inhibits chemical reactions and mitochondrial function, allowing more acetaldehyde to accumulate and causing further liver damage.

Chronic alcohol consumption leads to three types of alcoholic liver disease. **Fatty liver** is the accumulation of fat in liver cells. It occurs in almost all people who drink heavily due to increased synthesis and deposition of fat. If drinking continues, this condition may progress to **alcoholic hepatitis**. Both of these conditions are reversible if alcohol consumption is stopped and good nutritional and health practices are followed. If alcohol consumption continues, **cirrhosis** may develop (**Figure 12.25**).

Malnutrition and other health problems

Malnutrition is one of the complications of long-term excessive alcohol consumption. Alcohol contributes energy—7 Calories/g—but few nutrients; it may replace more nutrient-dense energy sources in the diet. Alcoholic beverages are also often consumed with high-sugar mixers, which add more empty calories to the diet. In addition to decreasing nutrient intake, alcohol interferes with nutrient absorption. Alcohol causes inflammation of the stomach, pancreas, and intestine, impairing digestion of food and absorption of nutrients into the blood. Deficiency of the B vitamin thiamin is a particular concern related to chronic alcohol consumption. Alcohol also contributes to malnutrition by altering the storage, metabolism, and excretion of other vitamins and some minerals.

In addition to causing liver disease and malnutrition, heavy drinking is associated with cancer of the oral cavity, pharynx, esophagus, larynx, breast, liver, colon, rectum, and stomach.[91] Even moderate alcohol consumption has been found to increase the risk of certain cancers in women.[92] Excess alcohol use also increases the risk of hypertension, heart disease, and stroke.[93] Some of this effect is related to the fact that calories consumed as alcohol are more likely to be deposited as fat in the abdominal region, and excess abdominal fat increases the risk of high blood pressure, heart disease, and diabetes.

---

alcoholic hepatitis An inflammation of the liver caused by alcohol consumption.

cirrhosis Chronic and irreversible liver disease characterized by loss of functioning liver cells and accumulation of fibrous connective tissue.

FIGURE 12.25 **Alcoholic cirrhosis** The liver on the left is normal. The one on the right has cirrhosis. This is an irreversible condition in which fibrous deposits scar the liver and interfere with its functioning. Because the liver is the primary site of many metabolic reactions, cirrhosis is often fatal.

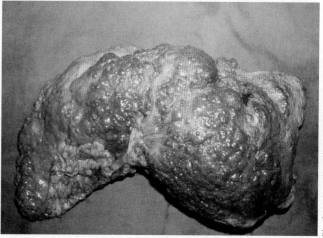

Benefits of Alcohol Consumption

For middle-aged and older adults, moderate alcohol consumption may have some benefits. Consuming alcoholic beverages can stimulate appetite, improve mood, and enhance social interactions. It can also reduce the risk of heart disease.[94] The primary mechanisms by which alcohol lowers cardiovascular risk are by raising blood levels of HDL cholesterol and by inhibiting the formation of blood clots. Although the benefits of low to moderate alcohol consumption are believed to be due to the alcohol itself, red wine is generally associated with the most benefit. The particular cardioprotective benefit of red wine is likely due not only to the alcohol but also to the phytochemicals in red wine.[95]

Whether or not the benefits of alcohol consumption outweigh the risks, drinking is a personal decision that must take into account medical and social considerations. If you do not drink, you should not begin drinking to reduce cardiovascular risk or achieve other potential health benefits. Anyone who chooses to drink should do so in moderation. Alcohol should be consumed slowly—at a rate of no more than one drink every 1.5 hours. Sipping, not gulping, allows the liver time to break down what has already been consumed. Alternating nonalcoholic and alcoholic drinks will also slow down the rate of alcohol intake and prevent dehydration. Alcohol absorption is most rapid on an empty stomach. Consuming alcohol with meals slows its absorption and may also enhance its protective effects on the cardiovascular system.

Concept Check

1. **How** much beer per day constitutes moderate drinking for a man?
2. **How** can alcohol metabolism lead to a fatty liver?
3. **What** are the symptoms of alcohol poisoning?
4. **How** does moderate alcohol intake reduce the risk of cardiovascular disease?

Summary

1 The Nutritional Health of America's Youth 372

- The diets of children and teens are generally too high in saturated fat, added sugars, and salt and are low in calcium and fiber because they do not meet recommendations for fruits, vegetables, dairy, and whole grains. An unhealthy food environment contributes to poor dietary patterns. Watching television contributes to an unhealthy food environment because it promotes snacking and poor food choices; many of the advertisements during children's programs are for foods high in calories, sodium, and sugar. Watching television also reduces the amount of exercise children get.

- Poor diets and inactivity contribute to the high incidence of obesity among children and teens, as shown in the graph. Obesity along with a poor diet and inactivity increase the risk of diabetes, high blood cholesterol, and high blood pressure.

Figure 12.4 Obesity and chronic disease among America's children

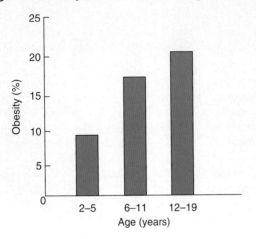

2 Nutrition for Children 376

- A child's diet should meet the needs for growth, development, and activity as well as reduce the risk of chronic disease later in life. Energy and protein needs per kilogram of body weight decrease as children grow, but total needs increase. The acceptable range of fat intake is higher for young children than for adults. Calcium, vitamin D, and iron intakes are often low in children's diets, putting them at risk for low bone density and anemia.
- To help children meet their nutrient needs and develop nutritious habits, caregivers should offer a variety of healthy foods at meals and snacks throughout the day. Children can then choose what and how much they consume. The **National School Lunch Program** provides low-cost school lunches, such as the one shown here, designed to meet nutrient needs and promote healthy diets.

Figure 12.8 Building a healthy tray

- A diet high in sugary foods can increase the risk of dental caries among children. Sugar intake has not been found to contribute to attention-deficit/hyperactivity disorder, but caffeine can contribute to hyperactivity.

- Children are at particular risk for lead toxicity. Reductions in the use of lead in paint and gasoline have decreased the incidence of high blood lead levels.

3 Nutrition for Adolescents 383

- During adolescence, the changes associated with **puberty** and the **adolescent growth spurt** have an impact on nutrient requirements. Body composition and the nutritional requirements of boys and girls diverge. Energy and protein requirements are higher during adolescence than during adulthood, and vitamin requirements increase to meet the needs of rapid growth. Calcium intake is often low in the adolescent diet, particularly if more sweetened beverages than milk are consumed, as shown. Iron deficiency anemia is common in adolescent girls due to low intake and iron losses through menstruation.

Figure 12.11 Beverage choices in children and teens

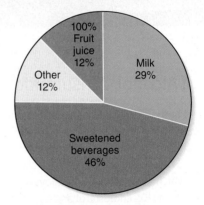

- The food choices of adolescents are usually determined more by social activities, peer pressure, and participation in athletics than by nutrient needs. Teens can improve their diets by making healthier fast-food choices. Poorly planned vegetarian diets can be low in iron, zinc, calcium, vitamin D, and vitamin B_{12} and high in saturated fat and cholesterol.
- Psychological and social changes occurring during the adolescent years make eating disorders more common than at any other time of life. Adolescent nutritional status may also be affected by weight-loss diets, use of supplements to enhance athletic performance, and the use of tobacco products or alcohol.

4 Nutrition for the Adult Years 388

- As a population, Americans are living longer but not necessarily healthier lives. Increasing **active life expectancy** is an important public health goal. The genes people inherit, as well as their diet, lifestyle, and other environmental factors, determine how long they live.
- Energy needs decrease with age, but the needs for protein, water, fiber, and most micronutrients remain the same. Decreases in nutrient intake and absorption and changes in the metabolism or absorption of certain micronutrients, including vitamin B_{12}, vitamin D, and calcium, put older adults at risk for deficiency. Iron requirements decrease in women after menopause, but many older adults are at risk for iron deficiency due to poor absorption or blood loss from disease or medications.

- Older adults are at risk for malnutrition due to the physiological changes that accompany **aging**, such as a decrease in muscle mass, illustrated here, and a decline in immune function. Chronic illnesses, which are more common in older adults than in younger people, may change nutrient requirements, decrease the appeal of food, and impair a person's ability to obtain and prepare an adequate diet. Medications to treat diseases may also affect nutritional status in older adults. Physical disabilities such as **arthritis** and **macular degeneration**, changes in mental status caused by **dementia** or depression, and social and economic factors increase the risk of **food insecurity** in older adults.

Figure 12.17b Causes and consequences of malnutrition

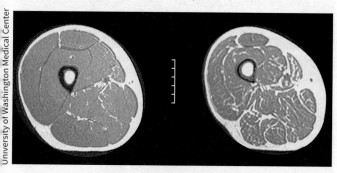

- A nutrient-dense diet and appropriate physical activity can prevent malnutrition, delay the chronic diseases associated with aging, and increase independence in older adults. The DETERMINE checklist helps identify older adults who are at risk for malnutrition. Economic or physical assistance may be required to meet nutritional needs.

5 The Impact of Alcohol Throughout Life 397

- **Alcohol**, which refers to **ethanol**, is absorbed rapidly, causing the blood alcohol level to rise and its effects to be felt almost immediately. Absorption is slowed when there is food in the stomach. Alcohol is metabolized primarily in the liver. Some is excreted in urine and exhaled in expired air.

- Moderate amounts of alcohol are broken down by the enzyme alcohol dehydrogenase (ADH). When greater amounts of alcohol are consumed, a second pathway in the liver, called the microsomal ethanol-oxidizing system (MEOS), also metabolizes alcohol. Alcohol metabolism increases fat synthesis in the liver and generates reactive oxygen molecules that can contribute to liver disease.

- In the short term, excess alcohol consumption causes **alcohol poisoning**, which interferes with brain function. Chronic alcohol consumption can lead to **alcohol use disorder**, or **alcoholism**, and can damage the liver, resulting in **fatty liver**, **alcoholic hepatitis**, and eventually **cirrhosis**, shown in the photo. Excess alcohol consumption also increases the risks of malnutrition and of developing hypertension, heart disease, and stroke, as well as certain types of cancer.

Figure 12.25 Alcoholic cirrhosis

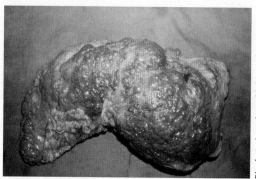

- Moderate alcohol consumption in adults can stimulate appetite, improve mood, and decrease the risk of heart disease.

Key Terms

- active life expectancy 388
- adolescent growth spurt 384
- aging 389
- alcohol 398
- alcohol poisoning 400
- alcoholic hepatitis 400
- alcohol use disorder or alcoholism 400
- Alzheimer's disease 394
- arthritis 393
- binge drinking 387
- blackout drinking 400
- cataracts 393
- cirrhosis 400
- dementia 393
- ethanol 398
- fatty liver 400
- food insecurity 394
- life expectancy 388
- life span 389
- macula 393
- macular degeneration 393
- National School Lunch Program 379
- physical frailty 393
- puberty 383

What is happening in this picture?

We all know that sugar-sweetened beverages add calories and little else to our diets, but that hasn't stopped us from buying them. Some cities have tried to change this by adding a special tax to soda, energy drinks, and other sweetened beverages.

Think Critically

1. Would an extra 2 cents per ounce influence your soda buying decisions?
2. What could you have to drink on a hot day that would not be taxed?
3. Name some potential health effects this tax would have if it did cause consumers to reduce their sweetened beverage intake.

CHAPTER 13

How Safe Is Our Food Supply?

Americans believe that the food they purchase is safe and that the facilities in which their food is prepared are sanitary and contaminant free. Of course, it would be impossible to inspect every product and every single facility every minute of every day, so sometimes food that reaches us is not safe.

The recent, widely publicized outbreaks of illness related to Chipotle Mexican Grill restaurants made consumers more aware and more cautious about the foods they choose. In these outbreaks, which occurred from Boston to Washington State, more than 200 people became ill from at least three different microorganisms. The outbreaks closed 40 restaurants, sent consumers scurrying, and caused the company's stock prices to plummet.

The outbreaks are over, but the specific foods that caused most of these illnesses have not been determined. The disease-causing microorganisms could have been present in the ingredients purchased by the restaurants, or they may have been introduced by unhygienic restaurant practices. Despite these frightening incidents, Chipotle has rebounded, we still go out to eat, and most food that arrives in our homes is safe and fit to eat. The modern U.S. food supply is perhaps the safest in human history, but danger still lurks if adequate food-handling practices are not in place from farm or feedlot to table.

CHAPTER OUTLINE

Keeping Food Safe 406
- The Role of Government
- The Role of Food Manufacturers and Retailers
- The Role of the Consumer

Pathogens in Food 410
- Bacteria
- Parasites
- Viruses
- Prions
- Molds

Preventing Microbial Food-Borne Illness 417
Thinking It Through: A Case Study on Tracking Food-Borne Illness

Agricultural and Industrial Chemicals in Food 421
- Pesticides: Risks and Benefits
Debate: Should You Go Organic?
- Industrial Contaminants
- Antibiotics and Hormones
- Choosing Wisely to Minimize Agricultural and Industrial Contaminants
What Should I Eat? Food Safety

Technology for Keeping Food Safe 427
- How Temperature Keeps Food Safe
- How Irradiation Preserves and Protects Food
- How Packaging Protects Food
- Food Additives
What a Scientist Sees: Processed Meat and Cancer

Biotechnology 432
- How Biotechnology Works
- Applications of Biotechnology
- Risks and Regulation of Biotechnology

13.1 Keeping Food Safe

LEARNING OBJECTIVES

1. **Describe** the causes of food-borne illness.
2. **Explain** why a contaminated food does not cause illness in everyone who eats it.
3. **Discuss** who is responsible for the safety of the U.S. food supply.
4. **Explain** how an HACCP system helps prevent food-borne illness.

Have you ever had food poisoning? Whether you know it or not, you probably have. Oftentimes what we call the 24-hour flu is actually food poisoning, also called **food-borne illness**. Most food-borne illness is caused by consuming food that has been contaminated by disease-causing **microbes**; occasionally, it is caused by toxic chemicals or other contaminants that find their way into food.

Whether or not you get sick from eating a contaminated food depends on how potent the contaminant is, how much of it you consume, and how often you consume it, as well as on your age, size, and health. Some food contaminants cause harm even when minute amounts are consumed, and almost any substance can be toxic if a large enough amount of it is consumed. How well a substance is absorbed and how it is metabolized by the body affect toxicity. Dietary factors and nutritional status can affect absorption. For example, mercury, which is extremely toxic, is not absorbed well if the diet is high in selenium, and lead absorption is decreased by the presence of iron and calcium in the diet. Contaminants that are stored in the body after being absorbed are more likely to be toxic because they accumulate over time, eventually causing symptoms of toxicity. Contaminants that are easily excreted from the body are less likely to cause toxicity.

An individual's size, overall health and nutritional status, and immune function affect his or her risk of food-borne illness. Infants and children are at greater risk than adults because their immune systems are immature, and their small size means that a given amount of contaminant represents a greater amount per unit of body weight than it would in an adult. Older adults, people with AIDS, and those receiving chemotherapy or other immunosuppressant drugs are at increased risk because their immune systems may be compromised. Pregnancy weakens the immune system, putting pregnant women and their unborn babies at risk. Poor nutritional status and chronic conditions such as diabetes and kidney disease may decrease the body's ability to detoxify harmful substances.

The Role of Government

Agencies at international, federal, state, and local levels monitor the safety of the food supply (**Table 13.1**). Federal agencies set standards and establish regulations for the safe handling of food and water and for the information included on food labels. They regulate the use of additives, packaging materials, and agricultural chemicals; inspect food processing and storage facilities; monitor domestic and imported foods for contamination; and investigate outbreaks of food-borne illness.

TABLE 13.1 Agencies that monitor the food supply

International organizations	
Food and Agriculture Organization of the United Nations (FAO)	Promotes and shares knowledge in all aspects of food quality and safety and in all stages of food production: harvest, postharvest handling, storage, transport, processing, and distribution.
World Health Organization (WHO)	Develops international food safety policies, food inspection programs, and standards for hygienic food preparation; promotes technologies that improve food safety and consumer education about safe food practices. Works closely with the FAO.

(continues)

---

food-borne illness An illness caused by consumption of contaminated food.

microbes Microscopic organisms, or microorganisms, including bacteria, viruses, and fungi.

TABLE 13.1 Agencies that monitor the food supply *(continued)*

Federal organizations

Agency	Role
U.S. Food and Drug Administration (FDA)	Ensures the safety and quality of all foods sold across state lines with the exception of red meat, poultry, and egg products; inspects food processing plants; inspects imported foods with the exception of red meat, poultry, and egg products; sets standards for food composition; oversees use of drugs and feed in food-producing animals; enforces regulations for food labeling, food and color additives, and food sanitation.
U.S. Department of Agriculture (USDA) Food Safety and Inspection Service (FSIS)	Enforces standards for the wholesomeness and quality of red meat, poultry, and egg products produced in the United States and imported from other countries. If an imported food is suspect, it can be tested for contamination and denied entry into the country.
U.S. Environmental Protection Agency (EPA)	Regulates pesticide levels and must approve all pesticides before they can be sold in the United States; establishes water quality standards.
National Marine Fisheries Service	Oversees the management of fisheries and fish harvesting; operates a voluntary program of inspection and grading of fish products.
National Oceanic and Atmospheric Administration (NOAA)	Oversees fish and seafood products. Its Seafood Inspection Program inspects and certifies fishing vessels, seafood processing plants, and retail facilities for compliance with federal sanitation standards.
Centers for Disease Control and Prevention (CDC)	Monitors and investigates the incidence and causes of food-borne illnesses.

State and local governments

Oversee all food within their jurisdiction; also inspect restaurants, grocery stores, and other retail food establishments, as well as dairy farms and milk processing plants, grain mills, and food manufacturing plants within local jurisdictions.

Each year, 1 in 6 Americans, or about 48 million people, get sick, 128,000 are hospitalized, and 3000 die from food-borne illnesses.[1] Media coverage of outbreaks of food-borne illness on cruise ships, deaths from *E. coli* infection, and recalls of products contaminated with *Salmonella* raise public concern about the safety of our food. The goal of current food safety policy in the United States is to prevent food-borne illness, not just react to food contamination after problems occur.[2] Because food can be contaminated anywhere in the supply chain—from where it is grown to when it is served in your home—keeping food safe requires attention at every step of food production and preparation (**Figure 13.1**).

The Role of Food Manufacturers and Retailers

The responsibility for providing safe food to the marketplace falls on the shoulders of food manufacturers, processors, and distributors. To meet this responsibility, they must establish and implement a **Hazard Analysis Critical Control Point (HACCP)** system. An HACCP system analyzes food production, processing, and transport, with the goal of identifying potential sources of contamination and points where measures can be taken to control contamination. Then, by monitoring these **critical control points**, such as **pasteurization** time and temperature, and establishing corrective actions, contamination can be prevented or eliminated (**Figure 13.2**). Unlike traditional methods of protecting the food supply, which use visual spot checks and random testing to catch contamination after it occurs, HACCP systems are designed to *prevent* contamination.

The Role of the Consumer

Although government agencies, manufacturers, and retailers are involved in creating a safe food supply, consumers also

---

Hazard Analysis Critical Control Point (HACCP) A food safety system that focuses on identifying and preventing hazards that could cause food-borne illness.

pasteurization The process of heating food products in order to kill disease-causing organisms.

408 CHAPTER 13 How Safe Is Our Food Supply?

PROCESS DIAGRAM

FIGURE 13.1 Keeping food safe from farm to table Keeping food safe involves identifying possible points of contamination along a food's journey from the farm to the dinner table and implementing controls to prevent or contain contamination.

1. **Farm:** Crops can be contaminated with bacteria before they are even harvested. Good agricultural practices help minimize contamination during growing, harvesting, sorting, packing, and storage.

2. **Processing:** Contamination of processing equipment can transfer microbes to food. To prevent contamination, processors must follow guidelines concerning cleanliness and training of workers; develop a protocol that anticipates how biological, chemical, or physical hazards are most likely to occur; and establish appropriate measures to prevent them from occurring.

3. **Transportation:** During transport, poor sanitation can contaminate food, and inadequate refrigeration can allow microbes to grow. Clean containers and vehicles, plus refrigeration, can keep food safe.

5. **Table:** Even a safe food can be contaminated in the home. Consumers can prevent food-borne illness at their table by carefully washing hands and food preparation equipment, as well as by handling, storing, and preparing food properly.

4. **Retail:** Food can become contaminated during handling or storage in grocery stores or during preparation in restaurants. The FDA's Food Code provides recommendations for the handling and service of food in an effort to help owners and employees at retail establishments prevent food-borne illness. Local health inspections ensure cleanliness and proper procedures.

Ask Yourself

Why does contamination that occurs during processing have the potential to make more people sick than contamination that occurs at home?

PROCESS DIAGRAM

FIGURE 13.2 HACCP in liquid egg production The scrambled eggs served in your cafeteria at school or work most likely came out of a carton rather than a shell. To produce this product, eggs are shelled, mixed together in large vats, pasteurized to kill microbial contaminants, packaged, and either refrigerated or frozen. A contaminated batch could sicken hundreds of people. This example shows how an HACCP system might be used to prevent contaminated eggs from reaching the consumer.

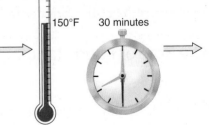

1. **Conduct a hazard analysis:** The manufacturer analyzes its processing steps for potential hazards and determines what preventive measures can be taken. Contamination of eggs with the bacterium *Salmonella* is a potential hazard. Adequate heating is a preventive measure that can eliminate this hazard.

2. **Identify the critical control points:** Critical control points are the steps in a food's processing at which the hazard can be eliminated. The critical control point in egg processing is pasteurization of the shelled egg mixture.

3. **Set critical limits:** Critical limits are the parameters that will prevent the hazard. The critical limits for egg pasteurization are sufficient heating time and temperature to ensure that *Salmonella* bacteria are killed.

(continues)

FIGURE 13.2 *(continued)*

Ask Yourself
What is the critical control point for the elimination of *Salmonella* in liquid egg production?

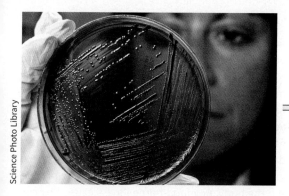
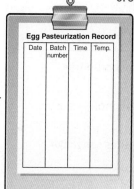

4 **Monitor the critical control points:** Procedures need to be in place to continually monitor the critical control points. Each batch of pasteurized eggs is tested for *Salmonella*. If the temperature is not hot enough or the heating is not continued long enough, *Salmonella* can survive, as shown here by the growing bacterial colonies.

5 **Establish corrective action:** If a critical limit is not met, corrective action is necessary. Batches of eggs that are contaminated with *Salmonella* are discarded, and the temperature of the heat chamber is adjusted to ensure that *Salmonella* in the next batch will be killed.

6 **Maintain record-keeping procedures:** Extensive records are kept, documenting the monitoring of critical control points and corrective actions taken. This enables the source of a problem to be traced in the event of an outbreak of food-borne illness.

7 **Institute verification procedures:** Plans and records are continuously reviewed to ensure that the HACCP plan is working and only safe eggs are reaching consumers.

TABLE 13.2 How to report food-related issues

Before reporting a suspected case of food-borne illness, get all the facts. Determine whether you have used the product as intended and according to the manufacturer's instruction. Check to see if the item is past its expiration date. After these steps have been taken, report the incident to the appropriate agency:

- **Problems related to any food except meat and poultry, including adverse reactions:** Report emergencies to the FDA's main emergency number, which is staffed 24 hours a day: 866-300-4374. Report nonemergencies to the FDA consumer complaint coordinator in your area, which you can find at www.fda.gov/Safety/ReportaProblem/.

- **Issues related to meat and poultry:** Report first to your state department of agriculture and then to the USDA Meat and Poultry Hotline (888-MPHotline or mphotline.fsis@usda.gov).

- **Restaurant food and sanitation problems:** Report directly to your local or state department of public health.

- **Issues related to alcoholic beverages:** Report to the U.S. Department of the Treasury's Bureau of Alcohol, Tobacco, Firearms, and Explosives.

- **Pesticide, air, and water pollution:** Report first to your state environmental protection department and then to the U.S. EPA.

- **Products purchased at the grocery store:** Return to the store. Grocery stores are concerned with the safety of the foods they sell, and they will take responsibility for tracking down and correcting the problem. They will either refund your money or replace the product.

---

need to assume responsibility for their food. Even a food that has been manufactured, packaged, and transported with great care can cause food-borne illness if it is not handled carefully at home. Consumers can prevent most food-borne illness through careful food handling, storage, and preparation (discussed in depth later in the chapter). They can also protect themselves and others by reporting incidents involving unsanitary, unsafe, deceptive, or mislabeled food to the appropriate agencies (**Table 13.2**).

Concept Check

1. **What** is the most common cause of food-borne illness?
2. **Why** might a contaminated food make a child sick but not affect an adult?
3. **What** agency monitors the safety of red meat, poultry, and eggs?
4. **How** does HACCP differ from traditional visual food inspection?

13.2 Pathogens in Food

LEARNING OBJECTIVES

1. **Distinguish** food-borne infection from food-borne intoxication.
2. **Discuss** three types of bacteria that commonly cause food-borne illness.
3. **Explain** how viruses, molds, and parasites can make us sick.

Most cases of food-borne illness in the United States are caused by food that has been contaminated with **pathogens**. The pathogens that most commonly affect the food supply include bacteria, viruses, molds, and parasites (**Table 13.3**). A typical case of food-borne illness causes a short bout of flulike symptoms, including abdominal pain, nausea, diarrhea, and vomiting. However, more severe symptoms, such as kidney failure, arthritis, paralysis, miscarriage, and even death, sometimes occur.

Any food-borne illness caused by pathogens that multiply in the human body is called a **food-borne infection**. Contracting a food-borne infection usually involves consumption of a large number of pathogens that infect the body or produce toxins within the body. Any food-borne illness caused by consuming a food that contains toxins produced by pathogens is referred to as **food-borne intoxication**. Even food that contains only a few pathogens can cause food-borne intoxication if the

TABLE 13.3 Summary of bacterial, viral, and parasitic food-borne illnesses[3]

Microbe	Sources	Symptoms	Onset (time after consumption)	Duration
Bacteria				
Campylobacter jejuni	Unpasteurized milk, untreated water, undercooked meat and poultry	Fever, headache, diarrhea, abdominal pain	2–5 days	2–10 days
Clostridium botulinum	Improperly canned foods, deep-dish casseroles, honey	Lassitude, weakness, vertigo, respiratory failure, paralysis	18–36 hours	10 days or longer (must administer antitoxin)
Clostridium perfringens	Fecal contamination, deep-dish casseroles	Nausea, diarrhea, abdominal pain	about 16 hours	12–24 hours
Escherichia coli O157:H7	Fecal contamination, undercooked ground beef	Abdominal pain, bloody diarrhea, kidney failure	1–9 days	2–9 days in uncomplicated cases
Listeria monocytogenes	Raw milk products; soft ripened cheeses; deli meats and cold cuts, raw and undercooked poultry; meats; raw and smoked fish; raw vegetables	Fever, headache, stiff neck, chills, nausea, vomiting. May cause spontaneous abortion or stillbirth in pregnant women and meningitis and blood infections in the fetus.	Hours to weeks	Days to weeks
Salmonella	Fecal contamination, raw or undercooked eggs and meat, especially poultry, contaminated produce	Nausea, abdominal pain, diarrhea, headache, fever	6–72 hours	4–7 days
Shigella	Fecal contamination of water or foods, especially salads such as chicken, tuna, shrimp, and potato salads	Diarrhea, abdominal pain, fever, vomiting	8–50 hours	5–7 days
Staphylococcus aureus	Human contamination from coughs and sneezes; eggs, meat, potato and macaroni salads	Severe nausea, vomiting, diarrhea	1–7 hours	hours–1 day
Vibrio parahaemolyticus	Raw seafood from contaminated water	Cramps, diarrhea, fever, nausea, vomiting	4–90 hours	2–6 days
Yersinia enterocolitica	Pork, unpasteurized milk, and oysters	Diarrhea, vomiting, fever, abdominal pain; often mistaken for appendicitis	1–11 days	A few days – 3 weeks

(continues)

pathogen A biological agent that causes disease.

TABLE 13.3 Summary of bacterial, viral, and parasitic food-borne illnesses[3] (continued)

Viruses				
Hepatitis A	Human fecal contamination of food or water, raw shellfish	Jaundice, liver inflammation, fatigue, fever, nausea, anorexia, abdominal discomfort	15–50 days	1–2 weeks to several months
Norovirus	Fecal contamination of water or foods, especially shellfish and salad ingredients	Diarrhea, nausea, vomiting	24–48 hours	12–60 hours
Parasites				
Anisakis simplex	Raw fish	Severe abdominal pain	24 hours–2 weeks	3 weeks
Cryptosporidium parvum	Fecal contamination of food or water	Severe watery diarrhea	7–10 days	2–14 days, but may become chronic
Giardia lamblia	Fecal contamination of water and uncooked foods	Diarrhea, abdominal pain, gas, anorexia, nausea, vomiting	1–2 weeks	2–6 weeks, but may be chronic
Toxoplasma gondii	Meat, primarily pork	Toxoplasmosis (can cause central nervous system disorders, flulike symptoms, and birth defects in the offspring of women exposed during pregnancy; see Chapter 11)	5–23 days	Several weeks–May become chronic
Trichinella spiralis	Undercooked pork, game meat	Muscle weakness, flulike symptoms	1–4 weeks	Several weeks

pathogens have produced enough toxin. Avoiding food-borne illness—both infection and intoxication—requires knowing how to handle and store food in ways that will prevent contamination and prevent or minimize the growth of pathogens that may already be present in the food. Even a food that is contaminated with pathogens can be safe if it is prepared in a manner that destroys any pathogens or toxins that are present.

Bacteria

Bacteria are present in the soil, on our skin, on most surfaces in our homes, and in the food we eat. Most are harmless, some are beneficial, and a few are pathogenic, causing food-borne infection or intoxication.[3] Some common causes of bacterial infections include *Salmonella*, *E. coli*, *Campylobacter jejuni*, *Listeria monocytogenes*, and *Vibrio*. *Staphylococcus aureus*, *Clostridium perfringens*, and *Clostridium botulinum* cause food-borne intoxication.

Salmonella
Salmonella is the most common cause of bacterial food-borne illness in the United States and is responsible for more deaths and hospitalizations than other food-borne pathogens.[4,5] It is found in animal and human feces and infects food through contaminated water or improper handwashing and food handling. Although poultry and eggs are the foods most often contaminated, *Salmonella* has also been found in a variety of foods ranging from peanut butter, ground meat, fruits, and vegetables to processed foods such as frozen pot pies (**Figure 13.3a** and **b**).[4] Because *Salmonella* is killed by heat, foods that are likely to be contaminated should be cooked thoroughly.

Campylobacter
Campylobacter jejuni is the leading cause of acute bacterial diarrhea in the United States, affecting about 845,000 people annually.[5] Common sources are undercooked chicken, unpasteurized milk, cheeses made from unpasteurized milk, and untreated water (**Figure 13.3b** and **c**). In 2011, *Campylobacter* was found on almost 50% of raw chicken sampled from grocery stores.[6] This organism grows slowly in cold temperatures and is killed by heat, so careful storage and thorough cooking are key to preventing infection.

Escherichia coli (*E. coli*)
E. coli is a bacterium that inhabits the gastrointestinal tracts of humans and other animals. It comes into contact with food through fecal contamination of water or unsanitary handling of food. Some strains of *E. coli* are harmless, but others can cause serious food-borne infection. One strain of *E. coli*, found in water contaminated by human or animal feces, is the cause of "travelers' diarrhea." Another strain, *E. coli* O157:H7, produces a toxin in the body that causes abdominal pain, bloody diarrhea, and in severe cases a form of kidney failure called **hemolytic-uremic syndrome**, which can be fatal.

E. coli O157:H7 entered the public spotlight in 1993, when it led to the deaths of several children who had consumed undercooked, contaminated hamburgers from a fast-food restaurant (**Figure 13.3d**).[7] Since then illness has been caused by *E. coli* contamination of alfalfa sprouts, lettuce, spinach, and green onions, as well as beef and even flour. In 2013, *E. coli* contamination of ready-to-eat salads sickened people in four states.[8] In 2015 food served at Chipotle restaurants in 11 states was linked with *E. coli* infection, and in 2017 a nine-state outbreak that sickened 16 people, 8 of whom were hospitalized, was linked to contaminated soynut butter.[9] Thorough cooking kills the bacteria; contaminated meat that is undercooked or contaminated produce that is eaten raw can cause illness (**Figure 13.3e**). In addition to concerns about *E. coli* O157:H7 in our meat and produce, a new,

NUTRITION INSIGHT

FIGURE 13.3 How bacteria contaminate our food Pathogenic bacteria can enter the food supply in a number of ways.

Pete Ryan / National Geographic Creative

a. Because poultry farms house large numbers of chickens in close proximity, one infected chicken can infect thousands of others. *Salmonella* can infect the ovaries of hens and contaminate the eggs before the shells are formed, so that the bacteria are present inside the shell when the eggs are laid. Therefore, eggs should never be eaten raw.

b. In processing plants, *Salmonella* and *Campylobacter* from infected birds can be transferred to the meat of healthy birds. Consumers should always handle raw chicken as if it contains pathogens.

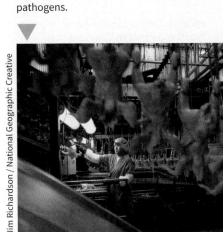
Jim Richardson / National Geographic Creative

David Arnold / National Geographic Creative

c. *Campylobacter* can be carried by healthy cattle and by flies on farms. As a result, the bacteria are commonly present in unpasteurized (raw) milk. Unpasteurized milk is also a common source of *Listeria*. During pasteurization, milk is heated to a temperature that is high enough to kill both *Campylobacter* and *Listeria*. Avoiding products made with unpasteurized milk will reduce the risk of *Campylobacter* and *Listeria* infection.

d. *E. coli* O157:H7 can live in the intestines of healthy cattle and contaminate the meat after slaughter. *E. coli*–contaminated meat that comes into contact with a grinder may contaminate hundreds of pounds of ground beef. Contaminated ground beef is a particular concern because the pathogens are mixed throughout during grinding rather than remaining on the surface, as they do on steaks and chops. The *E. coli* on the outside of the meat are quickly killed during cooking, but those in the interior survive if the meat is not cooked thoroughly.

Joel Sartore / NG Image Collection

e. *E. coli* and other bacteria can also contaminate fruits and vegetables if they are fertilized with raw or improperly composted manure or irrigated with water containing untreated sewage or manure. Produce may also be contaminated by wash water or through direct or indirect contact with cattle, deer, or sheep. Thoroughly washing fruits and vegetables can reduce the number of pathogens but does not make contaminated produce risk free.

Think Critically

Why should hamburger be cooked more thoroughly than steak to prevent *E coli* infection?

John A. Rizzo / Photodisc / Getty Images, Inc.

even more deadly strain of *E. coli* emerged in Europe in 2011, ultimately sickening more than 4000 people in 16 countries.[10]

Listeria Although food-borne *Listeria monocytogenes* sickens fewer people than *Salmonella* or *Campylobacter*, this bacterium is one of the leading causes of death from food-borne illness, with a fatality rate of 15 to 30%.[3,11] Almost all *Listeria* infections occur in people in high-risk groups, such as pregnant women, children, older adults, and people with compromised immune systems. It can cause flulike symptoms, or it can be more serious if the infection spreads through the bloodstream to the nervous system. Pregnant women are at particular risk; *Listeria* infection is about 20 times more common in pregnant women than in the general population. Infection during pregnancy is associated with an increased risk of spontaneous abortion and stillbirth, and it can be transmitted to the fetus, causing meningitis and serious blood infections.[11] *Listeria* is ubiquitous in the environment and can survive and grow at refrigerator temperatures. This bacterium can infect ready-to-eat foods such as hot dogs, lunchmeats, smoked fish, and cheeses even if they have been kept properly refrigerated. To prevent illness, ready-to-eat meats should be heated to the steaming point, and unpasteurized dairy products should be avoided.

Vibrio *Vibrio* bacteria can cause illness when contaminated seafood is eaten. *Vibrio* infection causes gastrointestinal upset, and it can be deadly in people with compromised immune systems. *Vibrio parahaemolyticus* sickens about 45,000 people per year. Another species, *Vibrio vulnificus*, infects fewer people but is more often fatal.[3] People most commonly become infected by eating raw or undercooked shellfish, particularly oysters. *Vibrio* bacteria grow in warm seawater. The incidence of *Vibrio* infection is higher during the summer months, when warm water favors growth.

Staphylococcus aureus *Staphylococcus aureus* is a common cause of bacterial food-borne intoxication. These bacteria live in human nasal passages and can be transferred to food through coughing or sneezing. They can then grow on the food, producing a toxin that causes symptoms including nausea, vomiting, diarrhea, abdominal cramps, and headache. Ham, salads, bakery products, and dairy products are common sources of *Staphylococcus aureus*.

Clostridium perfringens The bacterium *Clostridium perfringens* may cause illness by both infection and intoxication. It is often called the *cafeteria germ* because it grows in foods that are stored in large containers like those used in cafeterias. Little oxygen gets to the food at the center of a large container, thus providing an excellent growth environment for bacteria such as these, which thrive in low-oxygen environments. *C. perfringens* are difficult to kill because they form heat-resistant **spores**. Spores are a stage of bacterial life that remains dormant until environmental conditions favor their growth.

Clostridium botulinum Another strain of *Clostridium*, *Clostridium botulinum*, produces the deadliest of all bacterial food toxins. The heat-resistant spores of *C. botulinum* are found in soil, water, and the intestinal tracts of animals. The toxin is produced when the spores begin to grow and develop. When consumed, the toxin blocks nerve function, resulting in vomiting, abdominal pain, double vision, dizziness, and paralysis that leads to respiratory failure. If untreated, **botulism** is often fatal, but modern detection methods and rapid administration of antitoxin have reduced mortality rates. *C. botulinum* grows in low-oxygen, low-acid conditions, so improperly canned foods and foods such as potatoes or stew that are held in large containers where there is little exposure to oxygen provide optimal conditions for botulism spores to germinate (**Figure 13.4**).

Infant botulism is a type of botulism that is seen only in infants. Though rare, it occurs worldwide and is the most common form of botulism in the United States.[12] It is caused by ingestion of botulism spores. When ingested, the spores germinate in the infant's gastrointestinal tract, producing toxin. Some of the toxin is absorbed into the bloodstream, causing weakness, paralysis, and respiratory problems. Administration of an infant botulism antitoxin speeds recovery.[13] Only infants are affected because in adults, competing intestinal microflora prevent spores from germinating. Because honey can be contaminated with botulism spores, it should never be fed to infants under 1 year of age.

FIGURE 13.4 Canned foods and botulism Because acid prevents the germination of *Clostridium botulinum* spores, acidic canned foods such as most fruits, tomatoes, and pickles are less likely to cause botulism than low-acid canned foods, such as green beans, corn, peppers, asparagus, and mushrooms. To avoid botulism when home canning, follow canning instructions and use good hygiene. Discard bulging cans, as the bulge could indicate the presence of gas produced by the bacteria as they grow. Boiling canned foods for 10 minutes before eating destroys the toxin, but if the safety of a food is in question, it should be thrown away; even a taste of a food contaminated with botulism toxin can be deadly.

Viruses

Unlike bacteria, viruses that cause human diseases cannot grow and reproduce in foods. Human viruses can reproduce only inside human cells. They make us sick by turning our cells into virus-producing factories (**Figure 13.5**).

Norovirus   Noroviruses are a group of viruses that cause gastroenteritis, or what we commonly call "stomach flu." These viruses are the leading cause of infectious gastroenteritis among persons of all ages. Each year, they cause about 20 million people to become ill and lead to more than 55,000 hospitalizations and as many as 800 deaths.[14] Norovirus illness is contracted either by eating food that is contaminated with the virus or by touching a contaminated surface and then putting your fingers in your mouth. Shellfish can be contaminated with norovirus if the water in which they live is polluted with human or animal feces. Cooking destroys noroviruses, so water and uncooked foods such as leafy vegetables, fruits, and nuts are the most common causes of norovirus food-borne illness. Most outbreaks are caused by food contaminated during preparation and service; one study found that infected food handlers may have contributed to 82% of the outbreaks examined.[15] Norovirus infection can be spread from one infected person to another, so it spreads swiftly where many people congregate in a small area. Epidemics aboard cruise ships make headlines, but norovirus outbreaks are just as likely in nursing homes, restaurants, hotels, and dormitories as they are aboard cruise ships.

Hepatitis A   Hepatitis A is another viral infection that can be contracted from food or water that is contaminated with fecal matter. Hepatitis A infection causes liver inflammation, jaundice, fever, nausea, fatigue, and abdominal pain. The infection can require a recovery period of several months, but

PROCESS DIAGRAM   **FIGURE 13.5 How viruses make us sick**   Viruses make us sick by reproducing inside our cells. Viruses that cause food-borne illness enter the body through the gastrointestinal (GI) tract. Other types of viruses may enter the body through open cuts, the respiratory tract, or the genital tract.

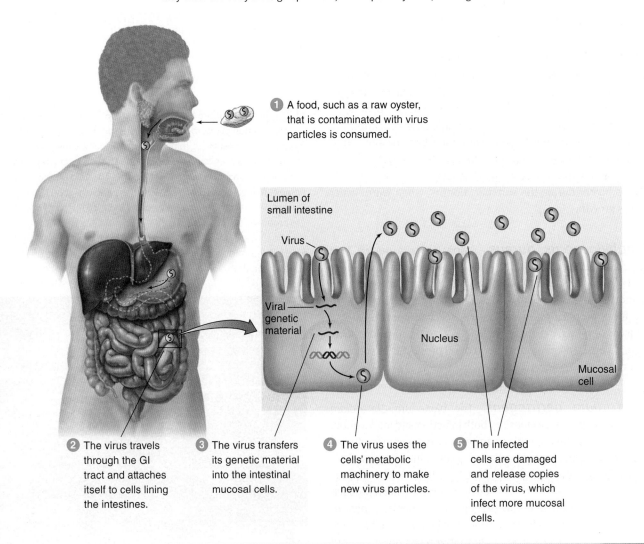

1. A food, such as a raw oyster, that is contaminated with virus particles is consumed.
2. The virus travels through the GI tract and attaches itself to cells lining the intestines.
3. The virus transfers its genetic material into the intestinal mucosal cells.
4. The virus uses the cells' metabolic machinery to make new virus particles.
5. The infected cells are damaged and release copies of the virus, which infect more mucosal cells.

FIGURE 13.6 **Mold toxin and liver cancer** The mold *Aspergillus flavus* produces *aflatoxin*, which is among the most potent carcinogens and mutagens known. The level of aflatoxin that may be present in foods in the United States is regulated, so there has never been an outbreak of illness caused by aflatoxin in the United States, but it is a problem in developing countries.

a. The filamentous growths seen in this electron micrograph belong to the mold *Aspergillus flavus*, which produces aflatoxin. This mold commonly grows on corn, rice, wheat, peanuts, almonds, walnuts, sunflower seeds, and spices such as black pepper and coriander.

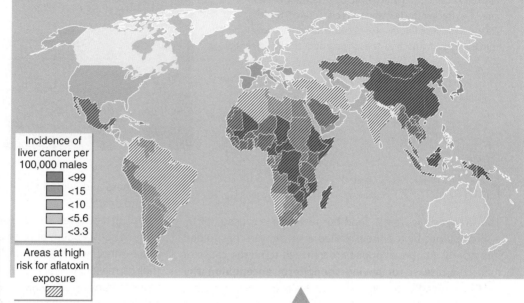

b. Exposure to aflatoxin can lead to liver cancer. Many regions with high rates of liver cancer also have high exposure to aflatoxin.[16]

Interpret the Data

Is the relationship between liver cancer and aflatoxin exposure, shown here, proof that aflatoxin causes liver cancer? Why or why not?

it does not require treatment and does not cause permanent liver damage. Hepatitis in drinking water is destroyed by chlorination. Cooking destroys the virus in food, and good sanitation can prevent its spread. A vaccine that protects against hepatitis A infection is available.

Molds

Many types of **mold** grow on foods such as bread, cheese, and fruit. Under certain conditions, these molds produce toxins (**Figure 13.6**). More than 250 different mold toxins have been identified. Cooking and freezing stop mold growth but do not destroy toxins that have already been produced. If a food is moldy, it should be discarded, the area where it was stored should be cleaned, and neighboring foods should be checked to see if they have also become contaminated.

Parasites

Some **parasites** are microscopic single-celled animals, and others are worms large enough to be seen with the naked eye.

Parasites that can be transmitted through consumption of contaminated food and water cause food-borne illness. *Giardia lamblia* is a single-celled parasite that causes diarrhea and other GI symptoms. It is often contracted by hikers who drink untreated water from streams contaminated with animal feces. *Giardia* infection is also becoming a problem in day-care centers when hands and surfaces are not thoroughly washed after diaper changes.[3] *Cryptosporidium parvum* is another single-celled parasite that causes diarrhea. It is commonly contracted from and spread by contaminated water, and the life stage of the parasite that causes infection is resistant to chlorine.[3]

Raw and undercooked pork and game meats can be a source of the parasite *Trichinella spiralis*. Once ingested, these small, wormlike organisms find their way to the muscles, where they grow, causing flulike symptoms. Fish are another common source of parasitic infections because they carry the larvae of parasites such as roundworms, flatworms, flukes, and tapeworms (**Figure 13.7**). As the popularity of eating raw fish has increased, so has the incidence of parasitic infections from fish. Parasites, including those in fish, are killed by thorough cooking. When consuming raw fish, parasitic infections can be avoided by eating fish that has been frozen.

---

mold Multicellular fungi that form filamentous branching growths.   **parasite** An organism that lives at the expense of another.

FIGURE 13.7 **Herring worm** The body cavity of this herring contains the larval form of the small roundworm *Anisakis simplex*, also called herring worm. When consumed in raw fish, these parasites burrow into the wall of the esophagus, stomach, or intestinal tract, causing *Anisakiasis*, which is characterized by severe abdominal pain.[3] The fresher the fish when it is eviscerated, the less likely it is to cause this disease because the larvae move from the fish's stomach to its flesh only after the fish dies.

Prions

The strangest, yet rarest, food-borne illness is caused not by a microbe but by a protein, called a **prion**, that has folded improperly. Abnormal prions are believed to be the cause of mad cow disease, or **bovine spongiform encephalopathy (BSE)**, a deadly degenerative neurological disease that affects cattle. The human form of this disease is **variant Creutzfeldt-Jakob Disease (vCJD)**. People are believed to contract it by eating tissue from cattle infected with BSE (**Figure 13.8**).[17] Symptoms of vCJD begin as mood swings and numbness and, within about 14 months, progress to dementia and death.

vCJD is believed to be transmitted by consumption of tissues likely to carry abnormal prions, such as the brain and nervous tissue, intestines, eyes, or tonsils of contaminated animals, but thus far meat (if free of high-risk tissue) and milk have not been found to transmit either BSE or vCJD.[3] Even though cooking does not destroy prions, the risk of acquiring vCJD is extremely small. Safeguards are in place to prevent cattle in the United States from contracting BSE. These include restrictions on the import of animals and animal products from countries where BSE has

PROCESS DIAGRAM **FIGURE 13.8** **How prions multiply** The abnormal prions that cause BSE differ from normal prions in the way these proteins are folded—that is, in their three-dimensional structure. After a person eats contaminated tissue, the improperly folded form of a prion is introduced into the brain, where it can reproduce by corrupting neighboring proteins, essentially changing their shape so that they, too, become abnormal prions. Because the abnormal prions are not degraded normally, they accumulate and form clumps called plaques. These plaques cause deadly nervous tissue damage.

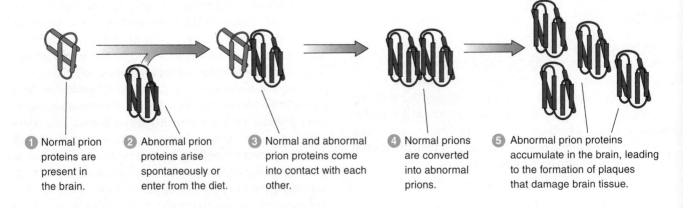

1. Normal prion proteins are present in the brain.
2. Abnormal prion proteins arise spontaneously or enter from the diet.
3. Normal and abnormal prion proteins come into contact with each other.
4. Normal prions are converted into abnormal prions.
5. Abnormal prion proteins accumulate in the brain, leading to the formation of plaques that damage brain tissue.

prion A pathogenic protein that is the cause of degenerative brain diseases called spongiform encephalopathies. *Prion* is short for *proteinaceous infectious particle*.

occurred and restrictions on what can be included in cattle feed. To prevent the transmission of vCJD to humans, cattle that can't walk or that show signs of neurological disease are banned from the food supply, and high-risk animal tissues are removed from the food supply.[18,19] On several occasions cattle with BSE have been identified in the United States, but meat from these animals did not enter the food supply. Thus far there has been no known instance of U.S. beef causing a case of vCJD.[17]

> **Concept Check**
>
> 1. **How** does *Clostridium botulinum* cause food-borne illness?
> 2. **What** pathogenic bacteria commonly contaminate chicken and eggs?
> 3. **How** do viruses make us sick?

13.3 Preventing Microbial Food-Borne Illness

LEARNING OBJECTIVES

1. **Discuss** how proper refrigeration can reduce food-borne illness.
2. **Describe** how safe handling can prevent cross-contamination.
3. **Explain** how proper cooking and storage temperatures prevent food-borne illness.
4. **List** three factors that could increase the risk of contracting food-borne illness from food served at a picnic.

The risk of microbial food-borne illness can be minimized by carefully choosing food to reduce the number of food-borne pathogens brought into the home (**Table 13.4**) and then, once in the home, storing food properly and handling and preparing it safely. Because microbes in food multiply when they are presented with the right conditions for growth, maintaining conditions that reduce growth can prevent illness.

Storing food at refrigerator or freezer temperatures either limits or stops microbial growth. Foods that are served cold should be kept cold until they are served. Frozen foods should be kept frozen and then thawed in the refrigerator or microwave before cooking, not thawed at room temperature, which favors microbial growth. Heating foods to the recommended temperature kills microbes and destroys toxins (**Figure 13.9**). Cooked food should be handled with care and kept hot until it is served, and it should be chilled quickly before storage.

Proper food handling can prevent the spread and growth of microbes. To prevent **cross-contamination**, foods that may contain live microbes should not be allowed to come into contact with cooked or ready-to-eat foods. Cooking surfaces and utensils should be clean; meats should be prepared using different utensils and cutting boards from those used for fruits and vegetables. Foods that are to be cooked may contain microbes so they should be kept separate from foods that will not be cooked. Cooked meat should never be returned to the same dish that held the raw meat, and sauces used to marinate uncooked foods should never be used as sauces on cooked food. Cooking utensils should be washed between steps in preparing food (see Figure 13.9).

Leftover cooked food should be refrigerated as soon as possible after it has been served. As a general rule, food should not be left unrefrigerated for more than 2 hours, or for more than an hour if the ambient temperature is above 90°F.[21] The temperature range that is most favorable for microbial growth is the range at which food usually sits between service and storage. Large portions of food should be divided before refrigeration so they will cool quickly. Most leftovers should be kept for only a few days. Leftovers should be thoroughly reheated to kill any microbes that have grown during storage. When in doubt about the safety of a food, throw it out.

Although much of the food-borne illness in the United States is caused by food prepared in homes, an outbreak in a commercial or institutional establishment usually involves more people and is more likely to be reported (see *Thinking It Through*). Food in retail establishments has many opportunities to be contaminated because of the large volume of food handled and the large number of people involved in its preparation. Consumers should have safety in mind when they choose restaurants. Restaurants should be clean, and cooked foods should be served hot. Cafeteria steam tables should be kept hot enough that the water is steaming and food is kept above

---

cross-contamination The transfer of contaminants from one food or object to another.

TABLE 13.4 Safe grocery choices

- Purchase food from reputable vendors.
- Choose produce that looks fresh and is not bruised or damaged.
- Choose meats that appear fresh and have been kept well refrigerated.
- Do not purchase food unless packaging is secure.
  - Select jars that are securely closed: Seals should not be broken, and safety "buttons" on jar lids should not be popped.
  - Avoid cans that are rusted, dented, or bulging.
- Choose frozen foods that are solidly frozen.
  - Select frozen foods from below the frost line in the freezer.
  - Avoid foods that contain frost or ice crystals.
- Check voluntary freshness dates and avoid foods with expired dates:[a]
  - **Sell-by or pull-by date:** Used by manufacturers to tell grocers when to remove their product from the shelves. You should buy the product before this date, but if the food has been handled and stored properly, it is usually still safe for consumption after the date. For example, milk is usually still good at least a week beyond its sell-by date if it has been properly refrigerated.
  - **Best if used by, use-by, quality assurance, or freshness date:** Used to specify the last date on which the product will retain maximum freshness, flavor, and texture. Beyond this date, the product's quality may diminish, but the food may still be safe if it has been handled and stored properly.
  - **Expiration date:** Used to specify the last day on which a product should be eaten. State governments regulate these dates for perishable items, such as milk and eggs. The FDA regulates only the expiration dates of infant formula.

Ask Yourself

Is the canned pineapple in this photo a safe choice? Why or why not?

[a] The Food Marketing Institute and the Grocery Manufacturers Association have advised food manufacturers and retailers to change current freshness and expiration labeling. The recommendation is to use two labels: the "Best if used by" date to indicate when the product should be consumed for peak flavor and the "Use by" date to indicate when perishable foods are no longer good.

140°F. Cold foods, such as salad bar items, should be kept either refrigerated or on ice to keep the food at 40°F or colder.

Picnics, potluck suppers, and other large events where food is served provide prime opportunities for bacteria to flourish because food is often left in the temperature danger zone for long periods before it is consumed (**Figure 13.10**). Whether you are at a picnic, potluck, taking your lunch to school or work, or eating at home, food should not be left unrefrigerated for more than an hour or two. When diners serve themselves at picnics and potlucks, cross-contamination from dirty hands or used plates and utensils is possible. Unlike the food in salad bars at restaurants, the food at a family picnic is not placed under a sneeze guard to prevent contamination from coughs and sneezes. All food transported out of the home or served outdoors at a picnic or other gathering should be approached with food safety in mind. Lunches should be transported to and from work or school in a cooler or an insulated bag and refrigerated upon arrival. Perishable foods that are brought home from work or school uneaten should be thrown out and not saved for another day.

NUTRITION INSIGHT

FIGURE 13.9 Safe food handling, storage, and preparation The Fight Bac! educational campaign recommends that consumers follow four steps—clean, separate, cook, and chill—to prevent food-borne illness (see www.fightbac.org).[20]

The Fight Bac! icon illustrates food-handling practices that will keep food safe from bacteria and prevent food-borne illness.

Ask Yourself

Why is a temperature between 40 and 140°F considered to be the danger zone?

Hands, countertops, cutting boards, and utensils should be washed with warm, soapy water before each step in food preparation.

Foods that are going to be cooked should not be prepared on the same surfaces as foods that are eaten raw.

FIGHT BAC!
Keep Food Safe From Bacteria

- **Clean** Wash hands and surfaces often.
- **Separate** Don't cross-contaminate.
- **Chill** Refrigerate promptly.
- **Cook** Cook to proper temperatures.

Minimum internal temperature for safety

- 165° Poultry, stuffing, casseroles, reheated leftovers
- 160° Ground meat, dishes containing egg
- 145° Beef, pork, lamb, veal roasts, steaks and chops* Fish**

*Allow meat to rest 3 minutes before carving or consuming
**Cook until flesh is opaque and separates easily with a fork

Cooked foods should be held at 140° or above.
Eggs should be cooked until the white and yolk are firm, not runny.

DANGER ZONE Temperatures in this zone allow rapid bacterial growth and production of bacterial toxins. Foods should only be allowed to remain in this temperature range for minimal amounts of time, generally less than 2 hrs. When cooling hot foods, temperature should be reduced to 40° within 2 hours

Refrigerator temperature: Stops the growth of all but a few cold-tolerant organisms

Freezer temperature: Prevents bacterial growth, but some bacteria are able to survive

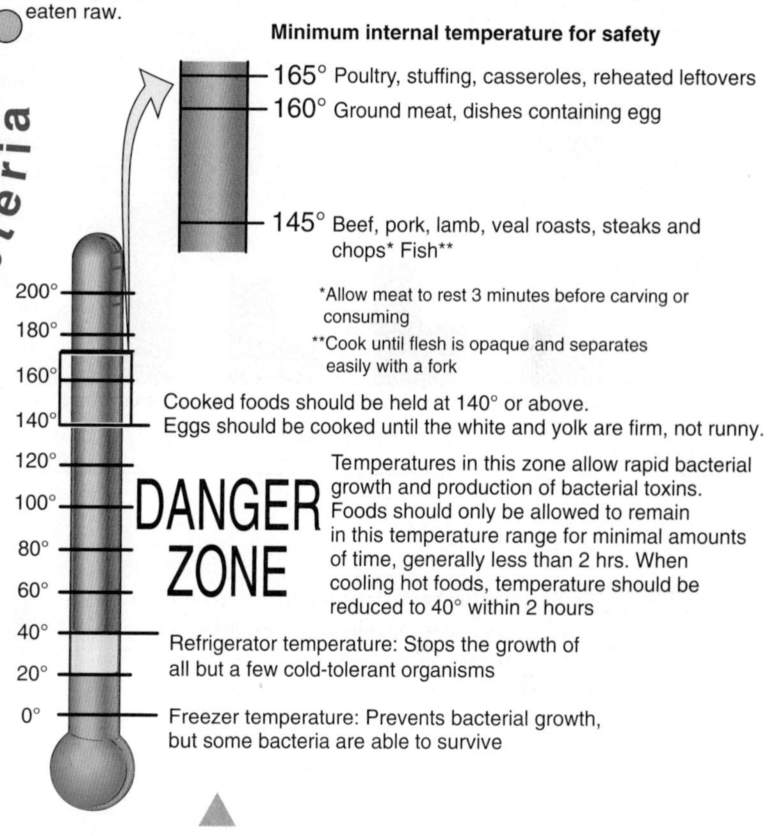

Temperature is one of the best weapons consumers have for preventing food-borne illness. Minimizing the time food temperatures remain in the danger zone—between 40 and 140°F—reduces bacterial growth and therefore the risk of food-borne illness. Use a food thermometer to make sure that the food is cooked to a safe internal temperature; color is not a good indicator of safety.

Fresh and frozen foods brought from the store should be refrigerated or frozen immediately. Fresh meat, poultry, and fish should be frozen if it will not be used within a day or two. Processed meats such as hot dogs and bologna should be refrigerated but can be kept longer than fresh meat. Freezers should be set to 0°F and refrigerators to less than 40°F.

Thinking It Through

A Case Study on Tracking Food-Borne Illness

On Friday, more than half of the 200 children enrolled at the local elementary school were either absent or went home sick sometime during the school day. They had symptoms that included nausea and vomiting, diarrhea, abdominal pain, and fever. Food-borne illness was suspected, and the local health department was notified. Inspectors were able to trace the source of the outbreak to the Spring Celebration held at the school on Thursday. For this event, the first-graders made cupcakes; the second-graders, cookies; the third-graders, fruit salad; and the fourth-graders, frozen custard. All the children were interviewed about which of these foods they had eaten that day, and the information was used to compile the following graph:

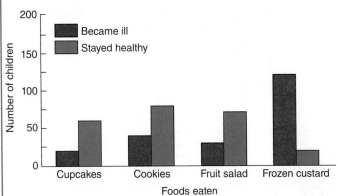

Based on the graph, which food is the most likely cause of the illness? Why?

Your answer:

The children's physicians found that their stool samples contained the bacterium *Salmonella enteritidis*. The health inspectors determined that the frozen custard recipe that the children used included raw eggs.

How can eggs become contaminated with *Salmonella*?

Your answer:

The cookie and cupcake recipes also called for eggs. Why are these unlikely to be the cause of the food-borne illness?

Your answer:

Suggest a reason 20 of the children who consumed the frozen custard did not get ill.

Your answer:

Suggest a food for next year's celebration that would be a safer choice.

Your answer:

(Check your answers in Appendix L.)

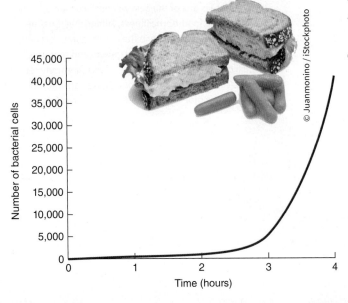

FIGURE 13.10 Exponential bacterial growth The number of bacterial cells doubles each time the cells divide, resulting in an exponential growth curve like the one shown here. If 10 bacterial cells contaminate an egg-salad sandwich during preparation and then it sits at room temperature for 4 hours, during which time the cells divide every 20 minutes, there will be 40,960 bacterial cells in the sandwich by the time you eat it.

Concept Check

1. **Why** would setting your refrigerator temperature too high increase your risk of food-borne illness?
2. **What** can be done to prevent raw chicken from contaminating a salad served at the same meal?
3. **How** does cooking prevent food-borne illness?
4. **What** factors should you consider when choosing foods that are safe to take to a picnic?

13.4 Agricultural and Industrial Chemicals in Food

LEARNING OBJECTIVES

1. **Illustrate** how contaminants move through the food chain and into our foods.
2. **Compare** the risks and benefits of using pesticides with those of growing food organically.
3. **Describe** how to minimize the risks of exposure to chemical contaminants.

Chemicals used in agricultural production and industrial wastes contaminate the environment and can find their way into the food supply. How harmful these chemicals are depends on whether they persist in the environment and whether they accumulate in the organisms that consume them or can be broken down and excreted by those organisms. Some contaminants are eliminated from the environment quickly because they are broken down by microorganisms or chemical reactions. Others remain in the environment for very long periods, and when taken up by plants and small animals, they are not metabolized or excreted. For example, when algae absorb mercury from seawater and small fish eat the algae, the mercury concentrates in the body fat of the small fish and cannot be readily excreted. When the small fish are consumed by larger fish, such as mackerel, and the mackerel are in turn eaten by still larger fish such as tuna, the mercury accumulates, reaching higher concentrations at each level of the food chain (**Figure 13.11**). This process is called

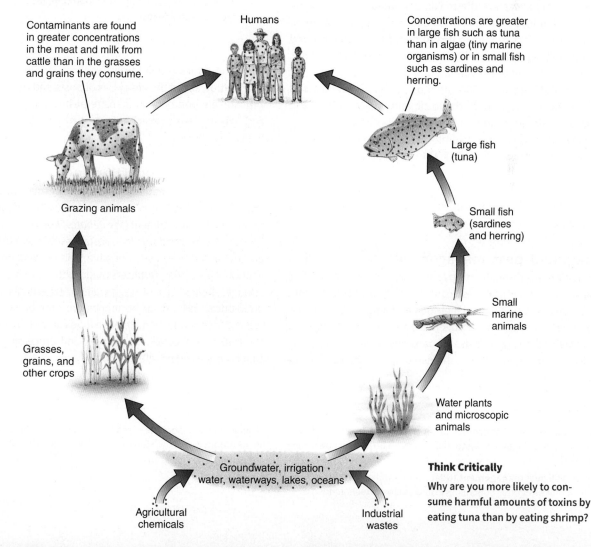

FIGURE 13.11 Contamination throughout the food chain Industrial pollutants and agricultural chemicals that contaminate the water supply enter the food chain and accumulate as they are passed through the chain. An animal at a higher level in the food chain has greater concentrations of these contaminants because it consumes all the contaminants that have been eaten by organisms at lower feeding levels.

Think Critically

Why are you more likely to consume harmful amounts of toxins by eating tuna than by eating shrimp?

bioaccumulation. Because the toxins are not eliminated from the body, the greater the amount consumed, the greater the amount present in the body.

Pesticides: Risks and Benefits

Pesticides are used to prevent plant diseases and insect infestations. They are applied both to crops in the fields and to harvested produce in order to prevent spoilage and extend shelf life. Crops that are grown using pesticides generally produce higher yields and look more appealing because they have less insect damage. Once pesticides have been applied, however, they can travel into water supplies, soil, and other parts of the environment. When pesticides enter the environment, pesticide residues are found not only on the treated plants but also in meat, poultry, fish, and dairy products (see Figure 13.11).

The potential risks of pesticides to consumers depend on the size, age, and health of the person who consumes the pesticides and on the type and amount consumed. To protect public health and the environment, the types of pesticides that may be used on food crops, the frequency of their use, and the amount of residue that may remain when foods reach consumers are regulated. The EPA approves and registers pesticides that are used in food production and establishes pesticide **tolerances**. Pesticide tolerances are the maximum amounts of pesticide residues that may remain in or on a food.[22] They are based on the potential risks to human health posed by a pesticide and are set well below the level that might cause harm to infants, children, adults, or the environment.

The FDA and the USDA monitor pesticide residues and enforce tolerances in both domestic and imported foods. In general, pesticide residue levels in the majority of domestic and imported foods have been found to be well below federally permitted limits (**Figure 13.12**).[23] Although repeated consumption of large doses of any one pesticide could be harmful, such a situation is unlikely because most people consume a variety of foods that have been produced in many different locations.

Integrated pest management One way to limit pesticide use is through **integrated pest management (IPM)**. IPM is an agricultural pest control method that combines chemical and nonchemical methods and emphasizes the use of natural toxins and more effective pesticide application. For example, increasing the use of naturally pest-resistant crop varieties that thrive without the use of pesticides can reduce

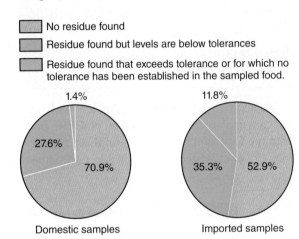

FIGURE 13.12 **Pesticide tolerances** As shown in these pie charts from an analysis of samples of domestically produced food and imported food, only a small percentage of foods exceed tolerances.[23] To further reduce pesticide risks and exposure in the United States, more effective, less toxic chemical pesticides are being developed; the use of older, more toxic products is decreasing; and production methods that result in low-pesticide and pesticide-free produce are being implemented.

Interpret the Data

What percentage of imported food samples was free of pesticide residues?

costs and do less environmental damage. IPM programs use information about the life cycles of pests and their interaction with the environment to manage pest damage economically and with the least possible hazard to people, property, and the environment.

Organic food production Organic food is produced using methods that minimize the use of synthetic pesticides and promote recycling of resources and conservation of soil and water to protect the environment (see *Debate: Should You Go Organic?*). The USDA sets standards for substances that can be used in or are prohibited from use in organic food production. Most conventional pesticides, fertilizers made with synthetic ingredients, sewage sludge, genetically modified ingredients, irradiation, antibiotics, and growth hormones are prohibited from use in organic food production. Before a food can be labeled "organic," the USDA must certify the farming and processing operations that produce and handle the food (**Figure 13.13**).

---

bioaccumulation The process by which compounds accumulate or build up in an organism faster than they can be broken down or excreted.

integrated pest management (IPM) A method of agricultural pest control that integrates nonchemical and chemical techniques.

organic food Food that is produced, processed, and handled in accordance with the standards of the USDA National Organic Program.

Debate

Should You Go Organic?

The Issue

As people become more and more concerned about food safety, nutritional health, and the environment, they are turning to organic food. Should you be choosing organic over conventionally produced food?

Organic foods account for over 4% of total U.S. food sales, and the top-selling category of organically grown food is fresh fruits and vegetables.[24] Consumers choose organic foods because they are often perceived as safer, more nutritious, and better for the environment. If safer means fewer synthetic pesticides, then organic produce is a good choice. It contains fewer pesticides and is significantly lower in nitrates than traditionally grown foods.[25] But it can be argued that there is little evidence that the current levels of pesticide exposure from conventional produce present risks to human health (see Figure 13.12). Moreover, other contamination risks are associated with organic foods. For example, manure is often used for fertilizer, and if the manure is not treated properly, it can contain pathogenic bacteria.[26]

Many consumers believe that organically produced food is not only safer but also more nutritious. *Nutritious* can mean that it contains more nutrients or is more effective at preventing nutrition-related diseases. Research data on the nutrient content of organically produced food are ambivalent. Some studies report that organically produced foods contain more nutrients than conventionally produced food, and others have found no consistent differences in nutrient content.[27] This confusion is not surprising because many factors—including growing conditions, season, the fertilizer regime, and the methods used for crop protection (for example, use of pesticides and herbicides)—affect the nutritional composition of fruits and vegetables. Nutrient content is also affected by how the food is stored, transported, and processed prior to consumption. There is little evidence that eating organic foods provides more nutritional or health benefits or disease protection than eating conventionally produced foods.[28]

What about the environment? It is hard to argue that organic farming is not better for the environment. Instead of synthetic pesticides and fertilizer, it relies on natural methods of pest control, crop rotation, compost, and cover crops to maintain the soil. The result is preservation of the soil, so crops can be grown far into the future, and reduction in the amounts of chemicals released into the environment. But organic growing still impacts the environment. As occurs with conventionally grown produce, runoff from manure can pollute waterways, and organic food, like conventional food, is often shipped long distances, using energy and generating pollution.

Is a diet based on organic foods safer, more nutritious, and better for the environment? We can assume that both conventional and organic foods sold in the United States are generally safe. Whether organic is more nutritious depends not only on individual foods but on the diet as a whole. If your choices of organic foods are limited by availability or cost, then choosing only organic may limit nutrient intake. Are organic foods better for the environment? They reduce pesticide and fertilizer use, but if organic foods are not available locally, the environmental cost of transporting them is no different than it is for conventional foods.

Think Critically

Are these organic onions that were shipped across the country a better environmental choice than conventionally grown onions from the farm across town? Why or why not?

FIGURE 13.13 Labeling organic foods Products that meet the definition of "100% organic" or "organic" may display the USDA "organic" seal shown here.

Labeling term	Meaning
100% organic	Contains 100% organically produced raw or processed ingredients.
Organic	Contains at least 95% organically produced raw or processed ingredients.
Made with organic ingredients	Contains at least 70% organically produced ingredients.

Organic farming techniques reduce farm workers' exposure to pesticides and decrease the quantity of pesticides introduced into the food supply and the environment. Organic foods, however, are not completely free of synthetic pesticides and other agricultural chemicals not approved for organic use because irrigation water, rain, and a variety of other sources can introduce trace amounts into organically grown foods. The threshold for pesticide residues in organic foods is set at 5% of the EPA's pesticide-residue tolerance.[29] Choosing organic food will reduce pesticide exposure, but it will not make your food risk free.

Industrial Contaminants

Industrial chemicals that contaminate the environment find their way into the food supply. Fish accumulate substances from the water in which they live and feed. Shellfish accumulate contaminants because they feed by passing large volumes of water through their bodies. Pollutants in the water can also contaminate crops and move through the food chain into meat and milk (see Figure 13.11).

One group of carcinogenic compounds that pollutes the environment is **polychlorinated biphenyls (PCBs)**. Prior to the 1970s, these chemicals were used in the manufacture of electrical capacitors and transformers, plasticizers, waxes, and paper. Runoff from manufacturing plants contained PCBs that contaminated water, particularly near the Great Lakes. PCBs are no longer produced, but because they do not degrade, they are still in the environment and accumulate in fish caught in contaminated waters. PCBs are a particular problem for pregnant and lactating women because prenatal exposure to PCBs and consumption of contaminated breast milk can damage the fetal and infant nervous system and cause learning deficits. Pregnant and breast-feeding women should check with their local health department for recommendations regarding fish consumption.

Other contaminants from manufacturing—such as chlordane (used to control termites); radioactive substances such as strontium-90; and toxic metals such as cadmium, lead, arsenic, and mercury—have found their way into fish and shellfish. Cadmium and lead can interfere with the absorption of other minerals. Cadmium can cause kidney damage, and lead can impair brain development. Arsenic is believed to increase the risk of cancer. Mercury, which has been found in large fish, particularly swordfish, king mackerel, tilefish, and shark, damages nerve cells (**Figure 13.14**). Because mercury is especially damaging during development, pregnant women and young children are advised to avoid certain types of fish and limit their consumption to one to three servings of fish per week, depending on the type of fish (see Chapter 11 and Appendix C).[30]

Antibiotics and Hormones

Antibiotics and hormones are administered to animals to improve health, increase growth, or otherwise enhance food production. To prevent these chemicals from being passed on to consumers, both the types of drugs used and when they can be administered are regulated, and animal tissues are monitored for drug residues.[31]

Animals are treated with antibiotics when they are sick, but for decades animals have also been given antibiotics to prevent disease and promote growth. This treatment increases the amount of meat produced and reduces costs, but if it is used improperly, antibiotic residues can remain in the meat. In addition, the overuse of antibiotics in animals can contribute to the development of antibiotic-resistant strains of bacteria. When exposed to an antibiotic, bacteria that are resistant to it survive and produce offspring that are also resistant. If these antibiotic-resistant bacteria infect humans, the resulting illness cannot be treated with that antibiotic. To avoid promoting antibiotic resistance to medications that are important for treating human disease,

FIGURE 13.14 **Mercury poisoning** This boy has Minamata disease, a neurological syndrome caused by mercury poisoning. The disease first appeared in Minamata, Japan, in 1956 and was caused by the release of mercury into the water by a local chemical factory. The mercury accumulated in the fat of fish and shellfish that were eaten by the local population.

FIGURE 13.15 **Bovine somatotropin** Consumer groups are concerned that the injection of cows with bST may cause health problems for the cows and for humans who consume the milk. The FDA has concluded that bST causes no serious long-term health problems in cows and is safe for humans. Milk from bST-treated cows does not require special labeling, but some dairies voluntarily label their products as bST free.

animals can now only be given antibiotics when approved or prescribed by a veterinarian.[32]

Hormones are used to increase weight gain in sheep and cattle and milk production in dairy cows. Before natural or synthetic hormones are approved for use, they must be shown to have the desired effect and to be safe for the treated animals, the environment, and the people who consume food from the animals.[33] Despite this, use of the synthetic, genetically engineered hormone **bovine somatotropin (bST)** has created public concern that consuming milk from bST-treated cows will cause harm to humans. Cows naturally produce bST, which stimulates milk production. Genetically engineered bST is produced by bacteria and injected into cows to further increase milk production (**Figure 13.15**). Milk from cows that have been treated with genetically engineered bST is indistinguishable from other milk.[34]

Choosing Wisely to Minimize Agricultural and Industrial Contaminants

Even though individual consumers cannot detect chemical contaminants in food, care in selection and preparation can reduce the amounts consumed. One of the easiest ways to reduce risk is to choose a wide variety of foods, thus avoiding excessive consumption of contaminants that may be present in any one food. Because chemicals accumulate through the food chain, eating lower on the food chain can minimize consumption of contaminants (see Figure 13.11).

To reduce exposure to pesticide residues in fruits and vegetables, consumers can choose organic foods or locally grown produce. Locally grown produce contains fewer pesticides because the pesticides used to prevent spoilage and extend the shelf life of shipped produce are not needed. Exposure to pesticide residues on conventionally grown produce can be minimized by washing and in some cases peeling (**Figure 13.16a**).

Intake of pesticides and other chemical pollutants from animal products can be minimized by choosing, preparing, and cooking wisely (**Figure 13.16b**). Pesticides and other toxins concentrate in fat, so intake can be reduced by trimming all fat from meat, removing the skin from poultry, and removing the skin, fatty material, and dark meat from fish. Broiling, poaching, boiling, and baking allow contaminants from the fatty portions of fish and other meats to drain out. Do

FIGURE 13.16 **Reducing exposure to pesticides and pollutants** Proper food selection and preparation can reduce exposure to pesticides and other pollutants.

a. Pesticide residues on fruits and vegetables can be removed or reduced by peeling or washing with tap water and scrubbing with a brush, if appropriate. In the case of leafy vegetables such as lettuce and cabbage, the outer leaves can be removed and discarded. Washing apples, cucumbers, eggplant, squash, and tomatoes may not remove all the pesticides because these fruits and vegetables are coated with wax in order to maintain freshness by sealing in moisture, but wax also seals in pesticides. The wax and pesticides can be removed by peeling, but removing the peel also eliminates fiber and some micronutrients.

George F. Mobley / National Geographic Creative

b. Exposure to chemical contaminants can be minimized by choosing saltwater fish caught well offshore, away from polluted coastal waters. Fish that live near the shore or spend part of their life cycle in fresh water are more likely to contain contaminants. Smaller species of fish are safer because they are lower in the food chain, and smaller fish within a species are safer because they are younger and hence have had less time to accumulate contaminants (see Figure 13.11).

What Should I Eat?

Food Safety

Avoid microbial food-borne illness

- Make sure your burger is well done.
- Skip the runny eggs and have them scrambled.
- Pass on the dough! Treat yourself to chocolate chip cookies only after they come out of the oven.
- Slice up some melon but make sure the knife, the cutting board, and the skin of the melon are clean before you slice it.

Reduce pesticides and pollutants in your food

- Buy locally grown produce.
- Look for the organic symbol.
- Try the small fish in the pond.
- Make a salad after you have washed the lettuce and peeled off the outer leaves.
- Trim the fat and don't eat the skin of poultry and fish.

 Use iProfile to compare the calories and grams of fat in a chicken breast with and without skin.

not eat the "tomalley" in lobster. The tomalley, a green paste inside the abdominal cavity of a cooked lobster, serves as the liver and pancreas and is the organ in which toxins accumulate. The analogous organ in blue crabs, called the "mustard" because of its yellow color, should also be avoided (see *What Should I Eat?*).

Concept Check

1. **How** can a pesticide used on broccoli plants end up in milk?
2. **What** are the benefits of organic foods?
3. **What** can you do to minimize PCBs in your diet?

13.5 Technology for Keeping Food Safe

LEARNING OBJECTIVES

1. **Describe** how temperature is used to prevent food spoilage.
2. **Discuss** how irradiation preserves food.
3. **Explain** how packaging protects food.
4. **Discuss** the risks and benefits of food additives.

Food spoilage occurs when the safety, taste, texture, or nutritional value of food changes as a result of either enzymes that are naturally present in the food or microbes that grow on the food. For thousands of years, humans have treated food in order to prevent spoilage. Techniques that preserve food work by destroying enzymes present in the food, by killing microbes, or by slowing microbial growth. As shown in **Table 13.5**, the acronym FAT TOM reminds us of the conditions required for microbial growth. Most food preservation techniques modify one or more of these factors to stop or slow microbial growth.

Most of the oldest methods of food preservation—heating, cooling, drying, smoking, and adding substances such as sugar or salt—are still used today. In addition, newer methods, such as irradiation and specialized packaging, have been developed. While all these technologies offer benefits, they can also create risks. Some risk arises when substances find their way into food, either accidentally or as a normal part of the production process. The FDA considers any substance that can be expected to become part of a food a **food additive** and regulates the types and amounts of food additives that may be present in a particular food. Unexpected substances that enter food are considered **accidental contaminants** and are not regulated.

How Temperature Keeps Food Safe

Cooking food is one of the oldest methods of ensuring that food is safe. Cooking kills disease-causing organisms and destroys most toxins. Other preservation techniques that rely on high temperature to kill microbes include pasteurization, sterilization, and **aseptic processing** (Figure 13.17). Lowering the temperature of food by means of refrigeration or freezing does not kill microbes but preserves the food and protects consumers because it slows or stops microbial growth.

Preservation techniques that rely on temperature benefit consumers by providing appealing, safe foods, but they are not risk free, particularly if used incorrectly. If foods are not heated long enough or to a high enough temperature, or if they are not kept cold enough, they could pose a risk of food-borne illness. In addition, some types of cooking can generate hazardous chemicals. Carcinogenic chemicals produced during the cooking of meats include **polycyclic aromatic hydrocarbons (PAHs)** and **heterocyclic amines (HCAs)**. PAHs are formed when fat drips on a grill and burns. They rise with the smoke and are deposited on the surface of the food. PAH formation can be minimized by selecting lower-fat meat and using a layer

TABLE 13.5 FAT TOM

Food	Food contains nutrients that promote bacterial growth.
Acidity	Most bacteria grow best at a pH near neutral. Some food additives, such as citric acid and ascorbic acid (vitamin C), are acids, which prevent microbial growth by lowering the pH of food.
Time	The longer a food sits at an optimum growth temperature, the more bacteria it will contain. Food should not be left in the temperature danger zone for more than 2 hours.
Temperature	The high temperatures of canning, cooking, and pasteurization kill microbes, and the low temperatures of freezing and refrigeration slow or stop microbial growth.
Oxygen	In order to grow, most bacteria need oxygen, so packaging that eliminates oxygen prevents their growth.
Moisture	Bacteria need water to grow, so preservation methods such as drying or use of high concentrations of salt or sugar, which draw water away by osmosis, prevent bacteria from growing.

food additive A substance that is either intentionally added to or can reasonably be expected to become a component of a food during processing.

accidental contaminant A substance not regulated by the FDA that unexpectedly enters the food supply.

aseptic processing The placement of sterilized food in a sterilized package using a sterile process.

FIGURE 13.17 **Aseptic processing** The juice boxes that fit so conveniently into school lunch bags are produced by aseptic processing. This technique heats foods to temperatures that result in sterilization. The sterilized foods are then placed in sterilized packages, using sterilized packaging equipment. If the package remains unopened, juice and other aseptically packaged foods can remain free of microbial growth at room temperature for years.

of aluminum foil to prevent fat from dripping on the coals. HCAs are produced by the burning of amino acids and other substances in meats and are formed during any type of high-temperature cooking. HCA formation can be reduced by precooking meat, marinating meat before cooking, cooking at lower temperatures, and reducing cooking time by using smaller pieces of meat and avoiding overcooking. The cooking temperatures recommended by the FDA are designed to prevent microbial food-borne illness and minimize the production of PAHs and HCAs. Because PAHs and HCAs are considered accidental contaminants, their amounts in food are not regulated by the FDA.

Another contaminant formed during food preparation is **acrylamide**. It is formed as a result of chemical reactions that occur during high-temperature baking or frying, particularly in carbohydrate-rich foods. The highest levels of acrylamide are found in French fries and snack chips. Smaller amounts are found in coffee and in foods made from grains, such as breakfast cereal and cookies. High doses can cause cancer and reproductive problems in experimental animals and act as neurotoxins in humans. Thus far, an association between dietary exposure to acrylamide and cancer in humans has not been consistently demonstrated.[35] Methods for reducing the amounts and potential toxicity of acrylamide in foods are being investigated.[36]

How Irradiation Preserves and Protects Food

Food **irradiation**, also called *cold pasteurization*, exposes food to high doses of X-rays, gamma radiation, or high-energy electrons in order to kill microbes and insects and inactivate enzymes that cause germination and ripening of fruits and vegetables. The FDA has approved irradiation for a variety of foods, including meat, poultry, shellfish, fruits, vegetables, flour, and spices and seasonings. (**Figure 13.18**).[37] Because irradiation produces compounds that are not present in the original foods, it is treated as a food additive, and the level of radiation that may be used is regulated. At the allowed levels of radiation, the

FIGURE 13.18 **Irradiated foods** When properly applied, irradiation increases the safety and shelf life of foods without compromising nutritional quality or changing taste, texture, or appearance.

a. Irradiated foods must be labeled with the radura symbol shown here and the statement "treated with radiation" or "treated by irradiation." This symbol is not required on the labels of foods that contain irradiated spices or other irradiated ingredients.

TREATED BY IRRADIATION

b. After two weeks in cold storage, the strawberries on the left, which were treated by irradiation, remain free of mold, whereas the untreated strawberries on the right, which were picked at the same time, are covered with mold.

irradiation A process that exposes foods to radiation in order to kill contaminating organisms and retard the ripening and spoilage of fruits and vegetables.

amounts of these compounds produced are almost negligible and have not been found to pose a risk to consumers.[38]

Food irradiation is a technique that is approved by the World Health Organization and used in more than 50 countries, yet there is still public fear and suspicion of this technology. The word *irradiation* fosters the belief that the food becomes radioactive. Opponents of food irradiation claim that it introduces carcinogens, depletes the nutritional value of food, and is used to allow the sale of previously contaminated foods. In fact, irradiated food is not radioactive, and scientific studies have found that the benefits of irradiation outweigh the potential risks.[39] Irradiation can decrease the amounts of certain nutrients in foods, but these nutrient losses are similar to those that occur with canning or refrigerated storage.[38] Because irradiation can be used in place of chemical treatments, it reduces consumers' exposure to chemical pesticides and preservatives.

How Packaging Protects Food

Packaging plays an important role in food preservation; it keeps molds and bacteria out, keeps moisture in, and protects food from physical damage. An open package of refrigerated cheddar cheese will be moldy in a few days, but an unopened package will stay fresh for weeks.

Food packaging is continually being improved. In the past two decades, for instance, consumer demand for fresh and easy-to-prepare foods has led manufacturers to offer partially cooked pasta, vegetables, seafood, fresh and cured meats, and dry products such as whole-bean and ground coffee in packaging that, if unopened, will keep perishable food fresh much longer than will conventional packaging. Vacuum packaging and **modified atmosphere packaging (MAP)** use plastics or other packaging materials that are impermeable to oxygen. In vacuum packaging, the air inside the package is removed prior to sealing in order to eliminate the oxygen. In modified atmosphere packaging, the air is flushed out and replaced with another gas, such as carbon dioxide or nitrogen. In both types of packaging, the low oxygen level prevents the growth of aerobic bacteria, slows the ripening of fruits and vegetables, and slows down oxidation reactions, which cause discoloration in fruits and vegetables and rancidity in fats.

Packaging can protect food from spoilage, but even the best packaging can introduce risk if it becomes part of the food. A variety of substances found in paper and plastic containers and packaging, and even dishes, can leach into food (**Figure 13.19**). Substances that are known to contaminate food are regulated by the EPA and the FDA. However, these regulations apply only to the intended use of the product. When a product is used improperly, substances from its packaging can migrate into food. For instance, some plastics migrate into food when heated in a microwave oven. Thus,

FIGURE 13.19 **Bisphenol A from plastics** Bisphenol A (BPA) is a chemical that has been used since the 1960s in the manufacture of many hard plastic bottles and the coating inside metal cans. The FDA has concluded that BPA is safe at the current levels occurring in foods, but to protect infants and young children, who may be more vulnerable to exposure, BPA-based materials are no longer used in baby bottles, sippy cups, and infant formula packaging.[40] To reduce BPA exposure, reduce your use of canned foods, avoid plastic containers marked with recycle codes 3 or 7, and don't microwave polycarbonate plastic food containers; over time, high temperatures may cause the polycarbonate to leach into the food.[41]

only containers designed for microwave cooking should be used for microwaving food.

Food Additives

What keeps bread from molding, gives margarine its yellow color, and keeps Parmesan cheese from clumping in the shaker? The answer to all these questions is food additives. Food additives ensure the availability of wholesome, appetizing, and affordable foods that meet consumer demands throughout the year. They are used to make food safer; maintain palatability; improve color, flavor, or texture; aid in processing; and enhance nutritional value (**Table 13.6**).[42] Many additives are familiar products such as sugar and spices. Others may sound like a chemical soup: calcium propionate in bread, disodium EDTA in kidney beans, and BHA in potato chips. Understanding what these chemicals are used for can help make the ingredient list a source of information rather than a cause for concern.

Substances that are intentionally added to foods are called **direct food additives**. Other substances that get into food

---

modified atmosphere packaging (MAP) A preservation technique used to prolong the shelf life of processed or fresh food by changing the gases surrounding the food in the package.

direct food additive A substance that is intentionally added to food. Direct food additives are regulated by the FDA.

TABLE 13.6 Common food additives[43]

Type of additive	What's on the label	What they do	Where they are used
Preservatives	Ascorbic acid, citric acid, sodium benzoate, calcium propionate, sodium erythorbate, sodium nitrite, calcium sorbate, potassium sorbate, BHA, BHT, EDTA, tocopherols	Maintain freshness; prevent spoilage caused by bacteria, molds, fungi, or yeast; slow or prevent changes in color, flavor, or texture; and delay rancidity	Jellies, beverages, baked goods, cured meats, oils and margarines, cereals, dressings, snack foods, fruits and vegetables
Sweeteners	Sucrose, glucose, fructose, sorbitol, mannitol, corn syrup, high-fructose corn syrup, saccharin, aspartame, sucralose, acesulfame potassium (acesulfame-K), neotame	Add sweetness	Beverages, baked goods, table-top sweeteners, many processed foods
Color additives	FD&C blue nos. 1 and 2, FD&C green no. 3, FD&C red nos. 3 and 40, FD&C yellow nos. 5 and 6, orange B, citrus red no. 2, annatto extract, beta-carotene, grape skin extract, cochineal extract or carmine, paprika oleoresin, caramel color, fruit and vegetable juices, saffron, colorings or color added	Prevent color loss due to exposure to light, air, temperature extremes, and moisture; enhance colors; give color to colorless and "fun" foods	Processed foods, candies, snack foods, margarine, cheese, soft drinks, jellies, puddings, and pie fillings
Flavors, spices, and flavor enhancers	Natural flavoring, artificial flavor, spices, monosodium glutamate (MSG), hydrolyzed soy protein, autolyzed yeast extract, disodium guanylate or inosinate	Add specific flavors or enhance flavors already present in foods	Many processed foods, puddings and pie fillings, gelatin mixes, cake mixes, salad dressings, candies, soft drinks, ice cream, BBQ sauce
Nutrients	Thiamine hydrochloride, riboflavin (vitamin B_2), niacin, niacinamide, folate or folic acid, beta-carotene, potassium iodide, iron or ferrous sulfate, alpha-tocopherols, ascorbic acid, vitamin D, amino acids (L-tryptophan, L-lysine, L-leucine, L-methionine)	Replace vitamins and minerals lost in processing; add nutrients that may be lacking in the diet	Flour, breads, cereals, rice, pasta, margarine, salt, milk, fruit beverages, energy bars, instant breakfast drinks
Emulsifiers	Soy lecithin, mono- and diglycerides, egg yolks, polysorbates, sorbitan monostearate	Allow smooth mixing and prevent separation; reduce stickiness; control crystallization; keep ingredients dispersed	Salad dressings, peanut butter, chocolate, margarine, frozen desserts
Stabilizers and thickeners, binders, and texturizers	Gelatin, pectin, guar gum, carrageenan, xanthan gum, whey	Produce uniform texture, improve "mouth-feel"	Frozen desserts, dairy products, cakes, pudding and gelatin mixes, dressings, jams and jellies, sauces
pH control agents and acidulants	Lactic acid, citric acid, ammonium hydroxide, sodium carbonate	Control acidity and alkalinity, prevent spoilage	Beverages, frozen desserts, chocolate, low-acid canned foods, baking powder
Leavening agents	Baking soda, monocalcium phosphate, calcium carbonate	Promote rising of baked goods	Breads and other baked goods
Anti-caking agents	Calcium silicate, iron ammonium citrate, silicon dioxide	Keep powdered foods free-flowing, prevent moisture absorption	Salt, baking powder, confectioners' sugar
Humectants	Glycerin, sorbitol	Retain moisture	Shredded coconut, marshmallows, soft candies, confections

unintentionally—such as the oil used to lubricate food processing machinery—are referred to as **indirect food additives**. The FDA regulates the amounts and types of direct and indirect food additives in food.

indirect food additive A substance that is expected to unintentionally enter food during manufacturing or from packaging. Indirect food additives are regulated by the FDA.

Regulating food additives Food additives improve food quality and help protect us from disease, but if the wrong additive is used or the wrong amount is added, it could do more harm than good. A manufacturer that wants to use a new food additive must submit to the FDA a petition that describes the chemical composition of the additive, how it is manufactured, and how it is detected in food. The manufacturer must prove that the additive will be effective for its intended purpose at the proposed levels, that it is safe for its intended use, and that its use is necessary. Additives may not be used to disguise inferior products or deceive consumers. They cannot be used if they significantly destroy nutrients or if the same effect can be achieved through sound manufacturing processes.

More than 600 chemicals defined as food additives were already in common use when legislation regulating food additives was passed. To accommodate substances that the FDA or the USDA had already determined to be safe, they were designated as **prior-sanctioned substances** and are exempt from regulation. For example, the nitrates and nitrites used to retard the growth of *Clostridium botulinum* in cured meats are prior-sanctioned substances. However, their use is still a concern because when consumed, they can form carcinogenic **nitrosamines** in the digestive tract. To minimize the risk posed by nitrosamines, the FDA limits the amount of nitrate and nitrite that can be added to food and requires the addition of antioxidants such as vitamin C, which reduce nitrosamine formation in the digestive tract. There is little evidence that nitrates and nitrites in the amounts consumed in the diet increase cancer risk.[44] However, cured and other processed meats have recently been classified as carcinogens (see *What a Scientist Sees: Processed Meat and Cancer*).[45]

A second category of additives that is not subject to food additive regulation consists of substances **generally recognized as safe (GRAS)**, based either on their history of

What a Scientist Sees

Processed Meat and Cancer

The World Health Organization has classified processed meats such as bacon, sausages, and hot dogs as carcinogens. A consumer might conclude from this that these should no longer be on the menu.

A scientist sees that there is strong scientific evidence that bacon and other processed meats are carcinogenic.[45] Because of this, they have been placed in the same carcinogenic classification group as tobacco, arsenic, and asbestos, but this does not mean that they are as dangerous. For example, smoking increases your risk of lung cancer by 2500%; that is, if you smoke, you are 25 times more likely to get lung cancer. In contrast, eating 50 grams (about 2 ounces) of processed meat, about 4 strips of bacon or one hot dog, every day increases the risk of colorectal cancer by 18%. This sounds like a lot, but the overall risk of colon cancer is low, so an increase of 18% is not that significant (see figure).

So should you pass on the bacon? You probably shouldn't have bacon and pastrami sandwiches every day, but occasionally

having bacon will not cause cancer. It can be part of a healthy dietary pattern that is based on whole grains, fruits, and vegetables with smaller amounts of lean meats and dairy products.

Think Critically

If tobacco and bacon are both carcinogens, why is smoking more of a cancer risk than eating bacon?

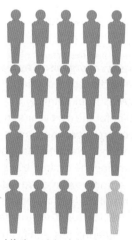

Lifetime risk of developing colon cancer is 1 in 20

18% increase →

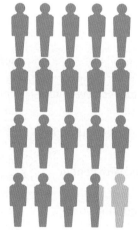

Lifetime risk of developing colon cancer if you eat 4 slice of bacon every day is about 1.2 in 20

The overall risk of developing colon cancer over a lifetime is 5%, or about 1 in 20. An 18% increase in this risk increases overall risk to only 5.9%.

generally recognized as safe (GRAS) A group of chemical additives that are considered safe, based on their long-standing presence in the food supply without harmful effects.

use in food before 1958 or on published scientific evidence. However, just because a substance is on the GRAS or prior-sanctioned list doesn't mean that it is safe or that it will stay on these lists. If new evidence suggests that a substance in either category is unsafe, the FDA may take action to require that the substance be removed from food products.

Substances that are toxic at some level of consumption may be harmless at a lower level. To ensure that additives are safe, most of those that are allowed in foods can be added only at levels 100 times below the highest level that has been shown to have no harmful effects. This is a greater margin of safety than exists for many vitamins and other naturally occurring substances.

The regulations for substances that cause cancer are far more rigid because of the **Delaney Clause**, part of the 1958 Food Additives Amendment. It states that a substance that induces cancer in either an animal species or humans, at any dosage, may not be added to food. Debate continues regarding whether the Delaney Clause should be liberalized to allow the use of substances that are added at a level so low that they would not represent a significant health risk.

Sensitivities to additives Some individuals are allergic or sensitive to certain food additives. For example, the flavor enhancer monosodium glutamate (MSG), commonly used in Chinese food, can cause adverse reactions known as *MSG symptom complex*, or *Chinese restaurant syndrome*, in sensitive individuals (see Chapter 6). Sulfites can cause symptoms ranging from stomachache and hives to severe asthma. Sulfites are used as preservatives in baked goods, canned foods, condiments, and dried fruits. Sensitive individuals can identify foods that contain sulfites by checking food labels. The forms of sulfites allowed in packaged foods include sulfur dioxide, sodium sulfite, sodium and potassium bisulfite, and sodium and potassium metabisulfite. Foods served in restaurants may also contain sulfites. For example, a potato dish served in a restaurant may be prepared using potatoes that were peeled and soaked in a sulfite solution before cooking.

Color additives can also cause adverse reactions. FD&C yellow no. 5, for instance, listed as tartrazine on medicine labels, may cause itching and hives in sensitive people. It is found in beverages, desserts, and processed vegetables. Color additives are listed in the ingredient list along with other food additives. Colors in foods are classified as certified or exempt. Certified colors are human-made, meet strict specifications for purity, and must be listed by name in the ingredient list. Colors that are exempt from certification include pigments from natural sources such as dehydrated beets and carotenoids; these may be listed collectively in the ingredient list as "artificial color."

Concept Check

1. **What** is pasteurization?
2. **How** does irradiation help extend the shelf life of fresh fruit?
3. **How** does modified atmosphere packaging prevent food spoilage?
4. **Why** are food additives regulated?

13.6 Biotechnology

LEARNING OBJECTIVES

1. **Explain** how genetic engineering introduces new traits into plants.
2. **List** ways in which genetic engineering is being used to enhance the food supply.
3. **Discuss** some potential risks associated with genetic engineering.
4. **Describe** how genetically modified foods and crops are regulated to ensure safety.

Biotechnology alters the characteristics of organisms by making selective changes in their DNA. The concept is not new. For centuries, farmers have selected seeds from plants with the most desirable characteristics to plant for the next year's crop, bred the animals that grew fastest or produced the most milk to improve the productivity of the next generation of animals, and crossbred plant varieties to combine the desired traits of each. However, these traditional methods may require many generations to produce the desired results. Biotechnology uses **genetic engineering**, or **genetic modification (GM)**, to select genes for specific traits. Genetic engineering has significantly

---

biotechnology The process of manipulating life forms via genetic engineering in order to provide desirable products for human use.

genetic engineering or **genetic modification (GM)** A set of techniques used to manipulate DNA for the purpose of changing the characteristics of an organism or creating a new product.

sped up the process of modifying the traits of organisms. Like all other new technologies, however, it may introduce new risks.

How Biotechnology Works

Genetically modified organisms (GMOs) are created through genetic engineering. To modify a plant such as corn, a piece of DNA containing the gene for a desired characteristic is taken from plant, animal, or bacterial cells and transferred to corn plant cells (**Figure 13.20**). The DNA is then referred to as **recombinant DNA** because the new DNA is a combination of the DNA from two organisms. The modified corn cells are then allowed to divide into more and more cells and eventually differentiate into the various types of cells that make up a whole corn plant. The new plant is a **transgenic** organism. Each cell in the new plant contains the transferred gene for the desired trait. This technique is used to introduce characteristics such as disease and drought resistance into plants. Genetic engineering is more difficult in animals because animal cells do not take up genes as easily as plant cells do, and making copies of these cells (clones) is also more difficult. However, these techniques have been used to produce cows that yield more milk, cattle and pigs that have more meat on them, and sheep that grow more wool.[46]

PROCESS DIAGRAM

FIGURE 13.20 **Engineering a genetically modified plant** Crops developed using the genetic engineering steps shown here are grown all over the world.

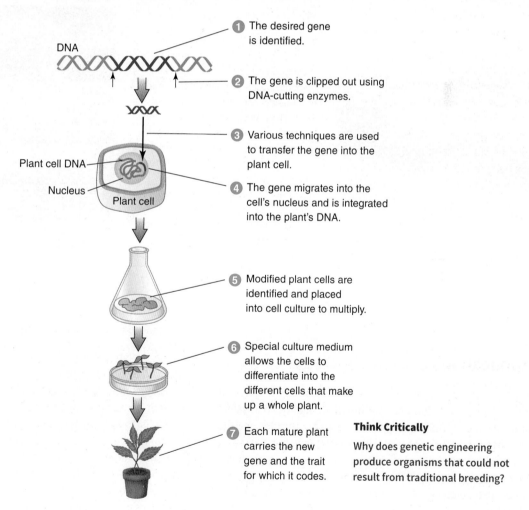

1. The desired gene is identified.
2. The gene is clipped out using DNA-cutting enzymes.
3. Various techniques are used to transfer the gene into the plant cell.
4. The gene migrates into the cell's nucleus and is integrated into the plant's DNA.
5. Modified plant cells are identified and placed into cell culture to multiply.
6. Special culture medium allows the cells to differentiate into the different cells that make up a whole plant.
7. Each mature plant carries the new gene and the trait for which it codes.

Think Critically

Why does genetic engineering produce organisms that could not result from traditional breeding?

recombinant DNA DNA that has been formed by joining DNA from different sources.

transgenic An organism with a gene or group of genes intentionally transferred from another species or breed.

NUTRITION INSIGHT

FIGURE 13.21 The potential of biotechnology Biotechnology has led to the creation of insect-resistant corn, virus-resistant papayas, and rice that is a source of β-carotene.

a. Insect-resistant corn is created by inserting a gene from the bacterium *Bacillus thuringiensis* (or Bt). The gene produces a protein called Bt toxin that is toxic to certain insects, such as this European corn borer, but safe for humans and other animals. The presence of the new gene improves the crop yield and also reduces the amounts of chemical pesticides that need to be applied.

Scott Camazine / PhotoResearchers // Getty Images, Inc.

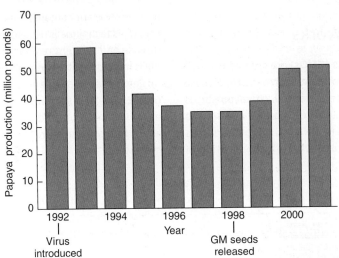

b. In the 1990s, the papaya crop in Hawaii was severely diminished by the papaya ring-spot virus (PRSV). In response, researchers developed papaya that is genetically modified to resist PRSV. The seeds were released for commercialization in 1998, allowing Hawaiian papaya production to rebound.[49] This was the first genetically enhanced fruit crop on the market. Genes for resistance to various viruses have also been used to create virus-resistant strains of potatoes, squash, cucumbers, and watermelons.[50]

c. Half the world's population depends on rice as a dietary staple, but rice is a poor source of vitamin A. Genetically modified rice, called Golden Rice (seen here compared with white rice) for the color imparted by the β-carotene pigment, has the potential to significantly increase vitamin A intake (discussed further in the *Debate* in Chapter 14). One variety contains enough provitamin A in ½ cup of cooked rice to meet the daily needs of a child.[47]

Courtesy Golden Rice Humanitarian Board / www.goldenrice.org

Applications of Biotechnology

The techniques of biotechnology can be used in a variety of ways in food production to alter quantity, quality, cost, safety, and shelf life. By making plants resistant to herbicides, insects, and various plant diseases, this technology has increased crop yields and reduced damage from insects and plant diseases (**Figure 13.21a** and **b**). By altering enzyme activity and other traits, biotechnology is being used to increase the shelf life of fresh fruits and vegetables and create products that have greater consumer appeal, such as seedless grapes and watermelons.

Biotechnology is also used in food processing. For example, many foods are produced with the help of enzymes. Rennet, an enzyme used in cheese production; enzymes used in the production of high-fructose corn syrup; and the enzyme lactase, used to reduce the lactose content of milk, are now all produced by GM microbes.

Biotechnology also has great potential for addressing the problem of world hunger and malnutrition. Although world hunger is rooted in political, economic, and cultural issues that cannot be resolved by agricultural technology alone, GM crops that target some of the major nutritional deficiencies worldwide are being developed. To address protein deficiency, varieties of corn, soybeans, and sweet potatoes with enhanced levels of essential amino acids are being developed. To address vitamin A deficiency, genes that code for the production of enzymes needed for the synthesis of the vitamin A precursor β-carotene have been inserted into rice (**Figure 13.21c**).[47] To address multiple nutrient deficiencies, the BioCassava Plus program has focused on using biotechnology to create cassava that will have higher levels of β-carotene, iron, and zinc.[48]

Risks and Regulation of Biotechnology

The rapid advancement of biotechnology during the past decade has created the potential for health problems and environmental damage. Regulations are in place to control the use of genetic engineering and GMOs. Despite these precautions, many consumers and scientists believe that the impact of this booming technology has not yet become apparent. They urge that this technology be used with caution to avoid health or environmental impacts that outweigh the benefits.

Consumer concerns Consumer safety concerns related to GM foods include the possibility that the nutrient content of a food may have been negatively affected or that an allergen or a toxin may have inadvertently been introduced into a food that was previously safe. For example, if DNA from fish or nuts—foods that commonly cause allergic reactions—were introduced into soybeans or corn, these foods would then be dangerous to individuals allergic to fish or nuts. To prevent this kind of situation from occurring unintentionally, biotechnology companies have established systems for monitoring the allergenic potential of proteins used for plant genetic engineering.

Environmental concerns An environmental concern about GM crops is that they will be used to the exclusion of other varieties, thereby reducing biodiversity. The ability of populations of organisms to adapt to new conditions, diseases, or other hazards depends on the presence of many different species and varieties that provide a diversity of genes. If farmers plant only GM insect-resistant, high-yielding crops, other species and varieties may eventually become extinct, and the genes for the traits they possess may be lost forever.

Another environmental issue is the possibility that GM crops will create "superweeds." This might occur, for example, if a trait such as increased rate of growth introduced into a domesticated plant species were passed on to a related wild species. This could produce a fast-growing weed, or superweed, that would compete with the domesticated species. As a safeguard, plant developers are avoiding introducing genes for traits that could increase a plant's competitiveness or other undesirable properties in weedy relatives.

There is also concern that crops that have been engineered to produce pesticides will promote the evolution of pesticide-resistant insects. An illustration is the case of insects that feed on plants modified to produce the Bt toxin (see Figure 13.21a). As more and more of the insects' food supply consists of plants that produce this pesticide, only insects that carry genes that make them resistant to Bt toxin survive and reproduce. This increases the number of Bt-resistant insects and therefore reduces the effectiveness of Bt toxin as a method of pest control. Although this is an important concern when growing GM crops, pesticide-resistant insects may also evolve when pesticides are sprayed on crops.

Regulation of GM food products The most common GM crops are soybeans, corn, cotton, and rapeseed (or canola) (**Figure 13.22**). Therefore, foods produced in the United States that contain corn or high-fructose corn syrup, such as many breakfast cereals, snack foods, and soft drinks; foods made with soybeans; and foods made with cottonseed and canola oils are likely to contain GM ingredients. You don't recognize these foods as genetically modified because they appear no different from other foods, and food manufacturers are not required to provide special labeling unless the food is known to pose a potential risk.

To ensure that GMOs cause no harm to consumers or to the environment, the FDA, the USDA, and the EPA are all involved in overseeing plant biotechnology. Safety and environmental issues are monitored at all stages of the process. The FDA regulates the safety and labeling of foods containing GM ingredients. Labeling of foods containing GM ingredients

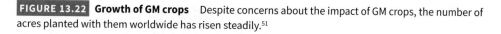

FIGURE 13.22 Growth of GM crops Despite concerns about the impact of GM crops, the number of acres planted with them worldwide has risen steadily.[51]

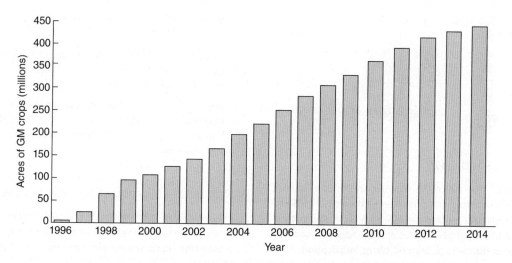

CHAPTER 13 How Safe Is Our Food Supply?

is required only if the nutritional composition of the food has been altered; if it contains potentially harmful allergens, toxins, pesticides, or herbicides; if it contains ingredients that are new to the food supply; or if it has been changed significantly enough that its traditional name no longer applies.[52] Premarket approval is required when the new food contains a substance that is not commonly found in foods or when it contains a substance that does not have a history of safe use in foods. To prevent material from a new plant variety intended for food use from inadvertently entering the food supply before its safety has been established, the FDA has asked developers to provide information about the safety of the new plants at a relatively early stage of development.

The USDA regulates agricultural products and research concerning the development of new plant varieties, including those developed through genetic engineering. The EPA regulates any pesticides that may be present in foods and sets tolerances for these pesticides. This includes GM plants containing proteins that protect them from insects or disease.

Concept Check

1. **Where** does the DNA introduced into GMOs come from?
2. **How** can biotechnology increase crop yields?
3. **Why** might a GM food cause an allergic reaction when the unmodified food does not?
4. **What** types of GM foods carry special labels?

Summary

1 Keeping Food Safe 406

- Most **food-borne illness** is caused by food contaminated with disease-causing **microbes**; occasionally, it can be caused by chemical contaminants in food. The harm caused by contaminants in the food supply depends on the type of toxin, dose, length of time over which the contaminant is consumed, how it is metabolized and excreted, and the size, age, and health of the consumer.
- The food supply is monitored for safety by food manufacturers and regulatory agencies at the international, federal, state, and local levels. Federal programs promote the use of **Hazard Analysis Critical Control Point (HACCP)** systems. HACCP systems monitor critical points in food handling and set limits on factors such as heating time and temperature, as illustrated here, to prevent and eliminate food contamination rather than catch it after it occurs. Consumers can prevent most cases of food-borne illness by following safe food-handling and preparation guidelines.

Figure 13.2 HACCP in liquid egg production

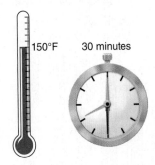

2 Pathogens in Food 410

- The **pathogens** that affect the food supply include bacteria, viruses, molds, parasites, and prions. Some bacteria cause **food-borne infection** because they are able to grow in the gastrointestinal tract when ingested. Others produce toxins in food, and consumption of the toxin causes **food-borne intoxication**.
- Viruses do not grow on food, but when consumed in food, they can reproduce in human cells, as shown here, and cause food-borne illness.

Figure 13.5 How viruses make us sick

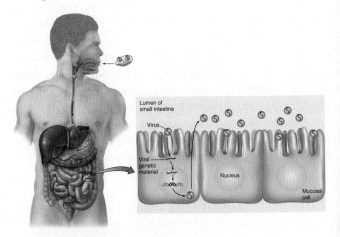

- **Molds** that grow on foods produce toxins that can harm consumers.
- **Parasites** include microscopic single-celled animals, as well as worms that can be seen with the naked eye. They are consumed in contaminated water and food.
- Improperly folded **prion** proteins cause **bovine spongiform encephalopathy (BSE)** in cattle. The risk of acquiring the human form of this deadly degenerative neurological disease is extremely low.

3 Preventing Microbial Food-Borne Illness 417

- The risk of food-borne illness can be decreased through proper food selection, storage, preparation, and cooking. These steps are emphasized by the Fight Bac! campaign, illustrated here.

Summary 437

Figure 13.9 Safe food handling, storage, and preparation

- Food should be refrigerated or frozen to slow or stop microbial growth and cooked to recommended temperatures to kill pathogens. Food should be held at temperatures that limit microbial growth. Careful food handling can prevent **cross-contamination**.

4 Agricultural and Industrial Chemicals in Food 421

- Contaminants in the environment can find their way into the food supply. Those that deposit in the fatty tissue of animals are not eliminated, leading to **bioaccumulation** as they pass through the food chain, as illustrated here.

Figure 13.11 Contamination throughout the food chain

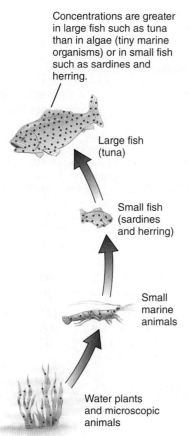

- Pesticides help increase crop yields and the quality of produce. To decrease the risk of pesticide toxicity, **tolerances** are established, and foods are monitored for pesticide residues. Safer pesticides are being developed, and U.S. farmers are reducing the amounts applied by using **integrated pest management (IPM)** and **organic food** production methods.
- Industrial pollutants such as **PCBs**, radioactive substances, and toxic metals have contaminated some waterways and the fish and shellfish that live in them. As these contaminants move through the food chain, their concentrations increase.
- Antibiotics and hormones are used to enhance animal food production. The amounts entering the food supply pose little risk, but the overuse of antibiotics contributes to the development of antibiotic-resistant strains of bacteria. Regulations now restrict the use of antibiotics in animals. Hormones can be used only if they are effective and safe for the treated animals, the environment, and consumers.
- Consumers can reduce the amounts of pesticides and other environmental contaminants in food by carefully selecting and handling produce; selecting saltwater varieties of fish caught far offshore in unpolluted waters; trimming fat from meat, poultry, and fish before cooking; and using cooking methods that allow contaminants from the fatty portions of fish and other meats to drain out.

5 Technology for Keeping Food Safe 427

- Heating foods to high temperatures, cooling them to low temperatures, and altering levels of acidity, moisture, and oxygen prevent food spoilage and lengthen shelf life by killing microbes or slowing their growth.
- **Irradiation** preserves food by exposing it to radiation. This process kills microbes, destroys insects, and slows the germination and ripening of fruits and vegetables. Irradiated foods can be identified by the radura symbol shown here.

Figure 13.18a Irradiated foods

- Packaging keeps molds and bacteria out of foods, keeps moisture in, and protects food from physical damage. Vacuum packaging and **modified atmosphere packaging (MAP)** reduce the oxygen available for microbial growth. The safety of packaging materials must be considered because components of packaging can leach into food.
- **Direct food additives** are used to preserve or enhance the appeal of food. **Indirect food additives** are substances known to find their way into food during cooking, processing, and packaging. Both are FDA regulated. **Accidental contaminants**, which enter food when it is handled or prepared incorrectly, are not regulated by the FDA.

6 Biotechnology 432

- **Biotechnology** produces **genetically modified organisms (GMOs)** by transferring a gene for a desired characteristic from one organism to another, as shown here. The result is a **transgenic** organism.

Figure 13.20 Engineering a genetically modified plant

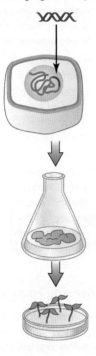

- Biotechnology has the potential to improve the quantity, safety, and quality of food, but it also has the potential to introduce allergens or toxins into foods or to negatively affect nutrient content. Environmental concerns about the use of GMOs include reduction of biologic diversity, creation of "superweeds," and evolution of pesticide-resistant insects. Regulations are in place to control the use of genetic engineering and GMOs.

Key Terms

- accidental contaminant 427
- acrylamide 428
- aseptic processing 427
- bioaccumulation 422
- biotechnology 432
- botulism 413
- bovine somatotropin (bST) 425
- bovine spongiform encephalopathy (BSE) 416
- critical control point 407
- cross-contamination 417
- Delaney Clause 432
- direct food additive 429
- food additive 427
- food-borne illness 406
- food-borne infection 410
- food-borne intoxication 410
- generally recognized as safe (GRAS) 431
- genetic engineering or genetic modification (GM) 432
- genetically modified organism (GMO) 433
- Hazard Analysis Critical Control Point (HACCP) 407
- hemolytic-uremic syndrome 411
- heterocyclic amines (HCAs) 427
- indirect food additive 430
- infant botulism 413
- integrated pest management (IPM) 422
- irradiation 428
- microbes 406
- modified atmosphere packaging (MAP) 429
- mold 415
- nitrosamine 431
- organic food 422
- parasite 415
- pasteurization 407
- pathogen 410
- polychlorinated biphenyls (PCBs) 424
- polycyclic aromatic hydrocarbons (PAHs) 427
- prion 416
- prior-sanctioned substance 431
- recombinant DNA 433
- spore 413
- tolerance 422
- transgenic 433
- variant Creutzfeldt-Jakob Disease (vCJD) 416

What is happening in this picture?

The Non-GMO Project label on this product indicates that it does not contain any genetically modified ingredients. Whether GMO labeling is voluntary, as in this case, or mandatory, designing an appropriate label is a complex issue. Simple labels such as this one may imply that the product is better than other products. However, a more detailed label that describes how a particular product was modified would be too lengthy and would not be understood by many consumers.

Think Critically

1. Do you think this is a safer choice than a similar product that includes GM ingredients? Why?
2. Should the FDA require labeling on all foods containing GM ingredients?
3. Why might food companies oppose this type of labeling?

CHAPTER 14

Feeding the World

Fantasists often consider how growing populations will face a world of limited food supplies. In the satirical 1973 film *Soylent Green*, which depicts a world of overpopulation and mass hunger, the titular line of dialogue, "Soylent Green is people!!!" reveals that the government-funded all-purpose food product is made from human remains. *Soylent Green* portrays an outrageous future, but it feels frighteningly possible.

Even though worldwide food shortage has been averted thus far, the production, distribution, safety, and sustainability of the world's food and water resources remain matters of great concern. The number of hungry people in the world exceeds the populations of the United States, Canada, and the European Union combined. Couple that with the millions of people who may have access to food yet continue to struggle to find safe, clean, and healthy options due to either a lack of access or class restrictions, and we have a legitimate global health crisis on our hands.

Thankfully, unlike in the reality presented in *Soylent Green*, most major governments actually *do* have the best interests of their people in mind when it comes to food production and distribution. But individual and national self-interest have often stymied more altruistic, public efforts to provide the less fortunate throughout the world with access to a more substantial, nutritious diet. Global, national, and local economic and political interests have the power to alleviate or exacerbate world nutrition conditions.

CHAPTER OUTLINE

The Two Faces of Malnutrition 441
- The Impact of Undernutrition
- Overnutrition: A World Issue
- Why Do Undernutrition and Overnutrition Exist Side by Side?

Causes of Hunger Around the World 443
- Food Shortages
- Poor-Quality Diets

Causes of Hunger in the United States 448
- Poverty and Food Insecurity
- Vulnerable Stages of Life

Eliminating World Hunger 450
- Providing Short-Term Food Aid
- Controlling Population Growth
- Increasing Food Production While Protecting the Environment
Thinking It Through: A Case Study on What One Person Can Do
- Increasing Food Availability Through Economic Development and Trade
- Ensuring a Nutritious Food Supply
Debate: Combating Vitamin A Deficiency with Golden Rice

Eliminating Food Insecurity in the United States 457
- The Nutrition Safety Net
What a Scientist Sees: The Cost of Wasted Food
- Nutrition Education
What Should I Eat? Make Your Meals Green

14.1 The Two Faces of Malnutrition

LEARNING OBJECTIVES

1. **Explain** what is meant by the two faces of malnutrition.
2. **Describe** the impact of undernutrition throughout the life cycle.
3. **Discuss** how nutrition transition affects dietary patterns and disease incidence.

For most of us, the image that comes to mind when we think of malnutrition around the world involves undernutrition: **hunger** and **starvation**. This image is certainly valid because about 795 million people around the world are chronically undernourished, and 45% of deaths in children under age 5 are related to undernutrition.[1,2] At the same time that health organizations are struggling with issues of undernutrition, however, rates of illness related to overnutrition are soaring.[3] The overweight and the undernourished both suffer from malnutrition and experience high levels of sickness and disability, shorter life expectancies, and reduced levels of productivity. These two faces of malnutrition complicate the goal of solving the problem of malnutrition worldwide.

The Impact of Undernutrition

In populations where hunger is a chronic problem, there is a **cycle of undernutrition** (**Figure 14.1a**). The cycle begins when women consume a nutrient-deficient diet during pregnancy. These women are more likely than others to give birth to low-birth-weight infants who are susceptible to illness and early death. The **infant mortality rate** and the number of low-birth-weight infants are indicators of a population's health and nutritional status (**Figure 14.1b**). In most industrialized countries, the infant mortality rate is less than 5 per 1000 live births; in developing countries, the rate is often over 50 per 1000 live births.[4] Low-birth-weight infants who survive require extra nutrients, which usually are not available. Malnutrition in infancy and childhood has a profound effect on growth and development

---

NUTRITION INSIGHT

FIGURE 14.1 The cycle of undernutrition Undernutrition impairs the health of individuals at every stage of life. Infants and children are at particular risk.

a. Undernutrition often begins in the womb, continues through infancy and childhood, and extends into adolescence and adulthood. Interruption of this cycle of undernutrition at any point can benefit both the individuals affected and their society.

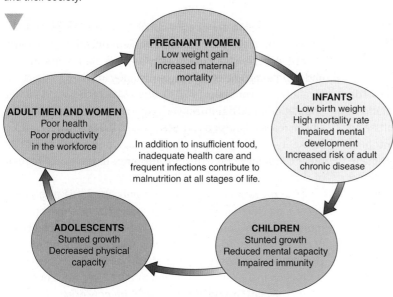

b. Low-birth-weight infants are at increased risk for complications, illness, and early death. Survivors often suffer lifelong physical and cognitive disabilities. A higher number of low-birth-weight infants means a higher infant mortality rate. Every year, more than 20 million low-birth-weight babies are born, 95% of them in developing countries.[5]

Karen Kasmauski / National Geographic Creative

(continues)

---

hunger Recurrent involuntary lack of food that over time may lead to malnutrition.

starvation A severe reduction in nutrient and energy intake that impairs health and eventually causes death.

infant mortality rate The number of deaths during the first year of life per 1000 live births.

FIGURE 14.1 *(continued)*

c. Well over half of all deaths in children under 5 years of age are due to infectious disease.[6] The rate of mortality from infections is increased among undernourished children. It is estimated that 45% of deaths in children under age 5 occur due to the presence of undernutrition.[2] Immunizations against infectious disease are less effective in undernourished children because their immune systems cannot respond normally.

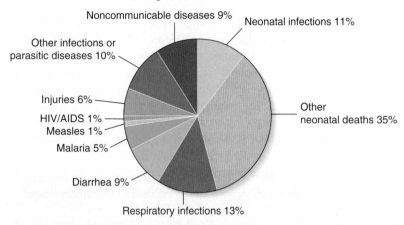

Interpret the Data

Undernutrition increases susceptibility to infections. What percentage of all deaths in children under age 5 is due to infectious diseases?

a. 9%   c. 50%
b. 17%   d. 63%

d. A nurse checks the growth of this child in Indonesia. Worldwide, stunting affects about one in four children under age 5.[7] Deficiencies of energy, protein, vitamin A, iron, and zinc, as well as prolonged infections, have been implicated as causes.

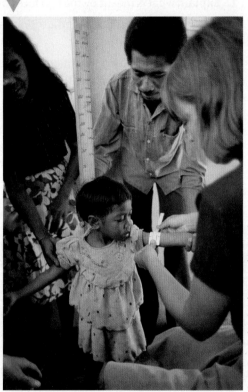

as well as on susceptibility to infectious disease. Infectious diseases are more common in undernourished children, and undernourished children may die of infectious diseases that would not be life threatening in well-nourished children (**Figure 14.1c**).

Undernutrition in children causes **stunting** (**Figure 14.1d**), which is an indicator of the nutritional health of a population's children. Stunting in childhood produces smaller adults who have a reduced work capacity and are unable to contribute optimally to their society's economic and social development. Stunted, undernourished women are more likely than others to give birth to low-birth-weight babies. In addition, those who have experienced lower birth weight and early-childhood stunting are more likely to develop diabetes and cardiovascular disease later in life.[8]

Overnutrition: A World Issue

For the first time in history, the number of overnourished people exceeds the number of undernourished people. Around the world, approximately 1.9 billion adults are overweight, and of these, 600 million are obese—about 13% of the world's adult population.[9] Once considered a problem only in high-income countries, overweight and obesity are now on the rise in low- and middle-income countries, particularly in urban settings.[10] If recent trends continue, by 2030, up to 51% of the world's adult population could be either overweight or obese.[11] Because obesity increases the risk of cardiovascular disease, hypertension, stroke, type 2 diabetes, certain cancers, and arthritis, among other conditions, it is a major contributor to the global burden of chronic disease and disability.

The prevalence of overweight and obesity is also growing among children worldwide. According to recent estimates, more than 41 million children under age 5 are overweight. Many of these children live in developing countries; the number of children who are overweight or obese has nearly doubled from 5.4 million in 1990 to 10.6 million in 2014.[9] In some countries, a high prevalence of overweight children now exists alongside a high prevalence of undernourished children. Overweight and obese children are likely to stay obese into adulthood and are more likely to develop diseases such as diabetes and cardiovascular disease at a younger age.

---

stunting A low height for age.

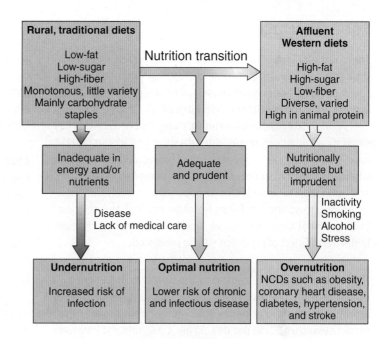

FIGURE 14.2 Nutrition transition This schematic shows the dietary changes and nutritional consequences of nutrition transition.[12] The traditional rural diet is often inadequate in energy, protein, or micronutrients. The affluent Western diet meets nutrient needs but increases the risk of nutrition-related noncommunicable diseases (NCDs) such as obesity, heart disease, and diabetes. A diet that falls somewhere between these extremes is optimal for health.

Why Do Undernutrition and Overnutrition Exist Side by Side?

We see the problems of undernutrition and overnutrition existing side by side because diets and lifestyles change as economic conditions improve. Traditional diets in developing countries are based on a limited number of foods—primarily starchy grains and root vegetables. As incomes increase and food availability improves, the diet becomes more varied, and energy intake increases with the addition of meat, milk, and other more calorie-dense foods (**Figure 14.2**). Along with these dietary changes, there is a decrease in activity due to occupations that are less physically demanding, greater access to transportation, more labor-saving technology, and more passive leisure time.

Some of the effects of this **nutrition transition** are positive: Life expectancy increases, and the frequencies of low birth weight, stunting, infectious diseases, and nutrient deficiencies decrease. However, as countries develop economically, there is an increase in the incidence of nutrition-related NCDs, such as obesity, heart disease, and diabetes (see Figure 14.2).[10] Three-quarters of the deaths from these diseases occur in low- and middle-income countries. Transition to a diet high in animal protein and refined foods also increases the use of natural resources and in the long term may deplete nonrenewable resources.

Concept Check

1. **How** has the prevalence of overnutrition changed in developing countries over the past few decades?
2. **What** is the impact of stunting on the health and productivity of a population?
3. **How** does nutrition transition affect the intake of animal products?

14.2 Causes of Hunger Around the World

LEARNING OBJECTIVES

1. **Explain** the concept of food insecurity.
2. **Discuss** the factors that cause food shortages for populations and individuals.
3. **Describe** the impact of nutrient deficiencies throughout the life cycle.

The specific reasons for hunger and **food insecurity** vary with time and location, but the underlying cause is that the available food is not distributed equitably. This inequitable distribution results in either a shortage of food or the wrong combination of foods to meet nutrient needs. This situation, in turn, results in protein-energy malnutrition and individual nutrient deficiencies.

nutrition transition A series of changes in diet, physical activity, health, and nutrition that occur as poor countries become more prosperous.

food insecurity A situation in which people lack adequate physical, social, or economic access to sufficient, safe, nutritious food that meets their dietary needs and food preferences for an active and healthy life.

Food Shortages

The most obvious example of a food shortage is **famine**. Drought, floods, earthquakes, and crop destruction due to diseases or pests are natural causes of famines. Human causes include wars and civil conflicts (**Figure 14.3**).

Food shortages due to famine are very visible because they cause many deaths in an area during a short period, but chronic food shortages take a greater toll. Chronic shortages occur when economic inequities result in lack of money, health care, and education for individuals or populations; when the population outgrows the food supply; when cultural and religious practices limit food choices; or when environmental damage limits the amount of food that can be produced.

Poverty Poverty is central to the problem of hunger and undernutrition (**Figure 14.4**). Today about 11% of people around the world live below the international poverty line, earning less than $1.90 per day.[13] This is a significant improvement from 1990, when 35% of the world lived below the poverty line. In addition to creating food insecurity, poverty reduces access to health care, increasing the prevalence of disease and disability. When diseases go untreated, nutrient needs are increased, a situation that further limits the ability to obtain an adequate diet and contributes to malnutrition.

FIGURE 14.3 Famine These young Somali girls are in line to get porridge at a feeding center. More than 20 million people in Somalia, Yemen, Sudan, and Nigeria face famine as the result of both natural and human-made disasters: Drought has killed crops and livestock, and the ongoing civil conflict has reduced access to food.[14] Regions that have barely enough food to survive under normal conditions are vulnerable to famine. This situation is analogous to a man standing in water up to his nostrils: If all is calm, he can breathe, but if there is a ripple, he will drown. A ripple such as a natural or civil disaster reduces the margin of survival and creates famine.

AMISOM / Alamy Stock Photo

Those who are poor also have less access to education, and this lack of access contributes to undernutrition and disease and reduces opportunities to escape poverty. Lack of education about food preparation and storage can affect food safety and the health of the household: Unsanitary food preparation increases the incidence of gastrointestinal diseases, which contribute to malnutrition.

Overpopulation Overpopulation exists when a region has more people than its natural resources can support. A fertile river valley can support more people per acre than can a desert environment. But even in fertile regions of the world, if the number of people increases excessively, resources are overwhelmed, and food shortages occur. At present, enough food is produced throughout the world to prevent hunger if that food is distributed equitably, but demand is rising. The human population is currently growing at a rate of more than 80 million persons per year (**Figure 14.5**).[15] This rate of growth could eventually outstrip the planet's ability to produce enough food to nourish the world's population.

Cultural practices In some cultures, access to food may be limited for certain individuals in households. For example, because they are viewed as less important, women and girls may receive less food than men and boys. How much food is available to an individual within a household depends on gender, control of income, education, age, birth order, and genetic endowments.

The cultural acceptability or unacceptability of foods also contributes to food shortages and malnutrition. If available foods are culturally unacceptable, a food shortage exists unless the population can be educated to accept the new food. For example, insects are eaten in some cultures and are an excellent source of protein, but in other cultures they are unacceptable as food.

Limited environmental resources The land and other resources available to produce food are limited. Some resources, such as minerals and fossil fuels, are present in finite amounts and are nonrenewable—that is, once they have been used, they cannot be replaced within a reasonable amount of time. Others, such as soil and water, are **renewable resources** because they will be available indefinitely if they are used at a rate at which the Earth can restore them. For example, when agricultural land is used wisely—that is, when crops are rotated, erosion is prevented, and contamination is limited—it can be reused almost endlessly. However, if land is not used carefully, damage caused by soil erosion, nutrient depletion, and accumulation of pollutants may reduce the amount of usable land over the long term.

---

famine A widespread lack of access to food due to a disaster that causes a collapse in food production and marketing systems.

renewable resource A resource that is restored and replaced by natural processes and can therefore be used forever.

Causes of Hunger Around the World **445**

NUTRITION INSIGHT **FIGURE 14.4** **The impact of poverty** Poverty increases malnutrition not only by limiting the availability of food but also by reducing access to health care and education.

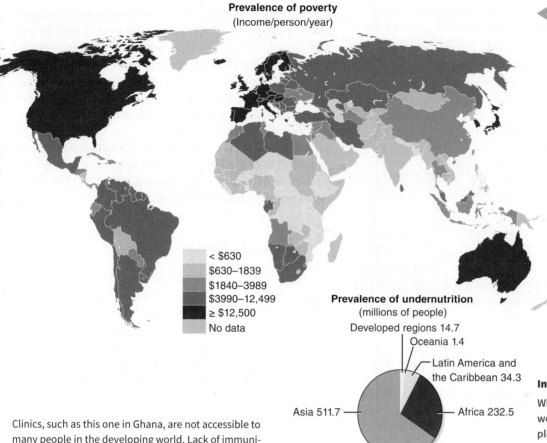

Prevalence of poverty
(Income/person/year)

< $630
$630–1839
$1840–3989
$3990–12,499
≥ $12,500
No data

Prevalence of undernutrition
(millions of people)

Developed regions 14.7
Oceania 1.4
Latin America and the Caribbean 34.3
Africa 232.5
Asia 511.7

Total = 795 million people

The regions of the world where poverty is the most prevalent (see map) correspond to the regions where there are the greatest numbers of undernourished people (see chart). Most extreme poverty is in developing countries, where 98% of the world's undernourished people live.[1,13] In wealthy countries, hungry people can usually obtain help to get food or money to buy food, but in poor countries, a family that cannot grow enough food or earn enough money to buy food may have nowhere to turn for help.

Interpret the Data

What percentage of the world's undernourished people live in Africa?

a. 29%   c. 62.5%
b. 50%   d. 4%

Clinics, such as this one in Ghana, are not accessible to many people in the developing world. Lack of immunizations and treatment for infections and other illnesses results in an increase in infectious disease and a decrease in survival rates from chronic diseases such as cancer. Lack of health care also increases infant mortality and the incidence of low-birth-weight births.

These young Indonesian girls are working in a textile factory rather than going to school. Lack of education prevents people from escaping poverty and contributes to undernutrition and disease because it leads to inadequate care for infants, children, and pregnant women.

Ask Yourself

How can lack of health care increase the incidence of undernutrition?

FIGURE 14.5 **World population growth** Most of the world's population growth is concentrated in the poorest countries.[16] Rapid population growth prevents developing countries from escaping poverty because their economies cannot keep pace. Efforts to produce enough food can damage the soil and deplete environmental resources, further reducing the capacity to produce food in the future.

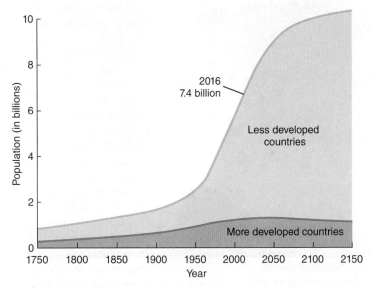

Modern mechanized agricultural methods have increased food production but use more energy and resources than more traditional labor-intensive farming. Large-scale farming can erode the soil and deplete its nutrients. Fertilizers and pesticides can contaminate groundwater and eventually pollute waterways. And if a product is shipped over long distances, requires refrigeration or freezing, or needs other types of processing, the environmental costs are increased even more.

Western food consumption patterns, which include more animal products than traditional diets, are also affecting our ability to feed the world's population.[17] In general, the environmental cost of producing plant-based foods is lower than that of producing animal-based foods.[18] Increases in animal production increase the use of land for growing grain to feed the animals. Raising cattle also creates both air and water pollution. The animals themselves produce methane, a greenhouse gas, in their gastrointestinal tracts. Large-scale "factory farming" makes the problem worse because more methane is produced when animal sewage is stored in ponds and heaps. In fact, livestock are responsible for a larger percentage of greenhouse gas emissions than all the cars in the world combined. It is not only the resources of the land that are at risk. Population growth has increased the demand for fish to the point that the Earth's oceans are being depleted (**Figure 14.6**).

Poor-Quality Diets

Even when there is enough food, malnutrition can occur if the quality of the diet is poor. The typical diet in developing countries is based on high-fiber grain products or root vegetables and has little variety. Adults who are able to consume a relatively large amount of this diet may be able to meet their nutrient needs. But individuals with high nutrient needs because they are ill or pregnant and those with limited capacity to consume this bulky grain diet, such as children and elderly individuals, are at risk for nutrient deficiencies. Deficiencies of protein, iron, iodine, and vitamin A are common with poor-quality diets (**Figure 14.7**).[19]

In addition to deficiencies of protein, iron, iodine, and vitamin A, deficiencies of folate, zinc, selenium, and calcium also cause problems throughout the world. Folate deficiency causes macrocytic anemia and during pregnancy increases the risk of neural tube defects. It has been associated with low

FIGURE 14.6 **Environmental impact on the oceans** Overfishing has severely reduced the numbers of many marine species. Pollution also threatens the world's fishing grounds. Oil spills and deliberate dumping occur offshore, and sewage, pesticides, organic pollutants, and sediments from erosion wash into coastal waters, where most fish spend at least part of their lives. Even aquaculture, designed to increase fish production, produces wastes that can pollute ocean water and harm other marine organisms.

FIGURE 14.7 Protein and micronutrient deficiencies
Deficiencies of protein and energy as well as iron, iodine, and vitamin A, are common throughout the developing world.

Protein-energy malnutrition is most common in children. When there is a general lack of food, the wasting associated with *marasmus* results, and when the diet is limited to starchy grains and vegetables, *kwashiorkor*, characterized by a bloated belly, can predominate (see Chapter 6). Other factors, such as metabolic changes caused by infection, may also play a role in the development of kwashiorkor.

Think Critically

Why are children who consume a starchy, low-protein diet less likely to meet their protein needs than adults consuming the same diet?

Marasmus

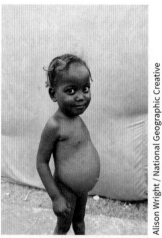
Kwashiorkor

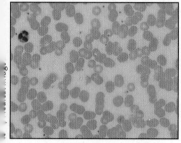
Normal red blood cells

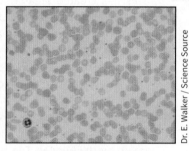
Iron deficiency anemia

Over 30% of the world's population suffers from iron deficiency anemia, which is characterized by small, pale red blood cells (see Chapter 8).[24] The lack of iron reduces the amount of hemoglobin produced, and the lack of hemoglobin lowers the blood's ability to deliver oxygen. In developing countries, intestinal parasites, which cause gastrointestinal blood loss, and acute and chronic infections, such as malaria, increase the risk and severity of dietary iron deficiency. Iron deficiency can have a major impact on the health and productivity of a population.

Although goiter, seen here, is a more visible manifestation of iodine deficiency, the subtle effects of deficiency on mental performance and work capacity may have a greater impact on the population as a whole. Iodine-deficient children have lower IQs and impaired school performance.[23] Iodine deficiency in children and adults is associated with apathy and decreased initiative and decision-making capabilities (see Chapter 8).

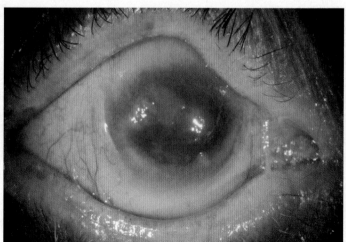

It is estimated that more than 250 million preschool children worldwide suffer from vitamin A deficiency.[25] Vitamin A deficiency leads to *xerophthalmia*, shown here. Vitamin A deficiency is the leading cause of preventable blindness among children. It also depresses immune function, thus increasing the risk of illness and death from infections, particularly measles and diarrheal diseases (see Chapter 7).

448 CHAPTER 14 Feeding the World

birth weight, preterm delivery, and fetal growth retardation.[20] It is estimated that about 17% of the world's population has inadequate zinc intake. Poor zinc status increases the risk of and mortality from infections and is associated with stunting.[20] Selenium deficiency is associated with an increased incidence of Keshan disease, a type of heart disease that affects mainly children and young women (see Chapter 8).[21] Inadequate calcium intake increases the risk of osteoporosis. Worldwide, one in three women and one in five men over age 50 will experience fractures due to osteoporosis.[22]

Concept Check

1. **What** causes food insecurity?
2. **How** can environmental damage lead to food shortages?
3. **Why** do children develop protein and micronutrient deficiencies more often than adults?

14.3 Causes of Hunger in the United States

LEARNING OBJECTIVES

1. **Discuss** the causes of food insecurity in the United States.
2. **Describe** how lack of access to health care, education, and transportation affect food security.
3. **List** the population groups that are at greatest risk for undernutrition in the United States.

Most of the nutritional problems in the United States are related to overnutrition. However, almost 13% of households in the United States experience food insecurity, putting this portion of the population at risk for undernutrition as well (**Figure 14.8**).[26] This situation is caused not by a general food shortage but by an inequitable distribution of food and money. The incidence of hunger and food insecurity is highest among women, infants, children, older adults, and those who are poor, homeless, ill, or disabled. However, a sudden decrease in income or increase in living expenses can put anyone at risk for food insecurity.

Poverty and Food Insecurity

In the United States, as elsewhere in the world, poverty is the main cause of food insecurity. Poverty reduces access to food, education, and health care. About 14% of Americans (43.1 million people) live at or below the poverty level.[27] These individuals have little money to spend on food and often have limited access to affordable food.

The high price of real estate in cities has driven supermarkets into the suburbs, and because many low-income city families do not own cars, they must shop at small, expensive corner stores or take public transportation if they wish to take advantage of lower prices at more distant, larger stores. This has created areas referred to as **food deserts**. Food deserts also exist in low-income rural communities where grocery stores may be great distances away. Lack of easy access to transportation and/or financial resources limits the ability to acquire affordable, healthy foods and increases the risk of nutrient deficiencies and nutrition-related chronic diseases.[28, 29]

Poverty also limits access to health care, leading to poorer health status. As in developing nations, poverty is reflected in infant mortality rates. The average infant mortality in the U.S. population is about 5.8 per 1000 live births.[4] However, there are groups within the population that have infant mortality rates as high as those in impoverished nations. Among African

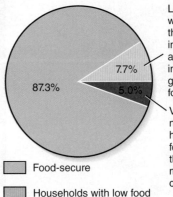

FIGURE 14.8 **Food insecurity in the United States** Current data show that 12.7% of U.S. households experienced food insecurity at some point during the year.[26]

Low food security means that families were able to avoid substantially disrupting their eating patterns or reducing food intake by using coping strategies, such as eating a less varied diet, participating in federal food assistance programs, or getting emergency food from community food pantries.

Very low food security means that the normal eating patterns of one or more household members were disrupted and food intake was reduced at times during the year because families had insufficient money or other resources to use for obtaining food.

food desert An area that lacks access to affordable fruits, vegetables, whole grains, low-fat milk, and other foods that make up a healthy diet.

Americans, the infant mortality rate is 11.1 per 1000 live births—over twice that of non-Hispanic whites.[30] This difference may reflect differences in infant mortality risk factors, such as poverty and lack of access to medical care.

Lack of education, which is both a cause and a consequence of poverty, also contributes to food insecurity. For people at or below the poverty level, educational opportunities are fewer and lower in quality than those for people with higher incomes. In the short term, lack of knowledge about food selection, food safety, and home economics can contribute to malnutrition. Too little food may cause the diet to be deficient in energy or particular nutrients, but poor food choices also allow food insecurity to coexist with obesity. Lack of education about food safety can also increase the incidence of food-borne illness. In the long term, lack of education prevents people from getting the higher-paying jobs that could allow them to escape poverty (**Figure 14.9**).

Poor families must use most of their income to pay for shelter, a situation that seriously reduces the chances that they will be adequately fed. The high cost of housing not only limits food budgets but also contributes to the growing problem of homelessness in the United States. About 550,000 Americans are homeless; two-thirds of these people stay in emergency shelters, transitional housing, or safe havens, and the remainder live in unsheltered locations such as in abandoned buildings, under bridges, or in their cars.[32] Homeless people are at high risk of food insecurity because they lack not only money but also cooking and food storage facilities.

Vulnerable Stages of Life

The high nutrient needs of pregnant and lactating women and small children put them at particular risk for undernutrition.

FIGURE 14.9 Education and poverty A lack of education limits the types of jobs available and hence the income level a person can attain.

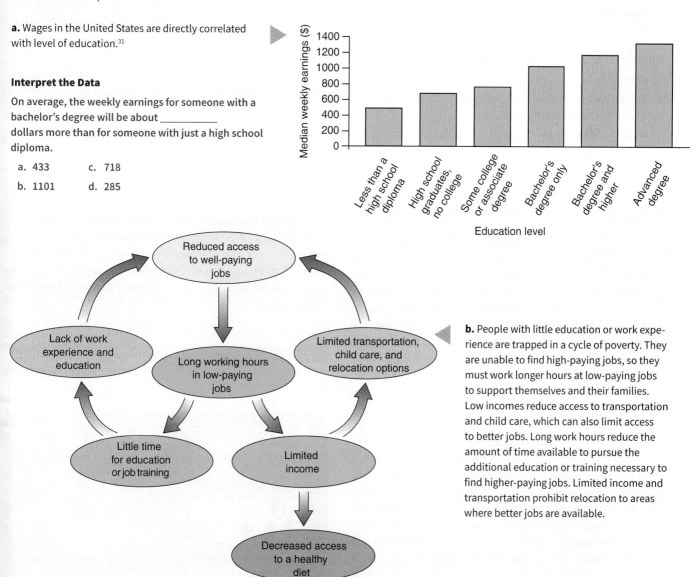

a. Wages in the United States are directly correlated with level of education.[31]

Interpret the Data

On average, the weekly earnings for someone with a bachelor's degree will be about _____ dollars more than for someone with just a high school diploma.

a. 433   c. 718
b. 1101   d. 285

b. People with little education or work experience are trapped in a cycle of poverty. They are unable to find high-paying jobs, so they must work longer hours at low-paying jobs to support themselves and their families. Low incomes reduce access to transportation and child care, which can also limit access to better jobs. Long work hours reduce the amount of time available to pursue the additional education or training necessary to find higher-paying jobs. Limited income and transportation prohibit relocation to areas where better jobs are available.

Almost one-third of households with children headed by single women live below the poverty line.[33] Poverty and food insecurity place these women and children at risk for malnutrition, and their special nutritional needs magnify this risk. Because of their increased need for some nutrients, malnutrition may occur in pregnant women, infants, and children even when the rest of the household is adequately fed. For example, the amount of iron in the family's diet may be enough to prevent anemia in all the family's members except a pregnant teenager.

Older adults are vulnerable to food insecurity and undernutrition due to the higher frequency of diseases and disabilities in this population group. Disease and disability may limit their ability to purchase, prepare, and consume food. Greater nutritional risk among older adults is associated with more hospital admissions and hence higher health-care costs. The number of individuals over age 85 could more than triple by 2060;[34] as the number of elderly people increases, so will the number of people at risk for food insecurity.

Concept Check

1. **Why** are some Americans hungry in a land of plenty?
2. **How** are education and poverty related to food insecurity?
3. **Why** are children at particular risk for undernutrition?

14.4 Eliminating World Hunger

LEARNING OBJECTIVES

1. **Discuss** strategies that can help reduce population growth.
2. **Describe** the role of sustainable agriculture in maintaining the food supply.
3. **Explain** how international trade can help eliminate hunger.
4. **Discuss** approaches used to ensure that the local food supply meets the nutrient needs of the population.

Solving the problem of world hunger is a daunting task that is being addressed on an international level. Beginning in 1996, the Food and Agriculture Organization of the World Health Organization began to hold World Food Summits to address the problem of widespread hunger. The initial goal was to cut world hunger in half by 2015. Although this goal was not reached, these efforts have had a huge impact. Estimates show that the percentage of undernourished people in the world decreased from 18.6% in 1990–1992 to 10.9% in 2014–2016. While this means there are 216 million fewer hungry people in the world than in the early 1900s, 795 million people are still undernourished.[1]

In 2000, the World Summit set Millennium Development Goals to address population growth, ensure that the nutrient needs of a large and diverse population are met with culturally acceptable foods, and increase food production without damaging the global ecosystem. In 2015 the goals were expanded into the Sustainable Development Goals, which are more comprehensive in their approach to eradicate extreme poverty and tackle other areas of importance to humanity and the planet (**Figure 14.10**).[35] Meeting these goals will require input from politicians, nutrition scientists, economists, and the food industry. Economic policies, technical advances, education, and legislative measures can create programs and policies to provide food in the short term; in the long term sustainable programs are needed to allow the continued production and distribution of acceptable foods.

Providing Short-Term Food Aid

When people are starving, short-term food and medical aid must be provided right away. The standard approach has been to bring food into stricken areas (**Figure 14.11**). This food generally consists of agricultural surpluses from other countries and often is not well planned in terms of its nutrient content. Although this type of relief is necessary for a population to survive an immediate crisis such as famine, it does little to prevent future hunger.

FIGURE 14.10 Sustainable development goals Ending poverty and hunger are two of the 17 Sustainable Development Goals set by the World Food Summit.

FIGURE 14.11 **Emergency food relief** Many organizations are working to combat world hunger. The American Red Cross and the High Commissioner for Refugees of the UN concentrate on famine relief. The Food and Agriculture Organization (FAO) works to improve the production, intake, and distribution of food worldwide. The World Health Organization (WHO) focuses on international health and emphasizes the prevention of nutrition problems, and the UN Children's Fund (UNICEF) targets education and vaccination and responds to crisis situations to improve the lives of children.

Controlling Population Growth

In the long term, solving the problem of world hunger requires balancing the number of people and the amount of food that can be produced. The world's population increased dramatically in the middle of the 20th century, but population growth has recently begun to slow. The birth rate worldwide has declined—from 5 children per woman in 1950 to 2.4 in 2014.[36] Continuing this downward trend will help to ensure that food production and natural resources can support the population. Changes in cultural and economic factors as well as family planning and government policies can be used to influence the birth rate.

Economic and cultural factors that affect birth rate

In many cultures, a large family is expected. A major reason for this expectation is high infant and child mortality rates. When infant mortality rates are high, people choose to have many children in order to ensure that some will survive. Higher birth rates in some developing countries are also due to the economic and societal roles of children. Children are needed to work farms, support the elders, and otherwise contribute to the economic survival of families. Programs that foster economic development and ensure access to food, shelter, and medical care have been shown to cause a decline in birth rates. Economic development reduces the need for children as workers.

Another cultural factor that influences birth rate is gender inequality. Girls are often kept at home to work rather than being sent to school. In most developing countries, the literacy rate is lower for women than for men, and fewer women attend primary and secondary school. This lack of education leaves women few options other than remaining home and having children. Providing education for girls has been shown to reduce birth rates (**Figure 14.12**).[37]

FIGURE 14.12 **Education and birth rate** Education increases the likelihood that women will have control over their fertility and gives them knowledge that can be used to improve the family's health and economic situation. Education builds job skills that allow women to join the workforce, marry later in life, and have fewer children. The graph shows that higher literacy among women is associated with lower birth rates. Women who are better educated have options other than having numerous children.

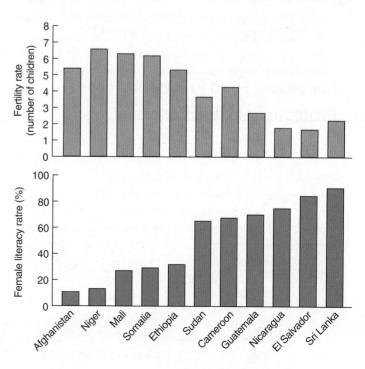

FIGURE 14.13 **Access to birth control** Access to birth control has reduced the fertility rate over the past several decades.

a. A birth control vendor explains condoms to women at a market in the Ivory Coast. The birth rate in this West African country declined from nearly 7 children per woman in 1988 to about 3.5 in 2016,[38] due in part to increased use of modern methods of birth control. Increased knowledge and availability of contraceptives is linked to a decrease in birth rate.

Karen Kasmauski / National Geographic Creative

b. As shown in the graph, the percentage of women in developing countries who are using contraception has been increasing but still lags behind use in the developed world. The availability of contraceptives gives women more control over the number of children they have and hence need to support.[39]

Family planning and government policies

Changes in cultural and economic factors may reduce the desire for large families, but reducing the birth rate also requires the availability of health and family-planning services. To be successful, family-planning efforts must be acceptable to the population and compatible with cultural and religious beliefs. Governments around the world have used a number of approaches, such as provision of contraceptives, education, and economic incentives, to decrease population growth (**Figure 14.13**).

Increasing Food Production While Protecting the Environment

Advances in agricultural technology have allowed food production to keep pace with population growth. However, the use of energy-intensive modern agricultural techniques has contributed to serious environmental problems. Commercial inorganic fertilizers and pesticides and modern farm machinery increase food production but at the same time pollute the air and water. Overuse of land causes deterioration of soil quality, which will limit food production in the future. For food production to continue to meet the needs of future generations, we need to increase food yields while conserving the world's natural resources (see *Thinking It Through*).

Sustainable agriculture uses food production methods that prevent damage to the environment and allow the land to restore itself so that food can be produced indefinitely. For example, contour plowing and terracing help prevent erosion, keeping the soil available for future crops. Rotating the crops grown in a field prevents the depletion of nutrients in the soil, reducing the need for fertilizers. Sustainable agriculture uses environmentally friendly chemicals that degrade quickly and do not persist as residues in the environment. It also relies on diversification. This approach to farming maximizes natural methods of pest control and fertilization and protects farmers from changes in the marketplace.

Sustainable agriculture is not a single program but involves choosing options that mesh well with local soil, climate, and farming techniques. In some cases, organic farming, which does not use synthetic pesticides, herbicides, and fertilizers (see Chapter 13), may be a more sustainable option. Organic techniques have a smaller environmental impact because they reduce the use of agricultural chemicals and the release of pollutants into the environment. Organic farming is also advantageous in terms of soil quality and biodiversity, but it has a disadvantage in terms of land use because crop yields are often

---

sustainable agriculture Agricultural methods that maintain soil productivity and a healthy ecological balance while having minimal long-term impacts.

Thinking It Through

A Case Study on What One Person Can Do

Keesha is concerned about the problems of hunger and malnutrition and the impact her choices have on the environment. Although she is a college student who cannot afford to make monetary contributions to relief organizations, she would like her everyday choices to have a minimal impact on the environment.

What are the advantages and disadvantages of the salad options shown in the photos in terms of convenience, food safety, waste, and environmental impact?

Your answer:

Keesha likes fish but has heard that some fish are endangered.

Go to the National Geographic Web site http://ocean.nationalgeographic.com/ocean/take-action/seafood-substitutions/ to find some ocean-friendly substitutes for her seafood choices.

Fish	Substitute variety
Atlantic cod	
Chilean sea bass	
Orange roughy	

The following are some inexpensive changes Keesha can make to reduce her impact on the environment.

What are the advantages and disadvantages of each?

Action	Advantages	Disadvantages
Bike instead of drive on short trips around town.		
Buy a canvas bag for carrying groceries.		
Carry a reusable water bottle rather than buy bottled water.		
Compost vegetable scraps.		
Buy locally grown produce.		
Buy organically grown produce.		

(Check your answers in Appendix L.)

---

lower. A combination of organic and conventional techniques, as is used with integrated pest management (see Chapter 13), might improve land use and protect the environment.

Other sustainable programs include agroforestry, in which techniques from forestry and agriculture are used together to restore degraded areas; natural systems agriculture, which attempts to develop agricultural systems that include many types of plants and therefore function like natural ecosystems; and the technique of reducing fertilizer use by matching nutrient resources with the demands of the particular crop being grown (**Figure 14.14**). One modern technology that may be integrated with sustainable systems is genetic engineering. As discussed in Chapter 13, genetic engineering can increase crop yields by inserting genes that improve the efficiency with which plants convert sunlight into food or genes that make plants resistant to herbicides, insects, and plant diseases.

Increasing Food Availability Through Economic Development and Trade

Hunger will exist as long as there is poverty. Even when food is plentiful, the poor do not have access to enough of the right foods to maintain their nutritional health. Economic development that leads to safe and sanitary housing, access to health care and education, and the resources to acquire enough food are essential if hunger is to be eliminated. Government policies can help reduce poverty and improve food security by increasing the population's income, lowering food prices, or funding food programs for those who are poor.

Some countries have the resources to grow enough food to feed their population, and others do not. When a country has few natural resources, access to international trade systems can

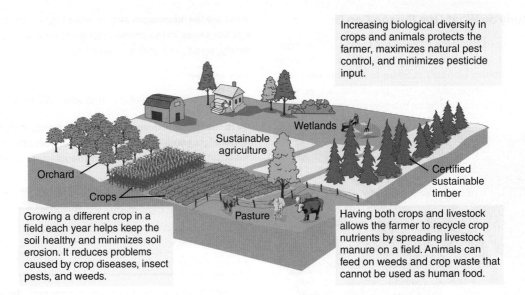

FIGURE 14.14 **A sustainable farm** A sustainable farm consists of a total agricultural ecosystem rather than a single crop. It may include field crops, fruit- and nut-bearing trees, herds of livestock, and forests.

help provide for their population. South Korea is an example of a country where an increase in food imports has helped feed the population. The mountainous terrain in South Korea means there is little arable land, and rice dominates areas where crops can be grown. Therefore, to increase food variety while keeping prices down, South Korea has turned to food imports. Today, South Korea imports about $25 billion in agricultural goods.[40] In general, countries around the world are becoming more dependent on food imports and on exports of food and other goods to pay for the food they import. This trade can increase the availability of food for the world's population as a whole.

Whether a country's agricultural emphasis is on producing **subsistence crops** or **cash crops** influences the availability of food for its people. Shifting to cash crops improves the country's cash flow but uses local resources to produce crops for export, reduces the diversity of crops grown, and limits the ability of its people to produce enough food to feed their families. For example, if a large portion of the arable land in a country is used to grow cash crops such as coffee and tea, little agricultural land remains for growing grains and vegetables that nourish the local population. If, however, the cash from the crop is used to purchase affordable nutritious foods from other countries, this decision may help alleviate undernutrition.

Ensuring a Nutritious Food Supply

To ensure the nutritional health of a population, the foods that are grown or imported must supply both sufficient energy and adequate amounts of all essential nutrients. To improve the quality of a diet that does not provide enough of all the essential nutrients, the dietary pattern needs to be changed, commonly consumed foods need to be fortified, or supplements containing the deficient nutrients need to be provided. For these changes to be effective, consumers must know how to choose foods that provide the missing nutrients and how to handle them safely.

Nutrition education Education can help improve nutrient intake by teaching consumers what foods to grow, which foods to choose, and how to prepare foods safely.

Education is particularly important when introducing a new crop. No matter how nutritious it may be, a new plant variety is not beneficial unless local farmers know how to grow it and the population accepts it as a food source and knows how to prepare it for consumption. For instance, white yams are common in some regions but are a poor source of β-carotene, which the body can use to make vitamin A. If sweet potatoes, which are rich in β-carotene, became an acceptable choice, the amount of vitamin A available to the population will increase.

Food safety is also a concern when changing traditional dietary practices. For example, introducing papaya to the diet as a source of vitamin A will not improve nutritional status if it is washed in unsanitary water and causes dysentery among the people it is meant to nourish.

---

subsistence crop A crop that is grown as food for a farmer's family, with little or nothing left over to sell.

cash crop A crop that is grown to be sold for monetary return rather than as food for the local population.

FIGURE 14.15 Nutritional and health benefits of breast-feeding Breast milk provides infants with optimal nutrition and immune factors that reduce the risk of infectious diseases. In developing nations, where infant mortality from infectious disease is high and nutritious alternatives to breast milk for the infant are not available, breast-feeding may even be recommended for women who are HIV positive who do not have access to HIV treatment. In such cases, the risk of the baby dying of malnutrition and other infections outweighs the risk of transmitting the virus in the milk.[42]

Education to encourage breast-feeding can also improve nutritional status and health (**Figure 14.15**). To achieve optimal growth, development, and health, the WHO recommends that infants be exclusively breast-fed for the first six months of life. After that, other foods should be offered, while breast-feeding continues for up to two years of age or beyond.[41]

Fortifying the food supply Although food fortification will not provide energy for a hungry population, it can increase the protein quality of the diet and eliminate micronutrient deficiencies. In order to solve a nutritional problem in a population, fortification must be implemented wisely. Fortification works only if vulnerable groups consume centrally processed foods. The foods selected for fortification should be among those that are consistently consumed by the at-risk population so that extensive promotion and reeducation are not needed to encourage their consumption. The nutrient should be added uniformly and in a form that optimizes its utilization.

Fortification has been used successfully in preventing health problems in the United States. Fortification of cow's milk to increase vitamin D intake was a major factor in the elimination of infantile rickets, and enrichment of grains with niacin helped eliminate pellagra. Fortification of salt with iodine is successfully eliminating iodine deficiency diseases in countries around the world (**Figure 14.16**).

An alternative to traditional fortification is biofortification, which involves breeding plants to increase the nutrient content of staple foods. For example, breeders have developed corn that provides higher levels of β-carotene than traditional varieties.[44] The challenge is to get local farmers and consumers to accept biofortified crops. (See *Debate: Combating Vitamin A Deficiency with Golden Rice*.)

FIGURE 14.16 Iodized salt The introduction of iodized salt into a population's diet helps eliminate iodine deficiency.

Over the past decade, the number of countries with salt iodization programs has increased dramatically. It is estimated that 70% of households worldwide now have access to iodized salt.[43]

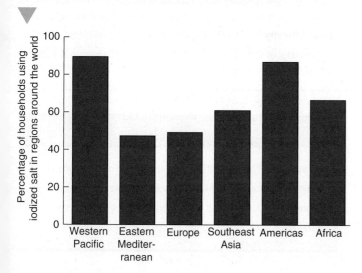

This logo is used around the world as an indicator of iodized salt.

Think Critically

The eastern Mediterranean has the lowest percentage of households with access to iodized salt of any region in the world. Based on your knowledge of dietary sources of iodine, why might this region not have a high incidence of iodine deficiency?

Debate

Combating Vitamin A Deficiency with Golden Rice

The Issue

Golden Rice is a genetically modified (GM) variety of rice developed to increase the vitamin A in the food supply of populations in which this deficiency is prevalent. Its development has stimulated debate about whether it is an effective way to prevent vitamin A deficiency and whether any genetically modified product is an appropriate and safe way to help alleviate malnutrition.

The β-carotene in Golden Rice gives it a yellow-orange color.

Erik De Castro / Reuters / Newscom

Each year vitamin A deficiency takes the sight and lives of hundreds of thousands of children worldwide. The deficiency is most common in impoverished populations where the dietary staple is deficient in vitamin A. Currently, this deficiency is addressed using vitamin A supplementation, fortification of the food supply, and interventions that increase the variety of the diet to include foods that are rich in vitamin A. A more controversial solution is growing rice that has been genetically engineered to synthesize β-carotene, a yellow-orange pigment that is a precursor to vitamin A. If it replaced white rice as a staple, Golden Rice could alleviate vitamin A deficiency. However, the controversy that ensued after it was developed has kept the preventive potential of Golden Rice from being fully assessed.

Initial concerns about Golden Rice focused on whether it would provide enough vitamin A to alleviate deficiency. The original variety provided so little β-carotene that a 2-year-old child would need to eat 3 kg of it each day to get enough vitamin A.[45] However, a newer variety contains enough β-carotene in a cup of cooked rice to provide about 430 µg of retinol, more than enough to meet the RDA for an 8-year-old child.[46] Critics of Golden Rice have expressed concern that the β-carotene in the rice might not provide usable vitamin A to the body. However, a study that compared the vitamin A value of the β-carotene in Golden Rice to that of a β-carotene supplement found that both resulted in similar blood levels of retinol.[46] Even though Golden Rice can provide enough vitamin A, many still argue that it may not necessarily be a solution to vitamin A malnutrition. Deficient populations typically suffer from other nutrient deficiencies in addition to vitamin A, and when protein, fat, or zinc is deficient, the body can't efficiently use vitamin A.[47, 48]

There is also concern that introducing Golden Rice, which is a GM organism, into the environment will cause harm. The worry is that its use will decrease the diversity of rice varieties grown. Reducing diversity increases the risk of crop destruction due to insects and disease.[47] Proponents of GM crops argue that this concern occurs whenever a new crop that is preferred by farmers is introduced and that the potential health benefits outweigh the environmental concerns. Opponents of Golden Rice also argue that it could exacerbate malnutrition by encouraging a diet based solely on rice.[49] Although the use of Golden Rice rather than white rice is unlikely to affect dietary diversity, switching to Golden Rice does little to address the underlying causes of malnutrition, which are poverty and lack of access to a varied diet.

While proponents debate opponents over the usefulness and safety of Golden Rice, vitamin A deficiency remains a major public health issue. Those who disagree with this approach argue that work on Golden Rice has diverted resources from proven programs that address multiple nutrient deficiencies. Advocates contend that the problem is not the rice but rather the regulatory climate that has prevented it from being introduced.[50] Golden Rice has been ready to be used by farmers since the beginning of the 21st century, but it is still not available for widespread use.[51] In efforts to move production forward, former Greenpeace leaders have now endorsed the Golden Rice NOW campaign in Europe to promote its use.[52, 53]

Although the debate continues, everyone would likely agree that the goal is to prevent malnutrition and, in the long term, eliminate poverty, which reduces access to a healthy diet. GM crops such as Golden Rice alone are not the solution to malnutrition. Supplementation may be necessary for those in immediate need, and fortification and supplementation may work better in urban settings. However, the question remains: Should Golden Rice be part of the arsenal available to eliminate vitamin A deficiency?

Think Critically

Do the risks of GM crops outweigh the risks of vitamin A deficiency? Defend your position.

Providing supplements Supplementing specific nutrients for at-risk segments of the population can help reduce the prevalence of malnutrition. In areas where deficiency is a problem, supplements of zinc and vitamin A can enhance child health and survival. The WHO recommends vitamin A supplementation in infants and children 6 months to 5 years of age in areas where vitamin A deficiency is a public health problem.[54] Appropriate supplementation of iron, folic acid, and calcium can improve maternal and newborn survival and health.[55]

Concept Check

1. **What** impact does education for women have on birth rates?
2. **How** does protecting the environment affect food security?
3. **When** can importing food ensure adequate food for a population?
4. **What** determines which foods are selected for fortification?

14.5 Eliminating Food Insecurity in the United States

LEARNING OBJECTIVES

1. **Describe** what is meant by a nutrition safety net.
2. **Explain** the role of food recovery in eliminating hunger.
3. **Discuss** the importance of nutrition education in ensuring a healthy population.

As with eliminating world hunger, eliminating hunger in the United States involves improving economic security, keeping food affordable, providing food aid to the hungry, and offering education about healthy diets that will meet nutrient needs and reduce diseases related to overconsumption.

The Nutrition Safety Net

Federal programs that provide access to affordable food and promote healthy eating have been referred to as a "nutrition safety net" for the American population (**Table 14.1**). The nutrition assistance programs include a combination of general nutrition assistance and specialized programs targeted to groups with particular nutritional risks: children, seniors, infants, women during and after pregnancy, Native Americans living on reservations, people with disabilities, and homeless people. About one of every four Americans receives some kind of food assistance, at a total cost of about $104.1 billion per year.[56]

The largest USDA program designed to make sure that all people have access to an adequate diet is the Supplemental Nutrition Assistance Program (SNAP) (previously known as the Food Stamp Program). SNAP accounts for over two-thirds of all federal food and nutrition assistance spending, and about 45.8 million people per month participate in the program.[56] SNAP provides monthly benefits in the form of coupons or debit cards that can be used to purchase food, thereby supplementing the food budgets of low-income individuals. SNAP, along with four other programs that target high-risk populations—the National School Lunch Program; the Special Supplemental Nutrition Program for Women, Infants, and Children (WIC); the Child and Adult Care Food Program; and the School Breakfast Program—account for 96% of the USDA's expenditure for food and nutrition assistance.[56]

In addition to federal nutrition assistance programs, church, community, and charitable emergency food shelters provide for the basic nutritional needs of many Americans. In the United States thousands of nonprofit food distribution programs help direct food to those in need. The leading hunger-relief charity in the United States is Feeding America, which provides more than 4 billion meals per year.[57] It includes a network of food banks across the country and supports thousands of local charitable organizations, such as food pantries and soup kitchens, which distribute food directly to hungry Americans.

Virtually all these food distribution programs use food obtained through **food recovery**, which involves collecting food that is wasted in fields, commercial kitchens, restaurants, and grocery stores and distributing it to those in need (**Figure 14.17**). It is estimated that 40% of America's food goes uneaten (See *What a Scientist Sees: The Cost of Wasted Food*).[58]

TABLE 14.1 Programs to prevent undernutrition in the United States

Program	Target population	Goals and methods
Supplemental Nutrition Assistance Program (SNAP)	Low-income individuals	Increases access to food by providing coupons or debit cards that can be used to purchase food at a grocery store
Commodity Supplemental Food Program (CSFP)	Low-income pregnant women, breast-feeding and non–breast-feeding postpartum women, infants and children under age 6, and elderly people	Provides food by distributing U.S. Department of Agriculture commodity foods
Special Supplemental Nutrition Program for Women, Infants, and Children (WIC)	Low-income pregnant women, breast-feeding and non–breast-feeding postpartum women, and infants and children under age 5	Provides debit cards for the purchase of foods (including infant formula and infant cereal) high in nutrients that are typically lacking in the program's target population; provides nutrition education and referrals for health care
WIC Farmers' Market Nutrition Program	WIC participants	Increases access to fresh produce by providing coupons that can be used to purchase produce at authorized local farmers' markets
School Breakfast Program	Low-income children	Provides free or low-cost breakfasts at school to improve the nutritional status of children
National School Lunch Program	Low-income children	Provides free or low-cost lunches at school to improve the nutritional status of children

(continues)

TABLE 14.1 Programs to prevent undernutrition in the United States (continued)

Program	Target population	Goals and methods
Special Milk Program	Low-income children	Provides milk for children in schools, camps, and child-care institutions that have no federally supported meal program
Summer Food Service Program	Low-income children	Provides free meals and snacks for children when school is not in session
Child and Adult Care Food Program	Children up to age 12 and elderly and disabled adults	Provides nutritious meals to children and adults in day-care settings
Team Nutrition	School-age children	Provides nutrition education, training and technical assistance, and resources to participating schools, with the goal of improving children's lifelong eating and physical activity habits
Head Start	Low-income preschool children and their families	Provides meals and education, including nutrition education
Nutrition Program for the Elderly	Individuals age 60 and over and their spouses	Provides free congregate meals in churches, schools, senior centers, or other facilities and delivers food to homebound people
Senior Farmers' Market Program	Low-income seniors	Provides coupons that can be exchanged for eligible foods at farmers' markets, roadside stands, and community-supported agricultural programs
Homeless Children Nutrition Program	Preschoolers living in shelters	Reimburses providers for meals served
Emergency Food Assistance Program	Low-income people	Provides commodities to soup kitchens, food banks, and individuals for home use
Healthy People 2020	U.S. population	Sets national health promotion objectives to improve the health of the U.S. population through the health-care system and industry involvement, as well as individual actions
Expanded Food and Nutrition Education Program (EFNEP)	Low-income families	Provides education on all aspects of food preparation and nutrition
Temporary Assistance for Needy Families (TANF)	Low-income households	Provides assistance and work opportunities to needy families by granting states federal funds to implement welfare programs
Food Distribution Program on Indian Reservations	Low-income households living on reservations and Native Americans living near reservations	Provides food by distributing USDA commodity foods

FIGURE 14.17 Field gleaning These oranges were harvested in California as part of a local gleaning program. Field gleaning is a type of food recovery that involves collecting crops that are not harvested because it is not economically profitable to harvest them or that remain in fields after mechanical harvesting. The word *gleaning* means "gathering after the harvest" and dates back at least as far as biblical times.

What a Scientist Sees

The Cost of Wasted Food

Ever go to the grocery store with the best nutrition intentions in mind and fill your cart with fresh fruits and vegetables? Then during the busy workweek, you forget that they are in your fridge or are too busy for all the required washing, chopping, and cooking. By the end of the week, most of the produce in the fridge ends up in the trash or on the compost heap.

What a scientist sees is that reducing this waste by just 15% could feed more than 25 million Americans every year. As shown in the graph, less than half of the fruits and vegetables on a farm end up getting eaten. Food losses occur at every step from farm to table. Losses on the farm occur when food is not harvested and postharvest when crops are discarded if they do not meet quality or appearance criteria. Losses continue during food processing, distribution, and at retailers. The USDA estimates that supermarkets lose $15 billion annually due to unsold fruits and vegetables.[58] The greatest losses occur in restaurants and households. Restaurants purchase more food than is ever cooked, and diners leave, on average, 17% of food uneaten. In homes, consumers throw out about 25% of the food and beverages they buy, resulting in a monetary loss of $1365 to $2275 annually for a family of four.[58] So the next time you are feeling health conscious in the grocery store, stock up on fresh produce, but only if you have your meals planned and the time to prepare them.

Think Critically

Suggest some things that could be done on your college campus to reduce food waste.

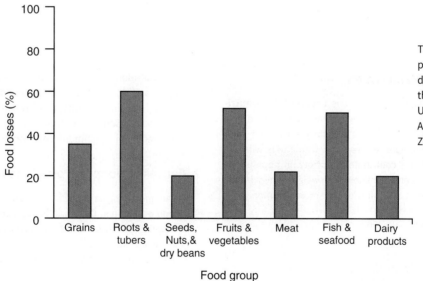

The graph shows the percentage of food from different categories that is wasted in the United States, Canada, Australia, and New Zealand.[59]

Nutrition Education

People who have more nutrition information and greater awareness of the relationship between diet and health consume healthier diets. Healthy diets not only improve current health by optimizing growth, productivity, and well-being but are essential for preventing chronic diseases. Increasing knowledge about nutrition can reduce medical costs and improve the quality of life.

Education can help individuals with lower incomes stretch their limited food dollars by making wise choices at the store and reducing food waste at home. It can promote community gardens to increase the availability of seasonal vegetables. It can teach people how to prepare foods received from

What Should I Eat?

Make Your Meals Green

Eat more plants
- Reduce the amount of meat in your meals.
- Eat vegetarian at least once a week.
- Eat lower on the food chain—more plant foods and small fish.
- Grow and eat some of your own vegetables.

Cut down on your contribution to pollution
- Buy in bulk in order to cut down on packaging.
- Use reusable bags to take your groceries home.

- Choose locally grown and organically produced foods.
- Cook from scratch and use fewer processed foods.

Reduce food waste
- Buy only what you know you will cook.
- Serve only portions you will eat.
- Freeze your leftovers for another day.
- Make soups and stews from vegetables that are past their prime.

 Use iProfile to compare the nutrients in a vegetarian meal versus a meat-based meal.

---

commodity distribution programs and food banks. It can explain safe food handling and preparation methods. Knowing which foods to choose and how to handle them safely is as important in preventing malnutrition as having the money to buy enough food. In addition to the programs described in Table 14.1, the *Dietary Guidelines for Americans*, MyPlate, and food labels educate the general public about making wise food choices (see *What Should I Eat?*).

Concept Check

1. **Which** federal nutrition programs help pregnant women and children?
2. **How** much of America's food goes uneaten?
3. **Why** does nutrition education save on food costs?

Summary

1 The Two Faces of Malnutrition 441

- In poorly nourished populations, there is a **cycle of undernutrition**, as illustrated in the figure. Poorly nourished women give birth to low-birth-weight infants at risk for disease and early death. If these children survive, they grow into adults who are physically unable to fully contribute to society. In populations where undernutrition is prevalent, low birth weight, a high **infant mortality rate**, **stunting**, and infections are more common.

Figure 14.1a Nutrition transition

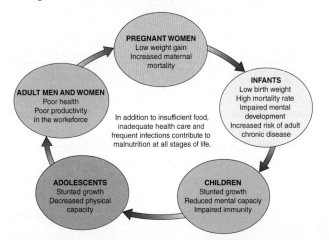

- Overnutrition coexists with **hunger** and **starvation** in both developed and developing nations around the world. As economic conditions improve, **nutrition transition** to more Western diet and lifestyle patterns contributes to the growing problem of nutrition-related noncommunicable diseases.

2 Causes of Hunger Around the World 443

- The underlying cause of hunger and **food insecurity** is that the food available in the world is not distributed equitably. **Famine** results from natural and human-caused disasters that temporarily disrupt food production and distribution. Chronic food shortage resulting in undernutrition is most common in the developing world, as shown in the chart. It occurs when economic inequities result in lack of money, health care, and education; when overpopulation and limited natural resources create a situation in which there are more people than food; when cultural practices limit food choices; and when **renewable resources** are misused, limiting the ability to continue to produce food.

Figure 14.4 The impact of poverty

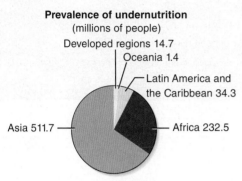

- Deficiencies of protein, iron, iodine, and vitamin A are common worldwide when the quality of the diet is poor. Pregnant women, children, older adults, and those who are ill may not be able to meet their nutrient needs with the available diet.

3 Causes of Hunger in the United States 448

- Both undernutrition and overnutrition are problems in the United States. As in developing nations, in the United States undernutrition and food insecurity are associated with poverty, which limits access to health care, adequate housing, and education, making it difficult to escape poverty, as depicted here.

Figure 14.9b Education and poverty

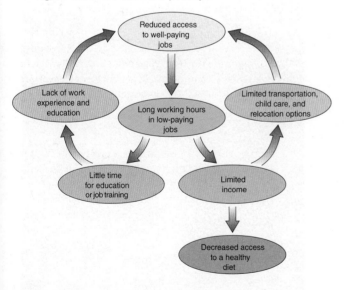

- High nutrient needs increase the risk of malnutrition in women and children, and disease and disability increase risk in older adults.

4 Eliminating World Hunger 450

- Short-term solutions to eliminating hunger provide food through relief at local, national, and international levels, as shown here.

Figure 14.11 Emergency food relief

- Eliminating world hunger in the long term requires controlling population growth. This can be addressed by improving economic conditions, providing education, particularly for women, and ensuring access to family planning services.
- **Sustainable agriculture** helps eliminate hunger by allowing food to be produced without damaging the environment.
- Economic development helps prevent hunger by eliminating poverty and ensuring access to health care and education. It also increases access to international trade, which can be used to import food or to export **cash crops** to bring more money into the country.
- Food fortification and dietary supplementation can be used to increase protein quality and eliminate micronutrient deficiencies, improving the overall quality of the diet.

5 Eliminating Food Insecurity in the United States 457

- Nutrition programs in the United States focus on maintaining a nutrition safety net that provides access to affordable food and education to promote healthy eating. This safety net includes federal nutrition assistance programs as well as church, community, and charitable emergency food shelters. Virtually all food distribution programs use food obtained through **food recovery**, which involves collecting and distributing food that is wasted in fields, as shown in the photo, as well as food left over in commercial kitchens, restaurants, and grocery stores.

- Nutrition education can help individuals with lower incomes make nutritious, economical choices at the store and reduce food waste at home. It can also help with meal planning skills and safe food handling and preparation methods.

Figure 14.17 Field gleaning

Key Terms

- cash crop 454
- cycle of undernutrition 441
- famine 444
- food desert 448
- food insecurity 443
- food recovery 457
- hunger 441
- infant mortality rate 441
- nutrition transition 443
- renewable resource 444
- starvation 441
- stunting 442
- subsistence crop 454
- sustainable agriculture 452

What is happening in this picture?

This 250-fold magnification shows the mouth of a hookworm, which the organism uses to attach to the lining of a human's small intestine and feed on blood. Hookworm larvae penetrate the skin, infecting people when they walk barefoot in contaminated soil. Hookworm infection affects about 576 to 740 million people worldwide.[60]

Think Critically

1. Why might you expect this infection to be more common in poor tropical and subtropical regions than elsewhere?
2. How would hookworm infection affect iron status? Why?
3. How would hookworm infection affect a population's productivity? Why?

Appendix A

Dietary Reference Intakes

TABLE A.1 Dietary Reference Intakes: Recommended Intakes for Individuals: Vitamins

Life Stage Group	Vitamin A (µg/day)[a]	Vitamin C (mg/day)	Vitamin D (µg/day)[b,c]	Vitamin E (mg/day)[d]	Vitamin K (µg/day)	Thiamin (mg/day)	Riboflavin (mg/day)	Niacin (mg/day)[e]	Vitamin B_6 (mg/day)	Folate (µg/day)[f]	Vitamin B_{12} (µg/day)	Pantothenic Acid (mg/day)	Biotin (µg/day)	Choline (mg/day)[g]
Infants														
0–6 mo	400*	40*	10*	4*	2.0*	0.2*	0.3*	2*	0.1*	65*	0.4*	1.7*	5*	125*
6–12 mo	500*	50*	10*	5*	2.5*	0.3*	0.4*	4*	0.3*	80*	0.5*	1.8*	6*	150*
Children														
1–3 y	300	15	15	6	30*	0.5	0.5	6	0.5	150	0.9	2*	8*	200*
4–8 y	400	25	15	7	55*	0.6	0.6	8	0.6	200	1.2	3*	12*	250*
Males														
9–13 y	600	45	15	11	60*	0.9	0.9	12	1.0	300	1.8	4*	20*	375*
14–18y	900	75	15	15	75*	1.2	1.3	16	1.3	400	2.4	5*	25*	550*
19–30y	900	90	15	15	120*	1.2	1.3	16	1.3	400	2.4	5*	30*	550*
31–50y	900	90	15	15	120*	1.2	1.3	16	1.3	400	2.4	5*	30*	550*
51–70y	900	90	15	15	120*	1.2	1.3	16	1.7	400	2.4[h]	5*	30*	550*
>70y	900	90	20	15	120*	1.2	1.3	16	1.7	400	2.4[h]	5*	30*	550*
Females														
9–13y	600	45	15	11	60*	0.9	0.9	12	1.0	300	1.8	4*	20*	375*
14–18y	700	65	15	15	75*	1.0	1.0	14	1.2	400[i]	2.4	5*	25*	400*
19–30y	700	75	15	15	90*	1.1	1.1	14	1.3	400[i]	2.4	5*	30*	425*
31–50y	700	75	15	15	90*	1.1	1.1	14	1.3	400[i]	2.4	5*	30*	425*
51–70y	700	75	15	15	90*	1.1	1.1	14	1.5	400	2.4[h]	5*	30*	425*
>70y	700	75	20	15	90*	1.1	1.1	14	1.5	400	2.4[h]	5*	30*	425*
Pregnancy														
14–18y	750	80	15	15	75*	1.4	1.4	18	1.9	600[j]	2.6	6*	30*	450*
19–30y	770	85	15	15	90*	1.4	1.4	18	1.9	600[j]	2.6	6*	30*	450*
31–50y	770	85	15	15	90*	1.4	1.4	18	1.9	600[j]	2.6	6*	30*	450*
Lactation														
14–18y	1200	115	15	19	75*	1.4	1.6	17	2.0	500	2.8	7*	35*	550*
19–30y	1300	120	15	19	90*	1.4	1.6	17	2.0	500	2.8	7*	35*	550*
31–50y	1300	120	15	19	90*	1.4	1.6	17	2.0	500	2.8	7*	35*	550*

Note: This table (taken from the DRI reports, see www.nap.edu) presents Recommended Dietary Allowances (RDAs) in **bold** type and Adequate Intakes (AIs) in ordinary type followed by an asterisk (*). RDAs and AIs may both be used as goals for individual intakes. RDAs are set to meet the needs of almost all (97 to 98 percent) healthy individuals in a group. For healthy breastfed infants, the AI is the mean intake. The AI for all other life stage and gender groups is believed to cover the needs of all healthy individuals in the groups, but lack of data or uncertainty in the data prevent being able to specify with confidence the percentage of individuals covered by this intake.

[a] As retinol activity equivalents (RAEs). 1 RAE = 1 µg retinol, 12 µg β-carotene, 24 µg α-carotene, or 24 µg β-cryptoxanthin. The RAE for dietary provitamin A carotenoids is two-fold greater than retinol equivalents (REs), whereas the RAE for preformed vitamin A is the same as RE.

[b] As cholecalciferol. 1 µg cholecalciferol = 40 IU vitamin D.

[c] Under the assumption of minimal sunlight.

[d] As α-tocopherol, which includes RRR-α-tocopherol, the only form of α-tocopherol that occurs naturally in foods, and the 2R-stereoisomeric forms of α-tocopherol (RRR-, RSR-, RRS, and RSS-α-tocopherol) that occur in fortified foods and supplements. It does not include the 2S-stereoisomeric forms of α-tocopherol (SRR-, SSR-, SRS-, and SSS-α-tocopherol), also found in fortified foods and supplements.

[e] As niacin equivalents (NEs). 1 mg niacin = 60 mg tryptophan; 0–6 months = pre-formed niacin (not NE).

[f] As dietary folate equivalents (DFE). 1 DFE = 1 µg food folate = 0.6 µg folic acid from fortified food or as a supplement consumed with food = 0.5 µg of a supplement taken on an empty stomach.

[g] Although AIs have been set for choline, there are few data to assess whether a dietary supply of choline is needed at all stages of the lifecycle, and it may be that the choline requirement can be met by endogenous synthesis at some of these stages.

[h] Because 10–30% of older people may malabsorb food-bound B_{12} it is advisable for those older than 50 years to meet their RDA mainly by consuming foods fortified with B_{12} or a supplement containing B_{12}.

[i] In view of evidence linking folate intake with neural tube defects in the fetus, it is recommended that all women capable of becoming pregnant consume 400 µg from supplements or fortified foods in addition to intake of food folate from a varied diet.

[j] It is assumed that women will consume 400 µg from supplements or fortified foods until their pregnancy is confirmed and they enter prenatal care, which ordinarily occurs after the end of the periconceptional period—the critical time for neural tube formation.

Source: Dietary Reference Intake Tables: The Complete Set. Institute of Medicine, National Academy of Sciences available online at www.nap.edu. Reprinted with permission from *Dietary Reference Intakes: The Essential Guide to Nutrient Requirements,* 2006, by the National Academy of Sciences, Washington, D.C. Institute of Medicine, Food and Nutrition Board Dietary Reference Intakes for Calcium and Vitamin D (2011), National Academies Press, Washington DC, 2011.

TABLE A.2 Dietary Reference Intakes: Recommended Intakes for Individuals: Minerals

Life Stage Group	Calcium (mg/day)	Chromium (μg/day)	Copper (μg/day)	Fluoride (mg/day)	Iodine (μg/day)	Iron (mg/day)	Magnesium (mg/day)	Manganese (mg/day)	Molybdenum (μg/day)	Phosphorous (mg/day)	Selenium (μg/day)	Zinc (mg/day)	Sodium (g/day)	Chloride (g/day)	Potassium (g/day)
Infants															
0–6 mo	200*	0.2*	200*	0.01*	110*	0.27*	30*	0.003*	2*	100*	15*	2*	0.12*	0.18*	0.4
6–12 mo	260*	5.5*	220*	0.5*	130*	11	75*	0.6*	3*	275*	20*	3	0.37*	0.57*	0.7
Children															
1–3 y	**700**	11*	**340**	0.7*	**90**	**7**	**80**	1.2*	**17**	**460**	**20**	**3**	1.0*	1.5*	3.0
4–8 y	**1,000**	15*	**440**	1*	**90**	**10**	**130**	1.5*	**22**	**500**	**30**	**5**	1.2*	1.9*	3.8
Males															
9–13 y	**1,300**	25*	**700**	2*	**120**	**8**	**240**	1.9*	**34**	**1,250**	**40**	**8**	1.5*	2.3*	4.5
14–18 y	**1,300**	35*	**890**	3*	**150**	**11**	**410**	2.2*	**43**	**1,250**	**55**	**11**	1.5*	2.3*	4.7
19–30 y	**1,000**	35*	**900**	4*	**150**	**8**	**400**	2.3*	**45**	**700**	**55**	**11**	1.5*	2.3*	4.7
31–50 y	**1,000**	35*	**900**	4*	**150**	**8**	**420**	2.3*	**45**	**700**	**55**	**11**	1.5*	2.3*	4.7
51–70 y	**1,000**	30*	**900**	4*	**150**	**8**	**420**	2.3*	**45**	**700**	**55**	**11**	1.3*	2.0*	4.7
>70 y	**1,200**	30*	**900**	4*	**150**	**8**	**420**	2.3*	**45**	**700**	**55**	**11**	1.2*	1.8*	4.7
Females															
9–13 y	**1,300**	21*	**700**	2*	**120**	**8**	**240**	1.6*	**34**	**1,250**	**40**	**8**	1.5*	2.3*	4.5
14–18 y	**1,300**	24*	**890**	3*	**150**	**15**	**360**	1.6*	**43**	**1,250**	**55**	**9**	1.5*	2.3*	4.7
19–30 y	**1,000**	25*	**900**	3*	**150**	**18**	**310**	1.8*	**45**	**700**	**55**	**8**	1.5*	2.3*	4.7
31–50 y	**1,000**	25*	**900**	3*	**150**	**18**	**320**	1.8*	**45**	**700**	**55**	**8**	1.5*	2.3*	4.
51–70 y	**1,200**	20*	**900**	3*	**150**	**8**	**320**	1.8*	**45**	**700**	**55**	**8**	1.3*	2.0*	4.7
>70 y	**1,200**	20*	**900**	3*	**150**	**8**	**320**	1.8*	**45**	**700**	**55**	**8**	1.2*	1.8*	4.
Pregnancy															
14–18 y	**1,300**	29*	**1,000**	3*	**220**	**27**	**400**	2.0*	**50**	**1,250**	**60**	**12**	1.5*	2.3*	4.
19–30 y	**1,000**	30*	**1,000**	3*	**220**	**27**	**350**	2.0*	**50**	**700**	**60**	**11**	1.5*	2.3*	4.
31–50 y	**1,000**	30*	**1,000**	3*	**220**	**27**	**360**	2.0*	**50**	**700**	**60**	**11**	1.5*	2.3*	4.
Lactation															
14–18 y	**1,300**	44*	**1,300**	3*	**290**	**10**	**360**	2.6*	**50**	**1,250**	**70**	**13**	1.5*	2.3*	5.
19–30 y	**1,000**	45*	**1,300**	3*	**290**	**9**	**310**	2.6*	**50**	**700**	**70**	**12**	1.5*	2.3*	5.
31–50 y	**1,000**	45*	**1,300**	3*	**290**	**9**	**320**	2.6*	**50**	**700**	**70**	**12**	1.5*	2.3*	5.

Note: This table (taken from the DRI reports, see www.nap.edu) presents Recommended Dietary Allowances (RDAs) in bold type and Adequate Intakes (AIs) in ordinary type followed by an asterisk (*). RDAs and AIs may both be used as goals for individual intakes. RDAs are set up to meet the needs of almost all (97–98%) healthy individuals in a group. It is calculated from an EAR. If sufficient scientific evidence is not available to establish an EAR, and thus calculate an RDA, an AI is usually developed. For healthy breastfed infants, the AI is the mean intake. The AI for all other life stage and gender groups is believed to cover the needs of all healthy individuals in the groups, but lack of data or uncertainty in the data prevent being able to specify with confidence the percentage of individuals covered by this intake.

Source: Dietary Reference Intake Tables: The Complete Set. Institute of Medicine, National Academy of Sciences. Available online at www.nap.edu. Reprinted with permission from *Dietary Reference Intakes: The Essential Guide to Nutrient Requirements,* 2006, by the National Academy of Sciences, Washington, D.C. Institute of Medicine, Food and Nutrition Board Dietary Reference Intakes for Calcium and Vitamin D (2011), National Academies Press, Washington DC, 2011.

TABLE A.3 Acceptable Macronutrient Distribution Ranges (AMDR) for Healthy Diets as a Percent of Energy

Age	Carbohydrate	Added Sugars	Total Fat	Linoleic Acid	α- Linolenic Acid	Protein
1–3 y	45–65	≤25	30–40	5–10	0.6–1.2	5–20
4–18 y	45–65	≤25	25–35	5–10	0.6–1.2	10–30
≥19 y	45–65	≤25	20–35	5–10	0.6–1.2	10–35

Source: Institute of Medicine, Food and Nutrition Board. "Dietary Reference Intakes for Energy, Carbohydrate, Fiber, Fat, Fatty Acids, Cholesterol, Protein, and Amino Acids." Washington, D.C.: National Academies Press, 2002, 2005.

TABLE A.4 Dietary Reference Intakes: Recommended Intakes for Individuals: Carbohydrates, Fiber, Fat, Fatty Acids, Protein, and Water

Life Stage Group	Carbohydrate (g/day)	Fiber (g/day)	Fat (g/day)	Linoleic Acid (g/day)	α-Linolenic Acid (g/day)	Protein (g/kg/day)[a]	Protein (g/day)	Water[b] (L/day)
Infants								
0–6 mo	60*	ND	31*	4.4*†	0.5*‡	1.52*	9.1*	0.7*
6–12 mo	95*	ND	30*	4.6*†	0.5*‡	**1.50**	**11.0**	0.8*
Children								
1–3 y	**130**	19*	ND	7*	0.7*	**1.10**	**13**	1.3*
4–8 y	**130**	25*	ND	10*	0.9*	**0.95**	**19**	1.7*
Males								
9–13 y	**130**	31*	ND	12*	1.2*	**0.95**	**34**	2.4*
14–18 y	**130**	38*	ND	16*	1.6*	**0.85**	**52**	3.3*
19–30 y	**130**	38*	ND	17*	1.6*	**0.80**	**56**	3.7*
31–50y	**130**	38*	ND	17*	1.6*	**0.80**	**56**	3.7*
51–70y	**130**	30*	ND	14*	1.6*	**0.80**	**56**	3.7*
>70 y	**130**	30*	ND	14*	1.6*	**0.80**	**56**	3.7*
Females								
9–13 y	**130**	26*	ND	10*	1.0*	**0.95**	**34**	2.1*
14–18 y	**130**	26*	ND	11*	1.1*	**0.85**	**46**	2.3*
19–30 y	**130**	25*	ND	12*	1.1*	**0.80**	**46**	2.7*
31–50y	**130**	25*	ND	12*	1.1*	**0.80**	**46**	2.7*
51–70y	**130**	21*	ND	11*	1.1*	**0.80**	**46**	2.7*
>70 y	**130**	21*	ND	11*	1.1*	**0.80**	**46**	2.7*
Pregnancy	**175**	28*	ND	13*	1.4*	**1.10**	**71**	3.0*
Lactation	**210**	29*	ND	13*	1.3*	**1.10**	**71**	3.8*

ND = not determined. *Values are AI (Adequate Intakes) (RDAs are in **bold** type), †Refers to all ω-6 polyunsaturated fatty acids, ‡Refers to all ω-3 polyunsaturated fatty acids.

Source: Institute of Medicine, Food and Nutrition Board. "Dietary Reference Intakes for Energy, Carbohydrate, Fiber, Fat, Fatty Acids, Cholesterol, Protein, and Amino Acids" (2002/2005); "Dietary Reference Intakes for Water, Potassium, Sodium, Chloride, and Sulfate" (2005) Washington, D.C.: National Academies Press.

[a] Based on g protein per kg of body weight for the reference body weight, e.g., for adults 0.8 g/kg body weight for the reference body weight.

[b] Total water includes all water contained in food, beverages, and drinking water.

TABLE A.5 Dietary Reference Intakes: Tolerable Upper Intake Levels (UL[a]): Vitamins

Life Stage Group	Vitamin A (µg/day)[b]	Vitamin C (mg/day)	Vitamin D (µg/day)	Vitamin E (mg/day)[c,d]	Vitamin K	Thiamin	Riboflavin	Niacin (mg/day)[d]	Vitamin B6 (mg/day)	Folate (µg/day)[d]	Vitamin B12	Pantothenic Acid	Biotin	Choline (g/day)	Carotenoids[e]
Infants															
0–6 mo	600	ND	25	ND	ND	ND	ND	ND	ND	ND	ND	ND	ND	ND	ND
6–12 mo	600	ND	38	ND	ND	ND	ND	ND	ND	ND	ND	ND	ND	ND	ND
Children															
1–3 y	600	400	63	200	ND	ND	ND	10	30	300	ND	ND	ND	1.0	ND
4–8 y	900	650	75	300	ND	ND	ND	15	40	400	ND	ND	ND	1.0	ND
Males, Females															
9–13 y	1,700	1,200	100	600	ND	ND	ND	20	60	600	ND	ND	ND	2.0	ND
14–18 y	2,800	1,800	100	800	ND	ND	ND	30	80	800	ND	ND	ND	3.0	ND
19–70 y	3,000	2,000	100	1,000	ND	ND	ND	35	100	1,000	ND	ND	ND	3.5	ND
>70 y	3,000	2,000	100	1,000	ND	ND	ND	35	100	1,000	ND	ND	ND	3.5	ND
Pregnancy															
14–18 y	2,800	1,800	100	800	ND	ND	ND	30	80	800	ND	ND	ND	3.0	ND
19–50 y	3,000	2,000	100	1,000	ND	ND	ND	35	100	1,000	ND	ND	ND	3.5	ND
Lactation															
14–18 y	2,800	1,800	100	800	ND	ND	ND	30	80	800	ND	ND	ND	3.0	ND
19–50 y	3,000	2,000	100	1,000	ND	ND	ND	35	100	1,000	ND	ND	ND	3.5	ND

[a]UL = The maximum level of daily nutrient intake that is likely to pose no risk of adverse effects. Unless otherwise specified, the UL represents total intake from food, water, and supplements. Due to lack of suitable data, ULs could not be established for vitamin K, thiamin, riboflavin, vitamin B_{12}, pantothenic acid, biotin, or carotenoids. In the absence of ULs, extra caution may be warranted in consuming levels above recommended intakes.

[b]As preformed vitamin A only.

[c]As α-tocopherol; applies to any form of supplemental α-tocopherol.

[d]The ULs for vitamin E, niacin, and folate apply to synthetic forms obtained from supplements, fortified foods, or a combination of the two.

[e]β-Carotene supplements are advised only to serve as a provitamin A source for individuals at risk of vitamin A deficiency.

ND = Not determinable due to lack of data of adverse effects in this age group and concern with regard to lack of ability to handle excess amounts. Source of intakes should be from food only to prevent high levels of intake.

Source: Dietary Reference Intake Tables: The Complete Set. Institute of Medicine, National Academy of Sciences. Available online at www.nap.edu. Reprinted with permission from 2006, by the National Academy of Sciences, Washington, D.C. Institute of Medicine, Food and Nutrition Board, Dietary Reference Intakes for Calcium and Vitamin D (2011), National Academies Press, Washington, DC, 2011.

TABLE A.6 Dietary Reference Intakes: Tolerable Upper Intake Levels (UL[a]): Minerals

Life Stage Group	Arsenic[b]	Boron (mg/day)	Calcium (g/day)	Chromium	Copper (μg/day)	Fluoride (mg/day)	Iodine (μg/day)	Iron (mg/day)	Magnesium (mg/day)[c]	Manganese (mg/day)	Molybdenum (μg/day)	Nickel (mg/day)	Phosphorus (g/day)	Selenium (μg/day)	Silicon[d]	Vanadium (mg/day)[e]	Zinc (mg/day)	Sodium (g/day)	Chloride (g/day)	Potassium
Infants																				
0–6 mo	ND	ND	1.0	ND	ND	0.7	ND	40	ND	ND	ND	ND	ND	45	ND	ND	4	ND	ND	ND
6–12 mo	ND	ND	1.5	ND	ND	0.9	ND	40	ND	ND	ND	ND	ND	60	ND	ND	5	ND	ND	ND
Children																				
1–3 y	ND	3	2.5	ND	1,000	1.3	200	40	65	2	300	0.2	3	90	ND	ND	7	1.5	2.3	ND
4–8 y	ND	6	2.5	ND	3,000	2.2	300	40	110	3	600	0.3	3	150	ND	ND	12	1.9	2.9	ND
Males, Females																				
9–13 y	ND	11	3.0	ND	5,000	10	600	40	350	6	1,100	0.6	4	280	ND	ND	23	2.2	3.4	ND
14–18 y	ND	17	3.0	ND	8,000	10	900	45	350	9	1,700	1.0	4	400	ND	ND	34	2.3	3.6	ND
19–30 y	ND	20	2.5	ND	10,000	10	1,100	45	350	11	2,000	1.0	4	400	ND	1.8	40	2.3	3.6	ND
31–50 y	ND	20	2.5	ND	10,000	10	1,100	45	350	11	2,000	1.0	4	400	ND	1.8	40	2.3	3.6	ND
51–70 y	ND	20	2.0	ND	10,000	10	1,100	45	350	11	2,000	1.0	4	400	ND	1.8	40	2.3	3.6	ND
>70 y	ND	20	2.0	ND	10,000	10	1,100	45	350	11	2,000	1.0	3	400	ND	1.8	40	2.3	3.6	ND
Pregnancy																				
14–18 y	ND	17	3.0	ND	8,000	10	900	45	350	9	1,700	1.0	3.5	400	ND	ND	34	2.3	3.6	ND
19–50 y	ND	20	2.5	ND	10,000	10	1,100	45	350	11	2,000	1.0	3.5	400	ND	ND	40	2.3	3.6	ND
Lactation																				
14–18 y	ND	17	3.0	ND	8,000	10	900	45	350	9	1,700	1.0	4	400	ND	ND	34	2.3	3.6	ND
19–50 y	ND	20	2.5	ND	10,000	10	1,100	45	350	11	2,000	1.0	4	400	ND	ND	40	2.3	3.6	ND

[a] UL = the maximum level of daily nutrient intake that is likely to pose no risk of adverse effects. Unless otherwise specified, the UL represents total intake from food, water, and supplements. Due to lack of suitable data, ULs could not be established for arsenic, chromium, silicon, and potassium. In the absence of ULs, extra caution may be warranted in consuming levels above recommended intakes.

[b] Although the UL was not determined for arsenic, there is no justification for adding arsenic to food or supplements.

[c] The ULs for magnesium represent intake from a pharmacological agent only and do not include intake from food and water.

[d] Although silicon has not been shown to cause adverse effects in humans, there is no justification for adding silicon to supplements.

[e] Although vanadium in food has not been shown to cause adverse effects in humans, there is no justification for adding vanadium to food and vanadium supplements should be used with caution. The UL is based on adverse effects in laboratory animals and this data could be used to set a UL for adults but not children and adolescents.

ND = Not determinable due to lack of data of adverse effects in this age group and concern with regard to lack of ability to handle excess amounts. Source of intake should be from food only to prevent high levels of intake.

Source: Dietary Reference Intake Tables: The Complete Set. Institute of Medicine, National Academy of Sciences. Available online at www.nap.edu. Reprinted with permission from. *Dietary Reference Intakes: The Essential Guide to Nutrient Requirements*, 2006, by the National Academy of Sciences, Washington, D.C. Institute of Medicine, Food and Nutrition Board. Dietary Reference Intakes for Calcium and Vitamin D (2011), National Academies Press, Washington, DC, 2011.

TABLE A.7 Dietary Reference Intake Values for Energy: Estimated Energy Requirement (EER) Equations and Values for Active Individuals by Life Stage Group

Life Stage Group	EER Prediction Equation	EER for Active Physical Activity Level (kCal\day)[a]	
		Male	Female
0–3 mo	EER = (89 × weight of infant in kg − 100) + 175	538	493 (2 mo)[c]
4–6 mo	EER = (89 × weight of infant in kg − 100) + 56	606	543 (5 mo)[c]
7–12 mo	EER = (89 × weight of infant in kg − 100) + 22	743	676 (9 mo)[c]
1–2 y	EER = (89 × weight of infant in kg − 100) + 20	1046	992 (2 y)[c]
3–8 y			
Male	EER = 88.5 − (61.9 × Age in yrs) + PA[b][(26.7 × Weight in kg) + (903 × Height in m)] + 20	1742 (6 y)[c]	
Female	EER = 135.3 − (30.8 × Age in yrs) + PA[b][(10.0 × Weight in kg) + (934 × Height in m)] + 20		1642 (6 y)[c]
9–13 y			
Male	EER = 88.5 − (61.9 × Age in yrs) + PA[b][(26.7 × Weight in kg) + (903 × Height in m)] + 25	2279 (11 y)[c]	
Female	EER = 135.3 − (30.8 × Age in yrs) + PA[b][(10.0 × Weight in kg) + (934 × Height in m)] + 25		2071 (11 y)[c]
14–18 y			
Male	EER = 88.5 − (61.9 − Age in yrs) + PA[b][(26.7 × Weight in kg) + (903 × Height in m)] + 25	3152 (16 y)[c]	
Female	EER = 135.3 − (30.8 × Age in yrs) + PA[b][(10.0 × Weight in kg) + (934 × Height in m)] + 25		2368 (16 y)[c]
19 and older			
Males	EER = 662 − (9.53 × Age in yrs) + PA[b][(15.91 × Weight in kg) + (539.6 × Height in m)]	3067 (19 y)[c]	
Females	EER = 354 − (6.91 × Age in yrs) + PA[b][(9.36 × Weight in kg) + (726 × Height in m)]		2403 (19 y)[c]
Pregnancy			
14–18 y			
1st trimester	Adolescent EER + 0		2368 (16 y)[c]
2nd trimester	Adolescent EER + 340		2708 (16 y)[c]
3rd trimester	Adolescent EER + 452		2820 (16 y)[c]
19–50 y			
1st trimester	Adult EER + 0		2403 (19 y)[c]
2nd trimester	Adult EER + 340		2743 (19 y)[c]
3rd trimester	Adult EER + 452		2855 (19 y)[c]
Lactation			
14–18 y			
1st 6 mo	Adolescent EER + 330		2698 (16 y)[c]
2nd 6 mo	Adolescent EER + 400		2768 (16 y)[c]
19–50 y			
1st 6 mo	Adult EER + 330		2733 (19 y)[c]
2nd 6 mo	Adult EER + 400		2803 (19 y)[c]

[a]The intake that meets the average energy expenditure of active individuals at a reference height, weight, and age.

[b]See table entitle "Physical Activity Coefficients (PA Values) for Use in EER Equations" to determine the PA value for various ages, genders, and activity levels.

[c]Value is calculated for an individual at the age in parentheses.

TABLE A.8 Physical Activity Coefficients (PA Values) for Use in EER Equations

Age and Gender	Sedentary	Low Active	Active	Very Active
3 to 18 y				
Boys	1.00	1.13	1.26	1.42
Girls	1.00	1.16	1.31	1.56
>19 y				
Men	1.00	1.11	1.25	1.48
Women	1.00	1.12	1.27	1.45

Source: Institute of Medicine, Food and Nutrition Board. "Dietary Reference Intakes for Energy, Carbohydrate, Fiber, Fat, Fatty Acids, Cholesterol, Protein, and Amino Acids." Washington, D.C.: National Academies Press, 2002, 2005.

TABLE A.9 Dietary Reference Intake Values for Energy: Total Energy Expenditure (TEE) Equations for Overweight and Obese Individuals

Life Stage Group	TEE Prediction Equation (Cal/day)	PA Values
Overweight boys aged 3–18 years	TEE = 114 − (50.9 × age in yrs) + PA[(19.5 × weight in kg) + (1161.4 × height in m)]	Sedentary = 1.00 Low active = 1.12 Active = 1.24 Very active = 1.45
Overweight girls aged 3–18 years	TEE = 389 − (41.2 × age in yrs) + PA[(15.0 × weight in kg) + (701.6 × height in m)]	Sedentary = 1.00 Low active = 1.18 Active = 1.35 Very active = 1.60
Overweight and obese men aged 19 years and older	TEE = 1086 − (10.1 × age in yrs) + PA[(13.7 × weight in kg) + (416 × height in m)]	Sedentary = 1.00 Low active = 1.12 Active = 1.29 Very active = 1.59
Overweight and obese women aged 19 years and older	TEE = 448 − (7.95 × age in yrs) + PA[(11.4 × weight in kg) + (619 × height in m)]	Sedentary = 1.00 Low active = 1.16 Active = 1.27 Very active = 1.44

Source: Institute of Medicine, Food and Nutrition Board "Dietary Reference Intakes for Energy, Carbohydrate, Fiber, Fat, Fatty Acids, Cholesterol, Protein, and Amino Acids," Washington, DC: National Academy Press, 2002, 2005.

Glossary

absorption The process of taking substances from the gastrointestinal tract into the interior of the body.

acceptable daily intake (ADI) The amount of a food additive or other substance that can be consumed daily over a lifetime without appreciable health risk to a person on the basis of all the known facts at the time of the evaluation.

Acceptable Macronutrient Distribution Ranges (AMDRs) Healthy ranges of intake for carbohydrate, fat, and protein, expressed as percentages of total energy intake.

accidental contaminant A substance not regulated by the FDA that unexpectedly enters the food supply.

acrylamide A chemical formed in foods when starches and other carbohydrates are overheated (over 120°C or 250°F) during cooking; may be a carcinogen.

active life expectancy The number of years a person is able to live in an independent state, without being limited by chronic conditions.

active transport The transport of substances across a cell membrane with the aid of a carrier molecule and the expenditure of energy.

added sugars Refined sugars and syrups that have been added to foods during processing or preparation.

adenosine triphosphate (ATP) A high-energy molecule that the body uses to power activities that require energy.

Adequate Intakes (AIs) Nutrient intakes that should be used as a goal when no RDA exists. AI values are an approximation of the nutrient intake that sustains health.

adipocyte A cell that stores fat.

adipose tissue Tissue found under the skin and around body organs that is composed of fat-storing cells.

adjustable gastric banding A surgical procedure in which an adjustable band is placed around the upper portion of the stomach to limit the volume that the stomach can hold and the rate of stomach emptying.

adolescent growth spurt An 18- to 24-month period of peak growth velocity that begins at about ages 10 to 13 in girls and 12 to 15 in boys.

aerobic activity Endurance activity that increases heart rate and uses oxygen to provide energy as ATP.

aerobic capacity The maximum amount of oxygen that can be consumed by the tissues during exercise. Also called maximal oxygen consumption, or VO_2 max.

aerobic metabolism Metabolism in the presence of oxygen. It can completely break down glucose to yield carbon dioxide, water, and energy in the form of ATP.

aerobic zone A level of activity that raises the heart rate to between 60 and 85% of maximum heart rate.

age-related bone loss Bone loss that occurs in both men and women as they advance in age.

aging The inevitable accumulation of changes associated with and responsible for an ever-increasing susceptibility to disease and death.

alcohol A molecule that contains 7 Calories per gram and is made by the fermentation of carbohydrates from plant products; the type of alcohol that is consumed in alcoholic beverages is called ethanol.

alcoholism See alcohol use disorder.

alcohol poisoning A potentially fatal condition characterized by mental confusion, stupor, vomiting, seizures, slow irregular breathing, and/or hypothermia caused by depression of the central nervous system due to excessive alcohol consumption.

alcohol use disorder or **alcoholism** A chronic disorder characterized by dependence on alcohol, with repeated excessive use of alcoholic beverages and development of withdrawal symptoms when alcohol intake is reduced.

alcoholic hepatitis An inflammation of the liver caused by alcohol consumption.

aldosterone A hormone secreted by the adrenal glands that increases sodium reabsorption and therefore enhances water reabsorption by the kidney.

allergen A substance that causes an allergic reaction.

alpha-carotene (α-carotene) A carotenoid, some of which can be converted into vitamin A, that is found in leafy green vegetables, carrots, and squash.

alpha-linolenic acid (α-linolenic acid) An 18-carbon omega-3 polyunsaturated fatty acid known to be essential in humans.

alpha-tocopherol (α-tocopherol) The form of tocopherol (vitamin E) active in humans.

Alzheimer's disease A disease that results in a relentless and irreversible loss of mental function.

amino acid pool All the amino acids in body tissues and fluids that are available for use by the body.

amino acids The building blocks of proteins. Each contains an amino group, an acid group, and a unique side chain.

anabolic steroids Synthetic fat-soluble hormones that mimic testosterone and are used to increase muscle strength and mass.

anaerobic metabolism Metabolism in the absence of oxygen.

anemia A condition in which the oxygen-carrying capacity of the blood is decreased by a reduced number of red blood cells or a reduced amount of hemoglobin in the cells.

anorexia nervosa An eating disorder characterized by self-starvation, a distorted body image, abnormally low body weight, and a pathological fear of becoming fat.

antibody A protein, released by a type of lymphocyte, that interacts with and neutralizes specific antigens.

antidiuretic hormone (ADH) A hormone secreted by the pituitary gland that increases the amount of water reabsorbed by the kidney and therefore retained in the body.

antigen A foreign substance that, when introduced into the body, stimulates an immune response.

antioxidant A substance that decreases the adverse effects of reactive molecules on normal physiological function.

appetite A desire to consume specific foods that is independent of hunger.

arachidonic acid A 20-carbon omega-6 polyunsaturated fatty acid that can be synthesized from linoleic acid.

arteriole A small artery that carries blood to the capillaries.

artery A blood vessel that carries blood away from the heart.

artificial sweetener See nonnutritive sweetener.

arthritis A disease characterized by inflammation of the joints, pain, and sometimes changes in structure.

GL-1

ascorbic acid The chemical term for vitamin C.

aseptic processing The placement of sterilized food in a sterilized package using a sterile process.

atherosclerosis A type of cardiovascular disease that involves the buildup of fatty material in the artery walls.

atherosclerotic plaque Cholesterol-rich material that is deposited in the arteries of individuals with atherosclerosis. It consists of cholesterol, smooth muscle cells, fibrous tissue, and eventually calcium.

atom The smallest unit of an element that retains the properties of the element.

atrophic gastritis An inflammation of the stomach lining that results in reduced secretion of stomach acid, microbial overgrowth, and, in severe cases, a reduction in the production of intrinsic factor.

atrophy Wasting or decrease in the size of a muscle or other tissue caused by lack of use.

autoimmune disease A disease that results from immune reactions that destroy normal body cells.

basal metabolic rate (BMR) The rate of energy expenditure under resting conditions. It is measured after 12 hours without food or exercise.

basal metabolism The energy expended to maintain an awake, resting body that is not digesting food.

behavior modification A process that is used to gradually and permanently change habitual behaviors.

beriberi A thiamin deficiency disease that manifests in one of two forms: dry beriberi, which causes weakness and nerve degeneration, or wet beriberi, which causes heart changes.

beta-carotene (β-carotene) A carotenoid found in many yellow and red-orange fruits and vegetables that is a precursor of vitamin A. It is also an antioxidant.

beta-cryptoxanthin (β-cryptoxanthin) A carotenoid found in corn, green peppers, and lemons that can provide some vitamin A activity.

bicarbonate A chemical released by the pancreas into the small intestine that neutralizes stomach acid.

bile A digestive fluid made in the liver and stored in the gallbladder that is released into the small intestine, where it aids in fat digestion and absorption.

binge drinking A pattern of drinking that brings a person's blood alcohol concentration to 80 mg/100 mL or above.

binge-eating disorder An eating disorder characterized by recurrent episodes of binge eating accompanied by a loss of control over eating in the absence of purging behavior.

bioaccumulation The process by which compounds accumulate or build up in an organism faster than they can be broken down or excreted.

bioavailability The extent to which the body can absorb and use a nutrient.

biotechnology The process of manipulating life forms via genetic engineering in order to provide desirable products for human use.

biotin One of the B vitamins, needed in energy metabolism.

blackout drinking Amnesia following a period of excessive alcohol consumption.

blood pressure The amount of force exerted by the blood against the walls of arteries.

body composition The term used to describe the different components (lean versus fat tissues) that when taken together make up an individual's body weight.

body image The way a person perceives and imagines his or her body.

body mass index (BMI) A measure of body weight relative to height that is used to compare body size with a standard.

bone remodeling The process whereby bone is continuously broken down and re-formed to allow for growth and maintenance.

bone resorption The process by which bone is broken down releasing calcium from bone into the blood.

botulism A severe food-borne intoxication that results from consuming the toxin produced by *Clostridium botulinum*.

bovine somatotropin (bST) A hormone naturally produced by cows that stimulates the production of milk. A synthetic version of this hormone is now being produced by genetic engineering.

bovine spongiform encephalopathy (BSE) A fatal neurological disease that affects cattle (mad cow disease). It is caused by a prion and may be transmitted to humans by consuming contaminated beef and beef-by-products.

bran The protective outer layers of whole grains. It is a concentrated source of dietary fiber.

brush border The microvilli surface of the intestinal mucosa, which contains some digestive enzymes.

bulimia nervosa An eating disorder characterized by the consumption of a large amount of food at one time (binge eating) followed by purging behaviors such as self-induced vomiting to prevent weight gain.

calcitonin A hormone produced by the thyroid gland that stimulates bone mineralization and inhibits bone breakdown, thus lowering blood calcium levels.

calorie A unit of measure used to express the amount of energy provided by food.

capillary A small, thin-walled blood vessel through which blood and the body's cells exchange gases and nutrients.

carbohydrate loading See glycogen supercompensation.

carbohydrates A class of nutrients that includes sugars, starches, and fibers. Chemically, they all contain carbon, along with hydrogen and oxygen, in the same proportions as in water (H_2O).

cardiorespiratory endurance The efficiency with which the body delivers to cells the oxygen and nutrients needed for muscular activity and transports waste products from cells.

cardiovascular disease Any disease affecting the heart and blood vessels.

cardiovascular system The organ system that includes the heart and blood vessels and circulates blood throughout the body.

carotenoids Yellow, orange, and red pigments synthesized by plants and many microorganisms. Some can be converted to vitamin A.

cash crop A crop that is grown to be sold for monetary return rather than as food for the local population.

cataracts A disease of the eye that results in cloudy spots on the lens (and sometimes the cornea), which obscure vision.

celiac disease A disorder that causes damage to the intestines when the protein gluten is eaten.

cell differentiation Structural and functional changes that cause cells to mature into specialized cells.

cell The basic structural and functional unit of living things.

cellular respiration The reactions that break down carbohydrates, fats, and proteins to produce carbon dioxide, water, and energy in the form of ATP.

cellulose An insoluble fiber that is the most prevalent structural material of plant cell walls.

cesarean section The surgical removal of the fetus from the uterus.

Chinese restaurant syndrome See MSG-symptom complex.

Choice Lists A new term for Exchange Lists; a system of grouping foods based on their carbohydrate, protein, fat, and energy content.

cholesterol A sterol, produced by the liver and consumed in the diet, that is needed to build cell membranes and make hormones and other essential molecules. High blood levels increase the risk of heart disease.

choline A compound needed for the synthesis of the phospholipid phosphatidylcholine, the neurotransmitter acetylcholine, and a number of other biochemical reactions. It is not a vitamin, but is considered an essential nutrient.

chylomicron A lipoprotein that transports lipids from the mucosal cells of the small intestine and delivers triglycerides to other body cells.

chyme A mixture of partially digested food and stomach secretions.

cirrhosis Chronic and irreversible liver disease characterized by loss of functioning liver cells and accumulation of fibrous connective tissue.

cobalamin The chemical term for vitamin B_{12}.

coenzyme An organic nonprotein substance that binds to an enzyme to promote its activity.

cofactor An inorganic ion or coenzyme that is required for enzyme activity.

collagen The major protein in connective tissue.

colostrum A substance, known as first milk, produced by the breast late in pregnancy and for up to a week after delivery. It contains more water, protein, immune factors, minerals, and vitamins and less fat than mature breast milk.

complete dietary protein See high-quality protein.

complex carbohydrates Carbohydrates composed of sugar molecules linked together in straight or branching chains. They include oligosaccharides, starches, and fibers.

conditionally essential amino acid An amino acid that is essential in the diet only under certain conditions or at certain times of life; also called semiessential amino acid.

constipation Infrequent or difficult defecation.

control group In a scientific experiment, the group of participants used as a basis of comparison. They are similar to the participants in the experimental group but do not receive the treatment being tested.

creatine A compound that can be converted into creatine phosphate, which replenishes muscle ATP during short bursts of activity. Creatine is a dietary supplement used by athletes to increase muscle mass and delay fatigue during short intense exercise.

creatine phosphate A compound stored in muscle that can be broken down quickly to make ATP.

cretinism A condition resulting from poor maternal iodine intake during pregnancy that impairs mental development and growth in the offspring.

crib death See sudden infant death syndrome (SIDS).

critical control point A possible point in food production, manufacturing, and transportation where contamination could occur or be prevented or eliminated.

critical period Time in growth and development when an organism is more susceptible to harm from poor nutrition or other environmental factors.

cross-contamination The transfer of contaminants from one food or object to another.

cycle of undernutrition A cycle in which undernutrition is perpetuated by an inability to meet nutrient needs at all life stages.

Daily Value A reference value for the intake of nutrients used on food labels to help consumers see how a given food fits into their overall diet.

DASH (Dietary Approaches to Stop Hypertension) eating plan A dietary pattern recommended to lower blood pressure. It is abundant in fruits and vegetables; includes low-fat dairy products, whole grains, legumes, and nuts; and incorporates moderate amounts of lean meat.

deamination The removal of the amino group from an amino acid.

dehydration A condition that results when not enough water is present to meet the body's needs.

Delaney Clause A clause added to the 1958 Food Additives Amendment of the Pure Food and Drug Act that prohibits the intentional addition to foods of any compound that has been shown to induce cancer in animals or humans at any dose.

dementia A deterioration of mental state that results in impaired memory, thinking, and/or judgment.

denaturation Alteration of a protein's three-dimensional structure.

dental caries Cavities, or decay of the tooth enamel, caused by acid produced when bacteria growing on the teeth metabolize carbohydrate.

designer food or **nutraceutical** A food or supplement thought to have health benefits in addition to its nutritive value.

diabetes mellitus A disease characterized by elevated blood glucose due to either insufficient production of insulin or decreased sensitivity of cells to insulin.

diarrhea An intestinal disorder characterized by frequent or watery stools.

dietary folate equivalent (DFE) The amount of folate equivalent to 1 µg of folate naturally occurring in food, 0.6 µg of synthetic folic acid from fortified food or supplements consumed with food, or 0.5 µg synthetic folic acid consumed on an empty stomach.

Dietary Guidelines for Americans A set of nutrition recommendations designed to promote population-wide dietary and lifestyle changes to reduce the incidence of nutrition-related chronic disease.

Dietary Reference Intakes (DRIs) A set of reference values for the intake of energy, nutrients, and food components that can be used for planning and assessing the diets of healthy people in the United States and Canada.

dietary supplement A product designed to supplement the diet; may include nutrients (vitamins, minerals, amino acids, fatty acids), enzymes, herbs, or other substances.

diet-induced thermogenesis See thermic effect of food (TEF).

diffusion The movement of molecules from an area of higher concentration to an area of lower concentration without the expenditure of energy.

digestion The process by which food is broken down into components small enough to be absorbed into the body.

dipeptide Two amino acids linked by a peptide bond.

direct food additive A substance that is intentionally added to food. Direct food additives are regulated by the FDA.

disaccharide A carbohydrate made up of two sugar units.

dispensable amino acid See nonessential amino acid.

diuretic A substance that increases the amount of urine passed from the body.

diverticula Sacs or pouches that protrude from the wall of the large intestine.

diverticulitis A condition in which diverticula in the large intestine become inflamed.

diverticulosis A condition in which out-pouches (or sacs) form in the wall of the large intestine.

docosahexaenoic acid (DHA) A 22-carbon omega-3 polyunsaturated fatty acid found in fish that may be needed in the diet of newborns. It can be synthesized from α-linolenic acid.

Down syndrome A disorder caused by a chromosomal abnormality that results in distinctive facial characteristics, mental impairment, and other abnormalities.

eating disorder A psychological illness characterized by specific abnormal eating behaviors, often intended to control weight.

eclampsia Convulsions or seizures that occur in association with preeclampsia. Untreated, it can lead to coma or death.

edema Swelling due to the buildup of extracellular fluid in the tissues.

eicosanoids Regulatory molecules that can be synthesized from omega-3 and omega-6 fatty acids.

eicosapentaenoic acid (EPA) A 20-carbon omega-3 polyunsaturated fatty acid found in fish that can be synthesized from α-linolenic acid but may be essential in humans under some conditions.

electrolyte A positively or negatively charged ion that conducts an electrical current in solution. Commonly refers to sodium, potassium, and chloride.

element A substance that cannot be broken down into products with different properties.

embryo A developing human from two through eight weeks after fertilization.

empty calories Calories from solid fats and/or added sugars, which add calories to the food but few nutrients.

emulsifier A substance with both water-soluble and fat-soluble portions that can break fat into tiny droplets and suspend it in a watery fluid.

endorphins Compounds that cause a natural euphoria and reduce the perception of pain under certain stressful conditions.

endosperm The largest portion of a kernel of grain. It is primarily starch and serves as a food supply for the sprouting seed.

energy balance The amount of energy consumed in the diet compared with the amount expended by the body over a given period.

energy-yielding nutrients Nutrients that can be metabolized to provide energy in the body. They include carbohydrates, fats, and proteins.

enrichment The addition of specific amounts of nutrients to replace those lost during processing.

enzyme A protein molecule that accelerates the rate of a chemical reaction without itself being changed.

epidemiology The branch of science that studies health and disease trends and patterns in populations.

epiglottis A piece of elastic connective tissue that covers the opening to the lungs during swallowing.

ergogenic aid A substance, appliance, or procedure that improves athletic performance.

essential fatty acid A fatty acid that must be consumed in the diet because it cannot be made by the body or cannot be made in sufficient quantities to meet the body's needs.

essential fatty acid deficiency A condition characterized by dry, scaly skin and poor growth that results when the diet does not supply sufficient amounts of linoleic acid and α-linolenic acid.

essential nutrient A nutrient that must be consumed in the diet because it cannot be made by the body or cannot be made in sufficient quantities to maintain body functions.

essential, or **indispensable, amino acid** An amino acid that cannot be synthesized by the body in sufficient amounts to meet its needs and therefore must be included in the diet.

Estimated Average Requirements (EARs) Nutrient intakes estimated to meet the needs of 50% of the healthy individuals in a given gender and life-stage group.

Estimated Energy Requirements (EERs) Energy intakes that are predicted to maintain body weight in healthy individuals.

ethanol The alcohol in alcoholic beverages; it is produced by yeast fermentation of sugar.

evidence-based practice Using the compiled evidence from all available well-controlled, peer-reviewed studies to develop recommendations and policies regarding nutrition and health care.

Exchange Lists A system of grouping foods based on their carbohydrate, protein, fat, and energy content. These have been revised as "Choice Lists".

experimental group In a scientific experiment, the group of participants who undergo the treatment being tested.

extreme obesity or **morbid obesity** A body mass index of 40 km/m² or greater.

facilitated diffusion Assisted diffusion of a substance across a cell membrane.

failure to thrive Inability of a child's growth to keep up with normal growth curves.

famine A widespread lack of access to food due to a disaster that causes a collapse in food production and marketing systems.

fasting hypoglycemia Low blood sugar that is not related to food intake; often caused by an insulin-secreting tumor.

fat-soluble vitamin A vitamin that does not dissolve in water; includes vitamins A, D, E, and K.

fatty acid A molecule made up of a chain of carbons linked to hydrogens, with an acid group at one end of the chain.

fatty liver The accumulation of fat in the liver. An early symptom of excess alcohol consumption.

feces Body waste, including unabsorbed food residue, bacteria, mucus, and dead cells, which is eliminated from the gastrointestinal tract by way of the anus.

female athlete triad The combination of energy restriction, changes in hormone levels that affect the menstrual cycle, and low bone mineral density that occurs in some female athletes, particularly those involved in sports in which low body weight and appearance are important; now recognized as part of relative energy deficiency in sport (RED-S).

fertilization The union of a sperm and an egg.

fetal alcohol spectrum disorders (FASDs) A range of physical and behavioral disorders or conditions and functional or mental impairments linked to prenatal alcohol exposure. One of the most severe FASDs is fetal alcohol syndrome.

fetal alcohol syndrome (FAS) A characteristic group of physical and mental abnormalities in an infant resulting from maternal alcohol consumption during pregnancy.

fetus A developing human from the ninth week after fertilization to birth.

fiber A type of carbohydrate that cannot be broken down by human digestive enzymes.

fibrin The protein produced during normal blood clotting that forms an interlacing fibrous network that is essential for formation of the clot.

fitness A set of attributes related to the ability to perform routine physical activities without undue fatigue.

fluorosis A condition caused by chronic overconsumption of fluoride, characterized by black and brown stains and cracking and pitting of the teeth.

foam cell A cholesterol-filled white blood cell.

folate A general term that refers to the many forms of this B vitamin, which is needed for the synthesis of DNA and the metabolism of some amino acids.

folic acid An easily absorbed form of the vitamin folate that is used in dietary supplements and fortified foods.

food additive A substance that is either intentionally added to or can reasonably be expected to become a component of a food during processing.

food allergy An adverse immune response to a specific food protein.

food desert An area that lacks access to affordable fruits, vegetables, whole grains, low-fat milk, and other foods that make up a healthy diet.

food environment The physical, economic, and social factors that affect eating habits and patterns.

food guide A food group system that suggests amounts of different types of foods needed to meet nutrient intake recommendations.

food insecurity A situation in which people lack adequate physical, social, or economic access to sufficient, safe, nutritious food that meets their dietary needs and food preferences for an active and healthy life.

food intolerance or **food sensitivity** An adverse reaction to a food that does not involve the production of antibodies by the immune system.

food recovery The collection of wholesome food for distribution to the poor and hungry, including collection of crops from farmers' fields that have already been mechanically harvested or from fields where it is not economically profitable to harvest.

food-borne illness An illness caused by consumption of contaminated food.

food-borne infection An illness caused by the ingestion of food containing microorganisms that can multiply inside the body and produce effects that are injurious.

food-borne intoxication An illness caused by consuming a food containing a toxin.

food sensitivity See food intolerance.

fortification The addition of nutrients to foods.

free radical A type of highly reactive atom or molecule that causes oxidative damage.

fructose A monosaccharide found in fruits and honey that is composed of six carbon atoms arranged in a ring structure; commonly called fruit sugar.

functional food A food that has health-promoting properties beyond basic nutritional functions.

galactose A monosaccharide composed of six carbon atoms arranged in a ring structure; when combined with glucose, it forms the disaccharide lactose.

gallstone A solid mass formed in the gallbladder or bile duct when substances in the bile harden.

gastric bypass A surgical procedure that reduces the size of the stomach and bypasses a portion of the small intestine.

gastric juice A substance produced by the gastric glands of the stomach that contains hydrochloric acid and an inactive form of pepsin.

gastric sleeve surgery A bariatric surgical procedure that removes part of the stomach leaving a thin vertical sleeve about the size of a banana. The smaller stomach restricts the amount of food that can be consumed at one session and therefore reduces caloric intake.

gastrin A hormone secreted by the mucosa of the stomach that stimulates the secretion of enzymes and acid in the stomach.

gastroesophageal reflux disease (GERD) A chronic condition in which acidic stomach contents leak into the esophagus, causing pain and damaging the esophagus.

gastrointestinal tract A hollow tube consisting of the mouth, pharynx, esophagus, stomach, small intestine, and large intestine in which digestion of food and absorption of nutrients occur; also called the alimentary canal, GI tract or digestive tract.

gene A length of DNA that contains the information needed to synthesize a polypeptide chain or a molecule of RNA; responsible for inherited traits.

gene expression The events of protein synthesis in which the information coded in a gene is used to synthesize a protein or a molecule of RNA.

generally recognized as safe (GRAS) A group of chemical additives that are considered safe, based on their long-standing presence in the food supply without harmful effects.

genetic engineering or **genetic modification (GM)** A set of techniques used to manipulate DNA for the purpose of changing the characteristics of an organism or creating a new product.

genetically modified organism (GMO) An organism whose genetic material has been altered using genetic engineering techniques.

germ The embryo or sprouting portion of a kernel of grain. It contains vegetable oil, vitamins, and minerals.

gestational diabetes A condition characterized by high blood glucose levels that develop during pregnancy.

gestational hypertension High blood pressure that develops after the 20th week of pregnancy and returns to normal after delivery. It may be an early sign of preeclampsia.

ghrelin A hormone produced by the stomach that stimulates appetite.

glucagon A hormone made in the pancreas that raises blood glucose levels by stimulating the breakdown of liver glycogen and the synthesis of glucose.

glucose A 6-carbon monosaccharide that is the primary form of carbohydrate used to provide energy in the body. Also known as blood sugar.

glutathione peroxidase A selenium-containing enzyme that protects cells from oxidative damage by neutralizing peroxides.

gluten-related disorders A term used to refer to all diseases triggered by the ingestion of gluten. Includes celiac disease, non-celiac gluten sensitivity, and wheat allergy.

glycemic index A ranking of the effect that the consumption of a single carbohydrate-containing food has on blood glucose in relation to consumption of a reference carbohydrate such as white bread or glucose.

glycemic load An index of the glycemic response that occurs after eating specific foods. It is calculated by multiplying a food's glycemic index by the amount of available carbohydrate in a serving of the food.

glycemic response The rate, magnitude, and duration of the rise in blood glucose that occurs after food is consumed.

glycogen The storage form of carbohydrate in animals, made up of many glucose molecules linked together in a highly branched structure.

glycogen supercompensation or **carbohydrate loading** A regimen designed to increase muscle glycogen stores beyond their usual level.

glycolysis An anaerobic metabolic pathway that splits glucose into two 3-carbon pyruvate molecules; the energy released from one glucose molecule is used to make two molecules of ATP.

goiter An enlargement of the thyroid gland caused by a deficiency of iodine.

goitrogen A substance that interferes with the utilization of iodine or the function of the thyroid gland.

Hazard Analysis Critical Control Point (HACCP) A food safety system that focuses on identifying and preventing hazards that could cause food-borne illness.

health claim A food label claim that describes the relationship between a nutrient or food and a disease or health condition. Only approved health claims may appear on food labels.

Healthy People A set of national health promotion and disease prevention objectives for the U.S. population.

heartburn A burning sensation in the chest or throat caused when acidic stomach contents leak back into the esophagus.

heat cramps Muscle cramps caused by an imbalance of sodium and potassium; may result from excessive exercise without adequate fluid and electrolyte replacement.

heat exhaustion Low blood pressure, rapid pulse, fainting, and sweating caused when dehydration decreases blood volume so much that blood can no longer both cool the body and provide oxygen to the muscles.

heat-related illnesses Conditions, including heat cramps, heat exhaustion, and heat stroke, that can occur due to an unfavorable combination of exercise, hydration status, and climatic conditions.

heat stroke Elevated body temperature as a result of fluid loss and the failure of the temperature regulatory center of the brain.

heme iron A readily absorbable form of iron found in meat, fish, and poultry that is chemically associated with certain proteins.

hemochromatosis An inherited disorder that results in increased iron absorption.

hemoglobin An iron-containing protein in red blood cells that binds and transports oxygen through the bloodstream to cells.

hemolytic anemia A condition in which there is an insufficient number of red blood cells because many have burst.

hemolytic-uremic syndrome A disorder, usually in children, that occurs when an infection in the digestive system produces toxic substances that destroy red blood cells causing them to block capillaries, eventually resulting in kidney failure.

hemorrhoid A swollen vein in the anal or rectal area.

hepatic portal vein The vein that transports blood from the gastrointestinal tract to the liver.

heterocyclic amines (HCAs) A class of mutagenic substances produced when there is incomplete combustion of amino acids during the cooking of meats—for example, when meat is charred.

high-density lipoprotein (HDL) A lipoprotein that picks up cholesterol from cells and transports it to the liver so that it can be eliminated from the body.

high-quality protein or **complete dietary protein** Easily digested protein that contains all of the amino acids needed for protein synthesis in amounts similar to those found in body proteins.

hormone A chemical messenger that is produced in one location in the body, released into the blood, and travels to other locations, where it elicits responses.

hunger A desire to consume food that is triggered by internal physiological signals. Also refers to recurrent involuntary lack of food that over time may lead to malnutrition.

hydrogenation A process used to make partially hydrogenated oils in which hydrogen atoms are added to the carbon–carbon double bonds of unsaturated fatty acids, making them more saturated. *Trans* fatty acids are formed during the process.

hydrolyzed protein or **protein hydrolysate** A mixture of amino acids or amino acids and polypeptides that results when a protein is completely or partially broken down by treatment with acid or enzymes.

hypercarotenemia A condition caused by the accumulation of carotenoids in the adipose tissue, causing the skin to appear yellow-orange.

hypertension Blood pressure that is consistently elevated to 140/90 millimeters (mm) mercury or greater.

hypertensive disorders of pregnancy A spectrum of conditions involving elevated blood pressure during pregnancy.

hypertrophy An increase in the size of a muscle or organ.

hypoglycemia Abnormally low blood glucose levels.

hyponatremia Abnormally low concentration of sodium in the blood.

hypothesis A proposed explanation for an observation or a scientific problem that can be tested through experimentation.

implantation The process through which a developing embryo embeds itself in the uterine lining.

incomplete dietary protein A protein that is deficient in one or more of the amino acids required for protein synthesis in humans.

indirect food additive A substance that is expected to unintentionally enter food during manufacturing or from packaging. Indirect food additives are regulated by the FDA.

indispensable amino acid See essential amino acid.

infancy The period of early childhood, generally from birth to 1 year of age.

infant botulism A potentially life-threatening disease in which the bacteria *Clostridium botulinum* grows in an infant's gastrointestinal tract.

infant mortality rate The number of deaths during the first year of life per 1000 live births.

inflammation A protective response to injury or destruction of tissues; signs of acute inflammation include pain, heat, redness, swelling, and loss of function.

insoluble fiber Fiber that does not dissolve in water and is less readily be broken down by bacteria in the large intestine. It includes cellulose, some hemicelluloses, and lignin.

insulin A hormone made in the pancreas that allows glucose to enter cells and stimulates the synthesis of protein, fat, and liver and muscle glycogen.

insulin resistance A condition in which the normal amount of insulin produces a subnormal effect in the body.

integrated pest management (IPM) A method of agricultural pest control that integrates nonchemical and chemical techniques.

intestinal microbiota The population of microorganisms associated with the gastrointestinal tract.

intrinsic factor A protein produced in the stomach that aids in the absorption of adequate amounts of vitamin B_{12}.

iodized salt Table salt to which a small amount of sodium iodide or potassium iodide has been added in order to supplement the iodine content of the diet.

ion An atom or a group of atoms that carries an electrical charge.

iron deficiency anemia An iron deficiency disease that occurs when the oxygen-carrying capacity of the blood is decreased because there is insufficient iron to make hemoglobin.

iron overload A condition in which iron accumulates in the tissues; characterized by bronzed skin, enlarged liver, diabetes mellitus, and abnormalities of the pancreas and the joints.

irradiation A process that exposes foods to radiation in order to kill contaminating organisms and retard the ripening and spoilage of fruits and vegetables.

Keshan disease A type of heart disease that occurs in an area of China where the soil is very low in selenium. It is believed to be cause by a combination of viral infection and selenium deficiency.

ketoacidosis A life-threatening condition in which ketone levels in the blood are high enough to increase blood acidity.

ketone or **ketone body** An acidic molecule formed when there is not sufficient carbohydrate to break down acetyl-CoA.

ketosis High levels of ketones in the blood.

kilocalorie (kcal) A unit of heat that is used to express the amount of energy provided by foods. It is the amount of heat required to raise the temperature of 1 kilogram of water 1 degree Celsius (1 kcalorie = 1000 calories. When Calorie is spelled with a capital C it denotes kilocalorie).

kwashiorkor A form of protein-energy malnutrition characterized by edema that is usually the result of a severe restriction in protein intake.

lactation Production and secretion of milk.

lacteal A lymph vessel in the villi of the small intestine that picks up particles containing the products of fat digestion.

lactic acid An end product of anaerobic metabolism and an additive used in food to maintain acidity.

lactose A disaccharide made of glucose linked to galactose that is found in milk.

lactose intolerance The inability to completely digest lactose due to a reduction in the levels of the enzyme lactase.

large for gestational age Refers to an infant weighing more than 4 kg (8.8 lb) at birth.

LDL receptor A protein on the surface of cells that binds to LDL particles and allows their contents to be taken up for use by the cell.

lean body mass Body mass attributed to non-fat body components such as bone, muscle, and internal organs; also called *fat-free mass*.

lecithin A phosphoglyceride composed of a glycerol backbone, two fatty acids, a phosphate group, and a molecule of choline; often used as an emulsifier in foods.

legume The starchy seed of a plant that produces bean pods; includes peas, peanuts, beans, soybeans, and lentils.

leptin A protein hormone produced by adipocytes that signals information about the amount of body fat.

let-down The release of milk from the milk-producing glands and its movement through the ducts and storage sinuses.

life expectancy The average length of life for a particular population of individuals.

life span The maximum age to which members of a species can live.

limiting amino acid The essential amino acid that is available in the lowest concentration relative to the body's needs.

linoleic acid An omega-6 essential fatty acid with 18 carbons and 2 carbon–carbon double bonds.

lipase A fat-digesting enzyme.

lipid bilayer Two layers of phosphoglyceride molecules oriented so that the fat-soluble fatty acid tails are sandwiched between the water-soluble phosphate-containing heads.

lipids A class of nutrients, commonly called fats, that includes fatty acids, triglycerides, phospholipids, and sterols; most do not dissolve in water.

lipoprotein A particle that transports lipids in the blood.

lipoprotein lipase An enzyme that breaks down triglycerides into free fatty acids and glycerol; attached to the outside of the cells that line the blood vessels.

liposuction A procedure that suctions out adipose tissue from under the skin; used to decrease the size of local fat deposits such as on the abdomen or hips.

low birth weight A birth weight less than 2.5 kg (5.5 lb).

low-density lipoprotein (LDL) A lipoprotein that transports cholesterol to cells.

lumen The inside cavity of a tube, such as the gastrointestinal tract.

lymphatic system The system of lymph vessels and other lymph organs and tissues that drains excess fluid from the spaces between cells and provides immune function.

lymphocyte A small white blood cell (leukocyte) that plays a large role in defending the body against disease.

macrocytic anemia A reduction in the blood's hemoglobin content and hence capacity to carry oxygen that is characterized by abnormally large immature and mature red blood cells.

macronutrients Nutrients needed by the body in large amounts. These include water and the energy-yielding nutrients carbohydrates, lipids, and proteins.

macrophage A type of white blood that ingests foreign material as part of the immune response to foreign invaders such as infectious microorganisms.

macula An oval, yellow-pigmented area on the central retina, which is the central point of sharpest vision.

macular degeneration Degeneration of a portion of the retina that results in loss of visual detail and eventually blindness.

major mineral A mineral required in the diet in an amount greater than 100 mg/day or present in the body in an amount greater than 0.01% of body weight.

malnutrition A condition resulting from an energy or nutrient intake either above or below that which is optimal.

maltose A disaccharide made of two glucose molecules linked together.

marasmus A form of protein-energy malnutrition characterized by wasting of both muscle and fat that is usually the result of a severe energy restriction.

maximum heart rate The maximum number of beats per minute that the heart can attain.

menopause The time in a woman's life when the menstrual cycle ends.

metabolic pathway A series of chemical reactions inside of a living organism that results in the transformation of one molecule into another.

metabolic syndrome A collection of health risks, including high blood pressure, altered blood lipids, high blood glucose, and a large waist circumference, that increase the chance of developing heart disease, stroke, and diabetes.

metabolism The sum of all the chemical reactions that take place in a living organism.

micelle A particle that is formed in the small intestine when the products of fat digestion are surrounded by bile. It facilitates the absorption of lipids.

microbes Microscopic organisms, or microorganisms, including bacteria, viruses, and fungi.

micronutrients Nutrients needed by the body in small amounts. These include vitamins and minerals.

microvillus (plural, **microvilli**) A minute projection on the mucosal cell membrane that increases the absorptive surface area in the small intestine.

mineral In nutrition, an element needed by the body to maintain structure and regulate chemical reactions and body processes.

mitochondrion (plural, **mitochondria**) A cellular organelle that is responsible for providing energy in the form of ATP via aerobic metabolism; the citric acid cycle and electron transport chain are located here.

modified atmosphere packaging (MAP) A preservation technique used to prolong the shelf life of processed or fresh food by changing the gases surrounding the food in the package.

mold Multicellular fungi that form filamentous branching growths.

molecule A group of two or more atoms of the same or different elements bonded together.

monoglyceride A glycerol molecule with one fatty acid attached.

monosaccharide A carbohydrate made up of a single sugar unit.

monosodium glutamate (MSG) A food additive used as a flavor enhancer, commonly used in Chinese food; made up of the amino acid glutamate bound to sodium.

monounsaturated fatty acid A fatty acid containing one carbon–carbon double bond.

morbid obesity See extreme obesity.

morning sickness Nausea and vomiting that affects many women during the first few months of pregnancy and that in some women can continue throughout the pregnancy.

MSG symptom complex or **Chinese restaurant syndrome** A group of symptoms including headache, flushing, tingling, burning sensations, and chest pain reported by some individuals after consuming monosodium glutamate (MSG).

mucosa The layer of tissue lining the gastrointestinal tract and other body cavities.

mucosal cell A type of epithelial cell that makes up mucus membranes such as those that line the gastrointestinal tract.

mucus A viscous fluid secreted by glands in the digestive tract and other parts of the body. It lubricates, moistens, and protects cells from harsh environments.

muscle endurance The ability of a muscle group to continue muscle movement over time.

muscle strength The amount of force that can be produced by a single contraction of a muscle.

muscle-strengthening exercise Activities that are specifically designed to increase muscle strength, endurance, and size; also called strength-training exercise or resistance-training exercise.

myelin A soft, white, fatty substance that covers nerve fibers and aids in nerve transmission.

myoglobin An iron-containing protein in muscle cells that binds oxygen.

MyPlate A plate-shaped food guide released in 2011 that suggests amounts and types of food from five food groups to meet the recommendations of the Dietary Guidelines.

National School Lunch Program A federally funded program designed to provide free or reduced cost lunches to school-age children.

neural tube defect An abnormality in the brain or spinal cord that results from errors that occur during prenatal development.

neurotransmitter A chemical substance produced by a nerve cell that can stimulate or inhibit another cell.

niacin A B vitamin needed in energy metabolism.

niacin equivalent (NE) A unit used to express the amount of niacin present in food, including that which can be made from its precursor, tryptophan. One NE is equal to 1 mg of niacin or 60 mg of tryptophan.

night blindness Inability to see clearly in dim light.

nitrogen balance The amount of nitrogen consumed in the diet compared with the amount excreted over a given period.

nitrosamine A carcinogenic compound produced by reactions between nitrites and amino acids.

nonessential, or dispensable, amino acid Amino acid that can be synthesized by the human body in sufficient amounts to meet needs.

nonexercise activity thermogenesis (NEAT) The energy expended for everything we do other than sleeping, eating, or sports-like exercise.

nonnutritive sweetener or artificial sweetener A substance used to sweeten food that provides few or no calories.

nursing bottle syndrome Extreme tooth decay in the upper teeth resulting from putting a child to bed with a bottle containing milk or other sweet liquids.

nutraceutical See designer food.

nutrient A substance in food that provides energy and structure to the body and regulates body processes.

nutrient content claim A claim on food labels used to describe the level of a nutrient in a food. The Nutrition Labeling and Education Act of 1990 defines these terms and regulates the circumstances under which they can be used.

nutrient density A measure of the nutrients provided by a food relative to its calorie content.

nutrigenetics The study of how the genes people inherit affect the impact their diet has on health.

nutrigenomics The study of how the nutrients and other food components people consume affect gene activity.

Nutrition Facts The portion of a food label that provides information about the nutritional composition of a food and how that food fits into the overall diet.

nutrition transition A series of changes in diet, physical activity, health, and nutrition that occur as poor countries become more prosperous.

nutritional genomics The study of how our genes affect the impact of nutrients or other food components on health (nutrigenetics) and how nutrients affect the activity of our genes (nutrigenomics).

nutritional programming The process through which nutritional conditions that exist while a baby is developing in the womb, and in infancy, affect future development and health.

nutritional status An individual's health, as it is influenced by the intake and utilization of nutrients.

obese Having excess body fat. Obesity is defined as having a body mass index (ratio of weight to height squared) of 30 kg/m^2 or greater.

obesity genes Genes that code for proteins involved in the regulation of body fat. When they are abnormal, the result is abnormal amounts of body fat.

obesogenic environment Surroundings, opportunities, and conditions of life that encourage weight gain by promoting excess calorie intake and a sedentary lifestyle.

oligosaccharide A carbohydrate made up of 3 to 10 sugar units.

omega-3 fatty acid A fatty acid containing a carbon–carbon double bond between the third and fourth carbons from the omega end; includes α-linolenic acid found in vegetable oils and eicosapentaenoic acid (EPA) and docosahexaenoic acid (DHA) found in fish oils.

omega-6 fatty acid A fatty acid containing a carbon–carbon double bond between the sixth and seventh carbons from the omega end; includes linoleic and arachidonic acid.

organ A discrete structure composed of more than one tissue that performs a specialized function.

organ system A group of cooperative organs.

organic compound A substance that contains carbon bonded to hydrogen.

organic food Food that is produced, processed, and handled in accordance with the standards of the USDA National Organic Program.

osmosis The unassisted diffusion of water across the cell membrane.

osteomalacia A vitamin D deficiency disease in adults, characterized by loss of minerals from bone, bone pain, muscle aches, and an increase in bone fractures.

osteopenia A reduction in bone density to below normal levels, but not low enough to be classified as osteoporosis.

osteoporosis A bone disorder characterized by reduced bone mass, increased bone fragility, and increased risk of fractures.

overload principle The concept that the body adapts to the stresses placed on it.

overnutrition Poor nutritional status resulting from a dietary intake in excess of that which is optimal for health.

overtraining syndrome A collection of emotional, behavioral, and physical symptoms that occurs when the amount and intensity of exercise exceeds an athlete's capacity to recover.

overweight Being too heavy for one's height, usually due to an excess of body fat. Overweight is defined as having a body mass index (ratio of weight to height squared) of 25 to 29.9 kilograms/meter^2 (kg/m^2).

oxidative stress An imbalance between reactive oxygen molecules and antioxidant defenses that results in damage.

oxidized LDL cholesterol A substance formed when the cholesterol in LDL particles is oxidized by reactive oxygen molecules. It is key in the development of atherosclerosis because it is taken up by macrophages, transforming them into foam cells.

oxytocin A hormone released by the posterior pituitary that stimulates the ejection or let-down of milk during lactation.

pancreatic amylase A starch-digesting enzyme found in pancreatic juice.

pancreatic juice Fluid secreted by the pancreas that contains bicarbonate to neutralize acid and enzymes that digest carbohydrates, fats, and proteins.

pantothenic acid One of the B vitamins, needed in energy metabolism.

parasite An organism that lives at the expense of another.

parathyroid hormone (PTH) A hormone released by the parathyroid gland that acts to increase blood calcium levels.

partially hydrogenated oils Liquid oils that have undergone a process that adds hydrogen atoms to the carbon–carbon double bonds of unsaturated fatty acids to make the oil more solid and/or increase the shelf life of the product.

pasteurization The process of heating food products in order to kill disease-causing organisms.

pathogen A biological agent that causes disease.

peak bone mass The maximum bone density attained at any time in life, usually occurring in young adulthood.

peer-review process The review of the design and validity of a research experiment by experts in the field of study who did not participate in the research.

pellagra A disease resulting from niacin deficiency, which causes dermatitis, diarrhea, dementia, and, if not treated, death.

pepsin A protein-digesting enzyme produced by the stomach. It is secreted in the gastric juice in an inactive form (pepsinogen) and activated by acid in the stomach.

peptic ulcer An open sore in the lining of the stomach, esophagus, or upper small intestine.

peptide Two or more amino acids joined by peptide bonds.

peptide bond The chemical linkage between the amino group of one amino acid and the acid group of another.

peristalsis Coordinated muscular contractions that move material through the GI tract.

pernicious anemia A macrocytic anemia resulting from vitamin B_{12} deficiency that occurs when dietary vitamin B_{12} cannot be absorbed due to a lack of intrinsic factor.

phagocyte A type of white blood cell that engulfs and consumes foreign particles such as bacteria.

pharynx A funnel-shaped opening that connects the nasal passages and mouth to the respiratory passages and esophagus. It is a common passageway for food and air and is responsible for swallowing.

phenylketonuria (PKU) A genetic disease in which the amino acid phenylalanine cannot be metabolized normally, causing it to build up in the blood. If untreated, the condition results in brain damage.

phosphate group A chemical group consisting of one phosphorus atom and four oxygen atoms.

phospholipid A type of lipid whose structure includes a phosphorus atom.

photosynthesis The metabolic process by which plants trap energy from the sun and use it to make sugars from carbon dioxide and water.

physical frailty Impairment in function and reduction in physiologic reserves severe enough to cause limitations in the basic activities of daily living.

phytochemical A substance found in plant foods that is not an essential nutrient but may have health-promoting properties.

pica An abnormal craving for and ingestion of nonfood substances that have little or no nutritional value.

placebo A harmless, inactive medicine or supplement that is indistinguishable in appearance from the real thing. It is used to disguise the control and experimental groups in an experiment.

placenta An organ produced from maternal and embryonic tissues. It secretes hormones, transfers nutrients and oxygen from the mother's blood to the fetus, and removes metabolic wastes.

plant sterol A compound found in plant cell membranes that resembles cholesterol in structure. It can lower blood cholesterol by competing with cholesterol for absorption in the gastrointestinal tract.

polychlorinated biphenyls (PCBs) Carcinogenic industrial compounds that have found their way into the environment and, subsequently, the food supply. Repeated exposure causes them to accumulate in biological tissues over time.

polycyclic aromatic hydrocarbons (PAHs) A class of mutagenic substances produced during cooking when there is incomplete combustion of organic materials—for example, when fat drips on a grill.

polypeptide A chain of amino acids linked by peptide bonds that is part of the structure of a protein.

polysaccharide A carbohydrate made up of many sugar units linked together.

polyunsaturated fatty acid A fatty acid that contains two or more carbon–carbon double bonds.

postmenopausal bone loss Accelerated bone loss that occurs in women for about 5 to 10 years surrounding menopause.

prebiotic A substance that passes undigested into the colon and stimulates the growth and/or activity of certain types of bacteria.

prediabetes A consistent elevation of blood glucose levels to between 100 and 125 mg/100mL of blood, a level above normal but not high enough to be diagnostic of diabetes but thought to increase the risk of developing diabetes.

preeclampsia A condition characterized by elevated blood pressure, a rapid increase in body weight, protein in the urine, and edema. Also called *toxemia*.

prehypertension Blood pressures of 120 to 139 millimeters of mercury systolic (top number) or 80 to 89 diastolic (bottom number). It increases the risk of developing hypertension as well as the risk of artery damage and heart disease.

preterm or **premature** Refers to an infant born before 37 weeks of gestation.

prion A pathogenic protein that is the cause of degenerative brain diseases called spongiform encephalopathies (such as bovine spongiform encephalopathy and variant Creutzfeldt-Jakob Disease). Prion is short for proteinaceous infectious particle.

prior-sanctioned substance A substance that the FDA or the USDA had determined was safe for use in a specific food prior to the 1958 Food Additives Amendment.

probiotic A product that contains live bacteria, which when consumed temporarily lives in the colon and confers health benefits on the host.

prolactin A hormone released from the anterior pituitary that stimulates the mammary glands to produce milk.

protease A protein-digesting enzyme.

protein A class of nutrients that includes molecules made up of one or more intertwining chains of amino acids. They contain carbon, hydrogen, oxygen, and nitrogen.

protein complementation The process of combining proteins from different sources so that they collectively provide the proportions of amino acids required to meet the body's needs.

protein-energy malnutrition (PEM) A nutritional deficiency resulting from inadequate energy and/or protein intake. It can be manifested as marasmus, kwashiokor, or both simultaneously.

protein quality A measure of how efficiently a protein in the diet can be used to make body proteins.

prothrombin A blood protein required for blood clotting.

provitamin or **vitamin precursor** A compound that can be converted into the active form of a vitamin in the body.

puberty A period of rapid growth and physical changes that ends in the attainment of sexual maturity.

pyridoxal phosphate The major coenzyme form of vitamin B_6 that functions in more than

100 enzymatic reactions, many of which involve amino acid metabolism.

qualified health claim A health claim on a food label that has been approved based on emerging but not well-established evidence of a relationship between a food, food component, or dietary supplement and reduced risk of a disease or health-related condition.

reactive hypoglycemia Low blood sugar that occurs an hour or so after the consumption of high-carbohydrate foods; results from an overproduction of insulin.

recombinant DNA DNA that has been formed by joining DNA from different sources.

Recommended Dietary Allowances (RDAs) Nutrient intakes that are sufficient to meet the needs of almost all healthy people in a specific gender and life-stage group.

refined Refers to foods that have undergone processing that changes or removes various components of the original food.

relative energy deficiency in sport (RED-S) Impaired physiological functioning caused by low energy intake relative to expenditure. It includes but is not limited to abnormalities in metabolic rate, menstrual function, bone health, protein synthesis, and cardiovascular health.

renewable resource A resource that is restored and replaced by natural processes and can therefore be used forever.

resistant starch Starch that escapes digestion in the small intestine of healthy people.

resting heart rate The number of times that the heart beats per minute while a person is at rest.

resting metabolic rate (RMR) The rate of energy expenditure at rest. It is measured after 5 to 6 hours without food or exercise.

retinoids The chemical forms of preformed vitamin A: retinol, retinal, and retinoic acid.

retinol activity equivalent (RAE) The amount of retinol, α-carotene, β-carotene, or β-crytoxanthin that must be consumed to equal the vitamin A activity of 1 μg of retinol.

retinol-binding protein A protein that is necessary to transport vitamin A from the liver to tissues in need.

rhodopsin A light-sensitive compound found in the retina of the eye that is composed of the protein opsin loosely bound to retinal.

riboflavin A B vitamin needed in energy metabolism.

rickets A vitamin D deficiency disease in children, characterized by poor bone development because of inadequate calcium absorption.

saliva A watery fluid that is produced and secreted into the mouth by the salivary glands. It contains lubricants, enzymes, and other substances.

salivary amylase An enzyme secreted by the salivary glands that breaks down starch into smaller units.

satiety The feeling of fullness and satisfaction caused by food consumption that eliminates the desire to eat.

saturated fat A type of lipid that is most abundant in solid animal fats and is associated with an increased risk of heart disease.

saturated fatty acid A fatty acid in which the carbon atoms are bonded to as many hydrogen atoms as possible; it therefore contains no carbon–carbon double bonds.

scientific method The general approach of science that is used to explain observations about the world around us.

scurvy A vitamin C deficiency disease characterized by bleeding gums, tooth loss, joint pain, bleeding into the skin and mucous membranes, and fatigue.

segmentation Rhythmic local constrictions of the intestine that mix food with digestive juices and speed absorption by repeatedly moving the food mass over the intestinal wall.

set point A level at which body fat or body weight seems to resist change despite changes in energy intake or output.

simple carbohydrates Carbohydrates known as sugars that include monosaccharides and disaccharides.

simple diffusion The unassisted diffusion of a substance across the cell membrane.

small for gestational age Refers to an infant born at term weighing less than 2.5 kg (5.5 lb).

sodium chloride The chemical formula of table salt.

soluble fiber Fiber that dissolves in water or absorbs water and is readily broken down by intestinal microbiota. It includes pectins, gums, and some hemicelluloses.

solute A dissolved substance.

solvent A fluid in which one or more substances dissolve.

Special Supplemental Nutrition Program for Women, Infants, and Children (WIC) A program funded by the federal government that provides pregnant and lactating women and their young children with nutrition education and financial assistance for purchasing nutritious foods.

sphincter A muscular valve that helps control the flow of materials in the gastrointestinal tract.

spina bifida A neural tube defect in which part of the spinal cord is exposed through a gap in the backbone, causing varying degrees of disability.

spore The dormant state of some bacteria that is resistant to heat but can germinate and produce a new organism when environmental conditions are favorable.

sports anemia A temporary decrease in hemoglobin concentration that occurs during exercise training. It occurs as an adaptation to training and does not impair delivery of oxygen to tissues.

starch A carbohydrate found in plants, made up of many glucose molecules linked in straight or branching chains.

starvation A severe reduction in nutrient and energy intake that impairs health and eventually causes death.

steroid precursor An androgenic hormone produced primarily by the adrenal glands and gonads that acts as precursor in the production of testosterone and estrogen.

sterol A type of lipid with a structure composed of multiple chemical rings.

structure/function claim A claim on a food label that describes the role of a nutrient or dietary ingredient in maintaining normal structure or function in humans.

stunting A low height for age.

subcutaneous fat Adipose tissue located under the skin, which is not associated with a great increase in the risk of chronic diseases.

subsistence crop A crop that is grown as food for a farmer's family, with little or nothing left over to sell.

sucrose A disaccharide commonly known as table sugar that is made of glucose linked to fructose.

sudden infant death syndrome (SIDS) or **crib death** The unexplained death of an infant, usually during sleep.

sugar unit A sugar molecule that cannot be broken down to yield other sugars.

Supplement Facts The portion of a dietary supplement label that includes information about serving size, ingredients, amount per serving (by weight), and percentage of Daily Value, if established.

sustainable agriculture Agricultural methods that maintain soil productivity and a healthy ecological balance while having minimal long-term impacts.

teratogen A chemical, biological, or physical agent that causes birth defects.

theory A formal explanation of an observed phenomenon made after a hypothesis has been tested and supported through extensive experimentation.

thermic effect of food (TEF) or **diet-induced thermogenesis** The energy required for the digestion of food and absorption, metabolism, and storage of nutrients.

thiamin A B vitamin needed in energy metabolism.

thirst A sensation of dryness in the mouth and throat associated with a desire for liquids.

thyroid gland A gland located in the neck that produces thyroid hormones and calcitonin.

thyroid hormones Hormones produced by the thyroid gland that regulate metabolic rate.

thyroid-stimulating hormone (TSH) A hormone that stimulates the synthesis and secretion of thyroid hormones from the thyroid gland.

tissue A collection of similar cells that together carry out a specific function.

tocopherol The chemical name for vitamin E.

Tolerable Upper Intake Levels (ULs) Maximum daily nutrient intake levels that are unlikely to pose risks of adverse health effects to almost all individuals in a given gender and life-stage group.

tolerances The maximum amount of pesticide residues that may legally remain in food, set by the EPA.

total energy expenditure The sum of basal energy expenditure, the thermic effect of food, and the energy used in physical activity, regulation of body temperature, deposition of new tissue, and production of milk.

trace mineral A mineral required in the diet in an amount of 100 mg or less per day or present in the body in an amount of 0.01% of body weight or less.

trans fat A term used to refer to triglycerides containing *trans* fatty acids. When consumed in the diet these fats increase the risk of heart disease.

trans fatty acid An unsaturated fatty acid in which the hydrogens are on opposite sides of the carbon–carbon double bond.

transamination The process by which an amino group from one amino acid is transferred to a carbon compound to form a new amino acid.

transcription The process of copying the information in DNA to a molecule of mRNA.

transgenic An organism with a gene or group of genes intentionally transferred from another species or breed.

transit time The time between the ingestion of food and the elimination of the solid waste from that food.

translation The process of translating the mRNA code into the amino acid sequence of a polypeptide chain.

triglyceride The major type of lipid in food and the body, consisting of three fatty acids attached to a glycerol molecule.

trimester A term used to describe each third or three-month period of a pregnancy.

tripeptide Three amino acids linked together by peptide bonds.

tropical oils A term used in the popular press to refer to the saturated plant oils—coconut, palm, and palm kernel oil—that are derived from plants grown in tropical regions.

type 1 diabetes The form of diabetes caused by autoimmune destruction of insulin-producing cells in the pancreas, usually leading to absolute insulin deficiency.

type 2 diabetes The form of diabetes characterized by insulin resistance and relative (rather than absolute) insulin deficiency.

undernutrition Poor nutritional status resulting from a dietary intake below that which meets nutritional needs.

underweight A body mass index of less than 18.5 kg/m^2, or a body weight 10% or more below the desirable body weight standard.

unrefined food A food eaten either just as it is found in nature or with only minimal processing.

unsaturated fat A type of lipid that is most abundant in plant oils and is associated with a reduced risk of heart disease.

unsaturated fatty acid A fatty acid that contains one or more carbon–carbon double bonds; may be either monounsaturated or polyunsaturated.

urea A nitrogen-containing waste product from the breakdown of proteins that is excreted in the urine.

variable A factor or condition that is changed in an experimental setting.

variant Creutzfeldt-Jakob Disease (vCJD) A rare, degenerative, fatal brain disorder in humans caused by a prion. It is believed that the persons who have developed vCJD became infected through their consumption of food products from cows with bovine spongiform encephalopathy.

vegan diet A plant-based diet that eliminates all animal products.

vegetarian diet A diet that includes plant-based foods and eliminates some or all foods of animal origin.

vein A vessel that carries blood toward the heart.

venule A small vein that drains blood from capillaries and passes it to larger veins for return to the heart.

very low birth weight A birth weight less than 1.5 kg (3.3 lb).

very-low-density lipoprotein (VLDL) A lipoprotein assembled by the liver that carries lipids from the liver and delivers triglycerides to body cells.

villus (plural, **villi**) A fingerlike protrusion of the lining of the small intestine that participates in the digestion and absorption of foodstuffs.

visceral fat Adipose tissue deposited in the abdominal cavity around the internal organs. High levels are associated with an increased risk of heart disease, high blood pressure, stroke, diabetes, and breast cancer.

vitamin An organic compound needed in the diet in small amounts to promote and regulate the chemical reactions and processes needed for growth, reproduction, and the maintenance of health.

vitamin A A fat-soluble vitamin needed in cell differentiation, reproduction, and vision.

vitamin B$_{12}$ One of the B vitamins, only found in animal products.

vitamin B$_6$ One of the B vitamins, needed in protein metabolism.

vitamin C A water-soluble vitamin needed for the maintenance of collagen.

vitamin D A fat-soluble vitamin needed for calcium absorption that can be made in the body when there is exposure to sunlight.

vitamin E A fat-soluble vitamin that functions as an antioxidant.

vitamin K A fat-soluble vitamin needed for blood clotting.

vitamin precursor See provitamin.

water intoxication A condition that occurs when a person drinks enough water to significantly lower the concentration of sodium in the blood.

water-soluble vitamin A vitamin that dissolves in water; includes the B vitamins and vitamin C.

Wernicke-Korsakoff syndrome A form of thiamin deficiency associated with alcohol abuse that is characterized by mental confusion, disorientation, loss of memory, and a staggering gait.

whole-grain product A product made from the entire kernel of a grain including the bran, endosperm, and germ.

xerophthalmia A spectrum of eye conditions resulting from vitamin A deficiency that may lead to blindness.

zoochemical A substance found in animal food (zoo means animal) that is not an essential nutrient but may have health-promoting properties.

References

Chapter 1

[1] Kantor, E.D., Rehm, C.D., Du, M., et al. Trends in dietary supplement use among US adults from 1999–2012. *JAMA* 316:1464–1474, 2016.

[2] American Dietetic Association. Position of the American Dietetic Association: Nutrient supplementation. *J Am Diet Assoc* 109:2073–2085, 2009.

[3] Zhou, X.F., Ding, Z.S., and Liu, N.B. Allium vegetables and risk of prostate cancer: Evidence from 132,192 subjects. *Asian Pac J Cancer Prev* 14:4131–4134, 2013.

[4] Antony, M.L., and Singh, S.V. Molecular mechanisms and targets of cancer chemoprevention by garlic-derived bioactive compound diallyl trisulfide. *Indian J Exp Biol* 49:805–816, 2011.

[5] Andres, S., Abraham, K., Appel, K.E., and Lampen, A. Risks and benefits of dietary isoflavones for cancer. *Crit Rev Toxicol* 41:463–506, 2011.

[6] Gencel, V.B., Benjamin, M.M., Bahou, S.N., and Khalil, R.A. Vascular effects of phytoestrogens and alternative menopausal hormone therapy in cardiovascular disease. *Mini Rev Med Chem* 12:149–174, 2012.

[7] Böhm, F., Edge, R., and Truscott, T.G. Interactions of dietary carotenoids with singlet oxygen (1O2) and free radicals: Potential effects for human health. *Acta Biochim Pol* 59:27–30, 2012.

[8] Nile, S.H., and Park, S.W. Edible berries: Review on bioactive components and their effect on human health. *Nutrition* 30:134–144, 2014.

[9] Toh, J.Y., Tan, V.M., Lim, P.C., et al. Flavonoids from fruit and vegetables: A focus on cardiovascular risk factors. *Curr Atheroscler Rep* 15:368, 2013.

[10] Rodriguez-Mateos, A., Heiss, C., Borges, G., and Crozier, A. Berry (poly)phenols and cardiovascular health. *J Agric Food Chem* 62:3842–3851, 2014.

[11] Kajla, P., Sharma, A., and Sood, D.R. Flaxseed—A potential functional food source. *J Food Sci Technol* 52:1857–1871, 2015.

[12] Nogueira Lde, P., Knibel, M.P., Torres, M.R., et al. Consumption of high-polyphenol dark chocolate improves endothelial function in individuals with stage 1 hypertension and excess body weight. *Int J Hypertens* 2012:147321, 2012.

[13] Khatua, T.N., Adela, R., and Banerjee, S.K. Garlic and cardioprotection: Insights into the molecular mechanisms. *Can J Physiol Pharmacol* 91:448–458, 2013.

[14] Ma, L., Dou, H.L., Wu, Y.Q., et al. Lutein and zeaxanthin intake and the risk of age-related macular degeneration: A systematic review and meta-analysis. *Br J Nutr* 107:350–359, 2012.

[15] Rosin, S., Ojansivu, I., Kopu, A., et al. Optimal use of plant stanol ester in the management of hypercholesterolemia. *Cholesterol* 2015:706970, 2015.

[16] Souza, R.G., Gomes, A.C., Naves, M.M., and Mota, J.F. Nuts and legume seeds for cardiovascular risk reduction: Scientific evidence and mechanisms of action. *Nutr Rev* 73:335–347, 2015.

[17] Zhang, J., Li, L., Song, P., et al. Randomized controlled trial of oatmeal consumption versus noodle consumption on blood lipids of urban Chinese adults with hypercholesterolemia. *Nutr J* 11:54, 2012.

[18] Calder, P.C. The role of marine omega-3 (n-3) fatty acids in inflammatory processes, atherosclerosis and plaque stability. *Mol Nutr Food Res* 56:1073–1080, 2012.

[19] Lambert, J.D. Does tea prevent cancer? Evidence from laboratory and human intervention studies. *Am J Clin Nutr* 98(6 Suppl):1667S–1675S, 2013.

[20] Borneo, R., and León, A.E. Whole grain cereals: Functional components and health benefits. *Food Funct* 3:110–119, 2012.

[21] Crowe, K.M., and Francis, C. Position of the Academy of Nutrition and Dietetics: Functional foods. *J Acad Nutr Diet* 113:1096–1103, 2013.

[22] Centers for Disease Control and Prevention. National Center for Health Statistics. *Overweight and Obesity*. Available online at www.cdc.gov/nchs/fastats/obesity-overweight.htm. Accessed January 14, 2017.

[23] U.S. Department of Agriculture and U.S. Department of Health and Human Services. *Dietary Guidelines for Americans, 2015–2020*, 8th ed. Available online at http://health.gov/dietary-guidelines/2015/. Accessed October 14, 2016.

[24] US Burden of Disease Collaborators. The state of US health, 1990–2010: Burden of diseases, injuries, and risk factors. *JAMA* 310:591–608, 2013.

[25] Ogden, C.L., Carroll, M.D., Lawman, H.G., et al. Trends in obesity prevalence among children and adolescents in the United States, 1988–1994 through 2013–2014. *JAMA* 315:2292–2299, 2016.

[26] Xu, J., Murphy, S.L., Kochanek, K.D., and Bastian, B.A. Deaths: Final data for 2013. *National Vital Statistics Reports* 64(2), Feb 2016. Available online at www.cdc.gov/nchs/data/nvsr/nvsr64/nvsr64_02.pdf. Accessed October 31, 2016.

[27] Fenech, M., El-Sohemy, A., Cahill, L., et al. Nutrigenetics and nutrigenomics: Viewpoints on the current status and applications in nutrition research and practice. *J Nutrigenet Nutrigenomics* 4:69–89, 2011.

[28] Bouchard, C., and Ordovas, J.M. Fundamentals of nutrigenetics and nutrigenomics. *Prog Mol Biol Transl Sci* 108:1–15, 2012.

[29] Hurlimann, T., Menuz, V., Graham, J., et al. Risks of nutrigenomics and nutrigenetics? What the scientists say. *Genes Nutr* 9: 370, 2014.

[30] Fallaize, R., Macready, A.L., Butler, L.T., et al. An insight into the public acceptance of nutrigenomic-based personalised nutrition. *Nutr Res Rev* 26:39–48, 2013.

[31] Saukko, P. State of play in direct-to-consumer genetic testing for lifestyle-related diseases: Market, marketing content, user experiences and regulation. *Proc Nutr Soc* 72:53–60, 2013.

[32] Livingstone, M.B., and Pourshahidi, L.K. Portion size and obesity. *Adv Nutr* 5:829–834, 2014.

Chapter 2

[1] Davis, C., and Saltos, E. Dietary recommendations and how they have changed over time. In *America's Eating Habits: Changes and Consequences*. Available online at www.ers.usda.gov/webdocs/publications/aib750/5831_aib750b_1_.pdf. Accessed November 14, 2016.

[2] U.S. Department of Agriculture and U.S. Department of Health and Human Services. *Dietary Guidelines for Americans, 2015–2020*, 8th ed. Available online at http://health.gov/dietaryguidelines/2015/. Accessed October 14, 2016.

[3] Finkelstein, E.A., Trogdon, J.G., Cohen, J.W., and Dietz, W. Annual medical spending attributable to obesity: Payer- and service-specific estimates. *Health Affairs* 28:w822–w831, 2009.

[4] Cawley, J., and Maclean, J.C. Report: *Unfit for Service: The Implications of Rising Obesity for U.S. Military Recruitment*, NBER Working Paper No. 16408, September 2010. Available online at www.nber.org/papers/w16408. Accessed November 2, 2016.

[5] Brownell, K.D., Kersh, R., Ludwig, D.S., et al. Personal responsibility and obesity: A constructive approach to a controversial issue. *Health Aff (Millwood)* 29:379–387, 2010.

[6] Stewart, H., Dong, D., and Carlson, A. Why are Americans consuming less fluid milk? A look at generational differences in intake frequency. *Economic Research Report*, No. 149. Economic Research Service, U.S. Department of Agriculture, May 2013.

[7] Hoy, M., and Goldman, J.D. Calcium intake of the U.S. population. *What We Eat in America*, NHANES 2009–2010. Food Surveys Research Group Dietary Data Brief No. 13, September 2014.

[8] HealthyPeople.gov. *About Healthy People*. Available online at www.healthypeople.gov/2020/about/default.aspx. Accessed October 15, 2016.

[9] Institute of Medicine, Food and Nutrition Board. *Dietary Reference Intakes for Energy, Carbohydrate, Fiber, Fat, Fatty Acids, Cholesterol, Protein, and Amino Acids*. Washington, DC: National Academies Press, 2002.

[10] U.S. Food and Drug Administration. *How to Understand and Use the Nutrition Facts Label*. Available online at www.fda.gov/Food/IngredientsPackagingLabeling/LabelingNutrition/ucm274593.htm. Accessed October 15, 2016.

[11] U.S. Department of Agriculture, Food Safety and Inspection Service. Nutrition labeling of single-ingredient products and ground or chopped meat and poultry products; final rule. *Fed Regist* 75:82147–82167, December 29, 2010.

[12] U.S. Food and Drug Administration. *Menu and Vending Machines Labeling Requirements*. Available online at www.fda.gov/Food/IngredientsPackagingLabeling/LabelingNutrition/ucm217762.htm. Accessed November 2, 2016.

[13] U.S. Food and Drug Administration. *Changes to the Nutrition Facts Label*. Available online at www.fda.gov/Food/GuidanceRegulation/GuidanceDocumentsRegulatoryInformation/LabelingNutrition/ucm385663.htm. Accessed October 17, 2016.

[14] U.S. Food and Drug Administration. *Guidance for Industry: A Food Labeling Guide*, January 2013. Available online at www.fda.gov/Food/GuidanceRegulation/GuidanceDocumentsRegulatoryInformation/LabelingNutrition/ucm2006828.htm. Accessed October 17, 2016.

[15] U.S. Department of Health and Human Services. U.S. Food and Drug Administration. Food Labeling: Revision of the nutrition and supplement facts labels. Final Rule. *Federal Register* 81(103):33742–33999, May 27, 2016. Available online at https://www.federalregister.gov/documents/2016/05/27/2016-11867/food-labeling-revision-of-the-nutrition-and-supplement-facts-labels. Accessed March 30, 2017.

[16] U.S. Food and Drug Administration. *Guidance for Industry: A Labeling Guide for Restaurants and Other Retail Establishments Selling Away-from-Home Foods*, April 2008. Available online at www.fda.gov/food/guidanceregulation/ucm053455.htm. Accessed October 17, 2016.

[17] U.S. Food and Drug Administration. *Guidance for Industry: A Dietary Supplement Labeling Guide*, April 2005. Available online at www.fda.gov/food/guidanceregulation/guidancedocumentsregulatoryinformation/dietarysupplements/ucm2006823.htm. Accessed October 17, 2016.

Chapter 3

[1] Li, D., Wang, P., Wang, P., et al. The gut microbiota: A treasure for human health. *Biotechnol Adv* 34:1210–1224, 2016.

[2] O'Connor, E.M. The role of gut microbiota in nutritional status. *Curr Opin Clin Nutr Metab Care* 16:509–516, 2013.

[3] de Moreno de LeBlanc, A., and LeBlanc, J.G. Effect of probiotic administration on the intestinal microbiota, current knowledge and potential applications. *World J Gastroenterol* 20:16518–16528, 2014.

[4] Power, S.E., O'Toole, P.W., Stanton, C., et al. Intestinal microbiota, diet and health. *Br J Nutr* 111:387–402, 2014.

[5] Morrison, D.J., and Preston, T. Formation of short chain fatty acids by the gut microbiota and their impact on human metabolism. *Gut Microbes* 7:189–200, 2016.

[6] Hempel, S., Newberry, S.J., Maher, A.R., et al. Probiotics for the prevention and treatment of antibiotic-associated diarrhea: A systematic review and meta-analysis *JAMA* 307:1959–1969, 2012.

[7] Salonen, A., and de Vos, W.M. Impact of diet on human intestinal microbiota and health. *Annu Rev Food Sci Technol* 5:239–262, 2014.

[8] Chan, Y.K., Estaki, M., and Gibson, D.L. Clinical consequences of diet-induced dysbiosis. *Ann Nutr Metab* 63(suppl 2):28–40, 2013.

[9] Sicherer, S.H., and Sampson, H.A. Food allergy: Epidemiology, pathogenesis, diagnosis, and treatment. *J Allergy Clin Immunol* 133:291–307, 2014.

[10] Beecher, C. Sesame gains traction in push for foods-labeling requirements. *Food Safety News* November 30, 2015.

[11] National Institute of Diabetes and Digestive and Kidney Diseases. *Celiac Disease*. Available online at http://digestive.niddk.nih.gov/ddiseases/pubs/celiac. Accessed April 5, 2017.

[12] van der Windt, D.A., Jellema, P., Mulder, C.J., et al. Diagnostic testing for celiac disease among patients with abdominal symptoms: A systematic review. *JAMA* 303:1738–1746, 2010.

[13] Rubio-Tapia, A., Ludvigsson, J.F., Brantner, T.L., et al. The prevalence of celiac disease in the United States. *Am J Gastroenterol* 107:1538–1545, 2012.

[14] Ludvigsson, J.F., Leffler, D.A., Bai, J.C., et al. The Oslo definitions for coeliac disease and related terms. *Gut* 62:43–52, 2013.

[15] Czaja-Bulsa, G. Non coeliac gluten sensitivity—A new disease with gluten intolerance. *Clin Nutr* 34:189–194, 2015.

[16] Mansueto, P., Seidita, A., D'Alcamo, A., and Carroccio, A. Non-celiac gluten sensitivity: Literature review. *J Am Coll Nutr* 33:39–54, 2014.

[17] Biesiekierski, J.R., Muir, J.G., and Gibson, P.R. Is gluten a cause of gastrointestinal symptoms in people without celiac disease? *Curr Allergy Asthma Rep* 13:631–638, 2013.

[18] Rubio-Tapia, A., Ludvigsson, J.F., Brantner, T.L., et al. The prevalence of celiac disease in the United States. *Am J Gastroenterol* 107:1538–1545, 2012.

[19] Beyond Celiac. *Nonceliac Gluten Sensitivity*. Available online at www.celiaccentral.org/non-celiac-gluten-sensitivity. Accessed April 5, 2017.

[20] Wu, J., Xu, S., and Zhu, Y. *Helicobacter pylori* CagA: A critical destroyer of the gastric epithelial barrier. *Dig Dis Sci* 58:1830–1837, 2013.

[21] National Institute of Diabetes and Digestive and Kidney Diseases. *Peptic Ulcers (Stomach Ulcers)*. Available online at www.niddk.nih.gov/health-information/health-topics/digestive-diseases/peptic-ulcer/Pages/overview.aspx. Accessed November 16, 2016.

[22] Marieb, E.N., and Hoehn, K. *Human Anatomy and Physiology*, 9th ed. Menlo Park, CA: Pearson, 2013.

Chapter 4

[1] U.S. Department of Agriculture and U.S. Department of Health and Human Services. *Dietary Guidelines for Americans, 2015–2020*, 8th

ed. Available online at http://health.gov/dietaryguidelines/2015/. Accessed October 15, 2016.
[2] National Institute of Diabetes and Digestive and Kidney Diseases. *Lactose Intolerance.* Available online at http://digestive.niddk.nih.gov/ddiseases/pubs/lactoseintolerance/. Accessed October 15, 2016.
[3] Li, D., Wang, P., Wang, P., et al. The gut microbiota: A treasure for human health. *Biotechnol Adv* 34:1210–1224, 2016.
[4] Kumar, V., Sinha, A.K., Makkar, H.P., et al. Dietary roles of non-starch polysaccharides in human nutrition: A review. *Crit Rev Food Sci Nutr* 52:899–935, 2012.
[5] Morrison, D.J., and Preston, T. Formation of short chain fatty acids by the gut microbiota and their impact on human metabolism. *Gut Microbes* 7:189–200, 2016.
[6] Burkitt, D.P., Walker, A.R.P., and Painter, N.S. Dietary fiber and disease. *JAMA* 229:1068–1074, 1974.
[7] Willett, W.C., and Stampfer, M.J. Current evidence on healthy eating. *Annu Rev Public Health* 34:77–95, 2013.
[8] Ravichandran, B. Sugar is the new tobacco. *The BMJ Blogs* March 13, 2015.
[9] Stanhope, K.L. Sugar consumption, metabolic disease and obesity: The state of the controversy. *Crit Rev Clin Lab Sci* 53:52–67, 2016.
[10] Bray, G.A. Energy and fructose from beverages sweetened with sugar or high-fructose corn syrup pose a health risk for some people. *Adv Nutr* 4:220–225, 2013.
[11] Bray, G.A., and Popkin, B.M. Dietary sugar and body weight: Have we reached a crisis in the epidemics of obesity and diabetes? Health be damned! Pour on the sugar. *Diabetes Care* 37:950–956, 2014.
[12] Te Morenga, L., Mallard, S., and Mann, J. Dietary sugars and body weight: Systematic review and meta-analyses of randomized controlled trials and cohort studies. *BMJ* 346:e7492, 2012.
[13] Aeberli, I., Hochuli, M., Gerber, P.A., et al. Moderate amounts of fructose consumption impair insulin sensitivity in healthy young men: A randomized control trial. *Diabetes Care* 36:150–156, 2013.
[14] Kahn, R., and Sievenpiper, J.L. Dietary sugar and body weight: Have we reached a crisis in the epidemics of obesity and diabetes? We have, but the pox on sugar is overwrought and overworked. *Diabetes Care* 37:957–962, 2014.
[15] Tappy, L., and Lê, K.A. Health effects of fructose and fructose-containing caloric sweeteners: Where do we stand 10 years after the initial whistle blowings? *Curr Diab Rep* 15:54, 2015.
[16] Sievenpiper, J.L. Sickeningly sweet: Does sugar cause chronic disease? No. *Can J Diabetes* 40:287–295, 2016.
[17] Rippe, J.M., and Angelopoulos, T.J. Added sugars and risk factors for obesity, diabetes and heart disease. *Int J Obes (Lond)* 40 (suppl 1):S22–S27, 2016.
[18] Centers for Disease Control and Prevention. *National Diabetes Statistics Report: Estimates of Diabetes and Its Burden in the United States, 2014.* Available online at www.cdc.gov/diabetes/pubs/statsreport14/national-diabetes-report-web.pdf. Accessed October 1, 2016.
[19] World Health Organization. *Global Report on Diabetes.* World Health Organization, 2016. Available online at http://apps.who.int/iris/bitstream/10665/204871/1/9789241565257_eng.pdf?ua=1. Accessed November 5, 2016.
[20] Aune, D., Norat, T., Romundstad, P., and Vatten, L.J. Whole grain and refined grain consumption and the risk of type 2 diabetes: A systematic review and dose-response meta-analysis of cohort studies. *Eur J Epidemiol.* 28:845–858, 2013.
[21] Ley, S.H., Ardisson Korat, A.V., Sun, Q., et al. Contribution of the Nurses' Health Studies to uncovering risk factors for type 2 diabetes: Diet, lifestyle, biomarkers, and genetics. *Am J Public Health* 106:1624–1630, 2016.
[22] Ley, S.H., Hamdy, O., Mohan, V., and Hu, F.B. Prevention and management of type 2 diabetes: Dietary components and nutritional strategies. *Lancet* 383:1999–2007, 2014.
[23] Salas-Salvadó, J., Martinez-González, M.Á., Bulló, M., and Ros, E. The role of diet in the prevention of type 2 diabetes. *Nutr Metab Cardiovasc Dis* 21(suppl 2):B32–B48, 2011.
[24] American Diabetes Association. Standards of Medical Care in Diabetes—2016. *Diabetes Care* 39 (suppl 1):S1–S112, 2016. Available online at http://care.diabetesjournals.org/content/suppl/2015/12/21/39.Supplement_1.DC2/2016-Standards-of-Care.pdf. Accessed November 1, 2016.
[25] Ebbeling, C.B. Sugar-sweetened beverages and body weight. *Curr Opin Lipidol* 25:1–7, 2014.
[26] Aller, E.E., Abete, I., Astrup, A., et al. Starches, sugars and obesity. *Nutrients* 3:341–369, 2011.
[27] Matarese, L.E., and Pories, W.J. Adult weight loss diets: Metabolic effects and outcomes. *Nutr Clin Pract* 29:759–767, 2014.
[28] Giacco, R., Della Pepa, G., Luongo, D., and Riccardi, G. Whole grain intake in relation to body weight: From epidemiological evidence to clinical trials. *Nutr Metab Cardiovasc Dis* 21:901–908, 2011.
[29] Rosset, R., Surowska, A., and Tappy, L. Pathogenesis of cardiovascular and metabolic disease: Are fructose-containing sugars more involved than other dietary calories? *Curr Hypertens Rep* 18:44, 2016.
[30] Cho, S.S., Qi, L., Fahey, G.C. Jr, and Klurfeld, D.M. Consumption of cereal fiber, mixtures of whole grains and bran, and whole grains and risk reduction in type 2 diabetes, obesity, and cardiovascular disease. *Am J Clin Nutr* 98:594–619, 2013.
[31] Threapleton, D.E., Greenwood, D.C., Evans, C.E.L., et al. Dietary fibre intake and risk of cardiovascular disease: Systematic review and meta-analysis. *BMJ* 347:f6879, 2013.
[32] Gunness, P., Michiels, J., Vanhaecke, L., et al. Reduction in circulating bile acid and restricted diffusion across the intestinal epithelium are associated with a decrease in blood cholesterol in the presence of oat β-glucan. *FASEB J* 30:4227–4238, 2016.
[33] Cohn, J.S., Kamili, A., Wat, E., et al. Reduction in intestinal cholesterol absorption by various food components: mechanisms and implications. *Atheroscler Suppl* 11:45–48, 2010.
[34] Centers for Disease Control and Prevention, National Center for Health Statistics. Dental caries and tooth loss in adults in the United States, 2011–2012. *NCHS Data Brief* No. 197, May 2015. Available online at www.cdc.gov/nchs/products/databriefs/db197.htm. Accessed November 21, 2016.
[35] Bradshaw, D.J., and Lynch, R.J. Diet and the microbial aetiology of dental caries: New paradigms. *Int Dent J* 63(suppl 2):64–72, 2013.
[36] Aune, D., Chan, D.S., Lau, R., et al. Dietary fibre, whole grains, and risk of colorectal cancer: Systematic review and dose-response meta-analysis of prospective studies. *BMJ* 343:d6617, 2011.
[37] Klement, R.J., and Kämmerer, U. Is there a role for carbohydrate restriction in the treatment and prevention of cancer? *Nutr Metab (Lond)* 8:75, 2011.
[38] Baena, R., and Salinas, P. Diet and colorectal cancer. *Maturitas* 80:258–264, 2015.
[39] Institute of Medicine, Food and Nutrition Board. *Dietary Reference Intakes for Energy, Carbohydrates, Fiber, Fat, Protein and Amino Acids.* Washington, DC: National Academies Press, 2002.
[40] Hoy, M.K., and Goldman, J.D. Fiber intake of the U.S. population. *What We Eat in America*, NHANES 2009–1010. Food Surveys Research Group Dietary Data Brief No. 12, September 2014.

[41] Food and Drug Administration. *Additional Information About High-Intensity Sweeteners Permitted for Use in Food in the United States.* Available online at www.fda.gov/Food/IngredientsPackagingLabeling/FoodAdditivesIngredients/ucm397725.htm. Accessed November 1, 2016.

[42] Fernstrom, J.D. Non-nutritive sweeteners and obesity. *Annu Rev Food Sci Technol* 6:119–136, 2015.

[43] Anderson, G.H., Foreyt, J. Sigman-Grant, M., and Allison, D.B. The use of low-calorie sweeteners by adults: Impact on weight management. *J Nutr* 142:1163S–1169S, 2013.

Chapter 5

[1] U.S. Department of Agriculture, Agricultural Research Service. 2016. Table 5. Energy intakes: Percentage of energy from protein, carbohydrate, fat, and alcohol, by gender and age. *What We Eat in America* (NHANES 2013–2014). Available online at www.ars.usda.gov/ARSUserFiles/80400530/pdf/1314/Table_5_EIN_GEN_13.pdf. Accessed November 25, 2016.

[2] U.S. Department of Agriculture and U.S. Department of Health and Human Services. *Dietary Guidelines for Americans, 2015-2020*, 8th ed., Washington, DC. Available online at https://health.gov/dietaryguidelines/2015/guidelines/. Accessed September 21, 2016.

[3] Bester, D., Esterhuyse, A.J., Truter, E.J., and van Rooyen, J. Cardiovascular effects of edible oils: A comparison between four popular edible oils. *Nutr Res Rev* 23:334–348, 2010.

[4] Feranil, A.B., Duazo, P.L., Kuzawa, C.W., and Adair, L.S. Coconut oil is associated with a beneficial lipid profile in pre-menopausal women in the Philippines. *Asia Pac J Clin Nutr* 20:190–195, 2011.

[5] Mensink, R.P., Zock, P.L., Kester, A.D., and Katan, M.B. Effects of dietary fatty acids and carbohydrates on the ratio of serum total to HDL cholesterol and on serum lipids and apolipoproteins: A meta-analysis of 60 controlled trials. *Am J Clin Nutr* 77:1146–1155, 2003.

[6] Assunção, M.L., Ferreira, H.S., dos Santos, A.F., et al. Effects of dietary coconut oil on the biochemical and anthropometric profiles of women presenting abdominal obesity. *Lipids* 44:593–601, 2009.

[7] Voon, P.T., Ng, T.K., Lee, V.K., and Nesaretnam, K. Diets high in palmitic acid (16:0), lauric and myristic acids (12:0 + 14:0), or oleic acid (18:1) do not alter postprandial or fasting plasma homocysteine and inflammatory markers in healthy Malaysian adults. *Am J Clin Nutr* 94:1451–1457, 2011.

[8] Han, J., Deng, B., Sun, J., et al. Effects of dietary medium-chain triglyceride on weight loss and insulin sensitivity in a group of moderately overweight free-living type 2 diabetic Chinese subjects. *Metabolism* 56: 985–991, 2007.

[9] St-Onge, M.P., and Bosarge, A. Weight-loss diet that includes consumption of medium-chain triacylglycerol oil leads to a greater rate of weight and fat mass loss than does olive oil. *Am J Clin Nutr* 87:621–626, 2008.

[10] St-Onge, M.P., Mayrsohn, B., O'Keeffe, M., et al. Impact of medium and long chain triglycerides consumption on appetite and food intake in overweight men. *Eur J Clin Nutr* 68:1134–1140, 2014.

[11] U.S. Food and Drug Administration. *Final Determination Regarding Partially Hydrogenated Oils (Removing Trans Fat)*. Available online at www.fda.gov/food/ingredientspackaginglabeling/foodadditivesingredients/ucm449162.htm. Accessed September 21, 2016.

[12] Mozaffarian, D., and Wu, J.H. Omega-3 fatty acids and cardiovascular disease: Effects on risk factors, molecular pathways, and clinical events. *J Am Coll Cardiol* 58:2047–2067, 2011.

[13] American Heart Association. *Diet and Lifestyle Recommendations.* Available online at www.heart.org/HEARTORG/GettingHealthy/Diet-and-Lifestyle-Recommendations_UCM_305855_Article.jsp. Accessed November 15, 2016.

[14] Mozaffarian, D., Benjamin, E.J., Go, A.S., et al; American Heart Association Statistics Committee; Stroke Statistics Subcommittee. Heart disease and stroke statistics—2016 update: A report from the American Heart Association. *Circulation* 133:e38–e360, 2016.

[15] Salisbury, D., and Bronas, U. Inflammation and immune system contribution to the etiology of atherosclerosis: Mechanisms and methods of assessment. *Nurs Res* 63:375–385, 2014.

[16] Stone, N.J., Robinson, J., Lichtenstein, A.H., et al. 2013 ACC/AHA guideline on the treatment of blood cholesterol to reduce atherosclerotic cardiovascular risk in adults: A report of the American College of Cardiology/American Heart Association Task Force on Practice Guidelines. *Circulation* 129(25, Suppl 2):S1–S45, 2014.

[17] Bhupathiraju, S.N., and Tucker, K.L. Coronary heart disease prevention: Nutrients, foods, and dietary patterns. *Clin Chim Acta* 412:1493–1514, 2011.

[18] Eckel, R.H., Jakicic, J.M., Ard, J.D., et al. 2013 AHA/ACC guideline on lifestyle management to reduce cardiovascular risk: A report of the American College of Cardiology/American Heart Association Task Force on Practice Guidelines. *Circulation* 129(25, Suppl 2):S76–S99, 2014.

[19] Chang, C.L., and Deckelbaum, R.J. Omega-3 fatty acids: Mechanisms underlying "protective effects" in atherosclerosis. *Curr Opin Lipidol* 24:345–350, 2013.

[20] Ros, E. Nuts and CVD. *Br J Nutr* 113(Suppl 2):S111–S120, 2015.

[21] Seal, C.J., and Brownlee, I.A. Whole-grain foods and chronic disease: Evidence from epidemiological and intervention studies. *Proc Nutr Soc* 74:313–319, 2015.

[22] AbuMweis, S.S., Marinangeli, C.P., Frohlich, J., and Jones, P.J. Implementing phytosterols into medical practice as a cholesterol-lowering strategy: Overview of efficacy, effectiveness, and safety. *Can J Cardiol* 30:1225–1232, 2014.

[23] Kerimi, A., and Williamson, G. The cardiovascular benefits of dark chocolate. *Vascul Pharmacol* 71:11–15, 2015.

[24] O'Keefe, J. H., Bhatti, S. K., Bajwa, A., et al. Alcohol and cardiovascular health: The dose makes the poison...or the remedy. *Mayo Clin Proc* 89:382–393, 2014.

[25] Nagarajan, S. Mechanisms of anti-atherosclerotic functions of soy-based diets. *J Nutr Biochem* 21:255–260, 2010.

[26] Debreceni, B., and Debreceni, L. The role of homocysteine-lowering B-vitamins in the primary prevention of cardiovascular disease. *Cardiovasc Ther* 32:130–138, 2014.

[27] Zhang, P.Y., Xu, X., and Li, X.C. Cardiovascular diseases: Oxidative damage and antioxidant protection. *Eur Rev Med Pharmacol Sci* 18:3091–3096, 2014.

[28] Schottenfeld, D., Beebe-Dimmer, J.L., Buffler, P.A., and Omenn, G.S. Current perspective on the global and United States cancer burden attributable to lifestyle and environmental risk factors. *Annu Rev Public Health* 34:97–117, 2013.

[29] Wolk, A. Potential health hazards of eating red meat. *J Intern Med* 281:106–122, 2017.

[30] Schwingshackl, L., and Hoffmann, G. Adherence to Mediterranean diet and risk of cancer: A systematic review and meta-analysis of observational studies. *Int J Cancer* 135:1884–1897, 2014.

[31] Albuquerque, R.C., Baltar, V.T., and Marchioni, D.M. Breast cancer and dietary patterns: A systematic review. *Nutr Rev* 72:1–17, 2014.

[32] Makarem, N., Chandran, U., Bandera, E.V., and Parekh, N. Dietary fat in breast cancer survival. *Annu Rev Nutr* 33:319–348, 2013.

[33] Devitt, A.A., and Mattes, R.D. Effects of food unit size and energy density on intake in humans. *Appetite* 42:213–220, 2004.

[34] Bes-Rastrollo, M., van Dam, R.M., Martinez-Gonzalez, M.A, et al. Prospective study of dietary energy density and weight gain in women. *Am J Clin Nutr* 88:769–777, 2008.

[35] Melanson, E.L., Astrup, A., and Donahoo, W.T. The relationship between dietary fat and fatty acid intake and body weight, diabetes, and the metabolic syndrome. *Ann Nutr Metab* 55:229–243, 2009.

[36] Institute of Medicine, Food and Nutrition Board. *Dietary Reference Intakes for Energy, Carbohydrates, Fiber, Fat, Protein, and Amino Acids*. Washington, DC: National Academies Press, 2002.

[37] Rong, Y., Chen, L., Zhu, T., et al. Egg consumption and risk of coronary heart disease and stroke: Dose-response meta-analysis of prospective cohort studies. *BMJ* 346:e8539, 2013.

[38] Calorie Control Council. *Glossary of Fat Replacers*. Available online at http://caloriecontrol.org/glossary-of-fat-replacers/. Accessed September 22, 2016.

Chapter 6

[1] Williams, C.D. Kwashiorkor: Nutritional disease of children associated with maize diet. *Lancet* 2:1151–1154, 1935.

[2] Marckmann, P., Osther, P., Pedersen, A. N., and Jespersen, B. High-protein diets and renal health. *J Ren Nutr* 25:1–5, 2015.

[3] Juraschek, S.P., Appel, L.J., Anderson, C.A., and Miller, E.R., 3rd. Effect of a high-protein diet on kidney function in healthy adults: Results from the OmniHeart trial. *Am J Kidney Dis* 61:547–554, 2013.

[4] Heaney, R.P., and Layman, D.K. Amount and type of protein influences bone health. *Am J Clin Nutr* 87:1567S–1570S, 2008.

[5] Mangano, K.M., Sahni, S., and Kerstetter, J.E. Dietary protein is beneficial to bone health under conditions of adequate calcium intake: An update on clinical research. *Curr Opin Clin Nutr Metab Care* 17:69–74, 2014.

[6] Tang, M., O'Connor, L.E., and Campbell, W.W. Diet-induced weight loss: The effect of dietary protein on bone. *J Acad Nutr Diet* 114:72–85, 2014.

[7] Remer, T., Krupp, D., and Shi, L. Dietary protein's and dietary acid load's influence on bone health. *Crit Rev Food Sci Nutr* 54:1140–1150, 2014.

[8] NIH, National Institute of Diabetes and Digestive and Kidney Diseases. *Diet and Kidney Stone Prevention* (NIH Publication No 13-6425), February 2013. Available online at https://www.niddk.nih.gov/health-information/health-topics/urologic-disease/diet-for-kidney-stone-prevention/Documents/Kidney_Stone_Diet_508.pdf. Accessed December 4, 2016.

[9] Gonzalez, C.A., and Riboli, E. Diet and cancer prevention: Contributions from the European Prospective Investigation into Cancer and Nutrition (EPIC) study. *Eur J Cancer* 46:2555–2562, 2010.

[10] Bouvard, V., Loomis, D., Guyton, K.Z., et al.; International Agency for Research on Cancer Monograph Working Group. Carcinogenicity of consumption of red and processed meat. *Lancet Oncol* 16:1599–1600, 2015.

[11] U.S. Department of Agriculture, Agricultural Research Service. 2016. Table 1. Nutrient Intakes from Food and Beverages: Mean amounts consumed per individual, by gender and age, in the United States, 2013–2014. *What We Eat in America* (NHANES 2013–2014). Available online at www.ars.usda.gov/ARSUserFiles/80400530/pdf/1314/Table_1_NIN_GEN_13.pdf. Accessed April 13, 2017.

[12] U.S. Department of Agriculture, Agricultural Research Service. 2016. Table 5. Energy Intakes: Percentage of energy from protein, carbohydrate, fat, and alcohol, by gender and age. *What We Eat in America* (NHANES 2013–2014). Available online at www.ars.usda.gov/ARSUserFiles/80400530/pdf/1314/Table_5_EIN_GEN_13.pdf. Accessed November 25, 2016.

[13] Pesta, D.H., and Samuel, V.T. A high-protein diet for reducing body fat: Mechanisms and possible caveats. *Nutr Metab* (Lond) 11:53–60, 2014.

[14] Martens, E.A., Gonnissen, H.K., Gatta-Cherifi, B., et al. Maintenance of energy expenditure on high-protein vs. high-carbohydrate diets at a constant body weight may prevent a positive energy balance. *Clin Nutr* 34:968–975, 2015.

[15] Matarese, L.E., and Pories, W.J. Adult weight loss diets: Metabolic effects and outcomes. *Nutr Clin Pract* 29:759–767, 2014.

[16] Soenen, S., Bonomi, A.G., Lemmens, S.G., et al. Relatively high-protein or "low-carb" energy-restricted diets for body weight loss and body weight maintenance? *Physiol Behav* 107:374–380, 2012.

[17] Neacsu, M., Fyfe, C., Horgan, G., and Johnstone, A.M. Appetite control and biomarkers of satiety with vegetarian (soy) and meat-based high-protein diets for weight loss in obese men: a randomized crossover trial. *Am J Clin Nutr* 100:548–558, 2014.

[18] U.S. Department of Agriculture and U.S. Department of Health and Human Services. *Dietary Guidelines for Americans, 2015–2020*, 8th ed., Washington, DC. Available online at https://health.gov/dietaryguidelines/2015/guidelines/. Accessed September 21, 2016.

[19] Santesso, N., Akl, E.A., Bianchi, M., et al. Effects of higher- versus lower-protein diets on health outcomes: a systematic review and meta-analysis. *Eur J Clin Nutr* 66:780–788, 2012.

[20] Schwingshackl, L., and Hoffmann, G. Long-term effects of low-fat diets either low or high in protein on cardiovascular and metabolic risk factors: A systematic review and meta-analysis. *Nutr J* 12:48–56, 2013.

[21] Lagiou, P., Sandin, S., Lof, M., et al. Low carbohydrate-high protein diet and incidence of cardiovascular diseases in Swedish women: Prospective cohort study. *BMJ* 344:e4026, 2012.

[22] Noto, H., Goto, A., Tsujimoto, T., and Noda, M. Low-carbohydrate diets and all-cause mortality: A systematic review and meta-analysis of observational studies. *PLoS One* 8:e55030, 2013.

[23] Bernstein, A.M., Sun, Q., Hu, F.B., et al. Major dietary protein sources and risk of coronary heart disease in women. *Circulation* 122:876–883, 2010.

[24] Levine, M.E., Suarez, J.A., Brandhorst, S., et al. Low protein intake is associated with a major reduction in IGF-1, cancer, and overall mortality in the 65 and younger but not older population. *Cell Metab* 19:407–417, 2014.

[25] Williams, A.N., and Woessner, K.M. Monosodium glutamate "allergy": Menace or myth? *Clin Exp Allergy* 39:640–646, 2009.

[26] National Institute of Diabetes and Digestive and Kidney Diseases. *Celiac Disease*. Available online at http://digestive.niddk.nih.gov/ddiseases/pubs/celiac/. Accessed November 28, 2016.

[27] Rubio-Tapia, A., Ludvigsson, J. F., Brantner, T. L., et al. The prevalence of celiac disease in the United States. *Am J Gastroenterol* 107:1538–1545, 2012.

[28] Institute of Medicine, Food and Nutrition Board. *Dietary Reference Intakes for Energy, Carbohydrates, Fiber, Fat, Protein and Amino Acids*. Washington, DC: National Academies Press, 2002.

[29] Baum, J.I., Kim, I.Y., and Wolfe, R.R. Protein consumption and the elderly: What is the optimal level of intake? *Nutrients* 8:359, 2016.

[30] Thomas, D.T., Erdman, K.A., and Burke, L.M. Position of the Academy of Nutrition and Dietetics, Dietitians of Canada, and the American College of Sports Medicine: Nutrition and athletic performance. *J Acad Nutr Diet* 116:501–528, 2016.

[31] Levesque, C.L., and Ball, R.O. Protein and amino acid requirements. In Stipanuk, M.H., and Caudill, M.A. (eds.), *Biochemical, Physiological and Molecular Aspects of Human Nutrition*, 3rd ed. St. Louis: Saunders Elsevier, 2013, pp. 331–356.

[32] Craig, W.J., Mangels, A.R., and American Dietetic Association. Position of the American Dietetic Association: Vegetarian diets. *J Am Diet Assoc* 109:1266–1282, 2009.

Chapter 7

[1] Fulgoni, V.L., 3rd, Keast, D.R., Bailey, R.L., and Dwyer, J. Foods, fortificants, and supplements: Where do Americans get their nutrients? *J Nutr* 141:1847–1854, 2011.

[2] Berner, L.A., Keast, D.R., Bailey, R.L., and Dwyer, J.T. Fortified foods are major contributors to nutrient intakes in diets of US children and adolescents. *J Acad Nutr Diet* 114:1009–1022, 2014.

[3] Valerie, T. Discretionary fortification—A public health perspective. *Nutrients* 6:4421–4433, 2014.

[4] Butte, N.F., Fox, M.K., Briefel, R.R., et al. Nutrient intakes of US infants, toddlers, and preschoolers meet or exceed dietary reference intakes. *J Am Diet Assoc* 110(12 Suppl.):S27–S37, 2010.

[5] Food and Drug Administration. Code of Federal Regulations, Title 21, volume 2. Section 101.45 Guidelines for voluntary nutrition labeling of raw fruits, vegetables, and fish, Revised April 1, 2016. Available online at www.accessdata.fda.gov/scripts/cdrh/cfdocs/cfcfr/cfrsearch.cfm?fr=101.45. Accessed January 25, 2017.

[6] Institute of Medicine, Food and Nutrition Board. *Dietary Reference Intakes for Thiamin, Riboflavin, Niacin, Vitamin B-6, Folate, Vitamin B-12, Pantothenic Acid, Biotin, and Choline*. Washington, DC: National Academies Press, 1998.

[7] Becker, D.A., Ingala, E.E., Martinez-Lage, M., et al. Dry beriberi and Wernicke's encephalopathy following gastric lap band surgery. *J Clin Neurosci* 19:1050–1052, 2012.

[8] Saffert, A., Pieper, G., and Jetten, J. Effect of package light transmittance on vitamin content of milk. *Packing Technology and Science* 21:47–55, 2008.

[9] Kavitha, B., Balasubramanian, R., and Kumar, T. Electrocardiographic enigma of a classical disease: pellagra. *Trop Doct* 42:211–213, 2012.

[10] Zeisel, S.H., and da Costa, K.A. Choline: An essential nutrient for public health. *Nutr Rev* 67:615–623, 2009.

[11] Patterson, K.Y., Bhagwat, S.A., Williams, J.R., et al. *USDA Database for the Choline Content of Common Foods, Release Two, January 2008*. Available online at www.ars.usda.gov/SP2UserFiles/Place/12354500/Data/Choline/Choln02.pdf. Accessed January 25, 2017.

[12] Debreceni, B., and Debreceni, L. Why do homocysteine-lowering B vitamin and antioxidant E vitamin supplementations appear to be ineffective in the prevention of cardiovascular diseases? *Cardiovasc Ther* 30:227–233, 2012.

[13] Talebi, M., Andalib, S., Bakhti, S., et al. Effect of vitamin B$_6$ on clinical symptoms and electrodiagnostic results of patients with carpal tunnel syndrome. *Adv Pharm Bull* 3:283–288, 2013.

[14] Biggs, W.S., and Demuth, R.H. Premenstrual syndrome and premenstrual dysphoric disorder. *Am Fam Physician* 84:918–924, 2011.

[15] Williams, J., Mai, C.T., Mulinare, J., et al. Updated estimates of neural tube defects prevented by mandatory folic acid fortification—United States, 1995–2011. *Morb Mortal Wkly Rep* 64:1–5, 2015.

[16] Crider, K.S., Bailey, L.B., and Berry, R.J. Folic acid food fortification—Its history, effect, concerns, and future directions. *Nutrients* 3:370–384, 2011.

[17] Craig, W.J., Mangels, A.R., and American Dietetic Association. Position of the American Dietetic Association: Vegetarian diets. *J Am Diet Assoc* 109:1266–1282, 2009.

[18] Willems, B.A., Vermeer, C., Reutelingsperger, C.P., and Schurgers, L.J. The realm of vitamin K dependent proteins: Shifting from coagulation toward calcification. *Mol Nutr Food Res* 58:1620–1635, 2014.

[19] DiNicolantonio, J.J., Bhutani, J., and O'Keefe, J.H. The health benefits of vitamin K. *Open Heart* 2:e000300, 2015.

[20] Theuwissen, E., Magdeleyns, E.J., Braam, L.A., et al. Vitamin K status in healthy volunteers. *Food Funct* 5:229–234, 2014.

[21] Food Labeling: Revision of the Nutrition and Supplement Facts Labels. *Federal Register*, 81 FR 33741–33999, May 27, 2016. Available online at www.federalregister.gov/documents/2016/05/27/2016-11867/food-labeling-revision-of-the-nutrition-and-supplement-facts-labels. Accessed December 28, 2016.

[22] Institute of Medicine, Food and Nutrition Board. *Dietary Reference Intakes for Vitamin C, Vitamin E, Selenium, and Carotenoids*. Washington, DC: National Academies Press, 2000.

[23] Hemilä, H., and Chalker, E. Vitamin C for preventing and treating the common cold. *Cochrane Database Syst Rev* 1:CD000980, 2013.

[24] Thomas, L.D., Elinder, C.G., Tiselius, H.G., et al. Ascorbic acid supplements and kidney stone incidence among men: a prospective study. *JAMA Intern Med* 173:386–388, 2013.

[25] Azzi, A., Meydani, S.N., Meydani, M., and Zingg, J.M. The rise, the fall and the renaissance of vitamin E. *Arch Biochem Biophys* 595:100–108, 2016.

[26] Galli, F., Azzi, A., Birringer, M., et al. Vitamin E: Emerging aspects and new directions. *Free Radic Biol Med* 102:16–36, 2016.

[27] U.S. Department of Agriculture, Agricultural Research Service. 2016. Table 1. Nutrient Intakes from Food and Beverages: Mean amounts consumed per individual, by gender and age, in the United States, 2013–2014. *What We Eat in America* (NHANES 2013–2014). Available online at www.ars.usda.gov/ARSUserFiles/80400530/pdf/1314/Table_1_NIN_GEN_13.pdf. Accessed January 25, 2017.

[28] Mathur, P., Ding, Z., Saldeen, T., and Mehta, J.L. Tocopherols in the prevention and treatment of atherosclerosis and related cardiovascular disease. *Clin Cardiol* 38:570–576, 2015.

[29] Fortmann, S.P., Burda, B.U., Senger, C.A., et al. Vitamin and mineral supplements in the primary prevention of cardiovascular disease and cancer: An updated systematic evidence review for the U.S. Preventive Services Task Force. *Ann Intern Med* 159:824–834, 2013.

[30] Vardi, M., Levy, N.S., and Levy, A.P. Vitamin E in the prevention of cardiovascular disease: The importance of proper patient selection. *J Lipid Res* 54:2307–2314, 2013.

[31] Institute of Medicine, Food and Nutrition Board. *Dietary Reference Intakes: Vitamin A, Vitamin K, Arsenic, Boron, Chromium, Copper, Iodine, Iron, Manganese, Molybdenum, Nickel, Silicon, Vanadium, and Zinc*. Washington, DC: National Academies Press, 2001.

[32] U.S. Department of Agriculture and U.S. Department of Health and Human Services. *Dietary Guidelines for Americans, 2015–2020*, 8th ed., Washington, DC. Available online at https://health.gov/dietaryguidelines/2015/guidelines/. Accessed September 21, 2016.

[33] World Health Organization. *Micronutrient Deficiencies: Vitamin A Deficiency: The Challenge*. Available online at www.who.int/nutrition/topics/vad/en/index.html. Accessed January 16, 2017.

[34] National Institutes of Health Office of Dietary Supplements. *Vitamin A*. Available online at https://ods.od.nih.gov/factsheets/VitaminA-HealthProfessional/. Accessed December 17, 2016.

[35] Druesne-Pecollo, N., Latino-Martel, P., Norat, T., et al. Beta-carotene supplementation and cancer risk: A systematic review and metaanalysis of randomized controlled trials. *Int J Cancer* 127:172–184, 2010.

[36] Gröber, U., Spitz, J., Reichrath, J., et al. Vitamin D: Update 2013: From rickets prophylaxis to general preventive healthcare. *Dermatoendocrinol* 5:331–347, 2013.

37. Wacker, M., and Holick, M.F. Vitamin D—Effects on skeletal and extra-skeletal health and the need for supplementation. *Nutrients* 5:111–148, 2013.
38. Institute of Medicine, Food and Nutrition Board. *Dietary Reference Intakes for Calcium and Vitamin D.* Washington DC: National Academies Press, 2011.
39. Holick, M.F. Vitamin D deficiency: What a pain it is. *Mayo Clin Proc* 78:1457–1459, 2003.
40. Holick, M.F., and Chen, T.C. Vitamin D deficiency: A worldwide problem with health consequences. *Am J Clin Nutr* 87(Suppl):1080S–1086S, 2008.
41. Dickinson, A., Blatman, J., El-Dash, N., and Francod, J. C. Consumer usage and reasons for using dietary supplements: Report of a series of surveys. *J Am Coll Nutr* 33:176–182, 2014.
42. Guallar, E., Stranges, S., Mulrow, C., et al. Enough is enough: Stop wasting money on vitamin and mineral supplements. *Ann Intern Med* 159:850–851, 2013.
43. American Dietetic Association. Position of the American Dietetic Association: Nutrient supplementation. *J Am Diet Assoc* 109:2073–2085, 2009.
44. Bruno, R.S., and Traber, M.G. Cigarette smoke alters human vitamin E requirements. *J Nutr* 135:671–674, 2005.
45. Gershwin, M.E., Borchers, A.T., Keen, C.L., et al. Public safety and dietary supplementation. *Ann N Y Acad Sci* 1190:104–117, 2010.
46. National Institutes of Health, National Center for Complementary and Integrative Health. *Herbs at a Glance.* Available online at https://nccih.nih.gov/health/herbsataglance.htm. Accessed January 30, 2017.
47. Laws, K.R., Sweetnam, H., and Kondel, T.K. Is *Ginkgo biloba* a cognitive enhancer in healthy individuals? A meta-analysis. *Hum Psychopharmacol* 27:527–533, 2012.
48. Tan, M.S., Yu, J.T., Tan, C.C., et al. Efficacy and adverse effects of ginkgo biloba for cognitive impairment and dementia: A systematic review and meta-analysis. *J Alzheimers Dis* 43:589–603, 2015.
49. National Institutes of Health, National Center for Complementary and Integrative Health. *Ginko.* Available online at https://nccih.nih.gov/health/ginkgo/ataglance.htm. Accessed February 1, 2017.
50. Diamond, B.J., and Bailey, M.R. Ginkgo biloba: Indications, mechanisms, and safety. *Psychiatr Clin North Am* 36:73–83, 2013.
51. Varteresian, T., and Lavretsky H. Natural products and supplements for geriatric depression and cognitive disorders: An evaluation of the research. *Curr Psychiatry Rep* 16:456, 2014.
52. National Institutes of Health, National Center for Complementary and Integrative Health. *St. John's Wort.* Available online at https://nccih.nih.gov/health/stjohnswort/ataglance.htm. Accessed February 1, 2017.
53. Shergis, J.L., Zhang, A.L., Zhou, W., and Xue, C.C. Panax ginseng in randomised controlled trials: A systematic review. *Phytother Res* 27:949–965, 2013.
54. Zeng, T., Guo, F.F., Zhang, C.L., et al. A meta-analysis of randomized, double-blind, placebo-controlled trials for the effects of garlic on serum lipid profiles. *J Sci Food Agric* 92:1892–1902, 2012.
55. National Institutes of Health, National Center for Complementary and Integrative Health. *Garlic.* Available online at https://nccih.nih.gov/health/garlic/ataglance.htm. Accessed February 1, 2017.
56. National Institutes of Health, National Center for Complementary and Integrative Health. *Echinachea* Available online at https://nccih.nih.gov/health/echinacea/ataglance.htm. Accessed February 1, 2017.
57. National Institutes of Health, National Center for Complementary and Integrative Health. *Saw Palmetto.* Available online at https://nccih.nih.gov/health/palmetto/ataglance.htm. Accessed December 19, 2016.
58. Food and Drug Administration. Current good manufacturing practice in manufacturing, packaging, labeling, or holding operations for dietary supplements; final rule. *Fed Regist* 72:34751–34958, June 25, 2007.

Chapter 8
1. World Health Organization. *Drinking Water.* Available online at www.who.int/mediacentre/factsheets/fs391/en/. Accessed January 1, 2017.
2. Shen, H-P. Body fluids and water balance. In M.H. Stipanuk and M.A. Caudill (eds.), *Biochemical, Physiological, and Molecular Aspects of Human Nutrition*, 3rd ed. St. Louis: Saunders Elsevier, 2013, pp. 781–800.
3. Institute of Medicine, Food and Nutrition Board. *Dietary Reference Intakes for Water, Potassium, Sodium, Chloride, and Sulfate.* Washington, DC: National Academies Press, 2004.
4. Drewnowski, A., Rehm, C.D., and Constant, F. Water and beverage consumption among adults in the United States: Cross-sectional study using data from NHANES 2005-2010. *BMC Public Health* 13:1068, 2013.
5. International Bottled Water Association. *Bottled Water Market.* Available online at www.bottledwater.org/economics/bottled-water-market. Accessed December 6, 2016.
6. U.S. Government Accountability Office. *Bottled Water: FDA Safety and Consumer Protections Are Often Less Stringent Than Comparable EPA Protections for Tap Water*, June 2009. Available online at www.gao.gov/new.items/d09610.pdf. Accessed December 5, 2016.
7. Environmental Working Group. *Bottled Water Quality Investigation: 10 Major Brands, 38 Pollutants*, October 2008. Available online at www.ewg.org/research/bottled-water-quality-investigation. Accessed December 6, 2016.
8. Environmental Working Group. *National Drinking Water Database.* Available online at www.ewg.org/tap-water/. Accessed December 5, 2016.
9. U.S. EPA. *Drinking Water Requirements for States and Public Water Systems.* Available online at www.epa.gov/dwreginfo/public-notification-rule. Accessed January 7, 2017.
10. Bosman, J. After water fiasco, trust of officials is in short supply in Flint. *New York Times*, October 8, 2016. Available online at www.nytimes.com/2016/10/09/us/after-water-fiasco-trust-of-officials-is-in-short-supply-in-flint.html?_r=0. Accessed January 7, 2017.
11. Ban the Bottle. *Bottled Water Facts.* Available online at www.banthebottle.net/bottled-water-facts/. Accessed December 6, 2016.
12. Go, A.S., Mozaffarian, D., Roger, V.L., et al.; American Heart Association Statistics Committee and Stroke Statistics Subcommittee. Heart disease and stroke statistics—2014 update: A report from the American Heart Association. *Circulation* 129:e28–e292, 2014.
13. American Heart Association. *Understanding Your Risk of Developing HBP.* Available online at www.heart.org/HEARTORG/Conditions/HighBloodPressure/UnderstandYourRiskforHighBloodPressure/Understand-Your-Risk-for-High-Blood-Pressure_UCM_002052_Article.jsp. Accessed January 1, 2014.
14. Carvalho, J.J., Baruzzi, R.G., Howard, P.F., et al. Blood pressure in four remote populations in the Intersalt study. *Hypertension* 14:238–246, 1989.
15. Appel, L.J., Moore, T.J., Obarzanek, E., et al. A clinical trial of the effects of dietary patterns on blood pressure. *N Engl J Med* 336:1117–1124, 1997.
16. Sacks, F.M., Svetkey, L.P., Vollmer, W.M., et al. Effects on blood pressure of reduced dietary sodium and the Dietary Approaches to Stop

Hypertension (DASH) diet. DASH-Sodium Collaborative Research Group. *N Engl J Med* 344:3–10, 2001.

17. Van Horn, L. Dietary sodium and blood pressure: How low should we go? *Prog Cardiovasc Dis* 58:61–68, 2015.

18. Houston, M.C., and Harper, K.J. Potassium, magnesium, and calcium: Their role in both the cause and treatment of hypertension. *J Clin Hypertens (Greenwich)* 10(7 Suppl. 2):3–11, 2008.

19. Greenland, P. Beating high blood pressure with low sodium DASH. *N Engl J Med* 344:53–55, 2001.

20. U.S. Department of Agriculture and U.S. Department of Health and Human Services. *Dietary Guidelines for Americans, 2015–2020*, 8th ed. Available online at https://health.gov/dietaryguidelines/2015/. Accessed January 21, 2017.

21. Cogswell, M.E., Zhang, Z., Carriquiry, A.L., et al. Sodium and potassium intakes among US adults: NHANES 2003–2008. *Am J Clin Nutr* 96:647–657, 2012.

22. U.S. Department of Health and Human Services and U.S. Department of Agriculture. *Dietary Guidelines for Americans 2005*. Available online at www.healthierus.gov/dietaryguidelines. Accessed May 19, 2005.

23. Centers for Disease Control and Prevention. *Ten Top Sources of Sodium*. Available online at www.cdc.gov/salt/sources.htm. Accessed February 1, 2017.

24. U.S. Food and Drug Administration. *Sodium Reduction*. Available online at www.fda.gov/Food/IngredientsPackagingLabeling/FoodAdditivesIngredients/ucm253316.htm. Accessed January 29, 2017.

25. Hamidi, M.S., Gajic-Veljanoski, O., and Cheung, A.M. Vitamin K and bone health. *J Clin Densitom* 16:409–413, 2013.

26. Wright, N.C., Looker, A.C., Saag, K.G., et al. The recent prevalence of osteoporosis and low bone mass in the United States based on bone mineral density at the femoral neck or lumbar spine. *J Bone Mineral Res* 29:2520–2526, 2014.

27. Dempster, D.W. Osteoporosis and the burden of osteoporosis-related fractures. *Am J Manag Care* 17(Suppl 6.):S164–S169, 2011.

28. Reid, I.R. Fat and bone. *Arch Biochem Biophys* 503:20–27, 2010.

29. Ilich, J.Z., Brownbill, R.A., and Coster, D.C. Higher habitual sodium intake is not detrimental for bones in older women with adequate calcium intake. *Eur J Appl Physiol* 109:745–755, 2010.

30. Mangano, K.M., Sahni, S., and Kerstetter, J.E. Dietary protein is beneficial to bone health under conditions of adequate calcium intake: An update on clinical research. *Curr Opin Clin Nutr Metab Care* 17:69–74, 2014.

31. Papapoulos, S.E. Use of bisphosphonates in the management of postmenopausal osteoporosis. *Ann N Y Acad Sci* 1218:15–32, 2011.

32. Patel, A. M., Adeseun, G. A., and Goldfarb, S. Calcium-alkali syndrome in the modern era. *Nutrients* 5:4880–4893, 2013.

33. Institute of Medicine, Food and Nutrition Board. *Dietary Reference Intakes for Calcium and Vitamin D*. Washington, DC: National Academies Press, 2011.

34. Park, S., Xu, F., Town, M., and Blanck, H.M. Prevalence of sugar-sweetened beverage intake among adults—23 states and the District of Columbia, 2013. *MMWR Morb Mortal Wkly Rep* 65:169–174, 2016.

35. U.S. Department of Agriculture and U.S. Department of Health and Human Services. *Scientific Report of the 2015 Dietary Guidelines Advisory Committee*. Available online at https://health.gov/dietaryguidelines/2015-scientific-report/. Accessed February 1, 2017.

36. Bauer, D.C. Clinical practice. Calcium supplements and fracture prevention. *N Engl J Med* 369:1537–1543, 2013.

37. Heaney, R.P., Weaver, C.M., and Recker, R.R. Calcium absorption from spinach. *Am J Clin Nutr* 47:707–709, 1988.

38. Calvo, M.S., and Uribarri, J. Public health impact of dietary phosphorus excess on bone and cardiovascular health in the general population. *Am J Clin Nutr* 98:6–15, 2013.

39. Chang, A.R., Lazo, M., Appel, L.J., et al. High dietary phosphorus intake is associated with all-cause mortality: Results from NHANES III. *Am J Clin Nutr* 99:320–327, 2014.

40. Menon, M.C., and Ix, J.H. Dietary phosphorus, serum phosphorus, and cardiovascular disease. *Ann N Y Acad Sci* 1301:21–26, 2013.

41. Institute of Medicine, Food and Nutrition Board. *Dietary Reference Intakes for Calcium, Phosphorus, Magnesium, Vitamin D, and Fluoride*. Washington, DC: National Academies Press, 1997.

42. Rosanoff, A., Weaver, C.M., and Rude, R.K. Suboptimal magnesium status in the United States: Are the health consequences underestimated? *Nutr Rev* 70:153–164, 2012.

43. Champagne, C.M. Magnesium in hypertension, cardiovascular disease, metabolic syndrome, and other conditions: A review. *Nutr Clin Pract* 23:142–151, 2008.

44. Rosanoff, A., Dai, Q., and Shapses, S.A. Essential nutrient interactions: Does low or suboptimal magnesium status interact with vitamin D and/or calcium status? *Adv Nutr* 7:25–43, 2016.

45. Everett, E.T. Fluoride's effects on the formation of teeth and bones, and the influence of genetics. *J Dent Res* 90:552–560, 2011.

46. Palmer, C.A., and Gilbert, J.A.; Academy of Nutrition and Dietetics. Position of the Academy of Nutrition and Dietetics: The impact of fluoride on health. *J Acad Nutr Diet* 112:1443–1453, 2012.

47. U.S. Department of Health and Human Services Federal Panel on Community Water Fluoridation. U.S. Public Health Service recommendation for fluoride concentration in drinking water for the prevention of dental caries. *Public Health Rep* 130:318–331, 2015.

48. Centers for Disease Control and Prevention. *Fluoridation Statistics*, 2014. Available online at www.cdc.gov/fluoridation/statistics/2014stats.htm. Accessed January 2, 2017.

49. Levy, S.M., Warren, J.J., Phipps, K., et al. Effects of life-long fluoride intake on bone measures of adolescents: A prospective cohort study. *J Dent Res* 93:353–359, 2014.

50. Beltrán-Aguilar, E.D., Barker, L., and Dye, B.A. Prevalence and severity of dental fluorosis in the United States, 1999–2004. *NCHS Data Brief* 53:1–8, 2010. Available online at www.cdc.gov/nchs/data/databriefs/db53.pdf. Accessed January 2, 2017.

51. Ganz, T. Systemic iron homeostasis. *Physiol Rev* 93:1721–1741, 2013.

52. Food and Nutrition Board, Institute of Medicine. *Dietary Reference Intakes: Vitamin A, Vitamin K, Arsenic, Boron, Chromium, Copper, Iodine, Iron, Manganese, Molybdenum, Nickel, Silicon, Vanadium, and Zinc*. Washington, DC: National Academies Press, 2001.

53. Miller, J.L. Iron deficiency anemia: A common and curable disease. *Cold Spring Harb Perspect Med* 3(7):a011866, 2013.

54. World Health Organization. *Micronutrient Deficiencies: Iron Deficiency Anemia*. Available online at www.who.int/nutrition/topics/ida/en/index.html. Accessed January 2, 2017.

55. Food and Drug Administration. *Dietary Supplement Labeling Guide: Chapter VIII. Other Labeling Information*. April 2005. Available online at www.fda.gov/Food/GuidanceRegulation/GuidanceDocumentsRegulatoryInformation/DietarySupplements/ucm070616.htm. Accessed January 2, 2017.

56. Camaschella, C., and Poggiali, E. Inherited disorders of iron metabolism. *Curr Opin Pediatr* 23:14–20, 2011.

57. Crownover, B.K., and Covey, C.J. Hereditary hemochromatosis. *Am Fam Physician* 87:183–190, 2013.

58. Fairweather-Tait, S.J., Bao, Y., Broadley, M.R., et al. Selenium in human health and disease. *Antioxid Redox Signal* 14:1337–1383, 2011.

[59] Food and Nutrition Board, Institute of Medicine. *Dietary Reference Intakes for Vitamin C, Vitamin E, Selenium, and Carotenoids*. Washington, DC: National Academies Press, 2000.

[60] Clark, L.C., Combs, G.F., Jr., Turnbull, B.W., et al. Effect of selenium supplementation for cancer prevention in patients with carcinoma of the skin. *JAMA* 276:1957–1968, 1996.

[61] Roman, M., Jitaru, P., and Barbante, C. Selenium biochemistry and its role for human health. *Metallomics* 6:25–54, 2014.

[62] Jarosz, M., Olbert, M., Wyszogrodzka, G., et al. Antioxidant and anti-inflammatory effects of zinc. Zinc-dependent NF-κB signaling. *Inflammopharmacology* 25:11–24, 2017.

[63] Maret, W. Metals on the move: Zinc ions in cellular regulation and in the coordination dynamics of zinc proteins. *Biometals* 24:411–418, 2011.

[64] Wong, C.P., and Ho, E. Zinc and its role in age-related inflammation and immune dysfunction. *Mol Nutr Food Res* 56:77–87, 2012.

[65] Chasapis, C.T., Loutsidou, A.C., Spiliopoulou, C.A., and Stefanidou, M.E. Zinc and human health: An update. *Arch Toxicol* 86:521–534, 2012.

[66] Prasad, A.S. Zinc: Role in immunity, oxidative stress and chronic inflammation. *Curr Opin Clin Nutr Metab Care* 12:646–652, 2009.

[67] Brown, K.H., Peerson, J.M., Baker, S.K., and Hess, S.Y. Preventive zinc supplementation among infants, preschoolers, and older prepubertal children. *Food Nutr Bull* 30(1 Suppl.):S12–S40, 2009.

[68] Singh, M., and Das, R.R. Zinc for the common cold. *Cochrane Database Syst Rev* 2:CD001364, 2011.

[69] World Health Organization. *Is It True That Lack of Iodine Really Causes Brain Damage?* Available online at www.who.int/features/qa/17/en. Accessed April 30, 2017.

[70] Pearce, E.N., Andersson, M., and Zimmermann, M.B. Global iodine nutrition: Where do we stand in 2013? *Thyroid* 23:523–528, 2013.

[71] Pearce, E.N., and Leung, A.M. The state of U.S. iodine nutrition: How can we ensure adequate iodine for all? *Thyroid* 23:924–925, 2013.

[72] National Institutes of Health. Office of Dietary Supplements. *Iodine: Fact Sheet for Professionals*. Available online at https://ods.od.nih.gov/factsheets/Iodine-HealthProfessional/#h6. Accessed January 27, 2017.

[73] Caldwell, K.L., Pan, Y., Mortensen, M.E., et al. Iodine status in pregnant women in the National Children's Study and in U.S. women (15–44 years), National Health and Nutrition Examination Survey 2005–2010. *Thyroid* 23:927–937, 2013.

[74] Hua, Y., Clark, S., Ren, J., and Sreejayan, N. Molecular mechanisms of chromium in alleviating insulin resistance. *J Nutr Biochem* 23:313–319, 2012.

[75] Di Luigi, L. Supplements and the endocrine system in athletes. *Clin Sports Med* 27:131–151, 2008.

Chapter 9

[1] The State of Obesity, Better Policies for a Healthier America. *The Healthcare Costs of Obesity*. Available online at http://stateofobesity.org/healthcare-costs-obesity/. Accessed February 23, 2017.

[2] Centers for Disease Control and Prevention, National Center for Health Statistics. *Overweight and Obesity*. Available online at www.cdc.gov/nchs/fastats/obesity-overweight.htm. Accessed January 14, 2017.

[3] Flegal, K.M., Kruszon-Moranm, D., Carroll, M.D., et al. Trends in obesity among adults in the United States, 2005 to 2014. *JAMA* 315:2284–2291, 2016.

[4] World Health Organization. *Obesity and Overweight*, Fact Sheet, June 2016. Available online at www.who.int/mediacentre/factsheets/fs311/en/. Accessed February 23, 2017.

[5] U.S. Department of Agriculture and U.S. Department of Health and Human Services. *Dietary Guidelines for Americans, 2015–2020*, 8th ed. Available online at http://health.gov/dietaryguidelines/2015/. Accessed January 14, 2017.

[6] Ogden, C.L., Carroll, M.D., Lawman, H.G., et al. Trends in obesity prevalence among children and adolescents in the United States, 1988–1994 through 2013–2014. *JAMA* 315:2292–2299, 2016.

[7] Flint, E., Cummins, S., and Sacker, A. Associations between active commuting, body fat, and body mass index: Population based, cross sectional study in the United Kingdom. *BMJ* 349:g4887, 2014.

[8] Faith, M.S., Butryn, M., Wadden, T.A., et al. Evidence for prospective associations among depression and obesity in population-based studies. *Obes Rev* 12:e438–e453, 2011.

[9] Trogdon, J.G., Finkelstein, E.A., Feagan, C.W., and Cohen, J.W. State- and payer-specific estimates of annual medical expenditures attributable to obesity. *Obesity (Silver Spring)* 20:214–220, 2012.

[10] Jensen, M.D., Ryan, D.H., Apovian, C.M., et al., 2013 AHA/ACC/TOS guideline for the management of overweight and obesity in adults, a report of the American College of Cardiology/American Heart Association Task Force on Practice Guidelines and The Obesity Society. *Circulation* 129(Suppl. 2):S102–S138, 2014.

[11] Flegal, K.M., Graubard, B.I., Williamson, D.F., and Gail, M.H. Excess deaths associated with underweight, overweight, and obesity. *JAMA* 293:1861–1867, 2005.

[12] Gallagher, D., Heymsfield, S., Heo, M., et al. Healthy percentage body fat ranges: An approach for developing guidelines based on body mass index. *Am J Clin Nutr* 72:694–701, 2000.

[13] Tchernof, A., and Després, J.P. Pathophysiology of human visceral obesity: An update. *Physiol Rev* 93:359–404, 2013.

[14] Byrne, N.M., Wood, R.E., Schutz, Y., and Hills, A.P. Does metabolic compensation explain the majority of less-than-expected weight loss in obese adults during a short-term severe diet and exercise intervention? *Int J Obes (Lond)* 36:1472–1478, 2012.

[15] Institute of Medicine, Food and Nutrition Board. *Dietary Reference Intakes for Energy, Carbohydrate, Fiber, Fat, Protein and Amino Acids*. Washington, DC: National Academies Press, 2002.

[16] Berryman, D.E., and Hulver, M.W. Cellular and whole-animal energetics. In M.H. Stipanuk and M.A. Caudill (eds.), *Biochemical, Physiological, and Molecular Aspects of Human Nutrition*, 3rd ed. St Louis: Saunders Elsevier, 2013, pp. 481–500.

[17] Xia, Q., and Grant, S.F. The genetics of human obesity. *Ann N Y Acad Sci* 1281:178–190, 2013.

[18] Waalen, J. The genetics of human obesity. *Transl Res* 164:293–301, 2014.

[19] Ravussin, E., Valencia, M.E., Esparza, J., et al. Effects of a traditional lifestyle on obesity in Pima Indians. *Diabetes Care* 17:1067–1074, 1994.

[20] Norman, R.A., Thompson, D.B., Foroud, T., et al. Genomewide search for genes influencing percent body fat in Pima Indians: Suggestive linkage at chromosome 11q21-q22. *Am J Human Genet* 60:166–173, 1997.

[21] Esparza, J., Fox, C., Harper, I.T., et al. Daily energy expenditure in Mexican and USA Pima Indians: Low physical activity as a possible cause of obesity. *Int J Obes Relat Metab Disord* 24:55–59, 2000.

[22] Speakman, J.R., Levitsky, D.A., Allison, D.B., et al. Set points, settling points and some alternative models: Theoretical options to understand how genes and environments combine to regulate body adiposity. *Dis Model Mech* 4:733–745, 2011.

[23] Hussain, S.S., De Silva, A., and Bloom, S.R. Controls of food intake and appetite. In A.C. Ross, B. Caballero, R.J. Cousins, et al. (eds.), *Modern*

24. Sáinz, N., Barrenetxe, J., Moreno-Aliaga, M.J., and Martínez, J.A. Leptin resistance and diet-induced obesity: Central and peripheral actions of leptin. *Metabolism* 64:35–46, 2015.
25. Hill, J.O., Wyatt, H.R., and Peters, J.C. Energy balance and obesity. *Circulation* 126:126–132, 2012.
26. Levine, J.A., Lanningham-Foster, L.M., McCrady, S.K., et al. Interindividual variation in posture allocation: Possible role in human obesity. *Science* 307:584–586, 2005.
27. Garvey, W.T., Garber, A.J., Mechanick. J.I., et al. American Association of Clinical Endocrinologists and American College of Endocrinology position statement on the 2014 advanced framework for a new diagnosis of obesity as a chronic disease. *Endocr Pract* 20:977–989, 2014.
28. Turk, M.W., Yang, K., Hravnak, M., et al. Randomized clinical trials of weight-loss maintenance: A review. *J Cardiovasc Nurs* 24:58–80, 2009.
29. American Association of Nutrition and Dietetics. Position of the Academy of Nutrition and Dietetics: Interventions for the treatment of overweight and obesity in adults. *J Acad Nutr Diet* 116:129–147, 2016.
30. Stevens, V.L., Jacobs, E.J., Sun, J., et al. Weight cycling and mortality in a large prospective US study. *Am J Epidemiol* 175:785–792, 2012.
31. Chaput, J.P., Klingenberg, L., Rosenkilde, M., et al. Physical activity plays an important role in body weight regulation. *J Obes* 2011, doi: 10.1155/2011/360257.
32. U.S. Department of Health and Human Services. *2008 Physical Activity Guidelines for Americans*. Available online at www.health.gov/paguidelines/pdf/paguide.pdf. Accessed February 15, 2017.
33. U.S. Department of Health and Human Services, Office of the Surgeon General. *The Surgeon General's Call to Action to Prevent and Decrease Overweight and Obesity*. Available online at www.ncbi.nlm.nih.gov/books/NBK44206/. Accessed June 12, 2014.
34. Martens, E.A., and Westerterp-Plantenga, M.S. Protein diets, body weight loss and weight maintenance. *Curr Opin Clin Nutr Metab Care* 17:75–79, 2014.
35. Acheson, K.J. Carbohydrate for weight and metabolic control: Where do we stand? *Nutrition* 26:141–145, 2010.
36. U.S. Food and Drug Administration. *Beware of Products Promising Miracle Weight Loss*. Available online at www.fda.gov/ForConsumers/ConsumerUpdates/ucm246742.htm. Accessed February 23, 2017.
37. Manore, M.M. Dietary supplements for improving body composition and reducing body weight: Where is the evidence? *Int J Sport Nutr Exerc Metab* 22:139–154, 2012.
38. Chan, T.Y. Potential risks associated with the use of herbal anti-obesity products. *Drug Saf* 32:453–456, 2009.
39. Lautz, D., Goebel-Fabbri, A., Halperin, F., and Goldfien, A.B. The great debate: Medicine or surgery. What is best for the patient with type 2 diabetes? *Diabetes Care* 34:763–770, 2011.
40. Buchwald, H., Estok, R., Fahrbach, K., et al. Weight and type 2 diabetes after bariatric surgery: Systematic review and meta-analysis. *Am J Med* 122:248–256, 2009.
41. Pontiroli, A.E., and Morabito, A. Long-term prevention of mortality in morbid obesity through bariatric surgery: A systematic review and meta-analysis of trials performed with gastric banding and gastric bypass. *Ann Surg* 253:484–487, 2011.
42. McEwen, L.N., Coelho, R.B., Baumann, L.M., et al. The cost, quality of life impact, and cost-utility of bariatric surgery in a managed care population. *Obes Surg.* 20:919–928, 2010.
43. Weiss, G. The true cost of my weight-loss surgery. *Time* January 30, 2014. Available online at http://time.com/money/2795119/the-true-cost-of-my-weight-loss-surgery/. Accessed June 25, 2017.
44. Chang, S.H., Stoll, C.R., Song, J., et al. The effectiveness and risks of bariatric surgery: An updated systematic review and meta-analysis, 2003–2012. *JAMA Surg* 149:275–287, 2014.
45. Quatromoni, P.A. A tale of two runners: A case report of athletes' experiences with eating disorders in college. *J Acad Nutr Diet* 117:21–31, 2017.
46. National Association of Anorexia Nervosa and Associated Disorders (ANAD). *Eating Disorders Statistics*. Available online at www.anad.org/get-information/about-eating-disorders/eating-disorders-statistics/. Accessed February 6, 2017.
47. American Psychiatric Association. *Diagnostic and Statistical Manual of Mental Disorders*, 5th ed. Arlington, VA: American Psychiatric Publishing, 2013.
48. Body image worries hit Zulu women. *BBC News*, April 16, 2004. Available online at http://news.bbc.co.uk/2/hi/health/3631359.stm. Accessed February 23, 2017.
49. Osad'an, R., and Hanna, R. The effects of the media on self-esteem of young girls. *Acta Technologica Dubnicae* 5:37–44, 2015.
50. Rancano, V. *Is It Time to Set Weight Minimums for the Fashion Industry?* Available online at www.npr.org/sections health-shots/2015/12/22/460682633/is-it-time-to-set-weight-minimums-for-the-fashion-industry/. Accessed February 15, 2017.
51. Jellinek, R.D., Myers, T.A., and Keller, K.L. The impact of doll style of dress and familiarity on body dissatisfaction in 6- to 8-year-old girls. *Body Image* 18:78–85, 2016.
52. Keel, P.K., and Forney, K.J. Psychosocial risk factors for eating disorders. *Int J Eat Disord* 46:433–439, 2013.
53. Wooldridge, T., and Lytle, P.P. An overview of anorexia nervosa in males. *Eat Disord* 20:368–378, 2012.
54. Arcelus, J., Mitchell, A.J., Wales, J., and Nielsen, S. Mortality rates in patients with anorexia nervosa and other eating disorders. A meta-analysis of 36 studies. *Arch Gen Psychiatry* 68:724–731, 2011.
55. American Dietetic Association. Position of the American Dietetic Association: Nutrition intervention and treatment of eating disorders. *J Am Diet Assoc* 111:1236–1241, 2011.
56. Vandereycken, W. History of anorexia nervosa and bulimia nervosa. In Fairburn, C.G., and Brownell, K.D. (eds.), *Eating Disorders and Obesity: A Comprehensive Handbook*, 2nd ed. New York: Guilford Press, 2002, pp. 151–154.
57. Bratland-Sanda, S., and Sundgot-Borgen, J. Eating disorders in athletes: overview of prevalence, risk factors and recommendations for prevention and treatment. *Eur J Sport Sci* 13:499–508, 2013.

Chapter 10

1. Centers for Disease Control and Prevention. *Physical Activity and Health. The Benefits of Physical Activity*. Available online at www.cdc.gov/physicalactivity/basics/pa-health/index.htm#LiveLonger. Accessed January 27, 2017.
2. Jacka, F.N., and Berk, M. Depression, diet and exercise. *Med J Aust* 199(6 Suppl.):S21–23, 2013.
3. Institute of Medicine, Food and Nutrition Board. *Dietary Reference Intakes for Energy, Carbohydrates, Fiber, Fat, Protein and Amino Acids*. Washington, DC: National Academies Press, 2002.
4. Strasser, B. Physical activity in obesity and metabolic syndrome. *Ann N Y Acad Sci* 1281:141–159, 2013.
5. Chaput, J.P., Klingenberg, L., Rosenkilde, M., et al. Physical activity plays an important role in body weight regulation. *J Obes* 2011, doi:10.1155/2011/360257.
6. Behm, D.G., and Chaouachi, A. A review of the acute effects of static and dynamic stretching on performance. *Eur J Appl Physiol* 111:2633–2651, 2011.

[7] Garber, C.E., Blissmer, B., Deschenes, M. R., et al. American College of Sports Medicine. American College of Sports Medicine position stand: Quantity and quality of exercise for developing and maintaining cardiorespiratory, musculoskeletal, and neuromotor fitness in apparently healthy adults: Guidance for prescribing exercise. *Med Sci Sports Exerc* 43:1334–1359, 2011.

[8] Behm, D.G, Blazevich, A.J., Kay, A.D., and McHugh, M. Acute effects of muscle stretching on physical performance, range of motion, and injury incidence in healthy active individuals: a systematic review. *Appl Physiol Nutr Metab* 41:1–11, 2016.

[9] Gallagher, D., Heymsfield, S., Heo, M., et al. Healthy percentage body fat ranges: An approach for developing guidelines based on body mass index. *Am J Clin Nutr* 72:694–701, 2000.

[10] U.S. Department of Agriculture and U.S. Department of Health and Human Services. *Dietary Guidelines for Americans, 2015-2020*, 8th ed., Washington, DC. Available online at https://health.gov/dietaryguidelines/2015/guidelines/. Accessed February 14, 2017.

[11] U.S. Department of Health and Human Services. *2008 Physical Activity Guidelines for Americans*. Available online at www.health.gov/paguidelines/pdf/paguide.pdf. Accessed February 14, 2017.

[12] Centers for Disease Control and Prevention, National Center for Health Statistics. *Exercise or Physical Activity*. Available online at www.cdc.gov/nchs/fastats/exercise.htm. Accessed February 14, 2017.

[13] Haskell, W.L., Lee, I-M., Pate, R.R., et al. Physical activity and public health: Updated recommendations for adults from the American College of Sports Medicine and the American Heart Association. *Circulation* 116:1081–1093, 2007.

[14] Gulati, M., Shaw, L.J., Thisted, R.A., et al. Heart rate response to exercise stress testing in asymptomatic women: The St. James Women Take Heart project. *Circulation* 122:130–137, 2010.

[15] Matthews, C.E., George, S.M., Moore, S.C., et al. Amount of time spent in sedentary behaviors and cause-specific mortality in US adults. *Am J Clin Nutr* 95: 437–445, 2012.

[16] Meeusen, R., Duclos, M., Foster, C., et al. Prevention, diagnosis, and treatment of the overtraining syndrome: Joint consensus statement of the European College of Sport Science and the American College of Sports Medicine. *Med Sci Sports Exerc* 45:186–205, 2013.

[17] Grassi, B., Rossiter, H.B., and Zoladz, J.A. Skeletal muscle fatigue and decreased efficiency: Two sides of the same coin? *Exerc Sport Sci Rev* 43:75–83, 2015.

[18] Finsterer, J. Biomarkers of peripheral muscle fatigue during exercise. *BMC Musculoskelet Disord* 13:218, 2012.

[19] Rivera-Brown, A.M., and Frontera, W.R. Principles of exercise physiology: Response to acute exercise and long-term adaptations to training. *PM R* 4:797–804, 2012.

[20] Thomas, D.T., Erdman, K.A., and Burke, L.M. Position of the Academy of Nutrition and Dietetics, Dietitians of Canada, and the American College of Sports Medicine: Nutrition and athletic performance. *J Acad Nutr Diet* 116:501–528, 2016.

[21] Mountjoy, M., Sundgot-Borgen, J., Burke, L., et al. The IOC consensus statement: Beyond the female athlete triad—Relative energy deficiency in sport (RED-S). *Br J Sports Med* 48:491–497, 2014.

[22] Bratland-Sanda, S., and Sundgot-Borgen, J. Eating disorders in athletes: Overview of prevalence, risk factors and recommendations for prevention and treatment. *Eur J Sport Sci* 13:499–508, 2013.

[23] Remick, D., Chancellor, K., Pederson, J., et al. Hyperthermia and dehydration-related deaths associated with intentional rapid weight loss in three collegiate wrestlers—North Carolina, Wisconsin, and Michigan, November–December, 1997. *MMWR Morb Mortal Wkly Rep* 47:105–108, 1998. Available online at www.cdc.gov/mmwr/preview/mmwrhtml/00051388.htm. Accessed June 15, 2014.

[24] Tipton, K.D. Efficacy and consequences of very-high-protein diets for athletes and exercisers. *Proc Nutr Soc* 7:1–10, 2011.

[25] Clénin, G., Cordes, M., Huber, A., et al. Iron deficiency in sports—Definition, influence on performance and therapy. *Swiss Med Wkly* 145:w14196, 2015.

[26] Latunde-Dada, G.O. Iron metabolism in athletes—Achieving a gold standard. *Eur J Haematol* 90:10–15, 2013.

[27] Sim, M., Dawson, B., Landers, G., et al. Iron regulation in athletes: Exploring the menstrual cycle and effects of different exercise modalities on hepcidin production. *Int J Sport Nutr Exerc Metab* 24:177–187, 2014.

[28] Food and Nutrition Board, Institute of Medicine. *Dietary Reference Intakes: Vitamin A, Vitamin K, Arsenic, Boron, Chromium, Copper, Iodine, Iron, Manganese, Molybdenum, Nickel, Silicon, Vanadium, and Zinc*. Washington, DC: National Academies Press, 2001.

[29] de Sousa, C.V., Sales, M.M., Rosa, T.S., et al. The antioxidant effect of exercise: A systematic review and meta-analysis. *Sports Med* 47: 277–293, 2017.

[30] NOAA's National Weather Service. *Heat Index*. Available online at www.weather.gov/os/heat/index.shtml. Accessed February 11, 2014.

[31] Knechtle, B., Gnädinger, M., Knechtle, P., et al. Prevalence of exercise-associated hyponatremia in male ultraendurance athletes. *Clin J Sport Med* 23:226–232, 2011.

[32] Bergstrom, J., Hermansen, L., Hultman, E., and Saltin, B. Diet, muscle glycogen and physical performance. *Acta Physiologica Scandinavica* 71:140–150, 1967.

[33] Burke, L.M., Hawley, J.A., Wong, S.H., and Jeukendrup, A.E. Carbohydrates for training and competition. *J Sports Sci* 29(Suppl. 1): S17–S27, 2011.

[34] Sedlock, D.A. The latest on carbohydrate loading: A practical approach. *Curr Sports Med Rep* 7:209–213, 2008.

[35] Howarth, K.R., Moreau, N.A., Phillips, S.M., and Gibala, M.J. Coingestion of protein with carbohydrate during recovery from endurance exercise stimulates skeletal muscle protein synthesis in humans. *J Appl Physiol* 106:1394–1402, 2009.

[36] Kanaley, J.A. Growth hormone, arginine and exercise. *Curr Opin Clin Nutr Metab Care* 11:50–54, 2008.

[37] Zajac, A., Poprzecki, S., Zebrowska, A., et al. Arginine and ornithine supplementation increases growth hormone and insulin-like growth factor-1 serum levels after heavy-resistance exercise in strength-trained athletes. *J Strength Cond Res* 24:1082–1090, 2010.

[38] Peternelj, T.T., and Coombes, J.S. Antioxidant supplementation during exercise training: Beneficial or detrimental? *Sports Med* 41:1043–1069, 2011.

[39] Domínguez, R., Cuenca, E., Maté-Muñoz, J.L., et al. Effects of beetroot juice supplementation on cardiorespiratory endurance in athletes. A systematic review. *Nutrients* 9(1), 2017.

[40] Clements, W.T., Lee, S.R., and Bloomer, R.J. Nitrate ingestion: A review of the health and physical performance effects. *Nutrients* 6: 5224–5264, 2014.

[41] Harris, R. C., and Stellingwerff, T. Effect of β-alanine supplementation on high-intensity exercise performance. *Nestle Nutr Inst Workshop Ser* 76:61–71, 2013.

[42] Quesnele, J. J., Laframboise, M. A., Wong, J. J., et al. The effects of beta-alanine supplementation on performance: A systematic review of the literature. *Int J Sport Nutr Exerc Metab* 24:14–27, 2014.

[43] Durkalec-Michalski, K., and Jeszka, J. The effect of β-hydroxy-β-methylbutyrate on aerobic capacity and body composition in trained athletes. *J Strength Cond Res* 30:2617–2626, 2016.

44. Baxter, J., Carlos, J.L., Thurmond, J., and Frost, D. Dietary toxicity of calcium β-hydroxy-β-methyl butyrate (CaHMB). *Food Chem Toxicol* 43:1731–1741, 2006.
45. Burke, L.M. Practical considerations for bicarbonate loading and sports performance. *Nestle Nutr Inst Workshop Ser* 75:15–26, 2013.
46. Meeusen, R., Roelands, B., and Spriet, L.L. Caffeine, exercise and the brain. *Nestle Nutr Inst Workshop Ser* 76:1–12, 2013.
47. Jeukendrup, A.E., and Randell, R. Fat burners: Nutrition supplements that increase fat metabolism. *Obes Rev* 12:841–851, 2011.
48. Di Luigi, L. Supplements and the endocrine system in athletes. *Clin Sports Med* 27:131–151, 2008.
49. Close, G.L., Hamilton, D.L., Philp, A., et al. New strategies in sport nutrition to increase exercise performance. *Free Radic Biol Med* 98:144–158, 2016.
50. Kim, H. J., Kim, C. K., Carpentier, A., and Poortmans, J. R. Studies on the safety of creatine supplementation. *Amino Acids* 40:1409–1418, 2011.
51. Andres, S., Ziegenhagen, R., Trefflich, I., et al. Creatine and creatine forms intended for sports nutrition. *Mol Nutr Food Res* 61(6), 2017.
52. Clegg, M.E. Medium-chain triglycerides are advantageous in promoting weight loss although not beneficial to exercise performance. *Int J Food Sci Nutr* 61:653–679, 2010.
53. Nissen, S.L., and Sharp, R.L. Effect of dietary supplements on lean mass and strength gains with resistance exercise: A meta-analysis. *J Appl Physiol* 94:651–659, 2003.
54. Pope, H.G. Jr, Wood, R.I., Rogol, A., et al. Adverse health consequences of performance-enhancing drugs: An Endocrine Society scientific statement. *Endocr Rev* 35:341–375, 2014.
55. El Osta, R., Almont, T., Diligent, C., et al. Anabolic steroids abuse and male infertility. *Basic Clin Androl* 26:2, 2016.
56. Brown, G.A., Vukovich, M., and King, D.S. Testosterone prohormone supplements. *Med Sci Sports Exerc* 38:1451–1461, 2006.
57. Birzniece, V., Nelson, A.E., and Ho, K.K. Growth hormone and physical performance. *Trends Endocrinol Metab* 22:171–178, 2011.
58. Meeusen, R., Roelands, B., and Spriet, L. L. Caffeine, exercise and the brain. *Nestle Nutr Inst Workshop Ser* 76:1–12, 2013.
59. Pesta, D. H., Angadi, S. S., Burtscher, M., and Roberts, C. K. The effects of caffeine, nicotine, ethanol, and tetrahydrocannabinol on exercise performance. *Nutr Metab (Lond)* 10:71, 2013.
60. Mora-Rodriguez, R., and Pallarés, J. G. Performance outcomes and unwanted side effects associated with energy drinks. *Nutr Rev* 72(Suppl. 1):108–120, 2014.
61. Berger, A.J., and Alford, K. Cardiac arrest in a young man following excess consumption of caffeinated "energy drinks" *Med J Aust* 190:41–43, 2009.
62. Clauson, K.A., Sheilds, K.M, McQueen, C.E., and Persad, N. Safety issues associated with commercially available energy drinks. *J Am Pharm Assoc* 48:e55–e63, 2008.
63. Ballard, S.L., Wellborn-Kim, J.J., and Clauson, K.A. Effects of commercial energy drink consumption on athletic performance and body composition. *Phys Sportsmen* 38:107–117, 2010.
64. Mora-Rodríguez, R., Pallarés, J. G., López-Gullón, J. M., et al. Improvements on neuromuscular performance with caffeine ingestion depend on the time-of-day. *J Sci Med Sport* 18:338–342, 2015.
65. Duchan, E., Patel, N.D., and Feucht, C. Energy drinks: A review of use and safety. *Phys Sportsmed* 38:171–179, 2010.
66. Heuberger, J.A., Cohen Tervaert, J.M., Schepers, F.M., et al. Erythropoietin doping in cycling: Lack of evidence for efficacy and a negative risk-benefit. *Br J Clin Pharmacol* 75:1406–1421, 2013.

Chapter 11
1. Committee to Reexamine IOM Pregnancy Weight Guidelines, Institute of Medicine, National Research Council. *Weight Gain During Pregnancy: Reexamining the Guidelines*. Washington, DC: National Academies Press, 2009.
2. Iacovidou, N., Varsami, M., and Syggellou, A. Neonatal outcome of preterm delivery. *Ann N Y Acad Sci* 1205:130–134, 2010.
3. Liu, P., Xu, L., Wang, Y., et al. Association between perinatal outcomes and maternal pre-pregnancy body mass index. *Obes Rev* 17:1091–1102, 2016.
4. Al-Hinai, M., Al-Muqbali, M., Al-Moqbali, A., et al. Effects of pre-pregnancy body mass index and gestational weight gain on low birth weight in Omani infants: A case-control study. *Sultan Qaboos Univ Med J* 13:386–391, 2013.
5. American Academy of Nutrition and Dietetics. Position of the Academy of Nutrition and Dietetics: Obesity, reproduction, and pregnancy outcomes. *J Acad Nutr Diet* 116:677–691, 2016.
6. Li, N., Liu, E., Guo, J., et al. Maternal prepregnancy body mass index and gestational weight gain on pregnancy outcomes. *PLoS One* 8:e82310, 2013.
7. Poston, L. Gestational weight gain: Influences on the long-term health of the child. *Curr Opin Clin Nutr Metab Care* 15:252–257, 2012.
8. ACOG Committee Opinion No. 650: Physical activity and exercise during pregnancy and the postpartum period. *Obstet Gynecol* 126:e135–e142, 2015.
9. U.S. Department of Health and Human Services. *2008 Physical Activity Guidelines for Americans*. Available online at www.health.gov/paguidelines. Accessed March 11, 2017.
10. Practice Bulletin No. 153: Nausea and vomiting of pregnancy. *Obstet Gynecol* 126:e12–e24, 2015.
11. Maltepe, C., and Koren, G. The management of nausea and vomiting of pregnancy and hyperemesis gravidarum—A 2013 update. *J Popul Ther Clin Pharmacol* 20:e184–e192, 2013.
12. National Center for Health Statistics, Centers for Disease Control and Prevention. *Infant Health*. Available online at www.cdc.gov/nchs/fastats/infant-health.htm. Accessed March 2, 2017.
13. Central Intelligence Agency. *The World Fact Book: Maternal Mortality Rate*. Available online at www.cia.gov/library/publications/the-world-factbook/rankorder/2223rank.html. Accessed March 2, 2017.
14. Leeman, L., Dresang, L.T., and Fontaine, P. Hypertensive disorders of pregnancy. *Am Fam Physician* 93:121–127, 2016.
15. Berg, C.J., Callaghan, W.M., Syverson, C., and Henderson, Z. Pregnancy-related mortality in the United States, 1998 to 2005. *Obstet Gynecol* 116:1302–1309, 2010.
16. Procter, S.B., and Campbell, C.G. American Academy of Nutrition and Dietetics. Position of American Academy of Nutrition and Dietetics: Nutrition and lifestyle for a healthy pregnancy. *J Acad Nutr Diet* 114:1099–1103, 2014.
17. American College of Obstetricians and Gynecologists, Task Force on Hypertension in Pregnancy. *Hypertension in Pregnancy, 2013*. Available online at www.acog.org/~/media/Task%20Force%20and%20Work%20Group%20Reports/HypertensioninPregnancy.pdf. Accessed March 2, 2017.
18. Food and Nutrition Board, Institute of Medicine. *Dietary Reference Intakes: Water, Potassium, Sodium, Chloride, and Sulfate*. Washington, DC: National Academies Press, 2004.
19. American Diabetes Association. Standards of Medical Care in Diabetes—2016. *Diabetes Care* 39(Suppl. 1):S1–S105, 2016.

[20] Food and Nutrition Board, Institute of Medicine. *Dietary Reference Intakes for Energy, Carbohydrates, Fiber, Fat, Protein and Amino Acids*. Washington, DC: National Academies Press, 2002.

[21] Scholtz, S.A., Colombo, J., and Carlson, S.E. Clinical overview of effects of dietary long-chain polyunsaturated fatty acids during the perinatal period. *Nestle Nutr Inst Workshop Ser* 77:145–154, 2013.

[22] Kaiser, L.L., Campbell, C.G. Academy Positions Committee Workgroup. Practice Paper of the Academy of Nutrition and Dietetics: Nutrition and lifestyle for a healthy pregnancy outcome. *J Acad Nutr Diet* 114:1447–1460, 2014.

[23] Food and Nutrition Board, Institute of Medicine. *Dietary Reference Intakes for Calcium and Vitamin D*. Washington, DC: National Academies Press, 2011.

[24] Lowensohn, R.I., Stadler, D.D., and Naze, C. Current concepts of maternal nutrition. *Obstet Gynecol Surv* 71:413–426, 2016.

[25] Center for Disease Control and Prevention. *Folic Acid: Birth Defects Count*. Available online at www.cdc.gov/ncbddd/birthdefectscount/data.html. Accessed March 2, 2017.

[26] Fekete, K., Berti, C., Cetin, I., et al. Perinatal folate supply: Relevance in health outcome parameters. *Matern Child Nutr* 6(Suppl. 2):23–38, 2010.

[27] Food and Nutrition Board, Institute of Medicine. *Dietary Reference Intakes for Thiamin, Riboflavin, Niacin, Vitamin B-6, Folate, Vitamin B-12, Pantothenic Acid, Biotin, and Choline*. Washington, DC: National Academies Press, 1998.

[28] Pepper, M.R., and Black, M.M. B12 in fetal development. *Semin Cell Dev Biol* 22:619–623, 2011.

[29] Food and Nutrition Board, Institute of Medicine. *Dietary Reference Intakes for Vitamin A, Vitamin K, Arsenic, Boron, Chromium, Copper, Iodine, Iron, Manganese, Molybdenum, Nickel, Silicon, Vanadium, and Zinc*. Washington, DC: National Academies Press, 2001.

[30] Cao, C., and O'Brien, K.O. Pregnancy and iron homeostasis: An update. *Nutr Rev* 71:35–51, 2013.

[31] Foster, M., Herulah, U.N., Prasad, A., et al. Zinc Status of vegetarians during pregnancy: A systematic review of observational studies and meta-analysis of zinc intake. *Nutrients* 7:4512–4525, 2015.

[32] Stagnaro-Green, A., Sullivan, S., and Pearce, E.N. Iodine supplementation during pregnancy and lactation. *JAMA* 308: 2463–2464, 2012.

[33] Caldwell, K.L., Pan, Y., Mortensen, M.E., et al. Iodine status in pregnant women in the National Children's Study and in U.S. women (15–44 years), National Health and Nutrition Examination Survey 2005–2010. *Thyroid* 23:927–937, 2013.

[34] Young, S.L. Pica in pregnancy: New ideas about an old condition. *Annu Rev Nutr* 30:403–422, 2010.

[35] Young, S.L., Sherman, P.W., Lucks, J.B., and Pelto, G.H. Why on earth? Evaluating hypotheses about the physiological functions of human geophagy. *Q Rev Biol* 86:97–120, 2011.

[36] Centers for Disease Control and Prevention. *Birth Defects*. Available online at www.cdc.gov/ncbddd/birthdefects/index.html. Accessed March 3, 2017.

[37] Stein, A.D., Zybert, P.A., van de Bor, M., and Lumey, L.H. Intrauterine famine exposure and body proportions at birth: The Dutch Hunger Winter. *Int J Epidemiol* 33:831–836, 2004.

[38] Langley-Evans, S.C. Nutrition in early life and the programming of adult disease: A review. *J Hum Nutr Diet* 28(Suppl. 1):1–14, 2015.

[39] Casanueva, E., Roselló-Soberón, M.E., and De-Regil, L.M. Adolescents with adequate birth weight newborns diminish energy expenditure and cease growth. *J Nutr* 136:2498–2501, 2006.

[40] National Center for Health Statistics, Centers for Disease Control and Prevention. *Teen Births*. Available online at www.cdc.gov/nchs/fastats/teen-births.htm. Accessed March 3, 2017.

[41] Carolan, M. Maternal age ≥45 years and maternal and perinatal outcomes: A review of the evidence. *Midwifery* 29:479–489, 2013.

[42] Kozuki, N., Lee, A.C., Silveira, M.F., et al; Child Health Epidemiology Reference Group, Small-for-Gestational-Age-Preterm Birth Working Group. The associations of birth intervals with small-for-gestational-age, preterm, and neonatal and infant mortality: A meta-analysis. *BMC Public Health* 13(Suppl. 3):S3, 2013.

[43] National Down Syndrome Society. *Down Syndrome Facts*. Available online at www.ndss.org/Down-Syndrome/Down-Syndrome-Facts/. Accessed March 3, 2017.

[44] de Graaf, J.P., Steegers, E.A., and Bonsel, G.J. Inequalities in perinatal and maternal health. *Curr Opin Obstet Gynecol* 25:98–108, 2013.

[45] American College of Obstetricians and Gynecologists. ACOG Committee Opinion No. 462: Moderate caffeine consumption during pregnancy. *Obstet Gynecol* 116(2 Pt. 1):467–468, 2010.

[46] U.S. Department of Agriculture and U.S. Department of Health and Human Services. *Dietary Guidelines for Americans, 2015–2020*, 8th ed. Available online at http://health.gov/dietaryguidelines/2015/. Accessed March 11, 2017.

[47] U.S. Food and Drug Administration. *Eating Fish: What Pregnant Women and Parents Should Know*. Available online at www.fda.gov/food/foodborneillnesscontaminants/metals/ucm393070.htm. Accessed March 3, 2017.

[48] Mayo Clinic Staff. *Pregnancy Nutrition: Foods to Avoid During Pregnancy*. Available online at www.mayoclinic.org/healthy-living/pregnancy-week-by-week/in-depth/pregnancy-nutrition/art-20043844. Accessed March 3, 2017.

[49] Lamont, R.F., Sobel, J., Mazaki-Tovi, S., et al. *Listeriosis* in human pregnancy: A systematic review. *J Perinat Med* 39:227–236, 2011.

[50] U.S. Food and Drug Administration, Center for Food Safety and Nutrition. *Bad Bug Book: Foodborne Pathogenic Microorganisms and Natural Toxins Handbook* (2nd ed.). Available online at www.fda.gov/food/foodborneillnesscontaminants/causesofillnessbadbugbook/default.htm. Accessed March 4, 2017.

[51] Centers for Disease Control and Prevention. *Alcohol and Pregnancy*. Available online at www.cdc.gov/vitalsigns/fasd/. Accessed March 4, 2017.

[52] Centers for Disease Control and Prevention. *Fetal Alcohol Spectrum Disorders (FADS), Data and Statistics*. Available online at www.cdc.gov/ncbddd/fasd/data.html#ref. Accessed March 4, 2017.

[53] Centers for Disease Control and Prevention, *Tobacco Use During Pregnancy*. Available online at www.cdc.gov/reproductivehealth/maternalinfanthealth/tobaccousepregnancy/index.htm. Accessed March 7, 2017.

[54] Clifford, A., Lang, L., and Chen, R. Effects of maternal cigarette smoking during pregnancy on cognitive parameters of children and young adults: A literature review. *Neurotoxicol Teratol* 34:560–570, 2012.

[55] Xiao, D., Huang, X., Yang, S., and Zhang, L. Direct effects of nicotine on contractility of the uterine artery in pregnancy. *J Pharmacol Exp Ther* 322:180–185, 2007.

[56] Task Force on Sudden Infant Death Syndrome; Moon, R.Y. SIDS and other sleep-related infant deaths: Expansion of recommendations for a safe infant sleeping environment. *Pediatrics* 128:e1341–e1367, 2011.

[57] Burke, H., Leonardi-Bee, J., Hashim, A., et al. Prenatal and passive smoke exposure and incidence of asthma and wheeze: Systematic review and meta-analysis. *Pediatrics* 129:735–744, 2012.

[58] Behnke, M., and Smith, V.C.; Committee on Substance Abuse; Committee on Fetus and Newborn. Prenatal substance abuse: Short- and long-term effects on the exposed fetus. *Pediatrics* 131:e1009–e1024, 2013.

[59] Fajemirokun-Odudeyi, O., and Lindow, S.W. Obstetric implications of cocaine use in pregnancy: A literature review. *Eur J Obstet Gynecol Reprod Biol* 112:2–8, 2004.

[60] Kocherlakota, P. Neonatal abstinence syndrome. *Pediatrics* 134: e547–e561, 2014.

[61] Innis, S.M. Impact of maternal diet on human milk composition and neurological development of infants. *Am J Clin Nutr* 99:734S–741S, 2014.

[62] Allen, L.H. B vitamins in breast milk: Relative importance of maternal status and intake, and effects on infant status and function. *Adv Nutr* 3: 362–369, 2012.

[63] Ballard, O., and Morrow, A.L. Human milk composition: Nutrients and bioactive factors. *Pediatr Clin North Am* 60:49–74, 2013.

[64] Centers for Disease Control and Prevention. *Growth Charts: WHO Growth Standards Are Recommended for Use in the U.S. for Infants and Children 0 to 2 Years of Age.* Available online at www.cdc.gov/growthcharts/who_charts.htm#The%20WHO%20Growth%20Charts. Accessed March 5, 2017.

[65] Adair, L.S. Long-term consequences of nutrition and growth in early childhood and possible preventive interventions. *Nestle Nutr Inst Workshop Ser* 78:111–120, 2014.

[66] DeBoer, M.D., Lima, A.A., Oría, R.B., et al. Early childhood growth failure and the developmental origins of adult disease: Do enteric infections and malnutrition increase risk for the metabolic syndrome? *Nutr Rev* 70:642–653, 2012.

[67] Wagner, C.L., and Greer, F.R. American Academy of Pediatrics Section on Breastfeeding, American Academy of Pediatrics Committee on Nutrition. Prevention of rickets and vitamin D deficiency in infants, children, and adolescents. *Pediatrics* 122:1142–1152, 2008.

[68] American Academy of Pediatrics, policy statement, Committee on Fetus and Newborn. Controversies concerning vitamin K and the newborn. *Pediatrics* 112:191–192, 2003.

[69] Lauritzen, L., Brambilla, P., Mazzocchi, A., et al. DHA effects in brain development and function. *Nutrients* 8(1), 2016.

[70] Tai, E.K., Wang, X.B., and Chen, Z.Y. An update on adding docosahexaenoic acid (DHA) and arachidonic acid (AA) to baby formula. *Food Funct* 4:1767–1775, 2013.

[71] Lapillonne, A., Groh-Wargo, S., Gonzalez, C.H., and Uauy, R. Lipid needs of preterm infants: Updated recommendations. *J Pediatr* 162 (3 Suppl.):S37–S47, 2013.

[72] Heaton, A.E., Meldrum, S.J., Foster, J.K., et al. Does docosahexaenoic acid supplementation in term infants enhance neurocognitive functioning in infancy? *Front Hum Neurosci* 7:774, 2013.

[73] Gale, C.R., Marriott, L.D., Martyn, C.N., et al. Group for Southampton Women's Survey Study. Breastfeeding, the use of docosahexaenoic acid-fortified formulas in infancy and neuropsychological function in childhood. *Arch Dis Child* 95:174–179, 2009.

[74] Makrides, M., Smithers, L.G., and Gibson, R.A. Role of long-chain polyunsaturated fatty acids in neurodevelopment and growth. *Nestle Nutr Workshop Ser Pediatr Program* 65:123–133, 2010.

[75] Section on Breastfeeding. Breastfeeding and the use of human milk. *Pediatrics* 129:e827–e841, 2012.

[76] Fleischer, D.M., Spergel, J.M., Assa'ad, A.H., and Pongracic, J.A. Primary prevention of allergic disease through nutritional interventions. *J Allergy Clin Immunol Pract* 1:29–36, 2013.

[77] Ziegler, E.E. Consumption of cow's milk as a cause of iron deficiency in infants and toddlers. *Nutr Rev* 69 (Suppl. 1):S37–S42, 2011.

[78] American Academy of Pediatrics. Policy statement: The use and misuse of fruit juice in pediatrics. *Pediatrics* 107:1210–1213, 2001. Reaffirmed August 2013.

Chapter 12

[1] Ogden, C.L., Carroll, M.D., Lawman, H.G., et al. Trends in obesity prevalence among children and adolescents in the United States, 1988–1994 Through 2013–2014. *JAMA* 315:2292–2299, 2016.

[2] The National Physical Activity Plan Alliance. *The 2016 United States Report Card on Physical Activity for Children and Youth.* Available online at www.physicalactivityplan.org/projects/reportcard.html. Accessed March 12, 2017.

[3] U.S. Department of Agriculture, Center for Nutrition Policy and Promotion. Diet quality of children age 2–17 years as measured by the Healthy Eating Index—2010. *Nutrition Insight* 52, July 2013. Available online at www.cnpp.usda.gov/sites/default/files/nutrition_insights_uploads/Insight52.pdf. Accessed March 20, 2017.

[4] Mancino, L., Todd, J.E., and Lin, B-H. *How Food Away from Home Affects Children's Diet Quality* (Economic Research Report no. ERR-104), October 2010. Available online at www.ers.usda.gov/publications/pub-details/?pubid=44756. Accessed March 12, 2017.

[5] Lin, B-H., and Guthrie, J. *Nutritional Quality of Food Prepared at Home and Away from Home, 1977–2008* (Economic Information Bulletin no. EIB-105), December 2012. Available online at www.ers.usda.gov/publications/pub-details/?pubid=43699. Accessed March 12, 2017.

[6] American Academy of Nutrition and Dietetics. Position of the Academy of Nutrition and Dietetics: Nutrition Guidance for Healthy Children Ages 2 to 11 Years. *J Acad Nutr Diet* 114: 1257–1276, 2014.

[7] Costa, S.M., Horta, P.M., and dos Santos, L.C. Food advertising and television exposure: Influence on eating behavior and nutritional status of children and adolescents. *Arch Latinoam Nutr* 62:53–59, 2012.

[8] Council on Communications and Media. Media and young minds. *Pediatrics* 138(5), November 2016.

[9] Council on Communications and Media. Media use in school-aged children and adolescents. *Pediatrics* 138(5), November 2016.

[10] American Academy of Pediatrics. *Family Media Plan.* Available online at www.healthychildren.org/English/media/Pages/default.aspx#home. Accessed March 20, 2017.

[11] Braithwaite, I., Stewart, A.W., Hancox, R.J., et al.; ISAAC Phase Three Study Group. The worldwide association between television viewing and obesity in children and adolescents: Cross sectional study. *PLoS One* 8:e74263, 2013.

[12] Batada, A., Seitz, M., Wotan, M., et al. Nine out of ten food advertisements shown during Saturday morning children's television programming are for food high in fat, sodium, or added sugars, or low in nutrients. *J Am Diet Assoc* 108:673–678, 2008.

[13] American Academy of Pediatrics. *Cholesterol Levels in Children and Adolescents.* Available online at www.healthychildren.org/English/healthy-living/nutrition/Pages/Cholesterol-Levels-in-Children-and-Adolescents.aspx. Accessed March 20, 2017.

[14] Expert Panel on Integrated Guidelines for Cardiovascular Health and Risk Reduction in Children and Adolescents, National Heart, Lung, and Blood Institute. Expert panel on integrated guidelines for cardiovascular health and risk reduction in children and adolescents: Summary report. *Pediatrics* 128(Suppl. 5):S213–S256, 2011.

[15] Centers for Disease Control and Prevention. *Basics About Childhood Obesity.* Available online at www.cdc.gov/obesity/childhood/defining.html. Accessed March 5, 2017.

[16] American Academy of Nutrition and Dietetics. Position of the Academy of Nutrition and Dietetics: Interventions for the prevention

and treatment of pediatric overweight and obesity. *J Acad Nutr Diet* 1375–1394, 2013.

[17] Barlow, S.E., Expert Committee. Expert Committee recommendations regarding the prevention, assessment, and treatment of child and adolescent overweight and obesity: Summary report. *Pediatrics* 120 (Suppl. 4):S164–S192, 2007.

[18] U.S. Department of Health and Human Services. *2008 Physical Activity Guidelines for Americans*. Available online at www.health.gov/paguidelines/pdf/paguide.pdf. Accessed March 6, 2017.

[19] National Diabetes Information Clearinghouse. *National Diabetes Statistics, 2014*. Available online at http://diabetes.niddk.nih.gov/dm/pubs/statistics/index.htm#13. Accessed March 6, 2017.

[20] U.S. Department of Agriculture, Agricultural Research Service. Table 5. Energy intakes: Percentages of energy from protein, carbohydrate, fat, and alcohol, by gender and age in the United States, 2013–2014. *What We Eat in America* (NHANES 2013–2014). Available online at www.ars.usda.gov/ARSUserFiles/80400530/pdf/1314/Table_5_EIN_GEN_13.pdf. Accessed March 17, 2017.

[21] American Heart Association. *Children and Cholesterol*. Available online at www.heart.org/HEARTORG/Conditions/Cholesterol/UnderstandYourRiskforHighCholesterol/Children-and-Cholesterol_UCM_305567_Article.jsp#.WMvK6BiZOjh. Accessed March 17, 2017.

[22] Luma, G.B., and Spiotta, R.T. Hypertension in children and adolescents. *Am Fam Physician* 73:1558–1568, 2006.

[23] U.S. Department of Health and Human Services, Centers for Disease Control and Prevention, National Center for Health Statistics. *Growth Charts*. Available online at www.cdc.gov/growthcharts. Accessed March 6, 2017.

[24] Institute of Medicine, Food and Nutrition Board. *Dietary Reference Intakes for Energy, Carbohydrate, Fiber, Fat, Protein, and Amino Acids*. Washington, DC: National Academies Press, 2002.

[25] Institute of Medicine, Food and Nutrition Board. *Dietary Reference Intakes for Water, Potassium, Sodium, Chloride, and Sulfate*. Washington, DC: National Academies Press, 2004.

[26] Butte, N.F., Fox, M.K., Briefel, R.R., et al. Nutrient intakes of US infants, toddlers, and preschoolers meet or exceed dietary reference intakes. *J Am Diet Assoc* 110(12 Suppl.):S27–S37, 2010.

[27] Fox, M.K., Condon, E., Briefel, R.R., et al. Food consumption patterns of young preschoolers: Are they starting off on the right path? *J Am Diet Assoc* 110(12 Suppl.):S52–S59, 2010.

[28] Berner, L.A., Keast, D.R., Bailey, R. L., Dwyer, J.T. Fortified foods are a major contributor to nutrient intakes in diets of US children and adolescents. *J Acad Nutr Diet* 114:1009–1022, 2014.

[29] Food and Nutrition Board, Institute of Medicine. *Dietary Reference Intakes for Calcium and Vitamin D*. Washington, DC: National Academies Press, 2011.

[30] Kumar, J., Muntner, P., Kaskel, F.J., et al. Prevalence and associations of 25-hydroxyvitamin D deficiency in US children: NHANES 2001–2004. *Pediatrics* 124:e362–e370, 2009.

[31] Madan, N., Rusia, U., Sikka, M., et al. Developmental and neurophysiologic deficits in iron deficiency in children. *Indian J Pediatr* 78:58–64, 2011.

[32] American Academy of Pediatrics. The use and misuse of fruit juice in pediatrics. *Pediatrics* 107:1210–1213, 2001.

[33] U.S. Department of Agriculture and U.S. Department of Health and Human Services. *Dietary Guidelines for Americans, 2015–2020*, 8th ed., Washington, DC. Available online at https://health.gov/dietaryguidelines/2015/guidelines/. Accessed March 6, 2017.

[34] Adolphus, K., Lawton, C.L., Champ, C.L., and Dye, L. The effects of breakfast and breakfast composition on cognition in children and adolescents: A systematic review. *Adv Nutr* 7:590S–612S, 2016.

[35] Alaimo, K., Olson, C., and Frongillo, E. Food insufficiency and American school-aged children's cognitive, academic, and psychosocial development. *Pediatrics* 108:44–53, 2001.

[36] Adolphus, K., Lawton, C.L., and Dye, L. The effects of breakfast on behavior and academic performance in children and adolescents. *Front Hum Neurosci* 7:425, 2013.

[37] Kleinman, R.E., Hall, S., Green, H., et al. Diet, breakfast, and academic performance in children. *Ann Nutr Metab* 46(Suppl. 1):24–30, 2002.

[38] American College of Pediatricians. *The Benefits of the Family Table*. May 2014. Available online at https://www.acpeds.org/the-college-speaks/position-statements/parenting-issues/the-benefits-of-the-family-table. Accessed May 11, 2017.

[39] Fisk, C.M., Crozier, S.R., Inskip, H.M., et al. Influences on the quality of young children's diets: The importance of maternal food choices. *Br J Nutr* 105:287–296, 2011.

[40] U.S. Department of Agriculture, Economic Research Service. *National School Lunch Program*. Available online at www.ers.usda.gov/topics/food-nutrition-assistance/child-nutrition-programs/national-school-lunch-program/. Accessed March 8, 2017.

[41] U.S. Department of Agriculture, Food and Nutrition Service. Nutrition standards in the national school lunch and school breakfast programs: Final rule. *Federal Register* 77(17):4087–4167, 2012.

[42] U.S. Department of Agriculture, Food and Nutrition Service. National school lunch and school breakfast program: Nutrition standards for all foods sold in school as required by the Healthy, Hunger-Free Kids Act of 2010: Interim Final Rule. *Federal Register* 78(125): 39068–39120, 2013.

[43] National Institute of Dental and Craniofacial Research. *Dental Caries (Tooth Decay) in Children (Age 2 to 11)*. Available online at www.nidcr.nih.gov/DataStatistics/FindDataByTopic/DentalCaries/DentalCariesChildren2to11.htm. Accessed March 8, 2017.

[44] Visser, S.N., Danielson, M.L., Wolraich, M.L., et al. Vital signs: National and state-specific patterns of attention deficit/hyperactivity disorder treatment among insured children aged 2–5 years—United States, 2008–2014. *MMWR Morb Mortal Wkly Rep* 65:443–450, 2016.

[45] Millichap, J.G., and Yee, M.M. The diet factor in attention-deficit/hyperactivity disorder. *Pediatrics* 129:330–337, 2012.

[46] Raymond, J., and Brown, M.J. Childhood blood lead levels in children aged <5 years—United States, 2009–2014. *MMWR Morb Mortal Wkly Rep* 66:1–10, 2017.

[47] Weng, F.L., Shults, J., Leonard, M.B., et al. Risk factors for low serum 25-hydroxyvitamin D concentrations in otherwise healthy children and adolescents. *Am J Clin Nutr* 86:150–158, 2007.

[48] Turer, C.B., Lin, H., and Flores, G. Prevalence of vitamin D deficiency among overweight and obese U.S. children. *Pediatrics* 131: e152–e161, 2013.

[49] Centers for Disease Control and Prevention. Iron deficiency—United States, 1999–2000. *MMWR Morb Mortal Wkly Rep* 51:897–899, 2002.

[50] National Institutes of Health, Office of Dietary Supplements. *Calcium Dietary Supplement Fact Sheet*. Available online at https://ods.od.nih.gov/factsheets/Calcium-HealthProfessional/. Accessed March 6, 2017.

[51] Risk Factor Monitoring and Methods Branch, National Cancer Institute. *Sources of Beverage Intakes Among the U.S. Population, 2005–06*. Available online at http://riskfactor.cancer.gov/diet/foodsources/beverages. Accessed March 6, 2017.

52. Poti, J.M., Slining, M.M., and Popkin, B.M. Where are kids getting their empty calories? Stores, schools, and fast-food restaurants each played an important role in empty calorie intake among U.S. children during 2009–2010. *J Acad Nutr Diet* 114:908–917, 2014.

53. Singh, T., Arrazola, R.A., Corey, C.G., et al. Tobacco use among middle and high school students—United States, 2011–2015. *MMWR Morb Mortal Wkly Rep* 65:361–367, 2016.

54. Centers for Disease Control and Prevention. *2014 Surgeon General's Report: The Health Consequences of Smoking—50 Years of Progress.* Available online at www.cdc.gov/tobacco/data_statistics/sgr/50th-anniversary/index.htm. Accessed March 9, 2017.

55. Cawley, J., Dragone, D., and Von Hinke Kessler Scholder, S. The demand for cigarettes as derived from the demand for weight loss: A theoretical and empirical investigation. *Health Econ* 25:8–23, 2016.

56. Palaniappan, U., Jacobs Starkey, L., O'Loughlin, J., and Gray-Donald, K. Fruit and vegetable consumption is lower and saturated fat intake is higher among Canadians reporting smoking. *J Nutr* 131:1952–1958, 2001.

57. Centers for Disease Control and Prevention. *Alcohol and Public Health. Fact Sheets—Underage drinking.* Available online at www.cdc.gov/alcohol/fact-sheets/underage-drinking.htm. Accessed March 6, 2017.

58. Kponee, K.Z., Siegel, M., and Jernigan, D.H. The use of caffeinated alcoholic beverages among underage drinkers: Results of a national survey. *Addict Behav* 39:253–258, 2014.

59. Kochanek, K.D., Murphy, S.L., Xu, J., and Tejada-Vera, B. Deaths: Final data for 2014. *National Vital Statistics Reports* 65(4), 2016. Available online at www.cdc.gov/nchs/data/nvsr/nvsr65/nvsr65_04.pdf. Accessed March 6, 2017.

60. Molla, M.T., Centers for Disease Control and Prevention (CDC). Expected years of life free of chronic condition-induced activity limitations—United States, 1999–2008. *MMWR Suppl* 62:87–92, 2013.

61. Administration on Aging, Administration for Community Living, Department of Health and Human Services. *A Profile of Older Americans 2016.* Available online at https://aoa.acl.gov/aging_statistics/profile/index.aspx. Accessed March 20, 2017.

62. Most, J., Tosti, V., Redman, L.M., and Fontana, L. Calorie restriction in humans: An update. *Ageing Res Rev* Aug 17, 2016. doi: 10.1016/j.arr.2016.08.005. [Epub ahead of print]

63. Kemnitz, J.W. Calorie restriction and aging in nonhuman primates. *ILAR J* 52:66–77, 2011.

64. Colman, R.J., Beasley, T.M., Kemnitz, J.W., et al. Caloric restriction reduces age-related and all-cause mortality in rhesus monkeys. *Nat Commun* 5:3557, 2014.

65. Mattison, J.A., Roth, G.S., Beasley, T.M., et al. Impact of caloric restriction on health and survival in rhesus monkeys from the NIA study. *Nature* 489:318–321, 2012.

66. Willcox, B.J., Willcox, D.C., Todoriki, H., et al. Caloric restriction, the traditional Okinawan diet, and healthy aging: The diet of the world's longest-lived people and its potential impact on morbidity and life span. *Ann N Y Acad Sci* 1114:434–455, 2007.

67. Willcox, D.C., Scapagnini, G., and Willcox, B.J. Healthy aging diets other than the Mediterranean: A focus on the Okinawan diet. *Mech Ageing Dev* 136–137:148–162, 2014.

68. Willcox, B.J., and Willcox, D.C. Caloric restriction, caloric restriction mimetics, and healthy aging in Okinawa: Controversies and clinical implications. *Curr Opin Clin Nutr Metab Care* 17:51–58, 2014.

69. Bendjilali, N., Hsueh, W.C., He, Q., et al. Who are the Okinawans? Ancestry, genome diversity, and implications for the genetic study of human longevity from a geographically isolated population. *J Gerontol A Biol Sci Med Sci.*, 69:1474–1484, 2014.

70. Robine, J.M., Herrmann, F.R., Arai, Y., et al. Exploring the impact of climate on human longevity. *Exp Gerontol* 47:660–671, 2012.

71. Cohn, S.H., Vartsky, D., Yasumura, A., et al. Compartmental body composition based on total-body nitrogen, potassium, and calcium. *Am J Physiol* 239:E524–E530, 1980.

72. Sechi, G., Sechi, E., Fois, C., and Kumar, N. Advances in clinical determinants and neurological manifestations of B vitamin deficiency in adults. *Nutr Rev* 74:281–300, 2016.

73. Institute of Medicine, Food and Nutrition Board. *Dietary Reference Intakes for Thiamin, Riboflavin, Niacin, Vitamin B_6, Folate, Vitamin B_{12}, Pantothenic Acid, Biotin, and Choline.* Washington, DC: National Academies Press, 1998.

74. National Institutes of Health, Office of Dietary Supplements. *Dietary Supplement Fact Sheet: Vitamin B_{12}.* Available online at http://ods.od.nih.gov/factsheets/vitaminb12.asp. Accessed March 6, 2017.

75. Bernstein, M., and Munoz, N., Academy of Nutrition and Dietetics. Position of the Academy of Nutrition and Dietetics: Food and nutrition for older adults: Promoting health and wellness. *J Acad Nutr Diet* 112:1255–1277, 2012.

76. U.S. Department of Health and Human Services. *Securing the Benefits of Medical Innovation for Seniors: The Role of Prescription Drugs and Drug Coverage.* Available online at https://aspe.hhs.gov/system/files/pdf/72831/innovation.pdf. Accessed March 9, 2017.

77. Montero-Fernández, N., and Serra-Rexach, J.A. Role of exercise on sarcopenia in the elderly. *Eur J Phys Rehabil Med* 49:131–143, 2013.

78. Centers for Disease Control and Prevention. *The State of Aging and Health in America 2013.* Available online at www.cdc.gov/aging/pdf/state-aging-health-in-america-2013.pdf. Accessed March 9, 2017.

79. Centers for Disease Control and Prevention. *Arthritis.* Available online at www.cdc.gov/arthritis/. Accessed March 5, 2017.

80. Henrotin, Y., and Lambert, C. Chondroitin and glucosamine in the management of osteoarthritis: An update. *Curr Rheumatol Rep* 15:361, 2013.

81. Hajjar, E.R., Cafiero, A.C., and Hanlon, J.T. Polypharmacy in elderly patients. *Am J Geriatr Pharmacother* 5:314–316, 2007.

82. Coleman-Jensen, A., Rabbitt, M.P., Gregory, C.A., and Singh, A. *Household Food Security in the United States in 2015* (Economic Research Report no 215), September, 2016. Available online at www.ers.usda.gov/webdocs/publications/err215/err-215.pdf. Accessed April 26, 2017.

83. Garber, C.E., Blissmer, B., Deschenes, M.R., et al. American College of Sports Medicine position stand: Quantity and quality of exercise for developing and maintaining cardiorespiratory, musculoskeletal, and neuromotor fitness in apparently healthy adults: Guidance for prescribing exercise. *Med Sci Sports Exerc* 43:1334–1359, 2011.

84. Dorner, B., Friedrich, E.K., and Posthauer, M.E. Position of the American Dietetic Association: Individualized nutrition approaches for older adults in health care communities. *J Am Diet Assoc* 110:1549–1553, 2010.

85. Silveri, M.M. Adolescent brain development and underage drinking in the United States: Identifying risks of alcohol use in college populations. *Harv Rev Psychiatry* 20:189–200, 2012.

86. Elamin, E.E., Masclee, A.A., Dekker, J., and Jonkers, D.M. Ethanol metabolism and its effects on the intestinal epithelial barrier. *Nutr Rev* 71:483–499, 2013.

87. National Institutes of Health Fact Sheets. *Alcohol-Related Traffic Deaths.* Available online at https://report.nih.gov/nihfactsheets/ViewFactSheet.aspx?csid=24. Accessed March 4, 2017.

[88] American Psychiatric Association. *Diagnostic and Statistical Manual of Mental Disorders*, 5th ed. Arlington, VA: American Psychiatric Publishing, 2013.

[89] National Institute on Alcohol Abuse and Alcoholism. *Alcohol Use Disorder*. Available online at www.niaaa.nih.gov/alcohol-health/overview-alcohol-consumption/alcohol-use-disorders. Accessed March 9, 2017.

[90] Enoch, M.A. Genetic influences on the development of alcoholism. *Curr Psychiatry Rep* 15:412, 2013.

[91] Boffetta, P., and Hashibe, M. Alcohol and cancer. *Lancet Oncol* 7:149–156, 2006.

[92] Cao, Y., Willett, W.C., Rimm, E.B., et al. Light to moderate intake of alcohol, drinking patterns, and risk of cancer: Results from two prospective US cohort studies. *BMJ* 351:h4238, 2015.

[93] Toma, A., Paré, G., and Leong, D.P. Alcohol and cardiovascular disease: How much is too much? *Curr Atheroscler Rep* 19:13, 2017.

[94] O'Keefe, J.H., Bhatti, S.K., Bajwa, A., et al. Alcohol and cardiovascular health: The dose makes the poison . . . or the remedy. *Mayo Clin Proc* 89:382–393, 2014.

[95] Saremi, A., and Arora, R. The cardiovascular implications of alcohol and red wine. *Am J Ther* 15:265–277, 2008.

Chapter 13

[1] Centers for Disease Control and Prevention. *CDC Estimates of Foodborne Illness in the United States*. Available online at www.cdc.gov/foodborneburden/estimates-overview.html. Accessed March 15, 2017.

[2] U.S. Food and Drug Administration. *FDA Food Safety Modernization Act (FSMA)*, Available online at www.fda.gov/Food/Guidance Regulation/FSMA/default.htm. Accessed April 4, 2017.

[3] U.S. Food and Drug Administration, Center for Food Safety and Nutrition. *Bad Bug Book: Foodborne Pathogenic Microorganisms and Natural Toxins Handbook* (2nd ed.). Available online at www.fda.gov/food/foodborneillnesscontaminants/causesofillnessbadbugbook/default.htm. Accessed April 4, 2017.

[4] Centers for Disease Control and Prevention. *Salmonella Is a Sneaky Germ: Seven Tips for Safer Eating*. Available online at www.cdc.gov/features/vitalsigns/foodsafety. Accessed March 18, 2017.

[5] Scallan, E., Hoekstra, R.M., Angulo, F.J., et al. Foodborne illness acquired in the United States—Major pathogens. *Emerg Infect Dis* 17:7–15, 2011.

[6] Centers for Disease Control and Prevention. *Campylobacter*. Available online at www.cdc.gov/foodsafety/diseases/campylobacter/index.html. Accessed March 23, 2017.

[7] Centers for Disease Control and Prevention. Case definitions for infectious conditions under public health surveillance. *MMWR Morb Mortal Wkly Rep* 46:17, 1997.

[8] Centers for Disease Control and Prevention. *Multistate Outbreak of Shiga Toxin-Producing Escherichia coli O157:H7 Infections Linked to Ready-to-Eat Salads (Final Update), December 11, 2013*. Available online at www.cdc.gov/ecoli/2013/O157H7-11-13/index.html. Accessed April 4, 2017.

[9] Centers for Disease Control and Prevention. *Reports of Selected of E. coli Outbreak Investigations*. Available online at www.cdc.gov/ecoli/outbreaks.html. Accessed March 18, 2017.

[10] Centers for Disease Control and Prevention (CDC). Outbreak of *Escherichia coli* O104:H4 infections associated with sprout consumption—Europe and North America, May–July 2011. *MMWR Morb Mortal Wkly Rep* 62:1029–1031, 2013.

[11] Mateus, T., Silva, J., Maia, R.L., and Teixeira, P. Listeriosis during pregnancy: A public health concern. *ISRN Obstet Gynecol* 2013:851712, 2013.

[12] Koepke, R., Sobel, J., and Arnon, S.S. Global occurrence of infant botulism, 1976–2006. *Pediatrics* 122:73–82, 2008.

[13] Barnes, M., Britton, P.N., and Singh-Grewal, D. Of war and sausages: A case-directed review of infant botulism. *J Paediatr Child Health* 49:E232–E234, 2013.

[14] Centers for Disease Control and Prevention. *Norovirus. U.S. Trends and Outbreaks*. Available online at www.cdc.gov/norovirus/trends-outbreaks.html. Accessed March 31, 2017.

[15] Hall, A.J., Eisenbart, V.G., Etingüe, A.L., et al. Epidemiology of foodborne norovirus outbreaks, United States, 2001–2008. *Emerg Infect Dis* 18:1566–1573, 2012.

[16] Liu, Y., and Wu, F. Global burden of aflatoxin-induced hepatocellular carcinoma: A risk assessment. *Environ Health Perspect* 118:818–824, 2010.

[17] Centers for Disease Control and Prevention. *vCJD (Variant Creutzfeldt-Jakob Disease)*. Available online at www.cdc.gov/prions/vcjd/about.html. Accessed March 18, 2017.

[18] Food and Drug Administration. *BSE (Bovine Spongiform Encephalopathy, or Mad Cow Disease)*. Available online at www.cdc.gov/prions/bse/index.html. Accessed March 18, 2017.

[19] U.S. Department of Agriculture. *Frequently Asked Questions on BSE (Bovine Spongiform Encephalopthy or Mad Cow Disease)*. Available online at www.usda.gov/topics/animals/bse-surveillance-information-center/bse-frequently-asked-questions. Accessed March 31, 2017.

[20] Partnership for Food Safety Education. *Fight Bac!* Available online at www.fightbac.org. Accessed March 19, 2017.

[21] U.S. Department of Agriculture. Food Safety and Inspection Service. *Food Safety Information: Danger Zone (40°–140°)*. Available online at www.fsis.usda.gov/wps/wcm/connect/8b705ede-f4dc-4b31-a745-836e66eeb0f4/Danger_Zone.pdf?MOD=AJPERES. Accessed April 4, 2017.

[22] U.S. Environmental Protection Agency. *Setting Tolerances for Pesticide Residues in Foods*. Available online at www.epa.gov/pesticide-tolerances/setting-tolerances-pesticide-residues-foods. Accessed March 19, 2017.

[23] U.S. Food and Drug Administration. *Pesticide Monitoring Program: Fiscal Year 2014 Pesticide Report*. Available online at www.fda.gov/downloads/Food/FoodborneIllnessContaminants/Pesticides/UCM546325.pdf. Accessed March 24, 2017.

[24] U.S. Department of Agriculture, Economic Research Service. *Organic Market Overview*. Available online at www.ers.usda.gov/topics/natural-resources-environment/organic-agriculture/organic-market-overview.aspx. Accessed March 19, 2017.

[25] Crinnion, W.J. Organic foods contain higher levels of certain nutrients, lower levels of pesticides, and may provide health benefits for the consumer. *Altern Med Rev* 1:4–12, 2010.

[26] Food and Agriculture Organization of the United Nations. *Organic Agriculture*. Available online at www.fao.org/organicag/oa-faq/oa-faq4/en/. Accessed March 27, 2017.

[27] Dangour, A.D., Dodhia, S.K., Hayter, A., et al. Nutritional quality of organic foods: A systematic review. *Am J Clin Nutr* 90:680–685, 2009.

[28] Forman, J., Silverstein, J.; Committee on Nutrition; Council on Environmental Health; American Academy of Pediatrics. Organic foods: Health and environmental advantages and disadvantages. *Pediatrics* 130:e1406–1415, 2012.

[29] U.S. Department of Agriculture, Agricultural Marketing Service. *National Organic Program*. Available online at www.ams.usda.gov/AMSv1.0/ams.fetchTemplateData.do?template=TemplateA&navID=NationalOrganicProgram&leftNav=NationalOrganicProgram&page=NOPNationalOrganicProgramHome&acct=AMSPW. Accessed March 19, 2017.

30. U.S. Food and Drug Administration. *Eating Fish: What Pregnant Women and Parents Should Know.* Available online at www.fda.gov/food/foodborneillnesscontaminants/metals/ucm393070.htm. Accessed March 3, 2017.
31. Animal Health Institute. *FDA Approval: About the FDA's Safety Assessment for Food Animals.* Available online at www.ahi.org/issues-advocacy/animal-antibiotics/fda-approval/#A. Accessed March 21, 2017.
32. U.S. Food and Drug Administration. *Fact Sheet: Veterinary Feed Directive Final Rule and Next Steps.* Available online at www.fda.gov/AnimalVeterinary/DevelopmentApprovalProcess/ucm449019.htm. Accessed March 21, 2017.
33. U.S. Food and Drug Administration. *Steroid Hormone Implants Used for Growth in Food-Producing Animals.* Available online at https://www.fda.gov/AnimalVeterinary/SafetyHealth/ProductSafetyInformation/ucm055436.htm. Accessed March 22, 2017.
34. U.S. Food and Drug Administration. *Report on the U.S. Food and Drug Administration's Review of the Safety of Recombinant Bovine Somatotropin. April 23, 2009.* Available online at http://www.fda.gov/animalveterinary/safetyhealth/productsafetyinformation/ucm130321.htm. Accessed March 22, 2017.
35. Riboldi, B.P., Vinhas, A.M., and Moreira, J.D. Risks of dietary acrylamide exposure: A systematic review. *Food Chem* 157C:310–322, 2014.
36. Pedreschi, F., Mariotti, M.S., and Granby, K. Current issues in dietary acrylamide: Formation, mitigation and risk assessment. *J Sci Food Agric* 94:9–20, 2014.
37. U.S. Food and Drug Administration. *Food Facts: Food Irradiation: What You Need to Know.* June 2016. Available online at www.fda.gov/downloads/Food/IngredientsPackagingLabeling/UCM262295.pdf. Accessed March 29, 2017.
38. U.S. Food and Drug Administration. *U.S. Regulatory Requirements for Irradiating Foods.* May 1999. Available online at www.fda.gov/Food/GuidanceRegulation/GuidanceDocumentsRegulatoryInformation/IngredientsAdditivesGRASPackaging/ucm110730.htm. Accessed March 22, 2017.
39. International Atomic Energy Agency. *Manual of Good Practice in Food Irradiation, 2015.* Available online at www-pub.iaea.org/MTCD/Publications/PDF/trs481web-98290059.pdf. Accessed March 29, 2017.
40. U.S. Food and Drug Administration, *Bisphenol A (BPA): Use in Food Contact Application.* Available online at www.fda.gov/Food/IngredientsPackagingLabeling/FoodAdditivesIngredients/ucm064437.htm. Accessed March 22, 2017.
41. National Institutes of Health, National Institute of Environmental Health Sciences. *Bisphenol A (BPA).* Available online at www.niehs.nih.gov/health/topics/agents/sya-bpa. Accessed March22, 2017.
42. U.S. Food and Drug Administration. *Everything Added to Food in the United States (EAFUS).* November 2011. Available online at www.fda.gov/Food/IngredientsPackagingLabeling/FoodAdditivesIngredients/ucm115326.htm. Accessed April 4, 2017.
43. U.S. Food and Drug Administration, International Food Information Center. *Overview of Food Ingredients, Additive, and Colors.* April 2010. Available online at www.fda.gov/Food/IngredientsPackagingLabeling/FoodAdditivesIngredients/ucm094211.htm. Accessed April 4, 2017.
44. Bedale, W., Sindelar, J.J., and Milkowski, A.L. Dietary nitrate and nitrite: Benefits, risks, and evolving perceptions. *Meat Sci* 120:85–92, 2016.
45. Bouvard, V., Loomis, D., Guyton, K.Z., et al., International Agency for Research on Cancer Monograph Working Group. Carcinogenicity of consumption of red and processed meat. *Lancet Oncol* 16:1599–1600, 2015.
46. Wheeler, M.B. Transgenic animals in agriculture. *Nature Education Knowledge* 4:1, 2013. Available online at www.nature.com/scitable/knowledge/library/transgenic-animals-in-agriculture-105646080. Accessed March 22, 2017.
47. Tang, G., Hu, Y., Yin, S.A., et al. β-carotene in golden rice is as good as β-carotene in oil at providing vitamin A to children. *Am J Clin Nutr* 96:658–664, 2012.
48. Donald Danforth Plant Science Center. *BioCassava Plus.* Available online at www.danforthcenter.org/scientists-research/research-institutes/institute-for-international-crop-improvement/crop-improvement-projects/biocassava-plus. Accessed April 4, 2017.
49. Gonsalves, D., and Ferreira, S. *Transgenic Papaya: A Case for Managing Risks of Papaya Ringspot Virus in Hawaii.* Available online at www.plantmanagementnetwork.org/pub/php/review/2003/papaya. Accessed March 21, 2017.
50. Smith, N. *Seeds of Opportunity: An Assessment of the Benefits, Safety and Oversight of Plant Genomics and Agricultural Biotechnology.* U.S. House of Representatives report, April 13, 2000. Available online at www.nicksmithconsulting.com/opportunity.pdf. Accessed March 21, 2017.
51. Clive, J. *Global Status of Commercialized Biotech/GM Crops: 2012* (ISAAA Brief No. 49). Available online at www.isaaa.org/resources/publications/briefs/49/executivesummary/pdf/B49-ExecSum-English.pdf. Accessed April 26, 2017.
52. U.S. Food and Drug Administration. *Biotechnology Guidance Documents & Regulatory Information.* Available online at www.fda.gov/Food/GuidanceRegulation/GuidanceDocumentsRegulatoryInformation/Biotechnology/default.htm. Accessed March 22, 2017.

Chapter 14

1. Food and Agriculture Organization of the United Nations. *The State of Food Insecurity in the World 2015, The Multiple Dimensions of Food Security.* Available online at www.fao.org/3/a-i4646e/i4646e01.pdf. Accessed March 24, 2017.
2. Black, R.E., Victora, C.G., Walker, S.P., et al. Maternal and Child Nutrition Study Group. Maternal and child undernutrition and overweight in low-income and middle-income countries. *Lancet* 382:427–451, 2013.
3. World Health Organization. *WHO Global Action Plan for the Prevention and Control of NCDs 2013–2020.* Available online at www.who.int/nmh/events/ncd_action_plan/en/. Accessed March 9, 2017.
4. Central Intelligence Agency. Country comparison: Infant mortality rate. *The World Fact Book.* Available online at www.cia.gov/library/publications/the-world-factbook/fields/2091.html#xx. Accessed March 9, 2017.
5. World Health Organization. *Low Birth Weight: Country, Regional and Global Estimates.* Available online at whqlibdoc.who.int/publications/2004/9280638327.pdf?ua=1. Accessed April 6, 2017.
6. World Health Organization. *Causes of Child Mortality, 2015.* Available online at www.who.int/gho/child_health/mortality/causes/en/. Accessed March 24, 2017.
7. United Nations Children's Fund, World Health Organization, The World Bank. *Joint Child Malnutrition Estimates, 2016.* Available online at http://datatopics.worldbank.org/child-malnutrition/. Accessed March 15, 2017.
8. Fall, C.H. Fetal programming and the risk of noncommunicable disease. *Indian J Pediatr* 80(Suppl. 1):S13–S20, 2013.
9. World Health Organization. *Obesity and Overweight.* June 2016. Available online at www.who.int/mediacentre/factsheets/fs311/en/. Accessed March 9, 2017.

[10] Food and Agriculture Organization of the United Nations. *The Nutrition Transition and Obesity*. Available online at www.fao.org/FOCUS/E/obesity/obes2.htm. Accessed March 9, 2017.

[11] Finkelstein, E.A., Khavjou, O. A., Thompson, et al. Obesity and severe obesity forecasts through 2030. *Am J Prev Med* 42:563–570, 2012.

[12] Vorster, H.H., Bourne, L.T., Venter, C.S., and Oosthuizen, W. Contribution of nutrition to the health transition in developing countries: A framework for research and intervention. *Nutr Rev* 57:341–349, 1999.

[13] World Food Bank. *Poverty Overview*. Available online at www.worldbank.org/en/topic/poverty/overview. Accessed March 15, 2017.

[14] Gettleman, J. Drought and war heighten threat of not just 1 famine, but 4. *NY Times* March 27, 2017. Available online at www.nytimes.com/2017/03/27/world/africa/famine-somalia-nigeria-south-sudan-yemen-water.html?_r=0. Accessed April 3, 2017.

[15] Population Reference Bureau. *2016 World Population Datasheet*. Available online at www.prb.org/pdf16/prb-wpds2016-web-2016.pdf. Accessed March 24, 2017.

[16] United Nations. *Sustainable Development Goals*. Available online at www.un.org/sustainabledevelopment/blog/2015/07/un-projects-world-population-to-reach-8-5-billion-by-2030-driven-by-growth-in-developing-countries/. Accessed March 22, 2017.

[17] Thornton, P., and Herrero, M. *The Inter-Linkages between Rapid Growth in Livestock Production, Climate Change, and the Impacts on Water Resources, Land Use, and Deforestation* (World Bank Policy Research Working Paper No. 5178). Available online at https://papers.ssrn.com/sol3/papers.cfm?abstract_id=1536991. Accessed March 24, 2017.

[18] Le Cotty, T., and Dorin, B. A global foresight on food crop needs for livestock. *Animal* 6:1528–1536, 2012.

[19] Centers for Disease Control and Prevention. *International Micronutrient Malnutrition Prevention and Control (IMMPaCt): Micronutrient Facts*. Available online at www.cdc.gov/immpact/micronutrients/. Accessed March 24, 2017.

[20] Baily, R.L., West, K.P., and Black, R.E. The epidemiology of global micronutrient deficiencies. *Ann Nutr Metab* 66:22–33, 2015.

[21] Loscalzo, J. Keshan disease, selenium deficiency, and the selenoproteome. *N Engl J Med* 370:1756–1760, 2014.

[22] International Osteoporosis Foundation. *Facts and Statistics*. Available online at www.iofbonehealth.org/facts-statistics. Accessed March 27, 2017.

[23] World Health Organization. *Is It True That Lack of Iodine Really Causes Brain Damage?* Available online at www.who.int/features/qa/17/en. Accessed March 27, 2017.

[24] World Health Organization. *Micronutrient Deficiencies: Iron Deficiency Anemia*. Available online at www.who.int/nutrition/topics/ida/en/. Accessed March 24, 2017.

[25] World Health Organization. *Micronutrient Deficiencies: Vitamin A Deficiency*. Available online at www.who.int/nutrition/topics/vad/en. Accessed March 24, 2017.

[26] Coleman-Jensen, A., Rabbitt, M., Gregory, C., and Singh, A. *Household Food Security in the United States in 2015* (Economic Research Report no. ERR-215). U.S. Department of Agriculture, Economic Research Service, September 2016. Available online at www.ers.usda.gov/publications/pub-details/?pubid=79760. Accessed March 23, 2017.

[27] U. S. Census Bureau. *Income and Poverty in the United States*. Available online at www.census.gov/library/publications/2016/demo/p60-256.html. Accessed March 23, 2017.

[28] Laraia, B.A. Food insecurity and chronic disease. *Adv Nutr* 4:203–212, 2013.

[29] Centers for Disease Control and Prevention. *A Look Inside Food Deserts*. Available online at www.cdc.gov/Features/fooddeserts. Accessed March 23, 2017.

[30] U.S. Department of Health and Human Services, Office of Minority Health. *Infant Mortality and African Americans*. Available online at https://minorityhealth.hhs.gov/omh/browse.aspx?lvl=4&lvlid=23. Accessed March 27, 2017.

[31] Bureau of Labor Statistics, U.S. Department of Labor. Median weekly earnings by educational attainment in 2014. *The Economics Daily*. Available online at www.bls.gov/opub/ted/2015/median-weekly-earnings-by-education-gender-race-and-ethnicity-in-2014.htm. Accessed April 2, 2017.

[32] U.S. Department of Housing and Urban Development. *The 2013 Annual Homeless Assessment Report (AHAR) to Congress*. November 2016. Available online at www.hudexchange.info/resources/documents/2016-AHAR-Part-1.pdf. Accessed March 27, 2017.

[33] University of Michigan, National Poverty Center. *Poverty Facts*. Available online at www.npc.umich.edu/poverty/. Accessed March 27, 2017.

[34] Administration on Aging, Administration for Community Living, Department of Health and Human Services. *A Profile of Older Americans, 2016*. Available online at https://aoa.acl.gov/aging_statistics/profile/index.aspx. Accessed March 20, 2017.

[35] United Nations. *Sustainable Development Goals*. Available online at www.un.org/sustainabledevelopment/sustainable-development-goals/. Accessed March 24, 2017.

[36] The World Bank. *Fertility Rate*. Available online at www.prb.org/pdf13/2013-population-data-sheet_eng.pdf. Accessed March 24, 2017.

[37] Berg, L.R., Hager, M.C., and Hassenzahl, D.M. *Visualizing Environmental Science*, 3rd ed. Hoboken, NJ: Wiley, 2011.

[38] Central Intelligence Agency. Africa: Cote D'Ivoire. *The World Fact Book*. Available online at www.cia.gov/library/publications/the-world-factbook/geos/iv.html. Accessed March 24, 2017.

[39] United Nations. *Millennium Development Goals Report*. 2012. Available online at www.un.org/millenniumgoals/pdf/MDG%20Report%202012.pdf. Accessed March 27, 2017.

[40] U.S. Department of Agriculture. Economic Research Service. *Trade*. Available online at www.ers.usda.gov/topics/international-markets-trade/countries-regions/south-korea/trade/. Accessed March 28, 2017.

[41] World Health Organization. *Global Strategy for Infant and Young Child Feeding*. Available online at www.who.int/nutrition/topics/global_strategy/en/index.html. Accessed March 28, 2017.

[42] World Health Organization. *Guidelines on Breast Feeding and HIV Infection 2010*. Available online at whqlibdoc.who.int/publications/2010/9789241599535_eng.pdf. Accessed March 28, 2017.

[43] World Health Organization. *Assessment of Iodine Deficiency Disorders and Monitoring Their Elimination*. Available online at www.who.int/nutrition/publications/micronutrients/iodine_deficiency/9789241595827/en/index.html. Accessed March 28, 2017.

[44] Li, S., Nugroho, A., Rocheford, T., and White, W.S. Vitamin A equivalence of the β-carotene in β-carotene-biofortified maize porridge consumed by women. *Am J Clin Nutr* 92:1105–1112, 2010.

[45] Enserink, M. Tough lessons from golden rice. *Science* 320:468–471, 2008.

[46] Tang, G., Hu, Y., Yin, S.A., et al. β-carotene in golden rice is as good as β-carotene in oil at providing vitamin A to children. *Am J Clin Nutr* 96:658–664, 2012.

[47] Greenpeace International. *Golden Rice's Lack of Luster: Addressing Vitamin A Deficiency Without Genetic Engineering.* November 9, 2010. Available online at www.greenpeace.org/international/en/publications/reports/Golden-rice-report-2010. Accessed March 28, 2017.

[48] Bienvenido, O. J. *Rice in Human Nutrition.* Available online at www.fao.org/docrep/t0567e/T0567E00.htm. Accessed March 28, 2017.

[49] Greenpeace International. *The Golden Rice Illusion.* Available online at www.greenpeace.org/international/en/campaigns/agriculture/problem/genetic-engineering/Greenpeace-and-Golden-Rice. Accessed April 17, 2014.

[50] Potrykus, I. Lessons from the "Humanitarian Golden Rice" project: Regulation prevents development of public good genetically engineered crop products. *N Biotechnol* 27:466–472, 2010.

[51] Moghissi, A. A., Pei, S., and Liu, Y. Golden rice: Scientific, regulatory and public information processes of a genetically modified organism. *Crit Rev Biotechnol* 36:535–541, 2016.

[52] Conner, S. *Former Greenpeace Leading Light Condemns Them for Opposing GM "Golden Rice" Crop That Could Save Two Million Children from Starvation per Year.* January 30, 2014. Available online at www.independent.co.uk/news/science/former-greenpeace-leading-light-condemns-them-for-opposing-gm-golden-rice-crop-that-could-save-two-million-children-from-starvation-per-year-9097170.html. Accessed March 28, 2017.

[53] *GoldenRiceNow!* Available online at www.allowgoldenricenow.org. Accessed March 28, 2017.

[54] World Health Organization. *Vitamin A Supplementation for Infant and Children 6-59 months of age.* Available online at www.who.int/nutrition/publications/micronutrients/guidelines/vas_6to59_months/en/. Accessed March 28, 2017.

[55] Micronutrient Initiative. *Sometimes Food Is Not Enough.* Available online at www.micronutrient.org/what-we-do/by-programs/supplementation/. Accessed March 28, 2017.

[56] Economic Research Service, U.S. Department of Agriculture. *The Food Assistance Landscape: FY 2015 Annual Report* (Economic Bulletin no. 150). March 2016. Available online at www.ers.usda.gov/publications/pub-details/?pubid=44062. Accessed March 27, 2017.

[57] Feeding America. *Food Bank Network.* Available online at feedingamerica.org/our-network/how-we-work.aspx. Accessed March 27, 2017.

[58] Gunders, D., Natural Resources Defense Council. *Wasted: How America Is Losing up to 40% of Its Food from Farm to Fork to Landfill* (NRDC Issue Paper IP:12-06B). August, 2012. Available online at www.nrdc.org/sites/default/files/wasted-food-IP.pdf. Accessed April 26, 2017.

[59] Food and Agricultural Association of the United Nations. *Global Food Losses and Food Waste—Extent, Causes and Prevention.* Rome, 2011.

[60] Centers for Disease Control and Prevention. *Parasites—Hookworm.* Available online at www.cdc.gov/parasites/Hookworm. Accessed March 28, 2017.

Index

Note: Page numbers with an "f" after them indicate the entry appears in a figure, and page numbers with a "t" after them indicate the entry appears in a table.

A

Absorption
  carbohydrates and, 89–91
  digestive system and, 55
  iron and, 249, 250f
  lipids and, 122–123, 122f
  proteins and, 147–148, 148f
  small intestine and, 63–64, 64f
  vitamin B_{12} and, 192, 192f, 1891
  vitamins and, 175, 175f
Acceptable Macronutrient Distribution Ranges (AMDRs), 33, 33f
Accidental contaminants, food safety and, 427
Acesulfame K, 107t
Acrylamide, 428
Active life expectancy, 388, 388f
Active lifestyle creation, 312–313, 312t
Active transport, 64
Activity, balancing calories with, 36–37
Added sugars
  characteristics of, 84, 85f
  disease and, 97, 97f
  refined grains and, 83, 84f
Additives. *See* Food additives
Adenosine diphosphate (ADP), 314
Adenosine triphosphate (ATP), 77–78, 77f
  exercise duration and, 313–314, 315f
  exercise intensity and, 316
  production of, 313
Adequate Intakes (AIs), 31, 32f
Adipocyte, 276
Adipose tissue, 125, 125f
Adjustable gastric banding, 291, 291f
Adolescent nutrition, 383–388
  alcohol use and, 387–388
  athletics impact and, 387
  beverage choices and, 384, 385f
  calcium needs and, 384
  eating disorders and, 387
  energy and nutrient needs and, 383–384
  fast food and, 385–386, 386t
  growth and development and, 384, 384f
  Healthy Eating Index and, 372, 372f
  iron needs and, 384
  meeting teen nutritional needs, 384–386
  MyPlate recommendations and, 385, 385f
  special concerns and, 387–388
  tobacco use and, 387
  vegetarian diets and, 386–387, 387f
  vitamin needs and, 383
Adult nutrition, 388–397. *See also* Older adults
  calorie restrictions and, 390
  declining muscle mass and, 389, 391f
  eating less and, 390, 390f
  energy and nutrient recommendations, 389, 391f
  health concerns and, 389–391
  identifying at risk adults, 394, 395t
  life expectancy and, 388, 388f
  malnutrition in older adults, 391–394
  meeting nutrient needs, 394, 396f
  overcoming economic and social issues, 396
  overcoming physical limitations, 396
  physical activity for older adults, 395–396, 397f
  successful aging and, 388–389, 388f
  vitamin and mineral deficiencies and, 391
  water needs and older adults, 389–391
Advantame, 107t
Advertising, food environment and, 372, 372f
Advertising claims, nutritional information and, 20, 21f
Aerobic activity, 307–309, 308f
Aerobic capacity, 309
Aerobic metabolism
  vs. anaerobic metabolism, 313, 314f
  fuels for, 316, 316f
  glucose and, 92
  long term energy and, 314–316
Aerobic training, 318–319, 318f
Aerobic zone, 310–311, 311f
Aging. *See also* Older adults
  age-related bone loss, 240
  factors in, 389, 389f
  heart disease and, 129t
  osteoporosis and, 242t
Agricultural chemicals, food safety and, 421–426
Agricultural contaminants, 425–426, 426f
Agroforestry, 453
Air displacement, body composition and, 271, 271f

Alcohol consumption, 397–401
  absorption, transport, excretion, 398, 399f
  adolescent nutrition and, 387–388
  adverse effects and, 398–400
  alcohol metabolism, 398–399, 399f
  alcohol poisoning, 400
  alcoholism and, 400
  benefits of, 401
  blackout drinking, 400
  cancers and, 400
  cirrhosis and, 400, 401f
  dietary supplements and, 211t
  heart disease and, 132f, 400
  hepatitis and, 400
  liver disease and, 400
  malnutrition and, 400
  osteoporosis and, 242t
  pregnancy and, 355, 355f
  short-term effects of, 400
Alcohol dehydrogenase (ADH), 398
Alcohol poisoning, 400
Alcohol use disorder, 400
Alcoholic cirrhosis, 400, 401f
Alcoholism, 400
Allergens, 67
Alpha-carotene, 201
Alpha-linolenic acid, 126
Alpha-tocopherol, 198
Alzheimer's disease, 394
Amino acids
  amino acid pools, 149, 149f
  as building blocks of proteins, 144
  deficiency of, 145
  as ergogenic supplement, 330t
  limiting of, 149–150, 150f
  peptide bonds and, 146
  polypeptides and, 146
  producing ATP from, 152f, 153
  proteins and, 146, 146f
  structure, 145, 145f
  supplements, 160, 160f
Anabolic steroids, 331–332, 332f
Anaerobic metabolism
  vs. aerobic metabolism, 313, 314f
  fuels for, 316, 316f
  glucose and, 92
  short-term energy and, 314
Anemia
  childhood nutrition and, 377
  folate and, 188, 189f
  iron deficiency anemia, 251–252, 251f
  sports anemia, 322, 322f
  vitamin B_6 and, 187
  vitamin B_{12} and, 190–191
Animal proteins, 144, 144f

Animal starch, 87
Animal studies, nutrition and, 17, 18f
Anisakis simplex, 411t
Anorexia athletica, 299, 299t
Anorexia nervosa, 296–297, 296f
Anti-caking agents, 430t
Antibiotics, food safety and, 424–425
Antibodies, 66
Anticoagulants, 194, 194f
Antidiuretic hormone (ADH), 222, 222f
Antigens, 66
Antioxidant minerals, 253–257
  manganese, 256
  molybdenum, 257
  selenium, 253–255
  sulfur, 256–257
  zinc, 255–256
Antioxidants, heart disease risk and, 131
Appetite stimulation, 267
Arachidonic acid (ARA), 126, 126f, 362
Arteries, 74
Arterioles, 74
Arthritis, older adults and, 393, 393f
Artificial sweeteners, 107–108, 107f–108f
Ascorbic acid. *See* Vitamin C
Aseptic processing, 427, 428f
Aspartame, 107t
Assisted living, 396–397
Atherosclerosis, 129–130, 130f
Atherosclerotic plaque, 129, 130f
Athletes. *See also* Exercise; Physical activity
  energy deficiency in sports, 320–321, 320f
  female athlete triad, 320–321, 320f
  precompetition meals, 326, 327f
  protein needs and, 159–160, 160f
  weight classes and, 321, 321f
Athletics, adolescent nutrition and, 387
Atoms, 52, 52f
Atrophic gastritis, 192
Atrophy, 308, 308f
Attention-deficit/hyperactivity disorder (ADHD), 382
Autoimmune disease, 96
Avoidant/restrictive food intake disorder, 299, 299t

I-1

B

Bacillus thuringiensis, 434, 434f
Bacteria, food safety and, 411–413, 412f
Balance, healthy diets and, 13–14, 14f
Bariatric surgery, 292
Basal metabolic rate (BMR), 273
Basal metabolism, 273
Beetroot juice, as ergogenic supplement, 330t
Behavior modification, body weight and, 284–285, 286f
Beriberi, 179
Beta-alanine, as ergogenic supplement, 330t
Beta-carotene, 201, 206, 206f
Beta-cryptoxanthin, 201–202
Beverage choices, children and adolescents and, 384, 385f
Beverages, calories from, 38, 38f
Bicarbonate, 61, 330t
Bifidobacterium, 67
Bile, 62–63
Binge drinking, adolescents and, 387–388
Binge eating, 294, 294f, 298, 298f
Bioaccumulation, 421–422, 421f
Bioavailablity, 174
Biochemistry, nutritional studies and, 17, 18f
Biofortification, 455
Biometric impedance, 271, 271f
Biotechnology, 432–436
  applications of, 434
  consumer concerns and, 435
  environmental concerns and, 435
  genetic engineering, 432–433
  genetic modification, 432
  genetically modified food regulations, 435–436, 435f7
  genetically modified organisms, 433, 433f
  potential of, 434, 434f
Biotin, 184–185
  deficiency and, 185
  raw eggs and, 185, 185f
  recommendations for, 177t
  sources of, 184
Birth rate, economic and cultural factors and, 451, 451f
Bispherol A, 429, 429f
Bisphosphonates, 242
Blackout drinking, 400
Blood alcohol levels, 398, 399f
Blood clotting, 194, 194f
Blood glucose, managing, 98–99, 100f
Blood lipid levels, heart disease and, 129t
Blood pressure. *See also* Hypertension
  diet effect on, 235, 235f
  electrolytes and, 233, 234f
  water distribution and, 220
Blood sugar, 85
Blood vessels, cardiovascular system and, 74, 74f
Body composition, 269, 271f, 309
Body fat location, 269–272, 272f
Body image, eating disorders and, 293–294, 294t
Body mass index (BMI)
  childhood obesity and, 373–374, 374f
  healthy weight and, 269, 270f
Body shape and size, 278–283
  food intake regulation and, 279–282, 280f
  genes *vs.* environment, 279, 279f
  leptin and, 280–281, 281f
  NEAT differences and, 282–283, 282f
  obesity genes and, 278–279, 278f
Body temperature, water and, 223, 223f
Body weight. *See also* Eating disorders; Energy balance; Obesity
  appetite stimulation and, 267
  behavior modification and, 284–285, 286f
  body composition and, 269, 271f
  body fat location and, 269–272, 272f
  body mass index and, 269, 270f
  carbohydrate-restricted diets, 289, 289f
  decreasing energy intake and, 284
  diets and fad diets and, 285–289, 288f–289f, 288t
  energy expenditure and, 267–268, 268f
  energy intake and, 320
  excess body fat and, 268–269, 268f
  food choices case study and, 287
  food intake regulation and, 279–282, 280f
  health and, 266–272
  healthy food patterns, 288, 288f
  hunger and, 267
  increasing physical activity and, 284
  liquid and prepackaged meals and, 289, 289f
  managing America's weight, 285
  moderate-to high-protein diets, 288, 288f
  modifying behavior and, 285
  NEAT differences and, 282–283, 282f
  obesity and, 266–267, 266f
  obesogenic environment and, 267
  osteoporosis and, 240, 242t
  portion distortion and, 267, 267f
  reduced fat diets, 289, 289f
  set point and, 279
  weight cycling and, 284, 284f
  weight gain suggestions, 285
  weight-loss goals and guidelines, 283–285, 283f
  weight-loss medications and supplements, 290–291, 290f
  weight-loss surgery, 291–293, 291f
Bone health
  age-related bone loss, 240
  bone remodeling, 240
  bone resorption, 242
  calcium and, 242–245
  childhood nutrition and, 377
  fluoride and, 246–249
  magnesium and, 246
  minerals and, 240–249
  osteoporosis and, 240–242, 241f
  peak bone mass, 240
  phosphorous and, 245–246
  postmenopausal bone loss, 240
Bone remodeling, 240
Bottle feeding
  do's and dont's of, 363, 363f
  nursing bottle syndrome, 363, 363f
Bottled water, 225
Botulism, 413, 413f
Bovine somatotropin, 425, 425f
Bovine spongiform encephalopathy (BSE), 416
Bowel health, carbohydrates and, 101–102
Bran, whole grains and, 84, 84f
Breakfast, school performance and, 379
Breast feeding
  food allergies and, 365–366
  health benefits of, 363–364, 365f
  nutritional status and health and, 455, 455f
Breast milk
  nutrients in, 363, 364f
  vitamin D and, 361
Brown rice, as soluble fiber source, 132f
Brush border, 62
Bulimia nervosa, 297–298, 297f

C

Cadmium, 424
Cafeteria germ, 413
Caffeine
  as ergogenic aid, 330t, 333
  hyperactivity in children and, 382
  pregnancy and, 354
  in sports drinks, 334, 334f
Calcitonin, 242
Calcium, 242–245
  adolescents and, 384
  bone resorption and, 242
  children and, 377
  food sources of, 243–244, 243f–244f
  in health and disease, 242–243, 243f
  high levels of, 243
  meeting needs for, 243–244
  older adults and, 391
  physical activity and, 322–323
  pregnancy and, 346
  recommendations and, 227t
  supplements, 244, 245f
  vegan diet and, 166t
  vitamin D and, 211
Calories
  adult restrictions and, 390
  balancing with activity, 13–14, 14f, 36–37
Campylobacter jejuni, 410t, 411, 412f
Cancer
  alcohol consumption and, 400
  dietary patterns and, 131
  processed meats and, 431, 431f
  selenium and, 254–255
Candy displays, 24, 24f
Capillaries, 73
Carbohydrate-restricted diets, 289, 289f
Carbohydrates
  absorption and, 89–91
  added sugar, 83–85, 84f, 85f
  bowel health and, 101–102
  case study on, 106, 106f
  cellular respiration and, 92, 94f
  characteristics of, 6, 7f
  children and, 376
  choosing wisely, 103–108
  complex carbohydrates, 87–88, 87f
  dental caries and, 101
  diabetes and, 96–99
  digestion and, 88–89, 88f
  food labels and, 104–105, 105f
  functions of, 92–95
  glucose as source of energy, 92–95
  glucose regulation and, 92, 93f
  glycemic response and, 91, 91f
  in health and disease, 96–102
  heart disease and, 100–101
  hypoglycemia and, 99–100
  lactose intolerance and, 89, 89f
  limited carbohydrates, 92–93, 95f
  MyPlate and, 104, 104f
  nonnutritive sweeteners and, 107–108, 107f–108f
  in our food, 83–85, 83f–84f
  physical activity and, 321–322, 321f
  protein breakdown and, 92–93
  recommendations for, 102–103, 103f
  simple carbohydrates, 85, 86f
  types of, 85–88
  weight management and, 100
  whole grains and, 83–84, 84f
Cardiorespiratory endurance, 307–309, 308f
Cardiovascular/circulatory system, 54t
Cardiovascular disease, lipids and, 128–129
Cardiovascular system
  blood circulation and, 73f
  heart and blood vessels and, 74, 74f
  hepatic portal circulation, 75, 75f
  nutrient delivery and, 73–75
Carnitine, as ergogenic aid, 330t, 333
Carotenoids, 201–202

INDEX   I-3

Cash crops, 454
Cataracts, older adults and, 393, 393f
Celiac disease, 67, 155–157
Cell differentiation, vitamin A and, 203
Cells, 52
Cellular respiration, 77, 77f, 92, 94f
Centers for Disease Control and Prevention (CDC), 407t
Cesarean section, 342
Chewing, 58
Child and Adult Care Food Program, 458t
Childhood nutrition, 376–383
  attention-deficit/hyperactivity disorder (ADHD), 382
  beverage choices and, 384, 385f
  breakfast and school performance, 379
  calcium, vitamin D, and bone health, 377
  chronic diseases and, 374–376, 375f
  day care and school and, 379–380
  dental caries and, 382
  developing healthy eating habits, 377–380
  dietary supplements and, 211t
  energy and nutrient needs and, 376–377, 376f
  food environment and, 372, 373f
  Healthy Eating Index and, 372, 372f
  iron and anemia, 377
  lead toxicity and, 382–383, 382f
  limiting juice consumption and, 378, 378f
  meals and snack patterns and, 378–379, 378t
  monitoring growth and, 376
  MyPlate recommendations and, 377–378, 377f
  obesity and, 373–374, 374f, 375f
  overnutrition and, 442
  special concerns and, 380–383
  under- and overnutrition and, 381, 381f, 442, 442f
Chinese restaurant syndrome, 155, 432
Chlordane, 424
Chloride, 227t, 232–239. *See also* Electrolytes
Choice (Exchange) Lists, 41
Cholesterol, 119–121, 121f
  childhood nutrition and, 374, 375f
  dietary fiber and, 101, 101f
  dietary recommendations, 133
  egg consumption and, 134, 134f
  elimination of, 124–125, 124f
Choline, 178t, 185
Chromium
  deficiency of, 260
  dietary sources of, 259–260
  as ergogenic supplement, 330t
  recommendations for, 228t
  role of, 259

Chronic diseases
  childhood nutrition and, 374–376, 375f
  older adults and, 393–394
  physical activity and, 305, 306f
Chylomicrons, 123–124
Chyme, 59, 60
Cigarette smokers, dietary supplements and, 211t
Cirrhosis, 400, 401f
Cis fatty acids, 118, 118f
Clinical trials, nutritional studies and, 17, 18f
Clostridium botulinum, 410t, 413, 413f
Clostridium perfringens, 410t, 413
Cocaine, 356
Coconut oil, 119, 119f
Coenzymes, 179, 180f
Collagen, 151, 151f, 197, 197f
Color additives, 430t, 432
Colostrum, 363
Commodity Supplemental Food Program (CSFP), 457t
Complementary proteins, 161–162, 161f
Complete dietary protein, 161
Complex carbohydrates, 87–88, 87f
Compounds, mineral absorption and, 229, 230f
Concealing clothing, vitamin D deficiency and, 210, 210f
Conditionally essential amino acids, 145, 145f
Constipation, 70–72, 71f
Consumer concerns, biotechnology and, 435
Consumers, food safety and, 407–409, 409t
Control groups, nutritional experiments and, 17
Copper, 228t, 253
Creatine, 331t, 332, 332f
Creatine phosphate, 314
Cretinism, 259
Crib death, 356
Critical control points, food safety and, 407, 407f
Crohn's disease, 65, 81, 195
Cross-contamination, food safety and, 417
Cryptosporidium parvum, 411t, 415
Cultural factors, birth rate and, 451, 451f
Cultural practices, food shortages and, 444
Cycle of undernutrition, 441–442, 441f–442f

D

Daily Value, 42, 43f
Dark chocolate, heart disease risk and, 132f
Dark skin pigmentation
  dietary supplements and, 211t
  vitamin D deficiency and, 210, 210f
DASH (Dietary Approaches to Stop Hypertension), 235–236, 236f

DASH Eating Plan, 35, 35f
Day care, nutrition and, 379–380
Deamination, 152, 152f
Dehydration, 224, 224f, 323–324, 323f
Delaney Clause, 432
Dementia, 393–394
Denaturation, protein, 147, 147f
Dental caries
  carbohydrates and, 101
  childhood nutrition and, 382
  fluoride and, 246
Dental pain, 70, 70f
Depression, older adults and, 394
DETERMINE checklist, 394, 395t
Developmentally appropriate foods, 366–367, 366f
Diabetes, 96–99
  blood glucose levels in, 96, 96f
  childhood nutrition and, 374, 375f
  diabetic ketoacidosis, 96
  gestational, 98, 345
  heart disease and, 129t
  insulin resistance and, 98
  managing blood glucose and, 98–99, 100f
  prediabetes, 98
  symptoms and complications, 98, 99f
  type 1 diabetes, 96
  type 2 diabetes, 98, 98f
  types of, 96–98
Diabetic ketoacidosis, 96
Diabulimia, 299, 299t
Diarrhea, 70–72
Diet-gene interactions, 10–12, 11f
Diet-induced thermogenesis, 275
Dietary factors, osteoporosis and, 240
Dietary folate equivalents (DFEs), 190
Dietary Guidelines for Americans, 33–38, 34t
  healthy eating patterns and, 35–36, 35f
  healthy weight and exercise recommendations, 37, 37f
  introduction of, 26
  limiting sugars, saturated fats, and sodium, 37–38
  meeting food group recommendations, 37, 38f
  shift to balance calories with activity, 36–37
  shift to healthier choices and, 36–38, 36f
  societal support and, 38
Dietary patterns, heart disease and, 129t, 131, 131f–132f
Dietary reference intakes (DRIs), 31–33
  for all population groups, 31, 31f
  for energy intake, 32–33
  for nutrient intake, 31–32
Dietary supplements
  calcium, 244, 245f
  checking label and, 213, 214f
  choosing with care, 213–214

ergogenic aids, 329–335, 330t–331t
groups requiring, 211, 211t
herbal supplements, 212, 212f–213f
labels, 45–47
as nutritional source, 3
structure/function claims and, 46–47, 47f
suggestions regarding, 213–214
supplement facts panel, 45–46
vitamins and, 211–214
weight loss, 290–291, 290f
zinc, 256
Diets, 285–289
  common dieting methods, 287, 288f–289f
  fad diets and, 287, 288t
  food choices case study, 287, 287f
  maintaining weight after, 286
Diffusion, 63–64
Digestion
  carbohydrates and, 88–89, 88f
  digestive system and, 55
  esophagus and, 58
  fiber and, 89
  large intestine and, 64
  lipids and, 122–123, 122f
  mouth and, 58
  pharynx and, 58
  proteins and, 147–148, 148f
  small intestine and, 60–63
  stomach and, 59–60
Digestive system, 54t, 55–57
  diarrhea and constipation and, 70–72
  digestion and absorption, 57–64
  gallstones and, 70, 71f
  gastroesophageal reflux disease and, 69–70, 70f
  gut immune function and, 66–68
  in health and disease, 65–72
  heartburn and, 69–70, 70f
  intestinal microbiota and, 65–66
  organs of, 55, 56f
  peptic ulcers and, 70, 71f
  secretions of, 55–57, 57f
Dipeptides, 146, 146f
Direct food additives, 429
Disaccharides, 85
Discretionary fortification, 172
Dispensable amino acids, 145, 145f
Diverticula, 101, 102f
Diverticulitis, 101
Docosahexaenoic acid (DHA), 126, 126f, 362
Down syndrome, 354, 354f
Dual-energy X-ray absorptiometry (DXA), 271, 271f

E

Eating disorders, 293–300
  adolescents and, 387
  anorexia nervosa, 296–297, 296f
  binge-eating disorder, 298, 298f
  bulimia nervosa, 297–298, 297f
  causes of, 293–296, 294f

Eating disorders (*Continued*)
  prevention and treatment, 299–300
  psychological issues and, 293–294
  sociocultural issues and, 294–296, 295f
  in special groups, 298–299, 299t
  types of, 293, 294t
Echinacea, 213, 213f
Eclampsia, pregnancy and, 344
Economic development, world hunger and, 453–454
Economic factors, birth rate and, 451, 451f
Economic issues, older adults and, 394, 396
Edema, 151, 151f, 343
Education, poverty and, 448, 449f
Egg production, critical control points in, 407, 408f–409f
Eggs, in diet, 134, 134f
Eicosapentaenoic acid (EPA), 126, 126f
Electrolytes, 232–239
  blood pressure regulation and, 233, 234f
  in the body, 232–233
  deficiency and toxicity, 233–234
  food labels and, 238, 239f
  functions of, 232, 233f
  hypertension and, 234–236, 235f
  meeting needs for, 236–238
  physical activity and, 323–325
  pregnancy and, 346
  regulating balance of, 232–233
Elements, 6
Embryo, nourishing, 341, 341f
Emergency Food Assistance Program, 458t
Emergency food relief, 451, 451f
Empty calories, 2, 2f
Emulsifiers, 119, 430t
Endocrine system, 54t
Endosperm, whole grains and, 84, 84f
Endurance, ergogenic supplements and, 333
Energy balance, 273–278
  basal metabolic rate, 273, 274f
  basal metabolism and, 273
  energy expenditure and, 273–275, 274f–275f
  energy intake and, 273, 274f
  estimated energy requirements, 276–278, 277f
  food labels and, 274, 274f
  physical activity value and, 276, 277f
  storing and retrieving energy, 276, 276f
  thermic effect of food and, 275, 275f
Energy drinks, 334, 334f
Energy expenditure, 273–275, 274f–275f
  basal metabolic rate and, 273
  basal metabolism and, 273
  body weight and, 267–268, 268f

diet-induced thermogenesis and, 375
  nonexercise activity thermogenesis (NEAT) and, 273
  thermic effect of food and, 375
  total energy expenditure, 273
Energy intake
  balancing with expenditure, 273, 274f
  body weight and, 284
  dietary reference intakes and, 32–33
Energy metabolism
  chromium and, 259–260
  iodine and, 257–259
  minerals and, 257–260
Energy needs
  adolescent nutrition and, 383–384
  adults and, 389
  body weight and composition and, 320
  female athlete triad, 320–321, 320f
  infant nutrition and, 359–360, 361f
  lactation and, 357–358, 357f, 358f
  physical activity and, 319–321, 319f
  pregnancy and, 345–346, 346f
  relative energy deficiency in sport (RED-S), 320–321, 320f
Energy-yielding nutrients, 7
Enrichment, 84, 84f
Environmental concerns, biotechnology and, 435
Enzymes
  activity of, 55, 57f
  as protein molecules, 151, 151f
Epidemiological studies, 17, 18f
Epiglottis, 58, 58f
Ergogenic supplements, 329–335, 330t–331t
  building muscle and, 331–332, 332f
  diet and performance and, 333–335, 335f
  endurance and, 333
  short, intense activities and, 332, 332f
  vitamin and mineral supplements, 329
Erythropoietin, 333
Escherichia coli (*E. coli*), 410t, 411–413, 412f
Esophagus, digestion and absorption and, 59, 59f
Essential amino acids, 145, 145f
Essential fatty acid deficiency, 126
Essential fatty acids, 116, 125–127, 126f
Estimated Average Requirements (EARs), 31, 32f
Estimated Energy Requirements (EERs), 32–33, 32f, 276–278, 277f

Evidence-based practice, nutritional experiments and, 17
Exercise. *See also* Physical activity
  blood flow at, 74, 74f
  duration and fuel use, 313–316
  eating and drinking after, 327–329
  eating and drinking before, 325–326
  eating and drinking during, 326–327
  fat-burning zone and, 317, 317f
  fatigue and, 316–318, 318f
  female athlete triad, 320–321, 320f
  fitness training and fuel use, 318–319, 318f
  intensity and fuel use, 316–318, 317f
  maximizing glycogen stores, 326, 326f
  osteoporosis and, 242t
  precompetition meal and, 326, 327t
  protein as fuel for, 316
  recommendations, 37, 37f
  relative energy deficiency in sport (RED-S), 320–321, 320f
  snacks case study, 328, 328f
  sports drinks and, 327, 329f
Expanded Food and Nutrition Education Program (EFNEP), 458t
Experimental groups, nutritional experiments and, 17
Exponential bacterial growth, 418, 420f

F
Facilitated diffusion, 63–64
Factory farming, 446
Fad diets, 285–289, 288f–289f, 288t
Failure to thrive, 359
Family history
  heart disease and, 129t
  osteoporosis and, 242t
Family planning, population growth and, 452, 452f
Famines, 444, 444f
Farmer's markets, 5
Fast foods, adolescent nutrition and, 385–386, 386t
Fasting, 128, 128f
Fasting hypoglycemia, 100
Fat burners, 291
Fat-burning zone, 317, 317f
Fat malabsorption, 202–203
Fat needs, physical activity and, 321–322, 321f
Fat replacers, 137–139, 139f
Fat-soluble vitamins, 122–123, 171, 178t
Fatigue, 316–318, 318f
Fats. *See also* Lipids
  choosing wisely, 133–135
  dietary recommendations, 133
  as energy source, 127–128
  healthy fats, 134

hidden fats, 114–115, 114f
  metabolism of, 127, 127f
  MyPlate choices, 135, 135f
  saturated fats, 115
  sources of, 114–115, 114f
  trans fats, 115
  understanding fat descriptors, 137, 138f
  unsaturated fats, 115
Fatty acids, 55, 115, 116–118, 117f, 118f
Fatty liver, 400
Feasting, 127–128, 128f
Feces, 55
Female athlete triad, 299, 299t, 320–321, 320f
Fertilization, 340–341, 340f
Fetal alcohol spectrum disorders (FASDs), 355
Fetal alcohol syndrome (FAS), 355, 355f
Fetus, nourishing, 341, 341f
Fiber, 87–88, 87f
  cholesterol and, 101, 101f
  digestion and, 89
  effects of, 90f
  insoluble fiber, 87–88
  soluble fiber, 87
Fibrin, 194, 194f
Field gleaning, 458, 458f
Fish
  omega-3 fatty acids and, 131f
  parasitic infections and, 415, 416f
Fitness, 307–309, 308f. *See also* Physical activity
  aerobic activity and, 307–309
  aerobic capacity and, 309
  body composition and, 309
  cardiorespiratory endurance and, 307–309
  flexibility and, 309
  fuel use and, 318–319, 318f
  muscle strength and endurance, 308
  overload principle and, 307
  programs, 309–313, 310f
  resting heart rate and, 309
Flavor enhancers, 430t
Flexibility exercises, 308, 308f, 309
Fluid balance, proteins and, 151, 151f
Fluid needs
  infants and, 360
  pregnancy and, 346
Fluoride, 246–249
  dental caries and, 246
  infants and, 361
  meeting needs of, 247
  recommendations and, 228t, 248, 248f
  water fluoridation, 248–249
Fluorosis, 248, 248f
Folate (Folic acid), 177t, 188–190
  dietary folate equivalents, 190
  fortification, 348, 348f
  homocysteine levels and, 190, 190f
  low folate status, 190

macrocytic anemia and, 188, 189f
neural tube defects and, 188–190, 189f
pregnancy and, 188, 346–347
spina bifida and, 189, 189f
toxicity and, 190
vitamin B$_{12}$ and, 191, 193f
Food additives, 429–432, 430t
classification of, 427
Delaney Clause and, 432
generally recognized as safe, 431–432
nitrosamines and, 431
prior-sanctioned substances and, 431
regulation of, 430–432
sensitivities to, 432
Food allergies
gut immune function and, 66–67, 68f
proteins and, 155–157, 157f
Food and Agriculture Organization of the United Nations (FAO), 406t
Food-borne illness, 354, 355t, 406
Food-borne infection, 410
Food-borne intoxication, 410
Food choices, determination of, 3–6
Food cravings and aversions, pregnancy and, 348–350
Food Distribution Program on Indian Reservations, 458t
Food environment, 3–6, 4f–5f
Food fortification, eliminating world hunger and, 455, 456f
Food group recommendations, meeting, 37, 38f
Food guides, 26, 26f
Food handling, 417, 419f
Food insecurity
older adults and, 394
United States and, 448–449, 448f
Food intake, regulation of, 279–282, 280f
Food intolerances, proteins and, 155–157, 157f
Food labels, 42–45, 43f–44f
carbohydrates and, 104–105, 105f
case study and, 45
electrolytes and, 238, 239f
energy balance and, 274, 274f
health claims and, 44, 46f
ingredient list and, 42
lipids and, 135–137, 137f
natural food claims and, 44, 46f
nutrient content claims and, 42–44
Nutritional Facts label and, 42
protein and allergen information, 157, 157f
qualified health claims and, 44
vitamins and, 173, 173f
Food manufacturers, food safety and, 407, 408f
Food marketing, 372
Food pathogens, 410–417, 410t–411t
bacteria, 411–413, 412f

food-borne infection and, 410
food-borne intoxication and, 410
molds, 415, 415f
parasites, 415, 416f
prions, 416–417, 416f
viruses, 414–415, 414f
Food portion sizes, 14
Food production, increasing, 452–453
Food recovery, 457
Food safety
agricultural and industrial chemicals and, 421–426
antibiotics and hormones and, 424–425
aseptic processing and, 427, 428f
biotechnology and, 432–436
consumer role and, 407–409, 409t
contamination through food chain, 421–422, 421f
cross-contamination and, 417
exponential bacterial growth and, 418, 420f
from farm to table, 407, 408f
food additives and, 429–432, 430t
food handling, storage, preparation, 417, 419f
food manufacturers and retailers and, 407, 408f
food pathogens and, 410–417, 410t–411t
government role and, 406–407, 406t–407t
industrial contaminants and, 424
irradiation and, 428–429, 428f
packaging and, 429, 429f
pesticides and, 422–424, 422f
preventing microbial illness, 417–426
reducing exposure risks, 425–426, 426f
restaurants and, 417–418
safe grocery choices and, 417, 418f
slowing microbial growth and, 427, 427t
technology for, 427–432
temperatures and, 427–428
tracking food-borne illness, 420, 420f
Food Safety and Inspection Service (FSIS), 407t
Food shortages, hunger and, 444–446
Food storage, 417, 419f
Formula, infant
advantages of, 364
DHA/ARA-fortified, 362
nutrients in, 363, 364f
Fortification, discretionary, 172
Fortified foods, 172
Free radicals, 196, 196f
Fructose, 85, 86f
Functional foods, 3, 4f

G

Galactose, 85, 86f
Gallbladder, 62–63
Gallstones, 70, 71f

Garlic, 212, 212f
Gastric bypass surgery, 291–293
Gastric juice, 59–60
Gastric sleeve surgery, 291, 291f
Gastroesophageal reflux disease (GERD), 69–70, 70f
Gastrointestinal tract, 55
Gastrointestinal tract crowding, pregnancy and, 343, 344f
Gatorade, 334
Gene expression
protein synthesis and, 149
vitamin A and, 203
zinc and, 255, 255f
Generally recognized as safe additives (GRAS), 431–432
Genes, diet interactions and, 10–12, 11f
Genetic engineering, 432–433, 453
Genetic modification, 432–433
Genetically modified organisms, 433, 433f
Genetics, body shape and size and, 278–279, 279f
Germ, whole grains and, 84, 84f
Gestational diabetes, 98, 345
Gestational hypertension, 344
Ghrelin, 280
Giardia lamblia, 411t, 415
Ginkgo biloba, 212, 212f
Ginseng, 212, 212f
Glucagon secretion, 92
Glucose
as energy source, 92–95
regulation of, 92, 93f
structure and sources of, 85, 86f
Glutathione, 257
Glutathione peroxidase, 253, 254f
Gluten-free foods, 68–69, 68f
Gluten-related disorders, 67–68
Glycemic index, 91
Glycemic load, 91
Glycemic response, 91, 91f
Glycogen, 87
Glycogen stores, 326, 326f
Glycolysis, 92
Goiter, 257, 258f, 447, 447f
Goitrogens, 259
Golden Rice, 434, 434f, 456
Government involvement, poor food choices and, 27, 27f
Government policies, population growth and, 452
Government role, food safety and, 406–407, 406t–407t
Green tea extract, 291
Growth charts, 360f
Growth hormone, 332
Growth spurt, adolescents and, 384, 384f
Gums, in baked goods, 139, 139f
Gut immune function, 66–68
celiac disease and, 67
food allergies and, 66–67, 68f
gluten-related disorders and, 67–68

H

Hazard Analysis Critical Control Point (HACCP) system, 407, 407f
Head Start, 458t
Health claims, food labels and, 44, 46f
Healthy blood
copper and, 253
iron and, 249–253
minerals and, 249–253
Healthy diets, 13–15
balance and, 13–14, 14f
case study on, 15, 15f
moderation and, 14–15
variety and, 13
Healthy Eating Index, 372, 372f
Healthy eating patterns, 35–36, 35f
Healthy food patterns, dieting methods and, 288, 288f
Healthy People 2020, 30, 458t
Heart, blood circulation and, 74, 74f
Heart disease
alcohol consumption and, 400
carbohydrates and, 100–101
dietary patterns and, 129t, 131, 131f–132f
lipids and, 128–131
risk factors and, 129, 129t
Heartburn, 69–70, 70f
Heat cramps, 324
Heat exhaustion, 324
Heat index, 324, 324f
Heat-related illnesses, 323–324, 323f
Heat stroke, 324
Heimlich maneuver, 58, 58f
Helicobacter pylori, 70
Heme iron, 249
Hemochromatosis, 242
Hemoglobin, 249
Hemolytic-uremic syndrome, 411
Hemorrhoids, 101
Hepatic portal circulation, 75, 75f
Hepatic portal vein, 75, 75f
Hepatitis, alcohol consumption and, 400
Hepatitis A, 411t, 414–415
Herring worm, 415, 416f
Heterocyclic amines, 427–428
High-density lipoproteins (HDLs), 125
High-protein diets, 155–156, 156f
High-quality protein, 161
Homeless Children Nutrition Program, 458t
Homocysteine levels, folate and, 190, 190f
Hormones, food safety and, 424–425
Humectants, 430t
Hunger. *See* World hunger
Hydrogenated oils, 135
Hydrogenation, 118
Hypercarotenemia, 206
Hypertension
childhood nutrition and, 374–376, 375f

Hypertension (Continued)
  DASH eating plan and, 235–236, 236f
  electrolytes and, 234–236, 235f
  heart disease and, 129t
  pregnancy and, 344–345
  preventing and treating, 235–236
  risk factors for, 235
Hypertrophy, 308, 308f
Hypoglycemia, 99–100
Hyponatremia, 224, 324–325, 325f

I

Illicit drug use, pregnancy and, 356
Implantation, 341
Incomplete dietary protein, 161
Indirect food additives, 430
Indispensable amino acids, 145, 145f
Individual testimonies, nutritional information and, 18, 20f
Industrial chemicals, food safety and, 421–426
Industrial contaminants
  food safety and, 424
  reducing exposure risks and, 425–426, 426f
Infant botulism, 413
Infant mortality rate, 441
Infant nutrition, 359–367
  avoiding bacterial contamination and, 365, 366f
  bottle feeding do's and dont's, 363, 363f
  breast-feeding health benefits, 363–364, 365f
  breast milk and formula nutrients and, 363, 364f
  developmentally appropriate foods, 366–367, 366f
  DHA/ARA-fortified formulas, 362
  dietary supplements and, 211t
  energy and macronutrient needs, 359–360, 361f
  failure to thrive and, 359
  fluid needs and, 360
  food allergies and, 365–366
  formula-feeding advantages, 364
  growth charts, 360f
  infant growth and development and, 359
  micronutrients at risk and, 360–361
  nursing bottle syndrome, 363, 363f
  safe feeding first year, 365–367, 366f
Inflammation, atherosclerosis and, 129
Ingredient lists, 42, 43f
Inner cities, food choices and, 4
Insect-resistant corn, 434, 434f
Insoluble fiber, 87–88
Insulin misuse, 299, 299t
Insulin resistance, 98
Integrated pest management, 422
Integumentary system, 54t

Intestinal microbiota, 64–66, 65f
Intoxication, water, 224
Iodine, 257–259
  cretinism and, 259
  deficiency disorders and, 258–259, 258f
  goiters and, 257, 258f
  goitrogens and, 259
  in health and disease, 257–259
  iodized salt, 259, 259f
  meeting needs for, 259
  recommendations and, 228t
  thyroid gland and, 257
  vegan diet and, 166t
Iodized salt, 259, 259f, 455, 455f
Iron, 249–253
  absorption and transport, 249, 250f
  adolescent nutrition and, 384
  children and, 377
  depletion, 9
  as ergogenic supplement, 331t
  in health and disease, 251–253
  heme iron, 249
  hemochromatosis and, 242
  infants and, 360
  iron deficiency, 251, 251f
  iron deficiency anemia, 251–252, 251f
  iron overload, 252
  meeting needs for, 249–251, 250f
  myoglobin and, 249
  older adults and, 391
  overdose, 10
  overload, 252
  physical activity and, 322, 322f
  pregnancy and, 347
  recommendations and, 227t
  toxicity and, 252, 252f
  vegan diet and, 166t
Iron deficiency anemia, 251–252, 251f
Irradiation, food safety and, 428–429, 428f

J

Juice consumption, children and, 378, 378f

K

Keshan disease, 254, 448
Ketoacidosis, 96
Ketone bodies, 94–95, 95f
Ketones, 128
Ketosis, 95
Kidney disease, high-protein diets and, 155
Kidney failure, 393
Kidney stones, 155
Kidneys, water loss and, 222, 222f
Kilocalories, 7
Kwashiorkor, 154, 154f, 447, 447f

L

Lactation, 356–358
  breast development and, 341
  energy and nutrient needs during, 357–358, 357f, 358f
  milk production and let-down, 356–357, 357f
  protein needs and, 159
Lacteal, 73
Lacto-ovo vegetarian diets, 164t
Lacto-vegetarian diets, 164t
Lactobacillus, 67
Lactose, 85, 86f
Lactose intolerance, 89, 89f
Large-for-gestational age babies, 343
Large intestine, digestion and absorption and, 64
LDL receptor, 124, 124f
Lead toxicity, childhood nutrition and, 382–383, 382f
Leavening agents, 430t
Lecithin, 119
Legumes, 144
Leptin, 280–281, 281f
Life expectancy, 388, 388f
Life span, 389
Linoleic acid, 125
Lipases, 62
Lipids
  cancer and, 131
  case study and, 136
  cholesterol elimination, 124–125, 124f
  choosing fats wisely, 133–135
  cis fatty acids, 118, 118f
  digestion and absorption of, 122–123, 122f
  essential fatty acids, 125–127, 126f
  fat and cholesterol recommendations, 133
  fat as source of energy, 127–128
  fat replacers, 137–139, 139f
  fats in our food, 114–115, 114f
  fatty acids, 116–118, 117f, 118f
  food labels and, 135–137, 137f
  functions of, 125–128
  heart disease and, 128–131
  hidden fats, 114–115, 115f
  MyPlate choices, 135, 135f
  obesity and, 133
  phospholipids, 118–119, 120f
  saturated fatty acids, 116–118
  sterols, 119–121, 121f
  trans fatty acids, 118, 118f
  transport and delivery, 123–125, 124f
  transport from liver, 124, 124f
  transport from small intestine, 123–124
  triglycerides, 116–118, 116f
  types of, 115–121
  unsaturated fatty acids, 116–118
  varying combinations of, 6, 7f
Lipoprotein lipase, 124, 124f
Lipoproteins, structure of, 123, 123f
Liposuction, 293
Liquid meals, 289, 289f
Listeria infection, 354
Listeria monocytogenes, 410t, 413
Liver, lipid transport and, 124, 124f
Liver disease, alcohol consumption and, 400
Low-birth weight infants, 341, 342f, 441, 441f

Low blood sodium, 324–325, 325f
Low-density lipoproteins (LDLs), 124, 124f
Low folate status, 190
Lumen, 55
Luo Han Guo, 107t
Lymphatic/immune system, 54t, 73, 75
Lymphocytes, 66

M

Macrocytic anemia, 188, 189f
Macronutrients
  definition, 6
  infant nutrition and, 359–360, 361f
  lactation and, 357–358, 357f
  pregnancy and, 345–346, 346f
Macular degeneration, older adults and, 393
Magnesium, 227t, 246, 247f
Malnutrition
  alcohol consumption and, 400
  causes and consequences and, 392–393, 392f
  older adults and, 391–394
  physiological changes and, 392–393, 392f
Maltose, 85, 86f
Manganese, 228t, 256
Marasmus, 154–154, 154f, 447, 447f
Maternal weight gain, 341–343, 342f
Maximum heart rate, 310
Meal patterns, children and, 378–379, 378t
Medication use, older adults and, 394
Medications, dietary supplements and, 211t
Mediterranean diet pyramid, 132f
Mediterranean Eating Pattern, 35, 35f
Medium-chain triglycerides, 119, 331t, 333
Menopause, 240
Mercury in fish, pregnancy and, 354, 355t
Mercury poisoning, 424, 425f
Metabolic pathways, 77
Metabolism, 77–78
Micelles, 122
Microbes, 406
Micronutrients
  children and, 376–377
  definition, 6
  food fortification and, 455
  infants and, 360–361
  lactation and, 357–358, 358f
  pregnancy and, 346, 347f
  world hunger and, 447, 447f
Microsomal ethanol-oxidizing system (MEOS), 398
Microvilli, 62
Milk consumption, 28, 28f
Milk production and let-down, 356–357, 357f
Minerals

INDEX I-7

antioxidant minerals, 253–257
bone health and, 240–249
electrolytes, 232–239
energy metabolism and, 257–260
healthy blood and, 249–253
mineral bioavailability, 229–230, 230f
mineral functions, 230–231, 231f
mineral ions, 230
older adults and, 391
osteoporosis and, 240–242, 241f
in our food, 228–229, 229f
overview of, 226–232
physical activity and, 322–323
pregnancy and, 346–348
recommendations regarding, 227t–228t
supplements, 329
understanding mineral needs, 230–232
vegetarian diets and, 165
Mitochondria, 77
Moderate-to-high protein diets, 288, 288f
Moderation, healthy diets and, 14–15
Modified atmosphere packaging (MAP), 429
Molds, food safety and, 415, 415f
Molecular biology, nutritional studies of, 17, 18f
Molecules, 52, 52f
Molybdenum, 228t, 257
Monoglycerides, 122
Monosaccharides, 85
Monosodium glutamate (MSG), 155, 432
Morning sickness, 343
Mouth, digestion and absorption and, 58
Movement, proteins and, 151, 151f
MSG symptom complex, 155
Mucosa, 55
Mucosal cells, 55
Mucus, 55
Muscle dysmorphia, 299, 299t
Muscle mass, declining, 389, 391f
Muscle strength and endurance, 308
Muscle strengthening, 311–312
Muscular system, 54t
Myoglobin, 249
MyPlate, 38–41
adolescents and, 385, 385f
building healthy eating style and, 39, 39f
carbohydrate choices and, 104, 104f
childhood recommendations, 377–378, 377f
Choice (Exchange) Lists compared, 41
fat in diet and, 135, 135f
food group choices and, 39–40, 40f
minerals and, 228–229, 229f
nutrient-dense foods and, 40–41, 41t

older adults and, 394, 396f
pregnancy and, 348, 349f
proteins and, 162–163, 162f
saturated fats, sodium, sugars and, 41
vegetarian diets and, 166, 167f

N

National Health and Nutrition Examination Survey (NHANES), 29
National Marine Fisheries Service, 407t
National Oceanic and Atmospheric Administration (NOAA), 407t
National School Lunch Program, 379–380, 380f, 457t
Natural foods, claims regarding, 44, 46f
Natural systems agriculture, 453
NEAT (nonexercise activity thermogenesis) differences, body weight and, 282–283, 282f
Neotame, 107t
Nervous system, 53t
Neural tube defects, 188, 348, 348f
Neural tube formation, 236
Neurotransmitters, 179
New molecule synthesis, 78
Niacin, 182–184
adverse effects and, 184
deficiency and, 183–184
equivalents, 183
meeting needs and, 183, 184f
pellagra and, 182–183, 183f
recommendations, 177t
Night blindness, 203–204
Night-eating syndrome, 299, 299t
Nitrogen balance, 158, 158f
Nitrosamines, 431
Non-celiac gluten sensitivity (NCGS), 67
Nonessential amino acids, 145, 145f
Nonexercise activity thermogenesis (NEAT), 273
Nonnutritive sweeteners, 107–108, 107f–108f
Nonvegetarian diets, 164t
Norovirus, 411t, 414
Nutraceuticals, 3
Nutrient-dense foods, 40–41, 41t
Nutrient intake
designer foods and, 3
dietary reference intakes and, 31, 31f
dietary supplements and, 3
empty calories and, 2, 2f
food choices and, 2–6
food fortification and, 3
nutraceuticals and, 3
phytochemicals and, 3, 3f
zoochemicals and, 3
Nutrients
adolescent nutrition and, 383–384

adult nutrition and, 389, 394, 396f
content claims, 42–44, 43f
density, 2, 2f
digestion and absorption of, 57–64
lactation and, 357–358, 357f, 358f
role of, 7–9, 8f
six classes of, 6–7
supplementation, world hunger and, 456
Nutrigenetics, 12
Nutrigenomics, 12
Nutrition, 26–30
education, 454–455, 455f, 459–460
food guides, 26, 26f
nutritional programming, 352
overnutrition, 10, 10f
past and present, 26
poor food choices, 27, 27f
study types, 17, 18f
undernutrition, 9–10, 9f
using recommendations regarding, 28–30
Nutrition Facts, 42
Nutrition Program for the Elderly, 458t
Nutrition transition, 443, 443f
Nutritional genomics, 11–12, 12f
Nutritional information
advertising claims and, 20, 21f
evaluation of, 15–21
individual testimonies and, 18, 20f
judging claims regarding, 17–21
science behind, 16, 16f
self-judgment and, 17–21
source of claims and, 18–20, 19f
study of, 16–17
time tested, 20–21
Nutritional status, evaluation of, 28–29, 29f–30f
Nuts, monosaturated fats and, 131f

O

Oatmeal, as soluble fiber source, 132f
Obesity. See also Body weight
childhood, 373–374, 374f, 375f
dietary fat and, 133
genetics and, 278–279, 278f
heart disease and, 129f
overnutrition and, 10, 10f, 442
surgery for, 292, 292f
Obesogenic environment, 267
Ocean depletion, world hunger and, 446, 446f
Older adults. See also Adult nutrition
Alzheimer's disease and, 394
arthritis and, 393, 393f
assisted living and, 396–397
cataracts and, 393, 393f
chronic illness and, 393–394
dementia and, 393–394

dietary supplements and, 211t
economic and social issues and, 394
food insecurity and, 394
macular degeneration and, 393
malnutrition and, 391–394
medication use and, 394
nutrient needs and, 394, 396f
physical activity and, 395–396, 397f
physiological changes and, 392–393, 392f
vitamin and mineral deficiencies and, 391
water needs and, 389–391
Olestra, 139, 139f
Oligosaccharides, 87
Omega-3 fatty acids, 126–127, 131f, 166t
Omega-6 fatty acids, 126–127
Organ systems, 52–55, 53t–54t
Organic compounds, 6
Organic food production, 422–424, 424f
Organization of life, 52–55, 52f
Organs, 52
Osmosis, 63, 220
Osteoarthritis, 393, 393f
Osteomalacia, 207
Osteoporosis, 240–242, 241f
factors affecting risk, 240, 242t
preventing and treating, 240–242
vitamin D and, 208
Overload principle, 307
Overnutrition, 10, 10f, 442
Overpopulation, world hunger and, 444, 446f
Overtraining syndrome, 312
Oxidative stress, 196
Oxygen, 313, 315f

P

Packaging, food safety and, 429, 429f
Pancreatic amylase, 62
Pancreatic juice, 61
Pantothenic acid, 177t, 185
Papaya ring-spot virus, 434, 434f
Parasites, food safety and, 415, 416f
Parathyroid hormone, 206
Partially hydrogenated oils, 118
Pasteurization, 407
Pathogens. See Food pathogens
Peak bone mass, 240
Pectins, 139, 139f
Peer-review process, nutritional experiments and, 17
Pellagra, 182–183, 183f
Pepsin, 59
Peptic ulcers, 70, 71f
Peptide bonds, 146
Percentage of Daily Value, 42, 43f
Peristalsis, 59
Pernicious anemia, 190–191
Pescetarian diets, 164t
Pesticide-resistant insects, 435

Pesticides, 422–424
  integrated pest management and, 422
  organic food production and, 422–424, 424f
  reducing exposure risks and, 425–426, 426f
  tolerances and, 422, 422f
pH control agents, 430t
Phagocytes, 66
Pharynx, digestion and absorption and, 58, 58f
Phenylketonuria (PKU), 145–146, 146f
Phospholipids, 115, 118–119, 120f
Phosphorus, 245–246
  deficiency of, 246
  in health and disease, 245–246
  meeting needs of, 246
  nonskeletal functions of, 245, 245f
  recommendations and, 227t
Photosynthesis, 87, 87f
Physical activity. See also Exercise; Fitness
  active lifestyle creation and, 312–313, 312t
  aerobic metabolism and, 314–316, 316f
  aerobic zone and, 310–311, 311f
  anaerobic metabolism and, 314, 316f
  body weight and, 284
  calcium and vitamin D needs and, 322–323
  carbohydrate, fat, and protein needs, 321–322, 321f
  chronic disease and, 305, 306f
  dehydration and heat-related illnesses, 323–324, 323f
  energy expenditure and, 273, 275f
  energy needs and, 319–321, 319f
  ergogenic aids and, 329–335, 330t–331t
  exercise duration and fuel use, 313–316
  exercise intensity and fuel use, 316–318, 317f
  fitness components and, 307–309, 308f
  fitness programs and, 310–312
  food and drink and, 305–307, 305f, 325–329
  fueling activity, 313–319
  heart disease and, 129t
  iron needs and, 322, 322f
  low blood sodium and, 324–325, 325f
  maximum heart rate and, 310
  muscle strengthening and stretching, 311–312
  older adults and, 395–396, 397f
  oxygen to muscle cells, 313, 315f
  pregnancy and, 343, 343t
  protein as fuel, 316
  recommendations, 309–313, 310f
  vitamin and mineral needs and, 322–323
  water and electrolytes and, 323–325
  weight management and, 305–307, 307f
Physical activity (PA) value, 276, 277f
Physical frailty, 393
Physical limitations, older adults and, 396–397
Physiological changes, malnutrition and, 392–393, 392f
Phytochemicals, 3, 3f
Pica, 299, 299t, 350, 351f
Pima Indians, 279, 279f
Placebo, nutritional experiments and, 17
Placenta, 341, 341f
Plant proteins, 144, 144f
Plant sterols, 121, 132f
Pollutants, reducing exposure risks and, 425–426, 426f
Polychlorinated biphenyls (PCBs), 424
Polycyclic aromatic hydrocarbons (PAHs), 427–428
Polypeptides, 146
Polysaccharides, 85
Poor-quality diets, world hunger and, 446–448, 447f
Population growth
  controlling of, 451–452
  economic and cultural factors, 451, 451f
  family planning and, 452, 452f
  government policies and, 452
Portion distortion, 267, 267f
Portion sizes, 14, 37, 37f
Postmenopausal bone loss, 240
Potassium, 232–239. See also Electrolytes
  on food labels, 238, 239f
  recommendations and, 227t
Potatoes, 13
Poverty
  education and, 448, 449f
  hunger and, 444–445, 445f
  pregnancy and, 354
  United States and, 448–449, 448f
Prebiotics, 66
Precompetition meals, 326, 327f
Prediabetes, 98
Preeclampsia, pregnancy and, 344
Pregnancy. See also Lactation
  alcohol intake and, 355, 355f
  body changes during, 340–345
  caffeine intake and, 354
  complications of, 343–345
  critical development periods and, 351–352, 352f
  dietary supplements and, 211t
  discomforts of, 343
  Down syndrome and, 354, 354f
  eclampsia and, 344
  edema and, 343
  energy and macronutrient needs, 345–346, 346f
  fluid and electrolyte needs, 346
  folate and, 188
  food-borne illness and, 354, 355t
  food cravings and aversions, 348–350
  gastrointestinal tract crowding and, 343, 344f
  gestational diabetes and, 345
  gestational hypertension and, 344
  high blood pressure and, 344–345
  hypertensive disorders of, 344
  legal and illicit drug use and, 356
  maternal age and health, 353–354, 353f
  maternal nutritional status and, 352
  maternal weight gain during, 341–343, 342f
  mercury in fish and, 354, 355t
  micronutrient needs and, 346, 347f
  morning sickness and, 343
  MyPlate recommendations, 348, 349f
  neural tube defects and, 348, 348f
  normal weight gain, 342
  nourishing embryo and fetus, 341, 341f
  nutrient needs case study, 350
  nutritional needs during, 345–351
  nutritional programming and, 352
  physical activity during, 343, 343t
  pica and, 350, 351f
  poverty and, 354
  preeclampsia and, 344
  prenatal development, 340–341, 341f
  prenatal supplements and, 348, 349f
  protein needs and, 159
  tobacco use and, 355–356
  toxic substances and, 354–356
  underweight at onset of, 342–343
  vitamin and mineral needs and, 346–348
Premature infants, 341
Prenatal development, 340–341, 341f
Prenatal supplements, 348, 349f
Prepackaged meals, 289, 289f
Preservatives, 430t
Preterm infants, 341
Prions, food safety and, 416–417, 416f
Prior-sanctioned substances, 431
Probiotics, 66
Processed meat, cancer and, 431, 431f
Proteases, 62
Protein-based fat replacers, 139, 139f
Protein-energy malnutrition, 153–154, 154f
Proteins
  amino acids and, 146–147, 146f
  athletic needs and, 159–160, 160f
  balancing intake and losses, 158
  breakdown, limited carbohydrates and, 92–93, 95f
  choosing protein wisely, 161–163
  combinations and, 6, 7f
  complementary proteins, 161–162, 161f
  contractile properties and, 151f, 153
  defense mechanisms and, 151f, 152
  deficiency, 153–155, 154f
  denaturation and, 147, 147f
  dietary guideline recommendations, 162–163
  digestion and absorption, 147–148, 148f
  as energy source, 152f, 153
  as ergogenic supplement, 331t
  fluid balance and, 151f, 153
  food allergies and intolerances, 155–157, 157f
  functions, 150–153, 151f
  gene expression and, 149
  growth and, 159, 159f
  high-protein diets, 155–156, 156f
  meeting protein needs, 157–163
  MyPlate recommendations, 162–163, 162f
  nitrogen balance and, 158, 158f
  in our food, 144, 144f
  physical activity and, 321–322, 321f
  pregnancy and lactation and, 159
  protein quality, 161
  recommended protein intake, 158–160
  structural, 150–152, 151f
  supplements, 160, 160f
  synthesis of, 149–150, 150f
  vegan diets and, 166t
  vegetarian diets and, 163–166, 164f
  vitamin K and, 195
  weight loss and, 156
  world hunger and, 447, 447f
Prothrombin, 194, 194f
Provitamins, 143
Psychological issues, eating disorders and, 293–294
Purging, eating disorders and, 294, 294f
Pyridoxal phosphate, 186

Q

Qualified health claims, food labels and, 44

R

Raw eggs, biotin availability and, 185
Reactive hypoglycemia, 100
Recombinant DNA, 433

Recommended Dietary Allowances (RDAs), 26, 31, 32f
Reduced fat diets, 289, 289f
Refined grains, 83
Rehydration, 224, 224f
Relative energy deficiency in sport (RED-S), 320–321, 320f
Renewable resources, hunger and, 444–446
Reproductive system, 53t
Resistant starch, 87
Respiratory system, 53t
Rest, blood flow at, 74, 74f
Resting heart rate, 309
Resting metabolic rate, 274, 274f
Restricted diets, dietary supplements and, 211t
Retailers, food safety and, 407, 408f
Retinoids, 201
Retinol activity equivalents, 202
Retinol-binding protein, 203
Rhodopsin, 203
Riboflavin, 180–182
   deficiency of, 182
   recommendations, 177t
   sources of, 182, 182f
Rickets, vitamin D and, 207, 208f
Rumination disorder, 299, 299t

S

Saccharin, 107t
Safe grocery choices, 417, 418f
Saliva, 58
Salivary amylase, 58
Salmonella, 410t, 411, 412f
Salt appetite, 233
Saturated fats, 37–38, 41, 115
Saturated fatty acids, 116–118
Saw palmetto, 213, 213f
School Breakfast Program, 380, 457t
School nutrition programs, 379–380
School performance, breakfast and, 379
Scientific method, nutritional information and, 16, 16f
Scurvy, 10, 197, 197f
Segmentation, 60
Selective eating disorder, 299, 299t
Selective estrogen receptor modulators (SERMS), 242
Selenium, 253–255
   cancer and, 254–255
   deficiency, 448
   glutathione peroxidase and, 253, 254f
   Keshan disease and, 254
   meeting needs for, 254
   recommendations and, 228t
   soil selenium, 254, 254f
   thyroid hormones and, 253
Self-esteem, eating disorders and, 293–294, 294t
Self-judgment, nutritional information and, 17–21
Semivegetarian diets, 164t

Senior Farmers' Market Program, 458t
Shigella, 410t
Sickle cell anemia, 198
Simple carbohydrates, 85, 86f
Simple diffusion, 63
Skeletal system, 54t
Skin, proteins and, 152
Skinfold thickness, body composition and, 271, 271f
Small for gestational age, 341
Small intestine
   digestion and absorption and, 60–64, 63f7
   lipid transport and, 123–125, 124f
   secretions and, 61–62
   structure of, 62f
Smoking
   heart disease and, 129t
   osteoporosis and, 242t
Snacking, children and, 378–379, 378t
Snacks, exercise and, 328, 328f
Social issues, older adults and, 394, 396
Sociocultural issues, eating disorders and, 294–296, 295f
Sodium, 232–239. See also Electrolytes; Hypertension
   on food labels, 238, 239f
   hidden sodium, 238
   limiting, 37–38, 41
   processing and, 236, 237f
   recommendations and, 227t
Sodium chloride, 232
Soil selenium, 254, 254f
Soluble fiber, 87
Solutes, water distribution and, 220
Solvent, water as, 223
South Korea, food imports and, 454
Soy-based diets, heart disease risk and, 132f
Special Milk Program, 458t
Special Supplemental Nutrition Program for Women, Infants, and Children (WIC), 354
Sphincter, 59
Spina bifida, 189, 189f
Spores, 413
Sports anemia, 322, 322f
Sports drinks, 327, 329f, 334, 334f
St. John's wort, 212, 212f
Staphylococcus aureus, 410t, 413
Starch, 87
Starvation. See World hunger
Steroid precursors, 332
Sterols, 115, 119–121, 121f
Stevia, 107t
Stomach
   activity regulation and, 60, 61f
   digestion and absorption and, 59–60
   gastric juices and, 59–60
   meal nutrient composition and, 60, 61f
   structure and function, 60f

Storage, food, 417, 419f
Stretching, 311–312
Structural proteins, 150–152
Structure/function claims, dietary supplement labels and, 46–47, 47f
Subcutaneous fat, 269
Subsistence crops, 454
Sucralose, 107t
Sucrose, 85, 86f, 139, 139f
Sudden infant death syndrome (SIDS), 356
Sugar, limiting, 37–38, 41
Sugar units, 85
Sugars. See also Added sugars
Sulfur, 227t, 256–257
Summer Food Service Program, 458t
Sunscreen, vitamin D deficiency and, 210, 210f
Superweeds, 435
Supplemental Nutrition Assistance Program (SNAP), 457t
Supplements. See Dietary supplements
Sustainable agriculture, 452–453, 454f
Sweeteners
   nonnutritive, 107–108, 107f–108f
   uses of, 430t
Synthesis, proteins, 149–150, 150f

T

Team Nutrition, 458t
Television viewing, diet quality and, 372, 372f
Temperatures, food safety and, 427–428
Temporary Assistance for Needy Family (TANF), 458t
Teratogens, 351
Thermic effect of food (TEF), 275, 275f
Thiamin, 179–180
   beriberi and, 179
   functions and deficiency, 180, 181f
   meeting needs and, 180, 181f
   neurotransmitter synthesis and, 179
   recommendations for, 177t
   Wernicke-Korsakoff syndrome and, 180
Thirst sensation, 221
Thyroid gland, 257
Thyroid hormones, 253
Thyroid-stimulating hormone, 258, 258f
Tobacco use
   adolescents and, 387
   pregnancy and, 355–356
Tocopherol. See Vitamin E
Tolerable Upper Intake Levels (ULs), 32, 32f
Tolerances, pesticide, 422, 422f
Tooth loss, 70, 70f
Total energy expenditure (TEE), 273

Toxic substances, pregnancy and, 354–356
Toxoplasma gondii, 411t
Toxoplasmosis, 354
Trans fat, 115
Trans fatty acids, 118, 118f
Transgenic organisms, 433
Transit time, food, 55
Transport, iron, 249, 250f
Trichinella spiralis, 411t, 415
Triglycerides
   fatty acids and, 116–118, 116f
   storage of, 125, 125f
   structure of, 115
Trimesters, pregnancy and, 342
Tripeptides, 146, 146f
Tropical oils, 116
Type 1 diabetes, 96
Type 2 diabetes, 98, 98f, 374, 375f

U

Undernutrition, 9–10, 9f, 441–442, 441f–442f
Underwater weighing, body composition and, 271, 271f
United States
   education and poverty, 449, 449f
   food recovery and, 457
   nutrition education and, 459–460
   nutrition safety net and, 457
   poverty and food insecurity and, 448–449, 448f
   programs preventing undernutrition, 457, 457t–458t
   vulnerable life stages and, 449–450
Unrefined foods, 83
Unsaturated fats, 115
Unsaturated fatty acids, 116–118
Urea, 152, 152f
Urinary system, 53t
U.S. Department of Agriculture (USDA), 407t
U.S. Environmental Protection Agency (EPA), 407t
U.S. Food and Drug Administration (FDA), 407t
USDA Food Patterns, 35, 35f
USDA Vegetarian Adaptations, 35, 35f

V

Vanadium, 331t
Variables, nutritional experiments and, 17
Variant Creutzfeldt-Jakob Disease, 416–417
Vegan diets
   animal products and, 163
   dietary supplements and, 211t
   exclusions and, 164t
   meeting nutrient needs with, 166t
Vegetarian diets, 163–166, 164f
   adolescents and, 386–387, 387f
   benefits of, 163–164
   case study, 165
   MyPlate recommendations and, 166, 167f

Vegetarian diets (*Continued*)
  planning of, 165–166
  risks of, 165
  vitamin and mineral deficiencies and, 165
Veins, 74
Venules, 74
Very-low-birth-weight infants, 341
Very-low-density lipoproteins (VLDLs), 124, 124f
Vibrio bacteria, 413
Vibrio parahaemolyticus, 410t
Villi, 62
Viruses, food safety and, 414–415, 414f
Visceral fat, 269–272
Visual cycle, 202f, 203
Vitamin A, 201–206
  carotenoids and, 201–202
  cell differentiation and, 203
  fat malabsorption and, 202–203
  functions and deficiency, 203–204, 447, 447f
  gene expression and, 203
  Golden Rice and, 456
  hypercarotenemia and, 206
  meeting needs and, 201–203, 201f
  night blindness and, 203–204
  recommendations and, 178t
  retinoids and, 201
  retinol activity equivalents (RAE) and, 202
  retinol-binding protein and, 203
  supplements and, 206, 206f
  toxicity, 204–206, 206f
  visual cycle and, 202f, 203
  xerophthalmia and, 204
Vitamin B_6, 186–188
  adolescents and, 383
  adverse effects and, 188
  anemia and, 187
  excessive, 10
  functions and deficiency, 186–187, 186f–187f
  older adults and, 391
  pyridoxal phosphate and, 186
  recommendations, 177t
  sources of, 187, 188f
  supplements and, 188
Vitamin B_{12}, 190–193
  absorption, 191, 192, 192f
  adolescents and, 383
  atrophic gastritis and, 192
  folate relationship and, 191, 193f
  intrinsic factor and, 191
  meeting needs and, 192, 193f
  older adults and, 391
  pernicious anemia and, 190–191
  pregnancy and, 346–347
  recommendations, 178t
  vegan diet and, 166t
Vitamin B complex, 185
Vitamin C, 196–198
  collagen and, 197, 197f
  deficiency, 10
  destruction of, 197
  meeting needs and, 197, 198f
  recommendations, 178t
  scurvy and, 197, 197f
  sickle cell anemia and, 198
  supplements and, 197–198
Vitamin D, 206–211
  activation and function, 206, 207f
  adolescents and, 383
  calcium and, 211, 244
  children and, 377
  concealing clothing and, 210, 210f
  dark skin pigmentation and, 210, 210f
  deficiency and, 206–208
  infants and, 361
  needs and, 208–211, 209f–210f
  older adults and, 391
  osteomalacia and, 207
  osteoporosis and, 208
  parathyroid hormone and, 206
  physical activity and, 322–323
  pregnancy and, 346
  recommendations, 178t
  rickets and, 207, 208f
  sunscreen and, 210, 210f
  vegan diet and, 166t
Vitamin E, 198–200
  antioxidant role of, 198, 199f
  deficiency and, 198
  hemolytic anemia and, 198
  longevity studies and, 21, 21f
  meeting needs and, 198–200, 199f
  recommendations for, 178t, 198
  supplements and, 200
Vitamin K, 194–195
  blood clotting and, 194, 194f
  deficiency and, 195
  infants and, 361
  meeting needs of, 195, 195f
  protein synthesis and, 195
  recommendations for, 178t
Vitamin precursor, 143
Vitamins. *See also* specific vitamin
  absorption of, 175, 175f
  adolescents and, 383
  antioxidant vitamins, 196–200, 196f
  bioavailability and, 174
  case study on modern diet, 205
  coenzymes and, 179, 180f
  deficiencies in older adults, 391
  dietary supplements and, 211–214
  discretionary fortification and, 172
  energy metabolism and, 179–185
  food labels and, 173, 173f
  free radicals and, 196, 196f
  functions of, 174, 176f
  gene expression and, 200–211, 200f
  healthy blood and, 186–195
  losses through preparation, 173, 174f
  in our food, 171–173, 171f
  oxidative stress and, 196
  physical activity and, 322–323, 329
  pregnancy and, 346–348
  primer on, 171–178, 171f
  recommended intakes of, 177t–178t
  tips for preserving, 173, 174f
  understanding needs and, 174–178
  vegetarian diets and, 165
Vulnerable life stages, hunger and, 449–450

W

Waste elimination, 75–76, 76f
Wasted food, 459
Water
  antidiuretic hormone and, 222, 222f
  balance, 221–222, 221f
  in the body, 220
  body temperature and, 223, 223f
  bottled water, 225
  dehydration and, 224, 224f
  distribution, 220, 220f
  fluoridation and, 248–249
  functions of, 222–223
  in health and disease, 223–224
  hyponatremia and, 224
  infants and, 376
  kidneys and, 222, 222f
  meeting water needs, 224–226
  in metabolism and transport, 223
  older adults and, 389–391
  physical activity and, 323–325
  as protection, 223
  regulating intake of, 221–222, 221f
  regulating loss of, 222, 222f
  water intoxication, 224
Water-soluble vitamins, 171, 177t–178t
Weight. *See* Body weight
Weight cycling, 284, 284f
Weight gain basics, 276
Weight gain suggestions, 285
Weight loss
  basics of, 276
  goals and guidelines, 283–285, 283f
  high-protein diets and, 156
  medications and supplements, 290–291, 290f
  surgery and, 291–293, 291f
Weight management, physical activity and, 305–307, 307f
Weightlifting, 308, 308f, 309
Wernicke-Korsakoff syndrome, 180
Western consumption patterns, 446
Wheat allergy, 67
Whole grains, 83–84, 84f
WIC Farmer's Market Nutrition Program, 457t
World Health Organization (WHO), 406t
World hunger
  breast feeding and, 455, 455f
  cash crops and, 454
  causes of, 443–448
  controlling population growth and, 451–452
  cultural practices and, 444
  economic development and trade and, 453–454
  elimination of, 450–456
  ensuring nutritious food supply and, 454–456
  famine and, 444, 444f
  food shortages and, 444–446
  food supply fortification and, 455, 456f
  increasing food production and, 452–453
  limited environmental resources and, 444–446
  malnutrition and, 441–443
  nutrition education and, 454–455, 455f
  nutrition transition and, 443, 443f
  ocean depletion and, 446, 446f
  overnutrition and, 442
  overpopulation and, 444, 446f
  poor-quality diets and, 446–448, 447f
  poverty and, 444–445, 445f
  providing short-term aid, 450–451, 451f
  renewable resources and, 444–446
  subsistence crops and, 454
  supplementation and, 456
  sustainable agriculture and, 452–453, 454f
  sustainable development goals and, 450, 450f
  undernutrition and, 441–442, 441f–442f
  United States and, 448–450, 449f, 450f, 457–460

X

Xerophthalmia, 204

Y

Yersinia enterocolitica, 410t
Yogurts, 67

Z

Zinc, 255–256
  deficiency and, 255, 448
  gene expression and, 255, 255f
  meeting needs for, 255, 256f
  pregnancy and, 347
  recommendations and, 228t
  supplements and, 256
  toxicity and, 255–256
  vegan diet and, 166t
Zoochemicals, 3